PRINCIPLES OF REFRIGERATION

PRINCIPLES OF REFRIGERATION

Second Edition

Roy J. Dossat
Professor of Mechanical Technology
University of Houston
Houston, Texas

John Wiley and Sons
New York
Santa Barbara
Chichester
Brisbane
Toronto

Library of Congress Cataloging in Publication Data:

Dossat, Roy J
Principles of refrigeration.

Includes index.
1. Refrigeration and refrigerating machinery.
I. Title.
TP492.D65 1978 621.5′6 78-2938
ISBN 0-471-03550-5

Printed in the United States of America

10 9 8 7 6 5 4 3 2 1

PREFACE

This textbook is a comprehensive, applications-oriented treatment of the mechanical refrigeration cycle and associated equipment. Although especially written for college-level courses in refrigeration, the arrangement and coverage of the material and the method of presentation are such that the text is also suitable for use in postsecondary occupational and adult evening programs, as well as for on-the-job training and self-instruction.

Despite a careful treatment of the thermodynamics of the refrigeration cycle, application of the calculus is not required, nor is a background in physics and thermodynamics presupposed. The first five chapters are introductory in nature and treat fundamental principles of physics and thermodynamics, including an introduction to psychrometrics, that are an essential prerequisite to any comprehensive study of mechanical refrigeration.

Throughout the text, emphasis is placed on the cyclic nature of the refrigeration system, and each part of the system is carefully examined with relation to the other parts and to the whole. Care is also taken continually to correlate theory with practice through the use of manufacturers' catalog data in many of the numerous examples and practice problems, a procedure that is made practical by the inclusion of certain pertinent manufacturers' equipment rating data as a part of the appendix.

In recognition of the change-over now in progress from the British system of measurement to the metric (SI) system, one section at the end of each chapter is devoted to repeating in metric units all the important relationships and calculations presented earlier in the chapter in British units. In addition, conversion factors for use in converting from one system of units to the other, along with frequently used values and constants for both systems of units, are conveniently listed in the appendix. Numerous examples and practice problems at the end of each chapter illustrate the use of these conversion factors and constants and provide interested students with the opportunity to become proficient not only in the use of both systems of measurement but also in converting from one system to the other.

Roy J. Dossat

ACKNOWLEDGMENTS

Most of the material in this textbook is based on information gathered from publications of the American Society of Heating, Refrigerating, and Air Conditioning Engineers and of the following equipment manufacturers:

Acme Industries, Inc.
Alco Valve Company
Anaconda Metal Hose Division, The American Brass Company
Bell & Gossett Company
Bohn Aluminum & Brass Corporation
Carrier Corporation
Controls Company of America
Dean Products, Inc.
Detroit Controls Division, American Radiator & Standard Sanitary Corporation
Detroit Ice Machine Company
Dole Refrigerating Company
Dunham-Bush, Inc.
Edwards Engineering Corporation
E. I. du Pont de Nemours & Company
Freezing Equipment Sales, Inc.
Frick Company
General Controls Company
General Electric Company
Halstead & Mitchell
Ingersoll-Rand Company
Kennard Division, American Air Filter Company, Inc.
Kramer Trenton Company
McQuay, Inc.
The Marley Company
Marsh Instrument Company
Mueller Brass Company
Penn Controls, Inc.
Prestcold, Ltd.
Recold Corporation
Singer Company, Controls Division
Sporlan Valve Company
Tecumseh Products Company
Tranter Manufacturing, Inc.
Tubular Exchanger Manufacturers Association, Inc.
Tyler Refrigeration Corporation
The Vilter Manufacturing Company
Worthington Corporation
York Corporation, Subsidiary of Borg-Warner Corporation

Appreciation is expressed to all these organizations for their contributions in the form of photographs and other art work, and for granting permission to reproduce proprietary data, without which this textbook would not have been possible.

R. J. D.

CONTENTS

PRINCIPLES OF REFRIGERATION

1

PRESSURE, WORK, POWER, ENERGY

1-1. Mass and Density

The *mass* (m) of a body, usually given in pounds (lb), and the *volume* (V), usually expressed in *cubic feet* (ft^3) or *cubic inches* ($in.^3$), are the primary measures of the quantity of matter in the body. *Density* (ρ) is defined as the mass per unit volume and *specific volume* (v), as the volume per unit mass; that is,

$$\rho = m/V \tag{1-1}$$

$$v = V/m \tag{1-2}$$

Density is usually expressed in pounds per cubic foot (lb/ft^3) and specific volume in cubic feet per pound (ft^3/lb). Notice that ρ equals $1/v$, and v equals $1/\rho$.

It will be shown later that the specific volume and density of a substance are not constant but vary with the temperature of the substance. However, for the temperature range normally encountered in refrigeration applications, a convenient value for the density of water, and one that is sufficiently accurate for most calculations, is 62.4 lb/ft^3. This is close to the maximum density for water and occurs at a temperature of approximately 39.2°F (4°C). The density of water decreases to approximately 59.8 lb/ft^3 at 212°F (100°C), the latter being the boiling point of water under normal barometric pressure. If more accurate values are required, they can be taken from the steam tables.

1-2. Specific Gravity

The specific gravity of a substance is the ratio of the density of the substance to that of some standard substance. In the case of liquids, the standard employed is usually water at its maximum density (62.4 lb/ft^3). If ρ_w is the density of water, then the specific gravity ρ_r of any substance is

$$\rho_r = \frac{\rho}{\rho_w} \tag{1-3}$$

Since specific gravity is a ratio, it has no dimensions.

1-3. Mass and Volume Flow Rates

Where mass quantities are expressed in pounds (lb), mass flow rates are expressed in pounds per second (lb/s), pounds per minute (lb/min), or pounds per hour (lb/hr). Similarly, where volume quantities are expressed in cubic feet (ft^3), volume flow rates are given in cubic feet per second (ft^3/s), cubic feet per minute (ft^3/min or cfm), and cubic feet per hour (ft^3/hr).

From Equations 1-1 and 1-2, and assuming consistent units, the following relationships exist between mass and volume, or mass flow rate and volume flow rate:

$$m = (V)(\rho) = \frac{V}{v} \qquad (1\text{-}4)$$

$$V = (m)(v) = \frac{m}{\rho} \qquad (1\text{-}5)$$

where m = the mass or mass flow rate
V = the volume or volume flow rate

Example 1-1 Water is flowing through a pipe at the rate of 5.6 ft^3/s. What is the mass flow rate in pounds per second (lb/s)?

Solution Assuming a water density of 62.4 lb/ft^3 and applying Equation 1-4,

$$m = (5.6 \text{ ft}^3/\text{s})(62.4 \text{ lb/ft}^3) = 349.44 \text{ lb/s}$$

1-4. Velocity and Speed

The simplest form of motion that a body can experience is to move at a constant speed in a straight line, in which case the body moves exactly the same distance in the same direction during each unit of time that the body is in motion. A body moving in such a manner is said to be moving at a constant velocity. Notice that the term *constant velocity* implies unchanging direction as well as unchanging speed, whereas the term *constant speed* implies only an unchanging rate of motion without regard for the direction of motion. Velocity, then, is a vector quantity, and the units of velocity properly should indicate direction as well as magnitude, while speed is a scalar quantity, and the units of speed indicate magnitude only. However, since the term *velocity* is so often used interchangeably with the term *speed*, and since the direction of motion will not be an important consideration in this book, no distinction will be made between these two terms. Both will be expressed as scalar quantities.

The speed (or velocity) of a moving body is the distance the body moves per unit of time. Whether or not the speed is constant, the *average speed* is the distance the body moves divided by the length of time required for the motion; that is,

$$v = \frac{s}{t} \qquad (1\text{-}6)$$

where v = the speed (or velocity) in units of distance per unit of time, usually feet per second (fps), feet per minute (fpm), or miles per hour (mph)
s = the distance traveled in feet (ft) or miles
t = the time in seconds (s), minutes (min), or hours (hr)

Example 1-2 A missile travels 1000 ft in 3.5 s. Compute the average velocity in feet per second.

Solution Applying Equation 1-6,

$$v = \frac{1000 \text{ ft}}{3.5 \text{ s}} = 285.7 \text{ fps}$$

1-5. Acceleration

Bodies in motion frequently experience changes in velocity. Motion in which the velocity is changing is known as *accelerated motion*, and the time rate of the change in velocity is known as *acceleration* (a). Acceleration may be either positive or negative, depending on whether the velocity is increasing or decreasing.

The simplest form of accelerated motion is *uniformly accelerated motion* wherein the motion occurs in a single direction and the speed changes at a constant rate. In such cases, if the initial velocity is v_o, the *instantaneous velocity* v_i at the end of t seconds will be

$$v_i = v_o + (a)(t) \qquad (1\text{-}7)$$

When a body starts from rest so that the initial velocity v_o is zero, Equation 1-7 may be simplified to

$$v_i = (a)(t) \qquad (1\text{-}8)$$

Then,

$$t = \frac{v_i}{a} \qquad (1\text{-}9)$$

and

$$a = \frac{v_i}{t} \qquad (1\text{-}10)$$

It can be shown that the average velocity v of a body uniformly accelerated from a position of rest is $v_i/2$. Then, in accordance with

Equation 1-6, the distance s in feet traveled by a body in t seconds will be

$$s = \frac{(v_i)(t)}{2} = \frac{(a)(t^2)}{2} \qquad (1\text{-}11)$$

1.6. Acceleration of Gravity

The most familiar example of uniformly accelerated motion is that of a freely falling body. A body falling freely toward the surface of the earth through the action of gravity alone will accelerate at a rate of 32.174 fps for each second that the body is falling. This value is known as the *standard acceleration of gravity*, or universal gravitational constant (g_c), and is the standard used for the *local acceleration of gravity* (g) at sea level. Since the effect of gravity on a body diminishes as the distance between the body and the center of earth increases, it follows that the local acceleration of gravity depends on the altitude, and since the earth tends to bulge at the equator, it will vary somewhat even at sea level at different latitudes. The value given is for a latitude of 45° but is sufficiently accurate at any latitude for routine calculations.

Example 1-3 A steel ball having a mass of 0.21 lb and starting from a position of rest is allowed to fall freely from the top of a building to the ground. If air friction is negligible and the elapsed time is 3.5 s, compute (a) the velocity of the ball at the instant of impact with the ground and (b) the height of the building in feet.

Solution Applying Equation 1-8,

$$v_i = (32.174 \text{ ft/s}^2)(3.5 \text{ s}) = 112.61 \text{ fps}$$

Applying Equation 1-11,

$$s = \frac{(112.61 \text{ fps})(3.5 \text{ s})}{2} = 197.1 \text{ ft}$$

1-7. Force

A force (F) is defined as a push or a pull. A force is anything that has a tendency to set a body in motion, to bring a moving body to rest, or to change the direction of motion. A force may also change the size or shape of a body. That is, the body can be twisted, bent, stretched, compressed, or otherwise distorted by the action of a force.

The most common unit of force is the pound (lb).* Mathematically, the magnitude of a force is proportional to the rate at which the force will cause a given mass to accelerate; that is,

$$F \cong (m)(a) \qquad (1\text{-}12)$$

where F = the force in pounds
m = the mass in pounds
a = the rate of acceleration in feet per second per second (ft/s^2)

In accordance with Equation 1-12, a force of 1 lb has been defined as the force that will cause a mass of 1 lb to be accelerated at a rate equal to the standard acceleration of gravity (32.174 ft/s^2). By definition, then, a 1-lb mass exerts a 1-lb gravitional force (weighs 1 lb) at any location where the local acceleration of gravity (g) is equal to the standard acceleration (g_c), an assumption that can be made for most routine calculations. The following equations further define the relationships among force, weight, and mass:

$$F = \frac{1}{g_c} ma \qquad (1\text{-}13)$$

$$W = \frac{1}{g_c} mg \qquad (1\text{-}14)$$

$$m = \frac{g_c}{g} W \qquad (1\text{-}15)$$

$$W = \frac{g}{g_c} m \qquad (1\text{-}16)$$

where F = the force in pounds
W = the weight (gravitational force) in pounds
a = the acceleration in feet per second per second
g = the local acceleration of gravity in feet per second per second
g_c = the universal gravitational constant (32.174 ft/s^2)

Notice again that where g is equal to g_c, the weight or gravitational force (W) in pounds is

* Although force, like velocity, is properly a vector quantity, for the purposes of this book, force will be treated as a scalar quantity.

numerically equal to the mass (m) in pounds. For this reason, mass and weight are frequently used interchangeably in the British system of measurements.

1-8. Pressure

Pressure is the force exerted per unit of area. It may be described as a measure of the intensity of a force at any given point on the contact surface. Whenever a force is evenly distributed over a given area, the pressure at any point on the contact surface is the same and can be calculated by dividing the total force exerted by the total area over which the force is applied. This relationship is expressed by the equation

$$p = \frac{F}{A} \tag{1-17}$$

where p = the pressure in units of F per units of A

F = the total force in any units of force

A = the total area in any units of area

As indicated in Equation 1-17, pressures are stated in units of force per unit of area, usually in pounds per square inch (psi) or pounds per square foot (psf). It will be shown later that pressures are stated also in terms of the height of a fluid column.

Example 1-4 A rectangular tank measuring 2 ft by 3 ft at the base is filled to a depth of 6 in. with water having a density of 62.4 lb/ft^3. Determine:

(a) the total gravitational force exerted on the base of the tank in pounds,

(b) the pressure exerted by the water on the base of the tank in pounds per square foot and pounds per square inch.

Solution

(a) Assuming that the local acceleration of gravity (g) is equal to the standard acceleration of gravity (g_c), the gravitational force (W) is numerically equal to the mass (m). The volume of the water is 3 ft^3 ($2 \times 3 \times 0.5$). Applying Equation 1-4,

$$m = (3 \text{ ft}^3)(62.4 \text{ lb/ft}^3) = 187.2 \text{ lb}$$

(b) Applying Equation 1-17,

$$p = \frac{187.2 \text{ lb}}{6 \text{ ft}^2} = 31.2 \text{ psf}$$

or

$$\frac{31.2 \text{ psf}}{144 \text{ in.}^2/\text{ft}^2} = 0.217 \text{ psi}$$

1-9. Atmospheric Pressure

The earth is surrounded by an envelope of atmosphere or air that extends upward from the surface of the earth to a distance of some 50 miles or more. Since air has mass and is subject to the action of gravity, it exerts a pressure that is known as the *atmospheric pressure*.

Imagine a column of air 1 in.^{2} in cross section extending from the surface of the earth at sea level through the upper limits of the atmosphere. Such a column of air supposedly would have a mass such that the gravitational force exerted at sea level (the base of the column) would be 14.696 lb. Since this total force is exerted on 1 in.^{2}, the pressure exerted by the atmosphere at sea level is 14.696 psi, usually rounded to 14.7 psi. This is the value given as the normal barometric (atmospheric) pressure at sea level and is sometimes referred to as a pressure of *one atmosphere*.

The atmospheric pressure does not remain constant but varies somewhat with temperature, humidity, and other conditions. The atmospheric pressure varies also with the altitude, decreasing as the altitude increases.

1-10. Barometers

Barometers are instruments used to measure the pressure of the atmosphere and are of several types. A simple barometer that measures pressure in terms of the height of a column of mercury can be constructed by filling with mercury a glass tube 30 or more inches in length and closed at one end. The mercury is held in the tube by placing the index finger over the open end of the tube while the tube is inverted in an open dish of mercury. When the finger is removed from the tube, the level of mercury in the tube will fall, leaving an almost perfect vacuum at the closed end (Fig. 1-1). The pressure exerted by the atmosphere

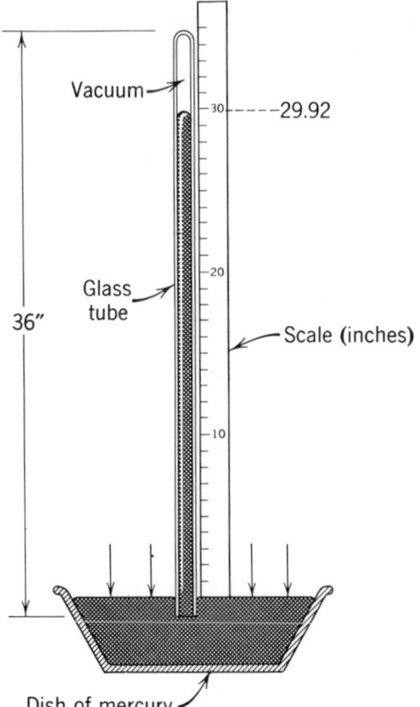

Fig. 1-1 The pressure exerted by the weight of the atmosphere on the open dish of mercury causes the mercury to stand up into the tube. The magnitude of the pressure determines the height of the mercury column.

on the mercury in the open dish will cause the mercury to stand up in the evacuated tube to a height depending on the amount of pressure exerted. The height of the mercury column in the tube is a measure of the pressure exerted by the atmosphere and is read in inches of mercury column (in. Hg). The normal pressure of the atmosphere at sea level (14.696 psi) exerted on the open dish will cause the mercury in the tube to stand to a height of 29.92 in. It follows, then, that a column of mercury 29.92 in. high is a measure of a pressure equivalent to 14.696 psi. By dividing 14.696 psi by 29.92 in. Hg, it is determined that 1 in. Hg is a measure of a pressure of approximately 0.491 psi, and the following relationships are established:

$$\text{in. Hg} = (\text{psi}/0.491) \qquad (1\text{-}18)$$

$$\text{psi} = (\text{in. Hg})(0.491) \qquad (1\text{-}19)$$

Example 1-5 What is the pressure of the atmosphere in pounds per square inch if a barometer reads 30.2 in. Hg?

Solution Applying Equation 1-19,

$$\text{psi} = (30.2 \text{ in. Hg})(0.491) = 14.83 \text{ psi}$$

Example 1-6 With reference to Fig. 1-1, how high will the mercury stand in the tube when the atmospheric pressure is 14.5 psi?

Solution Applying Equation 1-18,

$$\text{in. Hg} = \frac{14.5 \text{ psi}}{0.491} = 29.53 \text{ in. Hg}$$

1-11. Pressure Gages

Pressure gages are instruments used to measure fluid (gaseous or liquid) pressure in a closed vessel. Pressure gages commonly used in the refrigeration industry are of two principal types: manometer and bourdon tube.

1-12. Manometers

The *manometer* type of gage utilizes a column of liquid to measure the pressure, the height of the column indicating the magnitude of the pressure. The liquid used in manometers is usually either water or mercury. When mercury is used, the instrument is known as a mercury manometer or mercury gage, and when water is used, the instrument is called a water manometer or water gage. The simple barometer described previously is a manometer-type instrument.

A simple mercury manometer, illustrated in Figs. 1-2*a*, *b*, and *c*, consists of a U-shaped glass tube open at both ends and partially filled with mercury. When both legs of the U-tube are open to the atmosphere, atmospheric pressure is exerted on the mercury in

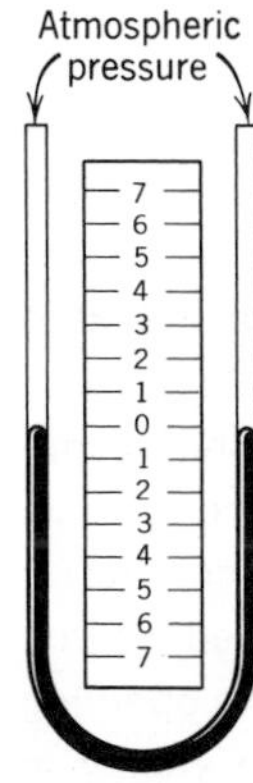

Fig. 1-2*a* Simple U-tube manometer. Since both legs of the manometer are open to the atmosphere and are at the same pressure, the level of the mercury is the same in both sides.

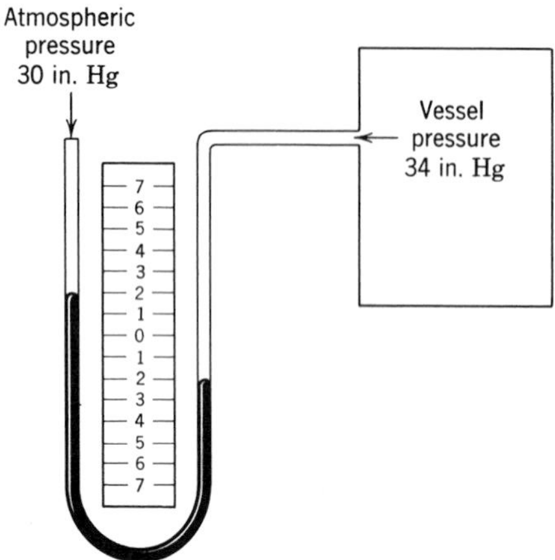

Fig. 1-2*b* Simple manometer indicates that the vessel pressure exceeds the atmospheric pressure by 4 in. Hg.

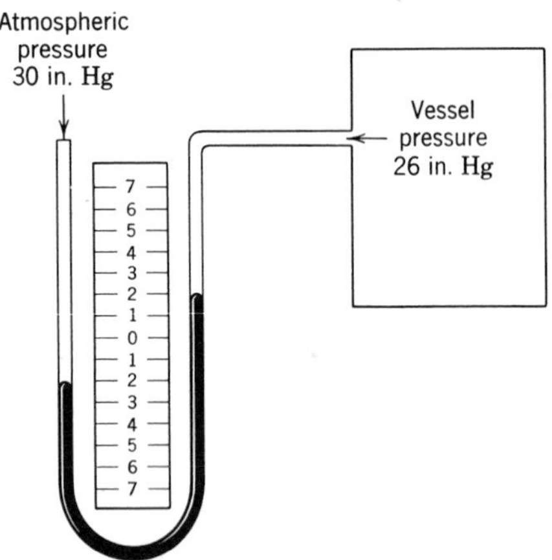

Fig. 1-2*c* Manometer indicates that the vessel pressure is 4 in. Hg less than the atmospheric pressure of 30 in. Hg.

both sides of the tube and the height of the two mercury columns is the same. The height of the two mercury columns at this position is marked as the zero point of the scale, and the scale is calibrated in inches to read the deviation of the mercury columns from the zero condition in either direction (Fig. 1-2*a*).

When in use, one side of the U-tube is connected to the vessel whose pressure is to be measured. The pressure in the vessel, acting on one leg of the tube, is opposed by the atmospheric pressure exerted on the open leg of the tube. If the pressure in the vessel is greater than that of the atmosphere, the level of the mercury on the vessel side of the U-tube is depressed while the level of the mercury on the open side of the tube is raised an equal amount (Fig. 1-2*b*). If the pressure in the vessel is less than that of the atmosphere, the level of the mercury in the open leg of the tube is depressed while the level of the mercury in the leg connected to the vessel is raised by an equal amount (Fig. 1-2*c*). In either case, the difference in the heights of the two mercury columns is a measure of the difference in pressure between the total pressure of the fluid in the vessel and the pressure of the atmosphere.

In Fig. 1-2*b*, the level of the mercury is 2 in. below the zero point in the side of the U-tube connected to the vessel and 2 in. above the zero point in the open side of the tube. This indicates that the pressure in the vessel exceeds the pressure of the atmosphere by 4 in. Hg (1.96 psi). In Fig. 1-2*c*, the level of the mercury is depressed 2 in. in the side of the tube open to the atmosphere and raised 2 in. in the side connected to the vessel, indicating that the pressure in the vessel is 4 in. Hg (1.96 psi) below (less than) atmospheric. Pressures below atmospheric are usually called "vacuum" pressures and may be read as "inches of mercury vacuum."

Manometers using water as the measuring fluid are particularly useful for measuring very small pressures. Because of the differences in the density of mercury and water, pressures that are too slight to visibly affect the height of a mercury column will produce easily detectable variations in the height of a water column. Atmospheric pressure, which will support a column of mercury only 29.92 in. high, will lift a column of water to a distance of approximately 34 ft. A pressure of 1 psi will raise a column of water 2.31 ft or 27.7 in. and a pressure of only 0.036 psi is sufficient to support a column of water 1 in. high. Hence, 1 in. of water column is equivalent to 0.036 psi.

Since all fluids are subject to thermal expansion and contraction with changes in temperature, where extreme accuracy is required, pressure readings from manometers must be corrected for temperature deviation.

TABLE 1-1 Units and Conversion Factors

To Convert	From	To	Multiply by*
Length	inch (in.)	meter	0.0254
	foot (ft)	meter	0.3048
Area (A)	ft^2	m^2	0.0929
	$in.^2$	cm^2	6.4516
Volume (V)	ft^3	m^3	0.0283
	gallon	m^3	0.003785
	gallon	ft^3	0.13368
Volume flow rate (V)	ft^3/min (cfm)	m^3/s	0.000472
	gpm	m^3/s	0.00006309
Mass (m)	lb	kg	0.45359
Mass flow rate (m)	lb/min	g/s	7.55987
Specific volume (v)	ft^3/lb	m^3/kg	0.062428
	ft^3/lb	cm^3/g	62.428
Density (ρ)	lb/ft^3	kg/m^3	16.0185
Velocity (v)	ft/min (fpm)	m/s	0.00508
	ft/sec (fps)	m/s	0.3048
	mph	m/s	0.447040
Pressure (p)	psi	Pa (N/m^2)	6894.76
	psf	Pa	47.8803
	mm Hg	Pa	133.322
	in. Hg (15.6°C)	Pa	3376,85
	in. H_2O (4°C)	Pa	249.082
	ft H_2O (4°C)	Pa	2988.98
	bar	Pa	100,000
	1 atmosphere	Pa	101,325
	in. Hg	psi	0.491
	psi	ft H_2O	2.31
Force (F)	lb	N	4.44822
Work and energy (w)	ft-lb	J	1.355818
	Btu	J	1055.06
	Btu	ft-lb	778
Power (P)	hp	W (J/s)	745.6999
	Btu/hr	W	0.293067
	ton (ref)	W	3516.8
Specific heat (c) or entropy (s)	Btu/lb R	J/g K	4.18682
Enthalpy (h)	Btu/lb	J/g	2.3244
Conductivity (k)	(Btu)(in.)/(hr)(ft^2)(R)	W/(m)(K)	0.1442285
Conductance (C) or coefficient of heat transfer (U)	Btu/(hr)(ft^2)(R)	W/(m^2)(K)	5.678286

* To convert in the opposite direction, *divide* by the factor in this column.

The relationship between the various units of pressure is shown in Table 1-1.

1-13. Bourdon Tube Gages

Because of the excessive length of tube required and other practical considerations, manometers are not usually employed to measure pressures much in excess of one atmosphere. Bourdon tube gages are widely used for measuring the higher pressures encountered in refrigeration practice. The actuating mechanism of the bourdon tube gage is

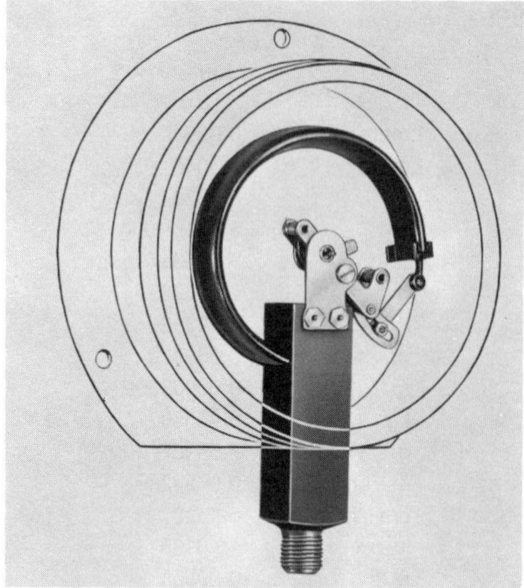

Fig. 1-3 Bourdon tube gage mechanism. (Courtesy Marsh Instrument Company.)

illustrated in Fig. 1-3. The bourdon tube, itself, is a curved, elliptical-shaped, metallic tube that tends to straighten as the fluid pressure in the tube increases and to curl tighter as the pressure decreases. Any change in the curvature of the tube is transmitted through a system of gears to the pointer. The direction and magnitude of the pointer movement depend on the direction and magnitude of the change in the curvature of the tube.

Bourdon tube gages are very rugged and will measure pressures either above or below atmospheric pressure. Those designed to measure pressures above atmospheric are known as "pressure" gages (Fig. 1-4*a*) and are generally calibrated in pounds per square inch, whereas those designed to read pressures below atmospheric are called "vacuum" gages and are usually calibrated in inches of mercury (Fig. 1-4*b*). In many cases, single gages, known as "compound" gages, are designed to measure pressures both above and below atmospheric (Fig. 1-4*c*). Such gages are calibrated to read in pounds per square inch above atmospheric and in inches of mercury below atmospheric.

1-14. Absolute and Gage Pressures

Absolute pressure is understood to be the "total" or "true" pressure of a fluid, whereas gage pressure is the pressure as indicated by a gage. It is important to understand that gages are calibrated to read zero at atmospheric pressure and that neither the manometer nor the bourdon tube gage measures the "total" or "true" pressure of the fluid in a vessel; both measure only the difference in pressure between the total pressure of the fluid in the vessel and the atmospheric pressure. When the fluid pressure is greater than the atmospheric pressure, the absolute pressure of the

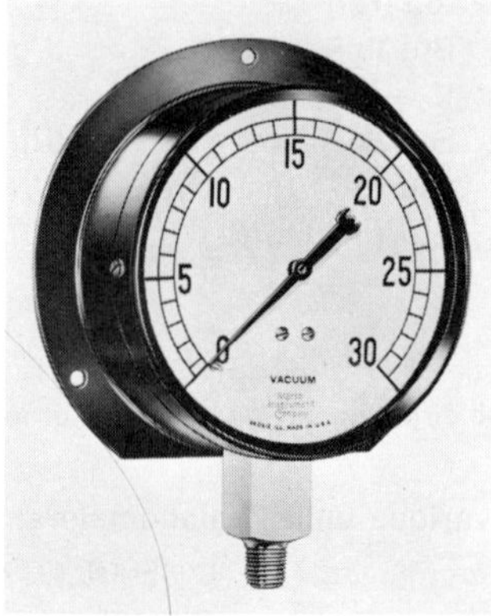

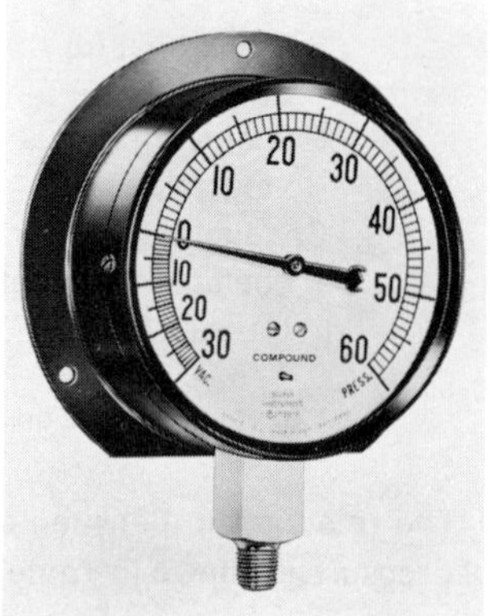

Fig. 1-4 Typical bourdon tube gages. (*a*) Pressure gage. (*b*) Vacuum gage. (*c*) Compound gage. (Courtesy Marsh Instrument Company.)

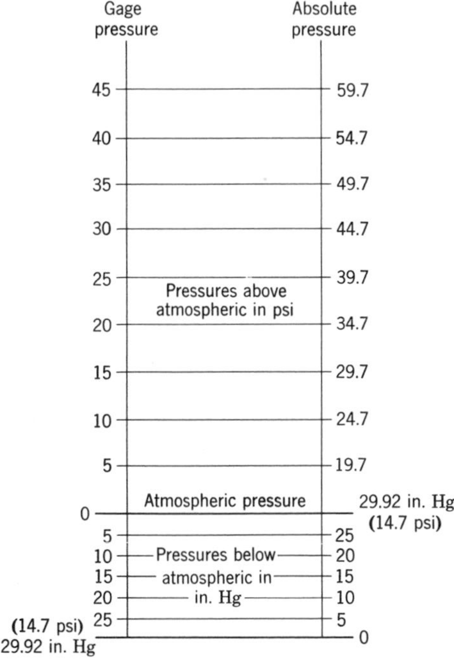

Fig. 1-5 Relationship between absolute and gage pressure, assuming standard barometric pressure.

fluid in the vessel is determined by adding the atmospheric pressure to the gage pressure, and when the fluid pressure is less than atmospheric, the absolute pressure of the fluid is found by subtracting the gage pressure from the atmospheric pressure. The relationship between absolute pressure and gage pressure is shown graphically in Fig. 1-5.

Example 1-7 A pressure gage on a refrigerant condenser reads 120 psi. What is the absolute pressure of the refrigerant in the condenser?

Solution Since the barometer reading is not given, it is assumed that the atmospheric pressure is normal at sea level (14.7 psi), and since the pressure of the refrigerant is above atmospheric, the atmospheric pressure of 14.7 psi is added to the gage pressure of 120 psi to determine that the absolute pressure of the refrigerant is 134.7 psia.

Example 1-8 A compound gage on the suction inlet of a vapor compressor reads 12 in. Hg, while a nearby barometer reads 29.6 in. Hg. Determine the absolute pressure of the vapor entering the compressor.

Solution Since the pressure of the vapor entering the compressor is less than atmospheric, the gage pressure of the vapor (12 in. Hg) is subtracted from the atmospheric pressure of 29.6 in. Hg to determine that the absolute pressure of the vapor is 17.6 in. Hg absolute or (17.6 in. Hg × 0.491 psi/in. Hg) 8.64 psia.

Example 1-9 During compression, the pressure of a vapor is increased from 10 in. Hg gage to 125 psig. Calculate the increase in pressure in pounds per square inch.

Solution From Equation 1-19, 10 in. Hg is equal to (10 in. Hg × 0.491) 4.91 psi. Since the initial pressure of the vapor is below atmospheric and the final pressure is above atmospheric, the increase in the pressure of the vapor is the sum of the two pressures and is equal to (4.91 + 125) 129.91 psi.

Notice that absolute pressures in pounds per square inch are abbreviated psia, whereas gage pressures in pounds per square inch are abbreviated psig.

1-15. Work

Mechanical *work* is done when a force acting on a body moves the body through a distance. Assuming that the action of the force is parallel to the direction of motion, the amount of work done (w) is equal to the force (F) multiplied by the distance (s) through which the force acts; that is,

$$w = (F)(s) \qquad (1\text{-}20)$$

where w = the work done in foot-pounds (ft-lb)
F = the force in pounds
s = the distance in feet

1-16. Power

Power is the time rate of doing work. The unit of mechanical power is the *horsepower* (hp). One horsepower has been defined as doing work at the rate of 33,000 ft-lb/min or 550 ft-lb/s.

The power required in horsepower can be determined from either of the following equations:

$$P = \frac{w}{(33{,}000)(t)} \tag{1-21}$$

where P = the power in horsepower
w = the work in foot-pounds
t = the time in minutes

or

$$P = \frac{w}{(550)(t)} \tag{1-22}$$

where t = the time in seconds

Example 1-10 A ventilating fan having a mass of 320 lb is hoisted 185 ft from the ground to the roof of a building. Neglecting friction and other losses, compute the work done.

Solution In this instance, the force F is the gravitational force W, which is numerically equal to the mass m. Applying Equation 1-20,

$$w = (320 \text{ lb})(185 \text{ ft}) = 59{,}200 \text{ ft-lb}$$

Example 1-11 A large crate is moved at constant velocity for a distance of 280 ft along a horizontal conveyer against a frictional force of 23.6 lb. Determine the work done.

Solution Applying Equation 1-20,

$$w = (23.6 \text{ lb})(280 \text{ ft}) = 6608 \text{ ft-lb}$$

Example 1-12 Neglecting friction and other losses, compute the power required to hoist the fan described in Example 1-10 if the time required is 5 min.

Solution From Example 1-10, the work done is 59,200 ft-lb and the time is given as 5 min. Applying Equation 1-21,

$$P = \frac{59{,}200 \text{ ft-lb}}{(33{,}000)(5 \text{ min})} = 0.36 \text{ hp.}$$

Example 1-13 Neglecting friction and other losses, determine the power required to pump 500 gallons per minute (gpm) of water up to a storage tank located a vertical distance of 130 ft above the ground if the water has a density of 8.33 lb/gal.

Solution Since the volume flow rate V is given in gallons per minute, the time t is fixed at 1 min. Applying Equation 1-21,

$$P = \frac{(500 \text{ gal})(8.33 \text{ lb/gal})(130 \text{ ft})}{(33{,}000)(1 \text{ min})}$$

$$= 16.4 \text{ hp}$$

1-17. Energy

Energy is described as the ability to do work. Energy is required for doing work, and a body is said to possess energy when it has the capacity to do work. The amount of energy required to do a given amount of work is always exactly equal to the amount of work done. Similarly, the amount of energy a body possesses is always equal to the amount of work the body can do in passing from one position or condition to another. Like work, mechanical energy is usually measured in foot-pounds.

1-18. Kinetic Energy

Energy may be possessed by a body in either one or both of two types: kinetic energy and potential energy. *Kinetic energy* is the energy a body possesses by virtue of its motion or velocity. For example, a falling body, a flowing fluid, and the moving parts of machinery all have kinetic energy because of their motion. The amount of kinetic energy (KE) a body possesses is a function of its mass m and its velocity v in the following relationship:

$$KE = \frac{(m)(v^2)}{(2)(g_c)} \tag{1-23}$$

where KE = the kinetic energy in foot-pounds
m = the mass in pounds
v = the velocity in feet per second (fps)
g_c = the universal gravitational constant (32.174 ft/s^2)

Example 1-14 An automobile having a mass of 3500 lb is moving at a rate of 50 mph. What is its kinetic energy?

Solution The given velocity of 50 mph is equal to 73.3 fps. Applying Equation 1-23,

$$KE = \frac{(3500 \text{ lb})(73.3 \text{ fps})^2}{(2)(32.174 \text{ ft/s}^2)} = 292{,}241 \text{ ft-lb}$$

1-19. Potential Energy

Potential energy is the energy a body possesses because of its position or configuration. The amount of work a body can do in passing from a given position or condition to some reference position or condition is a measure of the body's potential energy. For example, the driving head of a pile driver has potential energy of position when raised to some distance above the top of a piling. If released, the driving head can do the work of driving the piling. A compressed steel spring or a stretched rubber band possesses potential energy of configuration. Both the steel spring and the rubber band have the ability to do work because of their tendency to return to their normal condition.

The gravitational potential energy a body has by virtue of its position can be evaluated by the following equation:

$$PE = (m)(z) \tag{1-24}$$

where PE = the potential energy in foot-pounds
m = the mass in pounds,
z = the distance in feet above some datum or arbitrarily selected reference point

Example 1-15 Twelve hundred gallons of water are stored in a tank located 210 ft above the ground. Determine the potential energy of the water with relation to the ground if the density of the water is 8.33 lb/gal.

Solution Applying Equation 1-24,

$$\begin{aligned} PE &= (1200 \text{ gal})(8.33 \text{ lb/gal})(210 \text{ ft}) \\ &= 2{,}099{,}160 \text{ ft-lb} \end{aligned}$$

1-20. Total External Energy

The *total external energy* of a body is the sum of its kinetic and potential energies.

Example 1-16 Determine the total external energy per pound of water flowing at a rate of 50 fps in a raceway located 200 ft above a reference datum.

Solution Applying Equation 1-23,

$$KE = \frac{(1 \text{ lb})(50 \text{ fps})^2}{(2)(32.174 \text{ ft/s}^2)} = 38.85 \text{ ft-lb}$$

Applying Equation 1-24,

$$PE = (1 \text{ lb})(200 \text{ ft}) = 200 \text{ ft-lb}$$

$$\begin{aligned} \text{Total energy} &= KE + PE \\ &= 38.85 \text{ ft-lb} + 200 \text{ ft-lb} \\ &= 238.85 \text{ ft-lb} \end{aligned}$$

1-21. Law of Conservation of Energy

The first law of thermodynamics states, in effect, that the amount of energy in any thermodynamic system is constant. None is gained or expended except in the sense that it is converted from one form to another.

Energy is stored work. Before a body can possess energy, work must be done on the body. The work that is done on the body to give the body its motion, position, or configuration is stored in the body as energy. In all cases, the energy stored is equal to the work done. Whereas energy can be classified as being either kinetic or potential, energy may appear in any one of a number of different forms, such as mechanical energy, electrical energy, chemical energy, heat energy, and so forth and is readily converted from one form to another. Electrical energy, for instance, is converted into heat energy in an electric toaster or heater, and into mechanical energy in electric motors, solenoids, and other electrically operated mechanical devices. Mechanical energy, chemical energy, and heat energy are converted into electrical energy in the generator, battery, and thermocouple, respectively. Chemical energy is converted into heat energy in chemical reactions such as combustion and oxidation. These are only a few of the countless ways in which the transformation of energy can and does occur. There are many fundamental relationships that exist among the various forms of energy and their transformation, some of which are of particular importance in the study of refrigeration and are discussed later.

1-22. METRIC (SI) SYSTEM EQUIVALENTS

In the SI system of measurements, the basic unit of *length* is the *meter* (m). Other common units of length are the *centimeter* (cm), the *millimeter* (mm), and the *kilometer* (km). One kilometer is 1000 m, whereas 1 mm is equal to 1/1000th of a meter and 1 cm is equal to 1/100th of a meter.

Areas are usually expressed in *square meters* (m^2) or *square centimeters* (cm^2) and *volumes* in *cubic meters* (m^3) or *cubic centimeters* (cm^3). One cubic meter equals 1,000,000 cm^3. Other suggested measures of fluid volume are the liter (L) or the milliliter (mL). One liter is 1/1000th of a m^3 and 1 mL is equal to 1 cm^3.

The basic unit of *mass* is the *kilogram* (kg), which is equal to 1000 grams (g). *Density* is usually expressed in *kilograms per cubic meter* (kg/m^3) or in *grams per cubic centimeter* (g/cm^3) and *specific volume* in *cubic meters per kilogram* (m^3/kg) or *cubic centimeters per gram* (cm^3/g). An average density for water is 1000 kg/m^3 (see Table 1-2).

The unit of *force* is the *newton* (N), which is defined as the force that, when applied to a body having a mass of 1 kg, gives it an acceleration of one meter per second per second (1 m/s^2). Expressed as an equation, the relationship is

$$F = (m)(a) \tag{1-25}$$

where F = the force in newtons (N)
m = the mass in kilograms (kg)
a = the acceleration in meters per second per second (m/s^2)

Example 1-17 An unbalanced force acting on a body having a mass of 15 kg is accelerated at the rate of 10 m/s^2 in the direction of the net force. Determine the force.

Solution Applying Equation 1-25,

$$F = (15 \text{ kg})(10 \text{ m/s}^2) = 150 \text{ N}$$

The gravitational force exerted on (weight of) a given mass can be determined by substituting the local acceleration of gravity (g) for the acceleration factor (a) in Equation 1-25; that is,

$$F = (m)(g) \tag{1-26}$$

The standard value for the local acceleration of gravity at sea level (g_c) is 9.807 m/s^2 (Table 1-2). Notice that, in accordance with Equation 1-25, a 1-kg mass does not exert a 1-kg gravitational force (weigh 1 kg). Rather, assuming the local acceleration of gravity (g) to be equal to

TABLE 1-2 Frequently Used Values

	British Units	SI Units
Standard acceleration of gravity (g_c)	32.174 ft/s^2	9.80665 m/s^2
Normal atmospheric pressure	14.696 psi 29.921 in. Hg	101,325 Pa 760 mm Hg
Water		
Density	62.4 lb/ft^2 8.33 lb/gal	1000 kg/m^3
Specific heat	1 Btu/lb °F	4190 J/kg °K
Latent heat of fusion at 32°F (0°C)	144 Btu/lb	334.71 J/g
Latent heat of vaporization at 212°F (100°C)	970 Btu/lb	2255 J/g
Air		
Density (std)	0.075 lb/ft^3	1.2 kg/m^3
Specific heat (c_p)	0.24 Btu/lb °F	1 J/g °K

the standard acceleration of gravity (g_c), a 1-kg mass exerts a force of (1 kg × 9.807 m/s^2) 9.807 N.

Note Since the term "weight" has no significance when metric units are employed, the recommendation has been made that use of the term be discontinued with the move to metric units.

The unit of pressure is the *pascal* (Pa). A pressure of one pascal is exerted when a force of one newton is evenly applied over an area of one square meter. Some prefer to express pressure directly as units of force per unit area, that is, in newtons per square meter (N/m^2).

Another common unit of pressure measure is the *bar*. One bar is equal to 100,000 N/m^2 (Pa). Also, in terms of a column of fluid, pressures are measured in millimeters or centimeters of mercury (mm Hg or cm Hg) or in millimeters or centimeters of water (mm H_2O or cm H_2O). Conversion factors are listed in Table 1-1. As shown in Table 1-2, standard barometric pressure at sea level is 101,325 Pa. Notice that a pressure of 1 bar is equal approximately to a pressure of 1 standard atmosphere.

Example 1-18 A rectangular tank measuring 2 m by 3 m at the base is filled with water having a total mass of 18,000 kg. Determine:

(a) the gravitational force in newtons exerted on the base of the tank,

(b) the pressure exerted on the base of the tank in pascals.

Solution Applying Equation 1-26,

(a) The gravitational force

$$F = (18{,}000 \text{ kg})(9.807 \text{ m/s}^2)$$
$$= 176{,}526 \text{ N}$$

(b) The area of the base is 6 m^2. Applying Equation 1-17, the pressure, p,

$$= \frac{(18{,}000 \text{ kg})(9.807 \text{ m/s}^2)}{6 \text{ m}^2}$$
$$= 29{,}421 \text{ Pa or N/m}^2$$

All work and energy is measured in *joules* (J). When a force of one newton acts through a distance of one meter, one joule of work is done and one joule of energy is required to do the work. In accordance with Equation 1-20, repeated here, the amount of work done is

$$w = (F)(s) \qquad (1\text{-}20)$$

where w = the work done in joules (J)
F = the force in newtons (N)
s = the distance in meters (m)

Combining Equations 1-20 and 1-25, notice that

$$w = (m)(a)(s) \qquad (1\text{-}27)$$

and, in cases where a is equal to g,

$$w = (m)(g)(s) \qquad (1\text{-}28)$$

The unit of power is the *watt* (W). The watt is defined as doing work at the rate of one joule per second (J/s). The power required in watts can be determined from any of the following relationships:

$$P = \frac{w}{t} = \frac{(F)(s)}{t} = \frac{(m)(a)(s)}{t} = \frac{(m)(g)(s)}{t} \qquad (1\text{-}29)$$

where P = the power in watts (W)
w = the work in joules (J)
t = the time in seconds (s)

The kinetic energy of a body is given by the equation

$$KE = \frac{(m)(v^2)}{2} \qquad (1\text{-}30)$$

where KE = the kinetic energy in joules (J)
m = the mass in kilograms (kg)
v = the velocity in meters per second (m/s)

The gravitational potential energy of a body is defined by the equation

$$PE = (m)(g)(z) \qquad (1\text{-}31)$$

where PE = the potential energy in joules (J)
m = the mass in kilograms (kg)
g = the local acceleration of gravity in meters per second per second (m/s^2)
z = the distance in meters above some datum or reference point (m)

Example 1-19 A steady pressure of 1.5 bar is exerted on the top of a piston having an area of 0.02 m^2 while the piston moves through a distance of 0.08 m. Compute the work done in joules.

Solution Applying Equation 1-20, the work done, w,

$= (1.5 \text{ bar} \times 100{,}000 \text{ Pa/bar})(0.02 \text{ m}^2)(0.08 \text{ m})$

$= 240$ J

Example 1-20 A ventilating fan having a mass of 165 kg is hoisted 96 m from the ground to the roof of a building. Neglecting friction and other losses, compute the work done.

Solution In this instance, the force F is the gravitational force mg. Consequently, applying Equation 1-28,

$$w = (165 \text{ kg})(9.807 \text{ m/s}^2)(96 \text{ m}) = 155{,}343 \text{ J}$$

Example 1-21 Neglecting friction and other losses, compute the power required to hoist the fan described in Example 1-20 if the time required is 5 min.

Solution From Example 1-20, the work done is 155,343 J and the time is given as 5 min. (300 s). Applying Equation 1-29,

$$P = \frac{155,343 \text{ J}}{300 \text{ s}} = 517.81 \text{ W}$$

Example 1-22 An automobile having a mass of 1625 kg is moving at a rate of 50 km/hr. What is its kinetic energy?

Solution The given velocity of 50 km/hr is equal to 13.89 m/s. Applying Equation 1-30,

$$KE = \frac{(1625 \text{ kg})(13.89 \text{ m/s})^2}{2} = 156.757 \text{ kJ}$$

Example 1-23 Suppose that 150 m^3 of water are stored in a tank located 110 m above the ground. Determine the potential energy of the water with relation to the ground.

Solution Applying Equation 1-4,

$$m = (150 \text{ m}^3)(1000 \text{ kg/m}^3) = 150 \cdot 10^3 \text{ kg}$$

Applying Equation 1-31,

$$PE = (150 \cdot 10^3 \text{ kg})(9.807 \text{ m/s}^2)(110 \text{ m})$$
$$= 162 \cdot 10^6 \text{ J}$$

Customary Problems

1-1 A tank 2 ft by 8 ft at the base is filled to a depth of 18 in. with sodium chloride brine. If the total mass of the brine is 1782 lb, determine (a) the density, (b) the specific volume, and (c) the specific gravity of the brine solution.

1-2 A tank 3 ft by 4 ft is filled to a depth of 21 in. with a 60% solution of ethylene glycol having a specific gravity of 1.1. What is (a) the density, (b) the total mass, and (c) the specific volume of the glycol solution?

1-3 With reference to Problem 1-1, determine (a) the total force in pounds exerted by the sodium chloride on the base of the tank and (b) the resulting pressure in pounds per square foot and pounds per square inch exerted on the base of the tank.

1-4 A barometer reads 30.02 in. Hg. What is the atmospheric pressure in pounds per square inch?

1-5 A gage on a tank reads 122 psi. Assuming standard barometric pressure, what is the absolute pressure of the fluid in the tank?

1-6 A gage connected to a suction pipe reads 8 in. Hg. vacuum. If the atmospheric pressure is 30.04 in. Hg., what is the absolute pressure of the vapor in pounds per square inch?

1-7 A gage on the suction side of a refrigeration compressor reads 12 psi, while a gage on the discharge side reads 146 psi. What is the increase in pressure during the compression process?

1-8 Assume the suction gage in Problem 1-7 reads 12 in. Hg vacuum and compute the pressure increase in pounds per square inch.

1-9 A gas exerts a steady pressure of 56 psi on the top of a piston as the piston moves through a stroke of 3 in. If the diameter of the piston is 2 in., how much work is done by the gas in foot-pounds?

1-10 A pump delivers 2500 gpm of water through a vertical distance of 135 ft. Assuming a water density of 8.33 lb/gal and disregarding friction and other losses, compute (a) the work done in foot-pounds and (b) the power required in horsepower.

1-11 Two thousand gallons of water are stored at an average distance of 130 ft above the level of the water mains. What is the total potential energy of the water with respect to the mains?

1-12 Water leaves the nozzle of a hose at a velocity of 2400 fpm. Determine the kinetic energy per pound of water.

Metric Problems

1-13 A tank 0.85 m by 1.8 m at the base is filled to a depth of 68 cm with sodium chloride brine. If the total mass of the brine is 1238 kg, determine (a) the density, (b) the specific volume, and (c) the specific gravity of the brine solution.

1-14 A tank 1.5 m by 0.7 m is filled to a depth of 0.8 m with a 60% solution of ethylene glycol having a specific gravity of 1.1. What is (a) the density, (b) the total mass, and (c) the specific volume of the glycol solution?

1-15 A steady force of 630 N acts on a mass of 8 kg. What is the resulting acceleration?

1-16 Compute the force required to give a mass of 15 kg an acceleration of 20 m/s^2.

1-17 A certain mass of water exerts a gravitational force of 40 N. What is the mass of the water?

1-18 Calculate the gravitational force exerted by a mass of 3.73 kg.

1-19 With reference to Problem 1-13, determine (a) the total force in newtons exerted by the sodium chloride on the base of the tank and (b) the resulting pressure in pascals at the base of the tank.

1-20 A cylindrical tank 110 cm high has a base diameter of 30 cm. If the pressure at the base of the tank is 1240 Pa, what is the total force exerted on the base?

1-21 A barometer reads 756 mm Hg. What is the atmospheric pressure in pascals?

1-22 A mercury manometer reads 360 mm Hg vacuum. Determine the absolute pressure in pascals.

1-23 A gage on a refrigeration condenser reads 8.32 bar, while a barometer on the wall reads 770 mm Hg. What is the absolute pressure on the refrigerant in the condenser?

1-24 The gage on the suction side of a refrigeration compressor reads 150 mm Hg vacuum, while a gage on the discharge side of the compressor reads 6.8 bar. What is the increase in pressure in pascals suffered by the refrigerant vapor during the compression process?

1-25 The absolute pressure of the gas entering a compressor is 90,000 Pa, while the absolute pressure of the discharge gas is 850,000 Pa. Compute the increase in the pressure of the gas during the compression process.

1-26 Neglecting friction and other losses, determine the power in watts required to deliver 0.8 m^3/s of water through a vertical distance of 30 m.

1-27 Water flowing in a pipe has an average velocity of 2.3 m/s. What is the kinetic energy per kilogram of water?

1-28 An automobile having a mass of 8500 kg is moving at the rate of 105 km/hr. Compute the kinetic energy.

1-29 In a certain 630 km stretch of river, the river bed suffers a drop of 0.7 m/km. What is the total loss of potential energy per kilogram of water flowing down this stretch of the river?

1-30 Convert each of the following to equivalent metric units:
(a) a mass of 1595 lb
(b) a volume of 14.3 ft^3
(c) a volume of 16 gal

(d) a volume flow rate of 1500 cfm
(e) a volume flow rate of 32 gpm
(f) a density of 68 lb/ft^3
(g) a specific volume of 15 ft^3/lb
(h) a force of 2.4 lb
(i) a pressure of 100 psi
(j) a pressure of 22 in. Hg
(k) 1200 ft-lb of work
(l) 2.5 hp

1-31 Convert each of the following to equivalent British units:
(a) a mass of 65 kg
(b) a volume of 8.3 m^3
(c) a force of 23 N

1-32 A gage reads 750,000 Pa (7.5 bar). What is the gage pressure in pounds per square inch?

2

MATTER, INTERNAL ENERGY, HEAT, TEMPERATURE

2-1. Heat

Heat is a form of energy. This is evident from the fact that heat can be converted into other forms of energy and that other forms of energy can be converted into heat. Thermodynamically, heat is defined as *energy in transit from one body to another as the result of a temperature difference between the two bodies.* All other energy transfers occur as work.

2-2. Matter and Molecules

Everything that has mass or occupies space—all matter—is composed of molecules. The molecule is the smallest stable particle of matter into which a particular substance can be subdivided and still retain the identity of the original substance. For example, a grain of table salt (NaCl) may be broken down into individual molecules, and each molecule will be a molecule of salt, the original substance. However, all molecules are made up of atoms, so that it is possible to further subdivide a molecule of salt into its component atoms. A molecule of salt is made up of one atom of sodium and one atom of chlorine. Therefore, if a molecule of salt is divided into its atoms, the atoms will not be atoms of salt, the original substance, but atoms of two entirely different substances, one of sodium and one of chlorine.

There are some substances whose molecules are made up of only one kind of atom. A molecule of oxygen (O_2), for instance, is composed of two atoms of oxygen. If a molecule of oxygen is divided into its two component atoms, each atom will be an atom of oxygen, the original substance. However, the atoms of oxygen will not be stable in this condition. They will not remain as free and separate atoms of oxygen but, if permitted, will either join with atoms or molecules of another substance to form a new compound or rejoin each other to form again a molecule of oxygen.

It is assumed that the molecules that make up a substance are held together by forces of mutual attraction known as cohesion. These forces of attraction that the molecules have for each other may be likened to the attraction that exists between unlike electrical charges or between unlike magnetic poles. However, despite the mutual attraction that exists between the molecules and the resulting influence that each molecule has upon the others, the molecules are not tightly packed together. There is a certain amount of space between them, and they are relatively free to move about. The molecules are further assumed to be in a state of rapid and constant vibration or motion, the rate and extent of the vibration or movement being determined by the amount of energy they possess.

2-3. Internal Energy

It was shown in Chapter 1 that a body has external mechanical energy because of its velocity and/or its position or configuration in relation to some reference condition. A body also has *internal energy* resulting from the velocity and position or configuration of the molecules that make up the body.

The molecules of any material may possess both kinetic and potential energy. The total internal energy of a material is the sum of its internal kinetic and potential energy. This relationship is shown by the equation

$$U = K + P \tag{2-1}$$

where U = the total internal energy
K = the internal kinetic energy
P = the internal potential energy

2-4. Internal Kinetic Energy

Internal kinetic energy is the energy of molecular motion or velocity. When energy passing into a substance increases the motion or velocity of the molecules, the internal kinetic energy of the substance is increased, and this increase is reflected by an increase in the temperature of the substance. Conversely, if the internal kinetic energy of the substance is diminished by the loss of energy, the motion of the molecules will decrease, and the temperature will decrease accordingly.

It is evident from the foregoing that the temperature of a body is an index of the average velocity of the molecules that make up the body. According to the kinetic theory, if the loss of energy from a body continues until the internal kinetic energy is reduced to zero, the temperature of the body would drop to absolute zero (-459.7°F), and the molecules would become completely motionless.

2-5. States of Matter

Matter can exist in three different phases or states of aggregation: as a solid, as a liquid, or as a vapor or gas. For example, water is a liquid, but this same substance can exist as ice, which is a solid, or as steam, which is a vapor or a gas.

2-6. The Effect of Heat on the State of Aggregation

Many materials, under the proper conditions of pressure and temperature, can exist in any and all of the three physical states of matter. It will be shown presently that the amount of energy that molecules of a material have determines not only the temperature of the material but also which of the three physical states the material will assume at any particular time. In other words, the addition or removal of energy can bring about a change in the physical state of the material as well as a change in its temperature.

That energy can bring about a change in the physical state of a material is evident from the fact that many materials, such as metals, will become molten if sufficient heat is applied. The phenomenon of melting ice and boiling water is familiar to everyone. Each of these changes in the physical state is brought about by the addition of energy.

2-7. Internal Potential Energy

Internal potential energy is the energy of molecular separation or configuration. It is the energy that molecules have as a result of their position in relation to one another. The greater the degree of molecular separation, the greater is the internal potential energy.

When a material expands or changes its physical state with the addition of energy, a rearrangement of the molecules takes place which increases the distance between them. Inasmuch as the molecules are attracted to one another by forces that tend to pull them together, internal work must be done in order to further separate the molecules against these attractive forces. An amount of energy equal to the amount of internal work done must be supplied to the material. This energy is set up in the material as an increase in the internal potential energy, and it is this "stored" energy that is accounted for by the increase in the mean distance between the molecules.

It is important to understand that in this instance the energy passing into the material has no effect on molecular velocity (internal kinetic energy); only the degree of molecular separation (the internal potential energy) is affected.

2-8. The Solid Phase

A material in the solid phase has a relatively small amount of internal potential energy. The molecules of the material are rather closely bound together by each other's attractive forces and by the force of gravity. Material in the solid phase has a rather rigid molecular structure in which the position of each molecule is more or less fixed, and the motion of the molecules is limited to a vibratory type of movement which, depending on the amount of internal kinetic energy the molecules possess, may be either slow or rapid.

Because of its rigid molecular structure, a solid tends to retain both its size and its shape. A solid is practically noncompressible and will offer considerable resistance to any effort to change its shape.

2-9. The Liquid Phase

The molecules of a material in the liquid phase have more energy than those of a material in the solid phase, and they are not as closely bound together. Their greater energy allows them to overcome each other's attractive forces to some extent and to have more freedom to move about. They are free to move over and around one another in such a way that the material is said to "flow." Although a liquid is practically noncompressible and will retain its size, because of its fluid molecular structure, it will not retain its shape but will assume the shape of any containing vessel.

2-10. The Vapor or Gaseous Phase

The molecules of a material in the gaseous phase have an even greater amount of energy than those in the liquid phase. They have sufficient energy to overcome all restraining forces. They are no longer bound by each other's attractive forces, nor are they bound by the force of gravity. Consequently, they fly about at high velocities, continually colliding with each other and with the walls of the container. For this reason, a gas will retain neither its size nor its shape. It is readily compressible and completely fills any containing vessel. Furthermore, if the gas is not stored in a sealed container, it will escape from the container and be diffused into the surrounding air.

2-11. Temperature

Temperature is a property of matter. It is a measure of the level of the thermal pressure of a body. A high temperature indicates a high level of thermal pressure, and the body is said to be hot. Likewise, a low temperature indicates a low level of thermal pressure, and the body is said to be cold. It has already been shown that temperature is a function of the internal kinetic energy and, as such, is an index of the average molecular velocity.

2-12. Thermometers

The most frequently used instrument for measuring temperature is the thermometer. The operation of most thermometers depends upon the property of a liquid to expand or contract as its temperature is increased or decreased, respectively. Because of their low freezing temperatures and relatively constant coefficients of expansion, alcohol and mercury are the liquids most frequently used in thermometers. The mercury thermometer is the more accurate of the two because its coefficient of expansion is more constant through a greater temperature range than is that of alcohol. However, mercury thermometers have the disadvantage of being more expensive and more difficult to read. Alcohol is cheaper and can be colored for easy visibility.

Temperature scales in common use today are the Celsius and Fahrenheit scales. The point at which water freezes under standard barometric pressure is taken as the arbitrary zero point of the Celsius scale, and the point at which water boils under standard barometric pressure is designated as 100. The distance on the scale between these two points is divided into 100 equal units called degrees, so that the difference between the freezing and boiling points of water on the Celsius scale is 100°. Water freezes at 0°C and boils at 100°C.

Although there is some disagreement as to the actual method used by Fahrenheit in designing the first temperature scale, it was arrived at by means similar to those described for the Celsius scale. On the Fahrenheit scale, the point at which water freezes is marked as 32°, and the point at which water boils is 212°. Thus, there are 180 units between the freezing and boiling points of water. The zero or reference point on the Fahrenheit scale is placed

32 units or degrees below the freezing point of water and is assumed to represent the lowest temperature Fahrenheit could achieve with a mixture of ammonium chloride and snow.

Temperature readings on one scale can be converted to readings on the other scale by using the appropriate of the following equations:

$$°F = 9/5°C + 32 \tag{2-2}$$

$$°C = 5/9(°F - 32) \tag{2-3}$$

It should be noted that the difference between the freezing and boiling points of water on the Fahrenheit scale is 180°, whereas the difference between these two points on the Celsius scale is only 100°. Therefore, 100 Celsius degrees are equivalent to 180 Fahrenheit degrees. This establishes a relationship such that 1°C equals 9/5°F (1.8°F) and 1°F equals 5/9°C (0.555°C). This is shown graphically in Fig. 2-1. Since 0° on the Fahrenheit scale is 32°F below the freezing point of water, it is necessary to add 32°F to the Fahrenheit equivalent after converting from Celsius. Likewise, it is necessary to subtract 32°F from a Fahrenheit reading before converting to Celsius.

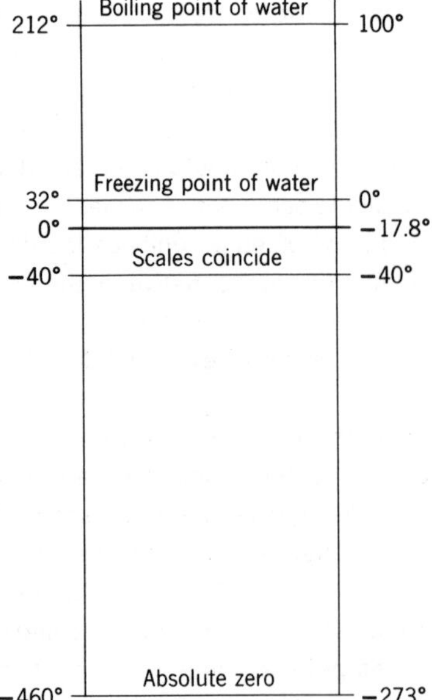

Fig. 2-1 Comparison of Fahrenheit and Centigrade temperature scales.

Example 2-1 Convert a temperature reading of 50°C to the equivalent Fahrenheit temperature.

Solution Applying Equation 2-2,

$$°F = 9/5(50°C) + 32$$
$$= 122°F$$

Example 2-2 A thermometer on the wall of a room reads 86°F. What is the room temperature in degrees Celsius?

Solution Applying Equation 2-3,

$$°C = 5/9(86°F - 32)$$
$$= 30°C$$

Example 2-3 A thermometer indicates that the temperature of a certain quantity of water is increased 45°F by the addition of heat. Compute the temperature rise in degrees Celsius.

Solution

Temperature rise in °F = 45°F

Temperature rise in °C 5/9(45°F) = 25°C

2-13. Absolute Temperature

Temperature readings taken from either the Fahrenheit or Celsius scales are based on arbitrarily selected zero points that, as has been shown, are not even the same for the two scales. When it is desired to know only the change in temperature that occurs during a process or the temperature of a substance in relation to some known reference point, such readings are adequate. However, when temperature readings are to be applied in equations dealing with certain fundamental laws, it is necessary to use temperature readings whose reference point is the true or absolute zero of temperature. Experiment has indicated that such a point, known as absolute zero, exists at approximately −460°F or −273°C (Fig. 2-1).

Temperature readings determined from absolute zero are designated as absolute temperatures and may be in either Fahrenheit or Celsius degrees. A temperature reading on the Fahrenheit scale can be converted to absolute temperature by adding 460° to the Fahrenheit reading. The resulting temperature is in degrees Rankine (°R).

Likewise, Celsius temperatures can be converted to absolute temperatures by adding 273° to the Celsius reading. The resulting temperature is stated in degrees Kelvin (°K).

In converting to and from absolute temperatures, the following relationships are used:

$$°R = °F + 460° \tag{2-4}$$

$$°F = °R - 460° \tag{2-5}$$

$$°K = °C + 273° \tag{2-6}$$

$$°C = °K - 273° \tag{2-7}$$

Example 2-4 A thermometer on the tank of an air compressor indicates that the temperature of the air in the tank is 95°F. Determine the absolute temperature in degrees Rankine.

Solution Applying Equation 2-4,

$$°R = 95°F + 460°$$
$$= 555°R$$

Example 2-5 The temperature of the vapor entering the suction of a refrigeration compressor is −20°F. Compute the temperature of the vapor in degrees Rankine.

Solution Applying Equation 2-4,

$$°R = -20°F + 460°$$
$$= 440°R$$

Example 2-6 If the temperature of a gas is 100°C, what is the temperature in degrees Kelvin?

Solution Applying Equation 2-6,

$$°K = 100°C + 273°$$
$$= 373°K$$

Example 2-7 The temperature of steam leaving a boiler is 610°R. What is the temperature of the steam on the Fahrenheit scale?

Solution Applying Equation 2-5,

$$°F = 610°R - 460°$$
$$= 150°F$$

2-14. Direction and Rate of Heat Transfer

Heat will pass from one body to another when and only when a difference in temperature exists between the two bodies. When a body is in thermal equilibrium with (i.e., at the same temperature as) its surroundings, there can be no transfer of energy as heat between the body and its surroundings.

Heat transfer is always from a region of high temperature to a region of lower temperature (from a warm body to a colder body) and never in the opposite direction. Since heat is energy and consequently is not destroyed or used up in any process, the heat energy that leaves one body must pass into and be absorbed by another body whose temperature is lower than that of the body losing the energy. The rate of heat transfer is always proportional to the difference in temperature causing the transfer.

2-15. Methods of Heat Transfer

The transfer of energy as heat occurs in three ways: (1) by conduction, (2) by convection, and (3) by radiation.

2-16. Conduction

Heat transfer by conduction occurs when energy is transmitted by direct contact between the molecules of a single body or between the molecules of two or more bodies in good thermal contact with each other. In either case, the heated molecules communicate their energy to the other molecules immediately adjacent to them. The transfer of energy from molecule to molecule by conduction is similar to that which takes place between the balls on a billiard table, wherein all or some part of the energy of motion of one ball is transmitted at the moment of impact to the other balls that are struck.

When one end of a metal rod is heated over a flame, some of the heat energy from the heated end of the rod will flow by conduction from molecule to molecule through the rod to the cooler end. As the molecules at the heated end of the rod absorb energy from the flame, their energy increases, and they move faster and through a greater distance. The increased energy of the heated molecules causes them to strike against the molecules immediately adjacent to them. At the moment of impact and because of it, the faster moving molecules transmit some of their energy to their slower moving neighbors, so that they too begin to move more rapidly. In this manner, energy passes from molecule to molecule from the heated end of the rod to the cooler end. However, in no case would it be possible for the molecules furthest from the heat source to have more energy than those at the heated end.

As heat passes through the metal rod, the air immediately surrounding the rod is also heated by conduction. The rapidly vibrating particles of the heated rod strike against the molecules of air that are in contact with the rod. The energy so imparted to the air molecules causes them to move about at a higher rate and communicate their energy to other nearby air molecules. Thus, some of the heat supplied to the metal rod is conducted to and carried away by the surrounding air.

If the heat supply to the rod is interrupted, heat will continue to be carried away from the rod by the air surrounding the rod until the temperature of the rod drops to that of the air. When this occurs, there will be no temperature differential, the system will be in equilibrium, and no heat will be transferred.

The rate of heat transfer by conduction, as previously stated, is in direct proportion to the difference in temperature between the high and low temperature parts. However, all materials do not conduct heat at the same rate. Some materials, such as metals, conduct heat very readily, whereas others, such as glass, wood, and cork, offer considerable resistance to the conduction of heat. Therefore, for any given temperature difference, the rate of heat flow by conduction through different materials of the same length and cross section will vary with the particular ability of the various materials to conduct heat. The relative capacity of a material to conduct heat is known as its conductivity. Materials that are good conductors of heat have a high conductivity, whereas materials that are poor conductors have a low conductivity and are used as heat insulators.

In general, solids are better conductors of heat than liquids, and liquids are better conductors than gases. This is accounted for by the difference in the molecular structure. Since the molecules of a gas are widely separated, the transfer of heat by conduction, that is, by direct contact between the molecules, is difficult.

2-17. Convection

Heat transfer by convection occurs when heat moves from one place to another by means of currents that are set up within some fluid medium. These currents are known as convection currents and result from the change in density that is brought about by the expansion of the heated portion of the fluid.

When any portion of a fluid is heated, it expands, and its volume per unit of mass increases. Thus the heated portion becomes lighter, rises to the top, and is immediately replaced by a cooler, heavier portion of the fluid. For example, assume that a tank of water is heated at the bottom in the center (Fig. 2-2). The heat from the flame is conducted through the metal bottom of the tank to the water inside. As the water adjacent to the heat source absorbs heat, its temperature increases, and it expands. The heated portion of the water, being lighter than the surrounding water, rises

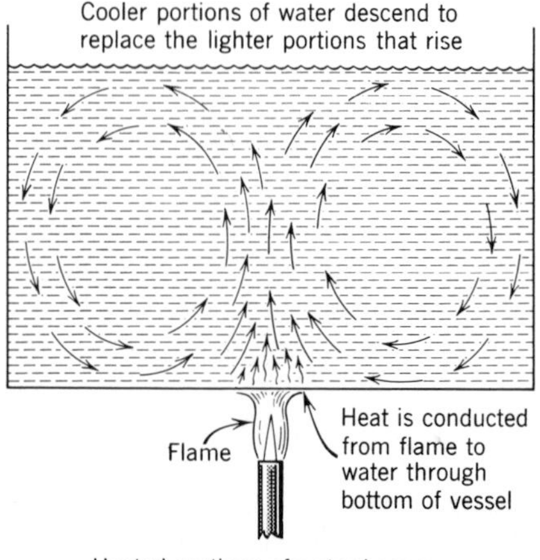

Fig. 2-2 Convection currents set up in a vessel of water when the vessel is heated at bottom center.

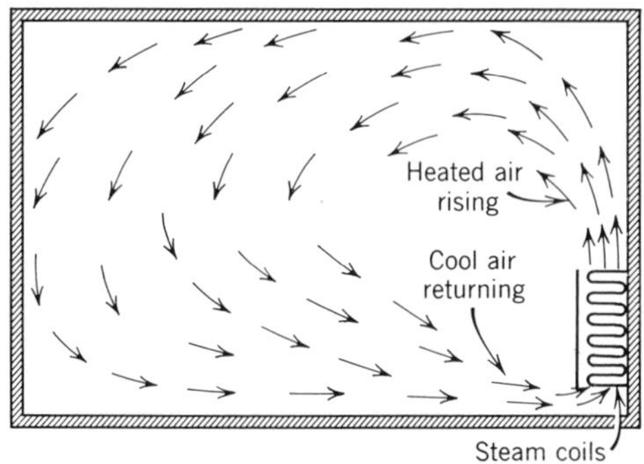

Fig. 2-3 Room heated by natural convection.

to the top and is replaced by cooler, denser water pushing in from the sides. As this new portion of water becomes heated, it too rises to the top and is replaced by cooler water from the sides. As this sequence continues, the heat is distributed throughout the entire mass of the water by means of the convection currents established within the mass.

Warm air currents, such as those that occur over stoves and other hot bodies, are familiar to everyone. Figure 2-3 illustrates how convection currents are utilized to carry heat to all parts of a heated space.

2-18. Radiation

Heat transfer by radiation occurs in the form of a wave motion similar to light waves wherein the energy is transmitted from one body to another without the need for intervening matter. Heat energy transmitted by wave motion is called radiant energy.

It is assumed that the molecules of a body are in rapid vibration and that this vibration sets up a wave motion in the space surrounding the body. Thus the internal molecular energy of the body is converted into radiant energy waves. When these energy waves are intercepted by another body of matter, they are absorbed by that body and are converted into its internal energy.

The earth receives heat from the sun by radiation. The energy of the sun's molecular vibration is imparted in the form of radiant energy waves to the space surrounding the sun. The energy waves travel across billions of miles of space and impress their energy upon the earth and upon any other material bodies that intercept their path. The radiant energy is absorbed and transformed into internal energy, so that the vibratory motion of the hot body (the sun) is reproduced in the cooler body (the earth).

All materials give off and absorb heat in the form of radiant energy. Any time the temperature of a body is greater than that of its surroundings, it will give off more heat by radiation than it absorbs. Therefore, it loses energy to its surroundings, and its internal energy decreases. If the temperature of the body is below that of its surroundings, it absorbs more radiant energy than it loses, and its internal energy increases. When no temperature difference exists, the energy exchange is in equilibrium, and the body neither gains nor loses energy.

Heat transfer through a vacuum is impossible by either conduction or convection, since these processes, by their very nature, require that matter be the transmitting media. Radiant energy, on the other hand, is not dependent upon matter as a medium of transfer and therefore can be transmitted through a vacuum. Furthermore, when radiant energy is transferred from a hot body to a cold through some intervening media such as air, the temperature of the intervening media is unaffected by the passage of the radiant energy. For example, heat is radiated from a "warm" wall to a "cold" wall through the intervening air without having any appreciable effect upon the temperature of the air. Since the molecules of the air are relatively few and widely separated, the waves of radiant energy can easily pass between them, so that only a very small part of the radiant energy is intercepted and absorbed by the molecules of the air. By far, the greater portion of the radiant energy impinges upon and is absorbed by the solid wall, whose molecular structure is much more compact and substantial.

Heat waves are very similar to light waves, differing from them only in length and frequency. Light waves are radiant energy waves of such lengths as to be visible to the human eye. Thus, light waves are visible heat waves. Whether heat waves are visible or invisible depends on the temperature of the radiating body. For example, when metal is heated to a sufficiently high temperature, it will "glow," that is, emit visible heat waves (light).

When radiant energy waves, either visible or invisible, strike a material body, they may be reflected, refracted, or absorbed by it, or they may pass through it to some other substance beyond.

The amount of radiant energy that will pass through a material depends on the degree of transparency. A highly transparent material, such as clear glass or air, will allow most of the radiant energy to pass through to the materials beyond, whereas opaque materials, such as wood, metal, and cork, cannot be penetrated by radiant energy waves, and none will pass through.

The amount of radiant energy that is either reflected or absorbed by a material depends on the nature of the material's surface, that is, its texture and its color. Materials with a light-colored, highly polished surface, such as a mirror, reflect a maximum of radiant energy, whereas materials with rough, dull, dark surfaces will absorb the maximum amount of radiant energy.

2-19. British Thermal Unit

The *quantity of heat energy* (Q) is measured in *British thermal units* (Btu). The British thermal unit is defined as the quantity of heat required to change the temperature of 1 lb of water 1°F. Added to 1 lb of water, 1 Btu will increase the temperature of the water 1°F. Similarly, removing 1 Btu from 1 lb of water will lower the temperature of the water 1°F.

Since the quantity of heat required to change the temperature of water 1°F varies slightly with the temperature range at which the change occurs, the Btu is more accurately defined as 1/180th of the quantity of heat required to raise the temperature of 1 lb of water from its freezing temperature (32°F) to its boiling temperature (212°F), a temperature change of 180°F.

Heat energy flow rates, also identified by the symbol Q, are usually given in Btu per minute or Btu per hour (Btuh). It is of interest to notice that energy flow rates are actually expressions of power, the rate of doing work.

2-20. Specific Heat

The *specific heat* (c) of any substance is the quantity of energy in Btu required to change the temperature of a 1 lb mass 1°F. For example, the specific heat of brass is 0.089 Btu/lb °F. This means that 0.089 Btu of heat energy must be supplied to 1 lb of brass to increase the temperature of the brass 1°F. Conversely, 0.089 Btu of energy must be given up by the brass to reduce the temperature of the brass 1°F. Notice that the specific heat of water, by definition of the Btu, is fixed at one Btu per pound per degree Fahrenheit (1 Btu/lb °F)

Although the specific heat of any substance varies with the temperature range, for most liquids and solids the change is small, and the specific heat can be assumed to be constant for most routine calculations. However, the specific heat of a substance changes significantly with a change in phase. For example, the specific heat of water is 1 Btu/lb °F, whereas the specific heat of ice is 0.5 Btu/lb °F.

The specific heat of any gas will take many different values depending on the conditions under which the temperature of the gas is caused or allowed to change. The specific heat of gases is discussed in Chapter 3.

2-21. Calculating Heat Quantity

From the definition of specific heat, it is evident that the quantity of heat energy supplied to, or given up by, any given mass of material to bring about a specified temperature change can be determined from the following relationship:

$$Q = (m)(c)(T_2 - T_1) \qquad (2\text{-}8)$$

where Q = the quantity of heat energy in British thermal units
m = the mass in pounds
c = the specific heat in Btu per pound per degree Fahrenheit (Btu/lb °F)
T_1 = the initial temperature in degrees Rankine or degrees Fahrenheit
T_2 = the final temperature in degrees Rankine or degrees Fahrenheit, consistent with T_1

Notice that mass flow rate can be substituted for the mass quantity (m) in Equation 2-8, in which case Q will be a heat energy flow rate rather than a heat energy quantity.

Example 2-8 Twenty pounds of water at an initial temperature of 80°F are heated until the temperature is increased to 190°F. Compute the quantity of heat energy supplied.

Solution Applying Equation 2-8,

$$Q = (20 \text{ lb})(1 \text{ Btu/lb °F})(190° - 80°)$$
$$= 2200 \text{ Btu}$$

Example 2-9 Suppose that 30 gpm of water are cooled from 60°F to 40°F. Calculate the heat energy removed in Btu/per hour.

Solution From Table 1-2, one gallon of water is approximately 8.33 lb, so that the mass flow rate is (30 gpm × 8.33 lb/gal × 60 min) 14,994 lb/hr.

Applying Equation 2-8,

$$Q = (14{,}994 \text{ lb/hr})(1 \text{ Btu/lb °F})(40° - 60°)$$
$$= -299{,}880 \text{ Btu/hr}$$

Notice in Example 2-9 that the result obtained from Equation 2-8 will be negative when T_2 is less than T_1, indicating that the heat energy is transferred from the material in question rather than to it. Except for problems that involve an energy balance, the negative sign is usually neglected.

2-22. Sensible Heat and Latent Heat

Heat energy transferred to or from a substance can bring about a change in the phase of the substance as well as a change in its temperature. As a matter of convenience, heat energy is divided into two types or categories, depending on which of these two effects the heat energy has on the substance that absorbs or gives up the energy. Heat energy that causes or accompanies a change in the temperature of a substance is called *sensible heat*, whereas heat energy that causes or accompanies a change in the phase of a substance is known as *latent heat*.

In progressing up the temperature scale, most materials will undergo two changes in the state of aggregation. First, they go from the solid phase to the liquid phase, and then, as the temperature of the liquid is further increased to a level beyond which it cannot exist as a liquid, the liquid will change into a vapor. When the change occurs between the solid and liquid phases in either direction, the latent heat involved is known as the *latent heat of fusion*. When the change occurs between the liquid and vapor phases in either direction, the latent heat involved is called the *latent heat of vaporization*.

2-23. Sensible Heat of the Solid

To obtain a better understanding of molecular (internal) energy, consider the progressive effects of energy as it is supplied to a solid substance whose initial thermodynamic condition is such that its temperature is absolute zero (0°R). Theoretically, at this condition the molecules of the substance have no energy and are completely at rest.

When energy is supplied to the solid, either as heat or as work, the molecules begin to vibrate slowly, and the temperature of the solid begins to climb. As more energy is supplied, molecular motion and the temperature of the solid continues to increase until the solid reaches its melting or fusion temperature. The total quantity of energy required to bring the temperature of the solid from the initial condition of absolute zero to the melting or fusion temperature is known as the *sensible heat of the solid* and can be calculated by applying Equation 2-8.

2-24. Latent Heat of Fusion

Upon reaching the fusion temperature, the molecules of the solid have the maximum motion possible within the limits of the rigid molecular structure of the solid phase. At this point, any additional energy supplied to the solid will cause some part of the solid to begin melting into the liquid phase, and if sufficient energy is supplied, the entire mass of the solid will pass into the liquid phase while the temperature remains constant. This applies literally only to crystalline solids. Noncrystalline solids, such as glass, have indefinite fusion temperatures. That is, the temperature will vary during the change of phase. However, for the purpose of calculating heat quantities, the temperature usually is assumed to remain constant during the phase change.

The exact temperature at which melting or fusion occurs varies with different materials and with the pressure. For instance, at normal atmospheric pressure, the fusion temperature of lead is approximately 600°F, whereas copper melts at approximately 2000°F and ice at 32°F. In general, the melting temperature decreases as the pressure increases except for non-crystalline solids, whose melting temperatures increase as the pressure increases. In most cases the effect of pressure on the fusion temperature is rather small, so that the fusion temperature remains practically constant over a relatively wide range of pressures.

The energy supplied to a substance during the change from the solid to the liquid phase is utilized by the molecules to partially overcome the attractive forces that bind them together, and they thereby separate themselves to the extent that they are able to move over and around one another so that the substance loses the rigidity of the solid phase and becomes fluid. In the liquid phase, the substance can no longer support itself independently and will assume the shape of any containing vessel.

The attraction that exists between the molecules of a solid is considerable, and a relatively large quantity of energy is required to overcome that attraction. For any substance, the latent heat of fusion is the quantity of energy required per unit mass to bring about the change of phase between the liquid and vapor phases. These values can be obtained from tabulated data and are given in Btu per pound.

It is important to emphasize that the change in phase occurs in either direction at the fusion temperature; that is, the temperature at which the solid will melt into a liquid is the same as that at which the liquid will freeze into a solid. Furthermore, the quantity of energy required to bring about the change will be the same in either case and can be determined from the following equation:

$$Q = (m)(h_{if}) \tag{2-9}$$

where Q = the quantity of latent heat in Btu
m = the mass in pounds
h_{if} = the latent heat of fusion in Btu per pound

Example 2-10 If the latent heat of fusion of water is 144 Btu/lb, determine the quantity of latent heat given up by 10 lb of water at 32°F when it freezes into ice at 32°F.

Solution Applying Equation 2-9,

$$Q_L = (10 \text{ lb})(144 \text{ Btu/lb}) = 1440 \text{ Btu}$$

Example 2-11 Suppose that 2000 Btu of energy are supplied to 25 lb of ice at 32°F. How many pounds of ice will be melted into water?

Solution Rearranging and applying Equation 2-9,

$$m = Q_L/h_{if} = \frac{2000 \text{ Btu}}{144 \text{ Btu/lb}} = 13.89 \text{ lb}$$

Example 2-12 Compute the cooling rate (energy flow rate in Btu/per hour) produced by ice melting at the rate of 150 lb/hr.

Solution Applying Equation 2-9,

$$Q_L = (150 \text{ lb/hr})(144 \text{ Btu/lb}) = 21{,}600 \text{ Btuh}$$

Notice in Example 2-12 that Q_L is an energy flow rate because m is a mass flow rate.

2-25. Sensible Heat of the Liquid

When a substance passes from the solid to the liquid phase, the resulting liquid is at the fusion temperature. If the liquid is removed from contact with the melting solid, the temperature of the liquid can be further increased by the addition of energy. The energy supplied to a liquid after the change of phase increases the internal kinetic energy (molecular velocity) of the fluid and the temperature of the fluid increases accordingly. Here again, a point is reached eventually where the molecules of the fluid have the maximum velocity possible within the limits of the liquid phase. When this point is reached the liquid will be at the maximum temperature it can have in the liquid phase at a given pressure, and any additional energy supplied to the liquid will cause some part of the liquid to begin vaporizing into the vapor phase. As more energy is supplied, the liquid will continue to vaporize and pass into the vapor phase while the temperature of the fluid remains constant, assuming that the pressure of the fluid remains constant.

The total quantity of energy supplied to a liquid to increase its temperature from the fusion temperature to the vaporizing temperature is known as the *sensible heat of the liquid* and can be calculated from Equation 2-8, which, for obvious reasons, is called the *sensible heat equation*.

2-26. Saturation Temperature

The temperature at which a fluid will change from the liquid phase to the vapor phase or, conversely, from the vapor phase to the liquid phase, is called the *saturation temperature*. A liquid at the saturation temperature is called a *saturated liquid*, and a vapor at the saturation temperature is called a *saturated vapor*. For any given pressure, the saturation temperature is the maximum temperature the liquid can have and the minimum temperature the vapor can have.

The saturation temperature is different for different fluids and, for a particular fluid, varies considerably with the pressure of the fluid. Iron, for example, vaporizes at approximately 4450°F, copper at 4250°F, and lead at 3000°F. At a pressure of one atmosphere, water vaporizes at 212°F and alcohol at 170°F. Some liquids vaporize at extremely low temperatures. A few of these are ammonia, oxygen, and helium, which vaporize at temperatures of −28°F, −296°F, and −452°F, respectively, under normal atmospheric pressure.

The effect of pressure on the saturation temperature of fluids will be discussed in detail in Chapter 4.

2-27. Latent Heat of Vaporization

Any energy supplied to a liquid after the liquid reaches the saturation temperature is utilized to increase the degree of molecular separation (increases the internal potential energy), and the fluid passes from the liquid to the vapor phase.* Since there is no increase in the internal kinetic energy (molecular velocity), the temperature of the fluid remains constant during the change of phase, and the vapor that results is at the vaporizing (saturation) temperature.

* Some of the energy supplied to the substance leaves the substance as external work and has no effect on the internal energy of the substance. When the pressure is constant, the amount of work done is proportional to the change in volume. External work is discussed in Section 2-29.

As the substance changes phase from a liquid to a vapor, the molecules of the substance acquire sufficient energy to substantially overcome all restraining forces, including the force of gravity. The amount of energy required to do the internal work necessary to overcome these restraining forces is very great. For this reason, the capacity of a substance to absorb energy while undergoing a change from the liquid to the vapor phase is enormous, many times greater even than its capacity to absorb energy in passing from the solid phase to the liquid phase.

The quantity of energy that a 1-lb mass of liquid will absorb in going from the liquid phase to the vapor phase, or give up in going from the vapor phase to the liquid phase, is called the *latent heat of vaporization*. The latent heat of vaporization is different for different fluids and, like the saturation temperature, varies significantly with the pressure for any particular fluid. In general, when the pressure increases, the saturation temperature of the fluid increases and the latent heat of vaporization decreases.

The quantity of energy required to vaporize or condense any given mass of fluid at the saturation temperature can be determined from the following relationship:

$$Q_L = (m)(h_{fg}) \qquad (2\text{-}10)$$

where Q_L = the quantity of latent heat energy in Btu

m = the mass in pounds

h_{fg} = the latent heat of vaporization in in Btu per pound

Example 2-13 Determine the quantity of energy required to vaporize 20 gallons of water at 212°F if the latent heat of vaporization of water at that temperature is 970 Btu/lb.

Solution From Table 1-2, 1 gal of water is approximately 8.33 lb, so that the total mass of the water is 166.6 lb. Applying Equation 2-10,

$$Q_L = (166.6 \text{ lb})(970 \text{ Btu/lb}) = 161{,}600 \text{ Btu}$$

Example 2-14 Suppose that 40,000 Btu of heat energy are supplied to 50 lb of water at 190°F. What part of the water in pounds will be vaporized?

Solution The temperature of the water must first be raised to the saturation temperature before any will be vaporized. Assuming that the water is in an open container under normal atmospheric pressure and applying Equation 2-8 to find the sensible heat required,

$$Q_s = (50 \text{ lb})(1 \text{ Btu/lb °F})(212\text{°F} - 190\text{°})$$
$$= 1100 \text{ Btu}$$

Subtracting the sensible heat required from the total heat supplied to find the available latent heat,

$$Q_L = Q_t - Q_s = 40{,}000 \text{ Btu} - 1100 \text{ Btu}$$
$$= 38{,}900 \text{ Btu}$$

Rearranging and applying Equation 2-10,

$$m = Q_L/h_{fg} = \frac{38{,}900 \text{ Btu}}{970 \text{ But/lb °F}} = 40.1 \text{ lb}$$

2-28. Superheat—the Sensible Heat of the Vapor

Once a liquid has been vaporized, the temperature of the resulting vapor can be further increased by the addition of heat. The heat added to a vapor after vaporization is the sensible heat of the vapor, more commonly called *superheat*. When the temperature of a vapor has been so increased above the saturation temperature, the vapor is said to be superheated and is called a *superheated vapor*. Superheated vapors are discussed at length in another chapter.

2-29. External Work

In the preceding sections, it has been shown that both the internal kinetic energy (molecular velocity) and the internal potential energy (degree of molecular separation) of a substance may be increased or decreased by the transfer of energy to or from the substance, respectively. It should be recognized, however, that not all the energy transferred to a substance is stored in the substance as an increase in the internal energy. Often, some or all the energy transferred to a substance passes through and leaves the substance as mechanical work. It should also be recognized that energy can be transferred to a substance in the form of mechanical energy (work) as well as in the form of heat energy. For example, the head of a nail struck by a hammer will become warm as part of the mechanical energy of the hammer blow is converted to the internal energy of the nail head. As the molecules of the metal that make up the nail head are jarred and agitated by the blow of the hammer, their motion or velocity is increased, and the temperature of the nail head increases. If a wire is bent rapidly back and forth, the bent portion of the wire becomes hot because of the agitation of the molecules. Also, everyone is familiar with the increase in temperature that is brought about by the friction of two surfaces rubbing together.

Often the external energy of a body is converted to internal energy and vice versa. For example, a bullet moving toward a target has kinetic energy because of its mass and velocity. At the time of impact with the target, the bullet loses its velocity and a part of its kinetic energy is imparted to the molecules of both the bullet and the target, so that the internal energy of each is increased.

2-30. Mechanical Energy Equivalent

It is frequently necessary to express work or mechanical energy in heat energy units. Experiments have established that 778 ft-lb of work or mechanical energy are equivalent to 1 Btu of heat energy. This value is known as the *mechanical energy equivalent* and the letter *J* traditionally has been used to represent this quantity in equations. The relationship of work or mechanical energy to heat energy is shown by the following equation:

$$Q = \frac{W}{J} \tag{2-11}$$

Example 2-15 Determine the heat energy equivalent of 36,000 ft-lb of mechanical energy.

Solution Applying Equation 2-11,

$$Q = \frac{36{,}000 \text{ ft-lb}}{778 \text{ ft-lb/Btu}} = 46.27 \text{ Btu}$$

Example 2-16 Express 15 Btu of heat energy as work in mechanical energy units.

Solution Applying Equation 2-11,

$$Q = (15 \text{ Btu})(778 \text{ ft-lb/Btu}) = 11{,}670 \text{ ft-lb}$$

2-31. METRIC SYSTEM EQUIVALENTS

In the metric (SI) system of measurements, heat energy, like all work and energy, is measured in joules. The joule has been previously defined as the quantity of energy or work required to cause a mass of 1 kg to accelerate at the rate of 1 m/s^2.

Equation 2-8 may be used to calculate the quantity of sensible heat transferred, in which case Q_s will be in joules when the mass (m) is in kilograms, the temperature is in degrees Celsius or degrees Kelvin, and the specific heat (c) is in joules per kilogram per degree Kelvin (J/kg °K). Where m is a mass flow rate in kilograms per second, Q will be an energy flow rate in joules per second, or watts. Again, notice that an energy flow rate is an expression of power.

Latent heat quantities in joules can be determined from Equations 2-9 and 2-10 when the latent heat of fusion (h_{if}) and the latent heat of vaporization (h_{fg}) are given in joules per kilogram. From Table 1-2, the latent heat of fusion (h_{if}) for water is 334.71 J/g, and the latent heat of vaporization (h_{fg}) under normal atmospheric pressure (saturation temperature of 100°C) is 2255 J/g.

Example 2-17 Twenty kilograms of water at an initial temperature of 25°C are heated until the temperature is increased to 80°C. Compute the quantity of heat energy supplied.

Solution From Table 1-2, the specific heat of water is 4190 J/kg °K. Applying Equation 2-8, the sensible heat supplied,

$$Q_s = (20\text{ kg})(4190\text{ J/kg °K})(80° - 25°)$$
$$= 4609\text{ kJ}$$

Example 2-18 One-tenth cubic meter (0.1 m^3) of water is cooled from 39°C to 2°C. Determine the quantity of heat energy rejected by the water.

Solution Assuming a water density of 1000 kg/m^3, the mass of water is (0.1 m^3 × 1000 kg/m^3) 100 kg. Applying Equation 2-8, the sensible heat rejected is

$$Q_s = (100\text{ kg})(4190\text{ J/kg °K})(39° - 2°)$$
$$= 15{,}500\text{ kJ}$$

Example 2-19 Suppose that 30 kg/s of water are cooled from 35°C to 10°C. Compute the required energy flow rate in joules per second (watts).

Solution Applying Equation 2-8, the energy flow rate,

$$Q_s = (30\text{ kg/s})(4190\text{ J/kg °K})(35° - 10°\text{C})$$
$$= 3142.5\text{ kJ/s or kW}$$

Example 2-20 Compute the cooling rate (energy flow rate in joules per second or watts) produced by ice melting at the rate of 150 kg/hr (2.5 kg/s).

Solution From Table 1-2, the latent heat of fusion of ice is 335 kJ/kg. Applying Equation 2-9, the latent cooling rate,

$$Q_L = (2.5\text{ kg/s})(335\text{ kJ/kg})$$
$$= 837.5\text{ kJ/s or kW}$$

Customary Problems

2-1 Convert the following Celsius temperature readings to equivalent temperatures on the Fahrenheit scale: (a) 25°C, (b) −40°C, and (c) 130°C.

2-2 Convert the following Fahrenheit temperature readings to equivalent temperatures on the Celsius scale: (a) −10°F, (b) 80°F, and (c) −215°F.

2-3 The temperature of air passing across a heating coil is increased from 20°F to 120°F. What is the temperature rise in degrees Celsius?

2-4 Convert the following Fahrenheit temperatures to equivalent temperatures in degrees Rankine: (a) 0°F, (b) −150°F, and (c) 32°F.

2-5 Convert the following temperatures in degrees Kelvin to equivalent temperatures in degrees Celsius: (a) 135°K, and (b) 310°K.

2-6 Thirty gallons of water are heated from 75°F to 180°F. Determine the quantity of heat in Btu required.

2-7 In a certain industrial process, 5000 gal of water are cooled from 90°F to 55°F each hour. Compute the required cooling rate in Btu per hour.

2-8 Calculate the quantity of heat energy in Btu that must be removed from 60 gal of water at 45°F and freeze it into ice at 32°F.

2-9 If 12,120 Btu are supplied to 3 gal of water at 198°F in an open container, what mass of the water in pounds will be vaporized?

2-10 A gas expanding in a cylinder does 25,000 ft-lb of work on a piston. Determine the quantity of heat energy in Btu required to do the work.

Metric Problems

2-11 The temperature of 150 kg of water is raised from 15°C to 85°C by the addition of heat. How much heat energy in joules is supplied?

2-12 Two cubic meters per second of water are cooled from 30°C to 2°C. Compute the rate of heat transfer in joules per second (watts).

2-13 What is the cooling rate in joules per second produced by the melting of ice at the rate of 12.36 kg/s?

2-14 Determine the quantity of heat in joules required to vaporize 50 kg of water at a saturation temperature of 100°C.

2-15 Twenty kilograms of water at 65°C are supplied with 5000 kJ of heat energy. What mass of the water will be vaporized?

2-16 Convert 10,000 J (10 kJ) of energy to heat energy in Btu.

2-17 Convert 500 J (0.5 kJ) to mechanical energy in foot-pounds.

2-18 Convert 150 Btu of heat energy into energy units in joules.

2-19 A gas expanding in a cylinder does 10,000 ft-lb of work on a piston. What is the energy required in joules?

2-20 Heat energy is transferred to a fluid at the rate of 15 W. What is the energy flow rate in Btu per hour?

2-21 A pump requires 2.85 hp. What is the power required in watts?

3

IDEAL GAS PROCESSES

3-1. The Effect of Heat on Volume

When either the velocity of the molecules or the degree of molecular separation is increased by the addition of energy, the mean distance between the molecules is increased, and the material expands so that a unit mass of the substance occupies a greater volume. This effect is in accordance with the theory of increased or decreased molecular activity described earlier. Consequently, when energy is supplied to or given up by an unconfined substance in any of the three physical states, the substance will expand or contract, respectively. That is, its volume will increase or decrease with the gain or loss of energy.

One of the few exceptions to this rule is water. If water is cooled, its volume will decrease normally until the temperature of the water drops to approximately 39.2°F (4°C). At this point, water is at its maximum density and, if further cooled, will begin to expand again and will continue to expand until the temperature drops to 32°F (0°C), at which point solidification will begin and will be accompanied by still further expansion. One cubic foot of water will freeze into 1.085 ft^3 of ice. This accounts for the tremendous expansive force created during the solidification process, which is sufficient to burst steel pipes or other restraining vessels.

Although the expansion of water as it solidifies appears to contradict the statement that the molecules of a substance are closer together in the solid phase than in the liquid phase, such is not the case. This apparent contradiction can be explained by the fact that ice is a crystalline solid, and although the molecules of the substance are actually closer together in the solid state, they are grouped together to form crystals. It is the relatively large spaces between the crystals of the solid, rather than any increase in the mean distance between the molecules, that accounts for the unusual increase in volume during solidification.

3-2. Expansion of Solids and Liquids

When a solid or a liquid is heated so that its temperature is increased, it will expand a given amount for each degree of temperature rise. As stated earlier, many temperature-measuring devices are based upon this principle. The amount of expansion that a material experiences with each degree of temperature rise is known as its coefficient of expansion. The coefficient of expansion is different for every material and will vary for any particular material, depending on the temperature range in which the change occurs.

Since solids and liquids are not readily compressible, if a solid or a liquid is restrained or confined so that its volume is not allowed to change normally with a change in temperature, tremendous pressures are created within the material itself and upon the restraining bodies, which are likely to cause buckling or rupturing of the material or the restraining bodies or both. To provide for the normal expansion and contraction occurring with temperature changes, expansion joints are built into highways, bridges, pipelines, and so forth. Likewise, liquid containers are never completely filled. Space must be allowed for the normal expansion. Otherwise the large expansive forces created by an increase in temperature will cause the containing vessel to rupture, sometimes with explosive force.

3-3. Pressure-Temperature-Volume Relationships of Gases

Because of its loose molecular structure, the change in the volume of a gas as it is heated or cooled is much greater than that which occurs in the case of a solid or liquid. In the following sections, it will be shown that a gas may change its condition in a number of different ways and that there are certain fundamental laws that govern the relationships among the pressure, temperature, and volume of the gas during these changes.

The relationships among these several properties are more readily examined when the gas is allowed to undergo a series of independent processes during which the gas passes from some initial state to some final state in such a way that only two of the three properties are allowed to change during any one process while the third property is held constant or unchanged.

In applying these gas laws, it is important to keep in mind that a gas always completely fills its containing vessel and that values for the several gas properties must be determined from the natural zero points.

3-4. Constant Pressure Process

If energy is supplied to a gas under such conditions that the pressure of the gas is kept constant, the volume of the gas will increase in direct proportion to the change in the absolute temperature of the gas. This is a statement of Charles' law for a constant pressure process. Written as an equation, with p constant,

$$T_1 V_2 = T_2 V_1 \tag{3-1}$$

where T_1 = the initial absolute temperature of the gas
T_2 = the final absolute temperature of the gas
V_1 = the initial volume of the gas
V_2 = the final volume of the gas

In order to better visualize a constant pressure change in condition, assume that a gas is confined in a cylinder equipped with a perfectly fitting, frictionless piston (Fig. 3-1*a*). The pressure of the gas is that which is exerted on the gas by the mass of the piston and by the air on top of the piston. Since the piston is free to move up and down in the cylinder, the gas is allowed to expand or contract (to change its volume) in such a way that the pressure of the gas remains constant. As the gas is heated, its temperature and volume increase, and the piston moves upward in the cylinder. As the gas is cooled, its temperature and volume decrease, and the piston moves downward in the cylinder. In either case, the pressure of the gas remains constant or unchanged.

Example 3-1 A gas having an initial volume of 5 ft^3 at a temperature of 520°R is heated under constant pressure until its volume increases to 10 ft^3. Determine the final temperature of the gas in degrees Rankine.

Solution Rearranging and applying Equation 3-1,

$$T_2 = \frac{T_1 V_2}{V_1} = \frac{(520°\text{R})(10\ \text{ft}^3)}{5\ \text{ft}^3} = 1040°\text{R}$$

Example 3-2 A gas having an initial temperature of 80°F is cooled at a constant pressure until its temperature is 40°F. If the initial volume of the gas is 8 ft^3, what is its final volume?

Solution Since the temperatures are given in degrees Fahrenheit, they must be converted to degrees Rankine. Rearranging and applying Equation 3-1,

$$V_2 = \frac{T_2 V_1}{T_1} = \frac{(40°\text{F} + 460°)(8\ \text{ft}^3)}{80°\text{F} + 460°} = 7.41\ \text{ft}^3$$

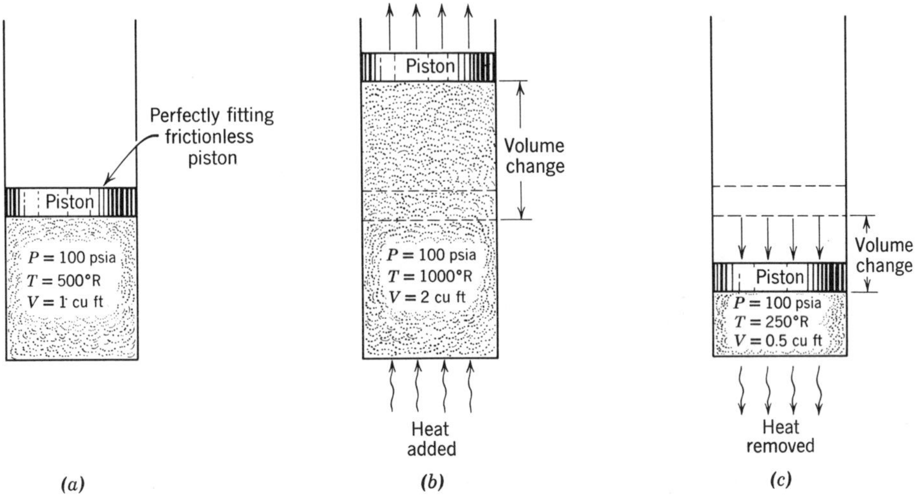

Fig. 3-1 Constant pressure process. (*a*) Gas confined in a cylinder with a perfectly fitting, frictionless piston. (*b*) As gas is heated, both the temperature and the volume of the gas increase. The increase in volume is exactly proportional to the increase in absolute temperature. (*c*) As gas is cooled, both the temperature and the volume of the gas decrease. The decrease in volume is exactly proportional to the decrease in absolute temperature.

3-5. Pressure-Volume Relationship at a Constant Temperature

When the volume of a gas is increased or decreased under such conditions that the temperature of the gas does not change, the absolute pressure will vary inversely with the volume. Thus, when a gas is compressed (volume decreased) while its temperature remains unchanged, its absolute pressure will increase in proportion to the decrease in volume. Similarly, when a gas is expanded at a constant temperature, its absolute pressure will decrease in proportion to the increase in volume. This is a statement of Boyle's law for a constant temperature process and is illustrated in Figs. 3-2*a*, *b*, and *c*.

It has been previously stated that the molecules of a gas fly about at random and at high velocities and that they frequently collide with one another and with the walls of the container. The pressure exerted by the gas is one manifestation of these molecular collisions. Billions and billions of gas molecules, traveling at high velocities, strike the walls of the container during each fraction of a second. It is this incessant molecular bombardment that produces the pressure that a gas exerts upon the walls of its container. The magnitude of the pressure exerted depends on the force and frequency of the molecular impacts on a given area. The greater the force and frequency of the impacts, the greater is the pressure. The number of molecules confined in a given space and their velocity will, of course, determine the force and the frequency of the impacts. That is, the greater the number of molecules (the greater the quantity of gas) and the higher the velocity of the molecules (the higher the temperature of the gas), the greater is the pressure. The force with which the molecules strike the container walls depends only on the velocity of the molecules. The higher the velocity, the greater is the force of impact. The greater the number of molecules in a given space and the higher the velocity, the more often the molecules will strike the walls.

When a gas is compressed at a constant temperature, the velocity of the molecules remains unchanged. The increase in pressure that occurs is accounted for by the fact that the volume of the gas is diminished and that a given number of gas molecules are confined in a smaller space, so that the frequency of impact is greater. The reverse of this holds true, of course, when the gas is expanded at a constant temperature.

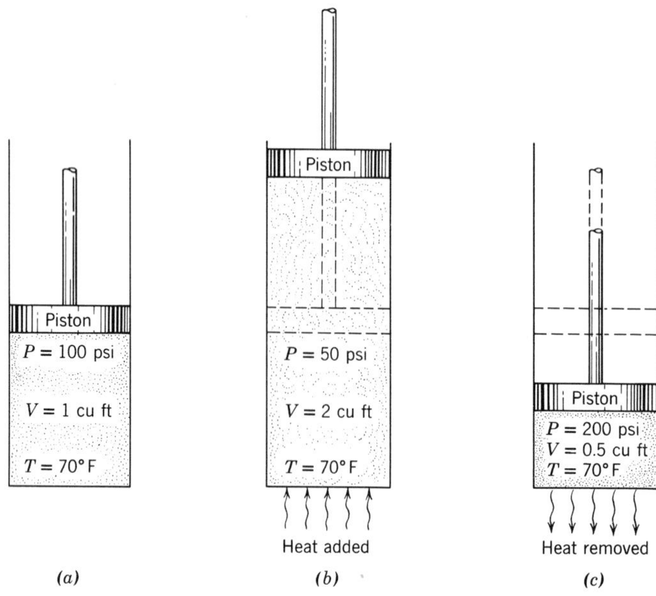

Fig. 3-2 Constant temperature process. (*a*) Initial condition. (*b*) Constant temperature expansion—volume change is inversely proportional to the change in absolute pressure. Heat must be added during expansion to keep temperature constant. (*c*) Constant temperature compression—volume change is inversely proportional to the change in absolute pressure. Heat must be removed during compression to keep temperature constant.

Any thermodynamic process that occurs in such a way that the temperature of the working substance does not change during the process is called an *isothermal* (constant temperature) *process*.

Boyle's law for a constant temperature process is represented by the following equation. With the temperature kept constant,

$$p_1V_1 = p_2V_2 \tag{3-2}$$

where p_1 = the initial absolute pressure
p_2 = the final absolute pressure
V_1 = the initial volume
V_2 = the final volume

Example 3-3 Three pounds of air are expanded at a constant temperature from an initial volume of 1.9 ft^3 to a final volume of 3.8 ft^3. If the initial pressure of the air is 36 psia, what is the final absolute pressure of the air in pounds per square inch?

Solution Rearranging and applying Equation 3-2,

$$p_2 = \frac{p_1V_1}{V_2} = \frac{(36 \text{ psia})(1.9 \text{ ft}^3)}{3.8 \text{ ft}^3} = 18 \text{ psia}$$

Example 3-4 Four cubic feet of gas are allowed to expand at a constant temperature from an initial pressure of 120 psia to a final pressure of 80 psia. Determine the final volume of the gas.

Solution Rearranging and applying Equation 3-2,

$$V_2 = \frac{p_1V_1}{p_2} = \frac{(120 \text{ psia})(4 \text{ ft}^3)}{80 \text{ psia}}$$
$$= 6 \text{ ft}^3$$

Example 3-5 A given mass of gas whose initial volume is 2 ft^3 is compressed at a constant temperature to a final volume of 0.4 ft^3. If the initial pressure of the gas is 30 psig, what is the final gage pressure of the gas in pounds per square inch?

Solution Adding the normal atmospheric pressure of 14.7 psi to the initial gage pressure of 30 psi gives an initial absolute pressure of 44.7 psi. Rearranging and applying Equation 3-2,

$$p_2 = \frac{p_1 V_1}{V_2} = \frac{(44.7 \text{ psia})(2 \text{ ft}^3)}{0.4 \text{ ft}^3}$$

$$= 223.5 \text{ psia}$$

Subtracting the atmospheric pressure,

$$223.5 \text{ psia} - 14.7 \text{ psi} = 208.8 \text{ psig}$$

3-6. Pressure-Temperature Relationship at a Constant Volume

Assume that a gas is confined in a closed cylinder so that its volume cannot change as it is heated or cooled (Fig. 3-3*a*). When the temperature of the gas is increased by the addition of heat, the absolute pressure will increase in direct proportion to the increase in absolute temperature (Fig. 3-3*b*). If the gas is cooled, the absolute pressure of the gas will decrease in direct proportion to the decrease in absolute temperature (Fig. 3-3*c*).

Whenever the temperature (velocity of the molecules) of a gas is increased while the volume of the gas (space in which the molecules are confined) remains the same, the magnitude of the pressure (the force and frequency of molecular impacts on the cylinder walls) increases. Likewise, when a gas is cooled at a constant volume, the force and frequency of molecular impingement on the walls of the container diminish, and the pressure of the gas will be reduced. The reduction in the force and the frequency of molecular impacts is accounted for by the reduction in molecular velocity.

The following equation is a statement of Charles' law for a constant volume process. With the volume constant,

$$T_1 p_2 = T_2 p_1 \qquad (3\text{-}3)$$

where T_1 = the initial absolute temperature
T_2 = the final absolute temperature
p_1 = the initial absolute pressure
p_2 = the final absolute pressure

Example 3-6 A certain mass of air confined in a tank has an initial temperature of 80°F and an initial pressure of 60 psia. If the gas is heated until the final pressure is 150 psia, what is the final temperature of the gas in degrees Fahrenheit?

Solution Rearranging and applying Equation 3-3,

$$T_2 = \frac{T_1 p_2}{p_1} = \frac{(80°\text{F} + 460°)(150 \text{ psia})}{60 \text{ psia}}$$

$$= 1350°\text{R} \quad \text{or} \quad 1350°\text{R} - 460° = 890°\text{F}$$

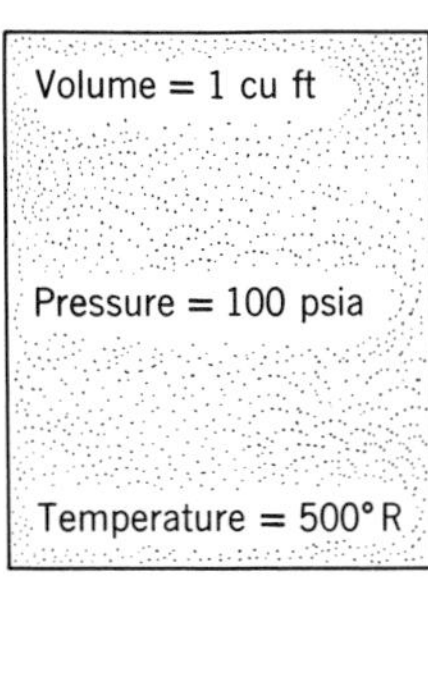

(*a*)

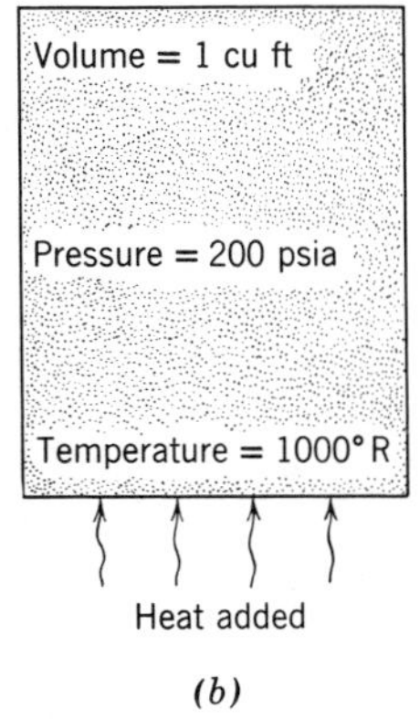

(*b*)

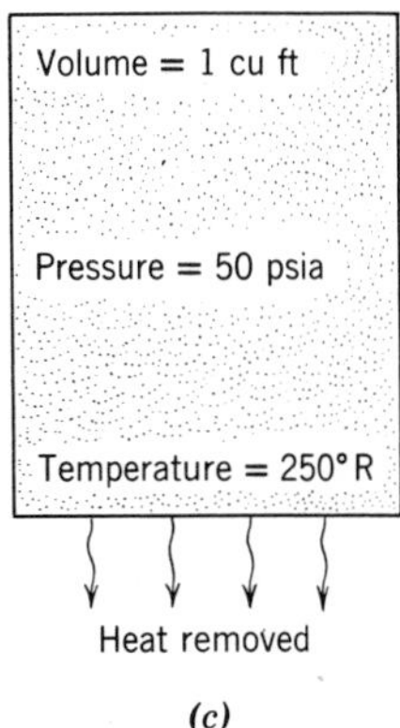

(*c*)

Fig. 3-3 Constant volume process. (*a*) Initial condition. (*b*) The absolute pressure increases in direct proportion to the increase in absolute temperature. (*c*) The absolute pressure decreases in direct proportion to the decrease in absolute temperature.

3-7. The General Gas Law

Combining Charles' and Boyle's laws produces the following equation:

$$\frac{p_1V_1}{T_1} = \frac{p_2V_2}{T_2} \qquad (3\text{-}4)$$

Equation 3-4 states that for any given mass of a gas, the product of the absolute pressure and the volume divided by the absolute temperature is always a constant; that is,

$$\frac{pV}{T} = \text{a constant} \qquad (3\text{-}5)$$

The value of the constant is different, of course, for different gases and, for any one particular gas, varies with the mass of the gas. However, when, for any one particular gas, a mass of 1 lb is used, the volume V becomes the specific volume v, and Equation 3-5 may be written

$$\frac{pv}{T} = R \qquad (3\text{-}6)$$

where R = the gas constant of the particular gas

The gas constant R is different for each gas and, for most common gases, can be found in tables. A few of these are listed in Table 3-1.

Multiplying both sides of Equation 3-6 by the mass m produces

$$pmv = mRT \qquad (3\text{-}7)$$

but since

$$mv = V \qquad (1\text{-}2)$$

then

$$pV = mRT \qquad (3\text{-}8)$$

where p = the absolute pressure in pounds per square foot
V = the volume in cubic feet
m = the mass in pounds
R = the gas constant (foot-pounds per pound per degree Rankine)
T = the absolute temperature in degrees Rankine

Equation 3-8 is known as the general gas law and is very useful in the solution of many problems involving gases. Since the value of R for most gases can be found in tables, if any three of the remaining four properties are known, the fourth can be found from Equation 3-8.

Example 3-7 The tank of an air compressor has a volume of 2 ft^2 and is filled with air at a temperature of 40°F. If a gage on the tank reads 150 psi, what is the mass of the air in the tank?

Solution From Table 3-1, R for air is 53.3 ft-lb/lb °R. Rearranging and applying Equation 3-8,

$$m = \frac{(150 \text{ psig} + 14.7 \text{ psi})(144)(2 \text{ ft}^3)}{(53.3)(40°\text{F} + 460°)}$$

$$= 1.78 \text{ lb}$$

Example 3-8 Two pounds of air have a volume of 5 ft^3. If the pressure of the air is 125 psia,

TABLE 3-1 Properties of Gases

	British Units*				SI Units*		
Gas	c_p	c_v	R	k	c_p	c_v	R
Air	0.2375	0.169	53.3	1.406	1.0000	0.711	287
Ammonia	0.508	0.399	90.5	1.273	2.1269	1.6705	487
Carbon dioxide	0.207	0.162	35.1	1.28	0.8709	0.6783	189
Carbon monoxide	0.243	0.173	55.1	1.403	1.0174	0.7243	297
Hydrogen	3.41	2.42	765.9	1.41	14.277	10.132	4124
Nitrogen	0.244	0.173	55.1	1.41	1.0216	0.7243	297
Oxygen	0.218	0.156	48.3	1.40	0.9127	0.6531	260
Sulfur dioxide	0.154	0.123	24.1	1.26	0.6448	0.5150	130

* c_p and c_v are in Btu/lb °F and J/g °K. R is in ft-lb/lb °R and J/kg °K

what is the temperature of the air in degrees Rankine?

Solution From Table 3-1, R for air is 53.3. Rearranging and applying Equation 3-8,

$$T = \frac{(125 \text{ psia})(144)(5 \text{ ft}^3)}{(2 \text{ lb})(53.3)}$$

$$= 844.3°\text{R}$$

3-8. External Work

Whenever a material undergoes a change in volume, work is done. If the volume of the material increases, work is done by the material. If the volume of the material decreases, work is done on the material. For example, consider a certain weight of gas confined in a cylinder equipped with a movable piston (Fig. 3-1*a*). As the gas is heated, its temperature increases and it expands, moving the piston upward in the cylinder against the pressure of the atmosphere. Work is done in that the weight of the piston is moved through a distance (Fig. 3-1*b*).* The agency doing the work is the expanding gas.

In order to do work, energy is required. In Fig. 3-1*b*, the energy required to do the work is supplied to the gas as the gas is heated by an external source. It is possible, however, for a gas to do external work without the addition of energy from an external source. In such cases, the gas does the work at the expense of its own energy. That is, as the gas expands and does work, its internal kinetic energy (and temperature) decreases by an amount equal to the amount of energy required to do the work.

When a gas is compressed (its volume decreased), a certain amount of work must be done on the gas in order to compress it, and an amount of energy equal to the amount of work done will be imparted to the molecules of the gas during the compression. That is, the mechanical energy of the piston motion will be transformed into the internal kinetic energy of the gas (molecular motion), and unless the gas is cooled during the compression, the temperature of the gas will increase in proportion to the amount of work done. The increase in the temperature of a gas as the gas is compressed is a common phenomenon and may be noted by feeling the valve stem of a tire being filled with a hand pump or by touching the head of an air compressor.

* Some work is done, also, in overcoming friction and in overcoming the pressure of the atmosphere.

3-9. The General Energy Equation

The law of conservation of energy clearly indicates that the energy transferred to a body must be accounted for in its entirety. It has been shown that some part (or all) of the energy taken in by a material may leave the material as work and that only the portion of the transferred energy that is not utilized to do external work remains in the body as "stored thermal energy." All the energy transferred to a body must be accounted for in one or some combination of the following three ways: (1) as an increase in the internal kinetic energy, (2) as an increase in the internal potential energy, or (3) as external work done. The general energy equation is a mathematical statement of this concept and may be written as

$$\Delta Q = \Delta K + \Delta P + \Delta W \qquad (3\text{-}9)$$

where ΔQ = the heat energy transferred to the body

ΔK = the part of the transferred energy that increases the internal kinetic energy

ΔP = the part of the transferred energy that increases the internal potential energy

ΔW = the part of the transferred energy that is utilized to do external work

The Greek letter Δ (delta) is used to identify a change of condition. For instance, where K represents the internal kinetic energy, ΔK represents the change in the internal kinetic energy.

3-10. External Work of a Solid or Liquid

When heat added to a material in either the solid or liquid state increases the temperature of the material, the material expands somewhat, and a small amount of work is done.

However, the increase in volume and the external work done is slight, and the portion of the transferred energy that is utilized to do external work or to increase the internal potential energy is negligible. For all practical purposes, it can be assumed that all the energy added to a solid or a liquid during a temperature change increases the internal kinetic energy. None leaves the material as work, and none is set up as an increase in the internal potential energy. In this instance, both ΔP and ΔW of Equation 3-9 are equal to zero, and therefore ΔQ is equal to ΔK.

When a solid melts into the liquid phase, the change in volume is again so small that the external work done may be neglected. Furthermore, since the temperature also remains constant during the phase change, none of the transferred energy increases the internal kinetic energy. All the energy taken in by the melting solid is set up as an increase in the internal potential energy. Therefore, ΔK and ΔW are both equal to zero, and ΔQ is equal to ΔP.

This is not true, however, when a liquid changes into a vapor. The change in volume that occurs, and therefore the external work done as the liquid changes into a vapor, is considerable. For example, when 1 lb of water at atmospheric pressure changes into vapor, its volume increases from 0.01671 ft^3 to 26.79 ft^3. Of the 970.4 Btu required to vaporize 1 lb of water, approximately 72 Btu of this energy are required to do the work of expanding against the pressure of the atmosphere. The remainder of the energy is set up in the vapor as an increase in the internal potential energy. In this instance, only ΔK is equal to zero, so that ΔQ is equal to ΔP plus ΔW.

3-11. "Ideal" or "Perfect" Gas

The various laws governing the pressure-volume-temperature relationships of gases as discussed in this chapter apply with accuracy only to a hypothetical "ideal" or "perfect" gas. A "perfect" gas is described as one in which there is no interaction between the molecules of the gas. The molecules of such a gas are entirely free and independent of each other's attractive forces. Hence, none of the energy transferred either to or from an ideal gas has any effect on the internal potential energy.

The concept of an ideal or perfect gas greatly simplifies the solution of problems concerning the changes in the condition of a gas. Many complex problems in mechanics are made simple by the assumption that no friction exists, the effects of friction being considered separately. The function of an ideal gas is the same as that of the frictionless surface. An ideal gas is assumed to undergo a change of condition without internal friction, that is, without the performance of internal work in overcoming intermolecular forces.

The idea of internal friction is not difficult to understand. A liquid such as oil will not flow readily at low temperatures because of the internal friction resulting from strong intermolecular forces within the liquid. However, as the liquid is heated and the molecules gain additional energy, the intermolecular forces are overcome somewhat, internal friction diminishes, and the liquid flows more easily.

Vaporization of the liquid, of course, causes a greater separation of the molecules and brings about a substantial reduction in internal friction, but some interaction between the molecules of the vapor still exists. In the gaseous state, intermolecular forces are greatest when the gas is near the liquid phase and diminish rapidly as the gas is heated and its temperature is raised farther and farther above the saturation temperature. A gas approaches the ideal state when it reaches a condition such that the interaction between the molecules, and hence, internal friction, is negligible.

Although no such thing as an ideal or perfect gas actually exists, many gases, such as air, nitrogen, hydrogen, and helium, so closely approach the ideal condition that any errors that may result from considering them to be ideal are of no consequence for most practical calculations.

Although it is important that the student of refrigeration understand and be able to apply the laws of perfect gases, it should be understood that gases as they normally occur in the mechanical refrigeration cycle are close to the saturation point (i.e., they are vapors) and do not approach the condition of an ideal or perfect gas.* They follow the gas laws in only a very general way, and therefore the use of the gas laws to determine the pressure-volume-temperature relationships of such vapors will result in considerable inaccuracy. In working with vapors, it is usually necessary to use values that have been determined experimentally or through rather complex calculations and that are made available in table and/or chart form. Several of these vapor tables and charts are included in this book and are discussed later.

3-12. Processes for Ideal Gases

A gas is said to undergo a process when it passes from some initial state or condition to some final state or condition. A change in the condition of a gas may occur in an infinite number of ways, only five of which are of interest. These are the (1) constant pressure (isobaric), (2) constant volume (isometric), (3) constant temperature (isothermal), (4) adiabatic, and (5) polytropic processes.

In describing an ideal gas, it has been said that the molecules of such a gas are so far apart that they have no attraction for one another and that none of the energy absorbed by an ideal gas has any effect on the internal potential energy. It is evident, then, that heat absorbed by an ideal gas will either increase the internal kinetic energy (temperature) of the gas or leave the gas as external work or both. Since the change in the internal potential energy, ΔP, will always be zero, the general energy equation for an ideal gas may be written as

$$\Delta Q = \Delta K + \Delta W \tag{3-10}$$

In order to better understand the energy changes that occur during the various processes, it should be kept in mind that a change in the temperature of the gas indicates a change in the initial kinetic energy of the gas, whereas a change in the volume of the gas indicates work done either by or on the gas.

* A vapor is sometimes defined as a gas at a condition close enough to the saturation point so that it does not follow the ideal gas laws even approximately. Some authorities define a vapor as a gas at any temperature below its critical temperature.

3-13. Constant Volume Process

When a gas is heated while it is so confined that its volume cannot change, its pressure and temperature will vary according to Charles' law (Fig. 3-3). Since the volume of the gas does not change, no external work is done, and ΔW is equal to zero. Therefore, for a constant volume process, indicated by the subscript v,

$$\Delta Q_v = \Delta K_v \tag{3-11}$$

Equation 3-11 states that during a constant volume process, all the energy transferred to the gas increases the internal kinetic energy of the gas. None of the energy leaves the gas as work.

When a gas is cooled (heat removed) while its volume remains constant, all the energy removed is effective in reducing the internal kinetic energy of the gas. It should be noted that in Equation 3-10, ΔQ represents heat transferred to the gas, ΔK represents an increase in the internal kinetic energy, and ΔW represents work done by the gas. Therefore, if heat is given up by the gas, ΔQ is negative. Likewise, if the internal kinetic energy of the gas decreases, ΔK is negative, and if work is done on the gas, rather than by it, ΔW is negative. Consequently, in Equation 3-11, when the gas is cooled, both ΔQ and ΔK are negative.

3-14. Constant Pressure Process

If the temperature of a gas is increased by the addition of heat while the gas is allowed to expand so that its pressure is kept constant, the volume of the gas will increase in accordance with Charles' law (Fig. 3-1). Since the volume of the gas increases during the process, work is done by the gas at the same time that its internal energy is increased. Hence, while one part of the transferred energy increases the initial kinetic energy of the gas, another part of the transferred energy leaves the gas as work. For a constant pressure process, identified by the subscript p, the energy equation may be written

$$\Delta Q_p = \Delta K_p + \Delta W_p \tag{3-12}$$

3-15. Specific Heat of Gases

The quantity of heat required to raise the temperature of 1 lb of a gas 1°F while the volume of the gas remains constant is known as the specific heat at a constant volume (c_v). Similarly, the quantity of heat required to raise the temperature of 1 lb of gas 1°F while the gas expands at a constant pressure is called the specific heat at a constant pressure (c_p). For any particular gas, the specific heat at a constant pressure is always greater than the specific heat at a constant volume. The reason for this is easily explained.

The quantity of energy required to increase the internal kinetic energy of a gas to the extent that the temperature of the gas is increased 1°F is exactly the same for all processes. Since no work is done during a constant volume process, the only energy required is that which increases the internal kinetic energy. However, during a constant pressure process, the gas expands a fixed amount for each degree of temperature rise, and a certain amount of external work is done. Therefore, during a constant pressure process, energy to do the work that is done must be supplied in addition to that which increases the internal kinetic energy. For example, the specific heat of air at a constant volume is 0.169 Btu/lb °F, whereas the specific heat of air at a constant pressure is 0.2375 Btu/lb °F. For either process, the increase in the internal energy of the air per degree of temperature rise is 0.169 Btu/lb. For the constant pressure process, the additional 0.0685 Btu/lb is the energy required to do the work resulting from the volume increase accompanying the temperature rise.

The specific heat of a gas may take any value, either positive or negative, depending on the amount of work the gas does, or has done on it, as its temperature changes.

3-16. The Change in Internal Kinetic Energy

During any process in which the temperature of the gas changes, there will be a change in the internal kinetic energy of the gas. Regardless of the process, when the temperature of a given mass of gas is increased or decreased, the change in the internal kinetic energy can be evaluated by the equation

$$\Delta K = (m)(c_v)(T_2 - T_1) \qquad (3\text{-}13)$$

where ΔK = the increase in the internal kinetic energy in British thermal units
m = the mass in pounds
c_v = the constant volume specific heat in Btu per pound per degree Fahrenheit
T_1 = the initial temperature in degrees Fahrenheit or degrees Rankine
T_2 = the final temperature in degrees Fahrenheit or degrees Rankine

Example 3-9 The temperature of 5 lb of air is increased by the addition of heat from an initial temperature of 75°F to a final temperature of 140°F. If c_v for air is 0.169 Btu/lb °F, what is the increase in the internal kinetic energy?

Solution Applying Equation 3-13,

$$\Delta K = (5\text{ lb})(0.169\text{ Btu/lb °F})(140° - 75°)$$
$$= 54.9\text{ Btu}$$

Example 3-10 Twelve pounds of air are cooled from an initial temperature of 95°F to a final temperature of 72°F. Compute the increase in the internal kinetic energy.

Solution Applying Equation 3-13,

$$\Delta K = (12\text{ lb})(0.169\text{ Btu/lb °F})(72° - 95°)$$
$$= -46.64\text{ Btu}$$

Notice in Example 3-10 that T_2 is less than T_1, so that ΔK is negative, indicating that the internal kinetic energy is decreased rather than increased.

3-17. Heat Transferred During a Constant Volume Process

For a constant volume process, since

$$\Delta Q_v = \Delta K_v$$

then

$$\Delta Q_v = (m)(c_v)(T_2 - T_1) \qquad (3\text{-}14)$$

Example 3-11 If, in Example 3-9, the gas is heated while its volume is kept constant, what is the quantity of heat transferred to the gas during the process?

Solution Applying Equation 3-14,

$$\Delta Q_v = (5\ \text{lb})(0.169\ \text{Btu/lb °F})(140° - 75°)$$
$$= 54.9\ \text{Btu}$$

Alternate Solution From Example 3-9, $\Delta K_v =$ 54.9 Btu. Since $\Delta Q_v = \Delta K_v$,

$$\Delta Q_v = 54.9\ \text{Btu}$$

Example 3-12 If, in Example 3-10, the air is cooled while its volume remains constant, what is the quantity of heat transferred to the air during the process?

Solution From Example 3-10, $\Delta K_v =$ -46.64 Btu. Since $\Delta Q_v = \Delta K_v$,

$$\Delta Q_v = -46.64\ \text{Btu}$$

In Example 3-12, notice that since ΔK_v is negative, indicating a decrease in the internal kinetic energy, ΔQ_v must of necessity also be negative, indicating that heat is transferred from the gas rather than to it.

3-18. External Work During a Constant Pressure Process

It will be shown that the work done during a constant pressure process can be evaluated by the equation

$$w_p = p(V_2 - V_1) \qquad (3\text{-}15)$$

where w_p = the work done in foot-pounds
p = the absolute pressure in pounds per square foot
V_1 = the initial volume in cubic feet
V_2 = the final volume in cubic feet

Assume that the piston in Figure 3-1*a* has an area of A square feet and that the absolute pressure of the gas in the cylinder is *p* pounds per square foot. Then, in accordance with Equation 1-17, the total force exerted on the top of the piston will be $(p)(A)$ pounds; that is,

$$F = (p)(A)$$

Assume now that the gas in the cylinder, having an initial volume V_1, is heated and allowed to expand to volume V_2 while its pressure is kept constant (Fig. 3-1*b*). In doing so, the force *pA* acts through a distance *s* and work is done (see Equation 1-19). The amount of work done may be written as

$$w = (p)(A)(s)$$

but since

$$(A)(s) = V_2 - V_1$$

then

$$w = p(V_2 - V_1)$$

Example 3-13 One pound of air having an initial volume of 13.34 ft^3 and an initial temperature of 70°F is heated and caused to expand at a constant pressure of one atmosphere (2116 psfa) to a final volume of 15 ft^3. Compute the external work done in foot-pounds.

Solution Applying Equation 3-15,

$$w_p = (2116\ \text{psfa})(15\ \text{ft}^3 - 13.34\ \text{ft}^3) = 3513\ \text{ft-lb}$$

In Equation 3-10, ΔW is always given in heat energy units. By application of the mechanical energy equivalent (Section 2-30), *w* in foot-pounds may be expressed as ΔW in Btu. The relationship is

$$\Delta W = \frac{w}{J} \qquad (3\text{-}16)$$

$$w = (\Delta W)(J) \qquad (3\text{-}17)$$

Example 3-14 Express the work done in Example 3-13 in terms of heat energy units.

Solution Applying Equation 3-16, the work in Btu, ΔW

$$= \frac{3513\ \text{ft-lb}}{778\ \text{ft-lb/Btu}}$$
$$= 4.52\ \text{Btu}$$

3-19. Heat Transferred During a Constant Pressure Process

According to Equation 3-12, ΔQ_p, the total heat transferred to a gas during a constant pressure process, is equal to the sum of ΔK_p, the increase in the internal kinetic energy of the gas, and ΔW_p, the external work done by the expansion of the gas.

Example 3-15 Compute the total heat energy transferred to the air during the constant pressure process described in Example 3-13.

Solution Applying Charles' law for a constant pressure process,

$$T_2 = \frac{T_1 V_2}{V_1} = \frac{(70°\text{F} + 460°)(15 \text{ ft}^3)}{(13.34 \text{ ft}^3)} = 596°\text{R}$$

Applying Equation 3-13,

$$\Delta K = (1 \text{ lb})(0.169 \text{ Btu/lb °F})(596° - 530°)$$
$$= 11.15 \text{ Btu}$$

From Example 3-14,

$$\Delta W_p = 4.52 \text{ Btu}$$

Applying Equation 3-12,

$$\Delta Q_p = 11.15 \text{ Btu} + 4.52 \text{ Btu} = 15.67 \text{ Btu}$$

Since the specific heat at constant pressure c_p takes into account not only the increase in the internal kinetic energy but also the work done per pound per degree of temperature change during a constant pressure process, for the constant pressure process only, the total heat transferred during the process can be determined by the following equation:

$$\Delta Q_p = (m)(c_p)(T_2 - T_1) \qquad (3\text{-}18)$$

Consequently, Equation 3-18 provides an alternate solution for Example 3-15. From Table 3-1, the specific heat or air at constant pressure, c_p, is 0.2375 Btu/lb °F. Applying Equation 3-18,

$$\Delta Q_p = (1 \text{ lb})(0.2375 \text{ Btu/lb °F})(596° - 530°)$$
$$= 15.67 \text{ Btu}$$

3-20. Pressure-Volume (pV) Diagram

Equation 3-4 is a statement that the thermodynamic state of a gas is adequately described by any two properties of the gas. Hence, using any two properties of the gas as mathematical coordinates, the thermodynamic state of a gas at any given instant may be shown as a point on a chart. Furthermore, when the conditions under which a gas passes from some initial state to some final state are known, the path that the process follows may be made to appear as a line on the chart.

The graphical representation of a process or cycle is called a process diagram or a cycle diagram, respectively, and is a very useful tool in the analysis and solution of cyclic problems.

Since work is a function of pressure and volume, when it is the work of a process or cycle that is of interest, the properties used as coordinates are usually the pressure and the volume. When the pressure and volume are used as coordinates to diagram a process or cycle, it is called a pressure-volume (pV) diagram.

To illustrate the use of the pV diagram, a pressure-volume diagram of the process described in Example 3-13 is shown in Figure 3-4. Notice that the absolute pressure in pounds per square foot (psfa) is used as the vertical coordinate, whereas the volume in cubic feet is used as the horizontal coordinate.

In Example 3-13, the initial condition of the gas is such that the pressure is one atmosphere and the volume is 13.34 ft^3. To establish the initial state of the gas on the pV chart, start at the origin and proceed upward along the vertical pressure axis to the given pressure, 2116 psfa. Draw a dotted line parallel to the base line through this point and across the chart. Next, from the point of origin proceed to the right along the horizontal volume axis to the given volume 13.34 ft^3. Through this point, draw a vertical dotted line across the chart. The intersection of the dotted lines at point 1 establishes the initial thermodynamic state of the gas.

According to Example 3-13, the gas is heated and allowed to expand at a constant pressure

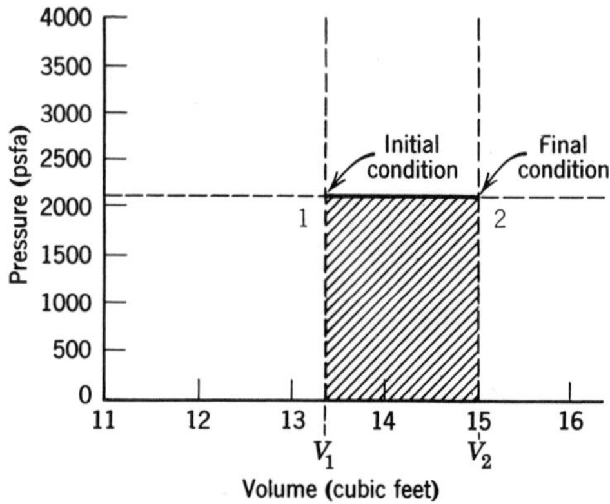

Fig. 3-4 Pressure-volume diagram of constant pressure process. Crosshatched area between process diagram and base line represents external work done during the process.

until its volume is 15 ft^3. Since the pressure remains the same during the process, the state point representing the final state of the gas must fall somewhere along the line of constant pressure already established. The exact point on the pressure line that represents the final state 2 is determined by the intersection of the line drawn through the point on the volume axis that identifies the final volume.

In passing from the initial state 1 to the final state 2, the air passes through a number of intermediate thermodynamic states, all of which can be represented by points that will fall along line 1 to 2. Line 1 to 2, then, represents the path that the process will follow as the thermodynamic state of the gas changes from 1 to 2 and is the pV diagram of the process described.

The area of a rectangle is the product of its two dimensions. In Fig. 3-4, the area of the rectangle under the process diagram (cross-hatched) is the product of its altitude p and its base $(V_2 - V_1)$. But according to Equation 3-15, the product $p(V_2 - V_1)$ is the external work done during a constant pressure process. It is evident, then, that the area between the process diagram and the volume axis is a measure of the external work done during the process in foot-pounds.

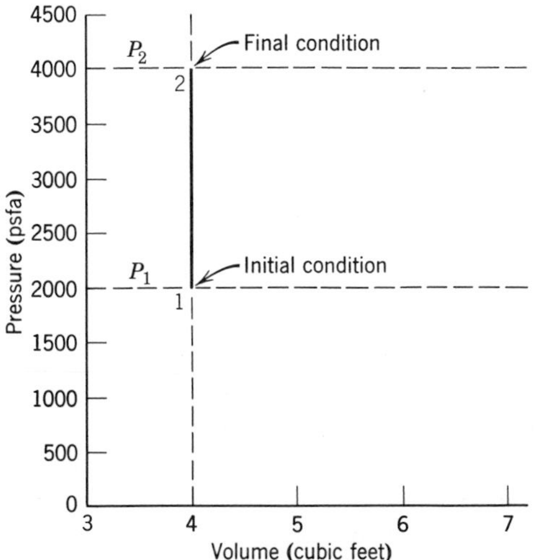

Fig. 3-5 Pressure-volume diagram of constant volume process. Since there is no area between the process diagram and the volume axis, there is no work done during a constant volume process.

Figure 3-5 is a pV diagram of a constant volume process. Assume that the initial condition of the gas at the start of the process is such that the pressure is 2000 psfa and the volume is 4 ft^3. The gas is heated while its volume is kept constant until the pressure increases to 4000 psfa. The process takes place along the constant volume line from the initial condition 1 to the final condition 2.

It has been stated that no work is done during a process unless the volume of the gas changes. Examination of the pV diagram in Fig. 3-5 will show that no work is indicated for the constant volume process. Since a line has only the dimension of length, there is no area between the process diagram and the base or volume axis. Consequently, no work is done.

3-21. Constant Temperature Process

According to Boyle's law, when a gas is compressed or expanded at a constant temperature, the pressure will vary inversely with the volume. That is, the pressure increases as the gas is compressed and decreases as the gas is expanded. Since the gas will do work as it expands, if the temperature is to remain constant, energy with which to do the work must be supplied from an external source (Fig. 3-2*b*). However, since the temperature of the gas remains constant, all the energy supplied to the gas during the process leaves the gas as work; none is stored in the gas as an increase in the internal energy.

When a gas is compressed, work is done on the gas, and if the gas is not cooled during the compression, the internal energy of the gas will be increased by an amount equal to the work of compression. Therefore, if the temperature of the gas is to remain constant during the compression, the gas must reject to the surroundings an amount of heat equal to the amount of work done on it during the compression (Fig. 3-2*c*).

Since there is no change in the internal kinetic energy during a constant temperature process, ΔK in Equation 3-10 is equal to zero, and the general energy equation for a constant temperature process may be written as

$$\Delta Q_t = \Delta W_t \qquad (3\text{-}19)$$

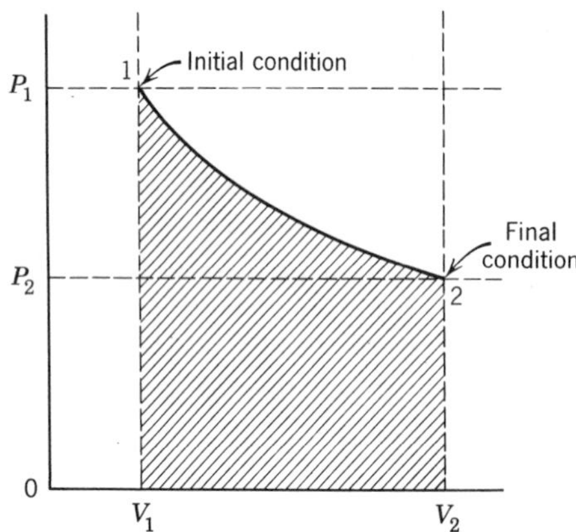

Fig. 3-6 Pressure-volume diagram of constant temperature process. Crosshatched area represents the work of the process.

3-22. Work of an Isothermal Process

A pV diagram of an isothermal expansion is shown in Fig. 3-6. In a constant temperature process, the pressure and volume both change in accordance with Boyle's law. The path followed by an isothermal expansion is indicated by line 1–2, and the work of the process, in foot-pounds, is represented by the area under the process diagram (crosshatched). This area, and therefore the work of the process, may be calculated by the equation

$$w_t = p_1 V_1 \times \ln\left(\frac{V_2}{V_1}\right) \tag{3-20}$$

where ln = natural logarithm ($\log_e$)

Example 3-16 A certain mass of gas having an initial pressure of 2500 psfa and an initial volume of 2 ft^3 is expanded isothermally to a volume of 4 ft^3. Determine:

(a) the final pressure of the gas in psfa,
(b) the work done by the gas in foot-pounds,
(c) the work done in Btu.

Solution

(a) Applying Equation 3-2,

$$p_2 = \frac{(2500 \text{ psfa})(2 \text{ ft}^3)}{4 \text{ ft}^3} = 1250 \text{ psfa}$$

(b) Applying Equation 3-20,

$$w_t = (2500 \text{ psfa})(2 \text{ ft}^3) \times \ln\left(\frac{4 \text{ ft}^3}{2 \text{ ft}^3}\right)$$

$$= 3466 \text{ ft-lb}$$

(c) Applying Equation 3-16,

$$\Delta W_t = \frac{3466 \text{ ft-lb}}{778 \text{ ft-lb/Btu}}$$

$$= 4.45 \text{ Btu}$$

Example 3-17 A certain mass of gas having an initial pressure of 1250 psfa and an initial volume of 4 ft^3 is compressed isothermally to a volume of 2 ft^3. Determine:

(a) the final pressure of the gas in psfa,
(b) the work done by the gas in foot-pounds,
(c) the work done in Btu.

Solution

(a) Applying Equation 3-2,

$$p_2 = \frac{p_1 V_1}{V_2}$$

$$= \frac{(1250 \text{ psfa})(4 \text{ ft}^3)}{2 \text{ ft}^3}$$

$$= 2500 \text{ psfa}$$

(b) Applying Equation 3-20,

$$w_t = (1250 \text{ psfa})(4 \text{ ft}^3) \times \ln\left(\frac{2 \text{ ft}^3}{4 \text{ ft}^3}\right)$$

$$= -3466 \text{ ft-lb}$$

(c) Applying Equation 3-16,

$$\Delta W_t = \frac{-3466 \text{ ft-lb}}{778 \text{ ft-lb/Btu}}$$

$$= -4.45 \text{ Btu}$$

Notice that the process in Example 3-17 is the exact reverse of that of Example 3-16. Where the process in Example 3-16 is an expansion, the process in Example 3-17 is a compression. Both processes occur between the same two conditions, except that the initial and the final conditions are reversed. Notice also that whereas work is done by the gas during the expansion process, work is done on the gas during the compression process. But since the change of condition takes place between the same limits in both cases, the amount of work done in each case is the same.

3-23. Heat Transferred During a Constant Temperature Process

Since there is no change in the temperature during an isothermal process, there is no change in the internal kinetic energy and ΔK equals zero. According to Equation 3-19, the heat energy transferred during a constant temperature process is exactly equal to the work done in Btu. During an isothermal expansion, heat is transferred to the gas to supply the energy to do the work that is done by the gas, whereas during an isothermal compression, heat is transferred from the gas so that the internal energy of the gas is not increased by the performance of work on the gas.

Example 3-18 Determine the quantity of heat transferred to the gas during the constant temperature expansion described in Example 3-16.

Solution From Example 3-16,

$$\Delta W_t = 4.45 \text{ Btu}$$

Since, in the isothermal process, ΔW_t equals ΔQ_t,

$$\Delta Q_t = 4.45 \text{ Btu}$$

Example 3-19 What is the quantity of heat transferred to the gas during the constant temperature process described in Example 3-17?

Solution From Example 3-17,

$$\Delta W_t = -4.55 \text{ Btu}$$

Since ΔW_t equals ΔQ_t,

$$\Delta Q_t = -4.55$$

Again, notice that a negative amount of heat is transferred to the gas, indicating that heat in this amount is actually given up by the gas during the process.

3-24. Adiabatic Process

An adiabatic process is one in which the gas is assumed to change its condition without the transfer of heat to or from the surroundings during the process. Furthermore, the pressure, volume, and temperature of the gas all vary during an adiabatic process, none remaining constant.

When a gas expands adiabatically, as in any other expansion, the gas does external work, and energy is required to do the work. In the process previously described, the gas absorbed the energy to do the work from an external source. Since, during an adiabatic process, no heat is transferred to the gas from an external source, the gas must do the external work at the expense of its own energy. An adiabatic expansion is always accompanied by a decrease in the temperature of the gas as the gas gives up its own internal energy to do the work (Fig. 3-7).

When a gas is compressed adiabatically, work is done on the gas by an external body. The energy of the gas is increased in an amount equal to the amount of work done, and since no heat energy is given up by the gas to an external body during the compression, the heat energy equivalent of the work done on the gas is set up as an increase in the internal energy of the gas, and the temperature of the gas increases.

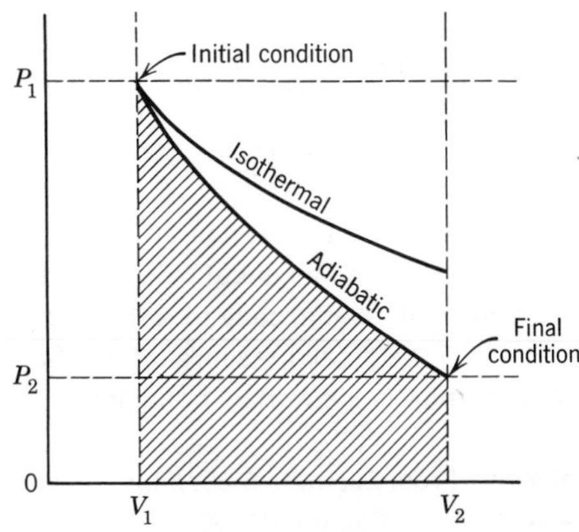

Fig. 3-7 Pressure-volume diagram of adiabatic process. An isothermal curve is drawn in for comparison.

Because no heat, as such, is transferred to or from the gas during an adiabatic process, ΔQ_a is always zero, and the energy equation for an adiabatic process can be written as

$$\Delta K_a + \Delta W_a = 0 \qquad (3\text{-}21)$$

Therefore,

$$\Delta W_a = -\Delta K_a \qquad \text{and} \qquad \Delta K_a = -\Delta W_a$$

3-25. Work of an Adiabatic Process

The work of an adiabatic process may be evaluated by the following equation:

$$w_a = \frac{P_1V_1 - P_1V_2}{K - 1} \qquad (3\text{-}22)$$

Example 3-20 A gas having an initial pressure of 2500 psfa and an initial volume of 2 ft^3 is expanded adiabatically to a final volume of 4 ft^3. If the final pressure is 945 psfa, determine the external work done in Btu.

Solution For air, the ratio of the specific heats,

$$k = \frac{c_p}{c_v} = \frac{0.2375}{0.169} = 1.406$$

Applying Equation 3-22,

$$w_a = \frac{(2500 \text{ psfa} \times 2 \text{ ft}^3) - (945 \text{ psfa} \times 4 \text{ ft}^3)}{1.406 - 1}$$

$$= 3005 \text{ ft-lb}$$

Applying Equation 3-16,

$$\Delta W_a = \frac{3005 \text{ ft-lb}}{778 \text{ ft-lb/Btu}}$$

$$= 3.86 \text{ Btu}$$

Example 3-21 A gas that has an initial pressure of 945 psfa and an initial volume of 4 cm^3 is compressed adiabatically to a final volume of 2 m^3. If the final pressure of the air is 2500 psfa, how much work is done in Btu?

Solution From Example 3-20, *k* for air equals 1.406. Applying Equation 3-22,

$$w_a = \frac{(945 \text{ psfa} \times 4 \text{ ft}^3) - (2500 \text{ psfa} \times 2 \text{ ft}^3)}{1.406 - 1}$$

$$= -3005 \text{ ft-lb}$$

Applying Equation 3-16,

$$\Delta W_a = \frac{-3005 \text{ ft-lb}}{778 \text{ ft-lb/Btu}}$$

$$= -3.86 \text{ Btu}$$

3-26. Comparison of the Isothermal and Adiabatic Processes

A comparison of the isothermal and adiabatic processes is of interest. Whenever a gas expands, work is done by the gas, and energy from some source is required to do the work. In an isothermal expansion, all the energy to do the work is supplied to the gas as heat from an external source. Since the energy is supplied to the gas from an external source at exactly the same rate that the gas is doing work, the internal energy of the gas neither increases nor decreases, and the temperature of the gas remains constant during the process. On the other hand, in an adiabatic expansion, there is no transfer of heat to the gas during the process, and all the work of expansion must be done at the expense of the internal energy of the gas. Therefore, the internal energy of the gas is always diminished by an amount equal to the amount of work done, and the temperature of the gas decreases accordingly.

Consider now isothermal and adiabatic compression processes. In any compression process, work is done on the gas by the compressing member, usually a piston, and an amount of energy equal to the amount of work done on the gas is transferred to the gas as work. During an isothermal compression process, energy is transferred as heat from the gas to an external sink at exactly the same rate that work is being done on the gas. Therefore, the internal energy of the gas neither increases nor decreases during the process, and the temperature of the gas remains constant. On the other hand, during an adiabatic compression, there is no transfer of energy as heat from the gas to an external sink. Therefore, an amount of energy equal to the amount of work done on the gas is set up in the gas as an increase in the internal energy, and the temperature of the gas increases accordingly.

3-27. The Polytropic Process

Perhaps the simplest way of defining a polytropic process is by comparison with the isothermal and adiabatic processes. The isothermal expansion, in which the energy to do the work of expansion is supplied entirely from an external source, and the adiabatic expansion, in which the energy to do the work of expansion is supplied entirely by the gas itself, may be thought of as the extreme limits between which all expansion processes will fall. Hence, any expansion process in which the energy to do the work of expansion is supplied partly from an external source and partly from the gas itself will follow a path that will fall somewhere between those of the isothermal and adiabatic processes (Fig. 3-8). Such a process is known as a polytropic process. If, during a polytropic expansion, most of the energy to do the work comes from an external source, the polytropic process will more nearly approach the isothermal. On the other hand, when the greater part of the energy to do the external work comes from the gas itself, the process more nearly approaches the adiabatic.

This is also true for the compression process. When a gas loses heat during a compression process, but not at a rate sufficient to maintain the temperature constant, the compression is polytropic. The greater the loss of heat, the closer the polytropic process approaches the isothermal. The smaller the loss of heat, the closer the polytropic process approaches the adiabatic. Of course, with no heat loss, the process becomes adiabatic.

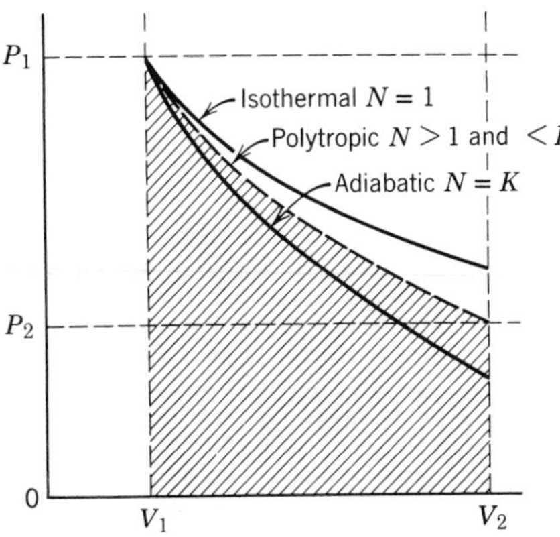

Fig. 3-8 Pressure-volume diagram of a polytropic process. Adiabatic and isothermal curves are drawn in for comparison.

The actual compression of a gas in a compressor will usually very nearly approach adiabatic compression. This is because the time of compression is normally very short, and there is not sufficient time for any significant amount of heat to be transferred from the gas through the cylinder walls to the surroundings. Water jacketing of the cylinder will usually increase the rate of heat rejection and move the path of the compression closer to the isothermal.

3-28. PVT Relationship During Adiabatic Processes

The temperature, pressure, and volume all change during an adiabatic process, and the relationships among these properties during an adiabatic process may be evaluated by the following equations:

$$T_2 = T_1 \times \left(\frac{V_1}{V_2}\right)^{k-1} \tag{3-23}$$

$$T_2 = T_1 \times \left(\frac{p_2}{p_1}\right)^{(k-1)/k} \tag{3-24}$$

$$p_2 = p_1 \times \left(\frac{V_1}{V_2}\right)^{k} \tag{3-25}$$

$$p_2 = p_1 \times \left(\frac{T_2}{T_1}\right)^{k/(k-1)} \tag{3-26}$$

$$V_2 = V_1 \times \left(\frac{T_1}{T_2}\right)^{1/(k-1)} \tag{3-27}$$

$$V_2 = V_1 \times \left(\frac{p_1}{p_2}\right)^{1/k} \tag{3-28}$$

Example 3-22 Air is expanded adiabatically from a volume of 2 ft^3 to a volume of 4 ft^3. If the initial pressure of the air is 4500 psfa, what is the final absolute pressure in pounds per square foot?

Solution From Table 3-1, k for air is 1.406. Applying Equation 3-25,

$$p_2 = p_1 \times \left(\frac{V_1}{V_2}\right)^k = (4500 \text{ psfa}) \times \left(\frac{2}{4}\right)^{1.406}$$

$$= (4500 \text{ psfa})(0.378) = 1698 \text{ psfa}$$

Example 3-23 Air is expanded adiabatically from a volume of 2 ft^3 to a volume of 4 ft^3. If the initial temperature of the air is 600°R, what is the final temperature in degrees Rankine?

Solution From Table 3-1, k for air is 1.406. Applying Equation 3-23,

$$T_2 = T_1 \times \left(\frac{V_1}{V_2}\right)^{k-1} = 600°\text{R} \times \left(\frac{2}{4}\right)^{1.406-1}$$

$$= (600°\text{R})(0.755) = 453°\text{R}$$

Example 3-24 Air is expanded adiabatically from an initial absolute pressure 4500 psfa to a final absolute pressure 1698 psfa. If the initial temperature is 600°R, what is the final temperature in degrees Rankine?

Solution Applying Equation 3-24,

$$T_2 = T_1 \times \left(\frac{p_2}{p_1}\right)^{k-1/k}$$

$$= 600°\text{R} \times \left(\frac{1698}{4500}\right)^{(1.406-1)/1.406}$$

$$= (600°\text{R})(0.378)^{0.289} = (600°\text{R})(0.755)$$

$$= 453°\text{R}$$

3-29. Exponent of Polytropic Expansion and Compression

The pressure-temperature-volume relationships for the polytropic process can be evaluated by Equations 3-23 through 3-28, except that the polytropic expansion or compression exponent n is substituted for k. Also, the work of a polytropic process can be determined by Equation 3-22 if n is substituted for k.

Usually, the value of n must be determined by actual test of the machine in which the expansion or compression occurs. In some instances, average values of n for some of the common gases undergoing changes under more or less standard conditions are given in tables. If the values of two properties are known for both initial and final conditions, the value of n may be calculated. The following sample equation shows the relationship:

$$n = \frac{\log(p_1/p_2)}{\log(V_1/V_2)} \tag{3-29}$$

The value of n depends on the specific heat of the gas during the process. Since the specific heat may take any value, it follows that, theoretically, n may have any value.

Broadly defined, a polytropic process is any process in which the specific heat remains constant. By this definition, all five processes discussed in this chapter are polytropic processes. It is sometimes convenient to restrict the term "polytropic" to mean only those processes that follow a path falling between those of the isothermal and adiabatic processes. The exponents of isothermal and adiabatic expansions or compressions are 1 and k, respectively. Hence, the value of n for the restricted polytropic process must fall somewhere between 1 and k. The closer the polytropic process approaches the adiabatic, the closer n will approach k.

Example 3-25 Air, having an initial absolute pressure of 4500 psfa and an initial temperature of 600°R, is expanded polytropically from a volume of 2 ft^3 to a volume of 4 ft^3. If the exponent of polytropic expansion is 1.2, determine the following:

(a) the mass of the air in pounds
(b) the final absolute pressure in pounds per square foot
(c) the final temperature in degrees Rankine
(d) the work done by the gas in Btu
(e) the increase in the internal energy in Btu
(f) the heat transferred to the gas in Btu

Solution

(a) From Table 3-1, R for air is 53.3. Rearranging and applying Equation 3-8,

$$m = \frac{pV}{RT} = \frac{(4500\text{ psfa})(2\text{ ft}^3)}{(53.3)(600°\text{R})} = 0.28\text{ lb}$$

(b) Applying Equation 3-25,

$$p_2 = p_1 \times \left(\frac{V_1}{V_2}\right)^n = (4500\text{ psfa}) \times \left(\frac{2}{4}\right)^{1.2}$$

$$= (4500\text{ psfa})(0.435) = 1959\text{ psfa}$$

(c) Applying Equation 3-23,

$$T_2 = T_1 \times \left(\frac{V_1}{V_2}\right)^{n-1} = (600°\text{R}) \times \left(\frac{2}{4}\right)^{1.2-}$$

$$= (600°\text{R})(0.5)^{0.2} = (600°\text{R})(0.87)$$

$$= 522°\text{R}$$

(d) Applying Equation 3-22,

$$w_n = \frac{(p_1V_1 - p_2V_2)}{n - 1}$$

$$\times \frac{(4500 \text{ psfa} \times 2 \text{ ft}^3) - (1959 \text{ psfa} \times 4 \text{ ft}^3)}{1.2 - 1}$$

$$= 5820 \text{ ft-lb}$$

Applying Equation 3-16,

$$\Delta W_n = \frac{(5820 \text{ ft-lb})}{(778 \text{ ft-lb/Btu})}$$

$$= 7.48 \text{ Btu}$$

(e) From Table 3-1, c_v for air equals 0.169 Btu/lb °F. Applying Equation 3-13,

$$\Delta K = (0.28 \text{ lb})(0.169 \text{ Btu/lb °F})(522° - 600°)$$

$$= -3.69 \text{ Btu}$$

(f) Applying Equation 3-10,

$$\Delta Q = \Delta K + \Delta W = -3.69 \text{ Btu} + 7.48 \text{ Btu}$$

$$= 3.79 \text{ Btu}$$

Notice in Example 3-25 that the work done by the air in the polytropic expansion is equal to 7.48 Btu. Of this amount, 3.79 Btu are supplied from an external source, while the other portion, 3.69 Btu, is supplied by the gas itself, thereby reducing the internal kinetic energy by this amount.

3-30. The Thermodynamic System

A *thermodynamic system* is any region or space, enclosed by real or imaginary *boundaries*, that is selected for the purpose of studying energy and its transformations. The space immediately adjacent to and outside the boundaries of the system is known as the *surroundings*. The boundaries of a system may be either fixed or elastic. A gas confined in a metal container is an example of a system with fixed boundaries. The gas in the container constitutes the system, and the walls of the container are taken as the boundaries of the system. Now assume that the gas is confined in a rubber balloon and that the walls of the balloon constitute the boundary of the system. If the gas is heated, the elastic walls of the balloon will stretch, and the boundary of the system will change (expand) as the gas expands.

The system chosen may be large or small. For example, it may be an entire refrigerating system or simply the gas in one cylinder of the compressor. The system may be a vacuum, or it may consist of one or more material components or even several phases of the same component. For instance, the system could consist of dry air and water vapor (two components) or water and water vapor (two phases of the same component). A *homogeneous* system is one that consists of a single substance or phase or a uniform mixture of several components or phases.

The system may be either *closed* or *open*. In a closed system, only energy crosses the boundary of the system, whereas in an open system, both energy and mass are exchanged between the system and the surroundings. Where the mass flow rate into and out of an open system is steady and uniform, the system is a *steady flow system*.

The *state* of a thermodynamic system is defined or described by the physical properties of the system, such as temperature, pressure, volume, and entropy. Since the state of a system is a fixed condition, it can be defined or described only when the properties of the system are constant or unchanging; that is, the state of a system can be described only when the system is in equilibrium.

All the thermodynamic systems discussed in this chapter are closed, homogeneous systems with real boundaries. For the constant volume process, the system boundaries are fixed. For the other processes, those occurring in a cylinder with a movable piston, the boundaries of the system were elastic.

3-31. Thermodynamic Processes

When a system changes from one state to another, it is said to undergo a *process*. Thermodynamic processes are either *reversible* or *irreversible*. A reversible process is one that theoretically is completely reversible in that it can be made to follow in reverse the exact path of the initial process and thereby return both the system and the surroundings to their initial states. All other processes are termed irreversible. Since a completely reversible process is not possible in practice, all actual or real processes are irreversible, although some can and do approach or approximate an ideal reversible process very closely.

There are a number of factors both internal and external that cause irreversibility in fluid processes. *Internal irreversibility* is brought about by internal fluid friction resulting from intermolecular forces and turbulence in the fluid. For example, assume that a gas at high pressure is confined behind a piston in a cylinder. If the piston is moved rapidly, a portion of the gas immediately adjacent to the piston will tend to expand into the void created by the receding piston, while another portion of the gas tends to remain at rest. This causes pressure and temperature differentials in the fluid, with the resulting turbulence and fluid friction accounting for some of the energy that otherwise would be delivered as useful work. For a process to be internally reversible, it must employ an ideal fluid (no intermolecular attraction) and it must take place very slowly so that, at any given instant, the properties of the system (pressure, temperature, etc.) are uniform throughout the system, thereby avoiding any fluid friction resulting from turbulence. All the ideal gas processes discussed in this chapter are considered to be internally reversible.

External irreversibility results from factors external to the system. One of the most frequent causes of external irreversibility is the mechanical friction encountered in rubbing surfaces such as bearings and cylinder walls. Another is heat transfer, which by its nature can occur in only one direction, from a higher temperature to a lower temperature. The frictionless, adiabatic (isentropic) process described in Section 3-24 is an externally (and internally) reversible process and one that has particular significance in the analysis of a vapor-compression refrigeration cycle.

Although ideal reversible processes are not possible in practice, they are nevertheless important in that they can be plotted graphically, as on the pV coordinates already described, and used in many ways to evaluate real processes or the results of real processes.

It can be demonstrated that any changes in the values of the various physical properties during any thermodynamic process depend only on the initial and final state of the system and are entirely independent of the path of the processes, that is, the manner in which the process takes place. On the other hand, the amount of heat energy transferred and/or the amount of work done during any thermodynamic process depends on the manner in which the process occurs, that is, the path of the process. Consequently, the physical properties of a system are known as *point* or *state* functions, whereas energy transfers, either as heat or as work, are recognized as *path* functions.

3-32. METRIC SYSTEM EQUIVALENTS

When employing metric units in the ideal gas formulas presented in this chapter, the mass (m) is in kilograms, the absolute temperature is in degrees Kelvin, the volume is in cubic meters, and the absolute pressure is in pascals. Metric system values for R, c_v, c_p, and k for a number of common gases are listed in Table 3-1. R has the units of joules per kilogram per degree Kelvin (J/kg °K), c_v and c_p are in joules per gram per degree Kelvin (J/g °K) and k is a dimensionless ratio that is the same for both measurements systems.

In all work equations, the work (w) is in joules, so that w is equal to ΔW.

Example 3-26 The tank of an air compressor has a volume of 0.2 m^3 and is filled with air at a temperature of 40°C. If the absolute pressure of the air is 7.5 bars, what is the mass of the air in the tank?

Solution From Table 3-1, R for air is 287 J/kg °K. Rearranging and applying Equation 3-8,

$$m = \frac{(7.5 \times 10^5 \text{ Pa})(0.2 \text{ m}^3)}{(287)(40°\text{C} + 273°)} = 1.67 \text{ kg}$$

Example 3-27 Heat is supplied to the air in Example 3-26 until the temperature of the air is

increased to 130°C while the volume is held constant. Determine:
(a) the final pressure of the air in bar,
(b) the increase in the internal energy,
(c) the quantity of heat transferred.

Solution
(a) Applying Equation 3-3, the final absolute pressure, p_2,

$$= \frac{(7.5 \text{ bar})(130°\text{C} + 273)}{40° + 273}$$

$$= 9.66 \text{ bar}$$

(b) From Table 3-1, c_v for air equals 711 J/kg °K. Applying Equation 3-13, the increase in internal energy,

$$\Delta K = (1.67 \text{ kg})(711 \text{ J/kg °K})(130° - 40°)$$
$$= 106.86 \text{ kJ}$$

(c) Since $\Delta Q_v = \Delta K_v$, $\Delta Q = 106.86$ kJ.

Example 3-28 Heat is supplied to the air in Example 3-26 until the temperature of the air is increased to 130°C while the pressure is held constant. Compute:
(a) the final volume of the air,
(b) the work done by the air,
(c) the increase in the internal kinetic energy,
(d) the quantity of heat transferred.

Solution From Table 3-1, c_v and c_p for air are 711 J/kg °K and 1000 J/kg °K, respectively.
(a) Applying Equation 3-1,

$$V_2 = \frac{(0.2 \text{ m}^3)(403 \text{ °K})}{313\text{°K}}$$

$$= 0.2575 \text{ m}^3$$

(b) Applying Equation 3-15,

$$w = (7.5 \times 10^5 \text{ Pa})(0.2575 \text{ m}^3 - 0.2 \text{ m}^3)$$
$$= 43{,}125 \text{ J} \quad \text{or} \quad 43.125 \text{ kJ}$$

(c) Applying Equation 3-13, the increase in internal energy,

$$\Delta K = (1.67 \text{ kg})(711 \text{ J/kg °K})(130° - 40°)$$
$$= 106.86 \text{ kJ}$$

(d) Applying Equation 3-18,

$$\Delta Q_p = (1.67 \text{ kg})(1000 \text{ J/kg °K})(130° - 40°)$$
$$= 150.3 \text{ kJ}$$

Example 3-29 Assume the air in Example 3-26 is expanded from its initial volume of 0.2 m^3 to a final volume of 0.4 m^3 while the temperature remains constant. Determine:
(a) the final pressure of the air in bar,
(b) the work done by the air,
(c) the quantity of heat transferred to the air.

Solution
(a) Applying Equation 3-2,

$$p_2 = \frac{(7.5 \text{ bar})(0.2 \text{ m}^3)}{0.4 \text{ m}^3}$$

$$= 3.75 \text{ bar}$$

(b) Applying Equation 3-20,

$$W_t = (7.5 \times 10^5 \text{ Pa})(0.2 \text{ m}^3) \times \ln\left(\frac{0.4 \text{ m}^3}{0.2 \text{ m}^3}\right)$$

$$= 103{,}972 \text{ J} \quad \text{or} \quad 104 \text{ kJ}$$

(c) Since $\Delta Q_t = \Delta W_t$, $\Delta Q = 104$ kJ.

Example 3-30 Assume the air in Example 3-29 is expanded adiabatically and compute:
(a) the final temperature in degrees Kelvin,
(b) the work done by the air,
(c) the increase in the internal kinetic energy.

Solution
(a) Applying Equation 3-23,

$$T_2 = (40°\text{C} + 273) \times \left(\frac{0.2 \text{ m}^3}{0.4 \text{ m}^3}\right)^{1.406-1}$$

$$= 236.2°\text{K}$$

(b) Applying Equation 3-25,

$$p_2 = (7.5 \times 10^5 \text{ Pa}) \times \left(\frac{0.2 \text{ m}^3}{0.4 \text{ m}^3}\right)^{1.406}$$

$$= 283{,}017 \text{ Pa}$$

Applying Equation 3-22,

$$W_a = \frac{(750{,}000 \text{ Pa})(0.2 \text{ m}^3) - (283{,}017 \text{ Pa})(0.4 \text{ m}^3)}{1.406 - 1}$$

$$= 90{,}623 \text{ J}$$

(c) Applying Equation 3-13,

$$\Delta K = (1.67 \text{ kg})(711 \text{ J/kg °K})(236.2° - 313°)$$
$$= 91{,}190 \text{ J}$$

Customary Problems

3-1 A gas having an initial volume of 4 ft^3 at a temperature of 1050°R is cooled under constant pressure until its volume decreases to 1.7 ft^3. Determine the final temperature of the gas in degrees Rankine.

3-2 A gas having an initial temperature of 60°F is heated under constant pressure until its temperature increases to 120°F. If the initial volume of the gas is 1.4 ft^3. what is its final volume?

3-3 Two pounds of air are compressed at constant temperature from an initial volume of 2.6 ft^3 to a final volume of 0.89 ft^3. If the initial absolute pressure of the air is 35 psia, what is the final absolute pressure in pounds per square inch?

3-4 Two cubic feet of a gas are allowed to expand from an initial pressure of 95 psia to a final pressure of 18 psia while the temperature remains constant. Determine the final volume of the gas.

3-5 A certain mass of gas is expanded isothermally from an initial volume of 0.62 ft^3 to a final volume of 3.16 ft^3. If the initial pressure of the gas is 115 psia, what is the final absolute pressure of the gas?

3-6 Air confined in a tank has an initial temperature of 170°F and an initial pressure of 60 psia. If the air is cooled until the pressure falls to 32 psia, what is the final temperature of the gas in degrees Fahrenheit?

3-7 The tank of an air compressor has a volume of 1.8 ft^3 and is filled with air at 95°F. If the pressure of the air is 150 psig, what is the mass of the air in the tank?

3-8 Determine the specific volume of air having a temperature of 70°F at normal barometric pressure.

3-9 Four pounds of oxygen have a volume of 3.2 ft^3. If the pressure of the oxygen is 250 psia, what is the temperature of the oxygen in degrees Rankine?

3-10 The volume of 0.1 lb of air at normal barometric pressure and a temperature of 70°F is 1.334 ft^3. If the volume of the air remains constant while the temperature of the air is increased to 135°F by the addition of heat, determine the following:
(a) the final absolute pressure of the air in pounds per square inch,
(b) the increase in the internal energy of the air in Btu,
(c) the work done by the air in foot-pounds,
(d) the heat transferred to the air in Btu.

3-11 A certain mass of air confined in a container has an initial temperature of 128°F and an initial pressure of 95 psia. If the air is cooled until the pressure of the air is reduced to 10 psia, what is the final temperature of the air in degrees Rankine?

3-12 The volume of 0.1 lb of air at normal barometric pressure and a temperature of 70°F is 1.334 ft^3. If the pressure of the air is kept constant while the temperature of the air is increased to 135°F by the addition of heat, determine the following:
(a) the final volume of the air,
(b) the increase in the internal kinetic energy of the air in Btu,
(c) the heat transferred to the air in Btu,
(d) the work done by the air in foot-pounds,
(e) the work done by the air in Btu.

3-13 The volume of 0.1 lb of air at normal barometric pressure and a temperature of 70°F is 1.334 ft^3. If the air is compressed isothermally to a final pressure of 135 psia, determine the following:
(a) the final volume of the air in cubic feet,
(b) the work done by the air in foot-pounds,
(c) the increase in the internal kinetic energy of the air,
(d) the heat transferred to the gas in Btu.

3-14 The volume of 0.1 lb of air at normal barometric pressure and a temperature of 70°F is 1.334 ft^3. If the air is compressed adiabatically to a final pressure of 135 psia, determine the following:

(a) the final temperature in degrees Rankine,
(b) the final volume in cubic feet,
(c) the work done by the air in foot-pounds,
(d) the increase in the internal energy of the air in Btu,
(e) the heat transferred to the air in Btu.

3-15 Assume that compression of the air in Problem 3-14 occurs polytropically in such a way that the compression exponent, n, equals 1.32. Compute the following:
(a) the final temperature of the air in degrees Rankine,
(b) the final volume of the air in cubic feet,
(c) the work done by the air in foot-pounds,
(d) the increase in the internal kinetic energy of the air in Btu,
(e) the heat transferred to the air in Btu.

Metric Problems

3-16 A gas having an initial volume of 4 m^3 at a temperature of 630°K is cooled under constant pressure until its volume decreases to 1.7 m^3. Determine the final temperature of the gas in degrees Kelvin.

3-17 A gas having an initial temperature of 16°C is heated under constant pressure until its temperature increases to 30°C. If the initial volume of the gas is 1.4 m^3, what is its final volume?

3-18 Two kilograms of air are compressed at constant temperature from an initial volume of 2.6 m^3 to a final volume of 0.89 m^3. If the initial absolute pressure of the air is 2.1 bar, what is the final absolute pressure in bars?

3-19 Two cubic meters of gas are allowed to expand from an initial absolute pressure of 6.2 bar to a final absolute pressure of 1.6 bar while the temperature remains constant. Determine the final volume of the gas.

3-20 A certain mass of gas is expanded isothermally from an initial volume of 0.62 m^3 to a final volume of 3.16 m^3. If the initial absolute pressure of the gas is 6.5 bar, what is the final absolute pressure of the gas?

3-21 Air confined in a tank has an initial temperature of 75°C and an initial absolute pressure of 4.6 bar. If the air is cooled until the absolute pressure falls to 2.3 bar, what is the final temperature of the gas in degrees Celsius?

3-22 The tank of an air compressor has a volume of 0.8 m^3 and is filled with air at 53°C. If the absolute pressure of the air is 6.25 bar, what is the mass of the air in the tank?

3-23 Determine the specific volume of air having a temperature of 21°C at normal barometric pressure.

3-24 Four kilograms of oxygen have a volume of 0.5 m^3. If the absolute pressure of the oxygen is 7.76 bar, what is the temperature of the oxygen in degrees Kelvin?

3-25 Two kilograms of air are heated at constant volume from 40°C to 92°C.
(a) How much work is done by the air?
(b) What is the increase in the internal kinetic energy?
(c) How much heat is supplied to the air?

3-26 A tank having a volume of 0.65 m^3 is filled with CO_2 at a temperature of 25°C and an absolute pressure of 3 bar. Heat is supplied to the tank until the temperature of the CO_2 increases to 150°C.
(a) What is the mass of the CO_2 in kilograms?
(b) What is the final absolute pressure in bars?
(c) What is the increase in the internal kinetic energy?
(d) How much heat is supplied?

3-27 Two kilograms of air at a pressure of 17.42 bar are caused to expand from an initial volume of 0.12 m^3 to a final volume of 0.2 m^3 by the addition of heat while the pressure of the air remains constant. If the initial temperature of the air is 363°K,
(a) What is the final temperature of the air in degrees Kelvin?
(b) How much heat is supplied?

(c) What is the increase in the internal kinetic energy?
(d) How much work is done by the gas?

3-28 Two cubic meters of CO_2 at an initial pressure of 8.5 bar are expanded isothermally (at constant temperature) to a volume of 4.5 m^3.
(a) What is the increase in the internal energy?
(b) How much work is done by the gas?
(c) How much heat energy is transferred to the gas?

3-29 One cubic meter of air having an initial temperature of 22°C and an initial pressure of 1 bar is compressed isothermally to a final pressure of 8 bar.
(a) How much work is done by the air?
(b) What is the increase in the internal kinetic energy?
(c) How much heat is transferred to the air?

3-30 Assume that the air in Problem 3-29 is compressed adiabatically rather than isothermally. Compute the following:
(a) the final temperature of the air,
(b) the final volume of the air,
(c) the heat transferred to the air,
(d) the work done by the air,
(e) the increase in the internal kinetic energy.

3-31 Assume that the air in Problem 3-29 is compressed polytropically rather than isothermally. If n is equal to 1.25, determine the following:
(a) the mass of the gas in the cylinder,
(b) the final temperature of the air,
(c) the final volume of the air,
(d) the work done by the air,
(e) the increase in the internal kinetic energy,
(f) the heat transferred to the air.

4

SATURATED AND SUPERHEATED VAPORS

4-1. Saturation Temperature

The temperature at which a fluid will change from the liquid phase to the vapor phase or, conversely, from the vapor phase to the liquid phase is called the *saturation temperature*. A liquid at the saturation temperature is called a *saturated liquid*, and a vapor at the saturation temperature is called a *saturated vapor*. It is important to recognize that the saturation temperature of the liquid (the temperature at which the liquid will vaporize) and the saturation temperature of the vapor (the temperature at which the vapor will condense) are the same for any given pressure.

For any given pressure, the saturation temperature is the maximum temperature the liquid can have and the minimum temperature that the vapor can have. Any attempt to raise the temperature of a liquid above the saturation temperature will only result in vaporizing some part of the liquid. Similarly, any attempt to reduce the temperature of a vapor below the saturation temperature will only result in condensing some part of the vapor.

4-2. Superheated Vapor

A vapor at any temperature above the saturation temperature corresponding to its pressure is known as a *superheated vapor*. Once a liquid has been vaporized, the temperature of the resulting vapor can be further increased by the addition of energy. When the temperature of a vapor has been so increased above the saturation temperature, the vapor is said to be superheated, and the energy supplied to superheat the vapor is commonly referred to as *superheat*.

Before a vapor can be superheated, the vapor must be removed from contact with the vaporizing liquid (Fig. 4-1). Also, before a superheated vapor can be condensed, it must first be cooled to the saturation temperature corresponding to its pressure.

4-3. Subcooled Liquid

If, after condensation, the resulting liquid is cooled so that its temperature is reduced below the saturation temperature, the liquid is said to be subcooled. Consequently, a liquid at any temperature below the saturation temperature is a *subcooled liquid*.

4-4. The Effect of Pressure on the Saturation Temperature

The saturation temperature of a fluid depends on the pressure of the fluid. Increasing the pressure raises the saturation temperature, while reducing the pressure lowers the saturation temperature.

To illustrate the effect of pressure on the saturation temperature of a liquid, assume that

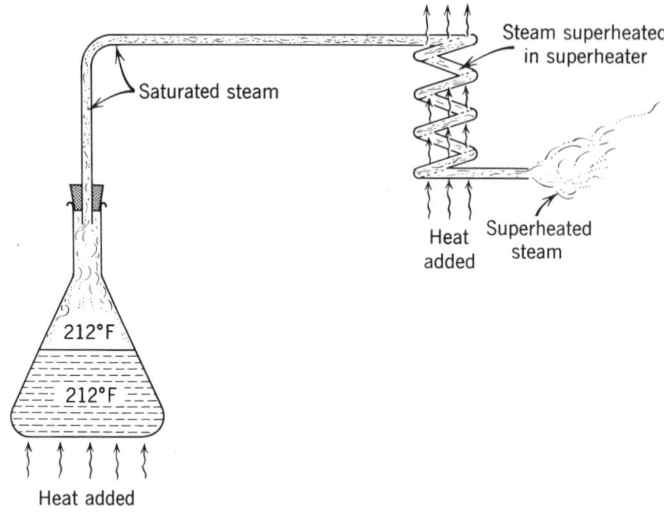

Fig. 4-1 Superheated vapor.

water is confined in a closed vessel that is equipped with a throttling valve at the top (Fig. 4-2*a*). A gage is used to determine the pressure exerted in the vessel, and two thermometers are installed so that one records the temperature of the vapor over the water. With the throttling valve wide open, the pressure exerted over the water is equal to one atmosphere. Since the saturation temperature of the water at normal atmospheric pressure is 212°F (100°C), the temperature of the water will rise as the water is heated until the temperature reaches 212°F. At this point, if the heating is continued, the water will begin to vaporize. Soon the space above the water will be filled with billons of water vapor molecules darting about at high velocities. Some of the vapor molecules will fall back into the water to become liquid molecules again, whereas others will escape through the opening to the outside

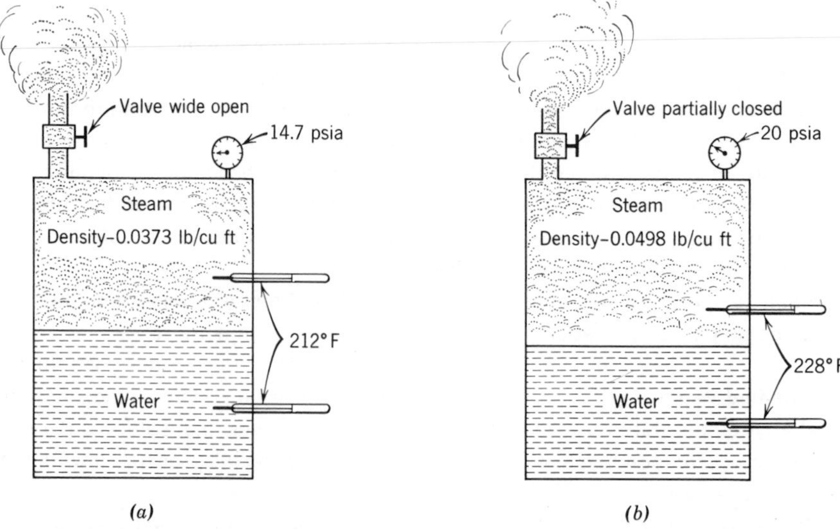

Fig. 4-2

and be carried away by air currents. If the opening at the top of the vessel is of sufficient size to allow the vapor to escape freely, the vapor will leave the vessel at the same rate that the liquid is vaporizing. That is, the number of molecules that are leaving the liquid to become vapor molecules will be exactly equal to the number of vapor molecules that are leaving the space, either by escaping to the outside or by falling back into the liquid. Thus the number of vapor molecules and the density of the vapor above the liquid will remain constant, and the pressure exerted by the vapor will be equal to that of the atmosphere outside the vessel.

Under this condition the water vapor from the vaporizing liquid will be saturated; that is, its temperature and pressure will be the same as that of the water, 212°F and 14.7 psia. The density of the water vapor at that temperature and pressure will be 0.0373 lb/ft^3, and its specific volume will be 1/0.0373 or 26.8 ft^3/lb.

Regardless of the rate at which the liquid is vaporizing, as long as the vapor is allowed to escape freely to the outside so that the pressure and density of the vapor over the liquid do not change, the liquid will continue to vaporize at 212°F.

Suppose that the throttling valve is partially closed so that the escape of the vapor from the vessel is somewhat impeded. For a time, the equilibrium will be disturbed in that the vapor will not be leaving the vessel at the same rate that it is being generated by the vaporizing liquid. Consequently, the number of vapor molecules in the space above the liquid will increase, thereby increasing the density and the pressure of the vapor over the liquid and raising the saturation temperature.

If it is assumed that the pressure of the vapor increases to 5.3 psig (20 psia) before equilibrium is again established, that is, before the rate at which the vapor is escaping to the outside is exactly equal to the rate at which the liquid is vaporizing, the saturation temperature will be 228°F, the density of the vapor will be 0.0498 lb/ft^3, and 1 lb of vapor will occupy a volume of 20.08 ft^3. This condition is illustrated in Fig. 4-2*b*.

By comparing the condition of the vapor in Fig. 4-2*b* with that of the vapor in Fig. 4-2*a*, it will be noted that the density of the vapor is greater at the higher pressure and saturation temperature. Furthermore, it is evident that the pressure and the saturation temperature of a liquid or vapor can be controlled by regulating the rate at which the vapor escapes from over the liquid.

In Fig. 4-2*a*, the rate of vaporization will have little or no effect on the pressure and saturation temperature because the vapor is allowed to escape freely so that the density and pressure of the vapor over the liquid will neither increase nor decrease as the rate of vaporization is changed. On the other hand, in Fig. 4-2*b*, any increase in the rate of vaporization will cause an increase in the density and pressure of the vapor and result in an increase in the saturation temperature. The reason is that any increase in the rate of vaporization will necessitate the escape of a greater quantity of vapor in a given length of time. Since the size of the vapor outlet is fixed by the throttling action of the valve, the pressure of the vapor in the vessel will increase until the pressure difference between the inside and outside of the vessel is sufficient to allow the vapor to escape at a rate equal to that at which the liquid is vaporizing. The increase in pressure, of course, results in an increase in the saturation temperature and in the density of the vapor. Likewise, any decrease in the rate of vaporization will have the opposite effect. The pressure and density of the vapor over the liquid will decrease, and the saturation temperature will be lower.

Assume now that the throttling valve on the container is again opened completely, as in Fig. 4-2*a*, so that the vapor is allowed to escape freely and unimpeded from over the liquid. The density and pressure of the vapor will decrease until the pressure of the vapor is again equal to that of the atmosphere outside the container. Since the saturation temperature of water at atmospheric pressure is 212°F and since a liquid cannot exist as a liquid at any temperature above its saturation temperature corresponding to its pressure, it is evident that the water must cool itself from 228°F to 212°F at the instant that the pressure drops to one atmosphere. To accomplish this cooling, a portion of the liquid will "flash" into the vapor state. The latent heat necessary to vaporize the portion of the liquid that flashes into the vapor state is supplied by the mass of the liquid, and as a result of supplying the vaporizing heat, the temperature of the mass of the liquid will be reduced to the new saturation temperature. Enough of the liquid will vaporize to provide the required amount of cooling.

4-5. Vaporization

The vaporization of a liquid can occur in two ways: (1) by evaporation and (2) by boiling or ebullition. The vaporization of a liquid by the process of evaporation occurs only at the free surface of the liquid and may take place at any temperature below the saturation temperature. Evaporation occurs without any visible disturbance of the liquid.

Boiling or ebullition, however, occurs only at the saturation temperature. Since the saturation temperature is the temperature at which the vapor pressure of the liquid is equal to the pressure exerted on the liquid, this type of vaporization occurs throughout the entire body of the liquid as well as at the free surface and is accompanied by considerable agitation of the liquid and the rapid formation of bubbles that expand, rise to the top of the liquid, and burst.

To this point, only the boiling type of vaporization has been considered.

4-6. Evaporation

Evaporation takes place continually, and the fact that water evaporates from lakes, rivers, ponds, clothes, and so forth is sufficient evidence that evaporation can and does occur at temperatures below the saturation temperature. Any liquid open to the atmosphere, regardless of its temperature, will gradually evaporate and be diffused into the air.

The vaporization of liquids at temperatures below their saturation temperature can be explained in this manner. The molecules of a liquid are in constant and rapid motion, their velocities being determined by the temperature of the liquid. In the course of their movements the molecules are continually colliding with one another, and as a result of these impacts, some of the molecules of the liquid momentarily attain velocities much higher than the average velocity of the other molecules of the mass. Thus, their energy is much greater than the average energy of the mass. If this occurs within the body of the liquid, the high-velocity molecules quickly lose their extra energy in subsequent collisions with other molecules. However, if the molecules attaining the higher than normal velocities are near the surface, they may project themselves from the surface of the liquid and escape into the air to become vapor molecules (Fig. 4-3). The molecules so escaping from the liquid are diffused throughout the air.

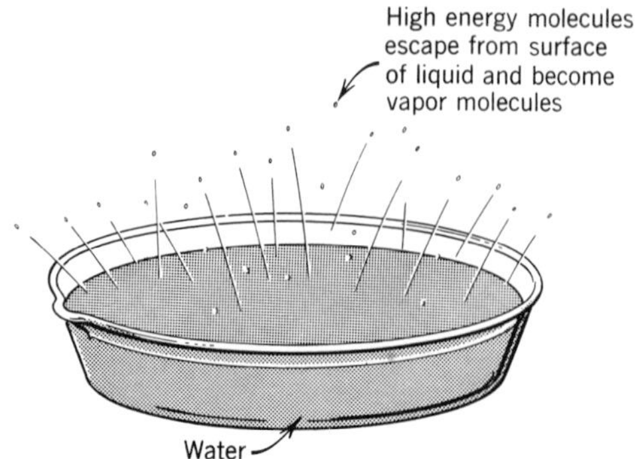

Fig. 4-3 Evaporation from surface of a liquid.

4-7. Rate of Vaporization

For any given temperature, some liquids will evaporate faster than others. Liquids having the lowest "boiling" points, that is, the lowest saturation temperature for a given pressure, evaporate at the highest rate. However, for any particular liquid, the rate of vaporization varies with a number of factors. In general, the rate of evaporation increases as the temperature of the liquid increases and as the pressure over the liquid decreases. Evaporation increases also with the amount of exposed surface. Furthermore, it will be shown later that the rate of evaporation is dependent on the degree of saturation of the vapor, which is always adjacent to and above the liquid.

4-8. The Cooling Effect of Evaporation

Since it is the higher velocity molecules (those having the most energy) that escape from the surface of an evaporating liquid, it follows that the average energy of the mass is thereby reduced and the temperature of the mass lowered. Whenever any portion of a liquid vaporizes, an amount of heat equal to the latent heat of vaporization must be absorbed by that portion, either from the mass of the liquid, from the surrounding air, or from adjacent objects. Thus the energy and temperature of the mass are reduced as it supplies the latent heat of vaporization to the portion of the liquid that vaporizes. The temperature of the mass is reduced to a point slightly below that of the surrounding media, and the temperature difference so established causes heat to flow from the surrounding media into the mass of the liquid. The energy lost by the mass during vaporization is thereby replenished, and evaporation becomes a continuous process as long as any of the liquid remains. The vapor resulting from evaporation is diffused into and carried away by the air.

4-9. Confined Liquid-Vapor Mixtures

When a vapor is confined in a container with a portion of its own liquid, both the vapor and the liquid will be saturated. To illustrate, assume that an open container is partially filled with water and is stored where the ambient temperature is 70°F (Fig. 4-4*a*). The water will be evaporating at 70°F, and as described in the previous section, the vapor molecules leaving the liquid will be diffused into the surrounding air so that evaporation will continue until all the liquid is evaporated. However, if a tightly fitting cover is placed over the container, the vapor molecules will be unable to escape to the outside, and they will collect above the liquid. Soon the space above the liquid will be so filled with vapor molecules that there will be as many molecules falling back into the liquid as there are leaving the liquid. A condition of equilibrium will be attained, the vapor will be saturated, and no further evaporation will occur. The energy of the liquid will be

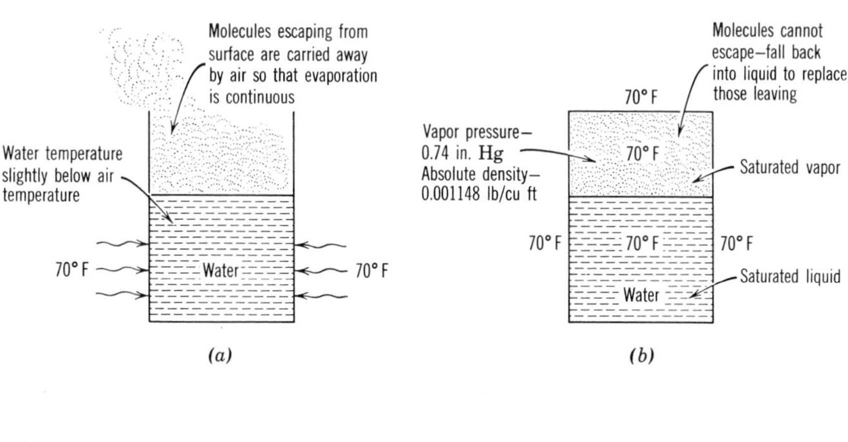

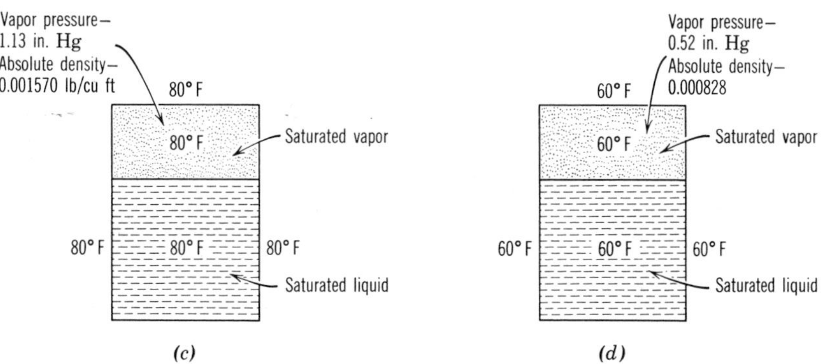

Fig. 4-4

increased by the vapor molecules that are returning to the liquid in exactly the same amount that it is diminished by the molecules that are leaving. Since no further cooling will take place by evaporation, the liquid will assume the temperature of the surrounding air, and heat transfer will cease.

If, at this point, the ambient temperature rises to, say 80°F, heat transfer will again take place between the surrounding air and the liquid. The temperature and average molecular velocity of the liquid will be increased, and evaporation will be resumed. The number of molecules leaving the liquid will again be greater than the number returning, and the density and pressure of the vapor above the liquid will be increased. As the density and pressure of the vapor increase, the saturation temperature of the liquid increases. Eventually, when the saturation temperature reaches 80°F and is equal to the ambient temperature, no further heat transfer will occur, and evaporation will cease. Equilibrium will have again been established. The density and pressure of the vapor will be greater than before, the saturation temperature of the liquid-vapor mixture will be higher, and there will be more vapor and less liquid in the container than previously (Fig. 4-4*c*).

Suppose now that the ambient temperature falls to 60°F. When this occurs, heat will flow from the 80°F liquid-vapor mixture to the cooler surrounding air. As the liquid-vapor mixture loses heat to the surrounding air, its temperature and average molecular velocity will be decreased, and many of the vapor molecules, lacking sufficient energy to remain in the vapor state, will fall back into the liquid and resume the molecular arrangement of the liquid state; that is, a part of the vapor will condense. The density and pressure of the vapor will be diminished, and the saturation temperature of the mixture will be reduced. When the saturation temperature of the mixture falls to 60°F, it will be the same as the ambient temperature and no further heat flow will occur. Equilibrium will have been established, and the number of molecules reentering the liquid will exactly equal those that are leaving. At this new condition, the density and pressure of the vapor will be less than before, the saturation temperature will be lower, and since a part of the vapor condensed into liquid, there will be more liquid and less vapor comprising the mixture than at the previous condition (Fig. 4-4*d*).

4-10. Sublimation

It is possible for a substance to go directly from the solid state to the vapor state without apparently passing through the liquid state. Any solid substance will sublime at any temperature below its fusion temperature. Sublimation takes place in a manner similar to evaporation, although much slower, in that the higher velocity molecules near the surface escape from the mass into the surrounding air and become vapor molecules. One of the most familiar examples of sublimation is that of solid CO_2 (dry ice), which, at normal temperatures and pressures, sublimes directly from the solid to the vapor state. Damp wash frozen on the line in winter will sublime dry. During freezing weather, ice and snow will sublime from streets and sidewalks.

4-11. Condensation

Condensation of a vapor may be accomplished in several ways: (1) by extracting heat from the vapor, (2) by increasing the pressure of the vapor, or (3) by some combination of these two methods.

4-12. Condensing by Extracting Heat from a Saturated Vapor

A saturated vapor has been previously described as one at a condition such that any further cooling will cause a part of the vapor to condense. This is because a vapor cannot exist as a vapor at any temperature below its saturation temperature. When the vapor is cooled, the vapor molecules cannot maintain sufficient energy and velocity to overcome the attractive forces of one another and remain as vapor molecules. Some of the molecules, overcome by the attractive forces, will revert to the molecular structure of the liquid state. When condensation occurs while the vapor is confined so that the volume remains constant, the density and pressure of the vapor will decrease, so that there is a decrease in the saturation temperature. If, as in a vapor condenser, more vapor is entering the vessel as the vapor condenses and drains from the vessel as a liquid, the density, pressure, and saturation temperature of the vapor will remain constant, and condensation will continue as long as heat is continuously extracted from the vapor.

4-13. Condensing by Increasing the Pressure of a Vapor

When a vapor is compressed at a constant temperature, its volume diminishes, and the density of the vapor increases as the molecules of the vapor are forced into a smaller volume. In the case of a superheated vapor, the saturation temperature of the vapor increases as the pressure increases until a point is reached at which the saturation temperature of the vapor is equal to the actual temperature of the vapor. When this occurs, the density of the vapor will be at a maximum value for that condition, and any further compression will cause a part of the vapor to assume the more restrained molecular structure of the liquid state. Thereafter, further compression will cause progressive increases in both the pressure and saturation temperature of the liquid-vapor mixture, accompanied by continuous and progressive condensation of the vapor into the liquid state. Assuming no heat transfer to the surroundings, the latent heat given up by the condensing vapor supplies the sensible heat necessary to provide the increase in the saturation temperature of the mixture which must accompany the increase in pressure. The process involved is essentially the reverse of the expansion process described in the last paragraph of Section 4-4.

4-14. Critical Temperature

The temperature of a gas may be raised to a point such that it cannot become saturated regardless of the amount of pressure applied. The critical temperature of any gas is the highest temperature the gas can have and still be condensable by the application of pressure. The critical temperature is different for every gas. Some gases have high critical temperatures, while the critical temperatures of others are relatively low. For example, the critical temperature of water vapor is approximately 706°F (375°C), whereas the critical temperature of air is approximately −225°F (−143°C).

4-15. Critical Pressure

Critical pressure is the lowest pressure at which a substance can exist in the liquid state at its critical temperature; that is, it is the saturation pressure at the critical temperature.

4-16. Important Properties of Gases and Vapors

Although a gas or vapor has many properties, only six are of particular importance in the study of refrigeration. These are pressure, temperature, volume, enthalpy, internal energy, and entropy. Pressure, temperature, and volume are called measurable properties because they can actually be measured. Enthalpy. internal energy, and entropy cannot be measured. They must be calculated and are therefore known as calculated properties.

Pressure, temperature, volume, and internal energy have already been discussed to some extent. A discussion of enthalpy and entropy follows.

4-17. Enthalpy

Enthalpy is a calculated property of matter that is sometimes very loosely defined as "total heat." More specifically, the enthalpy, H, of a given mass of material at any given thermodynamic condition is the summation of all the energy supplied to it to bring it to that condition from some initial condition arbitrarily taken as the zero point of enthalpy.

Whereas the total enthalpy, H, represents the enthalpy of m pounds, the specific enthalpy, h, is the enthalpy of one pound. Since it is usually the specific enthalpy rather than the total enthalpy that is of interest, hereinafter in this book the term "enthalpy" shall be used to mean specific enthalpy.

Mathematically, enthalpy is defined as

$$h = u + \frac{pv}{J} \qquad (4\text{-}1)$$

where h = the enthalpy in Btu per pound
u = the internal energy in Btu per pound
p = the absolute pressure in pounds per square foot
v = the specific volume in cubic feet per pound
J = the mechanical energy equivalent (778 ft-lb/Btu)

It was shown in Section 3-8 that not all the energy supplied to a fluid stays in the fluid as an increase in the internal energy of the fluid. In many cases, some part of or all the energy supplied to the fluid leaves the fluid as work. In Equation 4-1, the part of the transferred energy that is stored in the fluid as an increase in the internal energy is represented by the term u, whereas the part of the transferred energy that leaves the fluid as work is represented by the term pv/J. While the energy represented by the term pv/J is not stored in the fluid as an increase in the internal energy, it nevertheless represents energy that must be supplied to the fluid in order to bring the fluid to the given condition from the initial condition at the arbitrarily selected zero point of enthalpy. Also, this external work energy must pass back through and be given up by the fluid as the fluid returns to the initial condition.

Consider, for example, the vaporization of 1 lb of water into steam at 212°F under a pressure of one atmosphere. The volume of 1 lb of water at 212°F is 0.01672 ft^3/lb, whereas the volume of 1 lb of steam at 212°F is 26.80 ft^3/lb. Consequently, during the vaporization process the fluid expands at constant pressure from a volume of 0.01672 ft^3/lb to a volume of 26.80 ft^3/lb, thereby doing work in expanding against the pressure of the atmosphere.

The enthalpy of 1 lb of water at 212°F is 180.07 Btu/lb, while the enthalpy of 1 lb of steam at 212°F is 1150.4 Btu/lb, so that the enthalpy (latent heat) of vaporization is (1150.4 − 180.07) 970.3 Btu/lb, the latter being the total quantity of energy required to vaporize 1 lb of water at 212°F into steam at 212°F. Of the total energy supplied, the amount required to do the work of the constant pressure expansion is approximately

$$\begin{aligned} w_p &= p(v_2 - v_1) \\ &= (2116 \text{ psfa})(26.8 \text{ ft}^3/\text{lb} - 0.01672 \text{ ft}^3/\text{lb}) \\ &= 56{,}673 \text{ ft-lb} \end{aligned}$$

or

$$\frac{56{,}673 \text{ ft-lb}}{778 \text{ ft-lb/Btu}} = 72.8 \text{ Btu}$$

Consequently, the part of the energy supplied that is stored in the fluid as an increase in the internal energy is (970.3 − 72.8) = 897.5 Btu.

A pV diagram of the foregoing vaporization process is shown in Fig. 4-5. The values for enthalpy and specific volume were taken from

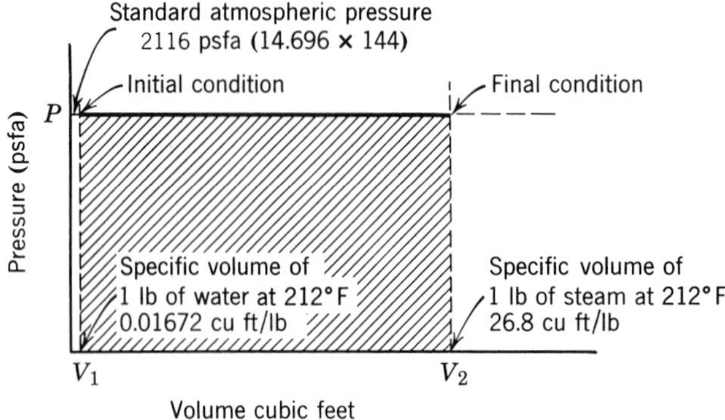

Fig. 4-5 Pressure-volume diagram showing the external work done by fluid expansion as 1 lb of water is vaporized at atmospheric pressure—approximately 56,700 ft-lb.

a vapor table. Vapor tables are discussed later in this chapter.

4-18. Entropy

Entropy, like enthalpy, is a calculated property of matter. The entropy S of a given mass of material at any given condition is an expression of the total energy transferred to the material per degree of absolute temperature to bring the material to that condition from some arbitrarily selected zero or reference point. For any one fluid, the reference point for entropy calculations is the same as that for the enthalpy calculations.

Again, as in the case of enthalpy, it is the specific entropy s rather than the total entropy S that is of interest. Therefore, the term "entropy" is used here to mean the specific entropy s unless otherwise indicated.

It has been shown (Section 3-18) that the work of a process can be expressed as the product of the change in volume ΔV and the average absolute pressure. Similarly, it is often convenient to express the heat energy transferred during a process as the product of two such factors. The concept of entropy makes this possible. The heat energy transferred during a process can be expressed as the product of the change in entropy and the average absolute temperature.* Mathematically the relationship is expressed by the following equations:

$$\Delta Q = (\Delta s)(T_m) \qquad (4\text{-}2)$$

$$\Delta s = \frac{\Delta Q}{T_m} \qquad (4\text{-}3)$$

$$T_m = \frac{\Delta Q}{\Delta s} \qquad (4\text{-}4)$$

where ΔQ = the heat energy transferred in Btu per pound
Δs = the change in entropy in Btu per pound per degree Fahrenheit
T_m = the average absolute temperature in degrees Rankine.

The relationships defined by the foregoing equations are for ideal reversible processes (see Section 3-31). However, since the change in entropy Δs between any two state points has a definite value, regardless of the path of the process, the change in entropy for an irreversible process can be determined by making the calculation for a reversible process which takes place between the same two state points.

On a pV diagram (Fig. 4-5), the area under the process line, which is the product of the change in volume ΔV and the average absolute pressure, represents the work of the process. Similarly, on a temperature-entropy (*Ts*) diagram (temperature plotted against entropy), the area under the process line, which is the product of the change in entropy ΔS and the average absolute temperature T_m, represents the heat transferred during the process. The vaporization process described in Section 4-17 is plotted on *Ts* coordinates in Fig. 4-6.

* The average absolute temperature is not merely the mean of the initial and final temperatures of the process but is the average of all the absolute temperatures through which the process passes.

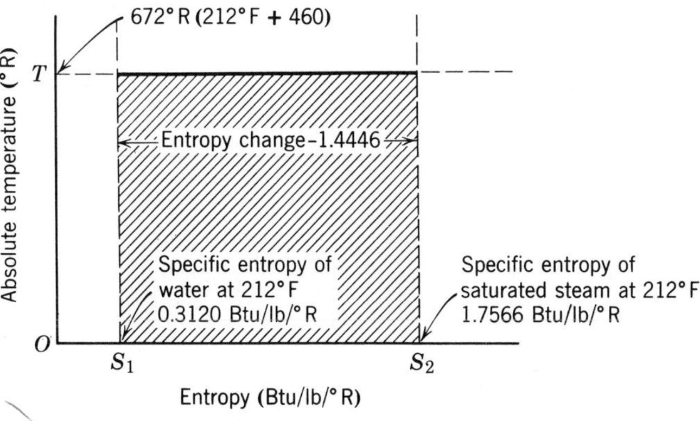

Fig. 4-6

Since the temperature remains constant at 212°F during the aforementioned vaporization process, the average absolute temperature T_m for the process is 672°R. From the steam table, the entropy of 1 lb of water at 212°F is 0.3120 Btu/lb °F, and the entropy of 1 lb of steam at 212°F is 1.7566 Btu/lb °F. Applying Equation 4-2, the heat transferred during the process is

$$Q = T_m(s_2 - s_1) = (672°\text{R})(1.7566 - 0.3120)$$
$$= 970.7 \text{ Btu/lb}$$

Entropy is frequently described as a measure of the unavailability of energy. Not all the energy of a system is available to do useful work. For example, consider a steam radiator being used to heat a room whose temperature is maintained at 75°F. Even under ideal conditions, the maximum amount of energy that would be available to the room from any steam delivered to the radiator is the heat given off by the steam as it condenses and by the resulting condensate as the latter cools to the room temperature of 75°F. Although the resultant 75°F condensate will still have a considerable amount of energy with reference to absolute zero, this energy is unavailable to the 75°F room because no temperature differential exists between the condensate and the room. Entropy is an index of this unavailability.

In all actual (irreversible) processes, the entropy of the system (or the surroundings) must of necessity increase, primarily because some of the available energy must be used to overcome friction (Section 3-31). On the other hand, in any ideal reversible adiabatic process (Section 3-24), frequently called a "frictionless adiabatic" or "isentropic" process, it is assumed that none of the available energy is used in overcoming friction, and since there is no heat transfer, the entropy (unavailable energy) is assumed to remain constant during the process.

For example, with reference to Fig. 4-7, as the gas in the cylinder expands isentropically, all the energy given up by the gas is transferred to the piston and subsequently stored in the flywheel. On the upstroke, the energy stored in the flywheel is returned to the piston and subsequently to the gas as the gas is compressed isentropically back to its initial state.

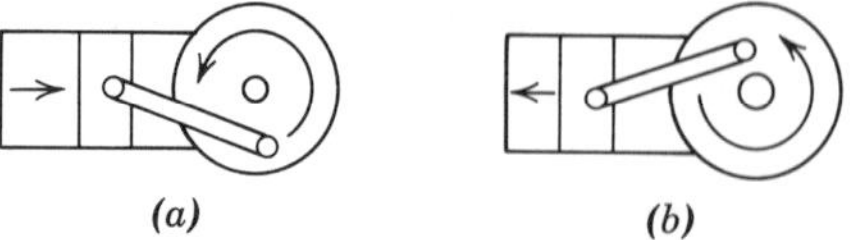

Fig. 4-7 (*a*) Expanding gas gives energy to flywheel during expansion process. (*b*) Flywheel returns stored energy to gas during compression process.

Notice that throughout the cycle, the available energy of the system remains unchanged so that the entropy (unavailability) also remains constant.

In contrast, assume now that the gas in the cylinder is expanded irreversibly in that some of the energy given up by the gas is used to overcome friction in the cylinder and other mechanical parts. It follows that only a part of the energy available from the gas is transferred to the piston and stored in the flywheel. Consequently, on the upstroke, the energy available from the flywheel will be somewhat less than that required to recompress the gas to its initial state. Moreover, during the recompression process, some of the available energy from the flywheel will be used in overcoming friction, so that the total energy of the gas at the end of the recompression process will be less than the total energy of the gas in the initial state. Because of the irreversible losses in energy suffered in both processes, it is evident that the available energy of the system has been decreased and that the entropy (unavailability) has been increased.

Ordinarily, the energy utilized to offset friction is converted into heat and dissipated to the surroundings. Consequently, if irreversible processes are to be repeated, energy to replace that lost to the surroundings must be supplied continuously to the system from some external source; otherwise the system would soon run out of energy.

4-19. Vapor Tables

Because of the relatively strong forces of mutual attraction that exist between the molecules of a vapor, internal friction is always present when a vapor undergoes a change in condition. Consequently, the behavior of vapors cannot be predicted by application of the ideal gas laws. For this reason, the properties of saturated fluids and superheated vapors have been determined by experiment for all common fluids, and these data are published in the form of tables and charts.

4-20. Properties of Saturated Fluids

The properties of saturated liquids and vapors at various conditions are given in *saturated vapor tables*, which usually list values for the following properties: (1) temperature, (2) pressure, (3) specific volume, (4) enthalpy, and (5) entropy (Fig. 4-8). The temperature given is the saturation temperature of the fluid and is usually listed in the extreme lefthand column as the entry property to the table. Normally, the corresponding saturation pressure is listed in the second column and is followed by the specific volume for both the liquid and the vapor states in the third and fourth columns, respectively. Some tables list the density in addition to, or in place of, the specific volume. If only the density is given and the specific volume is wanted, the specific volume is found by dividing the density into 1. Likewise, when the specific volume is given and the density is wanted, the density is found by taking the reciprocal of the specific volume.

Three values for enthalpy are listed in most tables: (1) the enthalpy of the saturated liquid (h_f), which is the summation of the energy required to raise the temperature of the liquid from the temperature at the assumed zero point of the enthalpy calculation to the saturation temperature listed in column one; (2) the enthalpy of vaporization (h_{fg}), which is the latent heat of vaporization at the given saturation temperature; and (3) the enthalpy of the saturated vapor (h_g), which is the sum of the enthalpy of the liquid (h_f) and the enthalpy of vaporization (h_{fg}). When the enthalpy of vaporization is not listed, it can be determined by subtracting h_f from h_g.

Two values of entropy are usually listed: s_f, the entropy of the saturated liquid, and s_g, the entropy of the saturated vapor. Some tables also list a value for s_{fg}, which is the change in entropy during the change in phase. When this value is not listed, it can be determined by subtracting s_f from s_g.

Some tables also list values for the internal energy of the fluid in both the liquid and vapor phases. When this is not the case, the value for the internal energy of the fluid in either phase can be computed by rearranging and applying Equation 4-1 to solve for the internal energy u.

It has been stated previously that the condition of a gas or a vapor can be determined when any two of its properties are known. However, for a saturated liquid or vapor at any one pressure, there is only one temperature that the fluid can have and still satisfy the conditions of saturation. This is true also for the other properties of a saturated liquid or vapor. Therefore, if either the temperature or the pressure of a saturated liquid or vapor is known, the value of the other properties can be read directly from the saturated vapor table.

Example 4-1 Determine the pressure, specific volume, enthalpy, entropy, and internal energy of 1 lb of R-12 saturated vapor at a temperature of 40°F.

Solution In Fig. 4-8 or Table 16-3 (Saturation Properties of R-12), locate the saturation temperature of 40°F in column one, and on the same line in column two, read the corresponding saturation pressure of 51.68 psia. Still on the same line, the specific volume of the vapor (v_g) is 0.792 ft³/lb, the enthalpy of the vapor (h_g) is 82.71 Btu/lb, and the entropy of the vapor (s_g) is 0.16833 Btu/lb °F.

Rearranging and applying Equation 4-1,

$$u = (82.71 \text{ Btu/lb}) - \frac{(51.68 \text{ psia} \times 144)(0.792 \text{ ft}^3\text{/lb})}{778 \text{ ft-lb/Btu}}$$

$$= 75.13 \text{ Btu/lb}$$

4-21. Superheated Vapor Tables

A superheated vapor table deals with the properties of a superheated vapor rather than those of a saturated vapor, and the arrangement is somewhat different from that of a saturated vapor table. One common form of the superheated vapor table is illustrated in Fig. 4-9.

Temp.	Pressure		Volume		Density		Heat Content from −40°			Entropy from −40°		Temp.
°F t	Abs. lb/in.^{2} p	Gage lb/in.^{2} p_d	Liquid ft^3/lb v_f	Vapor ft^3/lb v_g	Liquid lb/ft^3 $1/v_f$	Vapor lb/ft^3 $1/v_g$	Liquid Btu/lb h_f	Latent Btu/lb h	Vapor Btu/lb h_g	Liquid Btu/lb °F s_f	Vapor Btu/lb °F s_g	°F t
10	29.35	14.65	0.0112	1.351	89.45	0.7402	10.39	68.97	79.36	0.02328	0.17015	10
12	30.56	15.86	0.0112	1.301	89.24	0.7687	10.82	68.77	79.59	0.02419	0.17001	12
14	31.80	17.10	0.0112	1.253	89.03	0.7981	11.26	68.56	79.82	0.02510	0.16987	14
16	33.08	18.38	0.0112	1.207	88.81	0.8288	11.70	68.35	80.05	0.02601	0.16974	16
18	34.40	19.70	0.0113	1.163	88.58	0.8598	12.12	68.15	80.27	0.02692	0.16961	18
20	35.75	21.05	0.0113	1.121	88.37	0.8921	12.55	67.94	80.49	0.02783	0.16949	20
22	37.15	22.45	0.0113	1.081	88.13	0.9251	13.00	67.72	80.72	0.02873	0.16938	22
24	38.58	23.88	0.0113	1.043	87.91	0.9588	13.44	67.51	80.95	0.02963	0.16926	24
26	40.07	25.37	0.0114	1.007	87.68	0.9930	13.88	67.29	81.17	0.03053	0.16913	26
28	41.59	26.89	0.0114	0.973	87.47	1.028	14.32	67.07	81.39	0.03143	0.16900	28
30	43.16	28.46	0.0115	0.939	87.24	1.065	14.76	66.85	81.61	0.03233	0.16887	30
32	44.77	30.07	0.0115	0.908	87.02	1.102	15.21	66.62	81.83	0.03323	0.16876	32
34	46.42	31.72	0.0115	0.877	86.78	1.140	15.65	66.40	82.05	0.03413	0.16865	34
36	48.13	33.43	0.0116	0.848	86.55	1.180	16.10	66.17	82.27	0.03502	0.16854	36
38	49.88	35.18	0.0116	0.819	86.33	1.221	16.55	65.94	82.49	0.03591	0.16843	38
40	51.68	36.98	0.0116	0.792	86.10	1.263	17.00	65.71	82.71	0.03680	0.16833	40
42	53.51	38.81	0.0116	0.767	85.88	1.304	17.46	65.47	82.93	0.03770	0.16823	42
44	55.40	40.70	0.0117	0.742	85.66	1.349	17.91	65.24	83.15	0.03859	0.16813	44
46	57.35	42.65	0.0117	0.718	85.43	1.393	18.36	65.00	83.36	0.03948	0.16803	46
48	59.35	44.65	0.0117	0.695	85.19	1.438	18.82	64.74	83.57	0.04037	0.16794	48

Fig. 4-8 Dichlorodifluoromethane (Refrigerant-12) properties of saturated vapor.

Dichlorodifluoromethane (Refrigerant-12) Properties of Superheated Vapor

Temp. °F	Abs. Pressure 36 lb/in.² Gage Pressure 21.3 lb/in.² (Sat. Temp. 20.4° F)			Abs. Pressure 38 lb/in.² Gage Pressure 23.3 lb/in.² (Sat. Temp. 23.2° F)			Abs. Pressure 40 lb/in.² Gage Pressure 25.3 lb/in.² (Sat. Temp. 25.9° F)			Abs. Pressure 42 lb/in.² Gage Pressure 27.3 lb/in.² (Sat. Temp. 28.5° F)		
t	*V*	*H*	*S*	*V*	*H*	*S*	*V*	*H*	*S*	*V*	*H*	*S*
(at sat'n)	*(1.113)*	*(80.54)*	*(0.16947)*	*(1.058)*	*(80.86)*	*(0.16931)*	*(1.009)*	*(81.16)*	*(0.16914)*	*(0.963)*	*(81.44)*	*(0.16897)*
30	1.140	81.90	0.17227	1.076	81.82	0.17126	1.019	81.76	0.17030	0.967	81.65	0.16939
40	1.168	83.35	0.17518	1.103	83.27	0.17418	1.044	83.20	0.17322	0.991	83.10	0.17231
50	1.196	84.81	0.17806	1.129	84.72	0.17706	1.070	84.65	0.17612	1.016	84.56	0.17521
60	1.223	86.27	0.18089	1.156	86.19	0.17991	1.095	86.11	0.17896	1.040	86.03	0.17806
70	1.250	87.74	0.18369	1.182	87.67	0.18272	1.120	87.60	0.18178	1.063	87.51	0.18086
80	1.278	89.22	0.18647	1.208	89.16	0.18551	1.144	89.09	0.18455	1.087	89.00	0.18365
90	1.305	90.71	0.18921	1.234	90.66	0.18826	1.169	90.58	0.18731	1.110	90.50	0.18640
100	1.332	92.22	0.19193	1.260	92.17	0.19096	1.194	92.09	0.19004	1.134	92.01	0.18913
110	1.359	93.75	0.19462	1.285	93.69	0.19365	1.218	93.62	0.19272	1.158	93.54	0.19184
120	1.386	95.28	0.19729	1.310	95.22	0.19631	1.242	95.15	0.19538	1.181	95.09	0.19451
130	1.412	96.82	0.19991	1.336	96.76	0.19895	1.267	96.70	0.19803	1.204	96.64	0.19714
140	1.439	98.37	0.20254	1.361	98.32	0.20157	1.291	98.26	0.20066	1.227	98.20	0.19979
150	1.465	99.93	0.20512	1.387	99.89	0.20416	1.315	99.83	0.20325	1.250	99.77	0.20237
160	1.492	101.51	0.20770	1.412	101.47	0.20673	1.340	101.42	0.20583	1.274	101.36	0.20496
170	1.518	103.11	0.21024	1.437	103.07	0.20929	1.364	103.02	0.20838	1.297	102.96	0.20751
180	1.545	104.72	0.21278	1.462	104.67	0.21183	1.388	104.63	0.21092	1.320	104.57	0.21005
190	1.571	106.34	0.21528	1.487	106.29	0.21433	1.412	106.25	0.21343	1.343	106.19	0.21256
200	1.597	107.97	0.21778	1.512	107.93	0.21681	1.435	107.88	0.21592	1.365	107.82	0.21505
210	1.623	109.61	0.22024	1.537	109.57	0.21928	1.459	109.52	0.21840	1.388	109.47	0.21754
220	1.650	111.27	0.22270	1.562	111.22	0.22176	1.482	111.17	0.22085	1.411	111.12	0.22000
230	1.676	112.94	0.22513	1.587	112.89	0.22419	1.506	112.84	0.22329	1.434	112.80	0.22244
240	1.702	114.62	0.22756	1.612	114.58	0.22662	1.530	114.52	0.22572	1.457	114.49	0.22486
250	1.728	116.31	0.22996	1.637	116.28	0.22903	1.554	116.21	0.22813	1.480	116.19	0.22728
260	1.754	118.02	0.23235	1.662	117.99	0.23142	1.577	117.92	0.23052	1.502	117.90	0.22967
270	1.780	119.74	0.23472	1.687	119.71	0.23379	1.601	119.65	0.23289	1.524	119.62	0.23204
280	1.807	121.47	0.23708	1.712	121.45	0.23616	1.625	121.40	0.23526	1.547	121.36	0.23441
290	1.833	123.22	0.23942	1.737	123.20	0.23850	1.649	123.15	0.23760	1.570	123.11	0.23675
300	......	......	......	1.762	124.95	0.24083	1.673	124.92	0.23994	1.592	124.87	0.23909

Fig. 4-9 Abridged from typical superheated vapor table. (Copyright by E. I. du Pont de Nemours and Co., Inc. Reprinted by permission.)

Before examining the superheated vapor table, it is important to take note of one significant difference between a saturated and a superheated vapor. Whereas there is only one temperature that will satisfy the conditions of saturation, a superheated vapor may have any temperature above the saturation temperature corresponding to its pressure. The specific volume, enthalpy, and entropy of a superheated vapor at any one pressure will vary with the temperature. This does not mean that the properties of a superheated vapor are entirely independent of the pressure of the vapor but only that the properties of the superheated vapor at any one pressure will vary with the temperature.

Before the properties of a superheated vapor can be determined from a superheated vapor table, it is necessary to know either the saturation temperature or pressure of the vapor, as well as the actual superheated temperature.

Some superheated vapor tables use saturation temperature as the entry property (Table 16-2B) while others, including the one illustrated in Fig. 4-9, use pressure as the entry property. In either case, the corresponding pressure or saturation temperature is also listed. Notice in Fig. 4-9 that the saturation pressures are located at the top of the table and the corresponding saturation temperatures are shown in parenthesis immediately below the listed pressure. The superheated temperatures are listed on the right side of the table in 10°F increments. The body of the table provides values for specific volume (v), enthalpy (h), and entropy (s) for each of the superheated temperatures listed and also for the saturated condition. Similar values for superheated temperatures not listed can be determined by direct interpolation.

Example 4-2 One pound of superheated Refrigerant-12 vapor at a temperature of 50°F has a pressure of 40 psia. From the abbreviated table in Fig. 4-9, determine the following:

(a) the temperature, volume, enthalpy, and entropy of the vapor at saturation,
(b) the volume, enthalpy, and entropy of the vapor at the superheated condition,
(c) the degree of superheat of the vapor in degrees Fahrenheit,
(d) the amount of superheat in the vapor in Btu,
(e) the change in the volume during the superheating,
(f) the change in entropy during the superheating.

Solution

(a) From the head of the table, the saturation temperature corresponding to 40 psia

= 25.9°F

From the body of the table (first reading, italicized), the specific volume of the vapor at saturation

= 1.009 ft^3/lb

The enthalpy of the vapor at saturation

= 81.16 Btu/lb

The entropy of the vapor at saturation

= 0.16914 Btu/lb °F

(b) From the body of the table, the properties of the vapor superheated to 50°F (offset by heavy lines in Fig. 4-9) the specific volume

= 1.070 ft^3/lb

The enthalpy

= 84.65 Btu/lb

The entropy

= 0.17612 Btu/lb °F

(c) The superheated temperature

= 50.0°F

The temperature saturation

= 25.9°F

The degree of superheat of the vapor in degrees Fahrenheit

= 24.1°F

(d) The enthalpy of the superheated vapor

= 84.65 Btu/lb

The enthalpy of the vapor at saturation

= 81.16 Btu/lb

The amount of superheat in the vapor in Btu

= 3.49 Btu/lb

(e) The entropy of the superheated vapor

= 0.17612 Btu/lb °F

The entropy of the vapor at saturation

= 0.16914 Btu/lb °F

The change in entropy during the superheating

= 0.00698 Btu/lb °F

(f) The volume of the superheated vapor

= 1.070 ft^3/lb

The volume of the vapor at saturation

= 1.009 ft^3/lb

The change in volume during the superheating

= 0.061 ft^3/lb

4-22. Properties of Liquid-Vapor Mixtures

The so-called "steam" or "water vapor" that is usually seen rising from boiling water is not vapor at all but small drops of liquid that are entrained in the vapor. Because individual molecules are far too small to be seen, the vapor itself is not visible.

Dry saturated vapor is the term commonly used to describe a saturated vapor that is completely free of liquid particles, whereas the term *wet vapor* is used to denote vapor that has liquid in mixture with it. The *quality* of a vapor, designated as the x component, is an expression of the percent by weight vapor in any mixture of liquid and vapor, whereas the *moisture content*, designated as the y component, is an expression of the percent by weight liquid in the mixture.

Keeping in mind that any wet vapor (liquid-vapor mixture) must of necessity, be saturated, it follows that the temperature of any mixture will always be the saturation temperature corresponding to the pressure of the mixture. Other properties of the mixture can be determined by taking the appropriate values from the saturated vapor table and combining them in proportion to the percentages of liquid and vapor in the mixture, as follows:

$$v_m = (y)(v_f) + (x)(v_g) \tag{4-5}$$

$$h_m = (y)(h_f) + (x)(h_g) \tag{4-6}$$

$$s_m = (y)(s_f) + (x)(s_g) \tag{4-7}$$

where x is the quality of the vapor or the percent vapor in the mixture, and y is the percent moisture or liquid in the mixture. The subscript m identifies mixture quantities, and the subscripts f and g denote values for the liquid and vapor phases, respectively.

Example 4-3 Determine the specific volume, enthalpy, and entropy of R-12 vapor having a saturation temperature of 38°F and a quality of 15%.

Solution From Fig. 4-8 or Table 16-3, the specific volume, enthalpy, and entropy of liquid R-12 at 38°F are 0.0116 ft³/lb, 16.55 Btu/lb, and 0.03591 Btu/lb °F, respectively. In the same order, values for the vapor phase are 0.819 ft³/lb, 82.49 Btu/lb, and 0.16843 Btu/lb °F.

Applying Equation 4-5, the specific volume of the mixture, v_m

$$= (0.85)(0.0116) + (0.15)(0.819)$$
$$= 0.1327 \text{ ft}^3/\text{lb}$$

Applying Equation 4-6, the enthalpy of the mixture, h_m

$$= (0.85)(16.55) + (0.15)(82.49)$$
$$= 26.44 \text{ Btu/lb}$$

Applying Equation 4-7, the entropy of the mixture, s_m

$$= (0.85)(0.03591) + (0.15)(0.16843)$$
$$= 0.0558 \text{ Btu/lb °F}$$

4-23. METRIC SYSTEM EQUIVALENTS

The properties of Refrigerant-12 at saturation are listed in metric units in Table 16-2A, and the properties of superheated R-12 vapor are listed in Table 16-2B. Notice that enthalpy values are given in kilojoules per kilogram, which is also joules per gram, and those for entropy are given in kilojoules per kilogram per degree Kelvin.

Using metric units, the term pv/J in Equation 4-1 reduces to pv and has the units of J/kg, viz:

$$h = u + pv \tag{4-8}$$

where h = the enthalpy in J/kg
u = the internal energy in J/kg
p = the absolute pressure in Pa
v = the specific volume in m³/kg

Example 4-4 Compute the internal energy (u) of 1 kg of saturated R-12 vapor at 20°C.

Solution From Table 16-2A, the enthalpy of saturated vapor at 20°C is 359.729 J/g, the specific volume is 0.03078 m³/kg, and the absolute pressure is 567,290 Pa.

Applying Equation 4-8, the internal energy,

$$u = (359{,}729 \text{ J/kg}) - (567{,}290 \text{ Pa})(0.03078 \text{ m}^3/\text{kg})$$
$$= 342{,}260 \text{ J/kg or } 342.3 \text{ kJ/kg}$$

Example 4-5 Determine the specific volume, enthalpy, and entropy of 1 kg of Refrigerant-12 having a saturation temperature of −5°C and a quality of 32%.

Solution From Table 16-2A, the specific volume, enthalpy, and entropy of liquid R-12 are 0.70780 cm^3/g, 195.395 kJ/kg, and 0.98311 kJ/kg °K, respectively. In the same order, values for the vapor phase are 64.9629 cm^3/g, 349.321 kJ/kg, and 1.55710 kJ/kg °K.

(a) Applying Equation 4-5, the specific volume of the mixture, v_m

$$= (0.68)(0.70780) + (0.32)(64.9629)$$
$$= 21.27 \text{ cm}^3/\text{g} \quad \text{or} \quad 0.02127 \text{ m}^3/\text{kg}$$

(b) Applying Equation 4-6, the enthalpy of the mixture, h_m

$$= (0.68)(195.395) + (0.32)(349.321)$$
$$= 244.65 \text{ kJ/kg}$$

(c) Applying Equation 4-7, the entropy of the mixture, s_m

$$= (0.68)(0.98311) + (0.32)(1.55710)$$
$$= 1.17 \text{ kJ/kg °K}$$

Customary Problems

4-1 Saturated Refrigerant-12 liquid has a temperature of 50°F. What are the pressure, specific volume, enthalpy, entropy, and internal energy of the liquid?

4-2 Saturated Refrigerant-12 vapor has a temperature of 104°F. What are the pressure, specific volume, enthalpy, entropy, and internal energy of the vapor?

4-3 Saturated Refrigerant-12 vapor at 14°F has a quality of 18%. What are the pressure, specific volume, enthalpy, entropy, and internal energy of the mixture?

4-4 Refrigerant-12 vapor at a pressure of 36 psia has a temperature of 100°F. Using the abridged table in Fig. 4-9, determine the following:

(a) the temperature, specific volume, enthalpy, and entropy of the vapor at saturation,
(b) the specific volume, enthalpy, and entropy of the vapor at the superheated condition,
(c) the degree of superheat of the vapor in degrees Fahrenheit,
(d) the amount of superheat in the vapor in Btu per pound,
(e) the change in volume during the superheating,
(f) the change in entropy during the superheating.

4-5 Refrigerant-12 vapor at a pressure of 38 psia has a temperature of 72°F. Using the table in Fig. 4-9, determine the specific volume, enthalpy, and entropy of the vapor.

Metric Problems

4-6 Saturated Refrigerant-12 liquid has a temperature of 10°C. What are the pressure, specific volume, enthalpy, entropy, and internal energy of the liquid?

4-7 Saturated Refrigerant-12 vapor has a temperature of 40°C. What are the pressure, specific volume, density, enthalpy, entropy, and internal energy of the vapor?

4-8 Refrigerant-12 vapor has a saturation temperature of 40°C and is superheated to a temperature of 55°C. What are the pressure, specific volume, enthalpy, entropy, and internal energy of the vapor?

4-9 Refrigerant 12 vapor has a saturation temperature of 42°C and is superheated to a temperature of 72°C. What are the pressure, specific volume, enthalpy, entropy, and internal energy of the vapor?

4-10 Saturated Refrigerant-12 vapor at −10°C has a quality of 18%. What are the pressure, specific volume, enthalpy, entropy, and internal energy of the mixture?

5

PSYCHROMETRIC PROPERTIES OF AIR

5-1. Composition of Air

Air is a mechanical mixture of gases and water vapor. Dry air (air without water vapor) is composed chiefly of nitrogen (approximately 78% by volume) and oxygen (approximately 21%), the remaining 1% being made up of carbon dioxide and minute quantities of other gases, such as hydrogen, helium, neon, and argon. With regard to these dry air components, the composition of the air is practically the same everywhere. On the other hand, the amount of water vapor in the air varies greatly with the particular locality and with the weather conditions and normally is 1% to 3% by mass of the mixture. Since the water vapor in the air results primarily from the evaporation of water from the surface of various bodies of water, atmospheric humidity (water vapor content) is greatest in regions located near large bodies of water and is less in the more arid regions.

Since all air in the natural state contains a certain amount of water vapor, no such thing as "dry air" actually exists. Nevertheless, the concept of "dry air" is a very useful one in that it greatly simplifies psychrometric calculations. Hereinafter the term "dry air" will be used to denote air without water vapor, whereas the terms "air" and "moist air" will be used to mean the natural mixture of dry air and water vapor.

5-2. Dalton's Law of Partial Pressure

Dalton's law of partial pressure states, in effect, that in any mechanical mixture of gases and vapors (those that do not combine chemically) (1) each gas or vapor in the mixture exerts an individual partial pressure that is equal to the pressure that the gas would exert if it occupied the space alone, and (2) the total pressure of the gaseous mixture is equal to the sum of the partial pressures exerted by the individual gases or vapors.

Air, being a mechanical mixture of gases and water vapor, obeys Dalton's law. Therefore, the total barometric pressure is always equal to the sum of the partial pressures of the dry gases and the partial pressure of the water vapor. Since psychrometry is the study of the properties of air as affected by the water vapor content, the individual partial pressures exerted by the dry gases are unimportant, and, for all practical purposes, the total barometric pressure may be considered to be the sum of only two pressures: (1) the partial pressure exerted by the dry gases and (2) the partial pressure exerted by the water vapor.

5-3. Dew Point Temperature

It is important to recognize that the water vapor in the air is actually steam at low pressure and that this low-pressure steam, like high-pressure steam, will be in a saturated condition when its temperature is the saturation temperature corresponding to its pressure. Since all the components in a gaseous mixture occupy the same volume and are at the same temperature, it follows that when air is at any temperature above the saturation temperature corresponding to the partial pressure exerted by the water vapor, the water vapor in the air will be superheated. On the other hand, when air is at a temperature equal to the saturation temperature corresponding to the partial pressure of the water vapor, the water vapor in the air is saturated, and the air is said to be saturated (actually it is only the water vapor that is saturated). The temperature at which the water vapor in the air is saturated is known as the dew point (DP) temperature of the air. Obviously, then, the DP temperature of the air is always the saturation temperature corresponding to the partial pressure exerted by the water vapor. Consequently, when the partial pressure exerted by the water vapor is known, the DP temperature of the air can be determined from the steam tables. Likewise, when the DP temperature of the air is known, the partial pressure exerted by the water vapor can be determined from the steam tables.

Example 5-1 Assume that a certain quantity of air has a temperature of 80°F and that the partial pressure exerted by the water vapor in the air is 0.17811 psia. Determine the DP temperature of the air.

Solution From Table 5-1, the saturation temperature of steam corresponding to a pressure of 0.17811 psia is 50°F and is the DP temperature of the air.

Example 5-2 A certain quantity of air at a temperature of 80°F has a DP temperature of 60°F. Determine the partial pressure exerted by the water vapor in the air.

Solution From Table 5-1, the saturation pressure corresponding to 60°F is 0.2563 psia and is the partial pressure exerted by the water vapor.

It has been shown (Section 4-4) that the pressure exerted by any vapor is directly proportional to the density of the vapor. Since the DP temperature of the air depends only on the partial pressure exerted by the water vapor, it follows that for any given volume of air the DP temperature depends only on the mass of water vapor in the air. As long as the mass of water vapor per unit volume of air (vapor density) remains unchanged, the DP temperature of the air will also remain unchanged. Increasing the amount of water vapor in the air will increase the pressure exerted by the water vapor and raise the DP temperature. Likewise, reducing the amount of water vapor in the air will reduce the pressure exerted by the water vapor and lower the DP temperature.

5-4. Maximum Water Vapor Content

The maximum amount of vapor that can be mixed with any given volume of dry air depends only on the temperature of the air. Since the amount of water vapor in the air determines the partial pressure exerted by the water vapor, it is evident that the air will contain the maximum amount of water vapor when the water vapor in the air exerts the maximum possible pressure. Since the maximum pressure that can be exerted by any vapor is the saturation pressure corresponding to its temperature, the air will contain the maximum amount of water vapor (have maximum vapor density) when the pressure exerted by the water vapor is equal to the saturation pressure corresponding to the temperature of the air. At this condition, the temperature of the air and the DP temperature will be one and the same, and the air is said to be saturated. Notice that the higher the temperature of the air, the higher is the maximum possible vapor pressure and the greater is the possible water vapor content.

5-5. Absolute Humidity

The water vapor in the air is called humidity. The *absolute humidity* of the air at any given condition is the mass of water vapor per unit volume of air at that condition and, as such, is actually an expression of vapor density. Absolute humidity or vapor density is usually expressed in pounds per cubic foot.

TABLE 5-1

Temp, °F, t	Absolute pressure		Specific volume			Enthalpy			Entropy		
	Psi, p	In. Hg, p	Sat liquid, v_f	Evap, v_{fg}	Sat vapor, v_g	Sat liquid, h_f	Evap, h_{fg}	Sat vapor, h_g	Sat liquid, S_f	Evap, S_{fg}	Sat vapor, S_g
(1)	(2)	(3)	(4)	(5)	(6)	(7)	(8)	(9)	(10)	(11)	(12)
32	0.08854	0.1803	0.01602	3306	3306	0.00	1075.8	1075.8	0.0000	2.1877	2.1877
33	0.09223	0.1878	0.01602	3180	3180	1.01	1075.2	1076.2	0.0020	2.1821	2.1841
34	0.09603	0.1955	0.01602	3061	3061	2.02	1074.7	1076.7	0.0041	2.1764	2.1805
35	0.09995	0.2035	0.01602	2947	2947	3.02	1074.1	1077.1	0.0061	2.1709	2.1770
36	0.10401	0.2118	0.01602	2837	2837	4.03	1073.6	1077.6	0.0081	2.1654	2.1735
37	0.10821	0.2203	0.01602	2732	2732	5.04	1073.0	1078.0	0.0102	2.1598	2.1700
38	0.11256	0.2292	0.01602	2632	2632	6.04	1072.4	1078.4	0.0122	2.1544	2.1666
39	0.11705	0.2383	0.01602	2536	2536	7.04	1071.9	1078.9	0.0142	2.1489	2.1631
40	0.12170	0.2478	0.01602	2444	2444	8.05	1071.3	1079.3	0.0162	2.1435	2.1597
41	0.12652	0.2576	0.01602	2356	2356	9.05	1070.7	1079.7	0.0182	2.1381	2.1563
42	0.13150	0.2677	0.01602	2271	2271	10.05	1070.1	1080.2	0.0202	2.1327	2.1529
43	0.13665	0.2782	0.01602	2190	2190	11.06	1069.5	1080.6	0.0222	2.1274	2.1496
44	0.14199	0.2891	0.01602	2112	2112	12.06	1068.9	1081.0	0.0242	2.1220	2.1462
45	0.14752	0.3004	0.01602	2036.4	2036.4	13.06	1068.4	1081.5	0.0262	2.1167	2.1429
46	0.15323	0.3120	0.01602	1964.3	1964.3	14.06	1067.8	1081.9	0.0282	2.1113	2.1395
47	0.15914	0.3240	0.01603	1895.1	1895.1	15.07	1067.3	1082.4	0.0302	2.1060	2.1362
48	0.16525	0.3364	0.01603	1828.6	1828.6	16.07	1066.7	1082.8	0.0321	2.1008	2.1329
49	0.17157	0.3493	0.01603	1764.7	1764.7	17.07	1066.1	1083.2	0.0341	2.0956	2.1297
50	0.17811	0.3626	0.01603	1703.2	1703.2	18.07	1065.6	1083.7	0.0361	2.0903	2.1264
51	0.18486	0.3764	0.01603	1644.2	1644.2	19.07	1065.0	1084.1	0.0380	2.0852	2.1232
52	0.19182	0.3906	0.01603	1587.6	1587.6	20.07	1064.4	1084.5	0.0400	2.0799	2.1199
53	0.19900	0.4052	0.01603	1533.3	1533.3	21.07	1063.9	1085.0	0.0420	2.0747	2.1167
54	0.20642	0.4203	0.01603	1481.0	1481.0	22.07	1063.3	1085.4	0.0439	2.0697	2.1136
55	0.2141	0.4359	0.01603	1430.7	1430.7	23.07	1062.7	1085.8	0.0459	2.0645	2.1104
56	0.2220	0.4520	0.01603	1382.4	1382.4	24.06	1062.2	1086.3	0.0478	2.0594	2.1072
57	0.2302	0.4686	0.01603	1335.9	1335.9	25.06	1061.6	1086.7	0.0497	2.0544	2.1041
58	0.2386	0.4858	0.01604	1291.1	1291.1	26.06	1061.0	1087.1	0.0517	2.0493	2.1010
59	0.2473	0.5035	0.01604	1248.1	1248.1	27.06	1060.5	1087.6	0.0536	2.0443	2.0979
60	0.2563	0.5218	0.01604	1206.6	1206.7	28.06	1059.9	1088.0	0.0555	2.0393	2.0948
61	0.2655	0.5407	0.01604	1166.8	1166.8	29.06	1059.3	1088.4	0.0574	2.0343	2.0917
62	0.2751	0.5601	0.01604	1128.4	1128.4	30.05	1058.8	1088.9	0.0593	2.0293	2.0886
63	0.2850	0.5802	0.01604	1091.4	1091.4	31.05	1058.2	1089.3	0.0613	2.0243	2.0856
64	0.2951	0.6009	0.01605	1055.7	1055.7	32.05	1057.6	1089.7	0.0632	2.0194	2.0826
65	0.3056	0.6222	0.01605	1021.4	1021.4	33.05	1057.1	1090.2	0.0651	2.0145	0.0796
66	0.3164	0.6442	0.01605	988.4	988.4	34.05	1056.5	1090.6	0.0670	2.0096	2.0766
67	0.3276	0.6669	0.01605	956.6	956.6	35.05	1056.0	1091.0	0.0689	2.0047	2.0736
68	0.3390	0.6903	0.01605	925.9	925.9	36.04	1055.5	1091.5	0.0708	1.9998	2.0706
69	0.3509	0.7144	0.01605	896.3	896.3	37.04	1054.9	1091.9	0.0726	1.9950	2.0676
70	0.3631	0.7392	0.01606	867.8	867.9	38.04	1054.3	1092.3	0.0745	1.9902	2.0647
71	0.3756	0.7648	0.01606	840.4	840.4	39.04	1053.8	1092.8	0.0764	1.9854	2.0618
72	0.3886	0.7912	0.01606	813.9	813.9	40.04	1053.2	1093.2	0.0783	1.9805	2.0588
73	0.4019	0.8183	0.01606	788.3	788.4	41.03	1052.6	1093.6	0.0802	1.9757	2.0559
74	0.4156	0.8462	0.01606	763.7	763.8	42.03	1052.1	1094.1	0.0820	1.9710	2.0530
75	0.4298	0.8750	0.01607	740.0	740.0	43.03	1051.5	1094.5	0.0839	1.9663	2.0502
76	0.4443	0.9046	0.01607	717.1	717.1	44.03	1050.9	1094.9	0.0858	1.9615	2.0473
77	0.4593	0.9352	0.01607	694.9	694.9	45.02	1050.4	1095.4	0.0876	1.9569	2.0445
78	0.4747	0.9666	0.01607	673.6	673.6	46.02	1049.8	1095.8	0.0895	1.9521	2.0416
79	0.4906	0.9989	0.01608	653.0	653.0	47.02	1049.2	1096.2	0.0913	1.9475	2.0388
80	0.5069	1.0321	0.01608	633.1	633.1	48.02	1048.6	1096.6	0.0932	1.9428	2.0360
81	0.5237	1.0664	0.01608	613.9	613.9	49.02	1048.1	1097.1	0.0950	1.9382	2.0332
82	0.5410	1.1016	0.01608	595.3	595.3	50.01	1047.5	1097.5	0.0969	1.9335	2.0304
83	0.5588	1.1378	0.01609	577.4	577.4	51.01	1046.9	1097.9	0.0987	1.9290	2.0277
84	0.5771	1.1750	0.01609	560.1	560.2	52.01	1046.4	1098.4	0.1005	1.9244	2.0249
85	0.5959	1.2133	0.01609	543.4	543.5	53.00	1045.8	1098.8	0.1024	1.9198	2.0222
86	0.6152	1.2527	0.01609	527.3	526.3	54.00	1045.2	1099.2	0.1042	1.9153	2.0195
87	0.6351	1.2931	0.01610	511.7	511.7	55.00	1044.7	1099.7	0.1060	1.9108	2.0168
88	0.6556	1.3347	0.01610	496.6	496.7	56.00	1044.1	1100.1	0.1079	1.9062	2.0141
89	0.6766	1.3775	0.01610	482.1	482.1	56.99	1043.5	1100.5	0.1097	1.9017	2.0114

SOURCE: Reprinted by permission from KEENAN, J. H., and KEYES, F. G., "Thermodynamic Properties of Steam," John Wiley and Sons, Inc.

It was shown in Section 5-2 that the actual mass of water vapor per unit volume of air (vapor density) is solely a function of the DP temperature of the air. Because of this fixed relationship between the DP temperature and the absolute humidity of the air, when the value of one is known, the value of the other can readily be computed.

Since the vapor pressure exerted by the water vapor in the air is extremely low, the

TABLE 5-1 *(Continued)*

Temp, °F, t	Absolute pressure		Specific volume			Enthalpy			Entropy		
	Psi, p	In. Hg, p	Sat liquid, v_f	Evap, v_{fg}	Sat vapor, v_g	Sat liquid, h_f	Evap, h_{fg}	Sat vapor, h_g	Sat liquid, S_f	Evap, S_{fg}	Sat vapor, S_g
(1)	(2)	(3)	(4)	(5)	(6)	(7)	(8)	(9)	(10)	(11)	(12)
90	0.6982	1.4215	0.01610	468.0	468.0	57.99	1042.9	1100.9	0.1115	1.8972	2.0087
91	0.7204	1.4667	0.01611	454.4	454.4	58.99	1042.4	1101.4	0.1133	1.8927	2.0060
92	0.7432	1.5131	0.01611	441.2	441.3	59.99	1041.8	1101.8	0.1151	1.8883	2.0034
93	0.7666	1.5608	0.01611	428.5	428.5	60.98	1041.2	1102.2	0.1169	1.8838	2.0007
94	0.7906	1.6097	0.01612	416.2	416.2	61.98	1040.7	1102.6	0.1187	1.8794	1.9981
95	0.8153	1.6600	0.01612	404.3	404.3	62.98	1040.1	1103.1	0.1205	1.8750	1.9955
96	0.8407	1.7117	0.01612	392.8	392 8	63.98	1039.5	1103.5	0.1223	1.8706	1.9929
97	0.8668	1.7647	0.01612	381.7	381.7	64.97	1038.9	1103.9	0.1241	1.8662	1.9903
98	0.8935	1.8192	0.01613	370.9	370.9	65.97	1038.4	1104.4	0.1259	1.8618	1.9877
99	0.9210	1.8751	0.01613	360.4	360.5	66.97	1037.8	1104.8	0.1277	1.8575	1.9852
100	0.9492	1.9325	0.01613	350.3	350.4	67.97	1037.2	1105.2	0.1295	1.8531	1.9826
101	0.9781	1.9915	0.01614	340.6	340.6	68.96	1036.6	1105.6	0.1313	1.8488	1.9801
102	1.0078	2.0519	0.01614	331.1	331.1	69.96	1036.1	1106.1	0.1330	1.8445	1.9775
103	1.0382	2.1138	0.01614	321.9	321.9	70.96	1035.5	1106.5	0.1348	1.8402	1.9750
104	1.0695	2.1775	0.01615	313.1	313.1	71.96	1034.9	1106.9	0.1366	1.8359	1.9725
105	1.1016	2.2429	0.01615	304.5	304.5	72.95	1034.3	1107.3	0.1383	1.8317	1.9700
106	1.1345	2.3099	0.01615	296.1	296.2	73.95	1033.8	1107.8	0.1401	1.8274	1.9675
107	1.1683	2.3786	0.01616	288.1	288.1	74.95	1033.3	1108.2	0.1419	1.8232	1.9651
108	1.2029	2.4491	0.01616	280.3	280.3	75.95	1032.7	1108.6	0.1436	1.8190	1.9626
109	1.2384	2.5214	0.01616	272.7	272.7	76.94	1032.1	1109.0	0.1454	1.8147	1.9601
110	1.2748	2.5955	0.01617	265.3	265.4	77.94	1031.6	1109.5	0.1471	1.8106	1.9577
111	1.3121	2.6715	0.01617	258.2	258.3	78.94	1031.0	1109.9	0.1489	1.8064	1.9553
112	1.3504	2.7494	0.01617	251.3	251.4	79.94	1030.4	1110.3	0.1506	1.8023	1.9529
113	1.3896	2.8293	0.01618	244.6	244.7	80.94	1029.8	1110.7	0.1524	1.7981	1.9505
114	1.4298	2.9111	0.01618	238.2	238.2	81.93	1029.2	1111.1	0.1541	1.7940	1.9481
115	1.4709	2.9948	0.01618	231.9	231.9	82.93	1028.7	1111.6	0.1559	1.7898	1.9457
116	1.5130	3.0806	0.01619	225.8	225.8	83.93	1028.1	1112.0	0.1576	1.7857	1.9433
117	1.5563	3.1687	0.01619	219.9	219.9	84.93	1027.5	1112.4	0.1593	1.7816	1.9409
118	1.6006	3.2589	0.01620	214.2	214.2	85.92	1026.9	1112.8	0.1610	1.7776	1.9386
119	1.6459	3.3512	0.01620	208.6	208.7	86.92	1026.3	1113.2	0.1628	1.7735	1.9363
120	1.6924	3.4458	0.01620	203.25	203.27	87.92	1025.8	1113.7	0.1645	1.7694	1.9339
121	1.7400	3.5427	0.01621	198.02	198.03	88.92	1025.2	1114.1	0.1662	1.7654	1.9316
122	1.7888	3.6420	0.01621	192.93	192.95	89.92	1024.6	1114.5	0.1679	1.7614	1.9293
123	1.8387	3.7436	0.01622	188.01	188.02	90.91	1024.0	1114.9	0.1696	1.7574	1.9270
124	1.8897	3.8475	0.01622	183.23	183.25	91.91	1023.4	1115.3	0.1714	1.7533	1.9247
125	1.9420	3.9539	0.01622	178.59	178.61	92.91	1022.9	1115.8	0.1731	1.7493	1.9334
126	1.9955	4.0629	0.01623	174.09	174.10	93.91	1022.3	1116.2	0.1748	1.7454	1.9202
127	2.0503	4.1745	0.01623	169.71	169.72	94.91	1021.7	1116.6	0.1765	1.7414	1.9179
128	2.1064	4.2887	0.01624	165.46	165.47	95.91	1021.1	1117.0	0.1782	1.7374	1.9156
129	2.1638	4.4055	0.01624	161.33	161.35	96.90	1020.5	1117.4	0.1799	1.7335	1.9134
130	2.2225	4.5251	0.01625	157.32	157.34	97.90	1020.0	1117.9	0.1816	1.7296	1.9112
131	2.2826	4.6474	0.01625	153.43	153.44	98.90	1019.4	1118.3	0.1833	1.7257	1.9090
132	2.3440	4.7725	0.01626	149.65	149.66	99.90	1018.8	1118.7	0.1849	1.7218	1.9067
133	2.4069	4.9005	0.01626	145.97	145.99	100.90	1018.2	1119.1	0.1866	1.7179	1.9045
134	2.4712	5.0314	0.01626	142.40	142.42	101.90	1017.6	1119.5	0.1883	1.7141	1.9023
135	2.5370	5.1653	0.01627	138.93	138.95	102.90	1017.0	1119.9	0.1900	1.7102	1.9002
136	2.6042	5.3022	0.01627	135.56	135.58	103.90	1016.4	1120.3	0.1917	1.7063	1.8980
137	2.6729	5.4421	0.01628	132.29	132.30	104.89	1015.9	1120.8	0.1934	1.7024	1.8958
138	2.7432	5.5852	0.01628	129.10	129.12	105.89	1015.3	1121.2	0.1950	1.6987	1.8937
139	2.8151	5.7316	0.01629	126.00	126.02	106.89	1014.7	1121.6	0.1967	1.6948	1.8915
140	2.8886	5.8812	0.01629	122.99	123.01	107.89	1014.1	1122.0	0.1984	1.6910	1.8894
141	2.9637	6.0341	0.01630	120.06	120.08	108.89	1013.5	1122.4	0.2000	1.6873	1.8873
142	3.0404	6.1903	0.01630	117.22	117.23	109.89	1012.9	1122.8	0.2016	1.6835	1.8851
143	3.1188	6.3500	0.01631	114.45	114.46	110.89	1012.3	1123.2	0.2033	1.6797	1.8830
144	3.1990	6.5132	0.01631	111.75	111.77	111.89	1011.7	1123.6	0.2049	1.6760	1.8809

water vapor in the air approaches the condition of an ideal gas closely enough to permit application of the ideal gas laws. Consequently, the characteristic gas equation (Equation 3-8) can be used in conjunction with the steam table (Table 5-1) to determine the absolute humidity (vapor density) of any sample of air when either the DP temperature or the vapor pressure is known.

Example 5-3 Determine the absolute humidity of an air sample that has a DP temperature of 45°F, if the value of the gas constant R for low pressure water vapor is 85.66 ft-lb/lb °R.

Solution From the steam table (Table 5-1), the vapor pressure corresponding to a saturation temperature of 45°F is 0.14752 psia. Assuming a volume V of 1 ft^3 and applying

TABLE 5-1 *(Continued)*

Temp, °F, t	Absolute pressure		Specific volume			Enthalpy			Entropy		
	Psi, p	In. Hg, p	Sat liquid, v_f	Evap, v_{fg}	Sat vapor, v_g	Sat liquid, h_f	Evap, h_{fg}	Sat vapor, h_g	Sat liquid, S_f	Evap, S_{fg}	Sat vapor, S_g
(1)	(2)	(3)	(4)	(5)	(6)	(7)	(8)	(9)	(10)	(11)	(12)
145	3.281	6.680	0.01632	109.13	109.15	112.89	1011.2	1124.1	0.2066	1.6722	1.8788
146	3.365	6.850	0.01632	106.58	106.60	113.89	1010.6	1124.5	0.2083	1.6685	1.8768
147	3.450	7.024	0.01633	104.10	104.12	114.89	1010.0	1124.9	0.2099	1.6648	1.8747
148	3.537	7.202	0.01633	101.69	101.71	115.89	1009.4	1125.3	0.2116	1.6610	1.8726
149	3.627	7.384	0.01634	99.34	99.36	116.89	1008.8	1125.7	0.2133	1.6573	1.8706
150	3.718	7.569	0.01634	97.06	97.07	117.89	1008.2	1126.1	0.2149	1.6537	1.8685
151	3.811	7.759	0.01635	94.83	94.85	118.89	1007.6	1126.5	0.2165	1.6500	1.8665
152	3.906	7.952	0.01635	92.67	92.68	119.89	1007.0	1126.9	0.2182	1.6463	1.8645
153	4.003	8.150	0.01636	90.56	90.57	120.89	1006.4	1127.3	0.2198	1.6427	1.8624
154	4.102	8.351	0.01636	88.51	88.52	121.89	1005.8	1127.7	0.2214	1.6390	1.8604
155	4.203	8.557	0.01637	86.51	86.52	122.89	1005.2	1128.1	0.2230	1.6354	1.8584
156	4.306	8.767	0.01637	84.56	84.58	123.89	1004.7	1128.6	0.2246	1.6318	1.8564
157	4.411	8.981	0.01638	82.67	82.69	124.89	1004.1	1129.0	0.2263	1.6282	1.8545
158	4.519	9.200	0.01638	80.82	80.84	125.89	1003.5	1129.4	0.2279	1.6246	1.8525
159	4.629	9.424	0.01639	79.03	79.04	126.89	1002.9	1129.8	0.2295	1.6210	1.8505
160	4.741	9.652	0.01639	77.27	77.29	127.89	1002.3	1130.2	0.2311	1.6174	1.8485
161	4.855	9.885	0.01640	75.57	75.58	128.89	1001.7	1130.6	0.2327	1.6138	1.8466
162	4.971	10.122	0.01640	73.91	73.92	129.89	1001.1	1131.0	0.2343	1.6103	1.8446
163	5.090	10.364	0.01641	72.29	72.30	130.89	1000.5	1131.4	0.2360	1.6067	1.8427
164	5.212	10.611	0.01641	70.71	70.73	131.89	999.9	1131.8	0.2376	1.6032	1.8408
165	5.335	10.863	0.01642	69.17	69.19	132.89	999.3	1132.2	0.2392	1.5997	1.8388
166	5.461	11.120	0.01643	67.67	67.69	133.89	998.7	1132.6	0.2408	1.5961	1.8369
167	5.590	11 382	0.01643	66.21	66.23	134.89	998.1	1133.0	0.2424	1.5926	1.8350
168	5.721	11.649	0.01644	64.79	64.80	135.90	997.5	1133.4	0.2440	1.5891	1 8331
169	5.855	11.921	0.02644	63.40	63.41	136.90	996.9	1133.8	0.2455	1.5857	1.8312
170	5.992	12.199	0.01645	62.04	62.06	137.90	996.3	1134.2	0.2472	1.5822	1.8293
171	6.131	12.483	0.01645	60.72	60.74	138.90	995.7	1134.6	0.2488	1.5787	1.8275
172	6.273	12.772	0.01646	59.43	59.45	139.90	995.1	1135.0	0.2503	1.5753	1.8256
173	6.417	13.066	0.01647	58.18	58.20	140.90	994.5	1135.4	0.2519	1.5718	1.8237
174	6.565	13.366	0.01647	56.96	56.97	141.90	993.9	1135.8	0.2535	1.5684	1.8219
175	6.715	13.671	0.02648	55.76	55.78	142.91	993.3	1136.2	0.2551	1.5649	1.8200
176	6.868	13.983	0.01648	54.60	54.61	143.91	992.7	1136.6	0.2567	1.5615	1.8182
177	7.024	14.301	0.01649	53.46	53.48	144.91	992.1	1137.0	0.2583	1.5581	1.8164
178	7.183	14.625	0.01650	52.35	52.37	145.91	991.5	1137.4	0.2599	1.5547	1.8146
179	7.345	14.955	0.01650	51.27	51.29	146.92	990.8	1137.7	0.2614	1.5513	1.8127
180	7.510	15.291	0.01651	50.21	50.23	147.92	990.2	1138.1	0.2630	1.5480	1.8109
181	7.678	15.633	0.01651	49.18	49.20	148.92	989.6	1138.5	0.2645	1.5446	1.8091
182	7.850	15.982	0.01652	48.18	48.19	149.92	989.0	1138.9	0.2661	1.5412	1.8073
183	8.024	16.337	0.01653	47.19	47.21	150.93	988.4	1139.3	0.2676	1.5379	1.8055
184	8.202	16.699	0.01653	46.24	46.25	151.93	987.8	1139.7	0.2692	1.5346	1.8038
185	8.383	17.068	0.01654	45.29	45.31	152.93	987.2	1140.1	0.2708	1.5312	1.8020
186	8.567	17.443	0.01654	44.39	44.40	153.94	986.6	1140.5	0.2723	1.5279	1.8002
187	8.755	17.825	0.01655	43.50	43.51	154.94	986.0	1140.9	0.2739	1.5246	1.7985
188	8.946	18.214	0.01656	42.62	42.64	155.94	985.4	1141.3	0.2754	1.5213	1.7967
189	9.141	18.611	0.01656	41.77	41.79	156.95	984.8	1141.7	0.2770	1.5180	1.7950
190	9.339	19.014	0.01657	40.94	40.96	157.95	984.1	1142.0	0.2785	1.5147	1.7932
191	9.541	19.425	0.01658	40.13	40.15	158.95	983.4	1142.4	0.2801	1.5114	1.7915
192	9.746	19.843	0.01658	39.34	39.36	159.96	982.8	1142.8	0.2816	1.5082	1.7898
193	9.955	20.269	0.01659	38.57	38.58	160.96	982.2	1143.2	0.2831	1.5049	1.7880
194	10.168	20.703	0.01659	37.81	37.83	161.97	981.6	1143.6	0.2846	1.5017	1.7863
195	10.385	21.144	0.01660	37.07	37.09	162.97	981.0	1144.0	0.2862	1.4984	1.7846
196	10.605	21.593	0.01661	36.35	36.37	163.97	980.4	1144.4	0.2877	1.4952	1.7829
197	10.830	22.050	0.01661	35.64	35.66	164.98	979.7	1144.7	0.2892	1.4920	1.7812
198	11.058	22.515	0.01662	34.95	34.97	165.98	979.1	1145.1	0.2907	1.4888	1.7795
199	11.290	22.987	0.01663	34.28	34.30	166.99	978.5	1145.5	0.2923	1.4856	1.7779

Equation 3-8, the density of the water vapor is

$$\rho = m = \frac{pV}{RT}$$

$$= \frac{(0.14752 \text{ psia})(144)(1 \text{ ft}^3)}{(85.66 \text{ ft-lb/lb °R})(45° + 460°)}$$

$$= 0.000491 \text{ lb/ft}^3$$

The absolute humidity (vapor density) can be determined also directly from the steam table simply by taking the reciprocal of the vapor specific volume as listed in the table. With reference to Example 5-3, the specific volume of saturated vapor at the DP temperature of 45°F is given in Table 5-1 as 2036.4 ft^3/lb, so that the absolute humidity is (1/2036.4 ft^3/lb) 0.000491 lb/ft^3.

TABLE 5-1 *(Continued)*

Temp, °F, t	Absolute pressure		Specific volume			Enthalpy			Entropy		
	Psi, p	In. Hg, p	Sat liquid, v_f	Evap, v_{fg}	Sat vapor, v_g	Sat liquid, h_f	Evap, h_{fg}	Sat vapor, h_g	Sat liquid, S_f	Evap, S_{fg}	Sat vapor S_g
(1)	(2)	(3)	(4)	(5)	(6)	(7)	(8)	(9)	(10)	(11)	(12)
200	11.526	23.467	0.01663	33.62	33.64	167.99	977.9	1145.9	0.2938	1.4824	1.7762
202	12.011	24.455	0.01665	32.35	32.37	170.00	976.6	1146.6	0.2969	1.4760	1.7729
204	12.512	25.475	0.01666	31.14	31.15	172.02	975.4	1147.4	0.2999	1.4697	1.7696
206	13.031	26.531	0.01667	29.97	29.99	174.03	974.2	1148.2	0.3029	1.4634	1.7663
208	13.568	27.625	0.01669	28.86	28.88	176.04	972.9	1148.9	0.3059	1.4571	1.7630
210	14.123	28.755	0.01670	27.80	27.82	178.05	971.6	1149.7	0.3090	1.4508	1.7598
212	14.696	29.922	0.01672	26.78	26.80	180.07	970.3	1150.4	0.3120	1.4446	1.7566
214	15.289	31.129	0.01673	25.81	25.83	182.08	969.0	1151.1	0.3149	1.4385	1.7534
216	15.901	32.375	0.01674	24.88	24.90	184.10	967.8	1151.9	0.3179	1.4323	1.7502
218	16.533	33.662	0.01676	23.99	24.01	186.11	966.5	1152.6	0.3209	1.4262	1.7471
220	17.186	34.992	0.01677	23.13	23.15	188.13	965.2	1153.4	0.3239	1.4201	1.7440
222	17.861	36.365	0.01679	22.31	22.33	190.15	963.9	1154.1	0.3268	1.4141	1.7409
224	18.557	37.782	0.01680	21.53	21.55	192.17	962.6	1154.8	0.3298	1.4080	1.7378
226	19.275	39.244	0.01682	20.78	20.79	194.18	961.3	1155.5	0.3328	1.4020	1.7348
228	20.016	40.753	0.01683	20.06	20.07	196.20	960.1	1156.3	0.3357	1.3961	1.7318
230	20.780	42.308	0.01684	19.365	19.382	198.23	958.8	1157.0	0.3387	1.3901	1.7288
240	24.969	50.837	0.01692	16.306	16.323	208.34	952.2	1160.5	0.3531	1.3609	1.7140
250	29.825	60.725	0.01700	13.804	13.821	218.48	945.5	1164.0	0.3675	1.3323	1.6998
260	35.429	72.134	0.01709	11.746	11.763	228.64	938.7	1167.3	0.3817	1.3043	1.6860
270	41.858	85.225	0.01717	10.044	10.061	238.84	931.8	1170.6	0.3958	1.2769	1.6727
280	49.203	100.18	0.01726	8.628	8.645	249.06	924.7	1173.8	0.4096	1.2501	1.6597
290	57.556	117.19	0.01735	7.444	7.461	259.31	917.5	1176.8	0.4234	1.2238	1.6472
300	67.013	136.44	0.01745	6.449	6.466	269.59	910.1	1179.7	0.4369	1.1980	1.6350
310	77.68	.....	0.01755	5.609	5.626	279.92	902.6	1182.5	0.4504	1.1727	1.6231
320	89.66	.....	0.01765	4.896	4.914	290.28	894.9	1185.2	0.4637	1.1478	1.6115
330	103.06	.....	0.01776	4.289	4.307	300.68	887.0	1187.7	0.4769	1.1233	1.6002
340	118.01	.....	0.01787	3.770	3.788	311.13	879.0	1190.1	0.4900	1.0992	1.5891
350	134.63	.....	0.01799	3.324	3.342	321.63	870.7	1192.3	0.5029	1.0754	1.5783
360	153.04	.....	0.01811	2.939	2.957	332.18	862.2	1194.4	0.5158	1.0519	1.5677
370	173.37	.....	0.01823	2.606	2.625	342.79	853.5	1196.3	0.5286	1.0287	1.5573
380	195.77	.....	0.01836	2.317	2.335	353.45	844.6	1198.1	0.5413	1.0059	1.5471
390	220.37	.....	0.01850	2.0651	2.0836	364.17	835.4	1199.6	0.5539	0.9832	1.5371
400	247.31	.....	0.01864	1.8447	1.8633	374.97	826.0	1201.0	0.5664	0.9608	1.5272
425	325.92	.....	0.01902	1.4036	1.4226	402.27	801.2	1203.5	0.5974	0.9056	1.5030
450	422.6	.....	0.0194	1.0799	0.0993	430.1	774.5	1204.6	0.6280	0.8513	1.4793
475	539.9	.....	0.0199	0.8380	0.8579	458.6	745.4	1204.0	0.6584	0.7976	1.4560
500	680.8	.....	0.0204	0.6545	0.6749	487.8	713.9	1201.7	0.6887	0.7438	1.4325
525	848.1	.....	0.0210	0.5130	0.5340	518.0	679.1	1197.1	0.7191	0.6897	1.4088
550	1045.2	.....	0.0218	0.4022	0.4240	549.3	640.8	1190.0	0.7497	0.6346	1.3843
575	1275.4	.....	0.0226	0.3143	0.3369	582.1	597.7	1179.8	0.7809	0.5777	1.3585
600	1542.9	.....	0.0236	0.2432	0.2668	617.0	548.5	1165.5	0.8131	0.5176	1.3307
625	1852.0	.....	0.0250	0.1845	0.2095	654.4	491.4	1145.8	0.8467	0.4530	1.2997
650	2208.2	.....	0.0268	0.1348	0.1616	695.7	422.8	1118.5	0.8828	0.3809	1.2637
675	2618.7	.....	0.0297	0.0899	0.1196	745.4	332.6	1078.0	0.9251	0.2931	1.2183
700	3093.7	.....	0.0369	0.0392	0.0761	823.3	172.1	995.4	0.9905	0.1484	1.1389
702	3134.9	.....	0.0385	0.0325	0.0710	835.4	145.2	980.6	1.0006	0.1249	1.1256
704	3176.7	.....	0.0410	0.0234	0.0645	852.7	106.0	958.7	1.0152	0.0911	1.1063
705	3197.7	.....	0.0438	0.0152	0.0589	869.2	69.1	938.4	1.0293	0.0593	1.0886
705.4	3206.2	.....	0.0503	0	0.0503	902.7	0	902.7	1.0580	0	1.0580

5-6. Relative Humidity

Relative humidity (RH), expressed in percent, is the ratio of the actual partial pressure exerted by the water vapor in any volume of air to the partial pressure that would be exerted by the water vapor if the water vapor in the air is saturated at the temperature of the air; that is,

$$\text{RH} = \frac{\text{actual partial pressure}}{\text{partial pressure at saturation}} \times 100 \qquad (5\text{-}1)$$

RH is sometimes defined also as the ratio, expressed in percent, of the actual vapor density to the vapor density at saturation. The value obtained is essentially the same as that obtained from Equation 5-1.

Example 5-4 A certain sample of air has a temperature of 70°F and a DP temperature of 50°F. Determine the RH.

Solution From Table 5-1, the partial pressure of the water vapor corresponding to a 50°F DP temperature is 0.17811 psia and the partial pressure corresponding to a DP temperature of 70°F is 0.3631 psia. Applying Equation 5-4,

$$RH = \frac{0.17811 \text{ psia}}{0.3631 \text{ psia}} (100) = 49\%$$

Example 5-5 Compute the RH of the air in Example 5-4 if the temperature of the air is reduced to 60°F.

Solution Since the moisture content of the air remains the same, the DP temperature also remains the same. From Table 5-1, the partial pressures of water vapor corresponding to DP temperatures of 50°F and 60°F are 0.17811 psia and 0.2563 psia, respectively. Applying Equations 5-1,

$$RH = \frac{0.17811 \text{ psia}}{0.2563 \text{ psia}} (100) = 69.5\%$$

5-7. Humidity Ratio

The *humidity ratio* (w), sometimes called *specific humidity*, is an expression of the mass of water vapor per unit mass of dry air and is usually stated in grains per pound of dry air (gr/lb) or pounds per pound of dry air (lb/lb).* Seven thousand (7000) grains equals 1 lb.

For any given barometric pressure, the humidity ratio is a function of the DP temperature alone. However, the humidity ratio corresponding to any given DP temperature varies with the total barometric pressure, increasing as the total barometric pressure decreases. The reason for this is easily explained. In accordance with the ideal gas laws, the volume per unit mass of air will increase as the total barometric pressure decreases. Since the density of a vapor varies inversely with the volume, it follows that the mass of water vapor required to produce a given vapor density and vapor pressure increases as the volume of the air increases.

* Many sources use the terms "specific humidity" and "humidity ratio" interchangeably. However, the American Society of Heating, Refrigerating and Air Conditioning Engineers' *Handbook of Fundamentals*, 1972 edition, page 98, defines "specific humidity" as mass of water vapor per unit mass of moist air.

When the barometric pressure and the DP temperature are known, the humidity ratio is readily determined from the following relationship, which is derived from the characteristic gas equation (Equation 3-8):

$$w = \frac{0.622 p_w}{p - p_w} \tag{5-2}$$

where w = the humidity ratio, in pounds of water vapor per pound of dry air

p_w = the partial pressure of water vapor corresponding to DP temperature in pounds per square inch absolute

p = the barometric pressure in pounds per square inch

Example 5-6 Determine the humidity ratio for a sample of air at standard barometric pressure having a temperature of 70°F and a DP temperature of 50°F.

Solution From Table 5-1, the partial pressure of water vapor corresponding to a DP temperature of 50°F is 0.17811 psia and normal atmospheric pressure is 14,696 psi. Applying equation 5-2,

$$w = \frac{(0.622)(0.17811 \text{ psia})}{14.696 \text{ psi} - 0.17811 \text{ psia}}$$

$$= 0.00763 \text{ lb/lb}$$

Example 5-7 Determine the humidity ratio of the sample of air in Example 5-6 if the water vapor in the air is saturated at the temperature of the air.

Solution From Table 5-1, the partial pressure corresponding to a DP temperature of 70°F is 0.3631 psia. Substituting in Equation 5-2,

$$w = \frac{(0.622)(0.3631 \text{ psia})}{14.696 \text{ psi} - 0.3631 \text{ psia}}$$

$$= 0.01576 \text{ lb/lb}$$

5-8. Saturation Ratio

The saturation ratio, sometimes called percentage humidity, is the ratio of the mass of water vapor in the air per unit mass of dry air to the mass of water vapor required for saturation of the same air sample. The saturation ratio, like relative humidity, is expressed in percent:

$$\text{Saturation ratio} = \frac{w}{w_s}(100) \qquad (5\text{-}3)$$

where w = actual humidity ratio in pounds per pound of dry air

w_s = humidity ratio at saturation for same air temperature in pounds per pound of dry air

Example 5-8 Air at normal atmospheric pressure has a temperature of 70°F and a DP temperature of 50°F. Determine the saturation ratio of the air.

Solution From Examples 5-6 and 5-7, the humidity ratios corresponding to DP temperatures of 50°F and 70°F are 0.00763 lb/lb and 0.01576 lb/lb, respectively. Applying Equation 5-3,

$$\text{Saturation ratio} = \frac{0.00763}{0.01576}(100) = 48.4\%$$

5-9. Dry Bulb and Wet Bulb Temperatures

The dry bulb (DB) temperature of the air is the temperature as measured by an ordinary DB thermometer. When measuring the DB temperature of the air, the bulb of the thermometer should be shaded to reduce the effects of direct radiation.

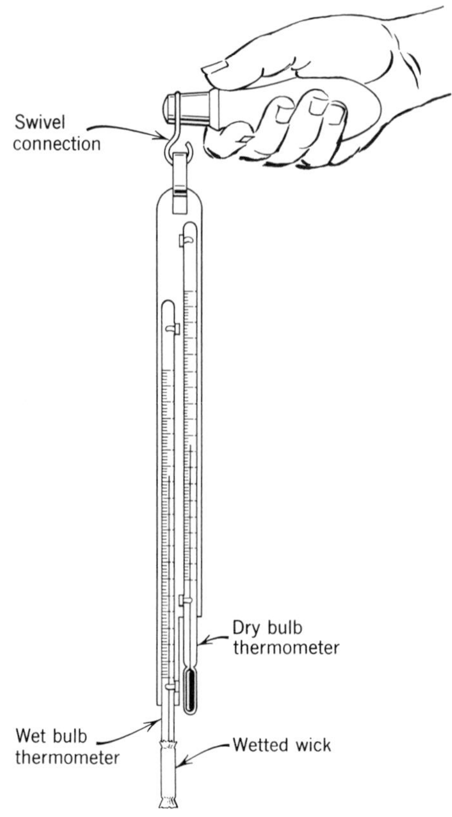

Fig. 5-1 Sling psychrometer.

The wet bulb (WB) temperature of the air is the temperature as measured by a wet bulb thermometer. A WB thermometer is an ordinary thermometer whose bulb is enclosed in a wetted cloth sac or wick. To obtain an accurate reading with a WB thermometer, the wick should be saturated with clean water at approximately the DB temperature of the air, and the air velocity around the wick should be maintained between 1000 and 2000 fpm. As a practical matter, this velocity can be simulated in still air by whirling the thermometer about on the end of a chain. An instrument especially designed for this purpose is the sling psychrometer (Fig. 5-1). The sling psychrometer is made up of two thermometers, one dry bulb and one wet bulb, mounted side by side in a protective case that is attached to a handle by a swivel connection so that the case can be easily rotated about the hand. After saturating the wick with clean water, the instrument is whirled rapidly in the air for approximately 1 minute, after which time readings can be taken from both the DB and WB thermometers. The process should be repeated several times to assure that the lowest possible WB temperature has been recorded.

Unless the air is 100% saturated, in which case the DB, WB, and DP temperatures of the air will be the same, the temperature recorded by a WB thermometer will always be lower than the DB temperature of the air. The amount by which the WB temperature is reduced below the DB temperature depends on the RH of the air and is called the WB depression.

Whereas a DB thermometer, being unaffected by humidity, measures only the actual temperature of the air, a WB thermometer, because of its wet wick, is greatly influenced by the moisture in the air; thus, a WB temperature is, in effect, a measure of the relationship between the DB temperature of the air and the moisture content of the air. In general, for any given DB temperature, the lower the moisture content of the air, the lower is the WB temperature. The reason for this is easily explained.

When unsaturated air is brought into contact with water, water will evaporate into the air at a rate proportional to the difference in pressure between the vapor pressure of the water and the vapor pressure of the water vapor in the air. Hence, when a WB thermometer is whirled rapidly about in unsaturated air, water will evaporate from the wick, thereby cooling the water remaining in the wick (and the thermometer bulb) to some temperature below the DB temperature of the air.

It is important to recognize that the WB temperature of the air is a measure of the relationship between the DB and DP temperatures of the air, and as such, it provides a convenient means of determining the DP temperature of the air when the DB temperature is known. Also, it will be shown later that the WB temperature is also an index of the total heat (enthalpy) of the air.

In order to understand why the WB temperature is a measure of the relationship between the DB and DP temperatures, a knowledge of the theory of the WB thermometer is required. When water evaporates from the wick of a WB thermometer, heat must be supplied to furnish the latent heat of vaporization. Before the temperature of the water in the wick is reduced below the DB temperature of the air, the source of the heat to vaporize the water is the water itself. Therefore, as water evaporates from the wick, the water remaining in the wick is cooled below the DB temperature of the air. When this occurs, a temperature differential is established and heat begins to flow from the air to the wick. Under this condition, one part of the vaporization heat is being supplied by the air, while the other part is supplied by the water in the wick. As the temperature of the wick continues to decrease, the temperature difference between the air and the wick increases progressively, so that more and more of the vaporization heat is supplied by the air and less and less is supplied by the water in the wick. When the temperature of the wick is reduced to the point where the temperature difference between the air and the wick is such that the flow of heat from the air is sufficient to supply all the vaporizing heat, the temperature of the wick will stabilize even though vaporization from the wick continues. The temperature at which the wick stabilizes is called the temperature of adiabatic saturation and is the WB temperature of the air.

Through careful analysis of the foregoing, it can be seen that the WB temperature depends on both the DB temperature and the amount of water vapor in the air. For example, the lower the RH of the air, the greater is the rate of evaporation from the wick and the greater is the amount of heat required for evaporation. Obviously, the greater the need for heat, the greater is the WB depression below the DB temperature. It follows also that the lower the DB temperature, the lower is the WB temperature for any given WB depression.

5-10. The Heat Content or Enthalpy of Air

Air has both sensible and latent heat. The total heat of the air at any condition is the sum of the sensible heat and latent heat contained therein.

It will be shown in the following sections that (1) the sensible heat of the air is a function of the DB temperature, (2) the latent heat of air is a function of the DP temperature, and (3) the total heat of the air is a function of the WB temperature.

5-11. Sensible Heat of the Air

For any given DB temperature, the sensible heat of air is taken as the enthalpy of dry air at that temperature as calculated from 0°F and can be computed from the sensible heat equation (Equation 3-16).

For this particular calculation, the term $(T_2 - T_1)$ in Equation 3-16 will be numerically equal to the DB temperature of the air, and assuming the mean specific heat of air at constant pressure to be 0.24 Btu/lb °F, Equation 3-16 may be written as

$$h_s = (0.24)(\text{DB}) \qquad (5\text{-}4)$$

where h_s = the specific enthalpy of dry air (sensible heat per pound of dry air) in Btu per pound.

To determine the total sensible heat of m pounds of air (enthalpy of dry air),

$$H_s = (m)(0.24)(\text{DB}) \quad (5\text{-}5)$$

or

$$H_s = (m)(h_s)$$

Example 5-9 Determine the sensible heat of 5 lb of air having a DB temperature of 70°F and a humidity ratio of 0.0092 lb/lb, the latter corresponding to a DP temperature of approximately 55°F.

Solution Applying Equation 5-5,

$$\begin{aligned} H_s &= (5\text{ lb})(0.24\text{ Btu/lb °F})(70°) \\ &= 84\text{ Btu} \end{aligned}$$

Notice in Example 5-9 that the sensible heat of the air is a function of the DB temperature alone and is unaffected by the DP temperature.

The quantity of sensible heat transferred when any given mass of air is heated or cooled between any initial and final DB temperatures is also a function of the sensible heat equation (Equation 3-18). Again, assuming the specific heat at constant pressure, c_p, for air to be 0.24 Btu/lb °F, Equation 3-18 can be written as

$$Q_s = (m)(0.24)(\Delta\text{DB}) \quad (5\text{-}6)$$

where Q_s = the sensible heat transferred in Btu
m = the mass of dry air in pounds
ΔDB = the difference between the initial and final DB temperatures in degrees Fahrenheit

The sensible heat transferred when m pounds of air are heated or cooled between any two DB temperatures can be determined also by multiplying the mass of the air by the difference in the specific enthalpy of dry air at the initial and final conditions:

$$Q_s = (m)(\Delta h_s) \quad (5\text{-}7)$$

Example 5-10 The temperature of 20 lb of air is increased from 55°F to 100°F by the addition of heat. Determine the quantity of sensible heat supplied.

Solution Applying Equation 5-6,

$$\begin{aligned} Q_s &= (20\text{ lb})(0.24)(100° - 55°) \\ &= 216\text{ Btu} \end{aligned}$$

5-12. Latent Heat of the Air

Since all the components of dry air are noncondensable at normal temperatures and pressures, for all practical purposes the only latent heat in the air is the latent heat of the water vapor in the air. Therefore, the amount of latent heat in any given quantity of air depends on the mass of water vapor in the air and on the latent heat of vaporization of water corresponding to the saturation temperature of the water vapor.

Inasmuch as the saturation temperature of the water vapor is also the DP temperature of the air, it follows that the DP temperature determines not only the mass of water vapor in the air but also the value of the latent heat of vaporization. Consequently, the latent heat of the air is a function of the DP temperature alone. As long as the DP temperature of the air remains unchanged, the latent heat of the air also remains unchanged.

The enthalpy of saturated water vapor (h_g) at various saturation (DP) temperatures as computed from 32°F is given in column nine of Table 5-1. Although the values listed include the sensible heat of the liquid above 32°F as well as the latent heat of vaporization at the given temperature, common practice is to treat the entire amount as latent heat.

Since the amount of sensible heat involved is relatively small, the error that accrues from assuming all the heat of the water vapor to be latent heat is of no particular consequence. As a practical matter, it is convenient to assume that the sensible heat of the air is the enthalpy of the dry air and that the latent heat of the air is the enthalpy of the water vapor, the sum of the two being the total heat or enthalpy of the air.

The latent heat of any given mass of air can be computed through the use of the following equation:

$$h_L = (m)(w \times h_w) \quad (5\text{-}8)$$

where h_L = the latent heat of any given mass of dry air having humidity ratio w in Btu
m = mass of dry air in pounds
w = the humidity ratio in pounds per pound

h_w = the specific enthalpy of the water vapor in the air, usually taken as the enthalpy of saturated water vapor (h_g) at a saturation temperature equal to the air DP temperature, in Btu per pound

For 1 lb of air, Equation 5-8 is shortened to

$$h_L = (w)(h_w) \tag{5-9}$$

where h_L = the latent heat per pound of dry air having humidity ratio w in Btu per pound

It is important to notice that in any mixture of dry air and water vapor, the mass of air (m) is taken as the mass of dry air, so that for m pounds of air, the actual mass of the air (dry air-water vapor mixture) is m pounds of dry air plus the mass of the water vapor in the mixture.

Example 5-11 Determine the latent heat of 10 lb of air having a DB temperature of 85°F and a DP temperature of 58°F.

Solution From Table 5-1, the saturation pressure corresponding to a saturation temperature of 58°F is 0.2386 psia and the enthalpy of saturated vapor (h_g) at that saturation temperature is 1087.1 Btu/lb. Assuming standard barometric pressure at sea level (14.696 psi) and substituting in Equation 5-2, the humidity ratio is

$$w = \frac{(0.622)(0.2386 \text{ psia})}{14.696 \text{ psi} - 0.2386 \text{ psia}}$$

$$= 0.01027 \text{ lb/lb}$$

Applying Equation 5-8,

$$H_L = (10 \text{ lb})(0.01027 \text{ lb/lb})(1087.1 \text{ Btu/lb})$$

$$= 111.65 \text{ Btu}$$

It should be noted that the calculations in the previous example, and all similar calculations, have some inherent inaccuracies:

1. The enthalpy of the water vapor (h_g) is not all latent heat; it is the sum of the sensible heat of the liquid above 32°F and the latent heat of vaporization.
2. The mass of the air is not 10 lb but actually 10 lb of dry air plus the mass of the water vapor mixed with the dry air, that is, 10 lb of dry air plus 0.01027 lb of water vapor (10.01027 lb).
3. The enthalpy of the water vapor (h_g) is the enthalpy at saturation and does not include the sensible heat of the vapor (superheat).

With reference to the third factor, recall that all the components in any mixture of gases and vapors are at the same temperature. Therefore, the water vapor in the air is always at the same temperature as the dry components (the DB temperature). For this reason, except in the case of saturated air (when DB and DP temperatures are the same), the water vapor in the air is in a superheated state, the degree of superheat being the difference between the DP and DB temperatures of the air. For instance, in Example 5-11, the water vapor in the air is superheated from 58°F (DP temperature) to 85°F (DB temperature). Assuming that the average specific heat of water vapor (c_p) is 0.45 Btu/lb °F and applying Equation 3-18, the quantity of superheat in the water vapor in Example 5-11 is 1.25 Btu (0.1027 lb)(0.45)(27°F) for the 10 lb of air.

From the foregoing, it is evident that the latent heat per pound of dry air (h_L) can be more accurately evaluated if h_w is corrected to take into account the superheat in the water vapor, that is,

$$h_w = h_g + (0.45)(\text{DB} - \text{DP}) \tag{5-10}$$

All the inaccuracies discussed in the preceding paragraphs are minor and can be ignored in most practical calculations.

When steam tables are not available, a convenient empirical equation for determining the enthalpy of low-pressure, low-temperature steam (such as that found in the air) is

$$h_w = 1060.8 + (0.45)(\text{DB}) \tag{5-11}$$

Equation 5-11 gives reasonably accurate total enthalpies for either saturated or superheated water vapor at temperatures between 32°F and 100°F.

Examination of the steam table (Table 5-1) shows that the enthalpy of saturated water vapor increases approximately 0.45 Btu/lb per degree of temperature rise above 32°F. Although the sensible heat of the liquid *increases* 1 Btu/lb °F, the latent heat of vaporization *decreases* approximately 0.55 Btu/lb per degree of increase in the saturation temperature, so that the net increase in the total enthalpy of saturated water vapor per degree of increase in saturation temperature above 32°F is 0.45 Btu/lb. Since the approximate increase in the total enthalpy of saturated water vapor per degree of temperature rise (0.45 Btu/lb) is equal to the average specific heat of water vapor (0.45 Btu/lb), the term 0.45, when added to the base value of 1060.8 Btu/lb, will give reasonably accurate enthalpies for either saturated or superheated water vapor.

The quantity of latent heat (Q_L) transferred to or from any given mass of air as water vapor is added or removed, respectively, can be determined by multiplying the mass of air (dry) by the difference in the latent heat per pound of air at the initial and final conditions; that is,

$$Q_L = (m)(h_{L,2} - h_{L,1}) \qquad (5\text{-}12)$$

or, since

$$h_{L,1} = (w_1)(h_{w,1}) \qquad \text{and} \qquad h_{L,2} = (w_2)(h_{w,2})$$

$$Q_L = (m)[(w_2 \times h_{w,2}) - (w_1 \times h_{w,1})] \qquad (5\text{-}13)$$

where the subscripts 1 and 2 indicate the initial and final conditions of the air, respectively. When the latent heat of the air at the initial condition is less than the latent heat at the final condition, the result obtained from Equations 5-12 and 5-13 will be negative, indicating that latent heat and water vapor are removed from the air rather than added to it. In practice, the negative sign is usually omitted.

Example 5-12 Fifty pounds of air per minute having DB and DP temperatures of 90°F and 60°F, respectively, are passed across a cooling coil and cooled to a final DB temperature of 50°F. Assuming that the air leaving the cooling coil is saturated (DB, WB, and DP temperatures are the same) and that the humidity ratios corresponding to DP temperatures of 60°F and 50°F are 0.01104 lb/lb and 0.00763 lb/lb, respectively, compute:

(a) the mass of water vapor condensed from the air in pounds per hour,
(b) the latent heat transferred in Btu per hour.

Solution From the steam table (Table 5-1), the enthalpy of saturated water vapor is 1088 Btu/lb at 60°F and 1083.7 Btu/lb at 50°F. Fifty pounds per minute is equivalent to 3000 lb/hr.

(a) The mass of water vapor $= (m)(w_1 - w_2)$ condensed per hour

$$m_w = (3000 \text{ lb/hr})(0.01104 \text{ lb/lb} - 0.00763 \text{ lb/lb})$$
$$= 10.23 \text{ lb/hr}$$

(b) Applying Equation 5-9,

$$h_{L,1} = (0.01104 \text{ lb/lb})(1088 \text{ Btu/lb}) = 12.01 \text{ Btu/lb}$$

$$h_{L,2} = (0.00763 \text{ lb/lb})(1083.7 \text{ Btu/lb}) = 8.27 \text{ Btu/lb}$$

Applying Equation 5-12,

$$Q_L = (3000 \text{ lb/hr})(12.01 \text{ Btu/lb} - 8.27 \text{ Btu/lb})$$
$$= 11{,}220 \text{ Btu/hr}$$

5-13. Total Heat of the Air

The total heat (enthalpy) of the air is the sum of the sensible heat of the air (the enthalpy of the dry air) and the latent heat of the air (the enthalpy of the water vapor), so that for 1 lb of air,

$$h_t = h_s + h_L \qquad (5\text{-}14)$$

where h_t = the enthalpy of the (moist) air in Btu per pound
h_s = the enthalpy of dry air in Btu per pound
h_L = the enthalpy of the water vapor in Btu per pound of dry air

For m pounds of air,

$$H_t = (m)(h_t) \qquad (5\text{-}15)$$

where H_t is the enthalpy of m pounds of air.

The total heat transferred (Q_t) to or from the air as the air is heated or cooled, respectively, may be determined from the following equation,

$$Q_t = (m)(h_{t,2} - h_{t,1}) \qquad (5\text{-}16)$$

where the subscripts 1 and 2 indicate the initial and final conditions of the air, respectively. When the enthalpy at condition 1 exceeds that at condition 2, the results obtained will be negative, indicating that heat is transferred from the air rather than to it. In practice, the negative sign is usually neglected.

Since the enthalpy of the air is the sum of the sensible and latent heat of the air, it follows that changes in either the sensible heat or the latent heat or both will cause corresponding changes in the enthalpy of the air. One notable exception is when the sensible heat and the latent heat of the air change in equal amounts but in opposite directions, in which case the change in one exactly offsets the change in the other, so that the enthalpy of the air remains unchanged during the process.

Adiabatic processes such as this are common and are usually referred to as either adiabatic humidification or evaporative cooling, depending on the purpose of the process.* In either case, the process is accomplished by bringing unsaturated air into intimate contact with water, usually by passing the air through a water spray (Fig. 5-2). Water is evaporated from the spray into the air, thereby increasing the humidity ratio, the DP temperature, and the latent heat of the air. Since the source of heat supplying the latent heat of vaporization to vaporize the water is the air itself, the sensible heat of the air decreases by an amount equal to the increase in the latent heat, and the DB

* Processes that occur without heat transfer to or from the working fluid are called adiabatic processes and are said to occur adiabatically.

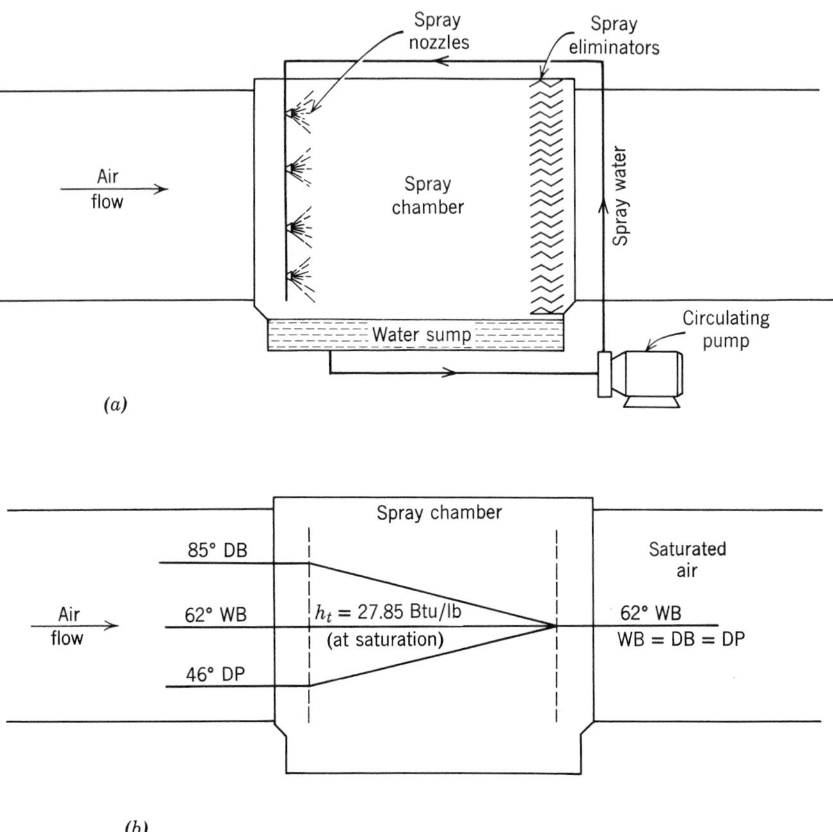

Fig. 5-2 Adiabatic humidification and/or evaporative cooling process. Air is brought into intimate contact with water spray in the spray chamber. In this instance, the spray water is recirculated without being either heated or cooled, in which case the temperature of the spray water will be the same as the WB temperature of the entering air.

temperature of the air is reduced accordingly.[†] If the air is maintained in contact with the water spray for a sufficient length of time, the air will become saturated at the entering WB temperature, since the latter remains constant throughout the entire process.

In the foregoing process, it is assumed that the temperature of the spray water in the spray chamber is the same as the WB temperature of the entering air, which is always the case when the spray water is continuously recirculated without being either heated or cooled by external means.

5-14. WB Temperature as an Index of Total Heat

It has been shown in preceding sections that the sensible heat of the air (the enthalpy of the dry air) is a function of the DB temperature of the air and that the latent heat of the air (the enthalpy of the water vapor mixed with the dry air) is a function of the DP temperature. Since for any given combination of DB and DP temperatures, the WB temperature of the air can have only one value, it follows that the WB temperature is an index of air total heat (enthalpy).

Although there is only one WB temperature that will satisfy any given combination of DB and DP temperatures, there are many combinations of DB and DP temperatures that will have the same WB temperature (Fig. 5-3). This means, in effect, that different samples of air having the same WB temperature have the same enthalpy, even though the ratio of sensible to latent heat may be different for the different samples.

Actually, the enthalpy of any given sample of air is a direct function of the WB temperature only when the air sample is saturated. For any sample of partially saturated air, the enthalpy deviates slightly from that of saturated air having the same WB temperature. Notice, in Fig. 5-3, that the difference between the enthalpy of partially saturated air and that of saturated air at the same WB temperature, called the enthalpy deviation, increases as the RH of the partially saturated air decreases (as the difference between the DB and DP temperatures increases). However, the error incurred in assuming that all samples of air having the same WB temperature have the same enthalpy is often small and in many cases

[†] Actually, the enthalpy of the air leaving an adiabatic saturator is slightly greater than that of the air entering the saturator. The increase in enthalpy is equal to the sensible heat of the liquid (32°F to the air WB temperature) of the water vapor added during the process.

Temperature of			Enthalpy, Btu/lb		
DB	DP	WB	Sensible	Latent	Total
60	60	60	14.39	11.98	*26.37*
65	57	60	15.59	10.75	26.34
70	53.7	60	16.79	9.53	26.32
75	50	60	17.99	8.30	26.29
80	45.5	60	19.19	7.05	26.24
85	40.7	60	20.39	5.81	26.20
90	34.7	60	21.59	4.58	26.17

Fig. 5-3 Some samples of air having the same WB temperature but different DB and DP temperatures. Notice that the true enthalpy (total heat) of the partially saturated samples differs slightly from that of saturated air (italicized) having the same WB temperature. The difference is known as the *enthalpy deviation*. The enthalpy deviation diminishes as the air sample approaches saturation.

can be ignored. Enthalpy deviation is discussed in more detail later.

5-15. Standard Air

In the foregoing sections the element of time has been largely ignored in all the basic mass-energy calculations. However, in practice, it is usually energy flow rates rather than energy quantities that are of interest, and for this reason, a time element is often introduced into the basic mass-energy equations. This is accomplished by simply substituting a mass flow rate, such as pounds per hour, for the mass *m* in pounds, in which case the results obtained from the equations will be energy flow rates (Btu per hour) rather than energy quantities (Btu).

In practice, it is also often convenient to express air quantities by volume rather than by mass. However, since the volume of any given mass of air varies with both the temperature and the pressure in accordance with the ideal gas laws, an air standard has been established so that an air quantity expressed as a volume or volume flow rate represents a known equivalent mass or mass flow rate. *Standard air* is defined as air having a density of 0.075 lb/ft^3 or a specific volume of (1/0.075) 13.34 ft^3/lb. Air at standard barometric pressure and 70°F substantially meets this condition.

The concept of standard air makes it possible to rate all air-handling equipment under equal conditions. Moreover, a standard air volume, when multiplied by the density of standard air (0.075 lb/ft^3), can be substituted for the mass *m* in any of the mass-energy equations.

Any given volume of air at any condition can be converted to an equivalent volume of standard air through the use of either of the following equations:

$$V_s = (V_a)\frac{v_s}{v_a} \qquad (5\text{-}17)$$

$$V_s = (V_a)\frac{T_s}{T_a} \qquad (5\text{-}18)$$

where V_s = the volume of standard air
V_a = the actual air volume
v_s = the specific volume of standard air (13.34 ft^3/lb)
v_a = the actual specific volume
T_s = the standard air temperature (430°R)
T_a = the actual air temperature in degrees Rankine

Example 5-13 Determine the equivalent volume of standard air for 1500 ft^3 of air having a DB temperature of 40°F.

Solution Applying Equation 5-18,

$$V_s = \frac{(1500\text{ ft}^3)(530°\text{R})}{40°\text{F} + 460°}$$

$$= 1590\text{ ft}^3\text{ (Std)}$$

5-16. Psychrometric Charts

Psychrometric charts are graphical presentations of the psychrometric properties of air. The use of such charts permits the graphical analysis of psychrometric data and processes and thereby facilitates the solution of many practical problems dealing with air that otherwise would require more tedious mathematical solutions. A typical psychrometric chart is illustrated in Fig. 5-4. The values given on the chart are for air at standard barometric pressure and should be corrected for use at other elevations.

The skeleton chart in Fig. 5-5 illustrates the general construction of the psychrometric chart and the relationships among some of the fundamental properties of air. Notice that the vertical lines on the chart are lines of constant DB temperature, whereas the horizontal lines are lines of constant DP temperature and humidity ratio. The closely spaced diagonal lines are lines of constant WB temperature and the more widely spaced lines are lines of constant specific volume.

The curved lines extending from lower left to upper right on the chart are lines of constant RH. The curved line bounding the chart on the left side is the line of 100% RH and is known as the saturation curve. Air at any condition such that its state can be plotted as a point falling anywhere along the saturation curve is saturated air. The other curved lines on the chart are enthalpy deviation lines.

The *enthalpy deviation* is the difference between the actual or true specific enthalpy

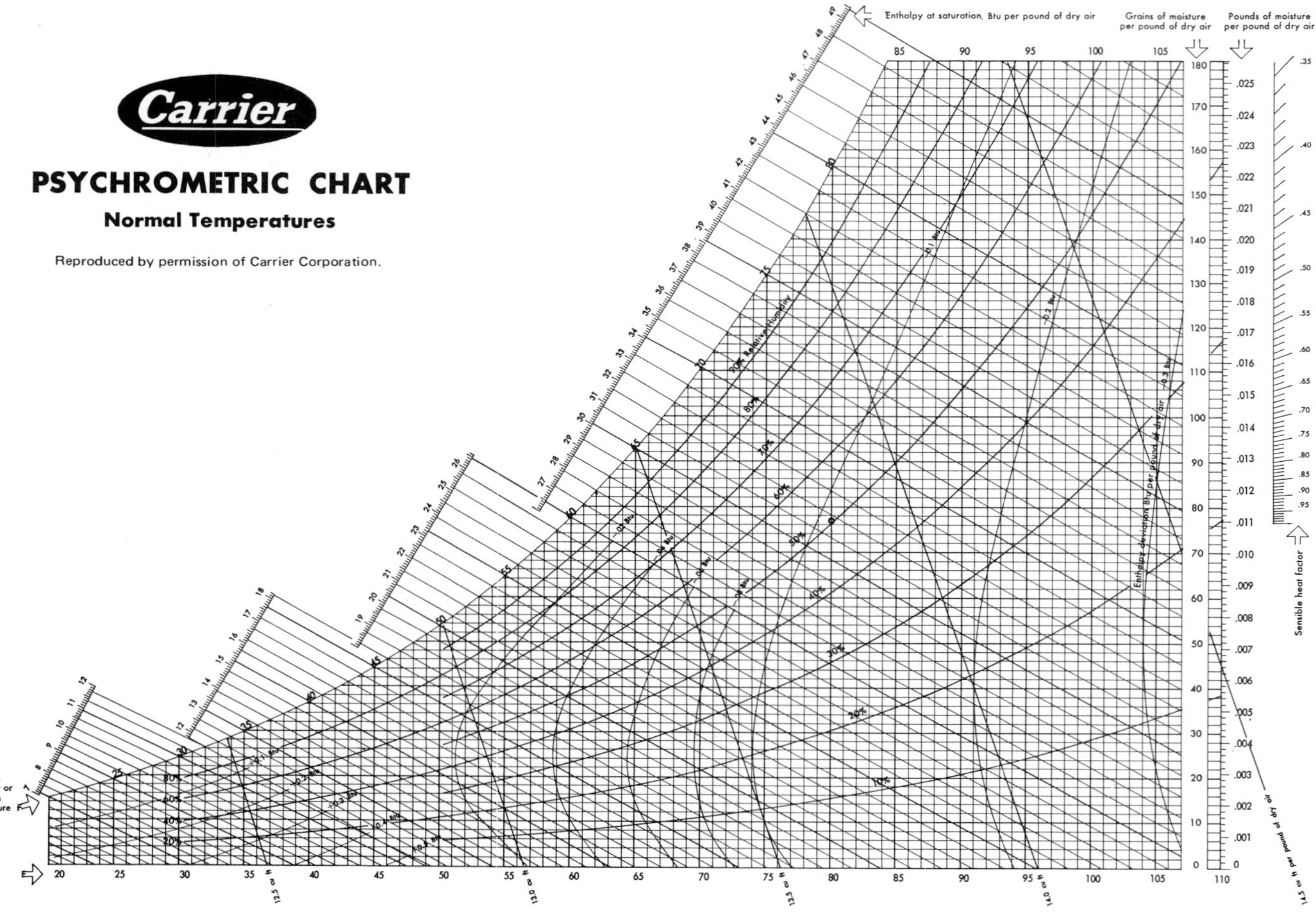
Carrier
PSYCHROMETRIC CHART
Normal Temperatures
Reproduced by permission of Carrier Corporation.
Enthalpy at saturation, Btu per pound of dry air
Grains of moisture per pound of dry air
Pounds of moisture per pound of dry air
Wet Bulb, Dew Point or Saturation Temperature F
Dry Bulb
90% Relative Humidity
Enthalpy deviation Btu per pound of dry air
Sensible heat factor
12.5 cu ft
13.0 cu ft
13.5 cu ft
14.0 cu ft
14.5 cu ft per pound of dry air
Below 32 F, properties and enthalpy deviation lines are for ice.
Copyright 1947 Carrier Corporation – Copyright 1959 Carrier Corporation – AC467 Printed in U.S.A.

Fig. 5-4

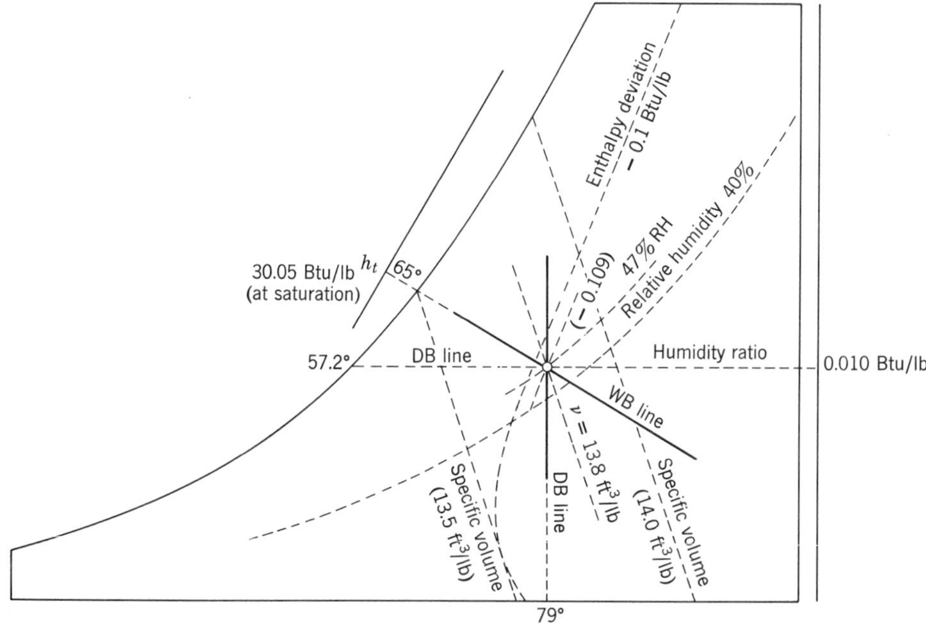

Fig. 5-5

of the air at any given condition and the specific enthalpy of saturated air at the same WB temperature (see Section 5-14).

Values for the DB and WB temperature lines are given at their intersection with the saturation curve. Values for DB temperature are given also at the base of the chart, as are the values for the specific volume lines.

Values for the DP temperature lines are read at the saturation curve and the corresponding humidity ratio is read on the same line from the scale on the right side of the chart.

The specific enthalpy of saturated air at any given WB temperature is found by following the WB line to the enthalpy scale above the saturation curve. Enthalpy deviations are determined by interpolation between the lines of enthalpy deviation. The true enthalpy of air at any condition is determined by adding (algebraically) the enthalpy deviation to the specific enthalpy at saturation.

The following example will serve to illustrate the use of the psychrometric chart.

Example 5-14 Readings from a sling psychrometer indicate that the air in a room has a DB temperature of 79°F and a WB temperature of 65°F. From the psychrometric chart, determine the humidity ratio, DP temperature, approximate specific volume, specific enthalpy at saturation, enthalpy deviation, actual specific enthalpy, and approximate relative humidity.

Solution Using the two known properties of the air as coordinates, the condition of the air can be established as a point on the chart. Once this point has been established, the other properties of the air at this condition can be read from the chart as follows (see Fig. 5-5):

Humidity ratio	= 0.010 lb/lb
DP temperature	= 57.2°F
Specific volume (by interpolation)	= 13.80 ft³/lb dry air
Specific enthalpy at saturation	= 30.05 Btu/lb
Enthalpy deviation	= −0.109 Btu/lb
Specific enthalpy (corrected)	= 30.05 − 0.109
	= 29.94 Btu/lb
Relative humidity	= 47%

Example 5-15 For the air sample described in Example 5-14, determine the sensible heat

per pound of air and the latent heat per pound of air.

Solution

(a) Applying Equation 5-4,

$$h_s = (0.24 \text{ Btu/lb °F})(79°\text{F}) = 18.96 \text{ Btu/lb}$$

(b) From Example 5-14,

$$h_t = 29.94 \text{ Btu/lb}$$

Applying Equation 5-14,

$$h = 29.94 - 18.96 = 10.98 \text{ Btu/lb}$$

Alternate Solution

(b) From Example 5-14,

$$w = 0.010 \text{ lb/lb}$$

Applying Equation 5-11,

$$h_w = 1060.8 + (0.45)(79°) = 1096.35 \text{ Btu/lb}$$

Applying Equation 5-9,

$$h_L = (0.010 \text{ lb/lb})(1096.35 \text{ Btu/lb}) = 10.96 \text{ Btu/lb}$$

5-17. Psychrometric Processes

Psychrometric processes can be illustrated and analyzed on a psychrometric chart. To illustrate the use of the psychrometric chart for this purpose, and also to introduce several important new concepts, a few of the more important psychrometric processes are analyzed in the following sections.

5-18. Air Mixtures

One of the psychrometric processes most frequently encountered is the mixing of two or more airstreams having different initial conditions. In such cases, the condition of the resulting mixture is readily determined through the use of a simple mass-energy balance. For example, with reference to Fig. 5-6*a*, two air quantities, one at condition A and one at condition B, are brought together to form a

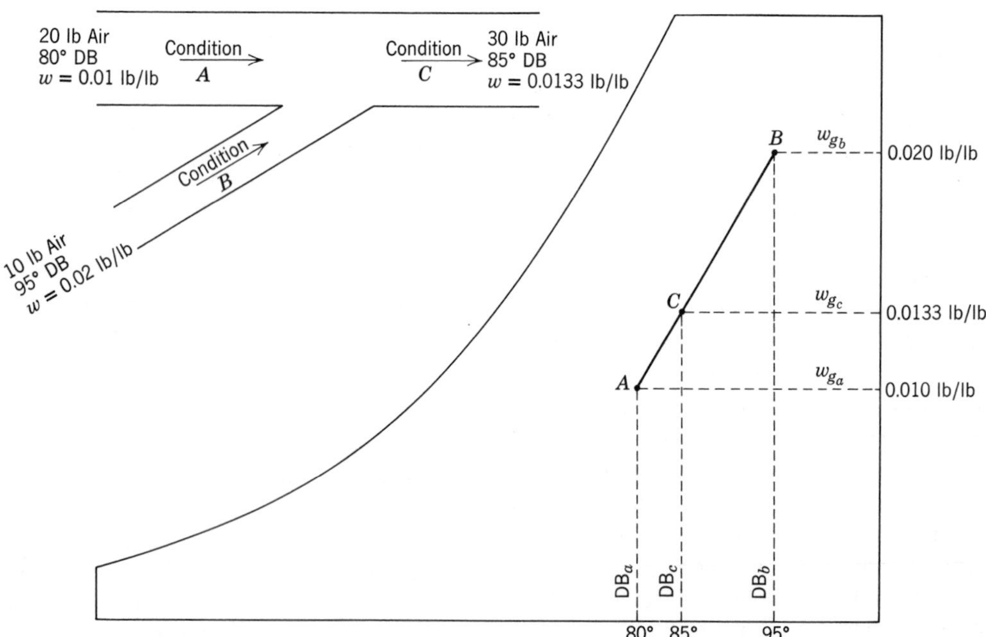

Fig. 5-6*a*;*b* Illustrates the mixing of two airstreams having initial conditions identified by states points A and *B*. State point *C* is the condition of the resulting mixture and divides line *AB* into two parts that are exactly proportional to the two air quantities being mixed. Line *AC* is proportional to the air quantity having initial condition *B*, while line *CB* is proportional to the air quantity having initial condition *A*.

mixture at some condition C. It is evident that the mass of the mixture (m_c) is equal to the sum of the masses of the two component air quantities, that is,

$$m_c = m_a + m_b \qquad (5\text{-}19)$$

Also, since the mixing of the two airstreams occurs adiabatically (without gain or loss of heat) and without gain or loss of moisture, it follows that the enthalpy of the mixture is the sum of the enthalpies of the components, that is,

$$H_{s,c} = H_{s,a} + H_{s,b} \qquad (5\text{-}20)$$

$$H_{L,c} = H_{L,a} + H_{L,b} \qquad (5\text{-}21)$$

$$H_{t,c} = H_{t,a} + H_{t,b} \qquad (5\text{-}22)$$

and

$$m_c w_c = m_a w_a = m_b w_b \qquad (5\text{-}23)$$

In accordance with Equation 5-5,

$$H_{s,c} = (m_c)(0.24)(T_c)$$

$$H_{s,a} = (m_a)(0.24)(T_a)$$

$$H_{s,b} = (m_b)(0.24)(T_b)$$

then

$$(m_c)(0.24)(T_c) = (m_a)(0.24)(T_a) + (m_b)(0.24)(T_b) \qquad (5\text{-}24)$$

Dividing both sides of Equation 5-24 by 0.24 and solving for T_c (DB temperature of mixture) gives

$$T_c = \frac{(m_a)(T_a) + (m_b)(T_b)}{m_c} \qquad (5\text{-}25)$$

Likewise, solving Equation 5-23 for the humidity ratio of the mixture results in

$$w_c = \frac{(m_a)(w_a) + (m_b)(w_b)}{m_c} \qquad (5\text{-}26)$$

Example 5-16 Twenty pounds of air having a DB temperature of 80°F and a humidity ratio of 0.010 lb/lb are mixed with 10 lb of air having a DB temperature of 95°F and a humidity ratio of 0.020 lb/lb (see Fig. 5-6*a*). Determine the DB temperature and humidity ratio of the mixture.

Solution (*by calculation*) Applying Equation 5-25,

$$T_c = \frac{(20\text{ lb})(80°\text{F}) + (10\text{ lb})(95°\text{F})}{30\text{ lb}} = 85°\text{F}$$

Applying Equation 5-26,

$$w_c = \frac{(20\text{ lb})(0.010\text{ lb/lb}) + (10\text{ lb})(0.020\text{ lb/lb})}{30\text{ lb}}$$

$$= 0.0133\text{ lb/lb}$$

Solution (*by psychrometric chart*) Plot the condition of each of the two components of the mixture as a point on the psychrometric chart (points *A* and *B* on the chart in Fig. 5-6*b*). Since the specific humidity and DB temperature scales are linear, the final state point, *C*, representing the condition of the mixture, will fall somewhere along a straight line connecting the two points representing the initial states of the two components before mixing. The exact point on line *AB* where state point *C* lies can be determined readily by calculating either the DB temperature or the specific humidity of the mixture, as shown in the preceding solution. Since point *C* will always divide the line *AB* into two parts that are exactly proportional to the masses (or standard air volumes) of the two components, the location of point *C* on line *AB* may also be determined graphically.

5-19. Sensible Heating

Sensible heating of the air occurs whenever the air is passed over a heating surface (dry), such as a steam or hot water coil, whose temperature is above the DB temperature of the air. In passing over the heating surface, the air absorbs heat (sensible) from the warmer surface, so that the DB temperature of the air increases and tends to approach the temperature of the heating surface (Fig. 5-7*a*). Since no moisture is added to or removed from the air during the heating process, the specific humidity, DP temperature, and latent heat content of the air do not change. For this reason, sensible heating (or cooling) processes always follow a straight line path along a line of constant DP temperature (constant humidity ratio) and are often referred to as *constant DP temperature processes*. Since, during the process, the total heat of the air increases by an amount equal to the increases in the sensible heat, the WB temperature also increases. The following example will serve to illustrate a typical sensible heating process.

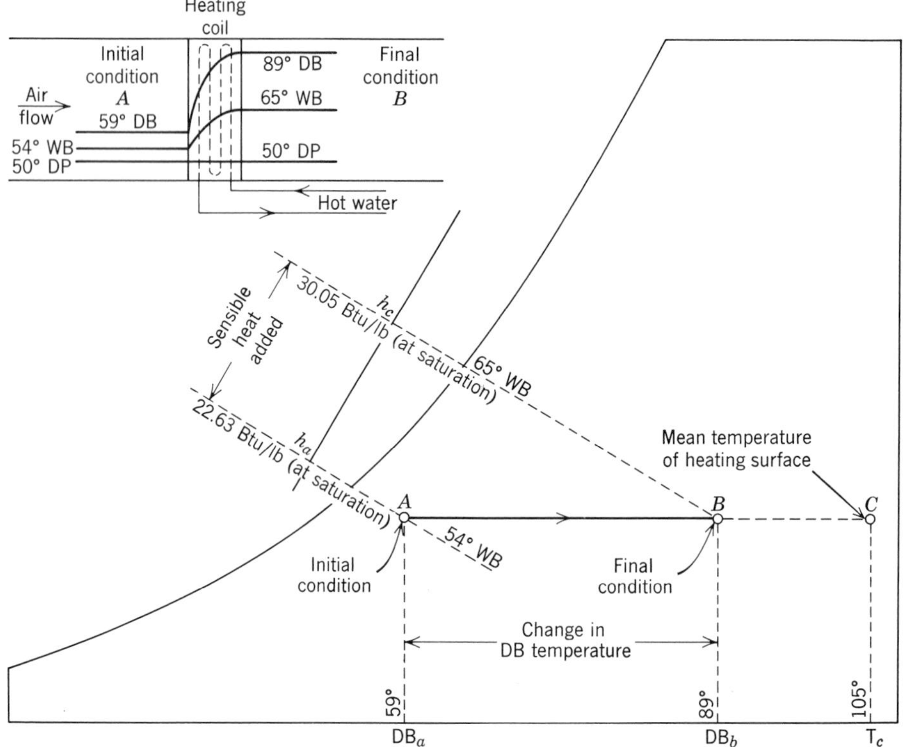

Fig. 5-7*a*;*b* Sensible heating process (see Example 5-17).

Example 5-17 One hundred pounds of air per minute having initial DB and WB temperatures of 59°F and 54°F are passed across a heating coil, and the DB temperature is raised to 89°F. Plot the process on a psychrometric chart and determine:

(a) the final WB temperature of the air,
(b) the sensible heat transferred in Btu per hour,
(c) the total heat transferred in Btu per hour.

Solution Using the initial DB and WB temperatures, which are given, the initial state of the air (the condition of the air entering the coil) can be located as point *A* on the chart in Fig. 5-7*b*. Since moisture is neither added nor removed during the process, the path of the process will be along a line of constant DP temperature (and constant humidity ratio) as the DB temperature increases. The final state of the air (the condition of the air leaving the coil) is indicated by point *B*, which is located at the intersection of the constant DP line with the constant DB line corresponding to the final DB temperature of the air.

(a) Once point B has been established on the chart, the final WB temperature can be read directly from the chart as 65°F.
(b) Since, in this process, the change in sensible heat equals the change in total heat, both can be determined by calculating either one, and therefore two equally simple solutions are possible. Both solutions follow.

Applying Equation 5-6,

$$Q_s = (100 \text{ lb/min} \times 60)(0.24)(89° - 59°)$$
$$= 43{,}200 \text{ Btu/hr}$$

From the psychrometric chart,

$$h_{t,a} = 22.63 - 0.026 = 22.60 \text{ Btu/lb}$$
$$h_{t,b} = (30.05 - 0.18) = 29.87 \text{ Btu/lb}$$

Applying Equation 5-16,

$$Q_t = (6000 \text{ lb/hr})(29.87 - 22.60)$$
$$= 43{,}620 \text{ Btu/hr}$$

5-20. Coil Bypass Factor

In the preceding section it was shown that the DB temperature of the air passing over a heating coil tends to approach the surface temperature of the coil (point *C* in Fig. 5-7*b*). If all the air passing through the coil came into intimate contact with the heating surface, and remained in contact with it for a sufficient length of time, the DB temperature of the leaving air could be raised to the temperature of the heating surface. However, as a practical matter, a certain portion of the total air quantity passing through any heating (or cooling) coil never comes into contact with the coil surface and therefore is unaffected by its passage through the coil. That is, it leaves the coil in the same condition that it enters the coil. The portion of the air that passes through the coil without contacting the coil surface is known as the *bypass air* and, when expressed as a ratio to the total air quantity, is called the *bypass factor* of the coil. It will be shown later that the bypass factor (BPF) of any coil is primarily a function of coil design.

From the foregoing, it can be assumed that the air leaving the coil is actually a mixture of two airstreams or components. One component is the portion of the air that comes into direct contact with the coil surface and is assumed to leave the coil at a DB temperature equal to the mean surface temperature of the coil. The other component is the bypass air, which does not contact the coil surface, and it is assumed to leave the coil at the same DB temperature as when entering.

Then, with reference to Fig. 5-7*b*, state point *B*, which represents the condition of the air leaving the heating coil, may also be thought of as being the condition of the mixture that results when two air quantities, having initial conditions represented by the state points *A* and *C*, are brought together. Since state point *B* will always divide a straight line (*AC*) drawn between the two initial state points (*A* and *C*) into two parts that are exactly proportional to the quantities of the two component airstreams (see Section 5-8), it follows that line *BC* represents one component air quantity—the bypass air—and that line *AB* represents the other component air quantity—that which actually contacts the coil surface. The line *AC*, then, represents the total quantity of the air mixture leaving the coil, which is the sum of the two component air quantities.

As previously stated, the BPF of any coil is the ratio of the bypass air quantity to the total air quantity. Accordingly, the ratio of line *BC* to line *AC* is equal to the bypass factor of the heating coil in this instance. Since, in terms of the DB temperature scale, the length of line *BC* is equal to $T_c - T_b$ and the length of line *AC* is equal to $T_c - T_a$, it follows that

$$\text{BPF} = \frac{T_c - T_b}{T_c - T_a} \qquad (5\text{-}27)$$

where T_a = the DB temperature of the air entering the coil
T_b = the DB temperature of the air leaving the coil
T_c = the mean effective temperature of the coil surface

It is important to understand that the bypass factor of any heating or cooling coil must always be found by dividing the *difference between the DB temperature of the leaving air and the coil surface temperature* by the *difference between the DB temperature of the entering air and the coil surface temperature.*

Example 5-18 Calculate the bypass factor of the heating coil used in the sensible heating process described in Example 5-17 if the mean effective temperature of the coil surface is 105°F.

Solution Applying Equation 5-27, the coil BPF

$$= \frac{105° - 89°}{105° - 59°}$$

$$= 0.348$$

The coil BPF of 0.348 means, in effect, that 34.8% of the total air quantity passes through the heating coil without contacting the coil surface and therefore leaves the coil at the entering air condition (A), while the other 65.2% of the total air quantity contacts the coil surface and has its temperature increased to the coil surface temperature (C). The mixing of these two air quantities produces the air mixture leaving the coil at condition (B).

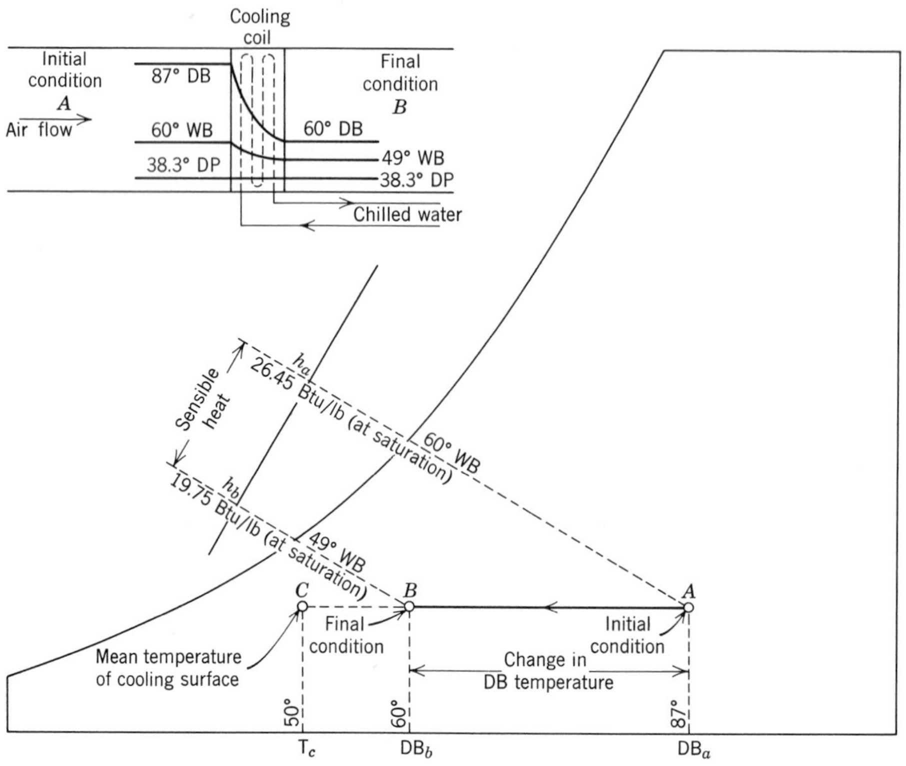

Fig. 5-8_a_;_b_ Sensible cooling process (see Example 5-19).

5-21. Sensible Cooling Processes

Sensible cooling of the air is accomplished by passing the air across a dry cooling surface whose temperature is below the DB temperature of the air but above the air DP temperature. As shown in Fig. 5-8, the sensible cooling process is similar to the sensible heating process in that no moisture is added or removed, and, therefore, the specific humidity, DP temperature, and latent heat content of the air remain the same or constant throughout the process. In addition, as in the case of sensible heating, the change in the total heat of the air is equal to the change in the sensible heat of the air.

Example 5-19 Four thousand pounds of air per hour, having initial DB and WB temperatures of 87°F and 60°F, respectively, are passed across a dry cooling surface and cooled to a final DB temperature of 60°F. If the temperature of the cooling surface is 50°F, determine:

(a) the sensible heat transferred in Btu per hour,
(b) the bypass factor of the coil,
(c) the bypass air quantity in pounds per hour.

Solution (See Fig. 5-8)

(a) Applying Equation 5-6,

$$Q_s = (4000\ \text{lb/hr})(0.24)(87^\circ - 60^\circ)$$
$$= 25{,}920\ \text{Btu/hr}$$

(b) Applying Equation 5-27,

$$\text{BPF} = \frac{60^\circ - 50^\circ}{87^\circ - 50^\circ}$$
$$= 0.27$$

(c) The bypass air quantity is

$$= (4000\ \text{lb/hr})(0.27)$$
$$= 1080\ \text{lb/hr}$$

5-22. Cooling and Dehumidification

Simultaneous cooling and dehumidification of the air will take place any time the air is passed across a cooling surface whose temperature is below the initial DP temperature of the air.

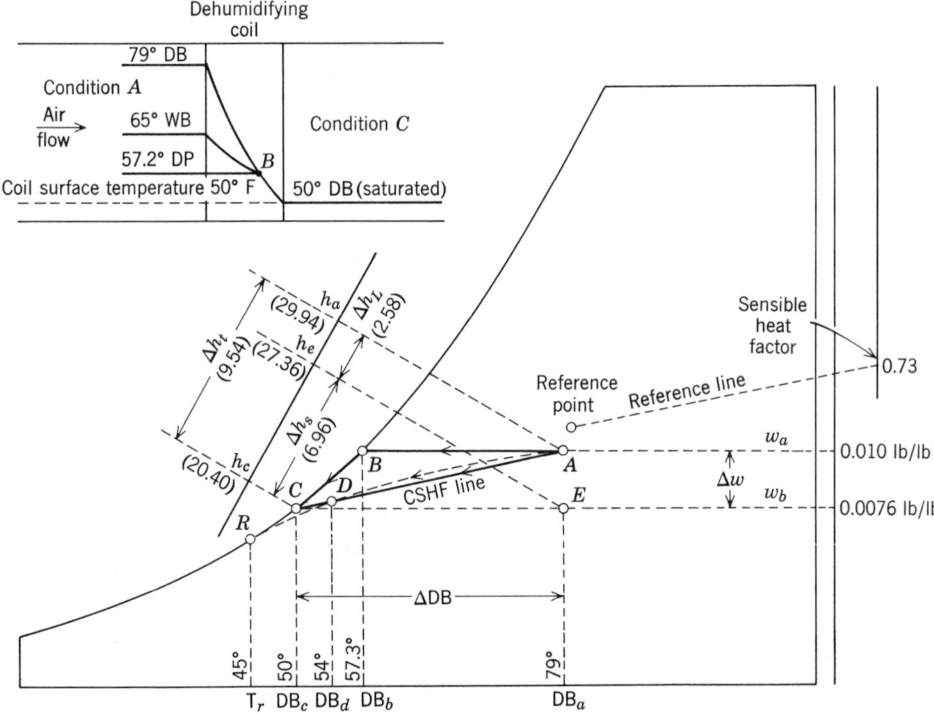

Fig. 5-9*a* Theoretical cooling and dehumidifying process (see Example 5-20). *b* Simultaneous cooling and dehumidification (see Sections 5-22 and 5-23).

This is true regardless of whether or not the DB temperature of the air leaving the cooling surface is below the initial DP temperature of the air. Although it may appear to do so, the latter statement does not contradict statements made previously that the temperature of the air must be reduced below the DP temperature before moisture can be condensed out of the air. The reason for this is made clear in Section 5-24 and is illustrated in Example 5-22. First, however, it is useful, for purposes of comparison, to examine the theoretical cooling and dehumidification process described in the following example.

Example 5-20 Two thousand pounds of air per hour, having initial DB and WB temperatures of 79°F and 65°F, respectively, are passed across a cooling coil whose mean effective surface temperature is 50°F. Assuming that all parts of the air contact the cooling surface so that the air leaves the coil saturated at the temperature of the coil surface, plot the path of the process on the psychrometric chart and determine:

(a) the total heat removed per pound of dry air,
(b) the sensible heat removed per pound of dry air,
(c) the latent heat removed per pound of dry air,
(d) the mass of water vapor condensed from the air in pounds per hour.

Solution Based on the premise that all parts of the air contact the cooling surface and are cooled simultaneously and progressively from the initial to the final condition, it is reasonable to assume that the air-conditioning process will follow the path *ABC*, as shown on the psychrometric chart in Fig. 5-9*b*, indicating that the air at initial condition A is first cooled with constant DP temperature (and humidity ratio) along line *AB* to saturation at the initial DP temperature of the air (point *B*) and then cooled and dehumidified along the saturation curve (line *BC*) to the leaving condition (point *C*). This theoretical process is illustrated in Fig. 5-9*a*.

(a) From the psychrometric chart,

$$h_{t,a} = 30.05 - 0.108 = 29.94 \text{ Btu/lb}$$

$$h_{t,c} = 20.40 \text{ Btu/lb}$$

Applying Equation 5-16,

$$Q_t = (1 \text{ lb})(29.94 - 20.40) = 9.54 \text{ Btu/lb}$$

(b) Applying Equation 5-6,

$$Q_s = (1 \text{ lb})(0.24)(79° - 50°) = 6.96 \text{ Btu/lb}$$

(c) The latent heat removed,

$$Q_L = Q_t - Q_s$$

$$= 9.54 - 6.96 = 2.58 \text{ Btu/lb}$$

(d) From the chart,

$$w_a = 0.010 \text{ lb/lb}$$

$$w_c = 0.0076 \text{ lb/lb}$$

Mass of water vapor condensed per hour, m

$$= (2000 \text{ lb/hr})(0.010 - 0.0076)$$

$$= 4.8 \text{ lb/hr}$$

An alternate solution to parts *a*, *b*, and *c* is illustrated on the chart in Fig. 5-9*b*.

5-23. Coil Sensible Heat Factor

In any cooling and dehumidifying process, both sensible and latent heat are removed at the cooling coil, the sum of the two being the total heat transferred. The ratio of sensible heat to total heat transferred is known as the *coil sensible heat factor* (CSHF), viz:

$$\text{CSHF} = \frac{Q_s}{Q_t} \qquad (5\text{-}28)$$

Example 5-21 Compute the CSHF for the coil in Example 5-20.

Solution Applying Equation 5-28,

$$\text{CSHF} = \frac{6.96 \text{ Btu/lb}}{9.54 \text{ Btu/lb}}$$

$$= 0.73$$

The CSHF can be determined directly from the psychrometric chart by employing the sensible heat factor scale on the right margin of the chart. A straight line (line *AC* in Fig. 5-9*b*) drawn between the entering air condition (A) and the coil surface temperature (C) is known as the CSHF line. Notice that the vertical component of this line is an index of the latent heat (moisture) removed during the process, as measured by the specific humidity scale, while the horizontal component is an index of the sensible heat removed, as measured by the DB temperature scale. Since the DB temperature and humidity ratio scales are linear, it is evident that the slope of this line will change in proportion to changes in the CSHF, becoming more vertical as the CSHF decreases and more horizontal as the CSHF increases. For a cooling process that is entirely sensible, the CSHF line is exactly horizontal and coincides with the lines of constant DP temperature.

To determine the CSHF directly from the chart, a straight line, originating at the chart reference point and extending to the sensible heat factor scale at the right margin of the chart, is drawn parallel to the CSHF line. The CSHF is read from the sensible heat factor scale at the point where the reference line intersects the scale. In Fig. 5-9*b*, read the CSHF as 0.73. Notice that this value corresponds to the result obtained in Example 5-21.

5-24. Practical Cooling and Dehumidifying Processes

The cooling and dehumidifying process described in Example 5-20 is purely hypothetical in that it can follow the path *ABC*, as shown in Fig. 5-9, only if all parts of the air passing through the coil actually contact the cooling surface and are cooled simultaneously from the initial condition A to the final condition C. Such an assumption ignores the existence of the bypass air (assumes coil BPF equals zero) and the fact that the temperature of all parts of the dehumidifying surface is not uniform and is therefore not practical.

In any actual cooling and dehumidifying process, there is always a certain amount of bypass air, so that the air leaving the coil is always a mixture of treated and untreated air. In such cases, it may be assumed that the treated portion of the air (that which contacts the cooling surface) follows the path *ABC* and leaves the coil saturated at the temperature of the coil surface (point *C*), while the untreated portion (the bypass air) leaves the coil at the entering condition (point *A*). Therefore, the air leaving the coil is actually a mixture of two airstreams whose initial conditions are represented, in this instance, by points *A* and *C*.

Since the condition of any mixture of two airstreams is such that it will plot as a point falling somewhere along a straight line connecting the state points that represent the initial conditions of the two airstreams, it follows that the condition of the air leaving any cooling and dehumidifying coil always will plot as some point, such as *D* in Fig. 5-9*b*, falling somewhere along a straight line (*AC*) drawn between the entering air condition (A) and the coil surface temperature (C). The exact condition of the leaving air, and therefore the exact location of point *D* on line *AC*, will depend on the percentage of the bypass air (the coil BPF). The greater the percentage of the bypass air (the higher the coil BPF), the closer the leaving air condition (D) will approach to the entering air condition (A).

On the psychrometric chart, the condition of the dehumidifying coil can be represented by a point plotted on the saturation curve at the temperature corresponding to the effective surface temperature of the coil, as illustrated by point *C* in Fig. 5-9*b*. This point is sometimes referred to as the *apparatus dew point* (ADP) of the coil.

It will be shown later (Chapter 11) that the surface temperature of a dehumidifying coil (or any heating or cooling coil) is not the same in all parts of the coil. For the dehumidifying coil illustrated in Fig. 5-9*b*, the actual surface temperatures in the various parts of the coil may range between the temperature limits roughly defined by points *B* and *R*, the former being the DP temperature of the entering air and the latter approaching the temperature of the refrigerant fluid in the tubes of the coil. The coil surface temperature (point *C*), then, is a theoretical equivalent or effective surface temperature, which may be considered as the uniform surface temperature that would produce the same leaving air conditions as would the varying surface temperature of an actual dehumidifying coil.

Line *AC* in Fig. 5-9*b* has been defined (Section 5-23) as the CSHF line, and as such is the locus of all possible leaving air conditions which will provide the desired ratio of sensible heat to total heat removal. However, this should not be interpreted to mean that the CSHF line represents the actual path of the dehumidifying process through a dehumidifying coil. For reasons discussed in Section 11-9, the air passing through a dehumidifying coil is actually a series of mixtures of treated and untreated air whose state points would plot along a curved path similar to the dashed line *AR* in Fig. 5-9*b*. However, the actual path of the process is unimportant, since for any given entering and leaving air conditions, the amount of sensible heat and moisture removed from the air will be the same regardless of the path of the progress. For this reason and because the CSHF line defines the apparatus dew point and is the locus of all possible leaving air conditions that will provide the desired sensible cooling to total cooling ratio, it is convenient to assume that the path of the dehumidifying process follows along the CSHF line.

Frequently, in air-conditioning practice, it is required to find the ADP, CSHF, and coil BPF that will provide a certain leaving air condition when the entering air condition is known. The following example serves to illustrate this procedure.

Example 5-22 A certain mass of air entering a dehumidifying coil has initial DB and WB temperatures of 85°F and 67°F, respectively. If it is desired that the air leaving the coil have a DB temperature of 63°F and WB temperature of 57°F, determine:

(a) the required ADP,
(b) the required BPF,
(c) the required CSHF.

Solution

(a) Plot the entering and leaving air conditions as state points *A* and *B* on the psychrometric chart (Fig. 5-10), and then draw a straight line from the entering air condition (A) through the leaving air condition (B) to intersection with the saturation curve at point *C*. This is the sensible heat factor line of the dehumidifying surface required to provide simultaneous cooling and dehumidification of the air in the proper proportions, and the required ADP can be read at the intersection of this line with the saturation curve (point *C*) as 50.3°F.

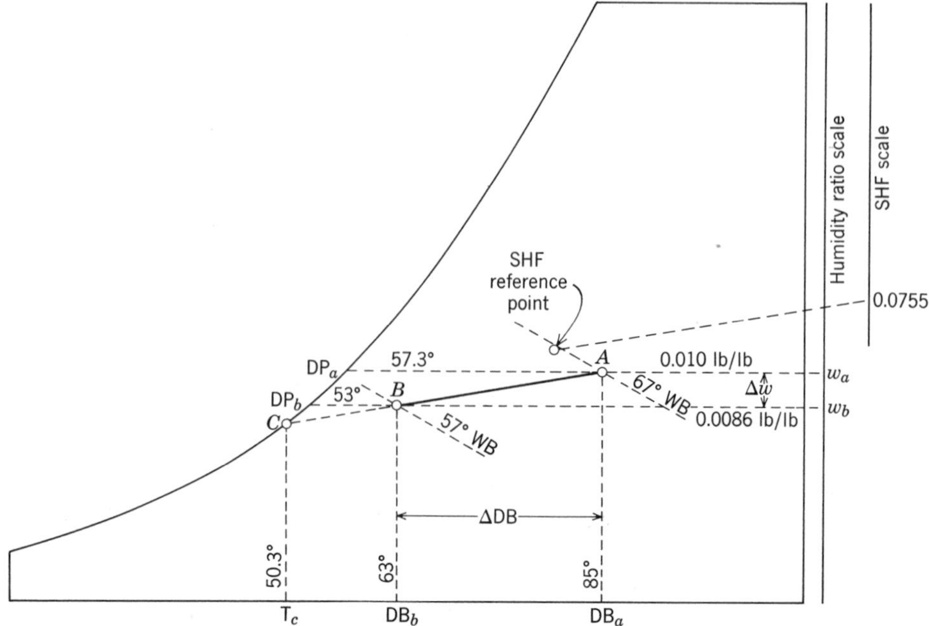

Fig. 5-10 Cooling and dehumidifying process. Notice that the leaving air DB temperature is above the entering air DP temperature (see Example 5-22).

(b) The required coil BPF is the ratio of the line *BC* to line *AC*, that is, applying Equation 5-27,

$$\text{BPF} = \frac{T_b - T_c}{T_a - T_c} = \frac{63° - 50.3°}{85° - 50.3°} = 0.366$$

(c) To determine the required CSHF, a reference line, originating at the chart reference point and extending to the sensible heat factor scale, is drawn parallel to line *AC*. The CSHF is read as the intersection of the reference line with the scale as 0.775.

It is of interest to notice, in Fig. 5-10, that moisture is removed from the air even though the leaving air DB temperature (63°F) is well above the DP temperature of the entering air (57.2°F).

5-25. METRIC SYSTEM EQUIVALENTS

The appropriate metric units may be substituted directly in all the equations developed in this chapter. However, since the enthalpy of dry air is calculated from 0°C and since the specific heat at constant pressure, c_p, for air is 1 kJ/kg °K, it follows that the specific enthalpy of dry air (sensible heat per kilogram of air) will be numerically equal to the DB temperature of the air in degrees Celsius, and that Equations 5-5 and 5-6 can be simplified to

$$H_s = (m)(\text{DB}) \tag{5-29}$$

$$Q_s = (m)(\Delta\text{DB}) \tag{5-30}$$

where H_s = the sensible heat in kilojoules
Q_s = the sensible heat transferred in kilojoules

Also, with converted constants, Equation 5-11 becomes

$$h_w = 2502 + (1.8)(\text{DB}) \tag{5-31}$$

where h_w = specific enthalpy of water vapor in kilojoules per kilogram

A psychrometric chart employing metric units is presented in Fig. 5-11. Notice that it is similar to the one in British units shown in Fig. 5-4 and may be used in the same manner.

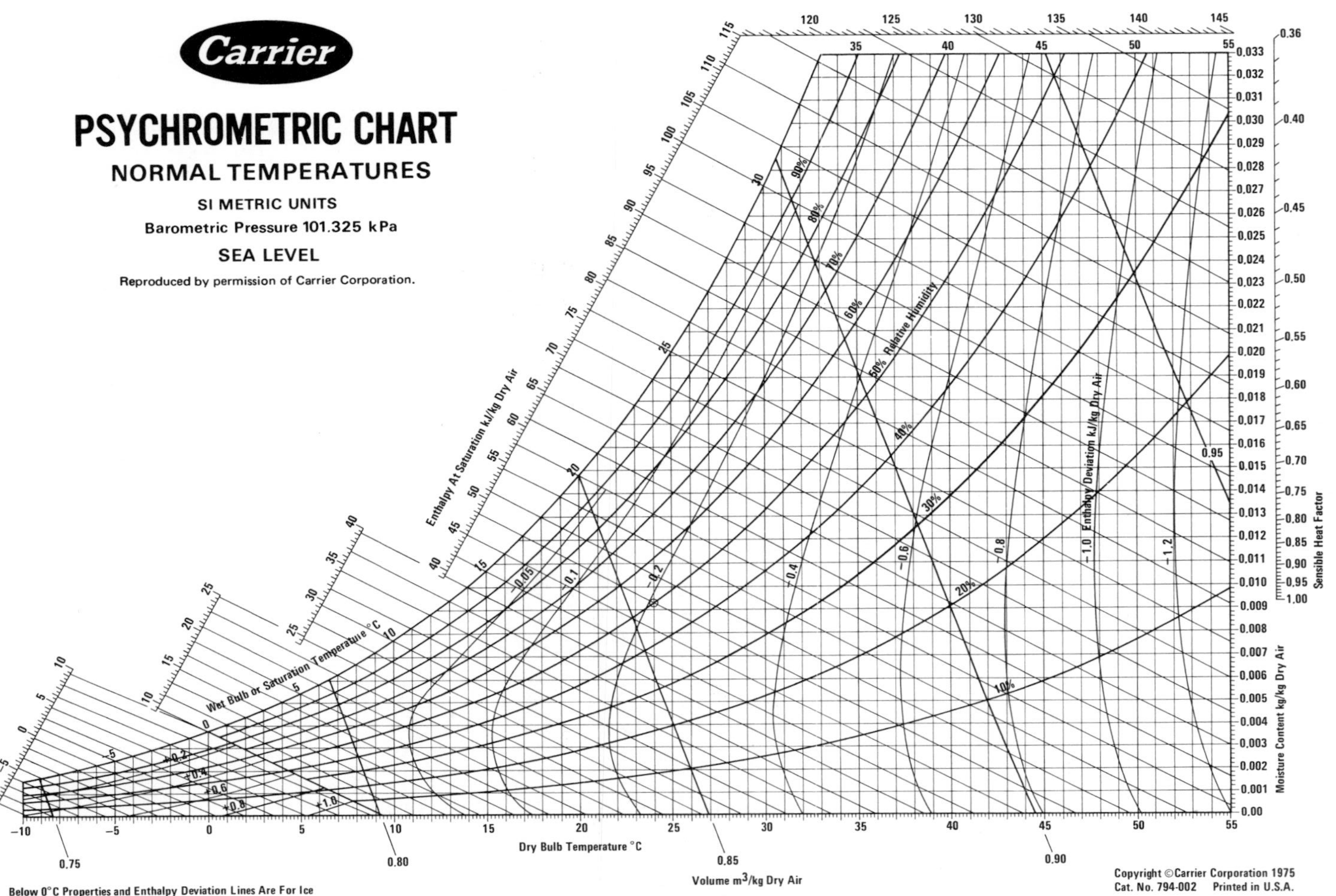
Carrier
PSYCHROMETRIC CHART
NORMAL TEMPERATURES
SI METRIC UNITS
Barometric Pressure 101.325 kPa
SEA LEVEL
Reproduced by permission of Carrier Corporation.
Enthalpy At Saturation kJ/kg Dry Air
Wet Bulb or Saturation Temperature °C
Relative Humidity
90%
80%
70%
60%
50%
40%
30%
20%
10%
Enthalpy Deviation kJ/kg Dry Air
Dry Bulb Temperature °C
Volume m³/kg Dry Air
Moisture Content kg/kg Dry Air
Sensible Heat Factor
Below 0°C Properties and Enthalpy Deviation Lines Are For Ice
Copyright ©Carrier Corporation 1975
Cat. No. 794-002 Printed in U.S.A.

Fig. 5-11

However, enthalpy values are not directly convertible, since the base for enthalpy calculations is not the same for the two charts. For the metric chart, the enthalpy of dry air and the enthalpy of water vapor are computed from the same base temperature, 0°C, whereas for the chart employing customary units, the enthalpy base for dry air is 0°F, while the enthalpy base for water vapor is 32°F.

In metric units *standard air* has a density of 1.2 kg/m^3 and a specific volume of 0.833 m^3/kg. Air at standard barometric pressure and a temperature of 21°C substantially meets this condition.

Example 5-23 A certain mass of air enters a dehumidifying coil at initial DB and WB temperatures of 30°C and 20°C, respectively. If it is desired that the air leaving the coil have a DB temperature of 17°C and a WB temperature of 14°C, determine the following:

(a) the mean effective surface temperature of the coil,
(b) the required coil bypass factor,
(c) the required coil sensible heat factor,
(d) the total heat removed per kilogram of air,
(e) the sensible heat removed per kilogram of air,
(f) the latent heat removed per kilogram of air,
(g) the mass of water vapor condensed per kilogram of air.

Solution

(a) Plot the entering and leaving air conditions as state points *A* and *B* on the psychrometric chart (Fig. 5-12), and then draw a straight line from the entering air condition (A) through the leaving air condition (B) to intersection with the saturation curve at point *C*. This is the sensible heat factor line of the dehumidifying surface required to provide simultaneous cooling and dehumidification of the air in the proper proportions, and the required coil mean effective surface temperature can be read at the intersection of this line with the saturation curve (point *C*) as 10°C.

(b) The required coil BPF is the ratio of line *BC* to line *AC*, that is, applying Equation 5-27,

$$\text{BPF} = \frac{T_b - T_c}{T_a - T_c} = \frac{17° - 10°}{30° - 10°} = 0.35$$

(c) To determine the required CSHF, a reference line, originating at the chart reference point and extending to the sensible heat

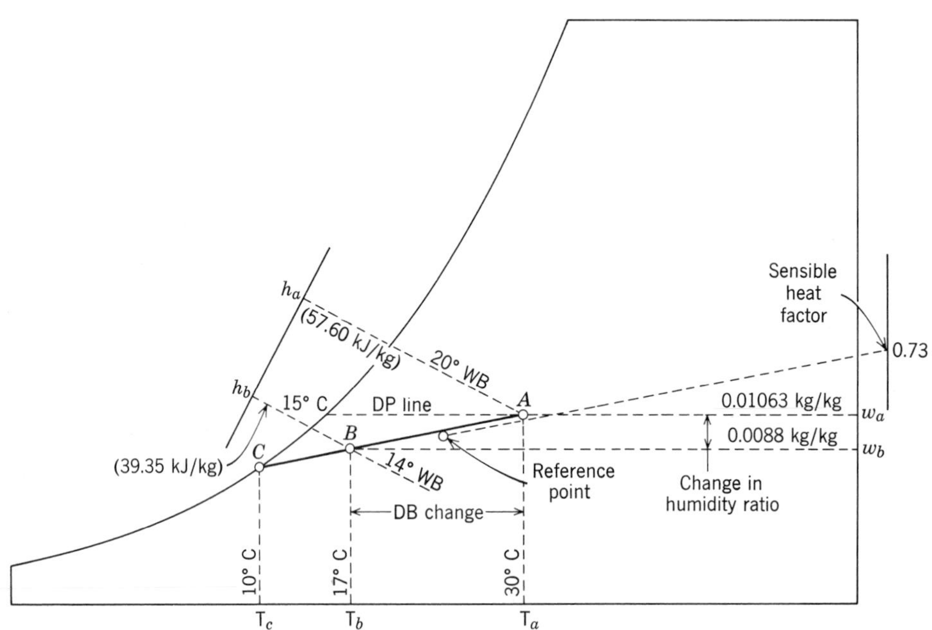

Fig. 15-12

factor scale, is drawn parallel to line *AC*. The CSHF is read as the intersection of the reference line with the scale as 0.73.

(d) From the psychrometric chart, h_a and h_b (corrected) are 57.25 kJ/kg (57.60 − 0.35) and 39.28 kJ/kg (39.35 − 0.075), respectively.
Applying Equation 5-16, the total heat removed,

$$Q_t = (1\text{ kg})(57.25 - 39.28) = 17.97\text{ kJ/kg}$$

(e) Applying Equation 5-30, the sensible heat removed,

$$Q_s = (1\text{ kg})(30° - 17°) = 13\text{ kJ/kg}$$

(f) The latent heat removed,

$$Q_L = Q_t - Q_s = 17.97\text{ kJ/kg} - 13\text{ kJ/kg} = 4.97\text{ kJ/kg}$$

(g) From the chart,

$$w_a = 0.01063\text{ kg/kg}$$

and

$$w_a = 0.0088\text{ kg/kg}$$

The mass of water vapor condensed per kilogram of air is

$$= (1\text{ kg})(0.01063 - 0.0088) = 0.00183\text{ kg/kg}$$

Example 5-24 Determine the equivalent standard air volume for 150 m³/s of air having a DB temperature of 15°C.

Solution Applying Equation 5-18,

$$V_s = \frac{(150\text{ m}^3/\text{s})(21°\text{C} + 273°)}{15°\text{C} + 273°} = 153\text{ m}^3/\text{s}$$

Customary Problems

5-1 Assume that a certain quantity of air has a temperature of 76°F and that the water vapor in the air exerts a partial pressure of 0.2386 psia. Determine the DP temperature of the air.

5-2 The partial pressure exerted by the water vapor in a certain sample of saturated air is 0.2563 psia. Determine (a) the air DP temperature and (b) the air DB temperature.

5-3 A certain quantity of air at a temperature of 90°F has a DP temperature of 56°F. What is the partial pressure exerted by the water vapor?

5-4 Determine the absolute humidity of the air in Problem 5-3 if the gas constant *R* for low pressure water vapor is 85.66 ft-lb/lb °R.

5-5 Compute the RH of the air samples in Problems 5-1, 5-2, and 5-3.

5-6 Determine the humidity ratio for the air samples in Problems 5-1, 5-2, and 5-3.

5-7 Compute the saturation ratio of the air samples in Problems 5-1, 5-2, and 5-3.

5-8 Calculate the sensible heat (enthalpy of dry air) of 5 lb of air having a DB temperature of 55°F and a DP temperature of 46°F.

5-9 The temperature of 150 lb of air is increased from 35°F to 120°F by the addition of heat. Calculate the quantity of heat supplied.

5-10 The temperature of 75 lb of air is reduced from 85°F to 40°F by passing the air across a cooling surface. Calculate the quantity of sensible heat removed from the air.

5-11 Determine the latent heat (enthalpy of the water vapor) of 10 lb of air having a DB temperature of 65°F and a DP temperature of 50°F.

5-12 Two thousand pounds of air per hour have DB and DP temperatures of 80°F and 60°F, respectively, and are passed across a cooling coil and cooled to a final DB temperature of 45°F. Assuming that the air leaving the coil is saturated, determine (a) the mass of the water vapor condensed from the air in pounds per hour and (b) the latent heat removed in Btu per hour.

5-13 Calculate the total heat in Btu per hour removed from the air in Problem 5-12.

5-14 What is the enthalpy (total heat) of 1 lb of air having a DB temperature of 70°F and a DP temperature of 55°F?

5-15 Determine the standard air volume of 500 ft^3 of air at 140°F.

5-16 What is the actual volume of air at a DB temperature of 0°F if the standard air volume is 750 cfm?

5-17 Readings from a sling psychrometer indicate that the air in a room has a DB temperature of 85°F and a WB temperature of 67°F. From the psychrometric chart, determine the humidity ratio, DP temperature, approximate specific volume, specific humidity at saturation, enthalpy deviation, actual enthalpy, and approximate relative humidity.

5-18 For the air sample described in Problem 5-17, determine (a) the sensible heat in Btu per pound of air and (b) the latent heat in Btu per pound of air.

5-19 Ten pounds of air having a DB temperature of 80°F and a humidity ratio of 0.010 lb/lb are mixed with 20 lb of air having a DB temperature of 95°F and a humidity ratio of 0.020 lb/lb. Determine the DB temperature and humidity ratio of the mixture.

5-20 Fifty pounds of air per minute have a DB temperature of 46°F and a WB temperature of 41°F and are passed across a heating surface, and the DB temperature is increased to 101°F. Plot the process on a psychrometric chart and determine (a) the final WB temperature of the air, (b) the sensible heat transferred in Btu per hour, and (c) the total heat transferred in Btu per hour.

5-21 Calculate the bypass factor of the heating coil used in Problem 5-20 if the mean temperature of the heating surface is 120°F.

5-22 Three thousand pounds of air per hour have initial DB and WB temperatures of 95°F and 67°F and are passed across a dry cooling coil, and the DB temperature is reduced to 64°F. If the temperature of the cooling surface is 55°F, determine (a) the sensible heat transferred in Btu per hour, (b) the bypass factor of the coil, and (c) the bypass air quantity in pounds per hour.

5-23 Three thousand pounds of air per hour have initial DB and WB temperatures of 86°F and 71°F and are cooled and dehumidified so that the leaving air DB and DP temperatures are 59°F and 53°F, respectively. Determine the following:

(a) the total heat removed in Btu per hour,
(b) the sensible heat removed in Btu per hour,
(c) the mass of water vapor condensed in pounds per hour,
(d) the latent heat removed in Btu per hour,
(e) the CSHF,
(f) the coil bypass factor,
(g) the mean effective surface temperature of the coil.

Metric Problems

5-24 Calculate the sensible heat (enthalpy of dry air) of 5 kg of air having a DB temperature of 15°C and a DP temperature of 3°C.

5-25 The temperature of 150 kg of air is increased from 2°C to 50°C by the addition of heat. Calculate the quantity of heat supplied in kilojoules.

5-26 Readings from a sling psychrometer indicate that the air in a room has a DB temperature of 46°C and a WB temperature of 26°C. From the psychrometric chart, determine the humidity ratio, DP temperature, approximate specific volume, specific humidity at saturation, enthalpy deviation, actual enthalpy, and approximate relative humidity.

5-27 For the air sample described in Problem 5-26, determine (a) the sensible heat in kilojoules per kilogram of air and (b) the latent heat in kilojoules per kilogram of air.

5-28 Thirty kilograms of air, with a DB temperature of 38°C and a humidity ratio of 0.020 kg/kg, are mixed with 20 kg of air having a DB temperature of 30°C and a humidity ratio of 0.011 kg/kg. Determine the DB temperature and humidity ratio of the mixture.

5-29 Two kilograms of air per second, with a DB temperature of 3°C and a WB temperature of 2°C, are passed across a heating surface and the DB temperature increased to 43°C. Plot the process on the chart and determine (a) final WB temperature of the air, (b) the sensible heat transferred in watts, and (c) the total heat transferred in watts.

5-30 Calculate the bypass factor of the heating coil used in Problem 5-29 if the mean temperature of the heating surface is 53°C.

5-31 Five kilograms of air per second, with initial DB and WB temperatures of 43°C and 24°C, respectively, are cooled and dehumidified so that the leaving air DB and WB temperatures are 14°C and 11°C, respectively. Determine the following:

(a) the total heat removed in watts,
(b) the sensible heat removed in watts,
(c) the mass of water vapor condensed in grams per second,
(d) the latent heat removed in watts,
(e) the coil sensible heat factor,
(f) the mean effective surface temperature of the coil,
(g) the coil bypass factor.

5-32 Determine the standard air volume of 30 m^3 of air at 65°C.

5-33 What is the actual air volume of air at a DB temperature of -20°C if the standard air volume is 3 m^3/s?

6

REFRIGERATION AND THE VAPOR COMPRESSION SYSTEM

6-1. Refrigeration

In general, refrigeration is defined as any process of heat removal. More specifically, refrigeration is defined as the branch of science that deals with the process of reducing and maintaining the temperature of a space or material below the temperature of the surroundings.

To accomplish this, heat must be removed from the body being refrigerated and transferred to another body whose temperature is below that of the refrigerated body. Since the heat removed from the refrigerated body is transferred to another body, it is evident that refrigerating and heating are actually opposite ends of the same process. Often only the desired result distinguishes one from the other.

6-2. Need for Thermal Insulation

Since heat will always migrate from a region of high temperature to a region of lower temperature, there is always a continuous flow of heat into the refrigerated region from the warmer surroundings. To limit the flow of heat into the refrigerated region to some practical minimum, it is usually necessary to isolate the region from its surroundings with a good heat-insulating material.

6-3. The Refrigeration Load

The rate at which heat must be removed from the refrigerated space or material in order to produce and maintain the desired temperature conditions is called the *refrigeration load*, the *cooling load*, or the *heat load*. In most refrigeration applications, the total cooling load on the refrigerating equipment is the sum of the heat gains from several different sources: (1) the heat transmitted by conduction through the insulated walls, (2) the heat that must be removed from the warm air that enters the space through opening and closing doors, (3) the heat that must be removed from the refrigerated product to reduce the temperature of the product to the storage temperature, and (4) the heat given off by people working in the space and by motors, lights, and other heat-producing equipment operating in the space.

Methods of calculating the cooling load are discussed in Chapter 10.

6-4. The Refrigerating Agent

In any refrigerating process, the substance employed as the heat absorber or cooling agent is called the refrigerant.

All cooling processes may be classified as either sensible or latent according to the effect

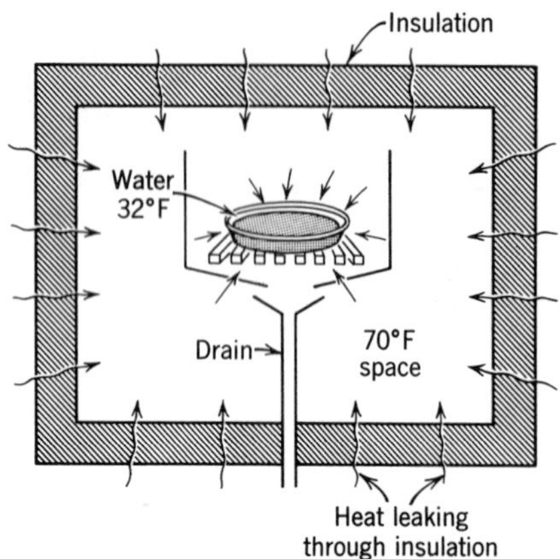

Fig. 6-1 Heat flows from warm space to cold water. Water temperature rises as space temperature decreases. Refrigeration will not be continuous.

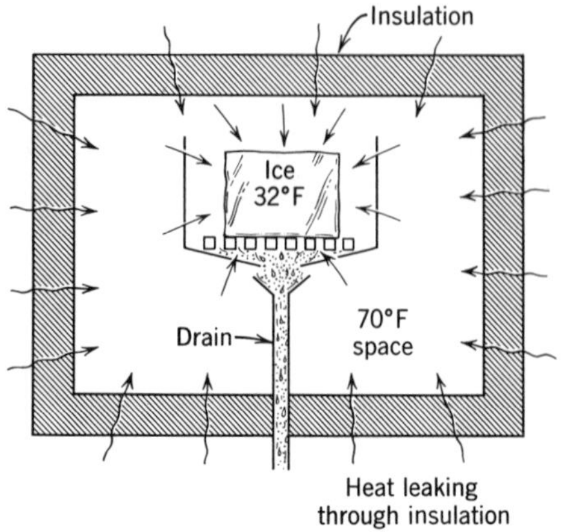

Fig. 6-2 Heat flows from warm space to cold ice. Temperature of space decreases as ice melts. Temperature of ice remains at 32°F. Heat absorbed by ice leaves space in water going out the drain.

the absorbed heat has upon the refrigerant. When the absorbed heat causes an increase in the temperature of the refrigerant, the cooling process is said to be sensible, whereas when the absorbed heat causes a change in the physical state of the refrigerant (either melting or vaporizing), the cooling process is said to be latent. With either process, if the refrigerating effect is to be continuous, the temperature of the refrigerant must be maintained continuously below that of the space or material being refrigerated.

To illustrate, assume that 1 lb of water at 32°F is placed in an open container inside an insulated space having an initial temperature of 70°F (Fig. 6-1). For a time, heat will flow from the 70°F space into the 32°F water and the temperature of the space will decrease. However, for each Btu of heat that the water absorbs from the space, the temperature of the water will increase 1°F, so that as the temperature of the space decreases, the temperature of the water increases. Soon the temperatures of the water and the space will be exactly the same and no heat transfer will take place. Refrigeration will not be continuous because the temperature of the refrigerant does not remain below the temperature of the space being refrigerated.

Now assume that 1 lb of ice, also at 32°F, is substituted for the water (Fig. 6-2). This time the temperature of the refrigerant does not change as it absorbs heat from the space. The ice merely changes from the solid to the liquid state while its temperature remains constant at 32°F. The heat absorbed by the ice leaves the space in the water going out the drain and the refrigerating effect will be continuous until all the ice has melted.

It is both possible and practical to achieve continuous refrigeration with a sensible cooling process provided that the refrigerant is continuously chilled and recirculated through the refrigerated space, as shown in Fig. 6-3.

Latent cooling may be accomplished with either solid or liquid refrigerants. The solid refrigerants most frequently employed are ice and solid carbon dioxide (dry ice). Ice, of course, melts into the liquid phase at 32°F, whereas solid carbon dioxide sublimes directly into the vapor phase at a temperature of −109°F under standard atmospheric pressure.

6-5. Ice Refrigeration

Melting ice has been used successfully for many years as a refrigerant. Not too many years ago ice was the only cooling agent available for use in domestic and small commercial refrigerators.

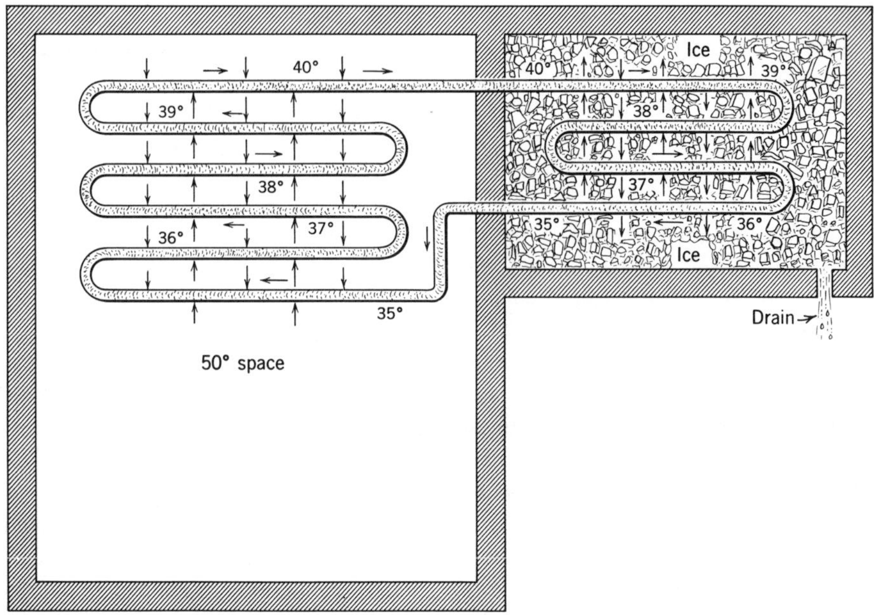

Fig. 6-3 Continuous sensible cooling. Heat taken in by the water in the space is given up to the ice.

In a typical ice refrigerator (Fig. 6-4) the heat entering the refrigerated space from all the various sources reaches the melting ice primarily by convection currents set up in the air of the refrigerated space. The air in contact with the warm product and walls of the space is heated by heat conducted to it from these materials. As the air is warmed it expands and rises to the top of the space carrying the heat with it to the ice compartment. In passing over the ice the air is cooled as heat is conducted from the air to the ice. On cooling, the air becomes more dense and falls back into the storage space, whereupon it absorbs more heat and the cycle is repeated.

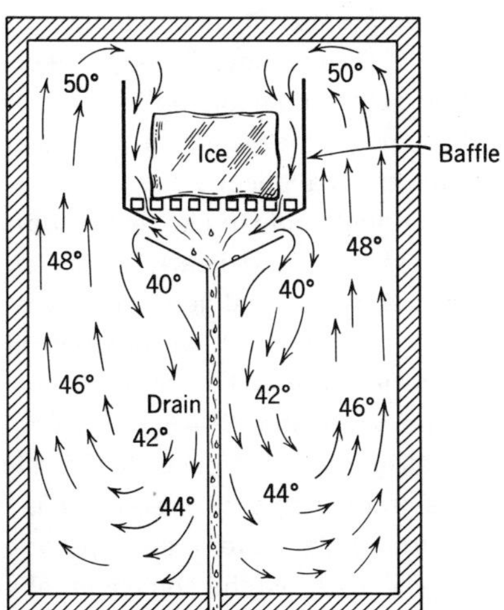

Fig. 6-4 Ice refrigerator. Heat is carried from warm walls and product to the ice by air circulation within the refrigerated space. Air circulation is by gravity.

The air in carrying the heat from the warm walls and stored product to the melting ice acts as a heat transfer agent. To ensure adequate air circulation within the refrigerated space, the ice should be located near the top of the refrigerator and proper baffling should be installed to provide direct and unrestricted paths of air flow. A drip pan must be located beneath the ice to collect the water that results from the melting.

Ice has certain disadvantages which tend to limit its usefulness as a refrigerant. For instance, with ice it is not possible to obtain the low temperatures required in many refrigeration applications. Ordinarily, 32°F is the minimum temperature obtainable through the melting of ice alone. In some cases, the melting temperature of the ice can be lowered to approximately 0°F by adding sodium chloride or calcium chloride to produce a freezing mixture.

Some of the other more obvious disadvantages of ice are the necessity of frequently replenishing the supply, a practice that is neither convenient nor economical, and the problem of disposing of the water resulting from the melting of the ice.

Another less obvious, but more important, disadvantage of employing ice as a refrigerant is the difficulty experienced in controlling the rate of refrigeration, which in turn makes it difficult to maintain the desired low temperature level within the refrigerated space. Since the rate at which the ice absorbs heat is directly proportional to the surface area of the ice and to the temperature difference between the space temperature and the melting temperature of the ice, the rate of heat absorption by the ice diminishes as the surface area of the ice is diminished by the melting process. Naturally, when the refrigerating rate diminishes to the point that the heat is not being removed at the same rate that it is accumulating in the space from the various heat sources, the temperature of the space will increase.

Despite its disadvantages, ice is preferable to mechanical refrigeration in some applications. Fresh vegetables, fish, and poultry are often packed and shipped in cracked ice to prevent dehydration and to preserve appearance. Also, ice has tremendous eye appeal and can be used to considerable advantage in the displaying and serving of certain foods such as salads and cocktails and in chilling beverages.

6-6. Liquid Refrigerants

The ability of liquids to absorb enormous quantities of heat as they vaporize is the basis of the modern mechanical refrigerating system. As refrigerants, vaporizing liquids have a number of advantages over melting solids in that the vaporizing process is more easily controlled; that is, the refrigerating effect can be started and stopped at will, the rate of cooling can be predetermined within small limits, and the vaporizing temperature of the liquid can be governed by controlling the pressure at which the liquid vaporizes. Moreover, the vapor can be readily collected and condensed back into the liquid state so that the same liquid can be used over and over again to provide a continuous supply of liquid for vaporization.

Until now, in discussing the various properties of fluids, water, because of its familiarity, has been used in most examples. However, because of its relatively high saturation temperature, and for other reasons, water is not suitable for use as a refrigerant in the vapor-compression cycle. In order to vaporize at temperatures low enough to satisfy most refrigeration requirements, water would have to vaporize under very low pressures, which are difficult to produce and maintain economically.

There are numerous other fluids which have lower saturation temperatures than water at the same pressure. However, many of these fluids have other properties that render them unsuitable for use as refrigerants. Actually, only a relatively few fluids have properties that make them desirable as refrigerants, and most of these have been compounded specially for that purpose.

There is no one refrigerant that is best suited for all the different applications and operating conditions. For any specific application the refrigerant selected should be the one whose properties most closely fit the particular requirements of the application.

Of all the fluids now in use as refrigerants, the one fluid that most nearly meets all the qualifications of the ideal general-purpose refrigerant is a fluorinated hydrocarbon of the methane series having the chemical name dichlorodifluoromethane (CCl_2F_2). It is one of a group of refrigerants introduced to the industry under the trade name of Freon but is now manufactured under several other proprietary designations. To avoid the confusion inherent in the use of proprietory or chemical names, this compound is now referred to as Refrigerant-12. Refrigerant-12 (R-12) has a saturation temperature of −21.6°F at standard atmospheric pressure. For this reason, R-12 can be stored as a liquid at ordinary temperatures only if confined under pressure in heavy steel cylinders.

Table 16-3 is a tabulation of the thermodynamic properties of R-12 saturated liquid and vapor. This table lists, among other things, the saturation temperature of R-12 corresponding to various pressures. Tables 16-4 through 16-6 list the thermodynamic properties of some of the other more commonly used refrigerants. These tables are similar to the saturated liquid and vapor tables previously discussed and are employed in the same manner.

6-7. Vaporizing the Refrigerant

An insulated space can be adequately refrigerated by merely allowing liquid R-12 to vaporize in a container vented to the outside as shown in Fig. 6-5. Since the R-12 is under standard atmospheric pressure, its saturation temperature is −21.6°F. Vaporizing at this low temperature, the R-12 readily absorbs heat from the 40°F space through the walls of the containing vessel. The heat absorbed by the vaporizing liquid leaves the space in the vapor escaping through the open vent. Since the temperature of the liquid remains constant during the vaporizing process, refrigeration will continue until all the liquid is vaporized.

Any container, such as the one in Fig. 6-5, in which a refrigerant is vaporized during a refrigerating process is called an evaporator and is one of the essential parts of any mechanical refrigerating system.

6-8 Controlling the Vaporizing Temperature

The temperature at which the liquid vaporizes in the evaporator can be controlled by controlling the pressure of the vapor over the liquid, which in turn is governed by regulating the rate at which the vapor escapes from the evaporator (Section 4-4). For example, if a hand valve is installed in the vent line and the vent is partially closed off so that the vapor cannot escape freely from the evaporator, vapor will collect over the liquid, causing the pressure in the evaporator to rise with a corresponding increase in the saturation temperature of the refrigerant (Fig. 6-6). By carefully adjusting the vent valve to regulate the flow of vapor from the evaporator, it is possible to control the pressure of the vapor over the liquid and cause the R-12 to vaporize at any desired temperature between −21.6°F and the space temperature. Should the vent valve be completely closed so

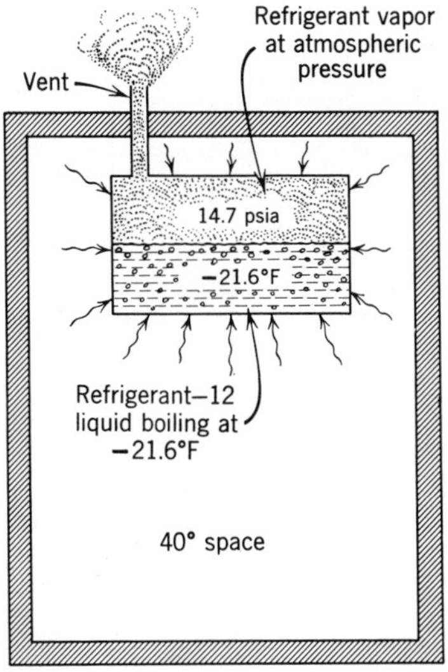

Refrigerant vapor at atmospheric pressure
Vent
14.7 psia
−21.6°F
Refrigerant–12 liquid boiling at −21.6°F
40° space

Fig. 6-5 The Refrigerant-12 liquid vaporizes as it takes in heat from the 40°F space. The heat taken in by the refrigerant leaves the space in the vapor escaping through the vent.

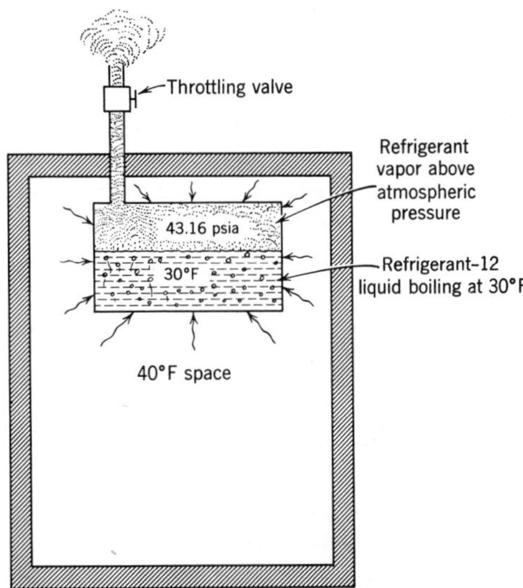

Fig. 6-6 The boiling temperature of the liquid refrigerant in the evaporator is controlled by controlling the pressure of the vapor over the liquid with the throttling valve in the vent.

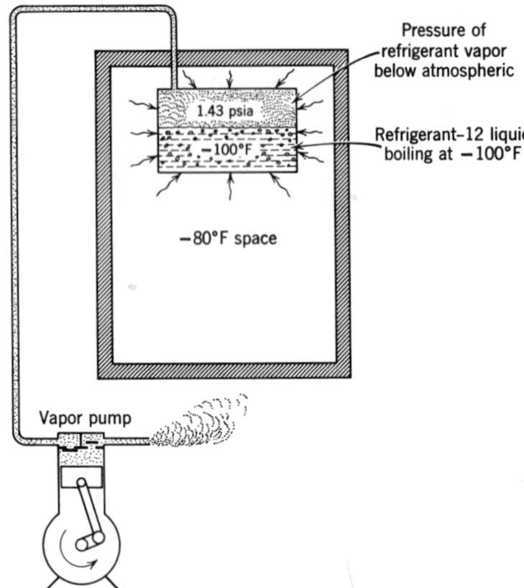

Fig. 6-7 Pressure of refrigerant in evaporator reduced below atmospheric by action of a vapor pump.

that no vapor is allowed to escape from the evaporator, the pressure in the evaporator will increase to a point such that the saturation temperature of the liquid will be equal to the space temperature, or 40°F. When this occurs, there will be no temperature differential and no heat will flow from the space to the refrigerant. Vaporization will cease and no further cooling will take place.

When vaporizing temperatures below the saturation temperature of the refrigerant corresponding to atmospheric pressure are required, it is necessary to reduce the pressure in the evaporator to some pressure below atmospheric. This can be accomplished through the use of a vapor pump as shown in Fig. 6-7. By this method, vaporization of the liquid R-12 can be brought about at very low temperatures in accordance with the pressure-temperature relationships given in Table 16-3.

6-9. Maintaining a Constant Amount of Liquid in the Evaporator

Continuous vaporization of the liquid in the evaporator requires that the supply of liquid be continuously replenished if the amount of liquid in the evaporator is to be maintained constant. One method of replenishing the supply of liquid in the evaporator is through the use of a float valve assembly, as illustrated in Fig. 6-8. The action of the float assembly is to maintain a constant level of liquid in the evaporator by allowing liquid to flow into the evaporator from the storage tank or cylinder at exactly the same rate that the supply of liquid in the evaporator is being depleted by vaporization. Any increase in the rate of vaporization causes the liquid level in the evaporator to drop slightly, thereby opening the needle valve wider and allowing liquid to flow into the evaporator at a higher rate. Likewise, any decrease in the rate of vaporization causes the liquid level to rise slightly, thereby moving the needle valve in the closing direction to reduce the flow of liquid into the evaporator. When vaporization ceases entirely, the rising liquid level will close the float valve tightly and stop the flow of liquid completely. When vaporization is resumed, the liquid level will fall allowing the float valve to open and admit liquid to the evaporator.

The liquid refrigerant does not vaporize in the storage cylinder and feed line because the pressure in the cylinder is such that the saturation temperature of the refrigerant is equal to the temperature of the surroundings (see

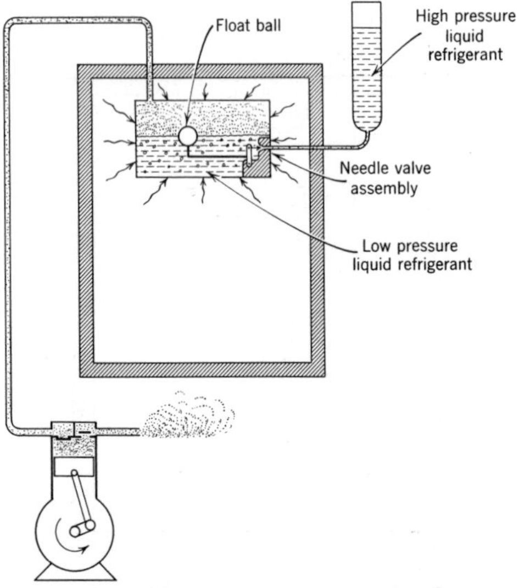

Fig. 6-8 Float valve assembly maintains constant liquid level in evaporator. The pressure of the refrigerant is reduced as the refrigerant passes through the needle valve.

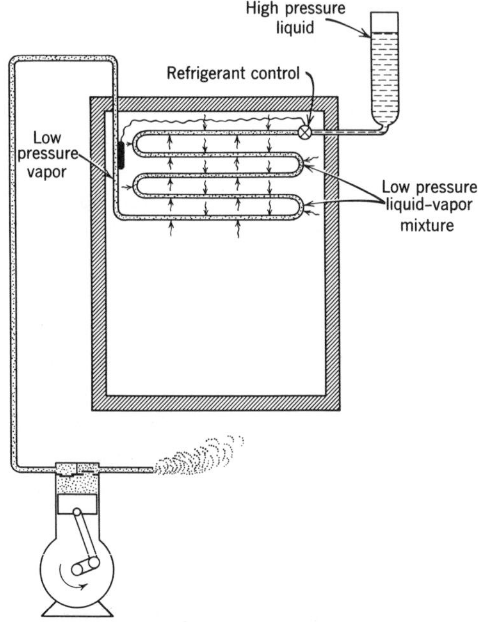

Fig. 6-9 Serpentine coil evaporator with thermostatic expansion valve refrigerant control.

Section 4-9). The high pressure existing in the cylinder forces the liquid to flow through the feed line and the float valve into the lower pressure evaporator. In passing through the float valve, the high pressure refrigerant undergoes a pressure drop that reduces its pressure to the evaporator pressure, thereby permitting the refrigerant liquid to vaporize in the evaporator at the desired low temperature.

Any device, such as the float valve illustrated in Fig. 6-8, used to regulate the flow of liquid refrigerant into the evaporator is called a refrigerant flow control. The refrigerant flow control is an essential part of every mechanical refrigerating system.

There are five different types of refrigerant flow controls, all of which are in use to some extent at the present time. Each of these distinct types is discussed at length in Chapter 17. The float type of control illustrated in Fig. 6-8 has some disadvantages, mainly bulkiness, which tend to limit its use to some few special applications. One widely used type of refrigerant flow control is the thermostatic expansion valve. A flow diagram illustrating the use of a thermostatic expansion valve to control the flow of refrigerant into a serpentine coil type evaporator is shown in Fig. 6-9.

6-10. Salvaging the Refrigerant

As a matter of convenience and economy, it is not practical to permit the refrigerant vapor to escape to the outside and be lost by diffusion into the air. The vapor must be collected continuously and condensed back into the liquid state so that the same refrigerant is used over and over again, thereby eliminating the need for ever replenishing the supply of refrigerant in the system. To provide some means of condensing the vapor, another piece of equipment, a condenser, must be added to the system (Fig. 6-10).

Since the refrigerant vaporizes in the evaporator because it absorbs the necessary latent heat from the refrigerated space, all that is required in order to condense the vapor back

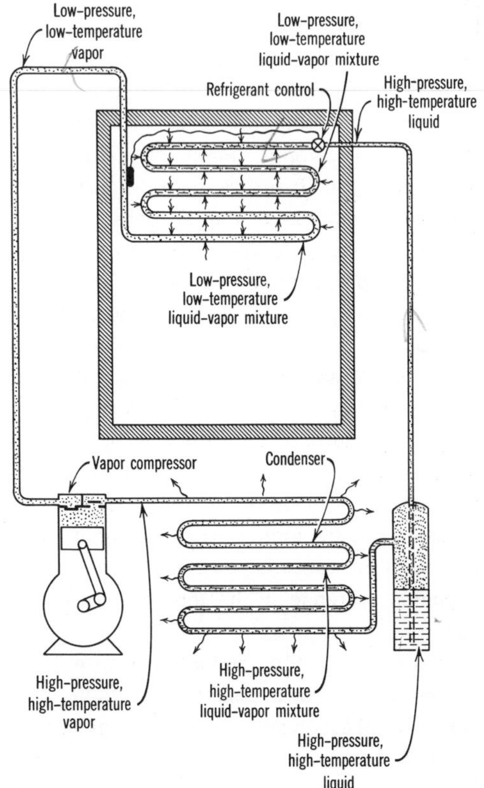

Fig. 6-10 Collecting and condensing the refrigerant vapor. Refrigerant absorbs heat in evaporator and gives off heat in the condenser.

into the liquid state is that the latent heat be caused to flow out of the vapor into another body. The body of material employed to absorb the latent heat from the vapor, thereby causing the vapor to condense, is called the condensing medium. The most common condensing media are air and water. The water used as a condensing medium is usually supplied from the city main or from a cooling tower. The air used as a condensing medium is ordinary outdoor air at normal temperatures.

For heat to flow out of the refrigerant vapor into the condensing medium the temperature of the condensing medium must be below that of the refrigerant vapor. However, since the pressure and temperature of the saturated vapor leaving the evaporator are the same as those of the vaporizing liquid, the temperature of the vapor will always be considerably below that of any normally available condensing medium. Therefore, heat will not flow out of the refrigerant vapor into the air or water used as the condensing medium until the saturation temperature of the refrigerant vapor has been increased by compression to some temperature above the temperature of the condensing medium. The vapor pump or compressor shown in Fig. 6-10 serves this purpose.

Before compression, the refrigerant vapor is at the vaporizing temperature and pressure. Since the pressure of the vapor is low, the corresponding saturation temperature is also low. During compression the pressure of the vapor is increased to a point such that the corresponding saturation temperature is above the temperature of the condensing medium being employed. At the same time, since mechanical work is done on the vapor in compressing it to the higher pressure, the internal energy of the vapor is increased with a corresponding increase in the temperature of the vapor.

After compression, the high-pressure, high-temperature vapor is discharged into the condenser where it gives up heat to the lower temperature condensing medium. Since a vapor cannot be cooled to a temperature below its saturation temperature, the continuous loss of heat by the refrigerant vapor in the condenser causes the vapor to condense into the liquid state at the new, higher pressure and saturation temperature. The heat given off by the vapor in the condenser is carried away by the condensing medium. The resulting condensed liquid, whose temperature and pressure will be the same as those of the condensing vapor, flows out of the condenser into the liquid storage tank and is then ready to be recirculated to the evaporator.

Notice that the refrigerant, sometimes called the working fluid, is merely a heat transfer agent which carries the heat from the refrigerated space to the outside. The refrigerant absorbs heat from the refrigerated space in the evaporator, carries it out of the space, and rejects it to the condensing medium in the condenser.

6-11. Typical Vapor-Compression System

A flow diagram of a simple vapor-compression system is shown in Fig. 6-11. The principal parts of the system are (1) an evaporator, whose function it is to provide a heat transfer surface through which heat can pass from the refrigerated space or product into the vaporizing refrigerant; (2) a suction line, which conveys the low pressure vapor from the evaporator to the suction inlet of the compressor; (3) a vapor compressor, whose function it is to remove the vapor from the evaporator, and to raise the temperature and pressure of the vapor to a point such that the vapor can be condensed with normally available condensing media; (4) a "hot-gas" or discharge line which delivers the high-pressure, high-temperature vapor from the discharge of the compressor to the condenser; (5) a condenser, whose purpose it is to provide a heat transfer surface through which heat passes from the hot refrigerant vapor to the condensing medium; (6) a receiver tank, which provides storage for the condensed liquid so that a constant supply of liquid is available to the evaporator as needed; (7) a liquid line, which carries the liquid refrigerant from the receiver tank to the refrigerant flow control; and (8) a refrigerant flow control, whose function it is to meter the proper amount of refrigerant to the evaporator and to reduce

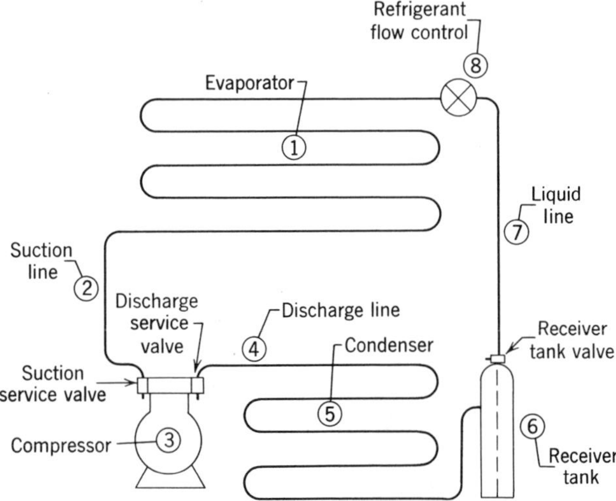

Fig. 6-11 Flow diagram of simple vapor compression system showing the principal parts.

the pressure of the liquid entering the evaporator so that the liquid will vaporize in the evaporator at the desired low temperature.

6-12. Service Valves

The suction and discharge sides of the compressor and the outlet of the receiver tank are usually equipped with manual shut-off valves for use during service operations. These valves are known as the "suction service valve," the discharge service valve," and the "receiver tank valve," respectively. Receiver tanks on large systems frequently have shut-off valves on both the inlet and the outlet.

6-13. Division of the System

A refrigerating system is divided into two parts according to the pressure exerted by the refrigerant in the two parts. The low pressure part of the system consists of the refrigerant flow control, the evaporator, and the suction line. The pressure exerted by the refrigerant in these parts is the low pressure under which the refrigerant is vaporizing in the evaporator. This pressure is known variously as the "low side pressure," the "evaporator pressure," the "suction pressure," or the "back pressure." During service operations this pressure is usually measured at the compressor by installing a compound gage on the gage port of the suction service valve.

The high pressure side or "high side" of the system consists of the compressor, the discharge or "hot gas" line, the condenser, the receiver tank, and the liquid line. The pressure exerted by the refrigerant in this part of the system is the high pressure under which the refrigerant is condensing in the condenser. This pressure is called the "condensing pressure," the "discharge pressure," or the "head pressure."

The dividing points between the high and low pressure sides of the system are the refrigerant flow control, where the pressure of the refrigerant is reduced from the condensing pressure to the vaporizing pressure, and the discharge valves in the compressor, through which the high pressure vapor is exhausted after compression.

Care should be taken not to confuse the suction and discharge valves in the compressor with the suction and discharge service valves. The suction and discharge valves in a reciprocating compressor perform the same function as the intake and exhaust valves in an automobile engine and are vital to the operation of the compressor, whereas the suction and discharge service valves serve no useful purpose insofar as the operation of the compressor is concerned. The latter valves are used only to facilitate service operations, as their nomenclature implies.

It should be noted that, although the compressor is considered to be a part of the high side of the system, the pressure on the suction side of the compressor and in the crankcase is the low side pressure. The change in pressure, of course, occurs in the cylinder during the compression process.

6-14. Condensing Units

The compressor, hot gas line, condenser, and receiver tank, along with the compressor driver (usually an electric motor), are often combined into one compact unit as shown in Fig. 6-12. Such an assembly is called a condensing unit because its function in the system is to reclaim the vapor and condense it back into the liquid state.

Condensing units are often classified according to the condensing medium used to con-

Fig. 6-12 Air-cooled condensing unit. Note fan mounted on motor shaft to circulate air over condenser.

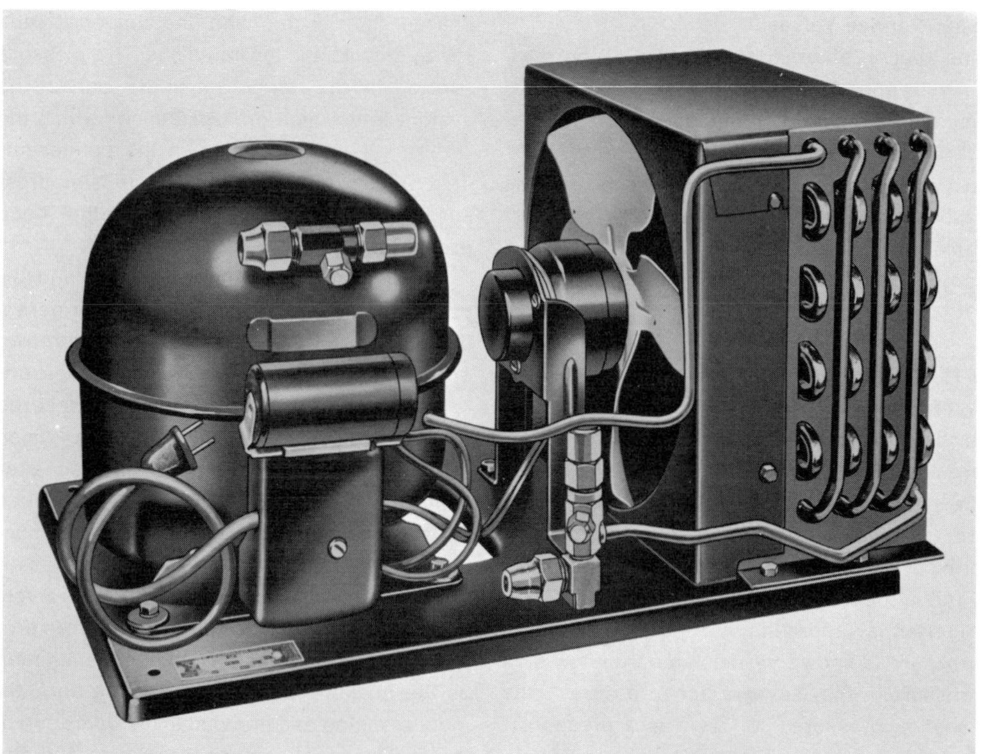

Fig. 6-13 Air-cooled condensing unit employing hermetic motor-compressor. Note separate fan to circulate air over condenser. (Courtesy Tecumseh Products Company.)

dense the refrigerant. A condensing unit employing air as the condensing medium (Fig. 6-12) is called an air-cooled condensing unit, whereas one employing water as the condensing medium is a water-cooled condensing unit.

6-15. Hermetic Motor-Compressor Assemblies

Condensing units of small ($\frac{1}{20}$ to $7\frac{1}{2}$) horsepower are often equipped with hermetically sealed motor-compressor assemblies. The assembly consists of a direct-driven compressor mounted on a common shaft with the motor rotor and the whole assembly hermetically sealed in a welded steel shell (Fig. 6-13).

Condensing units equipped with hermetically sealed motor-compressor assemblies are known as "hermetic condensing units" and are employed on a number of small commercial refrigerators and on almost all household refrigerators, home freezers, and window air conditioners. For reasons that will be shown later, many hermetic condensing units are not equipped with receiver tanks.

A variation of the hermetic motor-compressor assembly is the "accessible hermetic." It is similar to the full hermetic except that the shell enclosing the assembly is bolted together rather than seam welded (Fig. 6-14). The bolted construction permits the assemblies to be opened in the field for servicing. These units are available up through several hundred horsepower.

6-16. Definition of a Cycle

As the refrigerant circulates through the system, it passes through a number of changes in state or condition, each of which is called a process. The refrigerant starts at some initial state or condition, passes through a series of processes in a definite sequence, and returns to the initial condition. This series of processes is called a cycle. The simple vapor-compression refrigeration cycle is made up of four fundamental processes: (1) expansion, (2) vaporization, (3) compression, and (4) condensation.

To understand properly the refrigeration cycle it is necessary to consider each process in the cycle both separately and in relation to the complete cycle. Any change in any one process in the cycle will bring about changes in all the other processes in the cycle.

6-17. Typical Vapor-Compression Cycle

A typical vapor-compression cycle is shown in Fig. 6-15. Starting at the receiver tank, high-temperature, high-pressure liquid refrigerant flows from the receiver tank through the liquid line to the refrigerant flow control. The pressure of the liquid is reduced to the evaporator pressure as the liquid passes through the refrigerant flow control so that the saturation temperature of the refrigerant entering the evaporator will be below the temperature of the refrigerated space. It will be shown later that a part of the liquid vaporizes as it passes through the refrigerant control in order to reduce the temperature of the remaining liquid to the evaporating temperature.

In the evaporator, the liquid vaporizes at a constant pressure and temperature as heat to supply the latent heat of vaporization passes from the refrigerated space through the walls of the evaporator to the vaporizing liquid. By the action of the compressor, the vapor resulting from the vaporization is drawn from the evaporator through the suction line into the suction inlet of the compressor. The vapor leaving the evaporator is saturated and its temperature and pressure are the same as those of the vaporizing liquid. While flowing through the suction line from the evaporator to the compressor, the vapor usually absorbs heat from the air surrounding the suction line and becomes superheated. Although the temperature of the vapor increases somewhat in the suction line as the result of superheating, the pressure of the vapor does not change, so that the pressure of the vapor entering the compressor is the same as the vaporizing pressure.*

In the compressor, the temperature and pressure of the vapor are raised by compression and the high-temperature, high-pressure

* Actually, the pressure of the vapor decreases slightly between the evaporator and compressor because of the friction loss in the suction line resulting from the flow.

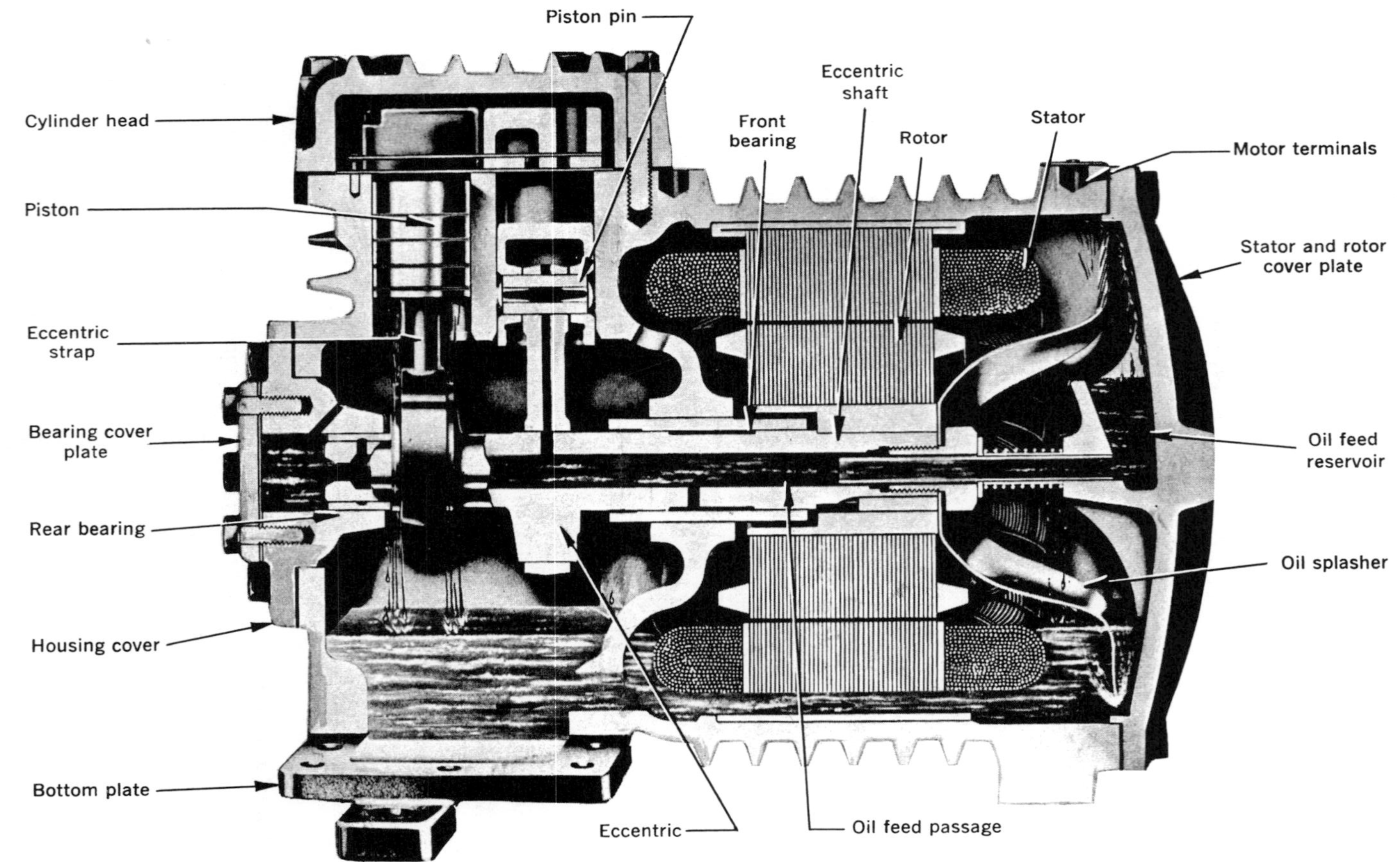

Fig. 6-14 Typical twin cylinder compressor—cutaway view illustrating bolted construction of "accessible hermetic" type motor-compressor unit. (Courtesy Carrier Corporation.)

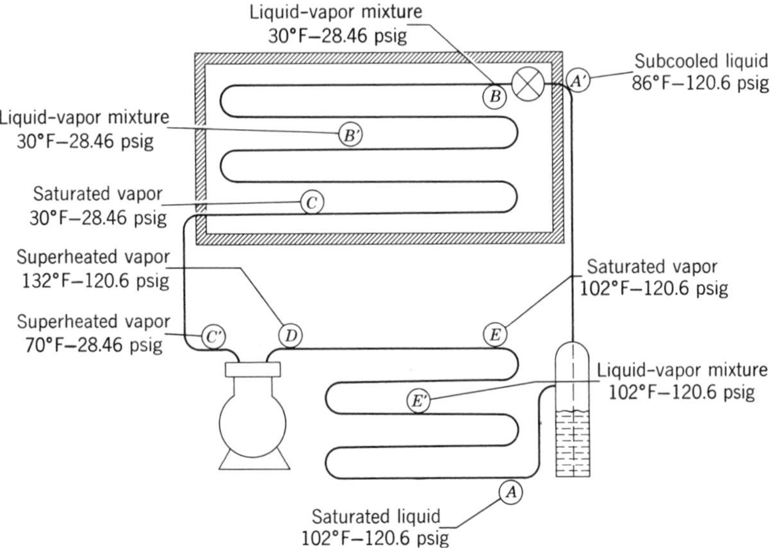

Fig. 6-15 Typical refrigeration system showing the condition of the refrigerant at various points.

vapor is discharged from the compressor into the hot-gas line. The vapor flows through the hot-gas line to the condenser, where it gives up heat to the relatively cool air being drawn across the condenser by the condenser fan. As the hot vapor gives off heat to the cooler air, its temperature is reduced to the saturation temperature corresponding to its new higher pressure and the vapor condenses back into the liquid state as additional heat is removed. By the time the refrigerant reaches the bottom of the condenser, all of the vapor is condensed and the liquid passes into the receiver tank, ready to be recirculated.

6-18. The Compression Process

In modern, high speed compressors, compression takes place very rapidly and the vapor is in contact with the compressor cylinder for only a short time. Because the time of compression is short and because the mean temperature differential between the refrigerant vapor and the cylinder wall is relatively small, the flow of heat either to or from the refrigerant during the compression process is usually negligible. Therefore, compression of the vapor in a refrigeration compressor is usually assumed to occur adiabatically.

Recall that during any adiabatic compression process, the internal energy of the gas is increased by an amount equal to the amount of work done on the gas to compress it (Section 3-24). Consequently, when the refrigerant vapor is compressed adiabatically in the refrigerant compressor, the temperature and enthalpy of the vapor are increased in proportion to the amount of work done on the vapor. The greater the work of compression, the greater is the increase in temperature and enthalpy.

The energy equivalent of the work done on the vapor to compress it is called the *heat of compression*, and the energy to do the work is supplied by the compressor driver. It will be shown later that the theoretical power required to drive the compressor can be determined directly from the heat of compression.

6-19. Discharge Temperature

Care should be taken not to confuse discharge temperature with condensing temperature. The discharge temperature is the temperature at which the vapor is discharged from the compressor, whereas the condensing temperature is the temperature at which the vapor condenses in the condenser and is the saturation temperature of the vapor corresponding to the pressure in the condenser.

The nature of the compression process in the compressor is such that the temperature of the vapor discharged from the compressor is always greater than the saturation temperature corresponding to the pressure of the vapor; that is, the vapor discharged from the compressor is always in a superheated state. The discharge vapor is cooled to the saturation (condensing) temperature as it flows through the hot-gas line and through the upper part of the condenser, whereupon the further removal of heat from the vapor causes the vapor to condense at the saturation temperature corresponding to the pressure in the condenser.

6-20. Condensing Temperature

To provide a continuous refrigerating effect the refrigerant vapor must be condensed in the condenser at the same rate that the refrigerant liquid is vaporized in the evaporator. This means that heat must leave the system at the condenser at the same rate that heat is taken into the system in the evaporator and suction line, and in the compressor as a result of the work of compression. Obviously, any increase in the rate of vaporization will increase the required rate of heat transfer at the condenser.

The rate at which heat will flow through the walls of the condenser from the refrigerant vapor to the condensing medium is the function of three factors: (1) the area of the condensing surface, (2) the coefficient of conductance of the condenser walls, and (3) the temperature difference between the refrigerant vapor and the condensing medium. For any given condenser, the area of the condensing surface and the coefficient of conductance are fixed so that the rate of heat transfer through the condenser walls depends only on the temperature difference between the refrigerant vapor and the condensing medium.

Since the condensing temperature is always equal to the temperature of the condensing medium plus the temperature difference between the condensing refrigerant and the condensing medium, it follows that the condensing temperature varies directly with the temperature of the condensing medium and with the required rate of heat transfer at the condenser.

6-21. Condensing Pressure

The condensing pressure is always the saturation pressure corresponding to the temperature of the liquid-vapor mixture in the condenser.

When the compressor is not running, the temperature of the refrigerant mixture in the condenser will be the same as that of the surrounding air, and the corresponding saturation pressure will be relatively low. Consequently, when the compressor is started, the vapor pumped over into the condenser will not begin to condense immediately because there is no temperature differential between the refrigerant and the condensing medium, and therefore no heat transfer between the two. Because of the throttling action of the refrigerant control, the condenser may be visualized as a closed container, and as more and more vapor is pumped into the condenser without condensing, the pressure in the condenser increases to a point where the saturation temperature of vapor is sufficiently high to permit the required rate of heat transfer between the refrigerant and the condensing medium. When the required rate of heat transfer is reached, the vapor will condense as fast as it is pumped into the condenser, whereupon the pressure in the condenser will stabilize and remain more or less constant during the balance of the running cycle.

6-22. Refrigerating Effect

The quantity of heat that each unit mass of refrigerant absorbs from the refrigerated space is known as the refrigerating effect. For example, when 1 lb of ice melts it will absorb from the surrounding air and from adjacent objects an amount of heat equal to its latent heat of fusion. If the ice melts at 32°F it will absorb 144 Btu/lb, so that the refrigerating effect of 1 lb of ice is 144 Btu.

Likewise, when a liquid refrigerant vaporizes as it flows through the evaporator it will absorb an amount of heat equal to that required to vaporize it; thus the refrigerating effect per unit mass of liquid refrigerant is potentially equal to its latent heat of vaporization. If the temperature of the liquid entering the refrigerant control from the liquid line is exactly equal to the vaporizing temperature in the evaporator, the entire mass of the liquid will vaporize in the evaporator and produce useful cooling, in which case the refrigerating effect per unit mass of refrigerant circulated will be equal to the total latent heat of vaporization. However, in an actual cycle the temperature of the liquid entering the refrigerant control is always considerably higher than the vaporizing temperature in the evaporator and must first be reduced to the evaporator temperature before the liquid can vaporize in the evaporator and absorb heat from the refrigerated space. For this reason, only a part of each pound of liquid actually vaporizes in the evaporator and produces useful cooling. Therefore, the refrigerating effect per unit mass of liquid circulated is always less than the total latent heat of vaporization.

With reference to Fig. 6-15, the pressure of the vapor condensing in the condenser is 120 psig and the condensing temperature (saturation temperature) of the R-12 vapor corresponding to this pressure is 102°F. Since condensation occurs at a constant temperature, the temperature of the liquid resulting from the condensation is also 102°F. After condensation, as the liquid flows through the lower part of the condenser it continues to give up heat to the cooler condensing medium, so that before the liquid leaves the condenser its temperature is usually reduced somewhat below the temperature at which it condensed. The liquid is then said to be subcooled. The temperature at which the liquid leaves the condenser depends upon the temperature of the condensing medium and upon how long the liquid remains in contact with the condensing medium after condensation.

The liquid may be further subcooled in the receiver tank and in the liquid line by surrendering heat to the surrounding air. In any case, because of the heat exchange between the refrigerant in the liquid line and the surrounding air, the temperature of the liquid approaching the refrigerant control is often reasonably close to the temperature of the air surrounding the liquid line. In Fig. 6-15, the liquid approaches the refrigerant control at a temperature of 86°F, whereas its pressure is still assumed to be the same as the condensing pressure, 120.6 psig. Since the saturation temperature corresponding to 120.6 psig is 102°F, the 86°F liquid at the refrigerant control is subcooled 16°F (102 − 86) below its saturation temperature.

Since the saturation pressure corresponding to 86°F is 93.2 psig, the R-12 can exist in the liquid state as long as its pressure is not reduced below 93.2 psig. However, as the liquid passes through the refrigerant control, its pressure is reduced from 120.6 psig to 28.46 psig, the saturation pressure corresponding to the 30° vaporizing temperature of the refrigerant in the evaporator. Since the R-12 cannot exist as a liquid at any temperature above the saturation temperature of 30°F when its pressure is 28.46 psig, the liquid must surrender enough heat to cool itself from 86°F to 30°F at the instant that its pressure is reduced in passing through the refrigerant control.

From Table 16-3, the enthalpy of liquid at 86°F and at 30°F is 27.72 Btu/lb and 14.76 Btu/lb, respectively, so that each pound of liquid must surrender 12.96 Btu (27.22 − 14.76) in order to cool from 86°F to 30°F.

Since the drop in pressure suffered by the refrigerant in the refrigerant control occurs instantaneously, there is no time for heat transfer to take place between the refrigerant and the surroundings. Consequently, the process is adiabatic, and the required reduction in the temperature of the fluid must be accomplished by energy transfers that take place solely and entirely within the fluid itself. For this reason a portion of each pound of liquid circulated will flash into the vapor phase as the refrigerant passes through the refrigerant control. The heat to supply the latent heat of vaporization for the portion of the liquid that vaporizes is drawn from the mass of the fluid, thereby reducing the temperature of the refrigerant fluid to the temperature of the evaporator. In this instance, enough of each pound of the refrigerant fluid circulated will vaporize as it passes through the orifice of the refrigerant control to absorb exactly the 12.96 Btu of sensible heat necessary to reduce the temperature of the refrigerant fluid from the liquid line temperature of 86°F to the evaporator temperature of 30°F.

As previously stated, the refrigerant is discharged from the refrigerant flow control in the form of a liquid-vapor mixture, so that only a portion of each pound of refrigerant circulated actually vaporizes in the evaporator and produces useful cooling. The other portion, that which vaporizes in the refrigerant control, produces no useful cooling and therefore represents a loss of refrigerating effect as compared to that which would be produced if the entire mass vaporized in the evaporator.

From the foregoing, it is evident that the refrigerating effect per pound of refrigerant circulated is equal to the total latent heat of vaporization at the vaporizing temperature minus the latent heat of the portion of the pound that vaporizes in the refrigerant control to reduce the temperature of the frigerant to the evaporator temperature.

From Table 16-3, the latent heat of vaporization of R-12 at 30°F is 66.85 Btu/lb. Since the loss of refrigerating effect is 12.96 Btu/lb, the refrigerating effect in this instance is (66.85 − 12.96) 53.89 Btu/lb.

The percentage of each pound of refrigerant that vaporizes in the refrigerant control can be determined by dividing the total latent heat of vaporization into the heat absorbed by that part of the pound that vaporizes in the control. In this instance the percentage of each pound vaporizing in the control is (12.96/66.85 × 100) 19.4%. Only 80.6% of each pound circulated actually vaporizes in the evaporator and produces useful cooling (66.85 × 0.806 = 53.89 Btu/lb).

Even though a portion of each unit mass circulated vaporizes as it passes through the refrigerant control, there is no heat transfer to or from the surroundings, and no external work is done. Consequently, there is no change in the enthalpy of the refrigerant during the expansion process, so that the enthalpy of the liquid-vapor mixture entering the evaporator is exactly the same as that of the refrigerant liquid approaching the refrigerant control. It follows, then, that the difference between the enthalpy of the refrigerant leaving the evaporator and the enthalpy of the liquid approaching the control is only the heat absorbed in the evaporator, which is, of course, the refrigerating effect. For this reason the refrigerating effect per unit mass can always be determined by simply subtracting the enthalpy of the liquid approaching the control from the enthalpy of the refrigerant vapor leaving the evaporator.

Example 6-1 Determine the refrigerating effect per pound if the temperature of the liquid R-12 approaching the refrigerant control is 86°F and the temperature of the saturated vapor leaving the evaporator is 30°F.

Solution From Table 16-3,

Enthalpy of R-12 saturated at 30°F	= 81.61 Btu/lb
Enthalpy of R-12 liquid at 86°F	= 27.72 Btu/lb
Refrigerating effect per pound	= 53.89 Btu/lb

Notice that, for any given refrigerant, the refrigerating effect produced per unit mass of refrigerant circulated depends on the difference between the evaporator temperature and the temperature of the liquid refrigerant approaching the refrigerant control. As this temperature difference increases, the refrigerating effect per unit mass is reduced because a greater portion of the liquid must vaporize in the control to provide the necessary drop in the temperature of the refrigerant. Conversely, reducing this differential will increase the refrigerating effect. For this reason, subcooling of the liquid approaching the control is always desirable in that it results in an increase in the refrigerating effect and an improvement in system efficiency.

6-23. System Capacity

The capacity of any refrigerating system is the rate at which it will remove heat from the refrigerated space. It has traditionally been expressed in Btu per hour or in terms of its ice-melting equivalent.

Before the era of mechanical refrigeration, ice was widely used as a cooling medium. With the development of mechanical refrigeration, it was only natural that the cooling capacity of mechanical refrigerators should be compared with an ice-melting equivalent. Hence, a refrigerating system having a capacity of 1 ton is one that has a cooling capacity equivalent to the melting of 1 ton of ice in a 24-hour period. Since 1 ton of ice will absorb 288,000 Btu (2000 lb × 144 Btu/lb) in melting, this is cooling at the rate of 12,000 Btu/hr or 200 Btu/min. In the metric system, 1 ton is a cooling rate of 3.517 kJ/s or kW.

Notice that the refrigerating capacity is actually an energy transfer rate and, as such, is an expression of power.

The capacity of a mechanical refrigerating system, that is, the rate at which the system will remove heat from the refrigerated space, depends on two factors: (1) the mass of refrigerant circulated per unit of time, and (2) the refrigerating effect per unit mass circulated. Expressed as an equation,

$$Q_e = (m)(q_e) \tag{6-1}$$

where Q_e = the refrigerating capacity in Btu per minute
m = the mass flow rate in pounds per minute
q_e = the refrigerating effect in Btu per pound

Example 6-2 A mechanical refrigerating system is operating under conditions such that the vaporizing temperature is 30°F while the temperature of the liquid approaching the refrigerant control is 86°F. If R-12 is circulated through the system at the rate of 5 lb/min, determine:

(a) the refrigerating capacity of the system in Btu per hour,
(b) the refrigerating capacity of the system in tons.

Solution

(a) From Example 6-1, refrigerating effect = 53.89 Btu/lb
Mass of refrigerant circulated per minute = 5 lb
Refrigerating capacity in Btu per minute = 5 × 53.89 = 269.45 Btu/min
Refrigerating capacity in Btu per hour = 269.45 × 60 = 16,167 Btu/hr

(b) Refrigerating capacity in tons $= \frac{269.45}{200}$ = 1.347 tons

6-24. Mass of Refrigerant Circulated per Minute per Ton

The mass of refrigerant which must be circulated per minute per ton of refrigerating capacity for any given operating conditions is found by dividing the refrigerating effect per pound at the given conditions into 200 Btu/min ton.

Example 6-3 An R-12 system is operating at conditions such that the vaporizing temperature is 20°F and the condensing temperature is 100°F. If it is assumed that no subcooling of the liquid occurs so that the temperature of the liquid at the refrigerant control is also 100°F, find:

(a) The refrigerating effect per pound,
(b) The mass of refrigerant circulated per minute per ton,
(c) The mass of refrigerant circulated per minute for a 10-ton system.

Solution

(a) From Table 16-3, enthalpy of R-12 saturated vapor at 20°F = 80.49 Btu/lb
Enthalpy of R-12 liquid at 100°F = 31.16 Btu/lb
Refrigerating effect, q_e = 49.33 Btu/lb

(b) Mass of refrigerant circulated per minute per ton $= \frac{200}{49.33}$ = 4.05 lb

(c) Mass of refrigerant circulated per minute for a 10-ton system = 10 × 4.05 = 40.5 lb

6-25. Volume Flow Rate of Vapor

When 1 lb of refrigerant vaporizes, the volume of saturated vapor produced depends on the refrigerant employed and on the vaporizing temperature. For any one refrigerant, the volume of vapor depends only on the vaporizing temperature and increases as the vaporizing temperature (and pressure) decreases. When the vaporizing temperature of the refrigerant is known, the volume of vapor produced per unit mass (specific volume) can be determined directly from the saturated vapor tables. Once the specific volume of the vapor is known, the total volume of vapor generated in the evaporator per unit time can be found by multiplying the refrigerant mass flow rate by the specific volume of the vapor, that is,

$$V = (m)(v) \qquad (6\text{-}2)$$

where V = the total volume of vapor generated in the evaporator in cubic feet per minute.

m = the mass flow rate of the refrigerant in pounds per minute

v = the specific volume of the vapor at the vaporizing temperature in cubic feet per pound

Example 6-4 With reference to Example 6-3, determine the volume of vapor generated in cubic feet per minute for each ton of refrigerating capacity.

Solution From Example 6-3, the mass flow rate of the R-12 is 4.05 lb/min ton, and from Table 16-3, the specific volume of the suction vapor at 20°F is 1.121 ft³/lb. Applying Equation 6-2,

$$\begin{aligned} V &= (4.05 \text{ lb/min ton})(1.121 \text{ ft}^3/\text{lb}) \\ &= 4.55 \text{ cfm/ton} \end{aligned}$$

6-26. Compressor Capacity

In any mechanical refrigerating system the capacity of the compressor must be such that vapor is withdrawn from the evaporator at the same rate that vapor is produced by the boiling action of the liquid refrigerant. If the refrigerant vaporizes faster than the compressor is able to remove the vapor, the excess vapor will accumulate in the evaporator and cause the pressure in the evaporator to increase, which in turn will result in raising the boiling temperature of the liquid. On the other hand, if the capacity of the compressor is such that the compressor removes the vapor from the evaporator too rapidly, the pressure in the evaporator will decrease and result in a decrease in the boiling temperature of the liquid. In either case, design conditions will not be maintained and the refrigerating system will be unsatisfactory.

Consequently, maintenance of the system design conditions, and therefore good refrigeration practice, requires the selection of a compressor whose capacity (displacement) is such that the compressor will displace in any given interval of time the same volume of vapor as that generated in the evaporator in that same time interval. For instance, in order to maintain equilibrium and produce a refrigerating capacity of 1 ton while operating at the conditions described in Example 6-4, the compressor must be selected to have a capacity (displacement) of 4.55 cfm.

6-27. METRIC CALCULATIONS

The following four examples are similar to Examples 6-1 through 6-4, except that metric units are employed.

Example 6-5 Determine the refrigerating effect per kilogram if the temperature of the R-12 liquid approaching the refrigerant control is 30°C (86°F) and the temperature of the saturated vapor leaving the evaporator is −5°C (23°F).

Solution From Table 16-2A,

Enthalpy of R-12 saturated vapor at −5°C	= 349.32 kJ/kg
Enthalpy of R-12 saturated liquid at 30°C	= 228.54 kJ/kg
Refrigerating effect per kilogram	= 120.78 kJ/kg

Example 6-6 A mechanical refrigerating system is operating under conditions such that

the vaporizing temperature is −5°C (23°F), while the temperature of the liquid approaching the refrigerant control is 30°C (86°F). If R-12 is circulated through the system at the rate of 0.3 kg/s, determine:

(a) the refrigerating capacity of the system in kilowatts,
(b) the refrigerating capacity of the system in tons.

Solution

(a) From Example 6-5, the refrigerating effect = 120.78 kJ(kg

Refrigerating capacity in kilowatts = (0.3 kg/s)(120.78 kJ/kg)
= 36.234 kJ/s (kW)

(b) System capacity in tons $= \dfrac{36.234 \text{ kW}}{3.517 \text{ kW/ton}}$
= 10.3 tons

Example 6-7 An R-12 system is operating at conditions such that the vaporizing temperature is −10°C (14°F) and the condensing temperature is 40°C (104°F). If it is assumed that no subcooling of the liquid occurs, so that the temperature of the liquid at the refrigerant control is also 40°C, find:

(a) the refrigerating effect per kilogram,
(b) the mass of refrigerant circulated in kilograms per second per kilowatt,
(c) the mass of refrigerant circulated per second for a 10-ton (35.17 kW) system.

Solution

(a) From Table 16-2A,

Enthalpy of R-12 saturated vapor at −10°C = 347.134 kJ/kg

Enthalpy of R-12 saturated liquid at 40°C = 238.535 kJ/kg

Refrigerating effect per kilogram = 108.599 kJ/kg

(b) Mass of refrigerant circulated per second per kilowatt $= \dfrac{1 \text{ kW (kJ/s)}}{108.599 \text{ kJ/kg}}$
= 0.0092 kg/s kW

(c) Mass of refrigerant circulated per second for a 10-ton (35.17 kW) system = (35.17 kW)(0.0092 kg/s kW)
= 0.3236 kg/s

Notice in Example 6-7 that expressing refrigerant mass flow rates in kilograms per second results in very cumbersome numerical values, particularly when the refrigerating capacity involved is relatively small. Consequently, it is usually more practical to determine refrigerant mass flow rates in grams per second (g/s) rather than in kilograms per second. Similarly, it is often more convenient to express volume flow rates in cubic centimeters per second (cm^3/s) rather than m^3/s and energy flow rates in joules per second (J/s, or W) rather than in kilojoules per second (kJ/s, or kW).

With reference to the refrigerant tables (Tables 16-2A and 16-2B), notice that the specific volume given in cubic meters per kilogram $\times 10^3$ is also numerically equal to cubic centimeters per gram and that the enthalpy values stated in kilojoules per kilogram are also numerically equal to joules per gram. By definition, of course, 1 kW is equal to 1000 W.

Example 6-8 With reference to Example 6-7, determine the volume of vapor generated in cubic centimeters per second per kilowatt of refrigerating capacity.

Solution From Example 6-7, the mass flow rate of R-12 is 9.2 g/s kW of refrigerating capacity, and from Table 16-2A, the specific volume of saturated vapor at −10°C is 76.65 cm^3/g. Applying Equation 6-2, the volume of vapor generated per second per kilowatt is

$$V = (9.2 \text{ g/s kW})(76.65 \text{ cm}^3/\text{g})$$
$$= 705.18 \text{ cm}^3/\text{s kW}$$

If required, the volume flow rate in cubic centimeters per second is converted to cubic meters per second by dividing by 1,000,000. In the foregoing example, the volume flow rate in cubic meters per second would be 0.00070518 (705.18×10^{-6}) m^3/s kW.

Customary Problems

6-1 Determine the refrigerating effect per pound of Refrigerant-12 if the temperature of the liquid approaching the control is 90°F and the temperature of the saturated vapor leaving the evaporator is −10°F.

6-2 Determine the refrigerating effect per pound of Refrigerant-22 if the evaporator temperature is 40°F and the liquid refrigerant enters the refrigerant control at a temperature of 100°F.

6-3 A mechanical refrigerating system employing R-12 is operating under such conditions that the evaporator temperature is −30°F and the temperature of the liquid refrigerant at the refrigerant control is 80°F. If the mass flow rate of refrigerant through the system is 5 lb/min, determine:
(a) the refrigerating effect per pound of refrigerant circulated,
(b) the refrigerating capacity of the system in Btu per minute,
(c) the refrigerating capacity of the system in tons.

6-4 A mechanical refrigerating system employing ammonia is operating under such conditions that the evaporator temperature is −10°F and the liquid approaching the refrigerant control is at a temperature of 105°F. If the system has a capacity of 18 tons, determine:
(a) the refrigerating effect per pound of refrigerant circulated,
(b) the mass flow rate in pounds per minute per ton
(c) the volume flow rate in cubic feet per minute per ton
(d) the total mass flow rate in pounds per minute
(e) the total volume flow rate in cubic feet per minute.

Metric Problems

6-5 Determine the refrigerating effect per kilogram of Refrigerant-12 if the temperature of the liquid approaching the refrigerant control is 32°C and the temperature of the saturated vapor leaving the evaporator is −20°C.

6-6 A mechanical refrigerating system employing Refrigerant-12 is operating under such conditions that the evaporator temperature is −15°C and the temperature of the liquid at the refrigerant control is 27°C. If the mass flow rate of refrigerant through the system is 0.78 kg/s, determine:
(a) the refrigerating effect per kilogram of refrigerant circulated,
(b) the refrigerating capacity of the system in kilowatts,
(c) the refrigerating capacity of the system in tons.

6-7 A mechanical refrigerating system employing R-12 is operating under such conditions that the evaporator temperature is −18°C and the liquid approaching the refrigerant control is at a temperature of 41°C. If the system has a capacity of 18 tons, determine:
(a) the refrigerating effect per kilogram of refrigerant circulated,
(b) the mass flow rate in grams per second per kilowatt,
(c) the volume flow rate in cubic centimeters per second per kilowatt,
(d) the total mass flow rate in grams per second,
(e) the total volume flow rate in cubic centimeters per second.

7

CYCLE DIAGRAMS AND THE SIMPLE SATURATED CYCLE

7-1. Cycle Diagrams

A good knowledge of the vapor-compression cycle requires an intensive study not only of the individual processes that make up the cycle but also of the relationships that exist among the several processes and of the effects that a change in any one process in the cycle has on all the other processes in the cycle. This is greatly simplified by the use of charts and diagrams upon which the complete cycle may be shown graphically. Graphical representation of the refrigeration cycle permits the desired simultaneous consideration of all the various changes in the condition of the refrigerant which occur during the cycle and the effect that these changes have on the cycle without the necessity of holding in mind all the different numerical values involved in cyclic problems.

The diagrams frequently used in the analysis of the refrigeration cycle are the pressure-enthalpy (*ph*) diagram and the temperature-entropy (*Ts*) diagram. Of the two, the pressure-enthalpy diagram seems to be the most useful and is the one that is used in this book.

7-2. The Pressure-Enthalpy Diagram

A pressure-enthalpy chart for R-12 is shown in Fig. 7-1. The condition of the refrigerant in any thermodynamic state can be represented as a point on the *ph* chart. The point on the *ph* chart that represents the condition of the refrigerant in any one particular thermodynamic state may be located if any two properties of the refrigerant at that state are known. Once the state point has been located on the chart, the other properties of the refrigerant for that state can be determined directly from the chart.

As shown by the skeleton *ph* chart in Fig. 7-2, the chart is divided into three areas that are separated from each other by the saturated liquid and saturated vapor lines. The area on the chart to the left of the saturated liquid line is called the subcooled region. At any point in the subcooled region the refrigerant is in the liquid phase and its temperature is below the saturation temperature corresponding to its pressure. The area to the right of the saturated vapor line is the superheated region and the refrigerant is in the form of a superheated vapor. The section of the chart between the saturated liquid and saturated vapor lines is the mixture region and represents the change in phase of the refrigerant between the liquid and vapor phases. At any point between the two saturation lines the refrigerant is in the form of a liquid-vapor mixture. The distance

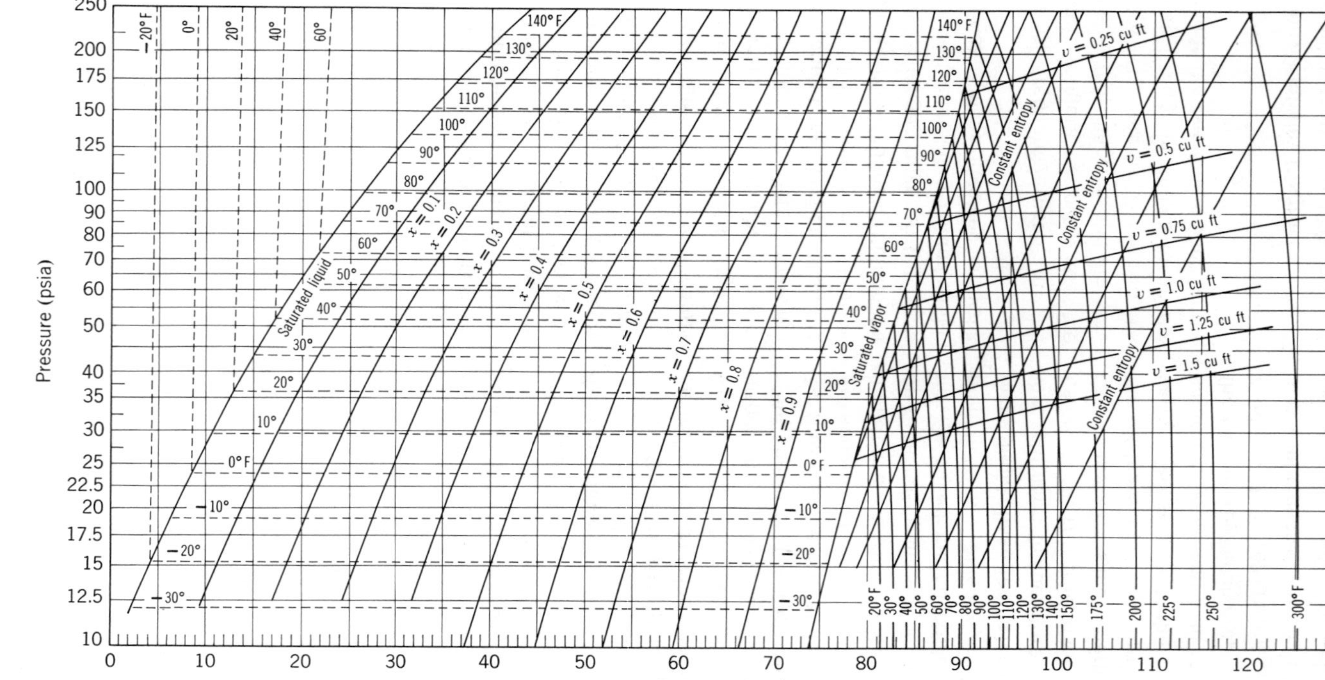

Fig. 7-1 Pressure-enthalpy diagram for Refrigerant-12.

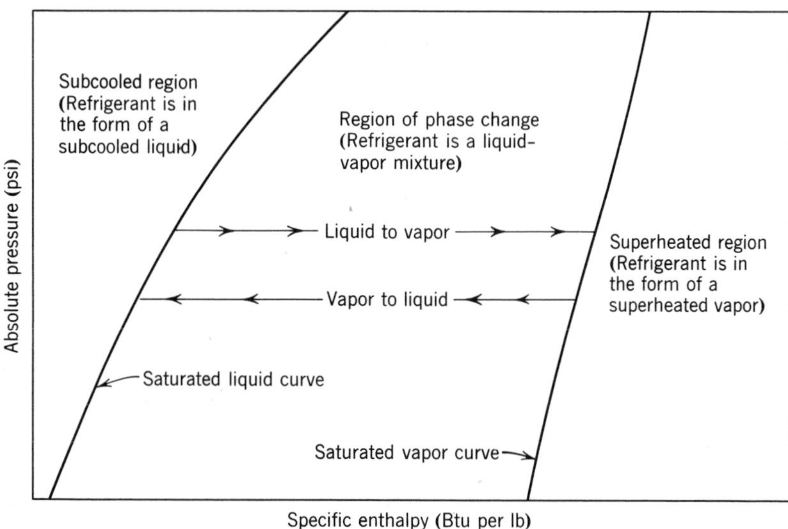

Fig. 7-2 Skeleton *ph* chart illustrating the three regions of the chart and the direction of phase changing.

between the two lines along any constant pressure line, as read on the enthalpy scale at the bottom of the chart, is the latent heat of vaporization of the refrigerant at that pressure. The saturated liquid and saturated vapor lines are not exactly parallel to each other because the latent heat of vaporization of the refrigerant varies with the pressure at which the change in phase occurs.

On the chart, the change in phase from the liquid to the vapor phase takes place progressively from left to right, whereas the change in phase from the vapor to the liquid phase occurs from right to left. Close to the saturated liquid line the liquid-vapor mixture is primarily liquid, whereas close to the saturated vapor line the liquid-vapor mixture is primarily vapor. The lines of constant quality (Fig. 7-3), extending from top to bottom through the center section of the chart and approximately parallel to the saturated liquid and vapor lines, indicate the percentage of vapor in the mixture in increments of 10%. For example, at any point on the constant quality line closest to the saturated liquid line the quality of the liquid-vapor mixture is 10%, which means that 10% (by mass) of the mixture is vapor. Similarly, the indicated quality of the mixture at any point along the constant quality line closest to the saturated vapor line is 90% and the amount of vapor in the liquid-vapor mixture is 90%. At any point on the saturated liquid line the refrigerant is a saturated liquid and at any point along the saturated vapor line the refrigerant is a saturated vapor.

The horizontal lines extending across the chart are lines of constant pressure and the vertical lines are lines of constant enthalpy.

The lines of constant temperature in the subcooled region are almost vertical on the chart and parallel to the lines of constant enthalpy. In the center section, since the refrigerant changes state at a constant temperature and pressure, the lines of constant temperature are parallel to and coincide with the lines of constant pressure. At the saturated vapor line the lines of constant temperature change direction again and, in the superheated vapor region, fall off sharply toward the bottom of the chart.

The straight lines which extend diagonally and almost vertically across the superheated vapor region are lines of constant entropy. The curved, nearly horizontal lines crossing the superheated vapor region are lines of constant volume.

The values of any of the various properties of the refrigerant which are of importance in the refrigerating cycle may be read directly from the *ph* chart at any point where the value of that

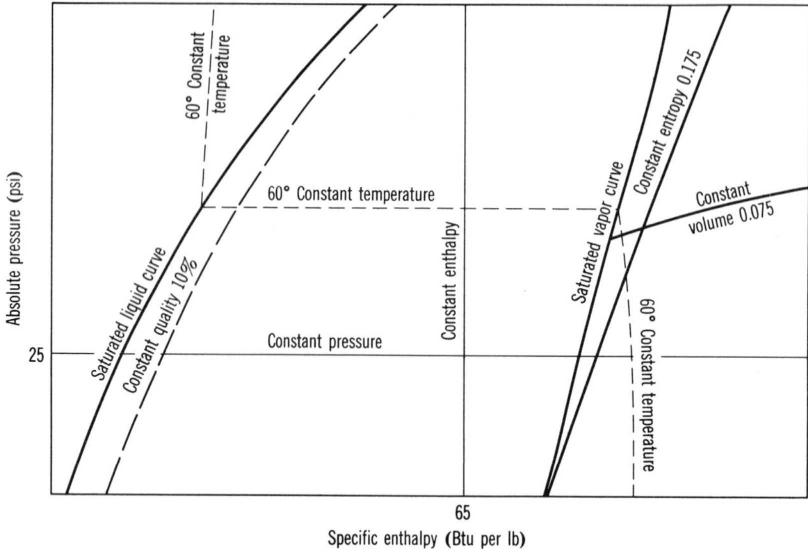

Fig. 7-3 Skeleton *ph* chart showing paths of constant pressure, constant temperature, constant volume, constant quality, constant enthalpy, and constant entropy. (Refrigerant-12.)

particular property is significant to the process occurring at that point. To simplify the chart, the number of lines on the chart is kept to a minimum. For this reason, the value of those properties of the refrigerant which have no real significance at some points in the cycle are omitted from the chart at these points. For example, in the liquid region and in the region of phase change (center section) the values of entropy and volume are of no particular interest and are therefore omitted from the chart in these sections.

Since the *ph* chart is based on a 1 lb mass of the refrigerant, the volume given is the specific volume, the enthalpy is in Btu per pound, and the entropy is in Btu per pound per degree of absolute temperature. Enthalpy values are found on the horizontal scale at the bottom of the chart and the values of entropy and volume are given adjacent to the entropy and volume lines, respectively. The values of both enthalpy and entropy are based on the arbitrarily selected zero point of −40°F.

The magnitude of the pressure in psia is read on the vertical scale at the left side of the chart. Temperature values in degrees Fahrenheit are found adjacent to the constant temperature lines in the subcooled and superheated regions of the chart and on both the saturated liquid and saturated vapor lines.

7-3. The Simple Saturated Refrigerating Cycle

A simple saturated refrigerating cycle is a theoretical cycle wherein it is assumed that the refrigerant vapor leaves the evaporator and enters the compressor as a saturated vapor at the vaporizing temperature and pressure and the liquid leaves the condenser and enters the refrigerant control as a saturated liquid at the condensing temperature and pressure. Although the refrigerating cycle of an actual refrigerating machine will deviate somewhat from the simple saturated cycle, the analysis of a simple saturated cycle is nonetheless worthwhile. In such a cycle, the fundamental processes which are the basis of every actual vapor compression refrigerating cycle are easily identified and understood. Furthermore, by using the simple saturated cycle as a standard against which actual cycles may be compared, the relative efficiency of actual refrigerating cycles at various operating conditions can be readily determined.

A simple saturated cycle for a R-12 system is plotted on a *ph* chart in Fig. 7-4. The system is

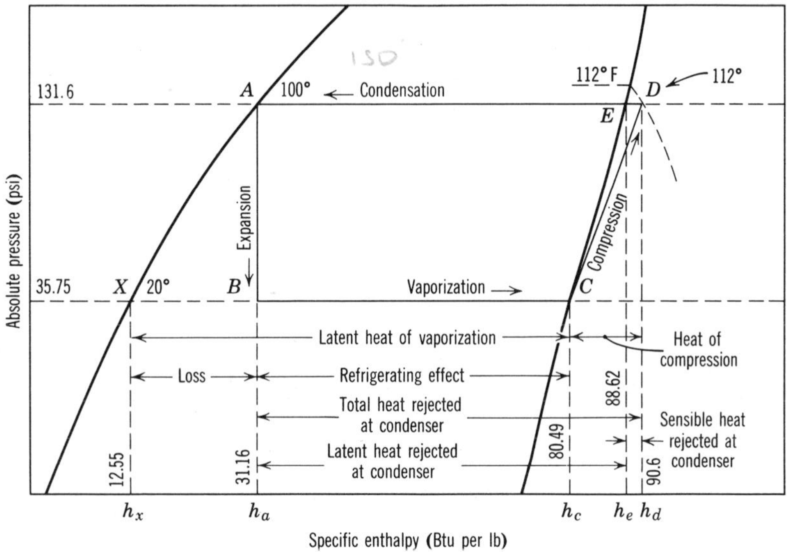

Fig. 7-4 Pressure-enthalpy diagram of a simple saturated cycle operating at a vaporizing temperature of 20°F and a condensing temperature of 100°F. (Refrigerant-12.)

assumed to be operating under such conditions that the vaporizing pressure in the evaporator is 35.75 psia and the condensing pressure in the condenser is 131.6 psia. The points *A*, *B*, *C*, *D*, and *E* on the *ph* diagram correspond to points in the refrigerating system as shown on the flow diagram in Fig. 7-5.

State point *A* can be described as some point near the bottom of the condenser where the condensing process has been completed and the refrigerant is a saturated liquid at the condensing temperature and pressure. The properties of the refrigerant at this point, as given in Table 16-3, are

$$p = 131.6 \text{ psia} \qquad T = 100°\text{F}$$

$$h = 31.16 \text{ Btu/lb} \qquad s = 0.06316 \text{ Btu/lb °F}$$

$$v = 0.0127 \text{ ft}^3/\text{lb}$$

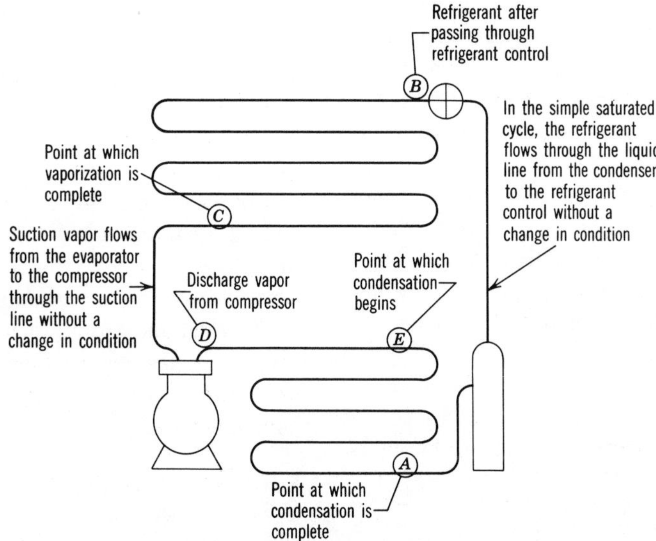

Fig. 7-5 Flow diagram of a simple saturated cycle.

At point *A*, the values of *p*, *T*, and *h* may be read directly from the *ph* chart. Since the refrigerant at point *A* is always a saturated liquid, point *A* will fall somewhere along the saturated liquid line and can be located on the *ph* chart if *p*, *T*, or *h* is known. In practice, *p* and *T* are readily measurable.

7-4. The Expansion Process

In the simple saturated cycle there is assumed to be no change in the properties (condition) of the refrigerant liquid as it flows through the liquid line from the condenser to the refrigerant control and the condition of the liquid approaching the refrigerant control is the same as its condition at point *A*. The process described by the initial and final state points *A–B* occurs in the refrigerant control when the pressure of the liquid is reduced from the condensing pressure to the evaporating pressure as the liquid passes through the control.* When the liquid is expanded into the evaporator through the orifice of the control, the temperature of the liquid is reduced from the condensing temperature to the evaporating temperature by the flashing into vapor of a portion of the liquid.

Process *A–B* is a throttling type of adiabatic expansion, frequently called "wire-drawing," in which the enthalpy of the working fluid does not change during the process. This type of expansion occurs whenever a fluid is expanded through an orifice from a high pressure to a lower pressure. It is assumed to take place without the gain or loss of heat through the piping or valves and without the performance of work.

Since the enthalpy of the refrigerant does not change during process *A–B*, point *B* is located on the *ph* chart by following the line of constant enthalpy from point *A* to the point where the constant enthalpy line intersects the line of constant pressure corresponding to the evaporating pressure. To locate point *B* on the *ph* chart, the evaporating pressure or temperature must be known.

As a result of the partial vaporization of the liquid refrigerant during process *A–B*, the refrigerant at point *B* is a liquid-vapor mixture whose properties are:

$p = 35.75$ psia
$T = 20°F$
$h = 31.16$ Btu/lb (same as at point *A*)
$v = 0.3154$ ft^3/lb
$s = 0.06657$ Btu/lb °F

Note The change in entropy during process *A–B* occurs as the result of allowing the fluid to expand from a higher pressure to a lower pressure without the performance of useful work, and as the result of energy transfers that take place within the fluid itself. A transfer of energy that takes place entirely within the working fluid does not affect the enthalpy of the fluid; only the entropy changes.

At point *B*, in addition to the values of *p*, *T*, and *h*, the approximate quality of the vapor can be determined from the *ph* chart by interpolating between the lines of constant quality. In this instance, the quality of the vapor as determined from the *ph* chart is approximately 27%.

Since the refrigerant at point *B* is a liquid-vapor mixture, only the values of *p* and *T* can be read directly from Table 16-3. However, because the enthalpy of the refrigerant at points *A* and *B* is the same, the enthalpy at point *B* may be read from Table 16-3 as the enthalpy at the conditions of point *A*. The quality of the vapor at point *B* can be determined as in Section 6-22, using enthalpy values taken either from Table 16-3 or from the *ph* chart directly.

The values of *s* and *v* at point *B* are usually of no interest and are not given either on the *ph* chart or in the vapor tables. If the values of *s* and *v* are desired, they must be calculated (see Section 4-22).

7-5. The Vaporizing Process

The process *B–C* is the vaporization of the refrigerant in the evaporator. Since vaporization takes place at a constant temperature and

* Process *A–B* is an irreversible adiabatic expansion during which the refrigerant passes through a series of state points in such a way that there is no uniform distribution of any of the properties. Hence, no true path can be drawn for the process and line *A-B* merely represents a process which begins at state point *A* and terminates at state point *B*.

pressure, process $B–C$ is both isothermal and isobaric. Therefore, point C is located on the *ph* chart by following the lines of constant pressure and constant temperature from point B to the point where they intersect the saturated vapor line. At point C the refrigerant is completely vaporized and is a saturated vapor at the vaporizing temperature and pressure. The properties of the refrigerant at point C, as given in Table 16-3 or as read from the *ph* chart, are:

p = 35.75 psia (same as at point B)
T = 20°F (same as at point B)
h = 80.49 Btu/lb
v = 1.121 ft^3/lb
s = 0.16949 Btu/lb °F

The enthalpy of the refrigerant increases during process $B–C$ as the refrigerant flows through the evaporator and absorbs heat from the refrigerated space. The quantity of heat absorbed by the refrigerant in the evaporator (refrigerating effect) is the difference between the enthalpy of the refrigerant at points B and C. Thus, if h_a, h_b, h_c, h_d, h_e, and h_x represent the enthalpies of the refrigerant at points A, B, C, D, E, and X, respectively, then

$$q_e = h_c - h_b \qquad (7\text{-}1)$$

where q_e = the refrigerating effect in Btu/lb.

But since h_b is equal to h_a, then

$$q_e = h_c - h_a \qquad (7\text{-}2)$$

When we substitute the appropriate values in Equation 7-2 for the example in question,

$$q_e = 80.49 - 31.16$$
$$= 49.33 \text{ Btu/lb}$$

On the *ph* diagram, the distance between point X and point C represents the total latent heat of vaporization of 1 lb of R-12 at the vaporizing pressure of 35.75 psia (h_{fg} in Table 16-3). Therefore, since the distance $B–C$ is the useful refrigerating effect, the difference between $X–C$ and $B–C$, which is the distance $X–B$, is the loss of refrigerating effect.

7-6. The Compression Process

In the simple saturated cycle, the refrigerant is assumed to undergo no change in condition while flowing through the suction line from the evaporator to the compressor. Process $C–D$ takes place in the compressor as the pressure of the vapor is increased by compression from the vaporizing pressure to the condensing pressure. For the simple saturated cycle, the compression process, $C–D$, is assumed to be isentropic.* An isentropic compression is a special type of adiabatic process which takes place without friction.† It is sometimes described as a "frictionless-adiabatic" or "constant-entropy" compression.

Since there is no change in the entropy of the vapor during process $C–D$, the entropy of the refrigerant at point D is the same as at point C. Therefore, point D can be located on the *ph* chart by following the line of constant entropy from point C to the point where the constant entropy line intersects the line of constant pressure corresponding to the condensing pressure.

At point D, the refrigerant is a superheated vapor whose properties are:

p = 131.6 psia
T = 112°F (approximate)
h = 90.6 Btu/lb (approximate)
v = 0.330 ft^3/lb (approximate)
s = 0.16949 Btu/lb°F (same as at point C)

All of the properties of the refrigerant at the condition of point D are taken from the *ph* chart. Since the values of T, h, and v require interpolation, they are only approximations.

Work is done on the refrigerant vapor during the compression process $C–D$ by the compressing member of the compressor, and the energy (enthalpy) of the vapor is increased by an amount exactly equal to the mechanical

* It will be shown later that compression of the vapor in an actual refrigerating compressor usually deviates somewhat from true isentropic compression.

† The term "adiabatic" is used to describe any number of processes that take place without the transfer of energy as heat to or from the working substance during the process. Thus, an isentropic process is only one of a number of different processes that may be termed adiabatic. For example, compare process $C–D$ with process $A–B$. Both are adiabatic, but $C–D$ is frictionless and reversible, whereas $A–B$ is a throttling type of process that involves friction and is irreversible.

work done on the vapor. The energy equivalent of the work done during the compression process is often referred to as the heat of compression and is equal to the difference in the enthalpy of the refrigerant at points D and C. Where q_w is the work (heat) of compression per pound of refrigerant circulated,

$$q_w = h_d - h_c \tag{7-3}$$

For the example in question,

$$q_w = 90.60 - 80.49 = 10.11 \text{ Btu/lb}$$

The mechanical work done on the vapor by the piston during the compression may be calculated from the heat of compression. If w is the work done in foot-pounds per pound of refrigerant circulated and J is the mechanical energy equivalent of heat, then

$$w = (q_w)(J) \tag{7-4}$$

or

$$w = J(h_d - h_c) \tag{7-5}$$

When we substitute in Equation 7-4,

$$w = (10.11 \text{ Btu/lb})(778 \text{ ft-lb/Btu})$$
$$= 7865.58 \text{ ft-lb/lb}$$

As a result of absorbing the heat of compression, the hot vapor discharged from the compressor is in a superheated condition, that is, its temperature is greater than the saturation temperature corresponding to its pressure. In this instance, the vapor leaves the compressor at a temperature of 112°F, whereas the saturation temperature corresponding to its pressure of 131.6 psia is 100°F. Before the vapor can be condensed, the superheat must be removed and the temperature of the vapor lowered from the discharge temperature to the saturation temperature corresponding to its pressure.

7-7. The Condensing Process

Usually, both processes D–E and E–A take place in the condenser as the hot gas discharged from the compressor is cooled to the condensing temperature and condensed. Process D–E occurs in the upper part of the condenser and to some extent in the hot gas line. It represents the cooling of the vapor from the discharge temperature to the condensing temperature as the vapor rejects heat to the condensing medium. During process D–E, the pressure of the vapor remains constant and point E is located on the *ph* chart by following a line of constant pressure from point D to the point where the constant pressure line intersects the saturated vapor curve.

At point E, the refrigerant is a saturated vapor at the condensing temperature and pressure. Its properties, as read from either the *ph* chart or from Table 16-3, are:

$p = 131.6$ psia (same as at point D)
$T = 100$°F
$h = 88.62$ Btu/lb
$s = 0.16584$ Btu/lb°F
$v = 0.319$ ft^3/lb

The quantity of sensible heat (superheat) removed per pound of vapor in the condenser in cooling the vapor from the discharge temperature to the condensing temperature is the difference between the enthalpy of the refrigerant at point D and the enthalpy at point $E (h_d - h_e)$.

Process E–A is the condensation of the vapor in the condenser. Since condensation takes place at a constant temperature and pressure, process E–A follows along lines of constant pressure and temperature from point E to point A. The heat rejected to the condensing medium during process E–A is the difference between the enthalpy of the refrigerant at points E and A $(h_e - h_a)$.

On returning to point A, the refrigerant has completed one cycle and its properties are the same as those previously described for point A.

Since both processes D–E and E–A occur in the condenser, the total amount of heat rejected by the refrigerant to the condensing medium in the condenser is the sum of the heat quantities rejected during processes D–E and E–A. The total heat given up by the refrigerant at the condenser is the difference between the enthalpy of the superheated vapor at point D and the saturated liquid at point A. Hence,

$$q_c = h_d - h_a \tag{7-6}$$

where q_c = the heat rejected at the condenser per pound of refrigerant circulated.

In this instance,

$$q_c = 90.60 - 31.16$$
$$= 59.44 \text{ Btu/lb}$$

If the refrigerant is to reach point *A* at the end of the cycle in the same condition as it left point *A* at the beginning of the cycle, the total heat rejected by the refrigerant to the condensing medium in the condenser must be exactly equal to the heat absorbed by the refrigerant at all other points in the cycle. In a simple saturated cycle, the energy of the refrigerant is increased at only two points in the cycle: (1) by the heat absorbed from the refrigerated space as the refrigerant vaporizes in the evaporator (q_e), and (2) by the energy equivalent of the mechanical work of compression in the compressor (q_w). Therefore,

$$q_c = q_e + q_w \tag{7-7}$$

In this instance,

$$q_c = 49.33 + 10.11 = 59.44 \text{ Btu/lb}$$

Where m is the mass flow rate of refrigerant circulated to produce the refrigerating capacity or rate, Q_e, of 1 ton,

$$m = \frac{Q_e}{q_e} \tag{7-8}$$

For the cycle in question,

$$m = \frac{200 \text{ Btu/min ton}}{49.33 \text{ Btu/lb}}$$
$$= 4.05 \text{ lb/min ton}$$

Then, where Q_c is the total quantity of heat rejected at the condenser per minute per ton,

$$Q_c = (m)(q_c) \tag{7-9}$$

or

$$Q_c = (m)(h_d - h_a) \tag{7-10}$$

For the cycle in question, applying Equation 7-9,

$$Q_c = (4.05)(59.44) = 240.73 \text{ Btu/min ton}$$

Where Q_w is the heat energy equivalent of the work of compression per minute per ton of refrigerating capacity,

$$Q_w = (m)(h_d - h_c) \tag{7-11}$$

or

$$Q_w = (m)(q_w) \tag{7-12}$$

Then, the work of compression per minute per ton in foot-pounds is

$$w = (J)(Q_w) \tag{7-13}$$

$$w = (J)(m)(h_d - h_c) \tag{7-14}$$

where J = the mechanical energy equivalent (see Section 2-30).

For the cycle in question, substituting in Equation 7-14,

$$w = (778)(4.05)(90.60 - 80.49)$$
$$w = 31{,}856 \text{ ft-lb/min ton}$$

7-8. Theoretical Power

The theoretical power (Thp) in horsepower required to drive the compressor per ton of refrigerating capacity may be found by applying the following equation (see Section 1-16):

$$\text{Thp/ton} = \frac{w}{33{,}000} \tag{7-15}$$

For the cycle in question,

$$\text{Thp/ton} = \frac{31{,}856 \text{ ft-lb/min ton}}{33{,}000 \text{ ft-lb/min hp}}$$
$$= 0.965 \text{ hp/ton}$$

A more convenient equation for determining the theoretical horsepower per ton is produced by combining Equations 7-14 and 7-15:

$$\text{Thp/ton} = \frac{(m)(h_d - h_c)}{42.42} \tag{7-16}$$

The compressor power as calculated from Equation 7-15 or 7-16 represents only the power required to compress the vapor. That is, it is the theoretical power which would be required per ton of refrigerating capacity by a 100% efficient system. It does not take into account the power required to overcome friction in the compressor and other power losses. The actual shaft power required per ton of refrigeration will usually be from 30% to 50% more than the theoretical power calculated, depending upon the efficiency of the compressor. The factors governing compressor efficiency are discussed later.

7-9. Coefficient of Performance

The coefficient of performance (c.o.p.) of a refrigerating cycle is an expression of the cycle efficiency and is stated as the ratio of the heat absorbed in the refrigerated space to the heat energy equivalent of the energy supplied to the compressor, that is,

$$\text{Coefficient of performance} = \frac{\text{Heat absorbed from the refrigerated space}}{\text{Heat energy equivalent of the energy supplied to the compressor}}$$

For the theoretical simple saturated cycle, this may be written as

$$\text{c.o.p.} = \frac{\text{Refrigerating effect}}{\text{Heat of compression}} \qquad (7\text{-}17)$$

$$= \frac{(h_c - h_a)}{(h_d - h_c)}$$

$$= \frac{q_e}{q_w}$$

Hence, for the cycle in question,

$$\text{c.o.p.} = \frac{49.33}{10.11}$$

$$= 4.88$$

7-10. Effect of Suction Temperature on Cycle Efficiency

The efficiency of the vapor-compression refrigerating cycle varies considerably with both the vaporizing and condensing temperatures. Of the two, the vaporizing temperature has by far the greater effect.

To illustrate the effect that varying the vaporizing temperature has on cycle efficiency, cycle diagrams of two simple saturated cycles operating at different vaporizing temperatures are drawn on the *ph* chart in Fig. 7-6. One cycle, identified by the points *A*, *B*, *C*, *D*, and *E*, is operating at a vaporizing temperature of 10°F and a condensing temperature of 100°F. A similar cycle having the same condensing temperature but operating at a vaporizing temperature of 40°F is set off by the points *A*, *B*′, *C*′, *D*′, and *E*.

To facilitate a comparison of the two cycles, the following values have been determined from the *ph* chart:

(a) For the 10°F cycle,

$$q_e = h_c - h_a = 79.36 - 31.16 = 48.20 \text{ Btu/lb}$$

$$q_w = h_d - h_c = 90.90 - 79.36 = 11.54 \text{ Btu/lb}$$

$$q_c = h_d - h_a = 90.90 - 31.16 = 59.74 \text{ Btu/lb}$$

(b) For the 40°F cycle,

$$q_e = h_{c'} - h_a = 82.71 - 31.16 = 51.55 \text{ Btu/lb}$$

$$q_w = h_{d'} - h_{c'} = 90.20 - 82.71 = 7.49 \text{ Btu/lb}$$

$$q_c = h_{d'} - h_a = 90.20 - 31.16 = 59.04 \text{ Btu/lb}$$

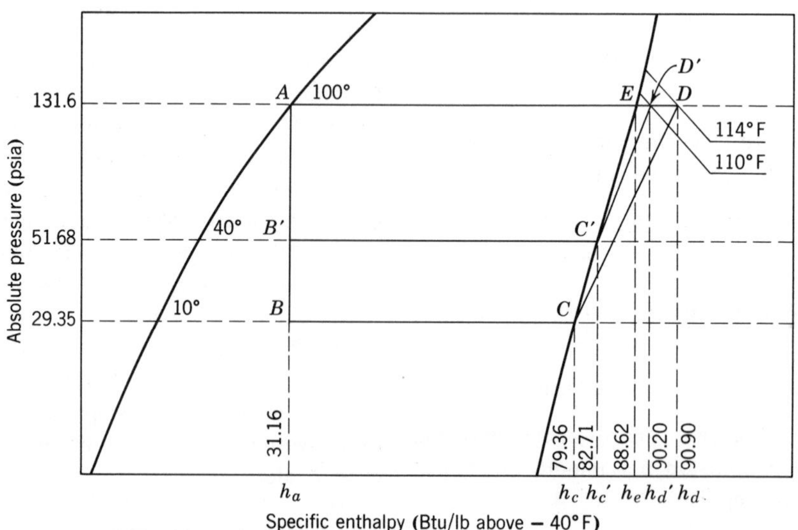

Fig. 7-6 Comparison of two simple saturated cycles operating at different vaporizing temperatures (figure distorted). (Refrigerant-12.)

In comparing the two cycles, note that the refrigerating effect per unit mass of refrigerant circulated is greater for the cycle having the higher vaporizing temperature. The refrigerating effect for the cycle having the 10°F vaporizing temperature is 48.20 Btu/lb. When the vaporizing temperature of the cycle is raised to 40°F, the refrigerating effect increases to 51.55 Btu/lb. This represents an increase in the refrigerating effect per pound of

$$\frac{(h_{c'} - h_a) - (h_c - h_a)}{h_c - h_a} \times 100$$

$$= \frac{51.55 - 48.20}{48.20} \times 100$$

$$= 6.95\%$$

The greater refrigerating effect per unit mass of refrigerant circulated obtained at the higher vaporizing temperature is accounted for by the fact that there is a smaller temperature differential between the vaporizing temperature and the temperature of the liquid approaching the refrigerant control. Consequently, at the higher suction temperature, a smaller fraction of the refrigerant vaporizes in the control and a greater portion vaporizes in the evaporator and produces useful cooling.

Since the refrigerating effect per unit mass is greater, the mass flow rate of refrigerant required to produce 1 ton of refrigerating capacity is less at the higher suction temperature than at the lower suction temperature. Whereas the mass flow rate of refrigerant required per ton for the 10°F cycle is

$$\frac{200}{h_c - h_a} = \frac{200}{48.20}$$

$$= 4.15 \text{ lb/min}$$

the mass flow rate required per ton for the 40°F cycle is only

$$\frac{200}{h_c - h_a} = \frac{200}{51.55}$$

$$= 3.88 \text{ lb/min}$$

The decrease in the mass flow rate at the higher suction temperature is

$$\frac{4.15 - 3.88}{4.15} \times 100 = 6.5\%$$

Since the difference between the vaporizing and condensing pressures is smaller for the cycle having the higher suction temperature, the work of compression per unit mass required to compress the vapor from the vaporizing pressure to the condensing pressure is less for the higher temperature cycle than for the lower temperature cycle. It follows then that the heat of compression per unit mass for the cycle having the higher vaporizing temperature is also less than that for the cycle having the lower vaporizing temperature. The heat of compression per pound for the 10°F cycle is 11.54 Btu, whereas the heat of compression for the 40°F cycle is only 7.49 Btu. This represents a decrease in the heat of compression per pound of

$$\frac{(h_d - h_c) - (h_{d'} - h_{c'})}{h_d - h_c}$$

$$= \frac{11.54 - 7.49}{11.54} \times 100$$

$$= 35.1\%$$

Because both the work of compression per unit mass and the mass flow rate of refrigerant required per ton of capacity are less at the higher suction temperature, the work of compression per ton and therefore the theoretical power required per ton will be smaller at the higher vaporizing temperature. The theoretical power required per ton of refrigerating capacity for the 10°F cycle is

$$\frac{m(h_d - h_c)}{42.42} = \frac{4.15 \times (90.90 - 79.36)}{42.42}$$

$$= 1.13 \text{ hp}$$

For the 40°F cycle, the theoretical power required per ton is

$$\frac{m(h_{d'} - h_{c'})}{42.42} = \frac{3.88 \times (90.20 - 82.71)}{42.42}$$

$$= 0.683 \text{ hp}$$

In this instance, increasing the vaporizing temperature of the cycle from 10°F to 40°F reduces the theoretical power required per ton by

$$\frac{1.13 - 0.683}{1.13} \times 100 = 39.5\%$$

Later, when the efficiency of the compressor is taken into consideration, it will be shown that the difference in the actual shaft horsepower required per unit capacity at the various vaporizing temperatures is even greater than that indicated by theoretical computations.

Since the coefficient of performance is an index of the power required per unit of refrigerating capacity and, as such, is an indication of cycle efficiency, the relative efficiency of the two cycles can be determined by comparing their coefficients of performance. The coefficient of performance for the 10°F cycle is

$$\frac{h_c - h_a}{h_d - h_c} = \frac{48.20}{11.54}$$
$$= 4.17$$

and the coefficient of performance for the 40°F cycle is

$$\frac{h_{c'} - h_a}{h_{d'} - h_{c'}} = \frac{51.55}{7.49}$$
$$= 6.88$$

It is evident that the coefficient of performance, and hence the efficiency of the cycle, improves considerably as the vaporizing temperature increases. In this instance, increasing the vaporizing temperature from 10°F to 40°F increases the efficiency of the cycle by

$$\frac{6.88 - 4.17}{4.17} \times 100 = 65\%$$

Although the difference in the mass flow rate per ton of refrigerating capacity at the various vaporizing temperatures is usually relatively small, the volume of vapor that the compressor must handle per minute per ton varies greatly with changes in the vaporizing temperature. This is probably one of the most important factors influencing the capacity and efficiency of a vapor-compression refrigerating system and is the one which is most likely to be overlooked by the student when studying cycle diagrams. The difference in the volume of vapor to be displaced per minute per ton at the various suction temperatures can be clearly illustrated by a comparison of the two cycles in question.

For the 10°F cycle, the volume of vapor to be displaced per minute per ton is

$$m(v) = 4.15 \times 1.351 = 5.6 \text{ ft}^3$$

whereas, at the 40°F suction temperature, the volume of vapor to be displaced per minute per ton is

$$m(v) = 3.88 \times 0.792 = 3.073 \text{ ft}^3$$

It is of interest to note that, whereas the decrease in the mass flow rate at the higher suction temperature is only 6.5%, the decrease in the volume of vapor handled by the compressor per minute per ton is

$$\frac{5.6 - 3.073}{5.6} \times 100 = 45\%$$

Obviously, the lower mass flow rate accounts for only a very small part of the reduction in the volume of vapor displaced per unit capacity at the higher suction temperature. To a far greater extent, the decrease in the volume flow rate per unit capacity can be attributed to the lower specific volume of the suction vapor which is coincident with the higher suction temperature and pressure (0.792 ft^3/lb at 40°F as compared to 1.351 ft^3/lb at 10°F). The effect of vaporizing temperature on compressor capacity and performance is discussed in more detail in Chapter 12.

The quantity of heat rejected at the condenser per minute per unit capacity is much smaller for the cycle having the higher vaporizing temperature. This is true even though the quantity of heat rejected at the condenser per pound of refrigerant circulated is nearly the same for both cycles. For the 10°F cycle, the quantity of heat rejected at the condenser per minute per ton is

$$m(h_d - h_a) = 4.15 \times 59.74 = 247.92$$

whereas for the 40°F cycle the heat rejected at the condenser per minute per ton is only

$$m(h_{d'} - h_a) = (3.88)(59.04) = 229.08 \text{ Btu}$$

The quantity of heat rejected per minute per ton at the condenser is less for the higher suction temperature because of (1) the lower mass flow rate and (2) the smaller heat of compression per unit mass.

It has been shown previously that the heat rejected at the condenser per unit mass of refrigerant circulated is the sum of the heat absorbed in the evaporator per unit mass (refrigerating effect) and the heat of compression per unit mass. Since increasing the vaporizing temperature of the cycle brings about an increase in the refrigerating effect as well as a decrease in the heat of compression, the quantity of heat rejected at the condenser per unit mass remains very nearly the same for both cycles (59.74 at 10°F as compared to 59.04 at 40°F). In general, this is true for all vaporizing temperatures because any increase or decrease in the heat of compression is usually accompanied by an offsetting increase or decrease in the refrigerating effect.

7-11. Effect of Condensing Temperature on Cycle Efficiency

Although the variations in cycle efficiency with changes in the condensing temperature are not as great as those brought about by changes in the vaporizing temperature, they are nonetheless important. In general, if the vaporizing temperature remains constant, the efficiency of the cycle decreases as the condensing temperature increases.

To illustrate the effect of condensing temperature on cycle efficiency, cycle diagrams of two saturated cycles operating at different condensing temperatures are drawn on the *ph* chart in Fig. 7-7. One cycle, *A*, *B*, *C*, *D*, and *E*, has a condensing temperature of 100°F, whereas the other cycle, *A′*, *B′*, *C*, *D′*, and *E′*, is operating at a condensing temperature of 120°F. The evaporating temperature of both cycles is 10°F. Values for cycle *A–B–C–D–E* have been determined in the previous section. Values for cycle *A′–B′–C–D′–E′* are as follows:

From the *ph* diagram,

$$q_e = h_c - h_{a'} = 79.36 - 36.16 = 43.20 \text{ Btu/lb}$$

$$q_w = h_{d'} - h_c = 93.20 - 79.36 = 13.84 \text{ Btu/lb}$$

$$q_c = h_{d'} - h_{a'} = 93.20 - 36.16 = 57.04 \text{ Btu/lb}$$

In a simple saturated cycle the liquid refrigerant reaches the refrigerant control at the condensing temperature. Therefore, as the

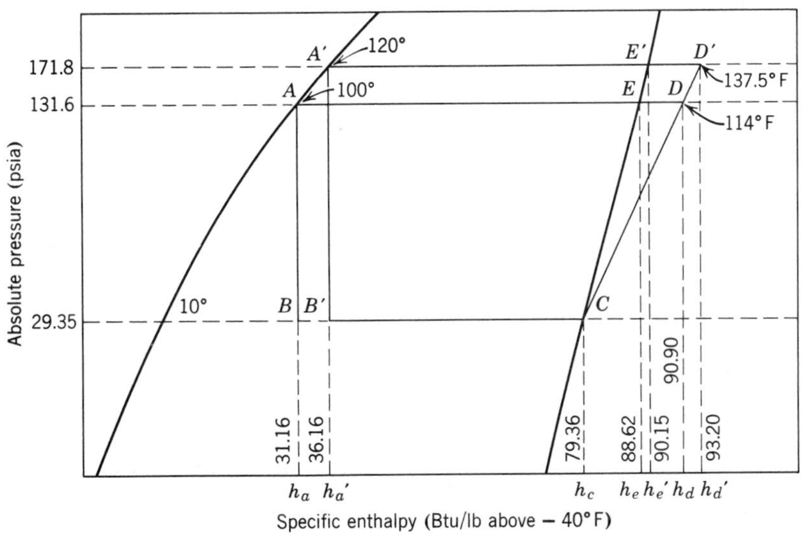

Fig. 7-7 Comparison of two simple saturated cycles operating at different condensing temperatures (figure distorted). (Refrigerant-12.)

condensing temperature is increased, the temperature of the liquid approaching the refrigerant control is increased and the refrigerating effect per pound is reduced. In this instance, the refrigerating effect is reduced from 48.20 Btu/lb to 43.20 Btu/lb when the condensing temperature is increased from 100°F to 120°F. This is a reduction of

$$\frac{48.20 - 43.20}{48.20} \times 100 = 10.37\%$$

Because the refrigerating effect per unit mass is less for the cycle having the higher condensing temperature, the mass flow rate of refrigerant per unit capacity must be greater. For the cycle having the 100°F condensing temperature the refrigerant mass flow rate is 4.15 lb/min ton. When the condensing temperature is increased to 120°F, the refrigerant mass flow rate increases to

$$\frac{200}{43.20} = 4.63 \text{ lb/min ton}$$

This is an increase of

$$\frac{4.63 - 4.15}{4.15} \times 100 = 11.57\%$$

Since the refrigerant mass flow rate (per unit capacity) is greater at the higher condensing temperature, it follows that the volume of vapor compressed per unit capacity is also greater. In a simple saturated cycle the specific volume of the suction vapor varies only with the vaporizing temperature. Since the vaporizing temperature is the same for both cycles, the specific volume of the vapor leaving the evaporator is also the same for both cycles, so that the difference in the volume of vapor handled per unit of capacity is in direct proportion to the difference in the mass flow rate per unit of capacity. At the 100°F condensing temperature the volume of vapor compressed per minute per ton is 5.6 ft^3, whereas at the 120°F condensing temperature the volume of vapor compressed per minute per ton increases to

$$4.63 \times 1.351 = 6.25 \text{ ft}^3$$

This represents an increase in the volume of vapor compressed per minute per ton of

$$\frac{6.25 - 5.6}{5.6} \times 100 = 11.57\%$$

Note that the percent increase in the volume of vapor handled by the compressor is exactly equal to the percent increase in the mass flow rate. Contrast this with what occurs when the vaporizing temperature is varied.

Since the difference between the vaporizing and condensing pressures is greater, the work of compression per unit mass of refrigerant circulated is also greater for the cycle having the higher condensing temperature. In this instance, the heat of compression increases from 11.54 Btu/lb for the 100°F condensing temperature to 13.84 Btu/lb for the 120°F condensing temperature. This is an increase of

$$\frac{13.84 - 11.54}{11.54} \times 100 = 20\%$$

As a result of the greater work of compression per unit mass and the greater mass flow rate per unit of capacity, the theoretical power required per unit of refrigerating capacity increases as the condensing temperature increases. Whereas the theoretical power required per ton at the 100°F condensing temperature is only 1.13 hp when the condensing temperature is increased to 120°F, the theoretical power per ton increases to

$$\frac{4.63 \times 13.84}{42.42} = 1.51 \text{ hp}$$

This is an increase in the power required per ton of

$$\frac{1.51 - 1.13}{1.13} \times 100 = 33.6\%$$

Notice that the percentage increase in the power required per ton at the higher condensing temperature is greater than the percentage increase in the work of compression per pound. This is accounted for by the fact that, in addition to the 20% increase in the work of compression per unit mass, there is also a

6.5% increase in the mass of refrigerant circulated per unit of capacity.

The coefficient of performance of the cycle at the 100°F condensing temperature is 4.17. When the condensing temperature is raised to 120°F, the coefficient of performance drops to

$$\frac{43.20}{13.84} = 3.12$$

Since the coefficient of performance is an index of the refrigerating capacity per unit of power, the decrease in refrigerating capacity per unit of power in this instance is

$$\frac{4.17 - 3.12}{3.12} \times 100 = 33.7\%$$

Obviously, the effect of raising the condensing temperature on cycle efficiency is the exact opposite of that of raising the evaporating temperature. Whereas raising the evaporating temperature increases the refrigerating effect per unit mass and reduces the work of compression so that the refrigerating capacity per unit of power increases, raising the condensing temperature reduces the refrigerating effect per unit mass and increases the work of compression so that the refrigerating capacity per unit of power decreases.

Although the quantity of heat rejected at the condenser per unit mass of refrigerant circulated varies only slightly with changes in the condensing temperature because any change in the heat of compression is accompanied by an offsetting change in the refrigerating effect, the total heat rejected at the condenser per unit of capacity varies considerably with changes in the condensing temperature because of the difference in the mass of refrigerant circulated per unit of capacity. It was shown in Section 7-7 that the total heat rejected at the condenser per unit capacity (Q_c) is always the sum of the heat absorbed in the evaporator per unit capacity (Q_e) and the heat of compression per unit capacity (Q_w). Since Q_e is a constant (200 Btu/min ton), Q_c will vary only with Q_w, the heat of compression per minute per ton. Furthermore, since Q_w always increases as the condensing temperature increases, it follows then that Q_c also increases as the condensing temperature increases.

For the two cycles in question, the heat rejected at the condenser per minute per ton at the 100°F condensing temperature is 247.92 Btu. For the 120°F condensing temperature, the quantity of heat rejected at the condenser per minute per ton increases to

$$4.63 \times (43.20 + 13.84) = 264.10 \text{ Btu}$$

The percent increase is

$$\frac{264.10 - 247.92}{247.92} \times 100 = 6.53\%$$

It is interesting to note also that the amount of sensible heat rejected at the condenser increases considerably at the higher condensing temperature, whereas the amount of latent heat rejected diminishes slightly. This indicates that, at the higher condensing temperature, a greater portion of the condenser surface is being used merely to reduce the temperature of the discharge vapor to the condensing temperature.

7-12. Summary

Since the capacity and efficiency of a refrigerating system improve significantly as the evaporator temperature is increased, it is evident that a refrigerating system should always be designed to operate at the highest practical evaporator temperature. Although the effect of the condensing temperature on the capacity and efficiency of the refrigeration cycle is considerably less than that of the evaporator temperature, it nonetheless should be kept as low as is practical.

In any event, the influence of the vaporizing and condensing temperatures on cycle efficiency is of sufficient importance to warrant a more intensive study. To aid in this, the relationship between the refrigerating effect per pound, the mass of refrigerant circulated per minute per ton, the specific volume of the suction vapor, the volume of vapor compressed per minute per ton, the horsepower required per ton, and the coefficient of performance of

Condensing Temperature, 100° F						Condensing Pressure, 136.16 Psia				
Suction temperature	50°	40°	30°	20°	10°	0°	−10°	−20°	−30°	−40°
Absolute suction pressure	61.39	51.68	43.16	35.75	29.35	23.87	19.20	15.28	12.02	9.32
Refrigerating effect per pound	52.62	51.55	50.45	49.33	48.20	47.05	45.89	44.71	43.54	42.34
Weight of refrigerant circulated per minute per ton	3.80	3.88	3.97	4.05	4.15	4.25	4.36	4.48	4.59	4.73
Specific volume of suction vapor	0.673	0.792	0.939	1.121	1.351	1.637	2.00	2.47	3.09	3.91
Volume of vapor compressed per minute per ton	2.56	3.08	3.77	4.55	5.60	6.96	8.72	11.10	14.20	18.50
Heat of compression per pound	6.01	7.49	8.79	10.11	11.54	13.29	14.85	16.73	18.50	20.40
Theoretical horsepower per ton	0.539	0.683	0.818	0.965	1.13	1.35	1.54	1.78	2.00	2.26
Coefficient of performance	8.76	6.88	5.74	4.88	4.17	3.54	3.09	2.67	2.36	2.07

Fig. 7-8

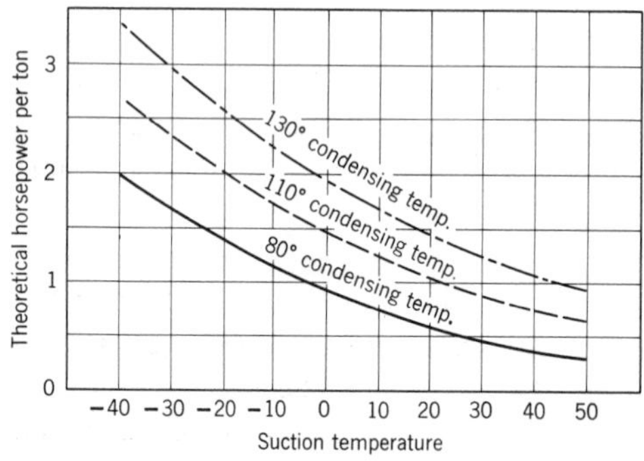

Fig. 7-9 The effect of condensing temperature on the horsepower per ton.

the cycle has been calculated for various suction temperatures. These values are given in tabular form in Fig. 7-8. In addition, the effect of condensing temperature on the horsepower required per ton of refrigerating capacity is shown for several condensing temperatures in Fig. 7-9.

Since the properties of the refrigerant at point *D* on the cycle diagram cannot readily be obtained from the refrigerant tables and since these properties are difficult to read accurately from the *ph* chart because of the size of the chart, the approximate isentropic discharge temperatures and the approximate enthalpy of the refrigerant vapor at point *D* have been compiled for a variety of vaporizing and condensing temperatures and are given in Table 7-1 to aid the student in arriving at more accurate solutions to the problems at the end of the chapter.*

* Since the new superheated vapor tables in SI units use saturation temperature, rather than pressure, as the entry property, the properties of the superheated vapor at point *D* are readily determined by interpolation from these tables.

TABLE 7-1 Temperature and Enthalpy of Discharge Vapor after Isentropic Compression

Saturated Suction Temperature	Condensing Temperature 80°		90°		100°	
	t	h	t	h	t	h
−40°	111.0°	91.6	121.0°	92.3	132.0°	93.9
−30°	105.0°	90.5	116.0°	92.0	127.5°	93.2
−20°	102.0°	90.2	112.5°	91.4	124.0°	92.6
−10°	97.5°	89.5	108.5°	90.7	119.9°	91.9
0°	95.0°	89.2	106.0°	90.3	117.0°	91.5
10°	92.0°	88.7	103.5°	89.9	114.0°	90.9
20°	90.0°	88.4	102.0°	89.6	112.0°	90.6
30°	88.0°	88.1	99.0°	89.1	110.8°	90.4
40°	86.0°	87.7	97.0°	88.8	109.5°	90.2
50°	84.0°	87.4	95.5°	88.6	107.0°	89.8

Saturated Suction Temperature	Condensing Temperature 110°		120°		130°	
	t	h	t	h	t	h
−40°	143.0°	95.1	155.0°	96.3	166.5°	97.3
−30°	138.0°	94.3	150.5°	95.5	161.5°	96.6
−20°	135.5°	93.7	147.0°	94.8	157.5°	95.8
−10°	131.6°	93.1	143.0°	94.2	154.0°	95.2
0°	128.5°	92.6	141.0°	93.7	152.0°	94.8
10°	126.5°	92.1	137.5°	93.2	148.5°	94.3
20°	124.0°	91.7	136.0°	92.8	147.2°	93.9
30°	122.0°	91.4	133.5°	92.5	146.0°	93.6
40°	120.0°	91.1	132.5°	92.2	143.5°	93.2
50°	118.0°	90.8	131.0°	92.0	142.0°	92.9

7-13. METRIC CALCULATIONS

The following example presents an analysis of a simple saturated refrigeration cycle in metric units.

Example 7-1 A refrigerating system, illustrated in Fig. 7-10, employs R-12 and is operating on a simple saturated cycle with an evaporator temperature of −5°C (23°F) and a condensing temperature of 40°C (104°F). Determine:

(a) the properties of p, T, v, h, and s for the refrigerant at state points A, B, C, D, and E,
(b) The compressor displacement in cubic centimeters per second required per kilowatt of refrigerant capacity,
(c) The theoretical power in watts required per kilowatt of refrigerant capacity,
(d) The total heat rejected at the condenser in kilowatts per kilowatt of refrigerating capacity,
(e) the coefficient of performance of the cycle.

Solution

(a) At state point A, the refrigerant is a saturated liquid at the condensing temperature of 40°C. From Table 16-2A,

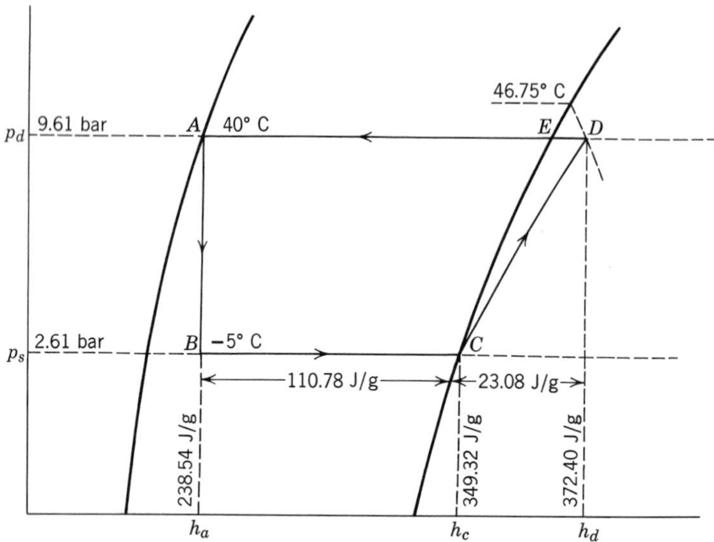

Fig. 7-10 Pressure-enthalpy diagram of simple saturated cycle (SI units).

$p = 9.6065$ bar, $v = 0.79802$ cm^3/g, $h =$ 238.535 J/g, and $s = 1.12984$ J/g K.

At state point *B* the refrigerant is a liquid-vapor mixture at the evaporator temperature and pressure of −5°C and 2.61 bar. The enthalpy (*h*) of the mixture is the same as that of the liquid at state point *A*, 238.535 J/g. By calculation (see Section 4-22), the specific volume (v) and the enthalpy (*s*) are 18.699 cm^3/g and 1.1438 J/g K, respectively.

At state point *C* the refrigerant is a saturated vapor at the vaporizing temperature of −5°C. From Table 16-2A, $p = 2.61$ bar, $v = 64.9629$ cm^3/g, $h =$ 349.32 J/g, and $s = 1.5571$ J/g K.

At state point *D* the refrigerant is a superheated vapor at the condensing pressure of 9.6065 bar, corresponding to a condensing (saturation) temperature of 40°C. Since the compression process is assumed to be isentropic (constant entropy), the entropy of the superheated vapor is the same as that of the saturated vapor at state point *C*, 1.5571 J/g K. By interpolation in the superheated vapor table (Table 16-2B), it can be determined that the temperature, enthalpy, and specific volume for superheated R-12 vapor, with a saturation temperature of 40°C and an entropy of 1.5571 J/g K, are 46.75°C, 372.40 J/g, and 18.9447 cm^3/g, respectively (see Fig. 7-11).

	v	h	s
Temp. °C	40°C (9.6065 bar)		
40	18.1706	367.146	1.5405
45	18.7495	371.054	1.5529
46.75	18.9447	372.403	1.5571
50	19.3072	374.904	1.5649
55	19.8469	378.708	1.5766
60	20.3711	382.475	1.5880

Fig. 7-11 See Example 7-1.

At state point *E* the refrigerant is a saturated vapor at the condensing temperature and pressure of 40°C and 9.6065 bar. From Table 16-2A, the specific volume, enthalpy, and entropy are 18.1706 cm^3/g, 367.146 J/g, and 1.54051 J/g K, respectively.

(b) Applying Equation 7-2,

$$q_e = 349.32 - 238.54$$
$$= 110.78 \text{ J/g}$$

Applying Equation 7-8,

$$m = \frac{1000 \text{ W}}{110.78 \text{ J/g}}$$

$$= 9.03 \text{ g/s kW}$$

Applying Equation 6-2,

$$V = (9.03 \text{ g/s kW})(64.96 \text{ cm}^3/\text{g})$$

$$= 586.59 \text{ cm}^3/\text{s kW}$$

(c) Applying Equation 7-3,

$$q_w = 372.40 - 349.32$$

$$= 23.08 \text{ J/g}$$

Applying Equation 7-12,

$$Q_w = (9.03 \text{ g/s kW})(23.08 \text{ J/g})$$

$$= 208.41 \text{ W/kW}$$

(d) Applying Equation 7-6,

$$q_c = 372.40 - 238.54$$

$$= 133.86 \text{ J/g}$$

Applying Equation 7-9,

$$Q_c = (9.03 \text{ g/s kW})(133.86 \text{ J/g})$$

$$= 1208.76 \text{ W/kW} \quad \text{or} \quad 1.208 \text{ kW/kW}$$

(e) Applying Equation 7-17,

$$\text{c.o.p.} = \frac{110.78 \text{ J/g}}{23.08 \text{ J/g}}$$

$$= 4.8$$

Customary Problems

7-1 A refrigerating system employing R-12 is operating on a simple saturated cycle with an evaporator temperature of 40°F and a condensing temperature of 120°F. Determine the following:

(a) the properties of p, T, v, h, and s for the refrigerant at state points A, B, C, D, and E as the latter are defined in Fig. 7-4,
(b) the refrigerating effect per pound of refrigerant circulated,
(c) the mass flow rate of refrigerant in pounds per minute per ton.
(d) the volume flow rate of refrigerant at the compressor inlet in cubic feet per minute per ton, that is, the required compressor displacement,
(e) the heat of compression per pound of refrigerant circulated,
(f) the heat of compression per minute per ton of refrigeration,
(g) the theoretical horsepower per ton of refrigeration,
(h) the coefficient of performance,
(i) the heat rejected per minute per ton at the condenser.

Metric Problems

7-2 A refrigeration system employing R-12 is operating on a simple saturated cycle with an evaporator temperature of 4°C and a condensing temperature of 42°C. Using metric units, determine the following:

(a) the properties of p, T, v, h, and s for the state points A, B, C, D, and E, as previously described.
(b) the compressor displacement in cubic centimeters per second required per kilowatt of refrigerating capacity,
(c) the theoretical power in watts required per kilowatt of refrigerating capacity,
(d) the total heat rejected at the condenser in kilowatts per kilowatt of refrigerating capacity,
(e) the coefficient of performance of the cycle.

8

ACTUAL REFRIGERATING CYCLES

8-1. Deviation from the Simple Saturated Cycle

Actual refrigerating cycles deviate somewhat from the simple saturated cycle. The reason for this is that certain assumptions are made for the simple saturated cycle which do not hold true for actual cycles. For example, in the simple saturated cycle, the drop in pressure in the lines and across the evaporator, condenser, etc., resulting from the flow of the refrigerant through these parts is neglected. Furthermore, the effects of subcooling the liquid and of superheating the suction vapor are not considered. Also, compression in the compressor is assumed to be true isentropic compression. In the following sections all these things are taken into account and their effect on the cycle is studied in detail.*

8-2. The Effect of Superheating the Suction Vapor

In the simple saturated cycle, the suction vapor is assumed to reach the suction inlet of the compressor as a saturated vapor at the vaporizing temperature and pressure. In practice, this is rarely true. After the liquid refrigerant has completely vaporized in the evaporator, the cold, saturated vapor usually will continue to absorb heat and thereby become superheated before it reaches the compressor (Fig. 8-1).

On the *ph* diagram in Fig. 8-2, a simple saturated cycle is compared to one in which the suction vapor is superheated from 20°F to 70°F. Points *A*, *B*, *C*, *D*, and *E* mark the saturated cycle, and points *A*, *B*, *C′*, *D′*, and *E* indicate the superheated cycle.

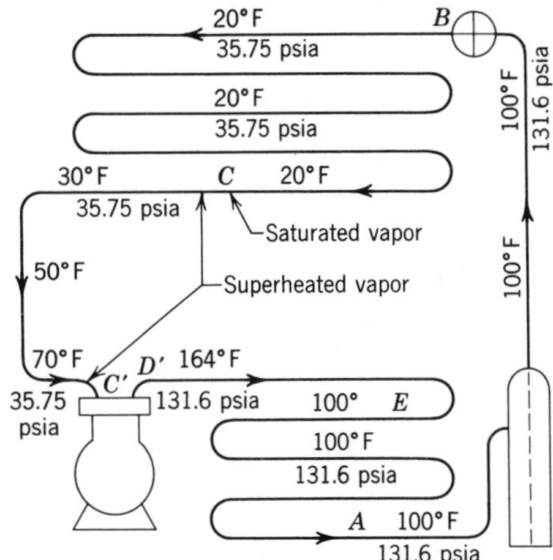

Fig. 8-1 Flow diagram of superheated cycle. Liquid completely vaporized at point *C*-saturated vapor continues to absorb heat while flowing from *C* to *C*—vapor reaches compressor in superheated condition. Notice the high discharge temperature. (Refrigerant-12.)

* The departure from true isentropic compression and the effect that it has on the cycle are discussed in Chapter 12.

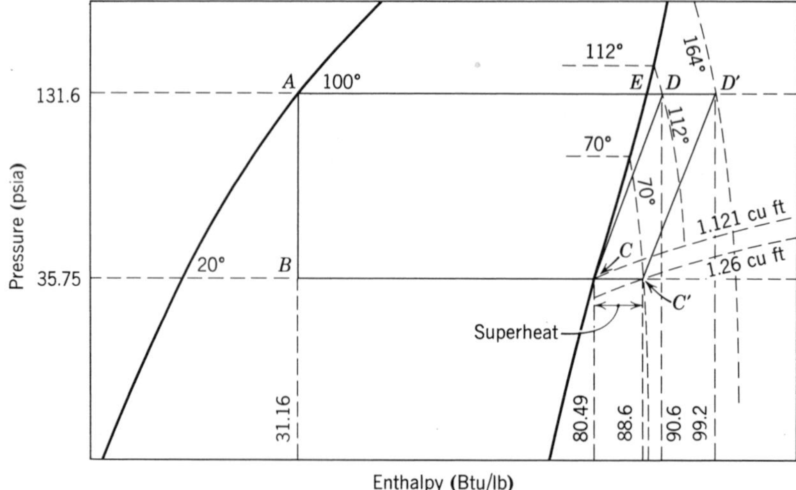

Fig. 8-2 *ph* diagram comparing simple saturated cycle to the superheated cycle. (Refrigerant-12).

If the slight pressure drop resulting from the flow of the vapor in the suction piping is neglected, it may be assumed that the pressure of the suction vapor remains constant during the superheating. That is, after the superheating, the pressure of the vapor at the suction inlet of the compressor is still the same as the vaporizing pressure in the evaporator. With this assumption, point C' can be located on the *ph* chart by following a line of constant pressure from point C to the point where the line of constant pressure intersects the 70°F constant temperature line. Point D' is found by following a line of constant entropy from point C' to the line of constant pressure corresponding to the condensing pressure.

In Fig. 8-2, the properties of the superheated vapor at points C' and D', as read from the *ph* chart, are as follows:

At point C',

$$p = 35.75 \text{ psia} \qquad T = 70°\text{F}$$
$$v = 1.260 \text{ ft}^3/\text{lb} \qquad h = 88.6 \text{ Btu/lb}$$
$$s = 0.1840 \text{ Btu/lb °R}$$

At point D',

$$p = 131.6 \text{ psia} \qquad T = 164°\text{F}$$
$$v = 0.380 \text{ ft}^3/\text{lb} \qquad h = 99.2 \text{ Btu/lb}$$
$$s = 0.1840 \text{ Btu/lb °R}$$

On the *ph* chart, process C–C' represents the superheating of the suction vapor from 20°F to 70°F at the vaporizing pressure, and the difference between the enthalpy of the vapor at these points is the amount of heat required to superheat each pound of refrigerant. In comparing the two cycles, the following observations are of interest:

1. The heat of compression per pound for the superheated cycle is slightly greater than that for the saturated cycle. For the superheated cycle, the heat of compression is

$$h_{d'} - h_{c'} = 99.2 - 88.6 = 10.6 \text{ Btu/lb}$$

whereas for the saturated cycle the heat of compression is

$$h_d - h_c = 90.6 - 80.49 = 10.11 \text{ Btu/lb}$$

In this instance, the heat of compression per pound is greater for the superheated cycle by

$$\frac{10.6 - 10.11}{10.11} \times 100 = 4.84\%$$

2. For the same condensing temperature and pressure, the temperature of the discharge vapor leaving the head of the compressor is considerably higher for the superheated cycle than for the saturated cycle—in this

case, 164°F for the superheated cycle as compared to 112°F for the saturated cycle.

3. For the superheated cycle, a greater quantity of heat must be rejected at the condenser per pound than for the saturated cycle. This is because of the additional heat absorbed by the vapor in becoming superheated and because of the small increase in the heat of compression per pound. For the superheated cycle, the heat rejected per pound is

$$h_{d'} - h_a = 99.2 - 31.16 = 68.04 \text{ Btu/lb}$$

and for the saturated cycle the heat rejected per pound is

$$h_d - h_a = 90.6 - 31.16 = 59.44 \text{ Btu/lb}$$

For the superheated cycle, this is an increase of

$$\frac{68.04 - 59.44}{59.44} \times 100 = 14.4\%$$

Notice that the additional heat rejected at the condenser in the superheated cycle is all sensible heat. The amount of latent heat rejected per pound is the same for both cycles. This means that in the superheated cycle a greater amount of sensible heat must be given up to the condensing medium before condensation begins and that a greater portion of the condenser will be used in cooling the discharge vapor to its saturation temperature.

Assuming that the pressure of the suction vapor remains constant during the superheating, the volume of the vapor increases with the temperature approximately in accordance with Charles' law.* Therefore, the specific volume of the superheated vapor always will be greater than that of saturated vapor at the same pressure. For example, in Fig. 8-2, the specific volume of the suction vapor increases from 1.121 ft^3/lb at saturation to 1.260 ft^3/lb when superheated to 70°F. This means that for each pound of refrigerant circulated, the compressor must compress a greater volume of vapor if the vapor is superheated than if the vapor is saturated. For this reason, in every instance where the vapor is allowed to become superheated before it reaches the compressor, the mass of refrigerant circulated by a compressor of any given displacement will always be less than when the suction vapor reaches the compressor in a saturated condition, provided the pressure is the same.

The effect that superheating of the suction vapor has on the capacity of the system and on the coefficient of performance depends entirely upon where and how the superheating of the vapor occurs and upon whether or not the heat absorbed by the vapor in becoming superheated produces useful cooling.

8-3. Superheating without Useful Cooling

Assume first that the superheating of the suction vapor occurs in the suction line in such a way that no useful cooling results. When this is true, the refrigerating effect per unit mass of refrigerant circulated is the same for the superheated cycle as for a saturated cycle operating at the same vaporizing and condensing temperatures, and therefore the mass flow rate of refrigerant required per unit refrigerating capacity will also be the same for both the superheated and saturated cycles. Then, for both cycles illustrated in Fig. 8-2,

the mass flow rate per ton m

$$= \frac{200}{h_c - h_a}$$

$$= \frac{200}{49.33}$$

$$= 4.05 \text{lb/min ton}$$

Since the mass flow rate of refrigerant is the same for both the superheated and saturated cycles and since the specific volume of the vapor at the compressor inlet is greater for the superheated cycle than for the saturated cycle, it follows that the volume flow rate of vapor that the compressor must handle per unit of refrigerating capacity is greater for the superheated cycle than for the saturated cycle.

* The temperature and volume of the vapor do not vary exactly in accordance with Charles' law because the refrigerant vapor is not a perfect gas.

For the saturated cycle, the specific volume of the suction vapor $v_{c'}$ $= 1.121\ ft^3/lb$

The volume of vapor compressed per minute per ton, V $= (m)(v)$ $= (4.05)(1.121)$ $= 4.55\ ft^3/min\ ton$

For the superheated cycle, the specific volume of the suction vapor $v_{c'}$ $= 1.260\ ft^3/lb$

The volume of vapor compressed per minute per ton, V $= (m)(v)$ $= (4.05)(1.26)$ $= 5.01\ ft^3/min\ ton$

The percentage increase in the volume flow rate of vapor for the superheated cycle is

$$\frac{5.01 - 4.55}{4.55} \times 100 = 10.3\%$$

This means that a compressor operating on the superheated cycle must have 10.3% more displacement than one operating on the saturated cycle.

Again, since the mass flow rate of refrigerant per unit of capacity is the same for both cycles and since the heat of compression per unit mass is greater for the superheated cycle, it follows that the power required per unit of refrigerating capacity is greater for the superheated cycle and the coefficient of performance is lower.

For the saturated cycle, the horsepower per ton

$$= \frac{m(h_d - h_c)}{42.42} = \frac{4.05 \times 10.11}{42.42} = 0.965\ \text{hp/ton}$$

The coefficient of performance

$$= \frac{h_c - h_a}{h_d - h_c} = \frac{49.33}{10.11} = 4.88$$

For the superheated cycle, the horsepower per ton

$$= \frac{m(h_{d'} - h_{c'})}{42.42} = \frac{4.05 \times 10.6}{42.42} = 1.01\ \text{hp/ton}$$

The coefficient of performance

$$= \frac{h_c - h_a}{h_{d'} - h_{c'}} = \frac{49.33}{10.60} = 4.65$$

In summary, when superheating of the vapor occurs without producing useful cooling, the volume flow rate of vapor per unit capacity, the power required per unit capacity, and the quantity of heat rejected at the condenser per unit capacity are all greater for the superheated cycle than for the saturated cycle. This means that the compressor, the compressor driver, and the condenser must all be larger for the superheated cycle than for the saturated cycle.

8.4. Superheating That Produces Useful Cooling

Assume, now, that all of the heat taken in by the suction vapor produces useful cooling. When this is true, the refrigerating effect per unit mass is greater by an amount equal to the amount of superheat. In Fig. 8-2, assuming that the superheating produces useful cooling, the refrigerating effect per pound for the superheated cycle is equal to

$$h_{c'} - h_a = 88.60 - 31.16 = 57.44\ \text{Btu/lb}$$

Since the refrigerating effect per unit mass is greater for the superheated cycle than for the saturated cycle, the mass flow rate of refrigerant per unit capacity is less for the superheated cycle than for the saturated cycle. Whereas the mass of refrigerant circulated per minute per ton for the saturated cycle is 4.05, the mass flow rate for the superheated cycle is

$$\frac{200}{h_{c'} - h_a} = \frac{200}{57.44} = 3.48\ \text{lb/min ton}$$

Even though the specific volume of the suction vapor and the heat of compression per unit mass are both greater for the superheated cycle than for the saturated cycle, the volume of vapor compressed per unit capacity and the power required per unit capacity are both less for the superheated cycle than for the saturated cycle. This is because of the reduction in the mass flow rate. The volume of vapor compressed per minute per ton and the power required per ton for the saturated cycle are 4.55 cu ft and 0.965 hp, respectively, whereas for the superheated cycle,

the volume of vapor compressed per minute per ton V

$$= m \times v_{c'} = 3.48 \times 1.260 = 4.38 \text{ cu ft/min ton}$$

The power required per ton

$$= \frac{m(h_{d'} - h_{c'})}{42.42} = \frac{3.48 \times 10.60}{42.42} = 0.870 \text{ hp/ton}$$

For the superheated cycle, both the refrigerating effect per pound and the heat of compression per pound are greater than for the saturated cycle. However, since the increase in the refrigerating effect is greater proportionally than the increase in the heat of compression, the coefficient of performance for the superheated cycle is higher than that of the saturated cycle. For the saturated cycle, the coefficient of performance is 4.88, whereas for the superheated cycle,

$$\text{the coefficient of performance} = \frac{(h_{c'} - h_a)}{(h_{d'} - h_{c'})} = \frac{57.44}{10.60} = 5.42$$

It will be shown in the following sections that the superheating of the suction vapor in an actual cycle usually occurs in such a way that a part of the heat taken in by the vapor in becoming superheated is absorbed from the refrigerated space and produces useful cooling, whereas another part is absorbed by the vapor after the vapor leaves the refrigerated space and therefore produces no useful cooling. The portion of the superheat which produces useful cooling will depend upon the individual application, and the effect of the superheating on the cycle will vary approximately in proportion to the useful cooling accomplished.

Regardless of the effect on capacity, except in some few special cases, a certain amount of superheating is usually unavoidable and, in most cases, desirable. When the suction vapor is drawn directly from the evaporator into the suction inlet of the compressor without at least a small amount of superheating, there is a good possibility that small particles of unvaporized liquid will be entrained in the vapor. Such a vapor is called a "wet" vapor. It will be shown later that "wet" suction vapor drawn into the cylinder of the compressor adversely affects the capacity of the compressor. Furthermore, since refrigeration compressors are designed as vapor pumps, if any appreciable amount of unvaporized liquid is allowed to enter the compressor from the suction line, serious mechanical damage to the compressor may result. Since superheating the suction vapor eliminates the possibility of "wet" suction vapor reaching the compressor inlet, a certain amount of superheating is usually desirable. Again, the extent to which the suction vapor should be allowed to become superheated in any particular instance depends upon where and how the superheating occurs and upon the refrigerant used.

Superheating of the suction vapor may take place in any one or in any combination of the following places:

1. In the end of the evaporator
2. In the suction piping installed inside the refrigerated space (usually referred to as a "drier loop")
3. In the suction piping located outside of the refrigerated space
4. In a liquid-suction heat exchanger.

8-5. Superheating in Suction Piping Outside the Refrigerated Space

When the cool refrigerant vapor from the evaporator is allowed to become superheated while flowing through suction piping located outside of the refrigerated space, the heat taken in by the vapor is absorbed from the surrounding air and no useful cooling results. It has already been demonstrated that superheating of the suction vapor which produces no useful cooling tends to reduce the efficiency of the cycle. Obviously, then, superheating of the vapor in the suction line outside of the refrigerated space should be eliminated whenever practical.

Superheating of the suction vapor in the suction line can be prevented for the most part by insulating the suction line. Whether or not the loss of cycle efficiency in any particular application is sufficient to warrant the additional expense of insulating the suction line depends primarily on the size of the system and on the operating suction temperature.

When the suction temperature is relatively high (35°F or 40°F), the amount of superheating will usually be small and the effect on the efficiency of the cycle will be negligible. The reverse is true, however, when the suction temperature is low. The amount of superheating is apt to be quite large.

Too, at low suction temperatures, when the efficiency of the cycle is already very low, each degree of superheat will cause a greater reduction in cycle efficiency percentagewise than when the suction temperature is high. It becomes immediately apparent that any appreciable amount of superheating in the suction line of systems operating at low suction temperatures will cause a significant reduction in the efficiency of the cycle and that, under these conditions, insulating of the suction line is necessary if the efficiency of the cycle is to be maintained at a reasonable level.

Aside from any considerations of capacity, even at the higher suction temperatures, insulating of the suction line is often required to prevent frosting or sweating of the suction line. In flowing through the suction piping, the cold suction vapor will usually lower the temperature of the piping below the dew point temperature of the surrounding air so that moisture will condense out of the air onto the surface of the piping, causing the suction piping to either frost or sweat, depending upon whether or not the temperature of the piping is below the freezing temperature of water. In any event, frosting or sweating of the suction piping is usually undesirable and should be eliminated by insulating the piping.

8-6. Superheating the Vapor Inside the Refrigerated Space

Superheating of the suction vapor inside the refrigerated space can take place either in the end of the evaporator or in suction piping located inside the refrigerated space, or both.

To assure the proper operation of the refrigerant control and to prevent liquid refrigerant from overflowing the evaporator and being carried back to the compressor, when certain types of refrigerant controls are used, it is necessary to adjust the control so that the liquid is completely evaporated before it reaches the end of the evaporator. In such cases, the cold vapor will continue to absorb heat and become superheated as it flows through the latter portion of the evaporator. Since the heat to superheat the vapor is drawn from the refrigerated space, useful cooling results and the refrigerating effect per unit mass of refrigerant is increased by an amount equal to the amount of heat absorbed in the superheating.

It has been shown that when the superheating of the suction vapor produces useful cooling the efficiency of the cycle is improved somewhat.* However, in spite of the increase in cycle efficiency, it must be emphasized that superheating the suction vapor in the evaporator is not economical and should always be limited to only that amount which is necessary to the proper operation of the refrigerant control. Since the transfer of heat through the walls of the evaporator per degree of temperature difference is not as great to a vapor as to a liquid, the capacity of the evaporator is always reduced in any portion of the evaporator where only vapor exists. Therefore, excessive superheating of the suction vapor in the evaporator will reduce the capacity of the evaporator unnecessarily and will require either that the evaporator be operated at a lower vaporizing

* Although this is true for systems using R-12 as a refrigerant, it will be shown later that this is not true for all refrigerants.

temperature or that a larger evaporator be used in order to provide the desired evaporator capacity. Since the space available for evaporator installation is often limited and since evaporator surface is expensive, the use of a larger evaporator is not usually practical. Because of the effect on cycle efficiency, the undesirability of lowering the vaporizing temperature is obvious.

Often, a certain amount of suction piping, usually called a drier loop, is installed inside the refrigerated space for the express purpose of superheating the suction vapor (Fig. 8-3). Use of a drier loop permits more complete flooding of the evaporator with liquid refrigerant without the danger of the liquid overflowing into the suction line and being drawn into the compressor. This not only provides a means of superheating the suction vapor inside the refrigerated space so that the efficiency of the cycle is increased without the sacrifice of expensive evaporator surface, but it actually makes possible more effective use of the existing evaporator surface. Also, in some instances, particularly where the suction temperature is high and the relative humidity of the outside air is reasonably low, superheating of the suction vapor inside the refrigerated space will raise the temperature of the suction piping and prevent the formation of moisture, thereby eliminating the need for suction line insulation. It should be noted, however, that the extent to which the suction vapor can be superheated inside the refrigerated space is limited by the space temperature. Ordinarily, if sufficient piping is used, the suction vapor can be heated to within 4°F to 5°F of the space temperature. Thus, for a 40°F space temperature, the suction vapor may be superheated to approximately 35°F.

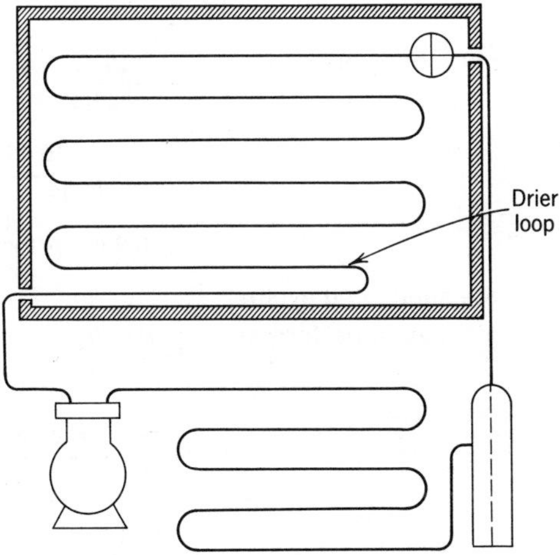

Fig. 8-3 Flow diagram showing drier loop for superheating suction vapor inside refrigerated space.

8-7. The Effects of Subcooling the Liquid

On the *ph* diagram in Fig. 8-4, a simple saturated cycle is compared to one in which the liquid is subcooled from 100°F to 80°F before it reaches the refrigerant control. Points *A*, *B*, *C*, *D*, and *E* designate the simple saturated cycle, whereas points *A′*, *B′*, *C*, *D*, and *E* designate the subcooled cycle.

It has been shown (Section 6-22) that when the liquid is subcooled before it reaches the refrigerant control the refrigerating effect per unit mass is increased. In Fig. 8-4, the increase in the refrigerating effect per pound resulting from the subcooling is the difference between h_b and $h_{b'}$, and is exactly equal to the difference between h_a and $h_{a'}$, which represents the heat removed per pound of liquid during the subcooling.

For the saturated cycle, the refrigerating effect per pound, q_e

$$= h_c - h_a = 80.49 - 31.16 = 49.33 \text{ Btu/lb}$$

For the subcooled cycle, the refrigerating effect per pound, q_e

$$= h_c - h_{a'} = 80.49 - 26.28 = 54.21 \text{ Btu/lb}$$

Because of the greater refrigerating effect per unit mass, the mass flow rate of refrigerant per unit capacity is less for the subcooled cycle than for the saturated cycle.

For the saturated cycle, the mass of refrigerant circulated per minute per ton m

$$= \frac{200}{49.33} = 4.05 \text{ lb}$$

For the subcooled cycle, the mass flow rate per ton m

$$= \frac{200}{54.21} = 3.69 \text{ lb}$$

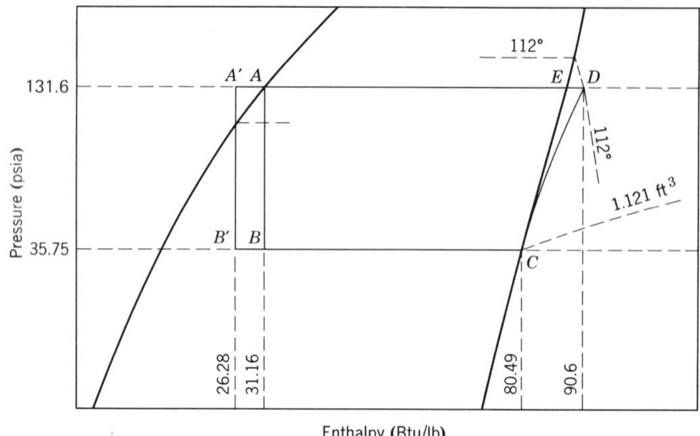

Fig. 8-4 *ph* diagrams comparing the subcooled cycle to the simple saturated cycle. (Refrigerant-12.)

Notice that the condition of the refrigerant vapor entering the suction inlet of the compressor is the same for both cycles. For this reason, the specific volume of the vapor entering the compressor will be the same for both the saturated and subcooled cycles and, since the mass flow rate per unit capacity is less for the subcooled cycle than for the saturated cycle, it follows that the volume of vapor which the compressor must handle per unit capacity will also be less for the subcooled cycle than for the saturated cycle.

For the saturated cycle, the specific volume of the suction vapor v_c $= 1.121$ ft³/lb

The volume of vapor compressed per minute per ton V $= (m)(v_c)$ $= 4.05 \times 1.121$ $= 4.55$ ft³/min

For the subcooled cycle, the specific volume of the suction vapor v_c $= 1.121$ ft³/lb

The volume of vapor compressed per minute per ton V $= (m)(v_c)$ $= 3.69 \times 1.121$ $= 4.14$ ft³/min

Because the volume of vapor compressed per unit capacity is less for the subcooled cycle, the compressor displacement required for the subcooled cycle is smaller than that required for the saturated cycle.

Notice also that the heat of compression per unit mass is the same for both the saturated and subcooled cycles. This means that the increase in refrigerating effect per unit mass resulting from the subcooling is accomplished without increasing the energy input to the compressor. Any change in the refrigerating cycle which increases the quantity of heat absorbed in the refrigerated space without causing an increase in the energy input to the compressor will increase the c.o.p. of the cycle and reduce the power required per unit capacity.

For the saturated cycle, the coefficient of performance

$$= \frac{h_c - h_a}{h_d - h_c} = \frac{80.49 - 31.16}{90.60 - 80.49} = 4.88$$

The horsepower per ton

$$= \frac{m(h_d - h_c)}{42.42} = \frac{4.05 \times 10.11}{42.42} = 0.965 \text{ hp/ton}$$

For the subcooled cycle, the coefficient of performance

$$= \frac{h_c - h_{a'}}{h_d - h_c} = \frac{80.49 - 26.28}{90.60 - 80.49} = \frac{54.21}{10.11} = 5.36$$

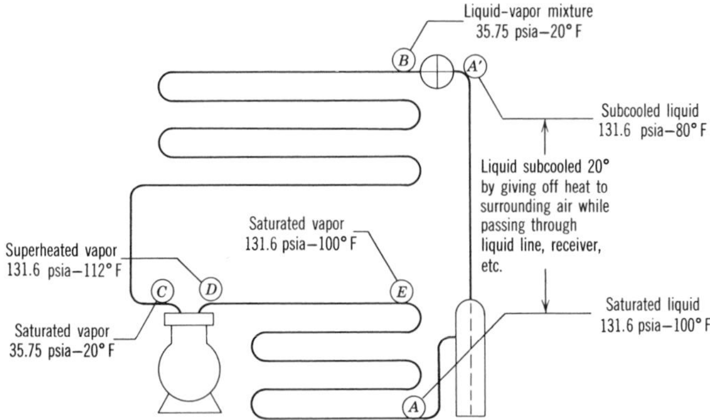

Fig. 8-5 Flow diagram illustrating subcooling of the liquid in the liquid line. (Refrigerant-12.)

$$\text{The horsepower per ton} = \frac{m(h_d - h_c)}{42.42}$$

$$= \frac{3.69 \times 10.11}{42.42}$$

$$= 0.879 \text{ hp/ton}$$

In this instance, the c.o.p. of the subcooled cycle is greater than that of the saturated cycle by

$$\frac{5.36 - 4.88}{4.88} \times 100 = 9.8\%$$

Subcooling of the liquid refrigerant can and does occur in several places and in several ways. Very often the liquid refrigerant becomes subcooled while stored in the liquid receiver tank or while passing through the liquid line by giving off heat to the surrounding air (Fig. 8-5). In some cases, a special liquid subcooler is used to subcool the liquid (Fig. 8-6). The gain in system capacity and efficiency resulting from the liquid subcooling is very often more than sufficient to offset the additional cost of the subcooler, particularly for low temperature applications.

Where a water-cooled condenser is used, the liquid subcooler may be piped either in series or in parallel with the condenser. When the subcooler is piped in series with the condenser, the cooling water passes through the subcooler first and then through the condenser, thereby bringing the coldest water into contact with the liquid being subcooled (Fig. 8-6). There is some doubt about the value of a subcooler piped in series with the condenser. Since the cooling water is warmed by the heat absorbed in the subcooler, it reaches the condenser at a higher temperature and the condensing temperature of the cycle is increased. Hence the increase in system efficiency resulting from the subcooling is offset to some extent by the rise in the condensing temperature.

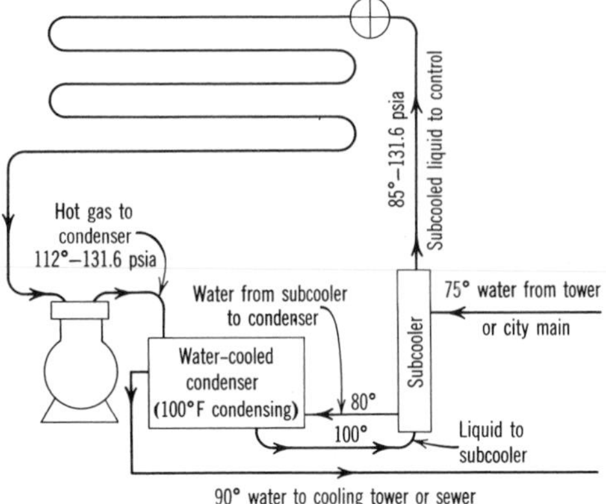

Fig. 8-6 Flow diagram illustrating subcooler piped in series with condenser.

When the subcooler is piped in parallel with the condenser (Fig. 8-7), the temperature of the water reaching the condenser is not affected by the subcooler. However, for either series or parallel piping, the size of the condenser water pump must be increased when a subcooler is added. If this is not done, the quantity of water

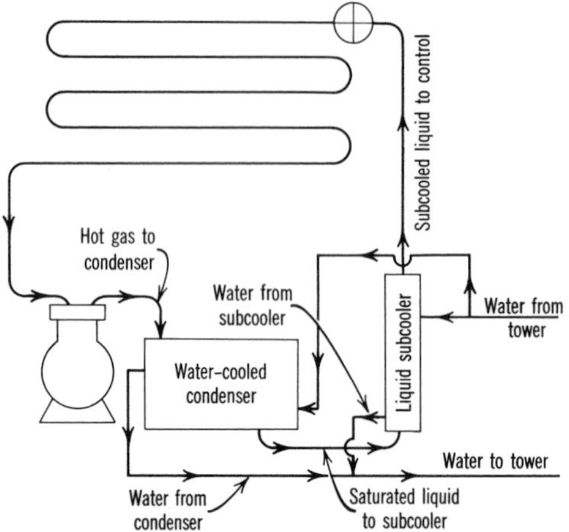

Fig. 8-7 Flow diagram showing parallel piping for condenser and subcooler.

circulated through the condenser will be diminished by the addition of the subcooler and the condensing temperature of the cycle will be increased, thus nullifying any benefit accruing from the subcooling.

When liquid subcoolers are employed in conjunction with air-cooled condensers, the subcooler is usually an intergal part of the condenser and the liquid is subcooled by giving up heat to the air circulated over the condenser.

8-8. Liquid-Suction Heat Exchangers

Another method of subcooling the liquid is to bring about an exchange of heat between the liquid and the cold suction vapor going back to the compressor. In a liquid-suction heat exchanger, the cold suction vapor is piped through the heat exchanger in counterflow to the warm liquid refrigerant flowing through the liquid line to the refrigerant control (Fig. 8-8). In flowing through the heat exchanger the cold suction vapor absorbs heat from the warm liquid so that the liquid is subcooled as the vapor is superheated, and, since the heat absorbed by the vapor in becoming superheated is drawn from the liquid, the heat of the liquid is diminished by an amount equal to the amount of heat taken in by the vapor. In each of the methods of subcooling discussed thus far, the heat given up by the liquid in becoming subcooled is given up to some medium external to the system and the heat then leaves the system. When a liquid-suction heat exchanger is used, the heat given up by the liquid in

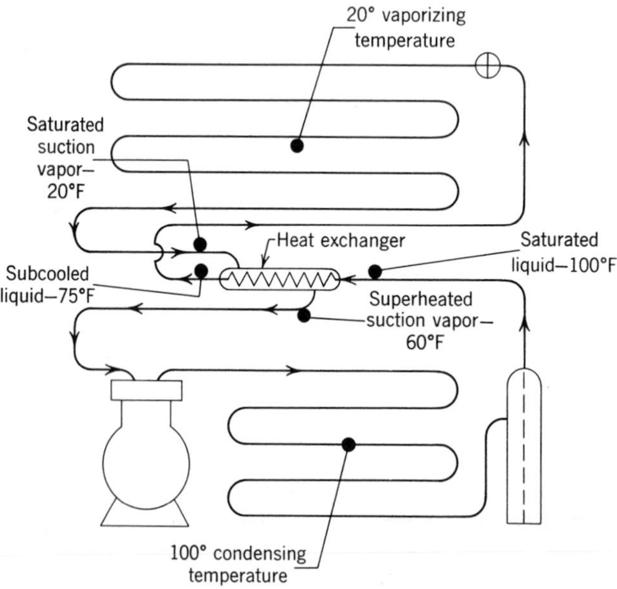

Fig. 8-8 Flow diagram of refrigeration cycle illustrating the use of a liquid-suction heat exchanger.

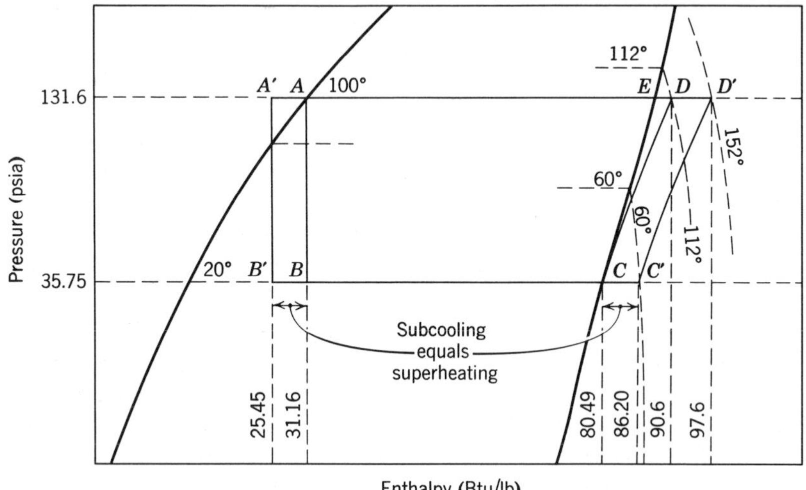

Fig. 8-9 *ph* Diagrams comparing simple saturated cycle to cycle employing a liquid-suction heat exchanger. The amount of subcooling is equal to the amount of superheating. (Refrigerant-12.)

becoming subcooled is absorbed by the suction vapor and remains in the system.

On the *ph* diagram in Fig. 8-9, a simple saturated cycle is compared to one in which a liquid-suction heat exchanger is employed. Points *A*, *B*, *C*, *D*, and *E* identify the saturated cycle and points *A*′, *B*′, *C*′, *D*′, *E* identify the cycle in which the heat exchanger is used. In the latter cycle, it is assumed that the suction vapor is superheated from 20°F to 60°F in the heat exchanger.

The heat absorbed per pound of vapor in the heat exchanger is

$$h_{c'} - h_c = 86.20 - 80.49 = 5.71 \text{ Btu/lb}$$

Since the heat given up by the liquid in the heat exchanger in becoming subcooled is exactly equal to the heat absorbed by the vapor in becoming superheated, $h_a - h_{a'}$ is equal to $h_{c'} - h_c$ and therefore is also equal to 5.71 Btu/lb. Since $h_a - h_{a'}$ represents an increase in the refrigerating effect, the refrigerating effect per pound for the heat exchanger cycle is

$$h_c - h_{a'} = 80.49 - 25.45 = 55.04$$

The heat of compression per pound for the heat exchanger cycle is

$$h_{d'} - h_{c'} = 97.60 - 86.20 = 11.40$$

Therefore, the coefficient of performance is

$$\frac{h_c - h_{a'}}{h_{d'} - h_{c'}} = \frac{55.04}{11.40} = 4.83$$

The coefficient of performance of the saturated cycle is 4.88. Therefore, the coefficient of performance of the heat exchanger cycle differs from than that of the saturated cycle by only

$$\frac{4.88 - 4.83}{4.88} \times 100 = 1.0\%$$

Depending upon the particular case, the coefficient of performance of a cycle employing a heat exchanger may be either greater than, less than, or the same as that of a saturated cycle operating between the same pressure limits. In any event, the difference is negligible, and it is evident that the advantages accruing from the subcooling of the liquid in the heat exchanger are approximately offset by the disadvantages of superheating the vapor. Theoretically, then, the use of a heat exchanger cannot be justified on the basis of an increase in system capacity and efficiency. However, since in actual practice a refrigerating system does not (cannot) operate on a simple saturated cycle, this does not represent a true appraisal of the practical value of the heat exchanger.

In an actual cycle, the suction vapor will always become superheated before the compression process begins because nothing can be done to prevent it. This is true even if no superheating takes place either in the evaporator or in the suction line and the vapor reaches the inlet of the compressor at the vaporizing temperature. As the cold suction vapor flows into the compressor, it will become superheated by absorbing heat from the hot cylinder walls. Since the superheating in the compressor cylinder will occur before the compression process begins, the effect of the superheating on cycle efficiency will be approximately the same as if the superheating occurred in the suction line without producing useful cooling.*

The disadvantages resulting from allowing the suction to become superheated without producing useful cooling have already been pointed out. Obviously, then, since superheating of the suction vapor is unavoidable in an actual cycle, whether or not a heat exchanger is used, any practical means of causing the vapor to become superheated in such a way that useful cooling results are worthwhile. Hence, the value of a heat exchanger lies in the fact that it provides a method of superheating the vapor so that useful cooling results. For this reason, the effect of a heat exchanger on cycle efficiency can be evaluated only by comparing the heat exchanger cycle to one in which the vapor is superheated without producing useful cooling.

The maximum amount of heat exchange which can take place between the liquid and the vapor in the heat exchanger depends on the initial temperatures of the liquid and the vapor as they enter the heat exchanger and on the length of time they are in contact with each other.

The greater the difference in temperature, the greater is the exchange of heat for any given period of contact. Thus, the lower the vaporizing temperature and the higher the condensing temperature, the greater is the possible heat exchange. Theoretically, if the two fluids remained in contact for a sufficient length of time, they would leave the heat exchanger at the same temperature. In practice, this is not possible. However, the longer the two fluids stay in contact, the more closely the two temperatures will approach one another. Since the specific heat of the vapor is less than that of the liquid, the rise in the temperature of the vapor is always greater than the reduction in the temperature of the liquid. For instance, the specific heat of R-12 liquid is approximately 0.24 Btu/lb, whereas the specific heat of the vapor is 0.15 Btu/lb. This means that the temperature reduction of the liquid will be approximately 62% (0.15/0.24) of the rise in the temperature of the vapor, or that for each 24°F rise in the temperature of the vapor, the temperature of the liquid will be reduced 15°F.

For the heat exchanger cycle in Fig. 8-9, the vapor absorbs 5.71 Btu/lb in superheating from 20°F to 60°F. Assuming that all of the superheating takes place in the heat exchanger, the heat given up by the liquid is 5.71 Btu, so that the temperature of the liquid is reduced 23.8°F (5.71/0.24) as the liquid passes through the heat exchanger.

8-9. The Effect of Pressure Losses Resulting from Friction

In overcoming friction, both internal (within the fluid) and external (surface), the refrigerant experiences a drop in pressure while flowing through the piping, evaporator, condenser, receiver, and through the valves and passages of the compressor (Fig. 8-10).

A *ph* diagram of an actual cycle, illustrating the loss in pressure occurring in the various parts of the system, is shown in Fig. 8-11. To simplify the diagram, no superheating or subcooling is shown and a simple saturated cycle is drawn in for comparison.

* It will be shown later that some advantages accrue from superheating which takes place in the compressor: (1) When the suction vapor absorbs heat from the cylinder walls, the cylinder wall temperature is lowered somewhat and this brings about a desirable change in the path of the compression process. However, the change is slight and is difficult to evaluate. (2) When hermetic motor-compressor assemblies are used, the suction vapor should reach the compressor at a relatively low temperature in order to help cool the motor windings.

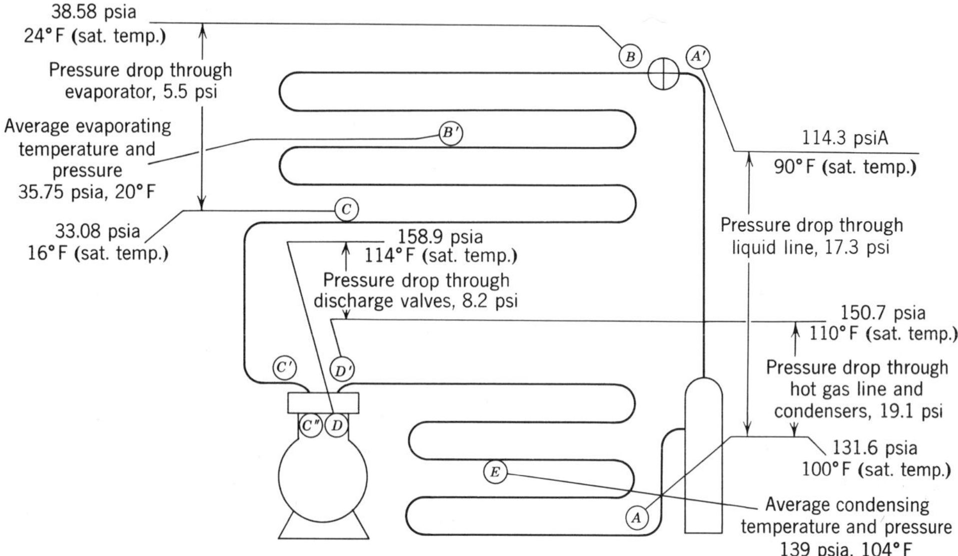

Fig. 8-10 Flow diagram illustrating the effect of pressure drop in various parts of the system. Pressure drops are exaggerated for clarity. (Refrigerant-12.)

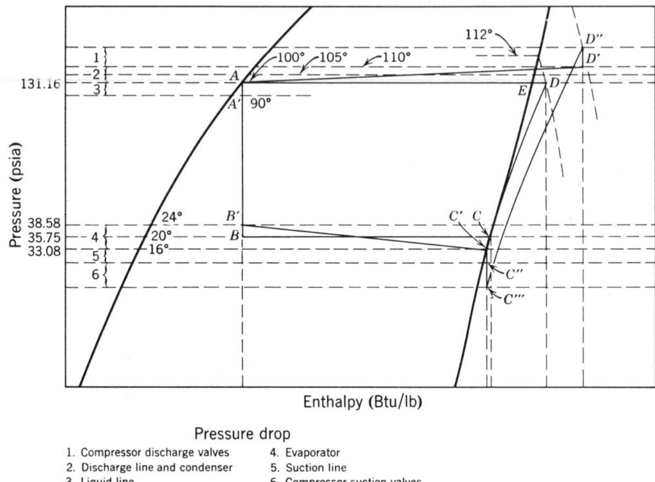

Fig. 8-11 *ph* diagram of refrigeration cycle illustrating the effect of pressure losses in the various parts of the system. A simple saturated cycle is drawn in for comparison. (Refrigerant-12.)

Line *B′–C′* represents the vaporizing process in the evaporator during which the refrigerant undergoes a drop in pressure of 5.5 psi. Whereas the pressure and saturation temperatures of the liquid-vapor mixture at the evaporator inlet are 38.58 psia and 24°F, respectively, the pressure of the saturated vapor leaving the evaporator is 33.08 psia, corresponding to a saturation temperature of 16°F. The average vaporizing temperature in the evaporator is 20°F, the same as that of the saturated cycle.

As a result of the drop in pressure in the evaporator, the vapor leaves the evaporator at a lower pressure and saturation temperature and with a greater specific volume than if no drop in pressure occurred.

The refrigerating effect per unit mass and the mass flow rate of refrigerant required per unit capacity are approximately the same for both cycles, but because of the greater specific volume the volume flow rate of vapor handled by the compressor per unit capacity is greater for the cycle experiencing the pressure drop. Also because of the lower pressure of the vapor leaving the evaporator, the vapor must be compressed through a greater pressure range during the compression process, so that the power required per unit capacity is also greater for the cycle undergoing the drop in pressure.

Line $C'-C''$ represents the drop in pressure experienced by the suction vapor in flowing through the suction line from the evaporator to the compressor inlet. Like pressure drop in the evaporator, pressure drop in the suction line causes the suction vapor to reach the compressor at a lower pressure and in an expanded condition so that the volume flow rate of vapor per unit capacity and the power required per unit capacity are both increased.

It is evident that the drop in pressure both in the evaporator and in the suction line should be kept to a minimum in order to obtain the best possible cycle efficiency. This applies also to heat exchangers or any other auxiliary device intended for installation in the suction line.

In Fig. 8-11, the pressure drops are exaggerated for clarity. Ordinarily, good evaporator design limits the pressure drop across the evaporator to 2 or 3 psi. Ideally, the suction line should be designed so that the pressure drop does not cause more than a 1°F or 2°F drop in saturation temperature.

Line $C''-C'''$ represents the drop in pressure that the suction vapor undergoes in flowing through the suction valves and passages of the compressor into the cylinder. The result of the drop in pressure through the valves and passages on the suction side of the compressor is the same as if the drop occurred in the suction line, and the effect on cycle efficiency is the same. Here again, good design requires that the drop in pressure be kept to a practical minimum.

Line $C'''-D''$ represents the compression process for the cycle undergoing the pressure drops. Notice that the vapor in the cylinder is compressed to a pressure considerably above the average condensing pressure. It is shown later that this is necessary in order to force the vapor out of the cylinder through the discharge valves against the condensing pressure and against the additional pressure occasioned by the spring-loading of the discharge valves.

Line $D''-D'$ represents the drop in pressure required to force the discharge valves open against the spring-loading and to force the vapor out through the discharge valves and passages of the compressor into the discharge line.

Line $D'-A$ represents the drop in pressure resulting from the flow of the refrigerant through the discharge line and condenser. That part of line $D'-A$ which represents the flow through the discharge line will vary with the particular case, since the discharge line may be either quite long or very short, depending upon the application. In any event, the result of the pressure drop will be the same. Any drop in pressure occurring on the discharge side of the compressor (in the discharge valves and passages, in the discharge line, and in the condenser) will have the effect of raising the discharge pressure and thereby increasing the work of compression and the horsepower per ton.

Line $A-A'$ represents the pressure drop resulting from the flow of the refrigerant through the receiver tank and liquid line. Since the refrigerant at A is a saturated liquid, the temperature of the liquid must decrease as the pressure decreases. If the liquid is not subcooled by giving up heat to an external sink as its pressure drops, a portion of the liquid must flash into a vapor in the liquid line in order to provide the required cooling of the liquid. Notice that point A' lies in the region of phase-change, indicating that a portion of the refrigerant is a vapor at this point.

Despite the flashing of the liquid and the drop in temperature coincident with the drop in pressure in the liquid line, the drop in pressure in the liquid line has no effect on cycle efficiency. The pressure and temperature of the liquid must be reduced to the vaporizing condition before it enters the evaporator in any case. The fact that a part of this takes place in the liquid line rather than in the refrigerant control has no direct effect on the efficiency of the system. It does, however, reduce the capacity of both the liquid line and the refrigerant control. Furthermore, passage of vapor through the refrigerant control will eventually cause damage to the refrigerant control by eroding the valve needle and seat.

Ordinarily, even without the use of a heat exchanger, sufficient subcooling of the liquid will occur in the liquid line to prevent the flashing of the liquid if the drop in pressure in the line is not excessive. Flashing of the liquid in the liquid line will usually not take place when the drop in the line does not exceed 5 psi.

The effect of pressure drop in the lines and in the other parts of the system is discussed more fully later in the appropriate chapters.

A *ph* diagram of a typical refrigeration cycle, which illustrates the combined effects of pressure drop, subcooling, and superheating, is compared to the *ph* diagram of the simple saturated cycle in Fig. 8-12.

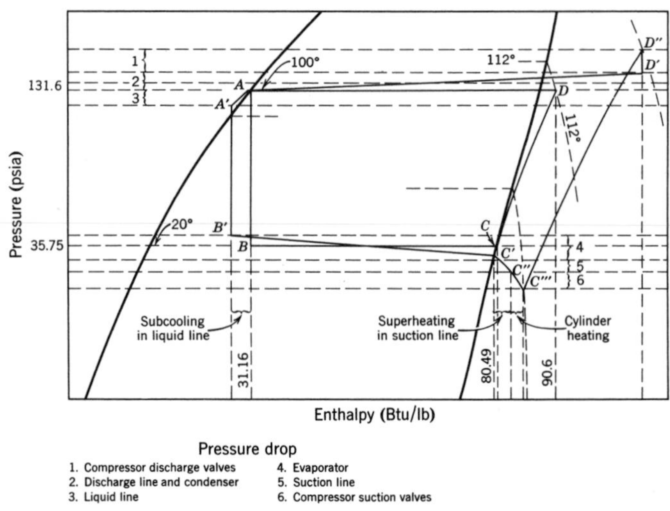

Fig. 8-12 *ph* diagram of actual refrigeration cycle illustrating effects of subcooling, superheating, and losses in pressure. A simple saturated cycle is drawn in for comparison. (Refrigerant-12).

9

SURVEY OF REFRIGERATION APPLICATIONS

9-1. History and Scope of the Industry

In the early days of mechanical refrigeration, the equipment available was bulky, expensive, and not too efficient. Also it was of such a nature as to require that a mechanic or operating engineer be on duty at all times. This limited the use of mechanical refrigeration to a few large applications such as ice plants, meat packing plants, and large storage warehouses.

In the span of only a few decades refrigeration has grown into the giant and rapidly expanding industry that it is today. This explosive growth came about as the result of several factors. First, with the development of precision manufacturing methods, it became possible to produce smaller, more efficient equipment. This, along with the development of "safe" refrigerants and the invention of the fractional horsepower electric motor, made possible the small refrigerating unit which is so widely used at the present time in such applications as domestic refrigerators and freezers, small air conditioners, and commercial fixtures. Today, there are few homes or business establishments in the United States that cannot boast of one or more mechanical refrigeration units of some sort.

Few people outside of those directly connected with the industry are aware of the significant part that refrigeration has played in the development of the highly technical society that America is today, nor do they realize the extent to which such a society is dependent upon mechanical refrigeration for its very existence. It would not be possible, for instance, to preserve food in sufficient quantities to feed the growing urban population without mechanical refrigeration. Too, many of the large buildings which house much of the nation's business and industry would become untenable in the summer months because of the heat if they were not air conditioned with mechanical refrigerating equipment.

In addition to the better known applications of refrigeration, such as comfort air conditioning and the processing, freezing, storage, transportation, and display of perishable products, mechanical refrigeration is used in the processing or manufacturing of almost every article or commodity on the market today. The list of processes or products made possible or improved through the use of mechanical refrigeration is almost endless. For example, refrigeration has made possible the building of huge dams which are vital to large-scale reclamation and hydroelectric projects. It has made possible the construction of roads and tunnels and the sinking of foundation and mining shafts through and across unstable ground formations. It has made possible the production of plastics, synthetic rubber, and many other new and useful materials and products. Because of mechanical refrigeration, bakers can get more loaves of bread from a barrel of flour, textile and paper manufacturers can speed up their machines and get more production, and better methods of hardening steels for machine tools are available. These represent only a few of the hundreds of ways in which mechanical refrigeration is now being used and many new uses are being found each year. In fact, the only thing slowing the growth of the refrigeration industry at the present time is the lack of an adequate supply of trained technical manpower.

9-2. Classification of Applications

For convenience of study, refrigeration applications may be grouped into six general categories: (1) domestic refrigeration, (2) commercial refrigeration, (3) industrial refrigeration, (4) marine and transportation refrigeration, (5) comfort air conditioning, and (6) industrial air conditioning. It will be apparent in the discussion which follows that the exact limits of these areas are not precisely defined and that there is considerable overlapping between the several areas.

9-3. Domestic Refrigeration

Domestic refrigeration is rather limited in scope, being concerned primarily with household refrigerators and home freezers. However, because the number of units in service is quite large, domestic refrigeration represents a significant portion of the refrigeration industry.

Domestic units are usually small in size, having horsepower ratings of between $\frac{1}{20}$ and $\frac{1}{2}$ hp, and are of the hermetically sealed type. Since these applications are familiar to everyone, they will not be described further here. However, the problems encountered in the design and maintenance of these units are discussed in appropriate places in the chapters which follow.

9-4. Commercial Refrigeration

Commercial refrigeration is concerned with the designing, installation, and maintenance of refrigerated fixtures of the type used by retail stores, restaurants, hotels, and institutions for the storing, displaying, processing, and dispensing of perishable commodities of all types. Commercial refrigeration fixtures are described in more detail later in this chapter.

9-5. Industrial Refrigeration

Industrial refrigeration is often confused with commercial refrigeration because the division between these two areas is not clearly defined. As a general rule, industrial applications are larger in size than commercial applications and have the distinguishing feature of requiring an attendant on duty, usually a licensed operating engineer. Typical industrial applications are ice plants, large food-packing plants (meat, fish, poultry, frozen foods, etc.), breweries, creameries, and industrial plants, such as oil refineries, chemical plants, rubber plants, etc. Industrial refrigeration includes also those applications concerned with the construction industry as described in Section 9-1.

9-6. Marine and Transportation Refrigeration

Applications falling into this category could be listed partly under commercial refrigeration and partly under industrial refrigeration. However, both these areas of specialization have grown to sufficient size to warrant special mention.

Marine refrigeration, of course, refers to refrigeration aboard marine vessels and includes, for example, refrigeration for fishing boats and for vessels transporting perishable cargo as well as refrigeration for the ship's stores on vessels of all kinds.

Transportation refrigeration is concerned with refrigeration equipment as it is applied to trucks, both long distance transports and local delivery, and to refrigerated railway cars. Typical refrigerated truck bodies are shown in Fig. 11-5.

9-7. Air Conditioning

As the name implies, air conditioning is concerned with the condition of the air in some designated area or space. This usually involves control not only of the space temperature but also of space humidity and air motion, along with the filtering and cleaning of the air.

Air conditioning applications are of two types, either comfort or industrial, according to their purpose. Any air conditioning which has as its primary function the conditioning of air for human comfort is called comfort air conditioning. Typical installations of comfort air conditioning are in homes, schools, offices, churches, hotels, retail stores, public buildings, factories, automobiles, buses, trains, planes, ships, etc.

On the other hand, any air conditioning which does not have as its primary purpose the conditioning of air for human comfort is called industrial air conditioning. This does not necessarily mean that industrial air conditioning systems cannot serve also as comfort air conditioning coincidentally with their primary function. Often this is the case, although not always so.

The applications of industrial air conditioning are almost without limit both in number and in variety. Generally speaking, the functions of industrial air conditioning systems are to (1) control the moisture content of hydroscopic materials; (2) govern the rate of chemical and biochemical reactions; (3) limit the variations in the size of precision manufactured articles because of thermal expansion and contraction; and (4) provide clean, filtered air which is often essential to trouble-free operation and to the production of quality products.

9-8. Food Preservation

The preservation of perishable commodities, particularly foodstuffs, is one of the most common uses of mechanical refrigeration. As such, it is a subject which should be given consideration in any comprehensive study of refrigeration.

At the present time, food preservation is more important than ever before in man's history. Today's large urban populations require tremendous quantities of food, which for the most part must be produced and processed in outlying areas. Naturally, these foodstuffs must be kept in a preserved condition during transit and subsequent storage until they are finally consumed. This may be a matter of hours, days, weeks, months, or even years in some cases. Too, many products, particularly fruit and vegetables, are seasonal. Since they are produced only during certain seasons of the year, they must be stored and preserved if they are to be made available the year round.

As a matter of life or death, the preservation of food has long been one of our most pressing problems. Almost from the very beginning of our existence on earth, it became necessary for us to find ways of preserving food during seasons of abundance in order to live through seasons of scarcity. It is only natural, then, that man should discover and develop such methods of food preservation as drying, smoking, pickling, and salting long before he had any knowledge of the causes of food spoilage. These rather primitive methods are still widely used today, not only in backward societies where no other means are available but also in the most modern societies where they serve to supplement the more modern methods of food preservation. For instance, millions of pounds of dehydrated (dried) fruit, milk, eggs, fish, meat, potatoes, etc., are consumed in the United States each year, along with huge quantities of smoked, pickled, and salted products, such as ham, bacon, and sausage, to name only a few. However, although these older methods are entirely adequate for the preservation of certain types of food, and often produce very unusual and tasty products which would not otherwise be available, they nonetheless have inherent disadvantages which limit their usefulness. Since by their very nature they usually bring about rather severe changes in appearance and taste, which in many cases are objectionable, they are not universally adaptable for the preservation of all types of food products. Furthermore, the keeping qualities of food preserved by such methods are limited as to time. Where a product is to be preserved indefinitely or for a long period of time, some other means of preservation ordinarily must be utilized.

The invention of the microscope and the subsequent discovery of microorganisms as a major cause of food spoilage led to the development of canning in France during the time of

Napoleon. With the invention of canning, man found a way to preserve food of all kinds in large quantities and for indefinite periods of time. Canned foods have the advantage of being almost entirely imperishable, easily processed, and convenient to handle and store. Today, more food is preserved by canning than by all other methods combined. The one big disadvantage of canning is that canned foods must be heat-sterilized, which frequently results in overcooking. Hence, although canned foods often have a distinctive and delicious flavor all their own, they usually differ greatly from the original fresh product.

The only means of preserving food in its original fresh state is by refrigeration. This, of course, is the principal advantage that refrigeration has over other methods of food preservation. However, refrigeration too has its disadvantages. For instance, when food is to be preserved by refrigeration, the refrigerating process must begin very soon after harvesting or killing and must be continuous until the food is finally consumed. Since this requires relatively expensive and bulky equipment, it is often both inconvenient and uneconomical.

Obviously, then, there is no one method of food preservation which is best in all cases and the particular method used in any one case will depend upon a number of factors, such as the type of product, the length of time the product is to be preserved, the purpose for which the product is to be used, and the availability of transportation and storage equipment. Very often it is necessary to employ several methods simultaneously in order to obtain the desired results.

9-9. Deterioration and Spoilage

Since the preservation of food is simply a matter of preventing or retarding deterioration and spoilage regardless of the method used, a good knowledge of the causes of deterioration and spoilage is a prerequisite to the study of preservation methods.

It should be recognized at the outset that there are degrees of quality and that all perishable foods pass through various stages of deterioration before becoming unfit for consumption. In most cases, the objective in the preservation of food is not only to preserve the foodstuff in an edible condition but also to preserve it as nearly as possible at the peak of its quality with respect to appearance, odor, taste, and vitamin content. Except for a few processed foods, this usually means maintaining the foodstuff as nearly as possible in its original fresh state.

Any deterioration sufficient to cause a detectable change in the appearance, odor, or taste of fresh foods immediately reduces the commercial value of the product and thereby represents an economic loss. Consider, for example, wilted vegetables or overripe fruit. Although their edibility is little impaired, an undesirable change in their appearance has been brought about which usually requires that they be disposed of at a reduced price. Too, since they are well on their way to eventual spoilage, their keeping qualities are greatly reduced and they must be consumed or processed immediately or become a total loss.

For obvious reasons, maintaining the vitamin content at the highest possible level is always an important factor in the processing and/or preservation of all food products. In fact, many food processors, such as bakers and dairyworkers, are now adding vitamins to their product to replace those which are lost during processing. Fresh vegetables, fruit, and fruit juices are some of the food products which suffer heavy losses in vitamin content very quickly if they are not handled and protected properly. Although the loss of vitamin content is not something which in itself is apparent, in many fresh foods it is usually accompanied by recognizable changes in appearance, odor, or taste, such as, for instance, wilting in leafy, green vegetables.

For the most part, the deterioration and eventual spoilage of perishable food are caused by a series of complex chemical changes which take place in the foodstuff after harvesting or killing. These chemical changes are brought about by both internal and external agents. The former are the natural enzymes which are inherent in all organic materials, whereas the latter are microorganisms which grow in and on the surface of the foodstuff. Although either agent alone is capable of bringing about the total destruction of a food product, both agents are involved in most cases of food spoilage. In any event, the activity of both of these spoilage agents must be either eliminated or effectively controlled if the foodstuff is to be adequately preserved.

9-10. Enzymes

Enzymes are complex, protein-like, chemical substances. Not yet fully understood, they are probably best described as chemical catalytic agents which are capable of bringing about chemical changes in organic materials. There are many different kinds of enzymes and each one is specialized in that it produces only one specific chemical reaction. In general, enzymes are identified either by the substance upon which they act or by the result of their action. For instance, the enzyme lactase is so known because it acts to convert lactose (milk sugar) to lactic acid. This particular process is called lactic acid fermentation and is the one principally responsible for the "souring" of milk. Enzymes associated with the various types of fermentation are sometimes called ferments.

Essential in the chemistry of all living processes, enzymes are normally present in all organic materials (the cell tissue of all plants and animals, both living and dead). They are manufactured by all living cells to help carry on the various living activities of the cell, such as respiration, digestion, growth, and reproduction, and they play an important part in such things as the sprouting of seeds, the growth of plants and animals, the ripening of fruit, and the digestive processes of animals, including man. However, enzymes are catabolic as well as anabolic. That is, they act to destroy dead cell tissue as well as to maintain live cell tissue. In fact, enzymes are the agents primarily responsible for the decay and decomposition of all organic materials, as, for example, the putrification of meat and fish and the rotting of fruit and vegetables.

Whether their action is catabolic or anabolic, enzymes are nearly always destructive to perishable foods. Therefore, except in those few special cases where fermentation or putrification is a part of the processing, enzymic action must be either eliminated entirely or severely inhibited if the product is to be preserved in good condition. Fortunately, enzymes are sensitive to the conditions of the surrounding media, particularly with regard to the temperature and the degree of acidity or alkalinity, which provides a practical means of controlling enzymic activity.

Enzymes are completely destroyed by high temperatures that alter the composition of the organic material in which they exist. Since most enzymes are eliminated at temperatures above 160°F, cooking a food substance completely destroys the enzymes contained therein. On the other hand, enzymes are very resistant to low temperatures and their activity may continue at a slow rate even at temperatures below 0°F. However, it is a well-known fact that the rate of chemical reaction decreases as the temperature decreases. Hence, although the enzymes are not destroyed, their activity is greatly reduced at low temperatures, particularly temperatures below the freezing point of water.

Enzymic action is greatest in the presence of free oxygen (as in the air) and decreases as the oxygen supply diminishes.

With regard to the degree of acidity or alkalinity, some enzymes require acid surroundings, whereas others prefer neutral or alkaline environments. Those requiring acidity are destroyed by alkalinity and those requiring alkalinity are likewise destroyed by acidity.

Although an organic substance can be completely destroyed and decomposed solely by the action of its own natural enzymes, a process known as autolysis (self-destruction), this seldom occurs. More often, the natural enzymes are aided in their destructive action by enzymes secreted by microorganisms.

9-11. Microorganisms

The term "microorganism" is used to cover a whole group of minute plants and animals of microscopic and submicroscopic size, of which only the following three are of particular interest in the study of food preservation: (1) bacteria, (2) yeasts, and (3) molds. These tiny organisms are found in large numbers everywhere—in the air, in the ground, in water, in and on the bodies of plants and animals, and in every other place where conditions are such that living organisms can survive.

Because they secrete enzymes which attack the organic materials upon which they grow, microorganisms are agents of fermentation, putrification, and decay. As such, they are both beneficial and harmful to human beings. Their growth in and on the surface of perishable foods causes complex chemical changes in the food substance which usually results in undesirable alterations in the taste, odor, and appearance of the food and which, if allowed to continue for any length of time, will render the food unfit for consumption. Too, some microorganisms secrete poisonous substances (toxins) which are extremely dangerous to health, causing poisoning, disease, and often death.

On the other hand, microorganisms have many useful and necessary functions. As a matter of fact, if it were not for the work of microorganisms, life of any kind would not be possible. Since decay and decomposition of all dead animal tissue are essential to make space available for new life and growth, the decaying action of microorganisms is indispensable to the life cycle.

Of all living things, only green plants (those containing chlorophyll) are capable of using inorganic materials as food for building their cell tissue. Through a process called photosynthesis, green plants are able to utilize the radiant energy of the sun to combine carbon dioxide from the air with water and mineral salts from the soil and thereby manufacture from inorganic materials the organic compounds which make up their cell tissue.

Conversely, all animals and all plants without chlorophyll (fungi) require organic materials (those containing carbon) for food to carry on their life activities. Consequently, they must of necessity feed upon the cell tissue of other plants and animals (either living or dead) and are, therefore, dependent either directly or indirectly on green plants as a source of the organic materials they need for life and growth.

It is evident, then, that should the supply of inorganic materials in the soil, which serve as food for green plants, ever become exhausted, all life would soon disappear from the earth. This is not likely to happen, however, since microorganisms, as a part of their own living process, are continuously replenishing the supply of inorganic materials in the soil.

With the exception of a few types of soil bacteria, all microorganisms need organic materials as food to carry on the living process. In most cases, they obtain these materials by decomposing animal wastes and the tissue of dead animals and plants. In the process of decomposition, the complex organic compounds which make up the tissue of animals and plants are broken down step by step and are eventually reduced to simple inorganic materials which are returned to the soil to be used as food by the green plants.

In addition to the important part they play in the "food chain" by helping to keep essential materials in circulation, microorganisms are necessary in the processing of certain fermented foods and other commodities. For example, bacteria are responsible for the lactic acid fermentation required in the processing of pickles, olives, cocoa, coffee, sauerkraut, ensilage, and certain sour milk products, such as butter, cheese, buttermilk, and yogurt, and for the acetic acid fermentation necessary in the production of vinegar from various alcohols. Bacterial action is useful also in the processing of certain other commodities such as leather, linen, hemp, and tobacco, and in the treatment of industrial wastes of organic composition.

Yeasts, because of their ability to produce alcoholic fermentation, are of immeasurable value to the brewing and wine-making industries and to the production of alcohols of all kinds. Too, everyone is aware of the importance of yeast in the baking industry.

The chief commercial uses of molds are in the processing of certain types of cheeses and, more important, in the production of antibiotics, such as penicillin and aureomycin.

Despite their many useful and necessary functions, the fact remains that microorganisms are destructive to perishable foods. Hence, their activity, like that of the natural enzymes, must be effectively controlled if deterioration and spoilage of the food substance are to be avoided.

Since each type of microorganism differs somewhat in both nature and behavior, it is worthwhile to examine each type separately.

9-12. Bacteria

Bacteria are a very simple form of plant life, being made up of one single living cell. Reproduction is accomplished by cell division. On reaching maturity, the bacterium divides into two separate and equal cells, each of which in turn grows to maturity and divides into two cells. Bacteria grow and reproduce at an enormous rate. Under ideal conditions, a bacterium can grow into maturity and reproduce in as little as 20 to 30 min. At this rate a single bacterium is capable of producing as many as 34 trillion descendants in a 24-hr period. Fortunately, however, the life cycle of bacteria is relatively short, being a matter of minutes or hours, so that even under ideal conditions they cannot multiply at anywhere near this rate.

The rate at which bacteria and other microorganisms grow and reproduce depends upon such environmental conditions as temperature, light, and the degree of acidity or alkalinity, and upon the availability of oxygen, moisture, and an adequate supply of soluble food. However, there are many species of bacteria, and they differ greatly both in their choice of environment and in the effect they have on their environment. Like the higher forms of plant life, all species of bacteria are not equally hardy with respect to surviving adverse conditions of environment, nor do all species thrive equally well under the same environmental conditions. Some species prefer conditions which are entirely fatal to others. Too, some bacteria are spore-formers. The spore is formed within the bacteria cell and is protected by a heavy covering or wall. In the spore state, which is actually a resting or dormant phase of the organism, bacteria are extremely resistant to unfavorable conditions of environment and can survive in this state almost indefinitely. The spore will usually germinate whenever conditions become favorable for the organism to carry on its living activities.

Most bacteria are saprophytes. That is, they are "free living" and feed only on animal wastes and on the dead tissue of animals and plants. Some, however, are parasites and require a living host. Most pathogenic bacteria (those causing infection and disease) are of the parasitic type. In the absence of a living host, some parasitic bacteria can live as saprophytes. Likewise, some saprophytes can live as parasites when the need arises.

Since bacteria are not able to digest insoluble food substances, they require food in a soluble form. For this reason, most bacteria secrete enzymes which are capable of rendering insoluble compounds into a soluble state, thereby making these materials available to the bacteria as food. The deterioration of perishable foods by bacteria growth is a direct result of the action of these bacterial enzymes.

Bacteria, like all other living things, require moisture as well as food to carry on their life activities. As in other things, bacteria vary considerably in their ability to resist drought. Although most species are readily destroyed by drying and will succumb within a few hours, the more hardy species are able to resist drought for several days. Bacterial spores can withstand drought almost indefinitely, but will remain dormant in the absence of moisture.

In their need for oxygen, bacteria fall into two groups: (1) those which require free oxygen (air) and (2) those which can exist without free oxygen. Some species, although having a preference for one condition or the other, can live in the presence of free oxygen or in the absence of it. Those bacteria living without free oxygen obtain the needed oxygen through chemical reaction which reduces one compound while oxidizing another. Decomposition which occurs in the presence of free oxygen is known as decay, whereas decomposition which takes place in the absence of free oxygen is called putrification. One of the products of putrification is hydrogen sulfide, a foul-smelling gas that is frequently noted arising from decomposing animal carcasses.

The Growth of Bacteria in Milk in Various Periods

Temp., °F	Time, hours 24	48	96	168
32	2,400	2,100	1,850	1,400
39	2,500	3,600	218,000	4,200,000
46	3,100	12,000	1,480,000	
50	11,600	540,000		
60	180,000	28,000,000		
86	1,400,000,000			

Fig. 9-1 From *ASRE Data Book*, Applications Volume, 1956–57. Reproduced by permission of the American Society of Heating, Refrigerating, and Air-Conditioning Engineers.

Bacteria are very sensitive to acidity or alkalinity and cannot survive in an either highly acid or highly alkaline environment. Most bacteria require either neutral or slightly alkaline surroundings, although some species prefer slightly acid conditions. Because bacteria prefer neutral or slightly alkaline surroundings, nonacid vegetables are especially subject to bacterial attack.

Light, particularly direct sunlight, is harmful to all bacteria. Whereas visible light only inhibits their growth, ultraviolet light is actually fatal to bacteria. Since light rays, ultraviolet or otherwise, have no power of penetration, they are effective only in controlling surface bacteria. However, ultraviolet radiation (usually from direct sunlight), when combined with drying, provides an excellent means of controlling bacteria growth.

For each species of bacteria there is an optimum temperature at which the bacteria will grow at the highest rate. Too, for each species there is a maximum and a minimum temperature which will permit growth. At temperatures above the maximum, the bacteria are destroyed. At temperatures below the minimum, the bacteria are rendered inactive or dormant. The optimum temperature for most saprophytes is usually between 75°F and 85°F, whereas the optimum temperature for parasites is around 99°F or 100°F. A few species grow best at temperatures near the boiling point of water, whereas a few other types thrive best at temperatures near the freezing point. However, most species are either killed off or severely inhibited at these temperatures. The effect of temperature on the growth rate of bacteria is illustrated by the chart in Fig. 9-1, which shows the growth rate of bacteria in milk at various temperatures. In general, the growth rate of bacteria is considerably reduced by lowering the temperature.

9-13. Yeasts

Yeasts are simple, one-cell plants of the fungus family. Of microscopic size, yeast cells are somewhat larger and more complex than the bacteria cells. Although a few yeasts reproduce by fission or by sexual process, reproduction is usually by budding. Starting as a small protrusion of the mature cell, the bud enlarges and finally separates from the mother cell. Under ideal conditions, budding is frequently so rapid that new buds are formed before separation occurs so that yeast clusters are formed.

Like bacteria, yeasts are agents of fermentation and decay. They secrete enzymes that bring about chemical changes in the food upon which they grow. Yeasts are noted for their ability to transform sugars into alcohol and carbon dioxide. Although destructive to fresh foods, particularly fruits and berries and their juices, the alcoholic fermentation produced by yeasts is essential to the baking, brewing, and wine-making industries.

Yeasts are spore-formers, with as many as eight spores being formed within a single yeast cell. Yeasts are widespread in nature and yeast spores are invariably found in the air and on the skin of fruit and berries, for which they have a particular affinity. They usually spend the winter in the soil and are carried to the new fruit in the spring by insects or by the wind.

Like bacteria, yeasts require air, food, and moisture for growth, and are sensitive to temperature and the degree of acidity or alkalinity in the environment. For the most part, yeasts prefer moderate temperatures and slight acidity. In general, yeasts are not as resistant to unfavorable conditions as are bacteria, although they can grow in acid surroundings which inhibit most bacteria. Yeast spores, like those of bacteria, are extremely hardy and can survive for long periods under adverse conditions.

9-14. Molds

Molds, like yeasts, are simple plants of the fungi family. However, they are much more complex in structure than either bacteria or yeasts. Whereas the individual bacteria or yeast plants consist of one single cell, an individual mold plant is made up of a number of cells which are positioned end to end to form long, branching, threadlike fibers called hypha. The network that is formed by a mass of these threadlike fibers is called the mycelium and is easily visible to the naked eye. The hyphae of the mold plant are of two general types. Some are vegetative fibers which grow under the surface and act as roots to gather food for the plant, whereas others, called aerial hyphae, grow on the surface and produce the fruiting bodies.

Molds reproduce by spore formation. The spores develop in three different ways, depending on the type of mold: (1) as round clusters within the fibrous hyphae network, (2) as a mass enclosed in a sac and attached to the end of aerial hyphae, and (3) as chainlike clusters on the end of aerial hyphae. In any case, a single mold plant is capable of producing thousands of spores which break free from the mother plant and float away with the slightest air motion.

Mold spores are actually seeds and, under the proper conditions, will germinate and produce mold growth on any food substance with which they come in contact. Since they are carried about by air currents, mold spores are found almost everywhere and are particularly abundant in the air.

Although molds are less resistant to high temperatures than are bacteria, they are more tolerant to low temperatures, growing freely at temperatures close to the freezing point of water. Mold growth is inhibited by temperatures below 32°F, more from the lack of free moisture than from the effect of low temperature. All mold growth ceases at temperatures of 10°F and below.

Molds flourish in dark, damp surroundings, particularly in still air. An abundant supply of oxygen is essential to mold growth, although a very few species can grow in the absence of oxygen. Conditions inside cold-storage rooms are often ideal for mold growth, especially in the wintertime. This problem can be overcome somewhat by maintaining good air circulation in the storage room, by the use of germicidal paints and ultraviolet radiation, and by frequent scrubbing.

Unlike bacteria, molds can thrive on foods containing relatively large amounts of sugars or acids. They are often found growing on acid fruits and on the surface of pickling vats, and are the most common cause of spoilage in apples and citrus fruits.

9-15. Control of Spoilage Agents

Despite complications arising from the differences in the reaction of the various types of spoilage agents to specific conditions in the environment, controlling these conditions provides the only means of controlling these spoilage agents. Thus, all methods of food preservation must of necessity involve manipulation of the environment in and around the preserved product in order to produce one or more conditions unfavorable to the continued activity of the spoilage agents. When the product is to be preserved for any length of time, the unfavorable conditions produced must be of sufficient severity to eliminate the spoilage agents entirely or at least render them ineffective or dormant.

All types of spoilage agents are destroyed when subjected to high temperatures over a period of time. This principle is used in the preservation of food by canning. The temperature of the product is raised to a level fatal to all spoilage agents and is maintained at this level until they are all destroyed. The product is then sealed in sterilized, air-tight containers to prevent recontamination. A product so processed will remain in a preserved state indefinitely.

The exposure time required for the destruction of all spoilage agents depends upon the temperature level. The higher the temperature level, the shorter is the exposure period required. In this regard, moist heat is more effective than dry heat because of its greater penetrating powers. When moist heat is used, the temperature level required is lower and the processing period is shorter. Enzymes and all living microorganisms are destroyed when exposed to the temperature of boiling water for approximately five minutes, but the more resistant bacteria spores may survive at this condition for several hours before succumbing. For this reason, some food products, particularly meats and nonacid vegetables, require long processing periods which frequently result in overcooking of the product. These products are usually processed under pressure so that the processing temperature is increased and the processing time shortened.

Another method of curtailing the activity of spoilage agents is to deprive them of the moisture and/or food which is necessary for their continued activity. Both enzymes and microorganisms require moisture to carry on their activities. Hence, removal of the free moisture from a product will severely limit their activities. The process of moisture removal is called drying (dehydration) and is one of the oldest methods of preserving foods. Drying is accomplished either naturally in the sun and air or artificially in ovens. Dried products which are stored in a cool, dry place will remain in good condition for long periods.

Pickling is essentially a fermentation process, the end result of which is the exhaustion of the substances which serve as food for yeasts and bacteria. The product to be preserved by pickling is immersed in a salt brine solution and fermentation is allowed to take place, during which the sugars contained in the food product are converted to lactic acid, primarily through the action of lactic acid bacteria.

Smoked products are preserved partially by the drying effect of the smoke and partially by antiseptics (primarily creosote) which are absorbed from the smoke.

Too, some products are "cured" with sugar or salt which act as preservatives in that they create conditions unfavorable to the activity of spoilage agents. Some other frequently used preservatives are vinegar, borax, saltpeter, benzoate of soda, and various spices. A few of the products preserved in this manner are sugarcured hams, salt pork, spiced fruits, certain beverages, jellies, jams, and preserves.

9-16. Preservation by Refrigeration

The preservation of perishables by refrigeration involves the use of low temperature as a means of eliminating or retarding the activity of spoilage agents. Although low temperatures are not as effective in bringing about the destruction of spoilage agents as are high temperatures, the storage of perishables at low temperatures greatly reduces the activity of both enzymes and microorganisms and thereby provides a practical means of preserving perishables in their original fresh state for varying periods of time. The degree of low temperature required for adequate preservation varies with the type of product stored and with the length of time the product is to be kept in storage.

For purposes of preservation, food products can be grouped into two general categories: (1) those that are alive at the time of distribution and storage and (2) those that are not. Nonliving food substances, such as meat, poultry, and fish, are much more susceptible to microbial contamination and spoilage than are living food substances, and they usually require more stringent preservation methods.

With nonliving food substances, the problem of preservation is one of protecting dead tissue from all the forces of putrification and decay, both enzymic and microbial. In the case of living food substances, such as fruit and vegetables, the fact of life itself affords considerable protection against microbial invasion, and the preservation problem is chiefly one of keeping the food substance alive while at the same time retarding natural enzymic activity in order to slow the rate of maturation or ripening.

Vegetables and fruit are as much alive after harvesting as they are during the growing period. Previous to harvesting they receive a continuous supply of food substances from the growing plant, some of which is stored in the vegetable or fruit. After harvesting, when the vegetable or fruit is cut off from its normal supply of food, the living processes continue through utilization of the previously stored food substances. This causes the vegetable or fruit to undergo changes which will eventually result in deterioration and complete decay of the product. The primary purpose of placing such products under refrigeration is to slow the living processes by retarding enzymic activity, thereby keeping the product in a preserved condition for a longer period.

Animal products (nonliving food substances) are also affected by the activity of natural enzymes. The enzymes causing the most trouble are those which catalyze hydrolysis and oxidation and are associated with the breakdown of animal fats. The principal factor limiting the storage life of animal products, in both the frozen and unfrozen states, is rancidity. Rancidity is caused by oxidation of animal fats and, since some types of animal fats are less stable than others, the storage life of animal products depends in part on fat composition. For example, because of the relative stability of beef fat, the storage life of beef is considerably greater than that of pork or fish, whose fatty tissues are much less stable.

Oxidation and hydrolysis are controlled by placing the product under refrigeration so that the activity of the natural enzymes is reduced. The rate of oxidation can be further reduced in the case of animal products by packaging the products in tight-fitting, gas-proof containers which prevent air (oxygen) from reaching the surface of the product. The packaging of fruit and vegetables in gas-proof containers, when stored in the unfrozen state, is not practical. Since these products are alive, packaging in gas-proof containers will cause suffocation and death. A dead fruit or vegetable decays very quickly.

As a general rule, the lower the storage temperature, the longer is the storage life of the product.

9-17. Refrigerated Storage

Refrigerated storage may be divided into three general categories: (1) short-term or temporary storage, (2) long-term storage, and (3) frozen storage. For short- and long-term storage, the product is chilled and stored at some temperature above its freezing point, whereas frozen storage requires freezing of the product and storage at some temperature between 10°F and −10°F, with 0°F being the temperature most frequently employed.

Short-term or temporary storage is usually associated with retail establishments where rapid turnover of the product is normally expected. Depending upon the product, short-term storage periods range from 1 or 2 days in some cases to a week or more in others, but seldom for more than 15 days.

Long-term storage is usually carried out by wholesalers and commercial storage warehouses. Again, the storage period depends on the type of product stored and upon the condition of the product on entering storage. Maximum storage periods for long-term storage range from seven to ten days for some sensitive products, such as ripe tomatoes, cantaloupes, and broccoli, and up to six or eight months for the more durable products, such as onions and some smoked meats. When perishable foods are to be stored for longer periods, they should be frozen and placed in frozen storage. Some fresh foods, however, such as tomatoes, are damaged by the freezing process and therefore cannot be successfully frozen. When such products are to be preserved for long periods, some other method of preservation should be used.

9-18. Storage Conditions

The optimum storage conditions for a product held in either short- or long-term storage depend upon the nature of the individual product, the length of time the product is to be held in storage, and whether the product is packaged or unpackaged. In general, the conditions required for short-term storage are more flexible than those required for long-term storage and, ordinarily, higher storage temperatures are permissible. Recommended storage conditions for both short- and long-term storage and the approximate storage life for various products are listed in Tables 10-9 through 10-12, along with other product data. These data are the result of both experiment and experience and should be followed closely, particularly for long-term storage, if product quality is to be maintained at a high level during the storage period.

9-19. Storage Temperature

Examination of the tables will show that the optimum storage temperature for most products is one slightly above the freezing point of the product. There are, however, notable exceptions.

Although the effect of incorrect storage temperatures generally is to lower product quality and shorten storage life, some fruits and vegetables are particularly sensitive to storage temperature and are susceptible to so-called cold storage diseases when stored at temperatures above or below their critical storage temperatures. For example, citrus fruits frequently develop rind pitting when stored at relatively high temperatures. On the other hand, they are subject to scald (browning of the rind) and watery breakdown when stored at temperatures below their critical temperature. Bananas suffer peel injury when stored below 56°F, whereas celery undergoes soggy breakdown when stored at temperatures above 34°F. Although onions tend to sprout at temperatures above 32°F, Irish potatoes tend to become sweet at storage temperatures below 40°F. Squash, green beans, and peppers develop pits on their surface when stored at or near 32°F. Too, whereas the best storage temperature for most varieties of apples is 30°F to 32°F, some varieties are subject to soft scald and soggy breakdown when stored below 35°F. Others develop brown core at temperatures below 36°F, and still others develop internal browning when stored below 40°F.

9-20. Humidity and Air Motion

The storage of all perishables in their natural state (unpackaged) requires close control not only of the space temperature but also of space humidity and air motion. One of the chief causes of the deterioration of unpackaged fresh foods, such as meat, poultry, fish, fruit, vegetables, cheese, and eggs, is the loss of moisture from the surface of the product by evaporation into the surrounding air. This process is known as dessication or dehydration. In fruit and vegetables, dessication is accompanied by shriveling and wilting and the product undergoes a considerable loss in both weight and vitamin content. In meats, cheese, etc., dessication causes discoloration, shrinkage, and heavy trim losses. It also increases the rate of oxidation. Eggs lose moisture through the porous shell, with a resulting loss of weight and general downgrading of the egg.

Dessication will occur whenever the vapor pressure of the product is greater than the vapor pressure of the surrounding air, the rate of moisture loss from the product being proportional to the difference in the vapor pressures and to the amount of exposed product surface.

The difference in vapor pressure between the product and the air is primarily a function of the relative humidity and the velocity of the air in the storage space. In general, the lower the relative humidity and the higher the air velocity, the greater will be the vapor pressure differential and the greater the rate of moisture loss from the product. Conversely, minimum moisture losses are experienced when the humidity in the storage space is maintained at a high level with low air velocity. Hence, 100% relative humidity and stagnant air are ideal conditions for preventing dehydration of the stored product. Unfortunately, these conditions are also conducive to rapid mold growth and the formation of slime (bacterial) on meats. Too, good circulation of the air in the refrigerated space and around the product is necessary for adequate refrigeration of the product. For these reasons, space humidity must be maintained at somewhat less than 100% and air velocities must be sufficient to provide adequate air circulation. The relative humidities and air velocities recommended for the storage of various products are listed in Tables 10-9 through 10-12.

When the product is stored in vapor-proof containers, space humidity and air velocity are not critical. Some products, such as dried fruits, tend to be hydroscopic and therefore require storage at low relative humidities.

9-21. Mixed Storage

Although the maintenance of optimum storage conditions requires separate storage facilities for most products, this is not usually economically feasible. Therefore, except when large quantities of product are involved, practical considerations often demand that a number of refrigerated products be placed in common storage. Naturally, the difference in the storage conditions required by the various products raises a problem with regard to the conditions to be maintained in a space designed for common storage.

As a general rule, storage conditions in such spaces represent a compromise and usually prescribe a storage temperature somewhat above the optimum for some of the products held in mixed storage. The higher storage temperatures are used in mixed storage in order to minimize the chances of damaging the more sensitive products which are subject to the aforementioned "cold storage diseases" when stored at temperatures below their critical temperature.

Although higher storage temperatures tend to shorten the storage life of some of the products held in mixed storage, this is not ordinarily a serious problem when the products are stored only for short periods as in temporary storage.

For long-term storage, most of the larger wholesale and commercial storage warehouses have a number of separate storage spaces available. General practice in such cases is to group the various products for storage, and only those products requiring approximately the same storage conditions are placed together in common storage.

Another problem associated with mixed storage is that of odor and flavor absorption. Some products absorb and/or give off odors while in storage. Care should be taken not to store such products together even for short periods. Dairy products in particular are very sensitive with regard to absorbing odors and flavors from other products held in mixed storage. On the other hand, potatoes are probably the worst offenders in imparting off-flavors to other products in storage and should never be stored with fruit, eggs, dairy products, or nuts.

9-22. Product Condition on Entering Storage

One of the principal factors determining the storage life of a refrigerated product is the condition of the product on entering storage. It must be recognized that refrigeration merely arrests or retards the natural processes of deterioration and therefore cannot restore to good condition a product which has already deteriorated. Neither can it make a high quality product out of one of initial poor quality. Hence, only vegetables and fruit in good condition should be accepted for storage. Those that have been bruised or otherwise damaged, particularly if the skin has been broken, have lost much of their natural protection against microbial invasion and are therefore subject to rapid spoilage by these agents. Too, as a general rule, since maturation and ripening continue after harvesting, vegetables and fruit intended for storage should be harvested before they are fully mature. The storage life of fully mature or damaged fruit and vegetables is extremely short even under the best storage conditions, and such products should be sent directly to market to avoid excessive losses.

Since a food product begins to deteriorate very quickly after harvesting or killing, it is imperative that preservation measures be taken immediately. To assure maximum storage life with minimum loss of quality, the product should be chilled to the storage temperature as soon as possible after harvesting or killing. When products are to be shipped over long distances to storage, they should be precooled and shipped by refrigerated transport.

9-23. Product Chilling or Precooling

Product chilling or precooling is distinguished from product storage in that the product enters the chilling room or precooler at a high temperature (usually either harvesting or killing temperature) and is chilled as rapidly as possible to the storage temperature, whereupon it is normally removed from the chilling room and placed in a holding cooler for storage. The handling of the product during the chilling period has a marked influence on the ultimate quality and storage life of the product.

The recommended conditions for product chilling rooms are given in Tables 10-9 through 10-12. Before the hot product is loaded into the chilling room, the chilling room temperature should be at the "chill finish" temperature. During loading and during the early part of the chilling period, the temperature and vapor pressure differential between the product and the chill room air will be quite large and the product will give off heat and moisture at a high rate. At this time, the temperature and humidity in the chill room will rise to a peak as indicated by the "chill start" conditions in the tables.* At the end of the cycle, the chill room temperature will again drop to the "chill finish" conditions. It is very important that the refrigerating equipment have sufficient capacity to prevent the chill room temperature from rising excessively during the peak chilling period.

9-24. Relative Humidity and Air Velocity in Chill Rooms

The importance of relative humidity in chilling rooms depends upon the product being chilled, particularly upon whether the product is packaged or not. Naturally, when the product is chilled in vapor-proof containers, the humidity in the chilling room is relatively unimportant. However, during loading and during the initial stages of chilling, chilling room humidity will be high if the containers are wet, but will drop rapidly once the free moisture has been evaporated.

Products chilled in their natural state (unpackaged) lose moisture very rapidly, often producing fog in the chilling room during the early stages of chilling when the product temperature and vapor pressure are high. Rapid chilling and high air velocity are desirable during this time so that the temperature and vapor pressure of the product are lowered as quickly as possible in order to avoid excessive moisture loss and shrinkage. High air velocity is needed also in order to carry away the vapor and thereby prevent condensation of moisture on the surface of the product.

Although high air velocity tends to increase the rate of evaporation of moisture from the product, it also greatly accelerates the chilling rate and results in a more rapid reduction in product temperature and vapor pressure. Since the reduction in vapor pressure resulting from the higher chilling rate more than offsets the increase in the rate of evaporation occasioned by the higher air velocity, the net effect of the higher air velocity during the early stages of chilling is to reduce the overall loss of moisture from the product. However, during the final stages of chilling, when the temperature and vapor pressure of the product are considerably lower, the effect of high air velocity in the chilling room is to increase the rate of moisture loss from the product. Therefore, the air velocity in the chilling room should be reduced during the final stages of chilling.

As a general rule, the humidity should be kept at a high level when products subject to dehydration are being chilled. Some extremely sensitive products, such as poultry and fish, are frequently chilled in ice slush to reduce moisture losses during chilling. For the same reason, eggs are sometimes dipped in a light mineral oil before chilling and storage. Too, poultry, fish, and some vegetables are often packed in ice for chilling and storage. When products packed in ice are placed in refrigerated storage, the slowly melting ice keeps the surface of the product moist and prevents excessive dehydration.

Some vegetables and fruit are precooled by a process known as hydrocooling, which involves flooding or spraying the product with chilled water or immersing it in an agitated bath of chilled water. Flooding is accomplished by showering the product with a liberal supply

* The temperatures listed in the tables as chill start temperatures are average values and are intended for use in selecting the refrigerating equipment. Actual temperatures in the chilling room during the peak chilling period are usually 3°F to 4°F higher than those listed.

of chilled water flowing under a gravity head from overhead flood pans, whereas spraying is accomplished by the use of overhead spray nozzles.

Also, some fresh vegetables, particularly those having a high ratio of surface area to volume, are precooled by causing the rapid evaporation of water from the surface of the product in a vacuum or flash chamber. Such a process is called vacuum cooling. Precooling is accomplished in a vacuum cooler (flash chamber) by reducing the pressure in the cooler until the corresponding saturation temperature of the water is below the temperature of the product being cooled, at which point vaporization of water from the surface of the product begins with the required latent heat being drawn from the product. As the pressure in the chamber is further reduced, cooling continues to the desired temperature level.

9-25. Combined Chilling and Storage

The use of the same room for both chilling and storage is not recommended for meats and similar products that are extremely sensitive to fluctuating temperature, relative humidity, and air motion. However, such limitations do not apply to chill rooms for fruit, such as apples and pears, since experience has shown that these can be handled in combined chilling and storage rooms without harmful effects to the product. This is because of the relative short loading-in periods and the fact that the product in storage provides a flywheel effect that permits little or no fluctuation in the space temperature. Freezing rooms in which the product to be frozen enters at temperatures up to 45°F are also exceptions.

9-26. Freezing and Frozen Storage

When a product is to be preserved in its original fresh state for relatively long periods, it is usually frozen and stored at approximately 0°F or below. The list of food products commonly frozen includes not only those which are preserved in their fresh state, such as vegetables, fruit, fruit juices, berries, meat, poultry, sea foods, and eggs (not in shell), but also many prepared foods, such as breads, pastries, ice cream, and a wide variety of specially prepared and precooked food products, including full dinners.

The following factors govern the ultimate quality and storage life of any frozen product:

1. The nature and composition of the product to be frozen
2. The care used in selecting, handling, and preparing the product for freezing
3. The freezing method
4. The storage conditions.

Only high quality products in good condition should be frozen. With vegetables and fruit, selecting the proper variety for freezing is very important. Some varieties are not suitable for freezing and will result in a low quality product or in one with limited keeping qualities

Vegetables and fruit to be frozen should be harvested at the peak of maturity and should be processed and frozen as quickly as possible after harvesting to avoid undesirable chemical changes through enzymic and microbial action.

Both vegetables and fruit require considerable processing before freezing. After cleaning and washing to remove foreign materials—leaves, dirt, insects, juices, etc.—from their surfaces, vegetables are "blanched" in hot water or steam at 212°F in order to destroy the natural enzymes. It will be remembered that enzymes are not destroyed by low temperature and, although greatly reduced, their activity continues at a slow rate even in food stored at 0°F and below. Hence, blanching, which destroys most of the enzymes, greatly increases the storage life of frozen vegetables. The time required for blanching varies with the type and variety of the vegetable and ranges from 1 to $1\frac{1}{2}$ min for green beans to 11 min for large ears of corn. Although much of the microbial population is destroyed along with the enzymes during the blanching process, many bacteria survive. To prevent spoilage by these viable bacteria, vegetables should be chilled to 50°F immediately after blanching and before they are packaged for the freezer.

Like vegetables, fruit must also be cleaned and washed to remove foreign materials and to reduce microbial contamination. Although fruit is perhaps even more subject to enzymic deterioration than are vegetables, it is never blanched to destroy the natural enzymes since to do so would destroy the natural fresh quality which is so desirable.

The enzymes causing the most concern with regard to frozen fruit are the ones which catalyze oxidation and result in rapid browning of the flesh. To control oxidation, fruit to be frozen is covered with a light sugar syrup. In some cases, ascorbic acid, citric acid, or sulfur dioxide are also used for this purpose.

As a general rule, meat products do not require any special processing prior to freezing. However, because of consumer demand, specially prepared meats and meat products are being frozen in increasing amounts. This is true also of poultry and seafood.

Because of the relative instability of their fatty tissue, pork and fish are usually frozen as soon after chilling as possible. On the other hand, beef is frequently "aged" in a chilling cooler for several days before freezing. During this time the beef is tenderized to some extent by enzymic activity. However, the aging of beef decreases its storage life, particularly if the aging period exceeds 6 or 7 days.

With poultry, experiments indicate that poultry frozen within 12 to 24 hr after killing is more tender than that frozen immediately after killing. However, delaying freezing beyond 24 hr tends to reduce storage life without appreciably increasing tenderness.

9-27. Freezing Methods

Food products may be either sharp (slow) frozen or quick frozen. Sharp freezing is accomplished by placing the product in a low temperature room and allowing it to freeze slowly, usually in still air. The temperature maintained in sharp freezers ranges from 0°F to −40°F. Since air circulation is usually by natural convection, heat transfer from the product ranges from 3 hr to 3 days, depending upon the bulk of the product and upon the conditions in the sharp freezer. Typical items which are sharp frozen are beef and pork half-carcasses, boxed poultry, panned and whole fish, fruit in barrels and other large containers, and eggs (whites, yolks, or whole) in 10 and 30 lb cans.

Quick freezing is accomplished in any one or in any combination of three ways: (1) immersion, (2) indirect contact, and (3) air blast.

9-28. Air Blast Freezing

Air blast freezing utilizes the combined effects of low temperature and high air velocity to produce a high rate of heat transfer from the product. Although the method employed varies considerably with the application, blast freezing is accomplished by circulating high-velocity, low-temperature air around the product. Regardless of the method used, it is important that the arrangement of the freezer is such that air can circulate freely around all parts of the product.

Packaged blast freezers are available in both suspended and floor-mounted models. Typical applications are shown in Figs. 9-2 through 9-4. Blast freezing is frequently carried out in insulated tunnels, particularly where large quantities of product are to be frozen (Figs. 9-5 and 9-6). In some instances, the product is carried through the freezing tunnel and frozen on slow-moving, mesh conveyor belts. The unfrozen product is placed on the conveyor at one end of the tunnel and is frozen by the time it reaches the other end. Another method is to load the product on tiered dollies.

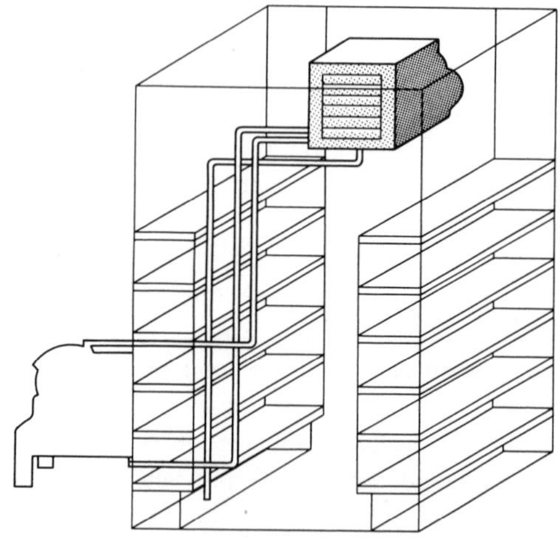

Fig. 9-2 Walk-in installation. Suspended blast freezer provides high-velocity air for fast freezing saving valuable floor space in small areas. (Courtesy Carrier Corporation.)

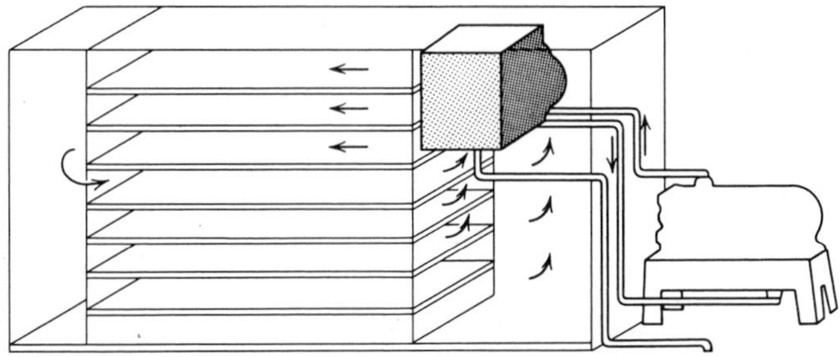

Fig. 9-3 Suspended blast freezer applied to reach-in cabinet distributes blast air through shelves. (Courtesy Carrier Corporation.)

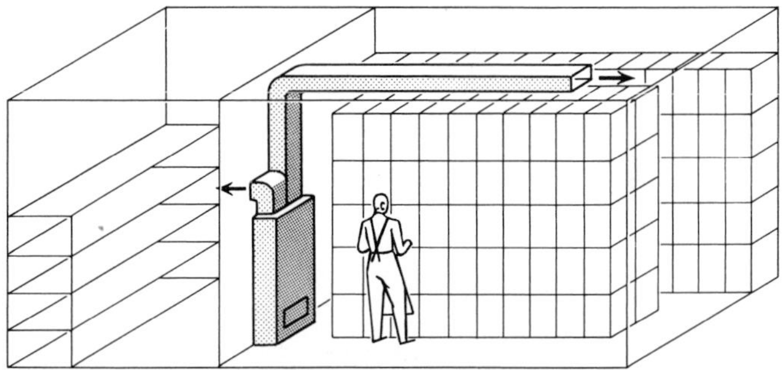

Fig. 9-4 Freezing in one room and storage in another is accomplished by single, floor-mounted blast freezers. (Courtesy Carrier Corporation.)

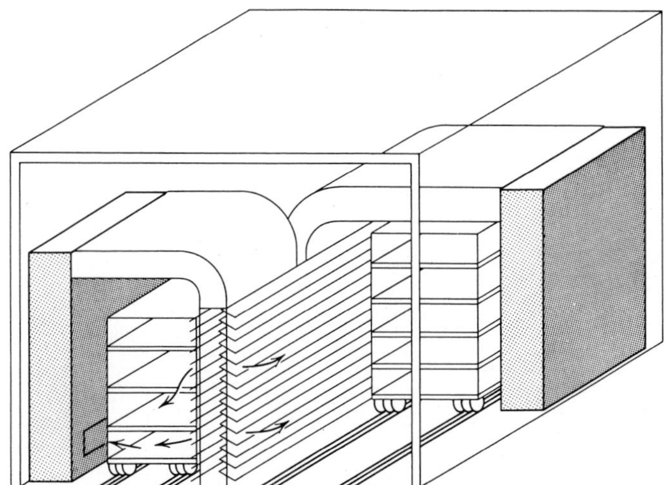

Fig. 9-5 Packaged blast freezers applied to freezing tunnel. High velocity, −15°F air is blasted through trucks. (Courtesy Carrier Corporation.)

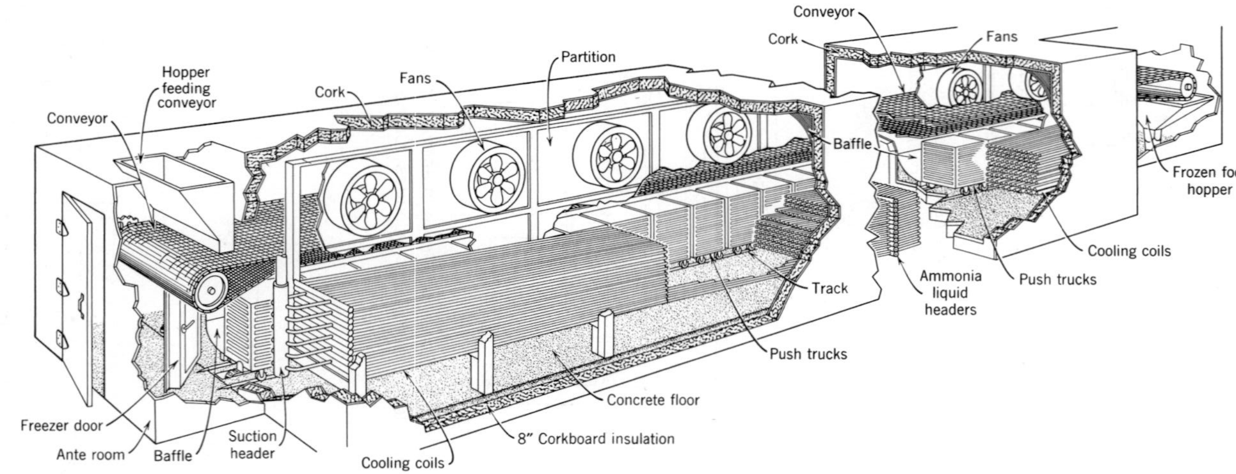

Fig. 9-6 Tunnel freezer for quick-freezing foods. (Courtesy Frick Company.)

The dollies are pushed into the tunnel and the product is frozen, whereupon they are pushed out of the freezing tunnel into a storage room (Fig. 9-5).

Although blast freezing is used to freeze nearly all types of products, it is particularly suitable for freezing products of nonuniform or irregular sizes and shapes, such as dressed poultry.

9-29. Indirect Contact Freezing

Indirect freezing is usually accomplished in plate freezers and involves placing the product on metal plates through which a refrigerant is circulated (Fig. 9-7). Since the product is in direct thermal contact with the refrigerated plate, heat transfer from the product occurs primarily by conduction so that the efficiency of the freezer depends, for the most part, on the amount of contact surface. This type of freezer is particularly useful when products are frozen in small quantities.

One type of plate freezer widely used by the larger commercial freezers to handle small, flat, rectangular, consumer-size packages is the multiplate freezer. The multiplate freezer consists of a series of horizontal, parallel, refrigerated plates which are actuated by hydraulic pressure so that they can be opened to receive the product between them and then closed on the product with any desired pressure. When the plates are closed, the packages are held tightly between the plates. Since both the top and the bottom of the packages are in good thermal contact with the refrigerated plates, the rate of heat transfer is high and the product is quickly frozen.

9-30. Immersion Freezing

Immersion freezing is accomplished by immersing the product in a low temperature brine solution, usually either sodium chloride or sugar. Since the refrigerated liquid is a good conductor and is in good thermal contact with

Fig. 9-7 Plate freezer for indirect contact freezing. (Courtesy Dole Refrigerating Company.)

all the product, heat transfer is rapid and the product is completely frozen in a very short time.

Another advantage of immersion freezing is that the product is frozen in individual units rather than fused together in a mass.

The principal disadvantage of immersion freezing is that juices tend to be extracted from the product by osmosis. This results in contamination and weakening of the freezing solution. Too, where a sodium chloride brine is used, salt penetration into the product may sometimes be excessive. On the other hand, when fruit is frozen in a sugar solution, sugar penetration into the fruit is entirely beneficial.

The products most frequently frozen by immersion are fish and shrimp. Immersion is particularly suitable for freezing fish and shrimp at sea, since the immersion freezer is relatively compact and space aboard ship is at a premium. In addition, immersion freezing produces a "glaze" (thin coating of ice) on the surface of the product which helps to prevent dehydration of unpackaged products during the storage period.

9-31. Quick Freezing Versus Sharp Freezing

Quick frozen products are nearly always superior to those which are sharp (slow) frozen. D. K. Tressler, in 1932, summarized the views of R. Plank, H. F. Taylor, C. Birdseye, and G. A. Fitzgerald, and stated the following as the main advantages of quick freezing over slow freezing:

1. The ice crystals formed are much smaller, and therefore cause much less damage to cells.
2. The freezing period being much shorter, less time is allowed for the diffusion of salts and the separation of water in the form of ice.
3. The product is quickly cooled below the temperature at which bacterial, mold, and yeast growth occurs, thus preventing decomposition during freezing.*

The principal difference between quick freezing and sharp freezing is in the size, number, and location of the ice crystals formed in the product as cellular fluids are solidified. When a product is slow frozen, large ice crystals are formed which result in serious damage to the tissue of some products through cellular breakdown. Quick freezing, on the other hand, produces smaller ice crystals which are formed almost entirely within the cell so that cellular breakdown is greatly reduced. Upon thawing, products which have experienced considerable cellular damage are prone to lose excessive amounts of fluids through "drip" or "bleed," with a resulting loss of quality.

Ice-crystal formation begins in most products at a temperature of approximately 30°F and, although some extremely concentrated fluids still remain unfrozen even at temperatures below −50°F, most of the fluids are solidified by the time the product temperature is lowered to 25°F. The temperature range between 30°F and 25°F is often referred to as the zone of maximum ice-crystal formation, and rapid heat removal through this zone is desirable from the standpoint of product quality. This is particularly true for fruits and vegetables because both undergo serious tissue damage when slow frozen.

Since animal tissue is much tougher and much more elastic than plant tissue, the freezing rate is not as critical in the freezing of meats and meat products as it is in fruits and vegetables. Recent experiment indicates that poultry and fish suffer little, if any, cellular damage when slow frozen. This does not mean, however, that quick frozen meats are not superior to those which are slow frozen, but only that, from the standpoint of cellular damage, quick freezing is not as important in the freezing of meats as it is in fruits and vegetables. For example, poultry that is slow frozen takes on a darkened appearance which makes it much less attractive to the consumer. This alone is enough to justify the quick freezing of poultry. Also, in all cases, quick freezing reduces the processing time and, consequently, the amount of bacterial deterioration. This is especially worthwhile in the processing of fish because of their tendency to rapid spoilage.

* *Air Conditioning Refrigerating Data Book*, Applications Volume, 5th Edition, American Society of Refrigerating Engineers, 1954–55, p. 1–2.

9-32. Packaging Materials

Dehydration, one of the principal factors limiting the storage life of frozen foods, is greatly reduced by proper packaging. Unpackaged products are subject to serious moisture losses not only during the freezing process but also during the storage period. While in storage, unpackaged frozen products lose moisture to the air continuously by sublimation. This eventually results in a condition known as "freezer burn," giving the product a white, leathery appearance. Freezer burn is usually accompanied by oxidation, flavor changes, and loss of vitamin content.

With few exceptions, all products are packaged before being placed in frozen storage. Although most products are packaged before freezing, some, such as loose frozen peas and lima beans, are packaged after the freezing process.

To provide adequate protection against dehydration and oxidation, the packaging material should be practically 100% gas and vapor proof and should fit tightly around the product to exclude as much air as possible. Too, air spaces in packages have an insulating effect which reduces the freezing rate and increases freezing costs.

The fact that frozen products are in competition to products preserved by other methods introduces several factors which must be taken into account when selecting packaging materials. When the product is to be sold directly to the consumer, the package must be attractive and convenient to use in order to stimulate sales. From a cost standpoint, the package should be relatively inexpensive and of such a nature that it permits efficient handling so as to reduce processing costs.

Some packaging materials in general use are aluminum foil, tin cans, impregnated paper-board cartons, paper-board cartons overwrapped with vapor-proof wrappers, wax paper, cellophane, polyethylene, and other sheet plastics.

Frozen fish are often given an ice glaze (a thin coating of ice) which provides an excellent protective covering. However, since the ice glaze is very brittle, glazed fish must be handled carefully to avoid breaking the glaze. Too, since the ice glaze gradually sublimes to the air, the fish must be reglazed approximately once a month by dipping into fresh water or by spraying.

9-33. Frozen Storage

The exact temperature required for frozen storage is not critical, provided that it is sufficiently low and that it does not fluctuate. Although 0°F is usually adequate for short-term (retail) storage, −5°F is the best temperature for all-around long-term (wholesale) storage. When products having unstable fats (oxidizable, free, fatty acids) are stored in any quantity, the storage temperature should be held at −10°F or below in order to realize the maximum storage life.

When products are stored above −20°F, which is normally the case, the temperature of the storage room should be maintained constant with a variation of not more than 1°F in either direction. Variations in storage temperature cause alternate thawing and refreezing of some of the juices in the product. This tends to increase the size of the ice crystals in the product and eventually results in the same type of cellular damage as occurs with slow freezing.

Since many packaging materials do not offer complete protection against dehydration, the relative humidity should be kept at a high level (85% to 90%) in frozen storage rooms, particularly for long-term storage.

Proper stacking of the product is also essential. Stacking should always be such that it permits adequate air circulation around the product. It is particularly important to leave a good size air space between the stored product and the walls of the storage room. In addition to permitting air circulation around the product, this eliminates the possibility of the product absorbing heat directly from the warm walls.

9-34. Commercial Refrigerators

The term "commercial refrigerator" is usually applied to the smaller, ready-built, refrigerated fixtures of the type used by retail stores and markets, hotels, restaurants, and institutions for the processing, storing, displaying, and dispensing of perishable commodities. The term is sometimes applied also to the larger, custom-built refrigerated fixtures and rooms used for these purposes.

Although there are a number of special purpose refrigerated fixtures which defy classification, in general, commercial fixtures can be grouped into three principal categories: (1) reach-in refrigerators, (2) walk-in coolers, and (3) display cases.

9-35. Reach-In Refrigerators

The reach-in refrigerator is probably the most versatile and the most widely used of all commercial fixtures. Typical users are grocery stores, meat markets, bakeries, drug stores, lunch counters, restaurants, florists, hotels, and institutions of all kinds. Whereas some reach-in refrigerators serve only a storage function, others are used for both storage and display (Fig. 9-8). Those serving only the storage function usually have solid doors, whereas those used for display have glazed doors.

9-36. Walk-In Coolers

Walk-in coolers are primarily storage fixtures and are available in a wide variety of sizes to fit every need. Nearly all retail stores, markets, hotels, restaurants, institutions, etc., of any size employ one or more walk-in coolers for the storage of perishables of all types. Some walk-in coolers are equipped with glazed reach-in doors. This feature is especially convenient for the storing, displaying, and dispensing of such items as dairy products, eggs, and beverages. Walk-in coolers with reach-in doors are widely used in grocery stores, particularly drive-in groceries, for handling such items.

9-37. Display Cases

The principal function of any kind of display fixture is to display the product or commodity as attractively as possible in order to stimulate sales. Therefore, in the design of refrigerated display fixtures, first consideration is given to the displaying of the product. In many cases, this is not necessarily compatible with providing the optimum storage conditions for the product being displayed. Hence, the storage life of a product in a display fixture is frequently very limited, ranging from a few hours in some

Fig. 9-8 Typical reach-in refrigerator. (Courtesy Tyler Refrigeration Corporation.)

instances to a week or more in others, depending upon the type of product and upon the type of fixture.

Display fixtures are of two general types: (1) the self-service case, from which the customer serves himself directly, and (2) the service case, from which the customer is usually served by an attendant. The former is very popular in supermarkets and other large, retail, self-service establishments, whereas the service case finds use in the smaller groceries, markets, bakeries, etc. Typical service cases are shown in Figs. 9-9 and 9-10.

Self-service cases are of two types, open and closed, with the open type gaining rapidly in popularity. With the advent of the supermarket, the trend has been increasingly toward the open type self-service case, and the older, closed type self-service cases are becoming obsolete. Several of the more popular types of open self-service cases are shown in Figs. 9-11 and 9-12. These are used to display meat, vegetables, fruit, frozen foods, ice cream, dairy products, delicatessen items, etc. The design of the case varies somewhat with the particular type of product being displayed. Too, designs are available for both wall and island installation. Although some provide additional storage space, others do not.

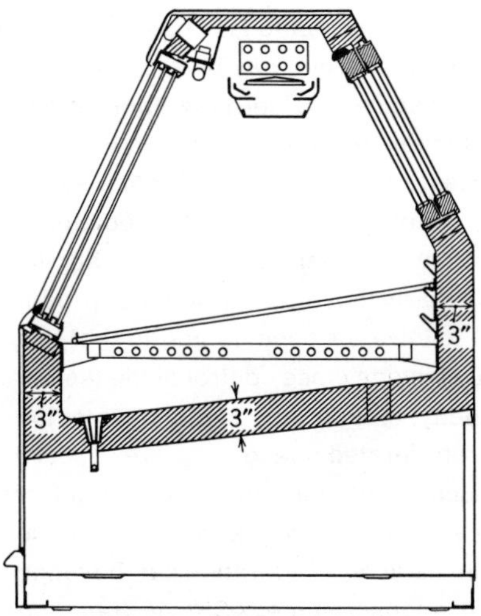

Fig. 9-9 Conventional single-duty service case for displaying meats. (Courtesy Tyler Refrigeration Corporation.)

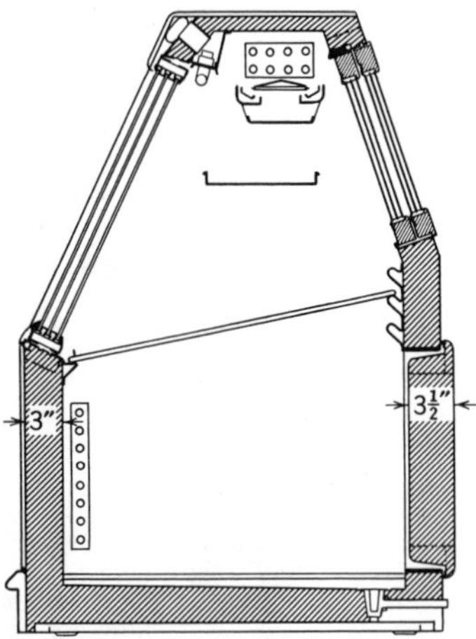

Fig. 9-10 Double-duty service case for displaying meats. (Courtesy Tyler Refrigeration Corporation.)

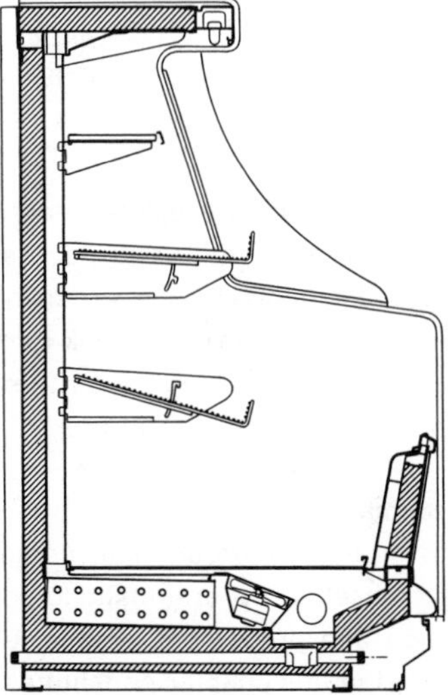

Fig. 9-11 High multishelf produce sales case. (Courtesy Tyler Refrigeration Corporation.)

9-38. Special Purpose Fixtures

Although all the refrigerated fixtures discussed in the preceding sections are available in a variety of designs in order to satisfy the specific requirements of individual products and applications, a number of special purpose fixtures are manufactured which may or may not fall into one of the three general categories already mentioned. Some of the more common special

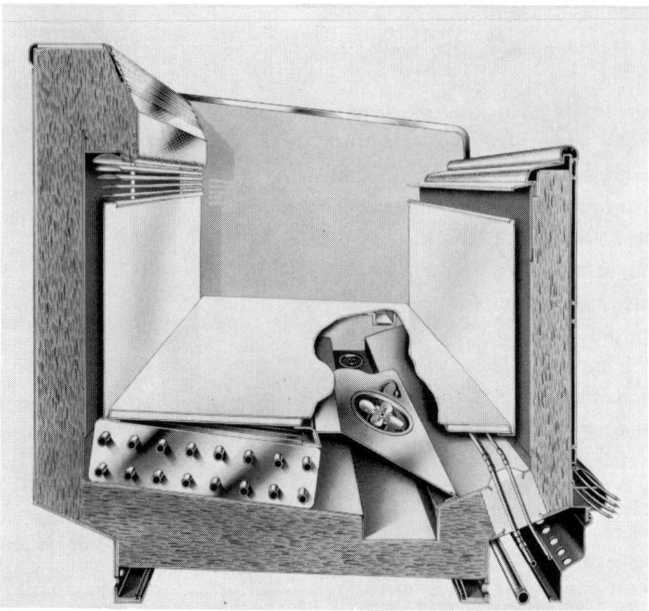

Fig. 9-12 Open-type display case for frozen foods and ice cream. (Courtesy Tyler Refrigeration Corporation.)

purpose fixtures are beverage coolers, milk coolers (dairy farm), milk and beverage dispensers, soda fountains, ice cream makers, water coolers, ice makers, back-bar refrigerators, florist boxes, dough retarders, candy cases, and mortuary refrigerators.

9-39. Summary

Recognizing that a thorough knowledge of the application itself is a prerequisite to good system design and proper equipment selection, the material in this chapter constitutes a brief survey of a few of the many applications of mechanical refrigeration, with special emphasis being given to the area of commercial refrigeration.

Obviously, the applications of mechanical refrigeration are too many and too varied to permit detailed consideration of each and every type. Fortunately, this is neither necessary nor desirable since methods of system designing and equipment selection are practically the same for all types of applications. Commercial refrigeration was selected for emphasis because this area embraces a wide range of applications and because the problems encountered in this area are representative of those in the other areas. Hence, even though the discussion in this chapter and in those which follow deals chiefly with commercial refrigeration, the principals of system design and the methods of equipment selection developed therein may be applied to all types of mechanical refrigeration applications.

Although no attempt is made in this book to discuss air conditioning as such except in a very general way, it should be pointed out that most commercial refrigeration applications, particularly those concerned with product storage, involve air conditioning in that they ordinarily include close control of the temperature, humidity, motion, and cleanliness of the air in the refrigerated space.

Much additional and detailed information concerning the many and varied applications of mechanical refrigeration can be found in the Applications Volume, *ASHRAE Handbook*, and Product Directory.

10

COOLING LOAD CALCULATIONS

10-1. The Cooling Load

The cooling load on refrigerating equipment seldom results from any one single source of heat. Rather, it is the summation of the heat which usually evolves from several different sources. Some of the more common sources of heat that supply the load on refrigerating equipment are:

1. Heat that leaks into the refrigerated space from the outside by conduction through the insulated walls.
2. Heat that enters the space by direct radiation through glass or other transparent materials.
3. Heat that is brought into the space by warm outside air entering the space through open doors or through cracks around windows and doors.
4. Heat given off by a warm product as its temperature is lowered to the desired level.
5. Heat given off by people occupying the refrigerated space.
6. Heat given off by any heat-producing equipment located inside the space, such as electric motors, lights, electronic equipment, steam tables, coffee urns, and hair driers.

It will be shown later that all of these sources of heat are not present in every application and that the importance of any one heat source with relation to the total cooling load will vary considerably with each application.

10-2. Equipment Running Time

Although refrigerating equipment capacities are normally given in Btu per hour, in refrigeration applications the total cooling load is usually calculated for a 24-hr period, that is, in Btu/24 hr. Then, to determine the required Btu per hour capacity of the equipment, the total load for the 24-hr period is divided by the desired running time for the equipment, that is,

$$\text{Required Btu/hr equipment capacity} = \frac{\text{Total cooling load, Btu/24 hr}}{\text{Desired running time (hrs)}} \tag{10-1}$$

Because of the necessity for defrosting the evaporator at frequent intervals, it is not practical to design the refrigerating system in such a way that the equipment must operate continuously in order to handle the load. In most cases, the air passing over the cooling coil is chilled to a temperature below its dew point and moisture is condensed out of the air onto the surface of the cooling coil. When the temperature of the coil surface is above the freezing temperature of water, the moisture condensed out of the air drains off the coil into the condensate pan and leaves the space through the condensate drain. However, when the temperature of the cooling coil is below the freezing temperature of water, the moisture condensed out of the air freezes into ice and adheres to the surface of the coil, thereby causing "frost" to accumulate on the coil surface. Since frost accumulation on the coil surface tends to insulate the coil and reduce the coil's capacity, the frost must be melted off periodically by raising the surface temperature of the coil above the freezing point of water and maintaining it at this level until the frost has melted off the coil and left the space through the condensate drain.

No matter how the defrosting is accomplished, the defrosting requires a certain amount of time, during which the refrigerating effect must be stopped in the coil being defrosted.*

One method of defrosting the coil is to stop the compressor and allow the evaporator to warm up to the space temperature and remain at this temperature for a sufficient length of time to allow the frost accumulation to melt off the coil. This method of defrosting is called "off-cycle" defrosting. Since the heat required to melt the frost in off-cycle defrosting must come from the air in the refrigerated space, defrosting occurs rather slowly and a considerable length of time is required to complete the process. Experience has shown that when off-cycle defrosting is used, the maximum allowable running time for the equipment is 16 hr out of each 24-hr period, the other 8 hr being allowed for the defrosting. This means, of course, that the refrigerating equipment must have sufficient capacity to accomplish the equivalent of 24 hr of cooling in 16 hr of actual running time. Hence, when off-cycle defrosting is used, the equipment running time used in Equation 10-1 is approximately 16 hr.

When the refrigerated space is to be maintained at a temperature below 34°F, off-cycle defrosting is not practical. The variation in space temperature which would be required in order to allow the cooling coil to attain a temperature sufficiently high to melt off the frost during every off cycle would be detrimental to the stored product. Therefore, where the space temperature is maintained below 34°F, some type of supplementary heat defrosting is ordinarily used. In such cases the surface of the coil is heated artificially, either with electric heating elements, with water, or with hot gas from the discharge of the compressor (see Chapter 20).

Defrosting by any of these means is accomplished much more quickly than when off-cycle defrosting is used. Hence, the shut-down time required is less for supplementary heat defrosting and the maximum allowable running time for the equipment can be greater than for the aforementioned off-cycle defrosting. For systems using supplementary heat defrosting the maximum allowable running time is usually from 18 to 20 hr out of each 24-hr period, depending upon how often defrosting is necessary for the application in question. As a general rule, the 18 hr running time is used.

It is of interest to note that since the temperature of the cooling coil in comfort air conditioning applications is normally around 40°F, no frost accumulates on the coil surface and, therefore, no down time is required for defrosting. For this reason, air conditioning systems are usually designed for continuous run and cooling loads for air conditioning applications are determined directly in Btu per hour. This holds true also for other applications where there is no frost accumulation on the cooling surface.

In cases where it is inconvenient to calculate the refrigeration load on a 24-hr basis, the load may be determined directly in Btu per hour provided that the result is multiplied by an appropriate factor which makes an allowance for the desired equipment operating time. Where the desired equipment operating time is 16 hr out of each 24 the refrigeration load in Btu per hour should be multiplied by (24/16) 1.5, which in effect increases the required equipment capacity by 50% so that the equipment selected will have the capacity to handle

* An exception to this is where a continuous brine spray is employed to keep the coil free of frost, in which case the equipment can be selected for continuous operation.

the 24-hr cooling load in the required 16 hr of operating time. Similar multipliers can be determined for other desired operating times.

10-3. Cooling Load Calculations

To simplify cooling load calculations, the total cooling load is divided into a number of individual loads according to the sources of heat supplying the load. The summation of these individual loads is the total cooling load on the equipment.

In commercial refrigeration, the total cooling load is divided into four separate loads, viz: (1) the wall gain load, (2) the air change load, (3) the product load, and (4) the miscellaneous or supplementary load.

10-4. The Wall Gain Load

The wall gain load, sometimes called the wall leakage load, is a measure of the heat flow rate by conduction through the walls of the refrigerated space from the outside to the inside. Since there is no perfect insulation, there is always a certain amount of heat passing from the outside to the inside whenever the inside temperature is below that of the outside. The wall gain load is common to all refrigeration applications and is ordinarily a considerable part of the total cooling load. Some exceptions to this are liquid chilling applications, where the outside area of the chiller is small and the walls of the chiller are well insulated. In such cases, the leakage of heat through the walls of the chiller is so small in relation to the total cooling load that its effect is negligible and it is usually neglected. On the other hand, commercial storage coolers and residential air conditioning applications are both examples of applications wherein the wall gain load often accounts for the greater portion of the total load.

10-5. The Air Change Load

When the door of a refrigerated space is opened, warm outside air enters the space to replace the more dense cold air which is lost from the refrigerated space through the open door. The heat which must be removed from this warm outside air to reduce its temperature and moisture content to the space design conditions becomes a part of the total cooling load on the equipment. This part of the total load is called the air change load.

The relationship of the air change load to the total cooling load varies with the application. Whereas in some applications the air change load is not a factor at all, in others it represents a considerable portion of the total load. For example, with liquid chillers, there are no doors or other openings through which air can pass and therefore the air change load is nonexistent. On the other hand, the reverse is true for air conditioning applications, where, in addition to the air changes brought about by door openings, there is also considerable leakage of air into the conditioned space through cracks around windows and doors and in other parts of the structure. Too, in many air conditioning applications outside air is purposely introduced into the conditioned space to meet ventilating requirements. When large numbers of people are in the conditioned space, the quantity of fresh air which must be brought in from the outside is quite large and the cooling load resulting from cooling and dehumidifying this air to the space design conditions is often a large part of the total cooling load in such applications.

In air conditioning applications, the air change load is called either the ventilating load or the infiltration load. The term ventilating load is used when the air changes in the conditioned space are the result of deliberate introduction of outside air into the space for ventilating purposes. The term infiltration load is used when the air changes are the result of the natural infiltration of air into the space through cracks around doors and other openings. Every air conditioning application will involve either an infiltration load or a ventilating load, but not both in the same application.

Since the doors on commercial refrigerators are equipped with well-fitted gaskets, the cracks around the doors are tightly sealed and there should be little, if any, leakage of air around the doors of a commercial fixture in good condition. Hence, in commercial refrigeration, the air changes are usually limited to those which are brought about by actual opening and closing of the door or doors.

The importance of minimizing or eliminating the leakage of air from the outside to the inside of coolers and freezers through the cracks around doors and through other openings cannot be overemphasized. Although such air leakage may or may not have an appreciable effect on the refrigeration load, the water vapor that condenses from the warm air in the affected crack spaces and that frequently freezes into ice in these openings can be very troublesome and should be prevented.

In addition to well-aligned and well-fitted doors and gaskets and the careful sealing of other wall openings to reduce air leakage, good design practice prescribes the use of a heater wire around the perimeter of the door to prevent condensation on these surfaces by maintaining their temperature above the DP temperature of the entering warm air. In addition, the installation of a small heated air vent to equalize the pressure between the inside and outside of the cooler or freezer is also important. Without such controlled venting, negative pressures can develop inside the cooler or freezer as a result of the air pressure drop that occurs in accordance with Charles' Law when the temperature of the warm air, which enters the space as the doors are opened and closed, is subsequently reduced to the space design temperature. Naturally, any air pressure differential between the inside and outside of the fixture greatly increases the tendency for air leakage around the door seals and through other wall openings, such as those through which refrigerant and water piping, drain lines, and electrical conduits pass to the outside.

10-6. The Product Load

The product load is made up of the heat that must be removed from the refrigerated product in order to reduce the temperature of the product to the desired level. The term "product" as used here is taken to mean any material whose temperature is reduced by the refrigerating equipment and includes not only perishable commodities, such as foodstuff, but also such items as welding electrodes, masses of concrete, plastic, rubber, and liquids of all kinds. In some instances, the product is frozen, in which case the latent heat removed is also a part of the product load.

The importance of the product load in relation to the total cooling load, like all others, varies with the application. Although it is nonexistent in some applications, in others it represents practically the entire cooling load. Where the refrigerated cooler is designed for product storage, the product is usually chilled to the storage temperature before being placed in the cooler and no product load need be considered since the product is already at the storage temperature.* However, in any instance where the product enters a storage cooler at a temperature above the storage temperature, the quantity of heat which must be removed from the product in order to reduce its temperature to the storage temperature must be considered as a part of the total load on the cooling equipment.

In some few instances, the product enters the storage fixture at a temperature below the normal storage temperature for the product. A case in point is ice cream, which is frequently chilled to a temperature of 0°F or −10°F during the hardening process, but is usually stored at about 10°F, which is the ideal dipping temperature. When such a product enters storage at a temperature below the space temperature, it will absorb heat from the storage space as it warms up to the storage temperature and thereby produce a certain amount of refrigerating effect of its own. In other words, it provides what might be termed a negative product load which could theoretically be subtracted from the total cooling load. This is never done, however, since the refrigerating effect produced is small and is not continuous in nature.

The cooling load on the refrigerating equipment resulting from product cooling may be either intermittent or continuous, depending on the application. The product load is a part of the total cooling load only while the temperature of the product is being reduced to the

* In the design of short-term storage coolers, general practice is to allow for 10°F of product cooling.

storage temperature, or while freezing is taking place. Once the product is cooled to the storage temperature, it is no longer a source of heat and the product load ceases to be a part of the load on the equipment. An exception to this is in the storage of fruit and vegetables which give off respiration heat for the entire time they are in storage at temperatures above the freezing temperature, even though there is no further decrease in their temperature (see Section 10-18).

There are, of course, a number of refrigeration applications where product cooling is more or less continuous, in which case the product load is a continuous load on the equipment. This is true, for instance, in chilling coolers where the primary function is to chill the warm product to the desired storage temperature.* When the product has been cooled to the storage temperature, it is usually moved out of the chilling room into a storage room and the chilling room is then reloaded with warm product. In such cases, the product load is continuous and is usually a large part of the total load on the equipment.

Liquid chilling is another application wherein the product provides a continuous load on the refrigerating equipment. The flow of the liquid being chilled through the chiller is continuous with warm liquid entering the chiller and cold liquid leaving. In this instance, the product load is practically the only load on the equipment since there is no air change load and the wall gain load is negligible, as is the miscellaneous load.

In air conditioning applications there is no product load as such, although there is sometimes a "pull-down load" which tends to disappear as steady-state design conditions are attained.

10-7. The Miscellaneous Load

The miscellaneous load, sometimes referred to as the supplementary load, takes into account all miscellaneous sources of heat. Chief among these are people working in or otherwise occupying the refrigerated space along with lights or other electrical equipment operating inside the space.

In most commercial refrigeration applications the miscellaneous load is relatively small, usually consisting only of the heat given off by lights and fan motors used inside the space.

In air conditioning applications, there is no miscellaneous load as such. This is not to say that human occupancy and equipment are not a part of the cooling load in air conditioning applications. On the contrary, people and equipment are often such large factors in the air conditioning load that they are considered as separate loads and are calculated as such. For example, in those air conditioning applications where large numbers of people occupy the conditioned space, such as churches, theaters, restaurants, etc., the cooling load resulting from human occupancy is frequently the largest single factor in the total load. Also, many air conditioning systems are installed for the sole purpose of cooling electrical, electronic, and other types of heat-producing equipment. In such cases, the equipment usually supplies the greater portion of the cooling load.

10-8. Factors Determining the Wall Gain Load

The quantity of heat transmitted through the walls of a refrigerated space per unit of time is the function of three factors whose relationship is expressed in the following equation:

$$Q = (A)(U)(D) \qquad (10\text{-}2)$$

where Q = the quantity of heat transferred in Btu per hour
A = the outside surface area of the wall (square feet)
U = the overall coefficient of heat transmission in Btu per hour per square foot per degree Fahrenheit
D = the temperature differential across the wall in degrees Fahrenheit

The coefficient of transmission or "U" factor is a measure of the rate at which heat will pass through a 1 ft^2 area of wall surface from the air on one side to the air on the other side

* In some instances, the product is not completely chilled to the final storage temperature in the chilling cooler, and some additional chilling is accomplished in the holding storage cooler.

for each 1°F of temperature difference across the wall. The value of the U factor is given in Btu per hour and depends on the thickness of the wall and on the materials used in the wall construction. Since it is desirable to prevent as much heat as possible from entering the space and becoming a load on the cooling equipment, the materials used in the construction of cold storage walls should be good thermal insulators so that the value of U is kept as low as is practical.

According to Equation 10-2, once the U factor is established for a wall, the rate of heat flow through the wall varies directly with the surface area of the wall and with the temperature differential across the wall. Since the value of U is given in Btu per hour per square foot per degree Fahrenheit, the total quantity of heat passing through any given wall in 1 hr can be determined by multiplying the U factor by the wall area in square feet and by the temperature difference across the wall in degrees Fahrenheit, that is, by application of Equation 10-2.

Example 10-1 Determine the total quantity of heat in Btu per hour which will pass through a wall 10 ft by 20 ft, if the U factor for the wall is 0.16 Btu/(hr)(ft^2)(°F) and the temperature on one side of the wall is 40°F while the temperature on the other side is 95°F.

Solution

Total wall area = (10 ft)(20 ft) = 200 ft^2

Temperature differential across wall, °F = 95° − 40° = 55°F

Applying Equation 10-2, the heat gain through the wall = (200)(0.16)(55) = 1760 Btu/hr

Since the value of U in Equation 10-2 is in Btu per hour, the result obtained from Equation 10-2 is in Btu per hour. To determine the wall gain load in Btu per 24 hr as required in refrigeration load calculations, the result of Equation 10-2 is multiplied by 24 hr. Hence, for calculation cooling loads in refrigeration applications, Equation 10-2 is written to include this multiplication, viz:

$$Q = A \times U \times D \times 24 \qquad (10\text{-}3)$$

10-9. Determination of the *U* Factor

Overall coefficients of transmission or U factors have been determined for various types of wall construction and these values are available in tabular form. Tables 10-1 through 10-3 list U values for various types of cold storage walls.

Example 10-2 From Table 10-1, determine the U factor for a wall constructed of 6-in. clay tile with 4 in. of corkboard insulation.

Solution From Table 10-1, first select the appropriate type of wall construction (third from top). In the next column select the desired thickness of clay tile (6 in.) and move to the right to the column listing values for 4 in. of insulation. Read the U factor of the wall, 0.064 Btu/(hr)(ft^2)(°F).

Should it be necessary, the U factor for any type of wall construction can be readily calculated provided that either the conductivity or the conductance of each of the materials used in the wall construction is known. The conductivity or conductance of most of the materials used in wall construction can be found in tables. Also, this information is usually available from the manufacturer or producer of the material. Table 10-4 lists the thermal conductivity or the conductance of some materials frequently used in the construction of cold storage walls.

The thermal conductivity or k factor of a material is the rate in Btu per hour at which heat passes through a 1 ft^2 cross section of the material 1 in. thick for each 1°F of temperature difference across the material and is given in Btu per hour per square foot per degree Fahrenheit per inch thickness.

Whereas the thermal conductivity or k factor is available only for homogeneous materials and the value given is always for a 1 in. thickness of the material, the thermal conductance or C factor is available for both homogeneous and nonhomogeneous materials and the value given in Btu per hour per square foot per degree Fahrenheit is for the specified thickness of the material.

TABLE 10-1 Heat Transmission Coefficients (U) for Cold Storage Rooms

Btu per hour per square foot per degree Fahrenheit difference between air on the two sides. Wind velocity 15 mph.

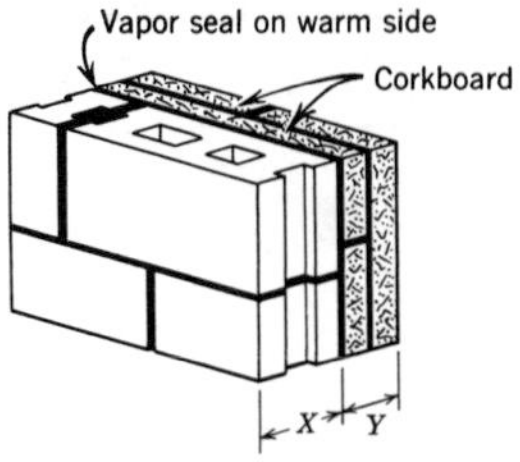

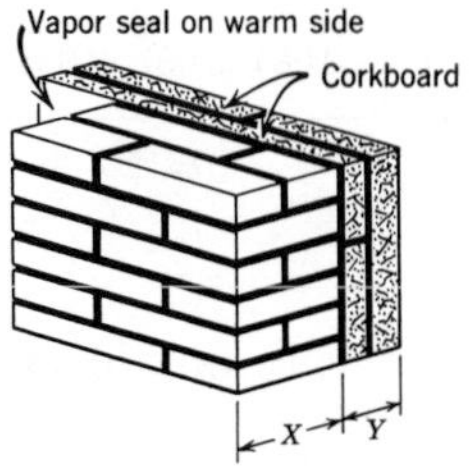

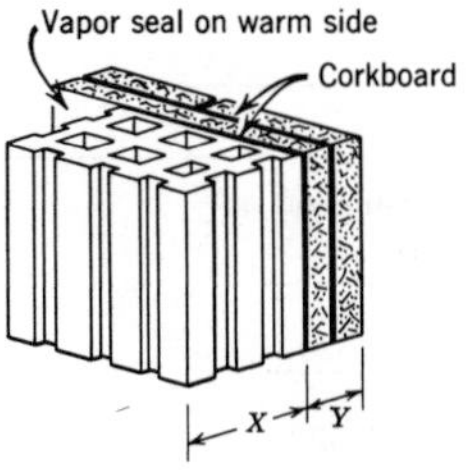

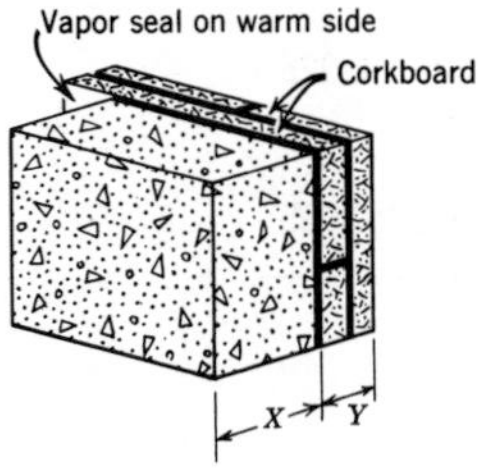

Wall Thickness X Inches	Thickness of Insulation, Y Inches						
	2	3	4	5	6	7	8
Concrete block 8	0.12	0.085	0.066	0.054	0.046	0.040	0.035
Concrete block 12	0.12	0.083	0.065	0.053	0.045	0.039	0.035
Cinder block 8	0.11	0.081	0.064	0.052	0.045	0.039	0.034
Cinder block 12	0.11	0.079	0.063	0.052	0.044	0.039	0.034
Common brick 8	0.11	0.081	0.064	0.053	0.045	0.039	0.034
Common brick 12	0.10	0.076	0.061	0.050	0.043	0.038	0.034
Clay tile 4	0.12	0.085	0.066	0.054	0.046	0.040	0.035
Clay tile 6	0.11	0.081	0.064	0.053	0.045	0.039	0.035
Clay tile 8	0.11	0.081	0.064	0.052	0.045	0.039	0.034
Concrete 6	0.13	0.089	0.069	0.056	0.047	0.041	0.036
Concrete 8	0.12	0.087	0.068	0.055	0.047	0.040	0.036
Concrete 10	0.12	0.086	0.067	0.055	0.046	0.040	0.035
Concrete 12	0.12	0.085	0.066	0.054	0.046	0.040	0.035

From *Carrier Design Data*. Reproduced by permission of Carrier Corporation.

TABLE 10-2 Heat Transmission Coefficients (*U*) for Cold Storage Rooms

Btu per hour per square foot per degree Fahrenheit difference between the air on the two sides. Outside wind velocity 15 mph.

Type of Construction	Insulating Material	Thickness of Insulation (Inches)						
		3⅝	5⅝	2	3	4	5	6
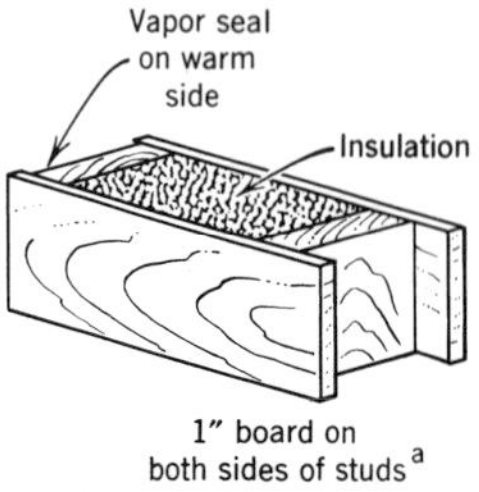 1″ board on both sides of studs[a] (Vapor seal on warm side; Insulation)	Granulated cork	0.079	0.055					
	Rock or palco wool	0.072	0.050					
	Sawdust	0.097	0.069					
	Corkboard	-----	-----	0.11	0.084	0.067	0.055	0.047
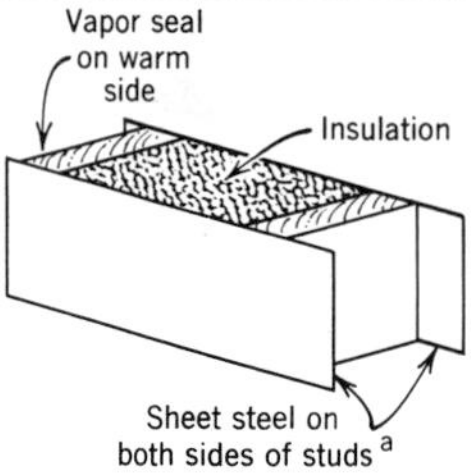 Sheet steel on both sides of studs[a] (Vapor seal on warm side; Insulation)	Glass or rock wool fill	0.084	0.055	-----	0.100	0.077	0.062	0.052

Type of Construction	Insulating Material	Thickness of Insulation (Inches)		
		8	10	12
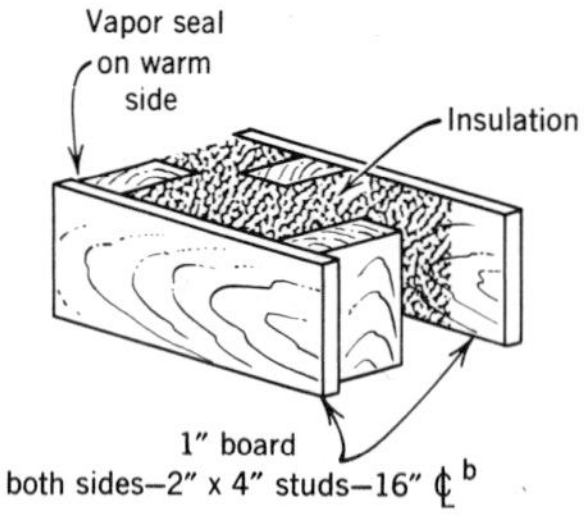 1″ board both sides—2″ x 4″ studs—16″ ℄[b] (Vapor seal on warm side; Insulation)	Granulated cork	0.040	0.033	0.027
	Palco or rock wool	0.036	0.029	0.025
	Sawdust	0.051	0.042	0.035

NOTES:

[a] Coefficients corrected for 2 x 4 or 2 x 6 studs, on 16 in. centers.

[b] Coefficients corrected for 2 x 4 studs

* Actual thickness 25/32 in.

From *Carrier Design Data*. Reproduced by permission of Carrier Corporation.

TABLE 10-3 Heat Transmission Coefficients (*U*) for Cold Storage Rooms

Btu per hour per square foot per degree Fahrenheit difference between air on the two sides. Wind velocity 15 mph.

	Wall, Floor or Ceiling Thickness X (Inches)	Thickness of Insulation, Y Inches						
		2	3	4	5	6	7	8
Self-supporting partition* 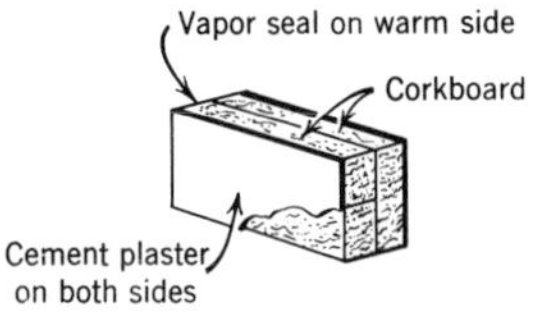 (Vapor seal on warm side; Corkboard; Cement plaster on both sides)	Cork partition	0.13	0.089	0.069	0.056	0.047	0.041	0.036
Floor* 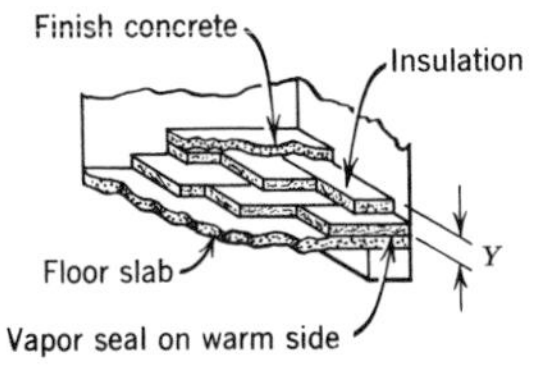 (Finish concrete; Insulation; Floor slab; Vapor seal on warm side)		Corkboard[a]						
	Slab 2 Finish 2	0.12	0.087	0.067	0.055	0.046	0.040	0.035
	Slab 5 Finish 3	0.12	0.084	0.066	0.054	0.046	0.040	0.035
	Slab 6 Finish 4	0.11	0.083	0.065	0.054	0.045	0.039	0.035
		Foamglas[a]						
	Slab 2 Finish 2	0.15	0.11	0.087	0.071	0.060	0.053	0.046
	Slab 5 Finish 3	0.15	0.11	0.084	0.070	0.059	0.052	0.046
	Slab 6 Finish 4	0.14	0.10	0.083	0.069	0.059	0.051	0.045
Ceiling* 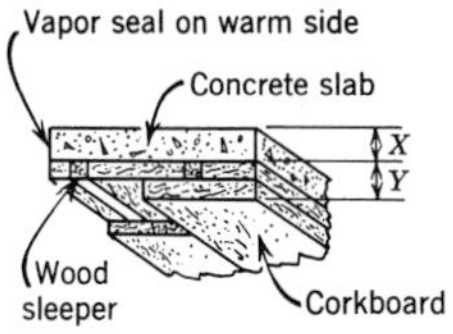 (Vapor seal on warm side; Concrete slab; Wood sleeper; Corkboard)	Concrete 4	0.12	0.089	0.069	0.056	0.048	0.042	0.036
	Concrete 8	0.12	0.086	0.067	0.055	0.047	0.041	0.036
Ceiling* 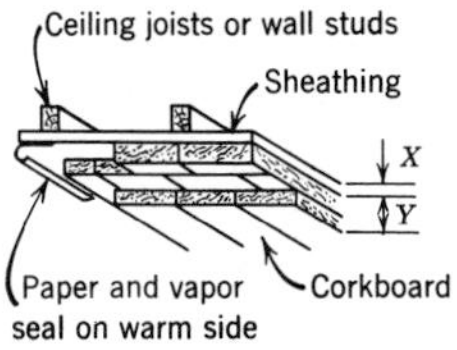 (Ceiling joists or wall studs; Sheathing; Paper and vapor seal on warm side; Corkboard)	Wood $^{25}/_{32}$ (actual)	0.11	0.082	0.064	0.053	0.045	0.039	0.035
Ceiling* 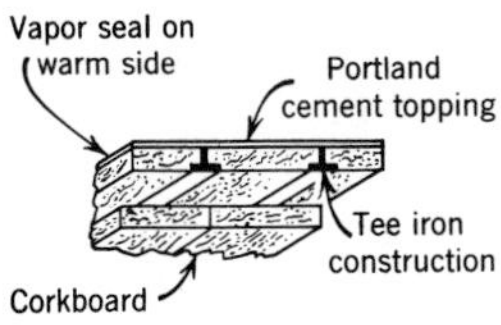 (Vapor seal on warm side; Portland cement topping; Tee iron construction; Corkboard)		0.13	0.092	0.072	0.059	0.050	0.043	0.038

[a] These values may also be used for floors on ground.

* Surface conductance for still air, 1.65, used on both sides

From *Carrier Design Data.* Reproduced by permission of Carrier Corporation.

TABLE 10-4 Thermal Conductivity of Materials Used in Cold Storage Walls

Material	Description	Thermal Conductivity (k)*	Thermal Conductance (C)*
Masonry	Brick, common	5.0	
	Brick, face	9.0	
	Concrete, mortar or plaster	5.0	
	Concrete, sand aggregate	12.0	
	Concrete block		
	Sand aggregate 4 in.		1.40
	Sand aggregate 8 in.		0.90
	Sand aggregate 12 in.		0.78
	Cinder aggregate 4 in.		0.90
	Cinder aggregate 8 in.		0.58
	Cinder aggregate 12 in.		0.53
	Gypsum plaster 1/2 in.		3.12
	Tile, hollow clay 4 in.		0.90
	Tile, hollow clay 6 in.		0.66
	Tile, hollow clay 8 in.		0.54
Woods	Maple, oak, similar hardwoods		1.10
	Fir, pine, similar softwoods		0.80
	Plywood $\frac{1}{2}$ in.		1.60
	Plywood $\frac{3}{4}$ in.		1.07
Roofing	Asphalt roll roofing	6.50	0.15
	Built-up roofing $\frac{3}{8}$ in.	3.00	0.33
Insulating materials	Blanket or batt, mineral or glass fiber	0.27	
	Board or slab		
	Cellular glass	0.40	
	Corkboard	0.30	
	Glass fiber	0.25	
	Expanded polystyrene	0.20	
	Expanded polyurethane	0.17	
	Loose fill		
	Milled paper or wood pulp	0.27	
	Sawdust or shavings	0.45	
	Mineral wool (rock, glass, slag)	0.27	
	Redwood bark	0.26	
	Wood fiber (soft woods)	0.30	
Surface conductance (convection coefficient)	Still air		1.65
	Moving air (7.5 mph)		4.00
	Moving air (15 mph)		6.00
Glass	Single pane		1.13
	Two pane		0.46
	Three pane		0.29
	Four pane		0.21

From *ASHRAE Data Book*, Fundamentals Volume, 1972 Edition, by permission of the American Society of Heating, Refrigerating, and Air-Conditioning Engineers.

For any homogeneous material, the thermal conductance can be determined for any given thickness of the material by dividing the *k* factor by the thickness in inches. Hence, for a homogeneous material,

$$C = \frac{k}{x} \tag{10-4}$$

where x = the thickness of material in inches.

Example 10-3 Determine the thermal conductance for a 5 in. thickness of corkboard.

Solution

From Table 10-4, *k* factor of corkboard $= 0.30\ (\text{Btu})(\text{in.})/(\text{hr})(\text{ft}^2)(°\text{F})$

Applying Equation 10-4, $C = \dfrac{0.30}{5}$

$= 0.06\ \text{Btu}/(\text{hr})(\text{ft}^2)(°\text{F})$

Since the rate of heat transmission through nonhomogeneous materials, such as the concrete building block in Fig. 10-1, will vary in the several parts of the material, the *C* factor from nonhomogeneous materials must be determined by experiment.

The resistance that a wall or a material offers to the flow of heat is inversely proportional to the ability of the wall or material to transmit heat. Hence, the overall thermal resistance of a wall can be expressed as the reciprocal of the overall coefficient of transmission, whereas the thermal resistance of an individual material can be expressed as the reciprocal of its conductivity or conductance, that is,

overall thermal resistance, $R = \dfrac{1}{U}$

Thermal resistance of an individual material $= \dfrac{1}{k}$ or $\dfrac{1}{C}$ or $\dfrac{x}{k}$

The terms $1/k$ and $1/C$ express the resistance to heat flow through a single material from surface to surface only and do not take into account the thermal resistance of the thin film of air which adheres to all exposed surfaces. In determining the overall thermal resistance to the flow of heat through a wall from the air on one side to the air on the other side, the resistance of the air on both sides of the wall should be considered. Air film coefficients or surface conductances for average wind velocities are given in Table 10-4.

When a wall is constructed of several layers of different materials the total thermal resistance of the wall is the sum of the resistances of the individual materials in the wall construction, including the air films, that is,

$$\frac{1}{U} = \frac{1}{f_i} + \frac{x}{k_1} + \frac{x}{k_2} + \frac{x}{k_n} + \frac{1}{f_0} \tag{10-5}$$

Therefore,

$$U = \frac{1}{\dfrac{1}{f_i} + \dfrac{x}{k_1} + \dfrac{x}{k_2} + \cdots \dfrac{x}{k_n} + \dfrac{1}{f_0}}$$

where $\dfrac{1}{f_i}$ = convection coefficient (surface conductance) of inside wall, floor, or ceiling

$\dfrac{1}{f_0}$ = convection coefficient (surface conductance) of outside wall, floor, or roof

Note When nonhomogeneous materials are used, $1/C$ is substituted for x/k.

Example 10-4 Assuming a wind velocity of 7.5 mph, calculate the value of *U* for a wall

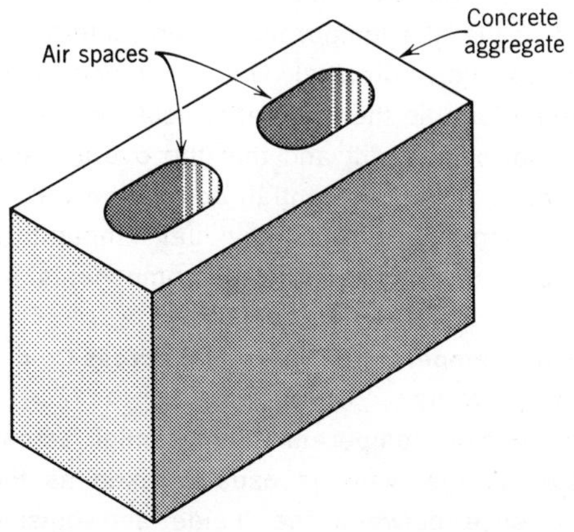

Fig. 10-1 Concrete aggregate building block.

constructed of 8 in. cinder aggregate building blocks, insulated with 4 in. of corkboard, and finished on the inside with 0.5 in. of cement plaster.

Solution

From Table 10-4,

8 in. cinder aggregate block	$C = 0.58$
Corkboard	$k = 0.30$
Cement plaster	$k = 5.00$
Inside convection coefficient	$f_i = 1.65$
Outside convection coefficient	$f_0 = 4.00$

Applying Equation 10-5, the over-all thermal resistance,

$$1/U = \frac{1}{4} + \frac{1}{0.58} + \frac{4}{0.3} + \frac{0.5}{5} + \frac{1}{1.65}$$

$$= 0.25 + 1.724 + 13.333$$
$$= + 0.1 + 0.607$$
$$= 16.01$$

Therefore, $U = 1/16.01$

$$= 0.062 \text{ Btu/(hr)(ft}^2\text{)(°F)}$$

For the most part, it is the insulating material used in the wall construction that determines the value of U for cold storage walls. The surface conductances and the conductances of the other materials in the wall have very little effect on the value of U because the thermal resistance of the insulating material is so large with relation to that of the air films and other materials. Therefore, for small coolers, it is sufficiently accurate to use the conductance of the insulating material alone as the wall U factor.

A realization of the full insulating values listed for the walls and materials described in the tables is dependent on proper installation and good workmanship. For example, many insulating materials are permeable to water vapor, and unless an adequate and unbroken vapor barrier is installed on the warm side of the insulating material, the higher vapor pressure on the warm side of the insulation will cause water vapor to migrate through the insulation toward the lower vapor pressure on the colder side. At some point in the travel of the vapor through the insulation, the temperature of the vapor will fall below the saturation temperature corresponding to its pressure, thereby causing the vapor to condense into water in the insulation, with the result that the insulation eventually becomes waterlogged. Moreover, in those cases where the temperature in the insulation is below the freezing temperature of water, the condensed vapor will subsequently freeze into ice. In either case, since both water and ice are good conductors, the insulating value of the insulating material will be completely destroyed.

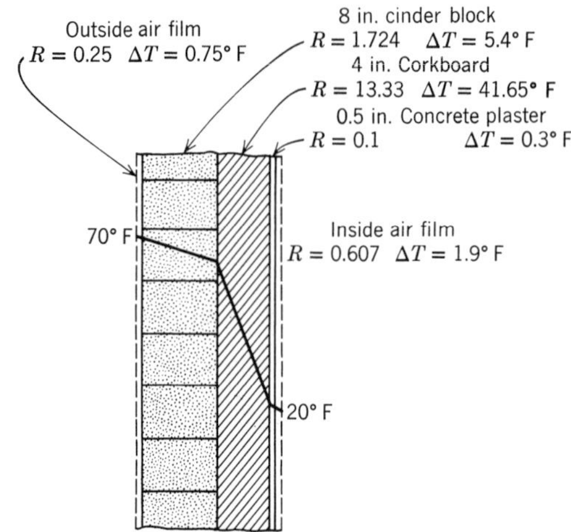

Fig. 10-2 Temperature gradiant through a typical cold storage (see Example 10-4).

The temperature gradient through a typical cold storage wall is illustrated in Fig. 10-2. Notice that the temperature drop (difference) across the individual wall components is proportional to the thermal resistance offered by that component and that the overall temperature drop (differential) across the wall is the summation of the individual temperature drops across the several wall components.

10-10. Temperature Differential across Cold Storage Walls

The design temperature differential across cold storage walls is usually taken as the difference between the inside and outside design temperatures.

The inside design temperature is that which is to be maintained inside the refrigerated space and usually depends upon the type of product to be stored and the length of time the product is to be kept in storage. The recommended storage temperatures for various products are given in Tables 10-9 through 10-12.

The outside design temperature depends on the location of the cooler. For cold storage walls located inside a building, the outside design temperature for the cooler wall is taken as the inside temperature of the building. When cold storage walls are exposed to the outdoors, the outdoor design temperature for the region (Table 10-5) is used as the outside design temperature. The outdoor design temperatures given in Table 10-5 are average outdoor temperatures and include an allowance for normal variations in the outdoor design dry bulb temperature during a 24-hr period. These temperatures should not be used for calculating air conditioning loads.

TABLE 10-5 Refrigeration Design Ambient Temperature Guide*

Location	Average Ambient Temp.	Maximum Ambient Temp.	Ground Temperature
Alabama			
Birmingham	88	99	70°F
Mobile	88	97	75
Arizona			
Flagstaff	75	90	60
Phoenix	100	113	80
Tucson	84	98	80
Arkansas			
Fort Smith	91	103	70
Little Rock	90	100	70
California			
Bakersfield	96	114	75
Fresno	94	111	80
Los Angeles	83	94	75
Oakland	75	89	65
Sacramento	90	108	80
San Diego	75	80	65
San Francisco	75	83	65
Colorado			
Colorado Springs	83	94	60
Denver	83	98	60
Grand Junction	88	102	60
Pueblo	83	100	55
Connecticut			
Hartford	83	94	65
New Haven	83	95	65
New London	83	93	65
Norwalk	83	96	65
Delaware			
Dover	87	96	65
Milford	87	98	65
Wilmington	87	94	65

TABLE 10-5 (*Continued*)

Location	Average Ambient Temp.	Maximum Ambient Temp.	Ground Temperature
District of Columbia			
Washington	89	98	65
Florida			
Jacksonville	88	96	80
Miami	88	90	80
Orlando	88	97	80
Tallahassee	88	100	80
Tampa	88	95	80
Georgia			
Atlanta	87	95	70
Savannah	89	99	75
Idaho			
Boise	89	105	60
Pocatello	83	100	60
Illinois			
Cairo	89	101	60
Chicago	87	98	60
Peoria	88	100	60
Quincy	90	103	60
Rockford	87	101	60
Springfield	90	102	60
Indiana			
Evansville	90	100	65
Fort Wayne	87	100	60
Indianapolis	89	99	60
South Bend	87	101	60
Terre Haute	90	100	65
Iowa			
Burlington	90	101	60
Davenport	90	100	60
Des Moines	90	102	60
Dubuque	90	99	60
Keokuk	90	101	60
Mason City	86	97	60
Sioux City	90	102	60
Kansas			
Concordia	93	108	60
Dodge City	92	106	60
Hutchinson	92	108	60
Salina	95	111	60
Topeka	92	105	60
Wichita	91	104	60
Kentucky			
Lexington	86	98	65
Louisville	88	99	65

TABLE 10-5 (*Continued*)

Location	Average Ambient Temp.	Maximum Ambient Temp.	Ground Temperature
Louisiana			
Baton Rouge	88	98	75
New Orleans	89	98	75
Shreveport	92	102	70
Maine			
Eastport	70	81	60
Portland	81	93	60
Maryland			
Baltimore	89	99	65
Cumberland	87	102	65
Massachusetts			
Boston	84	94	65
Fall River	81	90	60
Lawrence	81	94	60
Worcester	81	92	60
Michigan			
Alpena	82	95	60
Detroit	86	99	60
Grand Rapids	86	98	60
Jackson	86	99	60
Lansing	86	96	60
Marquette	81	96	60
Saginaw	88	101	60
Minnesota			
Duluth	79	92	50
Minneapolis	90	102	55
St. Cloud	88	101	55
Mississippi			
Jackson	90	99	75
Vicksburg	90	96	75
Missouri			
Hannibal	90	102	60
Kansas City	92	103	60
St. Joseph	92	103	60
St. Louis	92	103	60
Springfield	88	98	60
Montana			
Billings	85	104	55
Butte	75	96	55
Havre	82	99	50
Helena	82	102	55
Nebraska			
Lincoln	94	106	60
North Platte	89	103	55
Omaha	92	104	60

TABLE 10-5 (*Continued*)

Location	Average Ambient Temp.	Maximum Ambient Temp.	Ground Temperature
Nevada			
Reno	84	101	65
Tonopah	84	96	70
New Hampshire			
Concord	81	92	55
New Jersey			
Atlantic City	83	92	70
Paterson	85	95	70
Trenton	85	96	70
New Mexico			
Albuquerque	83	99	70
Santa Fe	81	90	65
New York			
Albany	83	96	60
Binghamton	83	94	60
Buffalo	80	89	65
Elmira	83	97	60
New York	85	93	65
Poughkeepsie	83	95	60
Rochester	83	95	60
Syracuse	83	96	60
Watertown	83	93	60
North Carolina			
Asheville	81	93	70
Charlotte	86	98	70
Raleigh	86	98	70
Wilmington	86	95	75
Winston-Salem	86	97	75
North Dakota			
Bismarck	87	103	50
Devils Lake	84	100	50
Ohio			
Akron	86	98	65
Canton	86	97	65
Cincinnati	88	100	65
Cleveland	83	95	65
Columbus	88	98	60
Dayton	88	99	65
Toledo	87	99	60
Youngstown	86	97	60
Oklahoma			
Oklahoma City	92	104	65
Tulsa	92	105	65
Oregon			
Portland	81	95	70

TABLE 10-5 (*Continued*)

Location	Average Ambient Temp.	Maximum Ambient Temp.	Ground Temperature
Pennsylvania			
Altoona	82	96	65
Erie	83	92	65
Harrisburg	85	97	70
Philadelphia	87	97	70
Pittsburgh	85	96	65
Scranton	82	95	65
Rhode Island			
Providence	83	94	65
South Carolina			
Charleston	88	98	75
Columbia	88	99	75
South Dakota			
Huron	93	107	55
Pierre	94	110	55
Rapid City	87	103	55
Sioux Falls	88	102	55
Tennessee			
Chattanooga	87	98	70
Knoxville	87	98	70
Memphis	89	99	70
Nashville	87	98	70
Texas			
Dallas	92	102	70
El Paso	92	102	70
Fort Worth	92	104	70
Houston	92	99	75
San Antonio	92	102	75
Utah			
Modena	80	97	60
Salt Lake City	88	101	60
Vermont			
Burlington	80	91	60
Virginia			
Lynchburg	87	99	75
Norfolk	87	95	75
Richmond	87	98	70
Washington			
Olympia	75	90	60
Seattle	75	86	75
Spokane	75	102	60
Walla Walla	87	105	60
West Virginia			
Charleston	87	102	65
Clarksburg	84	97	65

TABLE 10-5 (*Continued*)

Location	Average Ambient Temp.	Maximum Ambient Temp.	Ground Temperature
Huntington	87	100	65
Parkersburg	86	98	65
Wheeling	86	101	65
Wisconsin			
Green Bay	85	97	55
La Crosse	87	99	55
Madison	87	96	55
Milwaukee	87	99	55
Wyoming			
Cheyenne	79	94	55
Lander	80	98	55
Sheridan	86	102	55

* Do not use these temperatures for air conditioning design.

From *ASRE Data Book*, Design Volume, 1949 Edition, by permission of the American Society of Heating, Refrigerating, and Air-Conditioning Engineers.

10-11. Temperature Differential across Ceilings and Floors

When a cooler is located inside a building and there is adequate clearance between the top of the cooler and the ceiling of the building to allow free circulation of air over the top of the cooler, the ceiling of the cooler is treated the same as an inside wall. Likewise, when the top of the cooler is exposed to the outdoors, the ceiling is treated as an outdoor wall. The same holds true for floors except when the cooler floor is laid directly on a slab on the ground. As a general rule, the ground temperature under a slab varies only slightly the year round and is always considerably less than the outdoor design dry bulb temperature for the region in summer. Ground temperatures used in determining the temperature differential across the floor of cold storage rooms are given in Table 10-5 and are based on the regional outdoor design dry bulb temperature for winter.

Where the floor of a freezer is laid directly on a slab on the ground, some provision should be made to prevent the migration and eventual freezing of ground water under the freezer floor, since the expansive forces created by such freezing will eventually cause ground heaving and severe damage to the freezer structure. Preventive measures usually include some means of monitoring the ground temperature under the freezer, along with some means of supplying heat to the subsurface when ground temperatures under the freezer approach the freezing point. Warm air ducts, electric heating cables, and pipe coils for the circulation of brine or antifreeze solutions have all been employed to supply the heat necessary to maintain the ground temperature above the freezing point.

10-12. Effect of Solar Radiation

Whenever the walls of a refrigerator are so situated that they receive an excessive amount of heat by radiation, either from the sun or from some other hot body, the outside surface temperature of the wall will usually be considerably above the temperature of the ambient air. A familiar example of this phenomenon is the excessive surface temperature of an automobile parked in the sun. The temperature of the metal surface is much higher than that of the surrounding air. The amount by which the surface temperature exceeds the surrounding air temperature depends upon the amount of radiant energy striking the surface and upon the reflectivity of the surface. Recall (Section 2-18) that radiant energy waves are either reflected by or absorbed by any opaque material that they strike. Light-colored, smooth surfaces will tend to reflect more and absorb less radiant energy than dark, rough-textured surfaces. Hence, the surface temperature of smooth, light-colored walls will be somewhat lower than that of dark, rough-textured walls under the same conditions of solar radiation.

Since any increase in the outside surface temperature will increase the temperature differential across the wall, the temperature differential across sunlit walls must be corrected to compensate for the sun effect. Correction factors for sunlit walls are given in Table 10-6. These values are added to the normal temperature differential. For walls facing at angles to the directions listed in Table 10-1, average values can be used.

TABLE 10-6 Allowance for Solar Radiation

(Degrees Fahrenheit to be added to the normal temperature difference for heat leakage calculations to compensate for sun effect—not to be used for air-conditioning design)

Type of Surface	East Wall	South Wall	West Wall	Flat Roof
Dark-colored surfaces such as: Slate roofing, Tar roofing, Black paints	8	5	8	20
Medium-colored surfaces, such as: Unpainted wood, Brick, Red tile, Dark cement, Red, gray, or green paint	6	4	6	15
Light-colored surfaces, such as: White stone, Light-colored cement, White paint	4	2	4	9

From *ASRE Data Book*, Design Volume, 1957–1958 Edition, by permission of the American Society of Heating, Refrigerating, and Air-Conditioning Engineers.

10-13. Calculating the Wall Gain Load

In determining the wall gain load, the heat gain through all the walls, including the floor and ceiling, should be taken into account. When the several walls or parts of walls are of different construction and have different U factors, the heat leakage through the different parts is computed separately. Walls having identical U factors may be considered together, provided that the temperature differential across the walls is the same. Too, where the difference in the value of U is slight and/or the wall area involved is small, the difference in the U factor can be ignored and the walls or parts of walls can be grouped together for computation.

Example 10-5 A walk-in cooler 16 ft × 20 ft × 10 ft high is located in the southeast corner of a store building in Dallas, Texas (Fig. 10-3).

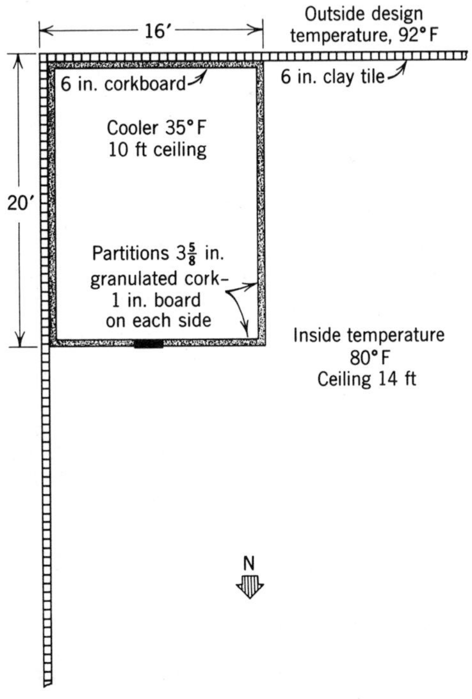

Fig. 10-3

The south and east walls of the cooler are adjacent to and a part of the south and east walls of the store building. The store has a 14 ft ceiling so that there is a 4 ft clearance between the top of the cooler and the ceiling of the store. The store is air conditioned and the temperature inside the store is maintained at approximately 80°F. The inside design temperature for the cooler is 35°F. Determine the wall gain load for the cooler if the walls of the cooler are of the following construction:

South and east (outside walls)	6 in. clay tile 6 in. corkboard 0.5 cement plaster finish on inside
North and west (inside walls)	1 in. board on both sides of 2 × 4 studs $3\frac{5}{8}$ in. granulated cork
Ceiling	Same as north and west walls
Floor	4 in. corkboard laid on 5 in. slab and finished with 3 in. of concrete

Solution

Wall surface area

North wall	10 × 16 = 160 ft^2
West wall	10 × 20 = 200 ft^2
South wall	10 × 16 = 160 ft^2
East wall	10 × 20 = 200 ft^2
Ceiling	16 × 20 = 320 ft^2
Floor	16 × 20 = 320 ft^2

Wall *U* factors (Tables 10-1, 10-2, and 10-3)

North and west walls	0.079 Btu/(hr)(ft^2)(°F)
South and east walls	0.045
Ceiling	0.079
Floor	0.066
From Table 10-5, outside design temperature for Dallas	92°F
From Table 10-5, design ground temperature for Dallas	70°F

Applying Equation 10-2,

North wall	160 × 0.079 × 45 =	569 Btu/hr
East wall	200 × 0.045 × 63 =	567
South wall	160 × 0.045 × 61 =	439
West wall	200 × 0.079 × 45 =	711
Ceiling	320 × 0.079 × 45 =	1,138
Floor	320 × 0.066 × 35 =	739
		4,163 Btu/hr

Total wall gain load

$$= 4{,}163 \times 24 = 99{,}912 \text{ Btu/24 hr}$$

A short method calculation may be used to determine the wall gain load for small coolers and for large coolers where the wall *U* factor and temperature difference are approximately the same for all the walls. Table 10-17 lists wall gain factors (Btu/24 hr ft^2) based on the thickness of the wall insulation and on the temperature differential across the wall. To compute the wall gain load in Btu/24 hr by the short method, multiply the total outside wall area (including floor and ceiling) by the appropriate wall gain factor from Table 10-17, that is,

$$\text{Wall gain load} = \text{Outside surface area} \times \text{wall gain factor}$$

	Outside Design Temp.	Inside Design Temp.	Normal Wall T.D.	Correction Factor from Table 10-6	Design Wall T.D.
North wall	80°F	35°F	45°F	0	45°F
South wall	92°F	35°F	57°F	4°F	61°F
East wall	92°F	35°F	57°F	6°F	63°F
West wall	80°F	35°F	45°F	0	45°F
Ceiling	80°F	35°F	45°F	0	45°F
Floor	70°F	35°F	35°F	0	35°F

To select the appropriate wall gain factor from Table 10-17, find the thickness of the wall insulation in the extreme left-hand column of the table, move right to the column headed by the design wall temperature difference, and read the wall gain factor in Btu per 24 hr per square foot. For example, assume that the walls of a cooler are insulated with the equivalent of 4 in. of cork-board and that the temperature difference across the walls is 55°F. From Table 10-17, read the wall gain factor of 99 Btu (24 hr)(ft^2) (see Example 10-18).

10-14. Calculating the Air Change Load

The space heat gain resulting from air changes in the refrigerated space is difficult to determine with any real accuracy except in those few cases where a known quantity of air is introduced into the space for ventilating purposes. When the mass of outside air entering the space in a 24-hr period is known, the space heat gain resulting from air changes depends upon the difference in the enthalpy of the air at the inside and outside conditions and can be calculated by applying the following equation:

$$\text{Air change load} = m(h_o - h_i) \qquad (10\text{-}6)$$

where m = mass of air entering space in 24 hr (lb/24 hr)

h_o = enthalpy of outside air (Btu/lb)

h_i = enthalpy of inside air (Btu/lb)

However, since air quantities are usually given in cubic feet rather than in pounds, to facilitate calculations the heat gain per cubic foot of outside air entering the space is listed in Tables 10-7A and 10-7B for various inside and outside air conditions. To determine the air change load in Btu per 24 hr, multiply the air quantity in cubic feet per 24 hr by the appropriate factor from Table 10-7A or 10-7B.

Where the ventilating air (air change) quantity is given in cubic feet per minute (cfm), convert cfm to cubic feet per 24 hr by multiplying by 60 min and by 24 hr.

Example 10-6 Three hundred cubic feet per minute of air are introduced into a refrigerated space for ventilation. If the inside of the cooler is maintained at 35°F and the outside dry bulb temperature and humidity are 85°F and 50%, respectively, determine the air change load in Btu/per 24 hr.

Solution

Cubic feet of air per 24 hr $= \text{cfm} \times 60 \times 24$
$= 300 \times 60 \times 24$
$= 432{,}000 \text{ ft}^3/24 \text{ hr}$

From Table 10-7A, air change factor $= 1.86 \text{ Btu/ft}^3$

Ventilating (air change) load $= 432{,}000 \text{ ft}^3/24 \text{ hr}$
1.86 Btu/ft^3
$= 803{,}520 \text{ Btu/24 hr}$

Except in those few cases where air is purposely introduced into the refrigerated space for ventilation, the air changes occurring in the space are brought about solely by infiltration through door openings. The quantity of outside air entering a space through door openings in a 24-hr period depends upon the number, size, and location of the door or doors, and upon the frequency and duration of the door openings. Since the combined effect of all these factors varies with the individual installation and is difficult to predict with reasonable accuracy, it is general practice to estimate the air change quantity on the basis of experience with similar applications. Experience has shown that, as a general rule, the frequency and duration of door openings and, hence, the air change quantity, depend on the inside volume of the cooler and the type of usage. Tables 10-8A and 10-8B list the approximate number of air changes per 24 hr for various cooler sizes. The values given are for average usage (see table footnotes). The ASRE *Data Book* defines average and heavy usage as follows:

Average usage includes installations not subject to extreme temperatures and where the quantity of food handled in the refrigerator is not abnormal. Refrigerators in delicatessens and clubs may generally be classified under this type of usage.

Heavy usage includes installations such as those in busy markets, restaurant and hotel kitchens where the room temperatures are likely to be high, where rush periods place

TABLE 10-7A Btu per Cubic Foot of Air Removed in Cooling to Storage Conditions above 30

Storage Room Temp., °F	Inlet Air Temperature, °F: 85			90			95		100	
	Inter. Air Relative Humidity, %: 50	60	70	50	60	70	50	60	50	60
65	0.65	0.85	1.12	0.93	1.17	1.44	1.24	1.54	1.58	1.95
60	0.85	1.03	1.26	1.13	1.37	1.64	1.44	1.74	1.78	2.15
55	1.12	1.34	1.57	1.41	1.66	1.93	1.72	2.01	2.06	2.44
50	1.32	1.54	1.78	1.62	1.87	2.15	1.93	2.22	2.28	2.65
45	1.50	1.73	1.97	1.80	2.06	2.34	2.12	2.42	2.47	2.85
40	1.69	1.92	2.16	2.00	2.26	2.54	2.31	2.62	2.67	3.06
35	1.86	2.09	2.34	2.17	2.43	2.72	2.49	2.79	2.85	3.24
30	2.00	2.24	2.49	2.26	2.53	2.82	2.64	2.94	2.95	3.35

Reprinted from *Refrigeration Engineering Data Book* by courtesy of American Society of Refrigerating Engineers.

TABLE 10-7B Btu per Cubic Foot Removed in Cooling to Storage Conditions below 30

Storage Room Temp., °F	Inlet Air Temperature, °F: 40		50		80		90		100	
	Inter. Air Relative Humidity, %: 70	80	70	80	50	60	50	60	50	60
30	0.24	0.29	0.58	0.66	1.69	1.87	2.26	2.53	2.95	3.35
25	0.41	0.45	0.75	0.83	1.86	2.05	2.44	2.71	3.14	3.54
20	0.56	0.61	0.91	0.99	2.04	2.22	2.62	2.90	3.33	3.73
15	0.71	0.75	1.06	1.14	2.20	2.39	2.80	3.07	3.51	3.92
10	0.85	0.89	1.19	1.27	2.38	2.52	2.93	3.20	3.64	4.04
5	0.98	1.03	1.34	1.42	2.51	2.71	3.12	3.40	3.84	4.27
0	1.12	1.17	1.48	1.56	2.68	2.86	3.28	3.56	4.01	4.43
−5	1.23	1.28	1.59	1.67	2.79	2.98	3.41	3.69	4.15	4.57
−10	1.35	1.41	1.73	1.81	2.93	3.13	3.56	3.85	4.31	4.74
−15	1.50	1.53	1.85	1.93	3.05	3.25	3.67	3.96	4.42	4.86
−20	1.63	1.68	2.01	2.09	3.24	3.44	3.88	4.18	4.66	5.10
−25	1.77	1.80	2.12	2.21	3.38	3.56	4.00	4.30	4.78	5.21
−30	1.90	1.95	2.29	2.38	3.55	3.76	4.21	4.51	5.00	5.44

Reprinted from *Refrigeration Engineering Data Book* by courtesy of American Society of Refrigerating Engineers.

TABLE 10-8A Average Air Changes per 24 Hours for Storage Rooms above 32°F due to Door Opening and Infiltration

(Does not apply to rooms using ventilating ducts or grilles)

Volume cu ft	Air Changes per 24 hr	Volume cu ft	Air Changes per 24 hr	Volume cu ft	Air Changes per 24 hr	Volume cu ft	Air Changes per 24 hr
250	38.0	1,000	17.5	6,000	6.5	30,000	2.7
300	34.5	1,500	14.0	8,000	5.5	40,000	2.3
400	29.5	2,000	12.0	10,000	4.9	50,000	2.0
500	26.0	3,000	9.5	15,000	3.9	75,000	1.6
600	23.0	4,000	8.2	20,000	3.5	100,000	1.4
800	20.0	5,000	7.2	25,000	3.0		

NOTE: For storage room with anterooms, reduce air changes to 50% of values in table.
For heavy duty usage, add 50% to values given in table.

From *ASRE Data Book*, Design Volume, 1949 Edition, by permission of the American Society of Heating, Refrigerating, and Air-Conditioning Engineers.

TABLE 10-8B Average Air Changes per 24 Hours for Storage Rooms below 32°F due to Door Opening and Infiltration

(Does not apply to rooms using ventilating ducts or grilles)

Volume cu ft	Air Changes per 24 hr	Volume cu ft	Air Changes per 24 hr	Volume cu ft	Air Changes per 24 hr	Volume cu ft	Air Changes per 24 hr
250	29.0	1,000	13.5	5,000	5.6	25,000	2.3
300	26.2	1,500	11.0	6,000	5.0	30,000	2.1
400	22.5	2,000	9.3	8,000	4.3	40,000	1.8
500	20.0	2,500	8.1	10,000	3.8	50,000	1.6
600	18.0	3,000	7.4	15,000	3.0	75,000	1.3
800	15.3	4,000	6.3	20,000	2.6	100,000	1.1

NOTE: (1) For storage rooms with anterooms, reduce air changes to 50% of values in table. For heavy duty usage, add 50% to values given in table.
(2) For locker plant rooms, double the above table values.

From *ASRE Data Book*, Design Volume, 1949 Edition, by permission of the American Society of Heating, Refrigerating, and Air-Conditioning Engineers.

heavy loads on the refrigerator, and where large quantities of warm foods are often placed in it.*

Example 10-7 A walk-in cooler 8 ft × 15 ft × 10 ft high is constructed of 4 in. of corkboard with 1 in. of wood on each side. The outside temperature is 95°F and the humidity is 50%. The cooler is maintained at 35°F and the usage is average. Determine the air change load in Btu per 24 hr.

Solution Since the walls of the cooler are approximately 6 in. thick, the inside dimensions of the cooler are 1 ft less than the outside dimensions (7 ft × 14 ft × 9 ft), so that the inside volume is 882 ft^3. From Table 10-8A by interpolation, the number of air changes per

* The *Refrigerating Data Book*, Basic Volume, The American Society of Refrigerating Engineers, 1949, New York, p. 327.

24 hr for a cooler volume of approximately 900 ft^3 is 19, and from Table 10-7A, the air change factor is 2.49 Btu/ft^3.

$$\begin{aligned}\text{Air change load} &= \text{inside volume}\\ &\quad\times \text{air changes}\\ &\quad\times \text{air change factor}\\ &= (882)(19)(2.49)\\ &= 41{,}700 \text{ Btu/24 hr}\end{aligned}$$

10-15. Calculating the Product Load

When a product enters a storage space at a temperature above the temperature of the space, the product will give off heat to the space until it cools to the space temperature. When the temperature of the storage space is maintained above the freezing temperature of the product, the amount of heat given off by the product in cooling to the space temperature depends upon the temperature of the space and upon the mass, specific heat, and entering temperature of the product. In such cases, the space heat gain from the product is computed by the following equation:

$$Q = (m)(c)(\Delta T) \qquad (10\text{-}7)$$

where Q = the quantity of heat in Btu
m = mass of the product (pounds)
c = the specific heat above freezing, Btu/(lb)(°F)
ΔT = the change in the product temperature (°F)

Example 10-8 Seventy-five hundred pounds of fresh beef enter a chilling cooler at 102°F and are chilled to 45°F each day. Compute the product load in Btu per 24 hr.

Solution From Table 10-11, the specific heat of beef above freezing is 0.75 Btu/lb °F. Applying Equation 10-7,

$$\begin{aligned}\text{product load} &= (7500)(0.75)(102 - 45)\\ &= 320{,}600 \text{ Btu/24 hr}\end{aligned}$$

Notice that no time element is inherent in Equation 10-7 and that the result obtained is merely the quantity of heat the product will give off in cooling to the space temperature. However, since in Example 10-8 the product is to be cooled over a 24-hr period, the resulting heat quantity represents the product load for a 24-hr period. When the desired cooling time is less than 24 hr, the equivalent product load for a 24-hr period is computed by dividing the heat quantity by the desired cooling time for the product to obtain the hourly cooling rate and then multiplying the result by 24 hr to determine the equivalent product load for a 24-hr period. When adjusted to include these two factors, Equation 10-7 is written

$$Q = \frac{(m)(c)(\text{TD})(24 \text{ hr})}{\text{desired cooling time (hr)}} \qquad (10\text{-}8)$$

Example 10-9 Determine the product load in Btu per 24 hr assuming that the beef described in Example 10-8 is chilled in 20 hr rather than over the entire 24-hr period.

Solution Applying Equation 10-8,

$$\begin{aligned}\text{product load} &= \frac{(7500)(0.75)(102 - 45)(24 \text{ hr})}{20 \text{ hr}}\\ &= 384{,}750 \text{ Btu/24 hr}\end{aligned}$$

10-16. Chilling Rate Factor

During the early part of the chilling period, the Btu per hour load on the equipment is considerably greater than the average hourly product load as calculated in the previous examples. Because of the high temperature difference which exists between the product and the space air at the start of the chilling period, the chilling rate is higher and the product load tends to concentrate in the early part of the chilling period (Section 9-23). Therefore, where the equipment selection is based on the assumption that the product load is evenly distributed over the entire chilling period, the equipment selected will usually have insufficient capacity to carry the load during the initial stages of chilling when the product load is at a peak.

To compensate for the uneven distribution of the chilling load, a chilling rate factor is sometimes introduced into the chilling load calculation. The effect of the chilling rate factor is to increase the product load calculation by an amount sufficient to make the average hourly cooling rate approximately equal to the hourly

TABLE 10-9 Design Data for Fruit Storage

FRUITS	TYPE OF STORAGE	DESIGN ROOM CONDITIONS: Temperature Recommended Deg F	Temperature Permissible Range Deg F	Relative Humidity Recommended %	Relative Humidity Permissible Range %	Grains per lb Air at Recommended Condition	Maximum Storage Period	CHILLING DATA: Product Temp. Deg F Start	Product Temp. Deg F Finish	Time Hr	Rate Factor	Est Prod Latent Heat Btu/lb 24 Hr (Ex. see Note f)	SPECIFIC HEAT Btu/lb/Deg F Before Freezing	SPECIFIC HEAT After Freezing	Latent Heat of Fusion Btu/lb	Water Content %	Freezing Point Deg F	Maximum Air Motion in Room Ft./Min
Apples	Short	35k	35-40	87	85.88	26.0						4.0	0.89	0.43	122	84	28.9	90
	Long	30k	30-32	87b	85-88	20.8	48 Mo g					0.2						60
	Chill Start	40		85		31.0		80	32	24	0.67	24.0f						150
	Chill Finish	30		85		20.4						0.3						60d
Apricots	Short	35	35-40	85	80-85	25.2						4.0	0.92	0.50	122	85	28.1	90
	Long	32	31-32	85a	80-85	22.3	7-14 Days					0.3						60
	Chill Start	40		85		31.0		80	33	20	0.67	20.0f						150
	Chill Finish	32		85		22.3						0.3						60d
Avocados	Short	40k	40-53	85b	85-90	31.0						4.5	0.91	0.49	136	94	27.2	90
	Long	38k	37-53	85b	85-90	28.8	10 Days					0.3						90
	Chill Start	40		85		31.0		80	39	22	0.67	22.0f						250
	Chill Finish	33		85		23.2						0.3						90d
Bananas i (See Document 2D-84)	Ripening	70	62-70	95	90-95	104.7		Heating 56°-70°				2.0	0.90		108	75	26-30	90
	Chill Start	70		95		104.7		68	56	12	0.1	11.0f						150
	Chill Finish	56		90		60	10 Days					1.0						90d
	Holding Green	56	56-60	92	90-95	61.3						1.0						90
	Holding Ripe	56	56-60	87	85-90	58.0						1.0						90
Berries (General)	Short	35	35-40	85	80-85	25.2						6.3	0.90	0.49	120	84	28-30	90
	Long	32	31-32	85b	80-85	22.3	3-10 Da. g					0.3						60
	Chill Start	40		85		31.0		80	34	20	0.67	20.0f						150
	Chill Finish	32		85		22.3						0.3						60d
Cranberries	Short	36	36-40	85	85-90	26.4						5.0	0.91	0.47	122	88	27.3	90
	Long	36	36-40	85b	85-90	26.4	1-3 Mo					0.2						90
	Chill Start	40		85		31.0		70	38	20	0.67	18.0f						150
	Chill Finish	36		85		26.2						0.2						90d
Dates (Cured)	Short	35k	35-40	70c	65-75	20.8						0.10	0.35		26	18	− 4	150
	Long	28k	28-32	70c	65-75	15.4	3-6 mo g					0.05						150
Dried Fruits	Short	35	35-40	70c	70-75	20.8						0.10	0.47	0.32	43	30		150
	Long	32 L	32-36	70c	70-75	18.6	9-12 Mo					0.07						150
Figs and Dates (Fresh)	Short	40	40-50	75	65-75	27.5						5.0	0.71	0.44	116	90	28.3	90
	Long	34	34-36	70	65-75	20.0	15 Days					0.4						90
Grapes (American Eastern)	Short	35	35-40	85	80-85	25.2						5.0	0.90	0.61	112	77	28.0	90
	Long	31	31-32	85b	80-85	21.3	3-8 Wk g					0.4						90
	Chill Start	40		85		31.0		70	34	20	0.80	14.0f						250
	Chill Finish	32		85		22.3						0.4						90d
Grapes (Vinifera California)	Short	35	35-40	85	85-90	25.2						5.0	0.85	0.59	112	79	24.3	90
	Long	30	30-31	85b	85-90	20.4	3-6 Mo					0.4						90
	Chill Start	40		85		31.0		70	34	20	0.80	14.0f						250
	Chill Finish	32		85		22.3						0.4						90d
Grapefruit	Short	40	40-45	90	85-90	32.8						2.0	0.91	0.49	128	88	28.4	90
	Long	32 j	32-34	85b	85-90	22.3	6-8 Wk					0.3						90
	Chill Start	40		85		31.0		75	34	22	0.70	19.0f						250
	Chill Finish	32		85		22.3						0.3						90d
Lemons	Short	55	55-60	85b	85-90	54.5						3.0	0.91	0.49	126	88	28.1	90
	Long	55	55-60	85b	85-90	54.5	1-4 Mo					0.3						90
	Chill Start	60		85		65.5		75	57	20	1.0	10.0f						250
	Chill Finish	55		85		54.5						0.3						90d
Limes	Short	45	45-50	85b	85-90	37.5						4.0	0.91	0.49	126	88	29.3	60
	Long	45	45-48	90b	85-90	39.6	6-8 Wk					0.2						60
	Chill Start	50		85		45.2		75	47	20	0.90	14.0f						150
	Chill Finish	45		85		37.5						0.2						60d
Oranges	Short	40	40-45	85	85-90	31.0						4.0	0.91	0.44	125	81	28.0	90
	Long	32	32-34	85b	85-90	22.3	8-10 Wk g					0.3						90
	Chill Start	40		85		31.0		75	32	22	0.70	19.0f						250
	Chill Finish	32		85		22.3						0.3						90d
Peaches	Short	35	35-40	85a	80-85	25.2						5.1	0.91	0.41	128	90	29.2	60
	Long	32	31-33	85b	80-85	22.3	2-4 Wk g					0.3						60
	Chill Start	40		85		31.0		85	34	24	0.62	23.0f						150
	Chill Finish	32		85		22.3						0.3						60d
Pears	Short	35	35-40	90a	85-90	26.8					0.80	6.0	0.91	0.49	122	84	27-28	60
	Long	31k	29-31	90b	85-90	22.7	1-7 Mo g					0.3						60
	Chill Start	40		85		31.0		70	34	24		17.0f						150
	Chill Finish	32		85		22.3						0.3						60d
Pineapples	Short	40	40-45	85	85-90	31.0						3.0	0.90	0.50	128	88		150
	Long Ripe	40	40-45	85b	85-90	31.0	2-4 Wk					0.1					29.9	150
	Green	50	50-60	90b	85-90	48.0	3-4 Wk					0.1					29.1	150
	Chill Start	45		85		37.5		85	40	3	0.67	24.0f						250
	Chill Finish	38		85		28.8						0.1						150d
Plums and Prunes (Fresh)	Short	35	35-40	85	80-85	25.2						4.0	0.88	0.48	116	80	28.0	90
	Long	32	31-32	85b	80-85	22.3	3-8 Wk g					0.3						90
	Chill Start	40		80		29.1		80	34	20	0.67	20.0f						250
	Chill Finish	32		80		21.1						0.3						90d
Quinces	Short	35	35-40	85	80-85	25.2						4.0	0.90	0.49	122	85	28.0	60
	Long	32	31-32	85b	80-85	22.3	2-3 Mo					0.3						60
	Chill Start	40		85		31.0		80	32	24	0.67	24.0f						150
	Chill Finish	32		85		22.3						0.3						60d

From *Carrier Design Data.* Reproduced by permission of Carrier Corporation.

TABLE 10-10 Design Data for Vegetable Storage

VEGETABLES	TYPE OF STORAGE	DESIGN ROOM CONDITIONS					Maximum Storage Period	CHILLING DATA				Est Prod Latent Heat Btu/lb 24 Hr (Ex. see Note f)	SPECIFIC HEAT Btu/lb/Deg F		Latent Heat of Fusion Btu/lb	Water Content %	Freezing Point Deg F	Maximum Air Motion in Room Ft./Min
		Temperature		Relative Humidity		Grains per lb Air at Recommended Condition		Product Temp. Deg F		Time Hr	Rate Factor		Before Freezing	After Freezing				
		Recommended Deg F	Permissible Range Deg F	Recommended %	Permissible Range %			Start	Finish									
Asparagus	Short	40	40-45	90	85-90	32.8						6.0	0.91	0.49	135	94.0	29.8	90
	Long	32	32-36	90a	85-90	23.7	30 Days					0.5						60
	Chill Start	40		85		31.0		60	34	24	0.90	13.0f						150
	Chill Finish	33		85		23.2						0.5						60d
Beans, Green	Short	40	40-45	90	85-90	32.8						3.0	0.87	0.47	119	83.0	29.7	90
	Long	33	32-40	90a	85-90	24.6	30 Days					0.7						60
	Chill Start	40		85		31.0		80	35	20	0.67	15.0f						150
	Chill Finish	33		85		23.2						0.7						60d
Beans (Lima)	Short	40	40-45	90	85-90	32.8	15 Days Shelled					3.0	0.78	0.36	99	68.5	28.4	90
	Long	33	32-40	90a	85-90	24.6	30 Days Unshelled					0.6						60
Beets, Tops Off	Short	40	40-45	90	85-90	32.8						2.0	0.90	0.48	129	90.0	26.9	90
	Long	32	32-36	95	95-98	25.0	1-3 Mo					0.3						60
Beets, Tops On	Short	40	40-45	90	85-90	32.8						3.0	0.90	0.48	129	90.0	31.0	90
	Long	32	32-36	90a	85-90	23.7	10-14 Days					0.4						60
	Chill Start	40		90		32.8		70	34	24	0.80	17.0f						150
	Chill Finish	32		90		23.7						0.4						60d
Broccoli	Short	40	40-45	90	90-95	32.8						4.0	0.90	0.48	135	93.0	29.2	90
	Long	32	32-35	90	90-95	23.7	7-10 Days					0.5						60
	Chill Start	40		90		32.8						14.0f						150
	Chill Finish	33		90		24.6		80	34	24	0.80	0.5						90d
Brussel Sprouts	Short	40	40-45	95	90-95	34.5						5.0	0.91	0.49	136	94.5	31.0	90
	Long	33	32-35	95b	90-95	26.0	3-4 Wk					0.5						60
	Chill Start	40		90		32.8		80	34	24	0.80	14.0f						150
	Chill Finish	33		90		24.6						0.5						90d
Cabbage	Short	35	35-40	95	90-95	28.2						7.0	0.93	0.47	132	91.5	31.2	90
	Long	32	32-36	95b	90-95	25.0	3-4 Mo					0.5						60
	Chill Start	40		90		32.8		70	34	24	0.80	17.0f						150
	Chill Finish	32		90		23.7						0.5						60d
Carrots, Tops Off	Short	40	40-45	90	85-90	32.8						2.0	0.93	0.45	126	88.0	30.4	90
	Long	32	32-36	95	95-98	25.0	4-5 Mo					0.3						60
Carrots, Tops On	Short	40	40-45	90	85-90	32.8						4.0	0.86	0.45	126	88.0	31.0	60
	Long	32	32-36	90b	85-90	23.7	10-14 Days					0.5						60
	Chill Start	40		90		32.8		70	34	24	0.80	17.0f						150
	Chill Finish	32		90		23.7						0.5						60d
Cauliflower	Short	35	35-40	90	85-90	26.8						4.0	0.90	0.46	133	92.5	30.1	90
	Long	32	32-36	90a	85-90	23.7	2-3 Wk					0.3						60
	Chill Start	40		90		32.8		70	34	24	0.80	17.0f						150
	Chill Finish	32		90		23.7						0.3						60d
Celery p	Short	35	35-40	90	90-95	26.8						4.0	0.91	0.46	136	94.5	29.7	90
	Long (Wetted)	32	31-32	90a	90-95	23.7	2-4 Mo					1.0						60
Corn (Green)	Short	35	35-40	90	85-90	26.8						7.0	0.86	0.38	108	75.5	28.9	90
	Long	32	31-32	90a	85-90	23.7	4-8 Days					0.5						60
	Chill Start	40		85		31.0		70	34	24	0.80	17.0f						150
	Chill Finish	32		85		22.3						0.5						60d
Cucumbers	Short	50	50-60	85	80-85	45.2						3.0	0.93	0.48	137	95.5	30.5	90
	Long	45	45-50	85	80-85	37.5	10-14 Days					0.2						90
	Chill Start	60		80		61.7		70	52	24	1.0	13.0f						250
	Chill Finish	50		80		42.6						0.2						150d
Endive p	Short	35	35-40	90	90-95	26.8						4.0	0.90	0.46	136	89.0	30.9	90
	Long (Iced)	35	32-36	90a	90-95	26.8	2-3 Wk					1.0						90

load at the peak condition. This results in the selection of larger equipment, having sufficient capacity to carry the load during the initial stages of chilling.

Chilling rate factors for various products are listed in Tables 10-9 through 10-12. The factors given in the tables are based on actual tests and on calculations and will vary with the ratio of the loading time to total chilling time. As an example, test results show that in typical beef and hog chilling operations the chilling rate is 50% greater during the first half of the chilling period than the average chilling rate for the entire period. The calculation without the chilling rate factor will, of course, show the average chilling rate for the entire period. To obtain this rate during the initial chilling period, it must be multiplied by 1.5. For convenience, the chilling rate factors are given in the tables in reciprocal form and are used in the denominator of the equation. Thus the chilling rate factor for beef as shown in the table is 0.67 (1/1.5).

TABLE 10-10 *(Continued)*

VEGETABLES	TYPE OF STORAGE	DESIGN ROOM CONDITIONS					Maximum Storage Period	CHILLING DATA				Est Prod Latent Heat Btu/lb 24 Hr (Ex. see Note f)	SPECIFIC HEAT Btu/lb/Deg F		Latent Heat of Fusion Btu/lb	Water Content %	Freezing Point Deg F	Maximum Air Motion in Room Ft./Min
		Temperature		Relative Humidity		Grains per lb Air at Recommended Condition		Product Temp. Deg F		Time Hr	Rate Factor		Before Freezing	After Freezing				
		Recommended Deg F	Permissible Range Deg F	Recommended %	Permissible Range %			Start	Finish									
Lettuce p	Short	35	35-40	90	90-95	26.8						7.0	0.90	0.46	136	89.0	31.2	90
	Long (Iced)	35	32-36	90a	90-95	26.8	2-3 Wk					1.0						60
Melons	Short	45	45-50	85	75-85	37.5						3.0	0.91	0.46	115	85.0	29.0	90
Watermelons Honeydews	Long	36	36-40	85a	75-85	26.2	2-4 Wk					0.2						150
Cantaloupes	Long	32	32-35	85	75-78	22.3	7-10 Days					0.2	0.91	0.47	128	89.0	29.0	90
	Chill Start	40		85		31.0		80	34	24	0.90	14.0f						250
	Chill Finish	32		85		22.3						.2						150d
Onions	Short	50	50-60	75	70-75	40.0						2.0	0.91	0.51	130	89.0	30.1	150
	Long	32	32-36	75	70-75	19.8	6-8 Mo					0.2						150
	Chill Start	40		75		27.5		70	34	24	0.80	10.0f						250
	Chill Finish	32		75		19.8						0.2						150d
Parsnips	Short	35	35-40	95	90-95	28.2						4.0	0.86	0.44	119	83.0	28.9	60
	Long	32	32-36	95b	90-95	25.0	2-4 Mo					0.5						60
	Chill Start	40		90		32.8		70	34	24	0.80	17.0f						150
	Chill Finish	32		90		23.7						0.5						90d
Peas (Green)	Short	35	35-40	90	85-90	26.8						3.0	0.82	0.45	107	80.0	28.9	90
	Long	32	32-36	90b	85-90	23.7	1-2 Wk					0.5						90
	Chill Start	40		85		31.0		80	34	20	0.67	14.0f						150
	Chill Finish	33		85		23.2						0.5						90d
Potatoes (Eating)		50r	50-70	85	85-90	45.2						3.0	0.86	0.47	113	78.5	28.9	150
Potatoes (Seed Stock)		36m	36-50	85a	85-90	26.4						0.5					28.9	150
Sauerkraut	Short	45	45-50	80	75-80	35.3						3.0	0.92	0.52	128	89.0	26.0	150
(In Kegs)	Long	30	30-32	80c	75-80	19.2	5 Mo					0.2						90
Spinach	Short	35	35-40	95a	90-95	28.2						7.0	0.92	0.51	129	90.0	30.3	90
	Long	32	32-36	95a	90-95	25.0	10-14 Days					0.5						60
Sweet	Short	55	55-60	85	80-85	54.5						3.0	0.86	0.42	102	78.0	28.5	150
Potato n	Long	55	55-60	85a	80-85	54.5	4-6 Mo					0.4						150
Tomatoes	Short	55	55-60	85	85-90	54.5						3.0	0.92	0.46	132	95.0	30.6	90
(Green)	Long	55	55-60	85	85-90	54.5	3-5 Wk					0.4						60
	Ripening	65	65-70	85	85-90	78.2						2.0						90
	Chill Start	70		85		93.3		80	52	34	1.0	14.0f						150
	Chill Finish	50		85		45.2						0.4						90d
(Ripe)	Long	45	40-50	85a	85-90	37.5	7-10 Days					3.0						90
Turnips	Short	35	35-40	95	95-98	28.2						4.0	0.90	0.45	128	89.5	30.5	90
	Long	32	32-36	95a	95-98	25.0	4-5 Mo					0.5						60
	Chill Start	40		95		34.5		70	34	24	0.80	17.0f						150
	Chill Finish	32		• 95		25.0						0.5						60d
Vegetables	Short	40	40-45	85b	85-90	31.0						5.0	0.90	0.45	130	90.0	30.0	90
(Wetted, Mixed)	Long	35	35-40	87b	85-90	26.0	2-4 Mo					1.2						90
	Chill Start	50		90		48.0		80	38	18	0.70	23.0f						150
	Chill Finish	35		90		26.8						1.2						90d

From *Carrier Design Data.* Reproduced by permission of Carrier Corporation.

Where a chilling rate factor is used, Equation 10-8 is written

$$Q = \frac{(m)(c)(\text{TD})(24 \text{ hr})}{(\text{chilling time})(\text{chilling rate factor})} \qquad (10\text{-}9)$$

Example 10-10 Recalculate the product load described in Example 10-9 employing the appropriate chilling rate factor.

Solution From Table 10-11, the chilling rate factor for beef chilling is 0.67. Applying Equation 10-9,

$$\text{product load} = \frac{(7500)(0.75)(102 - 45)(24 \text{ hr})}{(20 \text{ hr})(0.67)}$$

$$= 574{,}250 \text{ Btu/24 hr}$$

Chilling rate factors are usually applied to chilling rooms only and are not normally used in calculation of the product load for storage rooms. Since the product load for storage rooms usually represents only a small percentage of the total load, the uneven distribution of the product load over the cooling period will not ordinarily cause overloading of the equipment and/or unacceptable fluctuations in the

TABLE 10-11 Design Data for Meat Storage

MEATS	TYPE OF STORAGE	Design Room Conditions: Temperature Recommended Deg F	Design Room Conditions: Temperature Permissible Range Deg F	Design Room Conditions: Relative Humidity Recommended %	Design Room Conditions: Relative Humidity Permissible Range %	Grains per lb Air at Recommended Condition	Maximum Storage Period	Chilling Data: Product Temp. Deg F Start	Chilling Data: Product Temp. Deg F Finish	Chilling Data: Time Hr	Chilling Data: Rate Factor	Est Prod Latent Heat Btu/lb 24 Hr (Ex. see Note f)	Specific Heat Btu/lb/Deg F Before Freezing	Specific Heat Btu/lb/Deg F After Freezing	Latent Heat of Fusion Btu/lb	Water Content %	Freezing Point Deg F	Maximum Air Motion in Room Ft./Min
Bacon	Short	55	50-60	65	55-65	41.7	15 Days					2.5	0.50	0.30	29	20		150
	Hardening	28t	28-30	75	70-80	16.4						1.2						90
	Slicing Room	50	50-55	40	35-40	21.3												60
Beef- Combined Chill and Holding	Chill Start	38		85b		28.8		100	44	24	0.56	18.0f	0.75	0.40	98	72	31.3	250
	Chill Finish	33		85b		23.2						5.0						90d
Beef- Dried	Long	55	55-60	65	65-70	41.7	6 Mo					0.1	.22-.34	.19-.26	7-22	5-15		150
Beef- Fresh	Short	35	35-40	87b	85-90	26.0						5.0	0.75	0.40	98	72	31.3	60
	Long	30	30-32	87b	85-90	20.8	3 Wk					1.7						60
	Chill Start	45		87		38.3		100	44	18	0.67	22.0f						250
	Chill Finish	30		87		20.8						1.7						150d
Brined Meat	Short	40	40-45	85	80-85	31.0						1.0	0.75					150
	Long	31	31-32	85	80-85	21.3	6 Mo					0.8						150
Cut Meat	Short	34	34-38	87a	85-90	24.8	5 Days					5.6	0.72	0.40	95	65	29	60
Fish Frozen	Long	0	(−5)−0	85c	80-85	4.65	6 Mo					0.1	0.76	0.41	101	70	28	250
Iced	Short	34	34-38	85c	80-85	24.3						5.7						90
	Long	30	30-32	85a	80-85	20.4	15 Days					0.4						90
Hams, & Loins Fresh	Short	34	34-38	85	85-87	24.3						3.4	0.68	0.38	86.5	52	31.3	60
	Long	28	28-30	85b	85-87	18.5	3 Wk					1.8						60
Smoked	Short	55	50-60	65	55-65	41.7						1.3	0.60	0.32		57		150
	Chill Start s	60		70		53.9		105	57	8	1.00	5.0f						150
	Chill Finish s	55		70		44.8						.3						90d
Hog Chilling 18 Hrs	Chill Start	45		85		37.5		105	35	18	0.67	24.0f	0.68	0.38	86.5	60	27	250
	Chill Finish	30		85		20.4						1.9						150d
14 Hrs	Chill Start	38		90		30.1		105	35	14	0.67	23.0f						250
	Chill Finish	28		90		19.7						1.9						150d
Lamb	Short	34	34-38	90	85-90	25.8						3.4	0.67	0.30	83.5	58	29	60
	Long	28	28-30	90b	85.90	19.7	2 Wk					1.3						60
	Chill Start	45		90		39.6		100	40	5	0.75	19.0f						250
	Chill Finish	30		90		21.6						1.3						90d
Offal (Livers, Hearts, etc.)	Chill Start	40		85		31.0		90	35	18	0.70	21.0f	0.75	0.42	103	72		150
	Chill Finish	32		85		22.3						1.3						90d
Oysters Shell	Short	35	35-40	90c	85-90	26.8						4.2	0.83	0.44	116	80.4	27	90
	Long	32	32-38	90c	85-90	23.7	15 Days					0.5						90
Tub	Short	35	35-40	70	70-75	20.8						2.3	0.90	0.46	125	87	27	150
	Long	32	32-38	70	70-75	18.6	10 Days					0.2						150
Pork (Fresh)	Short	34	34-38	85	85-90	24.3	15 Days					3.4	0.68	0.38	86.5	60	28	90
Poultry Fresh	Long	28	28-30	87b	85-90	19.0	10 Days					0.4	0.79	0.37	106	74	27	60
Frozen	Long	0	(−5)−0	85	85-90	4.65	10 Mo					0.2						150
Wet Picked	Chill Start s	45		85		37.5		85	40	5	1.00	17.0f						150
	Chill Finish	32		85		22.3						0.4						90d
Sausage Casings (Salted)	Short	40	40-45	80c	75-80	29.1						0.2						150
	Long	31	31-32	80c	75-80	20.1	4 Mo					0.0	0.60					150
Franks and Smoked	Short	35	35-40	85a	80-90	25.2	48 Hr					4.3	0.86	0.56	86	60	29	60
	Chill Start	42		80		31.6		70	35	2	1.00	9.0f						150
	Chill Finish	32		80		21.1												60d
Fresh	Short	35	35-40	85a	85-90	25.2	7 Days					4.3	0.89	0.56	93	65	26	60
	Chill Start	42		85		33.6		70	35	2	1.00	9.0f						150
	Chill Finish	32		85		22.3												60d
Mfg. Room		55	55-60	40	35-40	25.5						0.0						60
Smoked Summer	Short	40	35 40	85	80-90	31.0	6 Mo					3.2	0.86	0.56	86	60	25	60
	Drying	50	48-56	70	65-80	37.2						5.0						60
	Long	32	32 34	70	70-75	18.6	6-8 Mo					2.0						60
Wrapping Room		45	45-50	85	80-85	37.5						0.0						60
Veal	Short	34	34-38	87b	85-90	24.8						3.6	0.71	0.39	91	63	29	60
	Long	28	28-30	87b	85-90	19.0	15 Days					1.3						60
	Chill Start	45		90		39.6		100	40	6	0.75	21.0f						90
	Chill Finish	30		90		21.6						1.3						60d

From *Carrier Design Data*. Reproduced by permission of Carrier Corporation.

TABLE 10-12 Design Data for Miscellaneous Storage

MISCEL-LANEOUS	TYPE OF STORAGE	Design Room Conditions: Temperature: Recommended Deg F	Temperature: Permissible Range Deg F	Relative Humidity: Recommended %	Relative Humidity: Permissible Range %	Grains, per lb Air at Recommended Condition	Chilling Data: Maximum Storage Period	Product Temp. Deg F: Start	Product Temp. Deg F: Finish	Time Hr	Rate Factor	Est Prod Latent Heat Btu/lb 24 Hr (Ex. see Note f)	Specific Heat Btu/lb/Deg F: Before Freezing	Specific Heat Btu/lb/Deg F: After Freezing	Latent Heat of Fusion Btu/lb	Water Content %	Freezing Point Deg F	Maximum Air Motion in Room Ft./Min
Beer (Whole-saler)																		
Wooden Keg	Short	35	35-40	85	80-85	25.2	6 Mo					7.0u	1.0			92.0	28	150
Metal Keg	Short	35	35-40	70c	65-70	20.8	6 Mo					0.4u						150
Butter or	Short	40	35-45	80c	75-80	29.1	10 Days					2.0	0.64	0.34	15	15.0	30.0w	150
Honey But'r	Long	0	(−5)−0	85	80-85	4.65	6 Mo					0.3						250
Candy	Long	65	60-70	55	50-55	50.3	6 Mo						0.93					60
Caviar	Short	40	40-45	85	80-85	31.0						2.0					20	150
(In Tubs)	Long	34	34-36	85b	80-85	24.3	15 Day					0.3						150
Cheese	Short	40	40-45	80b	75-80	29.1						2.3	0.64	0.36	79	55.0	17	90
American	Long	32	30-34	80b	75-80	21.1	15 Mo					0.5						90
Camembert	Short	40	40-45	85	80-85	31.0						2.5	0.70	0.40	86	60.0	18	90
	Long	40	34-34	85b	80-85	31.0	90 Days					0.2						90
Limburger	Short	40	40-45	85	80-85	31.0						2.5	0.70	0.40	86	60.0	19	90
	Long	31	30-34	85b	80-85	21.3	60 Days					0.3						90
Roquefort	Short	45	45-50	80	75-80	35.3						2.0	0.65	0.32	79	55.0	3	90
	Long	40	30-34	80b	75-80	29.1	60 Days					0.2						90
Swiss	Short	40	40-45	80	75-80	29.1						2.3	0.64	0.36	79	55.0	15	90
	Long	38	30-34	80b	75-80	27.0	60 Days					0.2						90
Chocolate (For Coating)	Long	60	60-70	55	50-55	42.1	6 Mo					0.1	0.56	0.30	40	0.5	85-95	60
Cream	Short	35	35-40	80c	75-80	23.8						2.0u	0.85	0.40	90	55.0	28	150d
(40%)	Long	5	(−5)−0	80	80-85	5.68	4 Mo					0.1u						150
Eggs Crated	Short	40	40-45	85b	80-85	31.0						3.6	0.85	0.45	100	74.2	31.6	90
(See Doc.	Long	30	30-31	85b	85-87	20.4	12 Mo					0.2						60
2D-85)	Chill Start	40		85b		31.0		45	30	10	0.85	7.0f						90
	Chill Finish	30		85b		20.4						0.2						60d
Eggs, Frozen	Long	5	(−5)−0	60		4.26	18 Mo					0.06u		0.45	100			250
10 lb cans	Chill Start	0		85c		4.65		40	5	24	0.67	9.0f						250
Doc. 2D-85	Chill Finish	0		85c		4.65						0.06						250
Fur, Woolens	Fumigated	35	35-40	65	60-65	19.3	6 Mo					0.1	0.40					150
(See Doc. 2D-83)	Ref only	15	15-18	70	65-70	8.2	6 Mo					0.1						150
Flour	Long	78	78-82	60	60-65	86.0	6 Mo						0.38	0.28		13.5		60
Flowers, Cut General		40k	33-40	85	85-90	31.0	3-14 Days					(120 Btu	0.92				27-31	60
Orchids Gardenias		45	45-50	85	85-90	37.5	1 Wk					Per Sq Ft Floor)					28-31	60
Hides, Curing		55	50-55	85	80-85	54.5						0.2	0.40					150
Storage	Long	36	32-40	75	70-75	23.1	5 Yr					0.1	0.40					150
Ice Cream	Hardening																	
5 Gal Cans	Start	0		85c		4.65		22	−10	8	0.75	1.1f,u	0.77		37	60.0	28.5-0	250
(See Doc.	Finish	−20		85c		1.55						0.1u						250
2D-86)	Start	0		85c		4.65		26	−10	8	0.75	1.3f,u			62			250
	Finish	−20		85c		1.55						0.1u						250
Lard	Short	45	45-50	80c	75-80	35.3						2.0	0.60		90		70	150
	Long	32	32-34	80c	75-80	21.1	6 Mo					0.3						150
Maple	Short	45	45-50	70c	65-70	29.9						0.7	0.24	0.21	7	5		250
Sugar	Long	31	31-32	70c	65-70	17.7	5 Mo					0.1						250
Maple	Short	45	45-50	70c	65-70	29.9						0.7	0.49	0.31	52	36.0		250
Syrup	Long	31	31-32	70c	65-70	17.7	5 Mo					0.1						250
Milk Bottled	Short	35	35-40	70c	65-75	20.8	5 Days					2.0	0.90	0.49	124	87.5	31	250
and Wet	Chill Start	40		80c		29.1		45	35	10	0.85	8.0f						250
Doc. 2D-59	Chill Finish	34		80c		23.0						0.1						250
Nuts, In	Short	40	40-45	70c	65-75	25.3						0.50	0.25	0.22	3-10	2-8		150
Shells	Long	32	32-40	70c	65-75	18.6	8-12 Mo					0.08						150
Nuts,	Short	40	40-45	70	65-75	25.3						0.50	0.30	0.24	4-14	3-10		150
Shelled	Long	32	32-40	70	65-75	18.6	6-10 Mo					0.08						150
Oleo	Short	45	45-50	80c	75-80	35.3						2.0	0.48					150
	Long	34	34-36	80c	75-80	23.0	90 Days					0.3						150
Vaccine Serum	Long	43	40-45	70	65-70	28.5	4 Mo					0.0						150
Shrubs	Long	28	24-29	70	60-80	15.4	6-8 Mo						0.60	0.35		50.0		

From *Carrier Design Data*. Reproduced by permission of Carrier Corporation.

space temperature, and therefore no allowance need be made for this condition.

10-17. Product Freezing and Storage

When a product is to be frozen and stored at some temperature below its freezing temperature, the product load is calculated in three parts:

1. The heat given off by the product in cooling from the entering temperature to its freezing temperature.
2. The heat given off by the product in solidifying or freezing.
3. The heat given off by the product in cooling from its freezing temperature to the final storage temperature.

The method of determining the product load resulting from temperature reduction (parts 1 and 3) has already been established. The product load resulting from freezing (part 2) is calculated from the following equation:

$$Q = (m)(h_{if}) \qquad (10\text{-}10)$$

where m = the mass of the product in pounds
h_{if} = the product latent heat in Btu per pound

When the chilling and freezing are accomplished over a 24-hr period, the summation of the three parts represents the product load for 24 hr. When the desired chilling and freezing time for the product are less than 24 hr, the summation of the three parts is divided by the desired processing time and then multiplied by 24 hr to determine the equivalent 24-hr product load.

Example 10-11 Five hundred pounds of poultry enter a chiller at 40°F and are frozen and chilled to a final temperature of −5°F for storage in 12 hr. Compute the product load in Btu per 24 hr.

Solution

From Table 10-11,

Specific heat above freezing = 0.79 Btu/lb °F

Specific heat below freezing = 0.37 Btu/lb °F

Latent heat = 106 Btu/lb

Freezing temperature = 27°F

To cool poultry from entering temperature to freezing temperature, applying Equation 10-7 = (500)(0.79)(40 − 27) = 5135 Btu

To freeze, applying Equation 10-10 = (500)(106) = 53,000 Btu

To cool from freezing temperature to final storage temperature, applying Equation 10-7 = 500 × 0.37 × [27 − (−5)] = 5920 Btu

Total heat given up by product (summation of 1, 2, and 3) = 64,000 Btu

Equivalent product load for 24-hr period Btu/24 hr $= \dfrac{64{,}000 \times 24 \text{ hr}}{12 \text{ hr}}$ = 128,000 Btu/24 hr

10-18. Respiration Heat

Fruits and vegetables are still alive after harvesting and continue to undergo changes while in storage. The more important of these changes are produced by respiration, a process during which oxygen from the air combines with the carbohydrates in the plant tissue and results in the release of carbon dioxide and heat. The heat released is called respiration heat and must be considered as a part of the product load where considerable quantities of fruit and/or vegetables are held in storage at temperatures above the freezing temperature. The amount of heat evolving from the respiration process depends upon the type and temperature of the product. Respiration heat for various fruits and vegetables is listed in Table 10-13.

Since respiration heat is given in Btu per pound per hr, the product load accruing from respiration heat is computed by multiplying the total mass of the product by the respiration

TABLE 10-13 Reaction Heat from Fruits and Vegetales

FRUITS		
Commodity	Temperature Deg F	Btu per hr per lb
Apples	32	.018
	40	.030
	60	.120
Apricots	32	.023
	40	.036
	60	.170
Bananas		
Holding	54	.069
Ripening	68	.190
Chilling	70-56	.500§
Berries	36	.115
	60	.345
Cherries	32	.032
	60	.250
Cranberries	32	.014
	40	.019
	50	.036
Dates, Fresh	32	.014
	40	.019
	50	.036
Grapefruit	32	.0096
	40	.022
	60	.058
Grapes	32	.0075
	40	.014
	60	.050
Lemons	32	.012
	40	.017
	60	.062
Limes	32	.012
	40	.017
	60	.062
Oranges	32	.017
	40	.029
	60	.104
Peaches	32	.023
	40	.036
	60	.170
Pears	32	.016
	60	.230
Plums	32	.032
	60	.250
Quinces	32	.018
	40	.030
	60	.120
Strawberries	32	.068
	40	.120
	60	.360

VEGETABLES		
Commodity	Temperature Deg F	Btu per hr per lb
Asparagus	32	.035
	40	.170
Beans, Lima	32	.170
	60	.820
Beans, String	32	.099
	40	.140
	60	.470
Beets	32	.055
	40	.085
	60	.150
Brussel Sprouts	32	.059
	40	.095
	60	.280
Cabbage	32	.059
	40	.095
	60	.280
Cauliflower	32	.059
	40	.095
	60	.280
Carrots	32	.045
	40	.073
	60	.170
Celery	32	.059
	40	.095
	60	.280
Corn, Sweet	32	.035
	40	.170
Cucumber	32	.028
	40	.041
	60	.175
Endive	40	.200
Lettuce	32	.240
	40	.330
	60	.960
Melons (Except Watermelons)	32	.028
	40	.041
	60	.175
Mushrooms	32	.130
	50	.460
Onions	32	.018
	50	.039
	70	.075
Parsnips	32	.045
	40	.073
	60	.170
Peas	32	.170
	60	.820
Peppers	32	.057
	60	.180
Potatoes	32	.014
	40	.030
	70	.060
Spinach	40	.200
Sweet Potatoes	40	.070
Tomatoes		
(Green)	60	.130
(Ripe)	40	.027
Turnips	32	.040
	40	.050

From *Carrier Design Data*. Reproduced by permission of Carrier Corporation.

heat as given in Table 10-13, that is,

$$Q\ (\text{Btu/24 hr}) = \text{Mass of product (lb)} \times \text{respiration heat (Btu/lb hr)} \times 24\ \text{hr} \qquad (10\text{-}11)$$

10-19. Containers and Packing Materials
When a product is chilled in containers, such as milk in bottles or cartons, eggs in crates, and fruit and vegetables in baskets and lugs, the heat given off by the containers and packing materials in cooling from the entering temperature to the space temperature must be considered as a part of the product load (see Example 10-15).

10-20. Calculating the Miscellaneous Load
The miscellaneous load consists primarily of the heat given off by lights and electric motors operating in the space and by people working in the space. The heat given off by lights is 3.42 Btu/watt hr. The heat given off by electric

TABLE 10-14 Heat Equivalent of Electric Motors

Motor hp	Connected Load in Refr. Space[1]	Motor Losses Outside Refr. Space[2]	Connected Load Outside Refr. Space[3]
	Btu/hp-hr		
$\frac{1}{8}$ to $\frac{1}{2}$	4250	2545	1700
$\frac{1}{2}$ to 3	3700	2545	1150
3 to 20	2950	2545	400

[1] For use when both useful output and motor losses are dissipated within refrigerated space; motors driving fans for forced circulation unit coolers.

[2] For use when motor losses are dissipated outside refrigerated space and useful work of motor is expended within refrigerated space; pump on a circulating brine or chilled water system, fan motor outside refrigerated space driving fan circulating air within refrigerated space.

[3] For use when motor heat losses are dissipated within refrigerated space and useful work expended outside of refrigerated space; motor in refrigerated space driving pump or fan located outside of space.

From *ASRE Data Book*, Design Volume, 1949 Edition, by permission of the American Society of Heating, Refrigerating, and Air-Conditioning Engineers.

TABLE 10-15 Heat Equivalent of Occupancy

Cooler Temperature, F	Heat Equivalent/Person Btu/hr
50	720
40	840
30	950
20	1050
10	1200
0	1300
−10	1400

From *ASRE Data Book*, Design Volume, 1949 Edition, by permission of the American Society of Heating, Refrigerating, and Air-Conditioning Engineers.

motors and by people working in the space is listed in Tables 10-14 and 10-15, respectively. The following calculations are made to determine the heat gain from miscellaneous:

Lights: wattage × 3.42 Btu/watt hr × 24 hr

Electric motors: factor (Table 10-14) × horsepower × 24 hr

People: factor (Table 10-15) × number of people × 24 hr

10-21. Use of Safety Factor

The total cooling load for a 24-hr period is the summation of the heat gains as calculated in the foregoing sections. It is common practice to add 5% to 10% to this value as a safety factor. The percentage used depends upon the reliability of the information used in calculating the cooling load. As a general rule 10% is used.

After the safety factor has been added, the 24-hr load is divided by the desired operating time for the equipment to determine the average load in Btu per hour (see Section 10-2). The average hourly load is used as a basis for equipment selection.

10-22. Short Method Load Calculations

Whenever possible the cooling load should be determined by using the procedures set forth in the preceding sections of this chapter. However, when coolers are used for general-purpose storage, the product load is frequently unknown and/or varies somewhat from day to day so that it is not possible to compute the product load with any real accuracy. In such cases, a short method of load calculation can be employed which involves the use of load factors which have been determined by experience. When the short method of calculation is employed, the entire cooling load is divided into two parts: (1) the wall gain load and (2) the usage or service load.

The wall gain load is calculated as outlined in Section 10-13. The usage load is computed by the following equation:

$$\text{Usage load} = \text{interior volume} \times \text{usage factor} \qquad (10\text{-}12)$$

Notice that the usage factors listed in Table 10-16 vary with the interior volume of the cooler

TABLE 10-16 Usage Heat Gain

Volume cu ft	Service*	Temperature difference (ambient temp minus storage room temp), F deg										
		1	40	50	55	60	65	70	75	80	90	100
20	Average	4.68	187	234	258	281	305	328	351	374	421	468
	Heavy	5.51	220	276	303	331	358	386	413	441	496	551
30	Average	3.30	132	165	182	198	215	231	248	264	297	330
	Heavy	4.56	182	228	251	274	297	319	342	365	410	456
50	Average	2.28	91	114	126	137	148	160	171	182	205	228
	Heavy	3.55	142	177	196	213	231	249	267	284	320	355
75	Average	1.85	74	93	102	111	120	130	139	148	167	185
	Heavy	2.88	115	144	158	173	188	202	216	230	259	288
100	Average	1.61	64	81	84	97	105	113	121	129	145	161
	Heavy	2.52	101	126	139	151	164	176	189	202	227	252
200	Average	1.38	55	69	76	83	90	97	103	110	124	138
	Heavy	2.22	90	111	122	133	144	155	166	178	200	222
300	Average	1.30	52.0	65	71.5	78	84.5	91	97.5	104	117	130
	Heavy	2.08	83.2	104	114	125	135	146	156	166	187	208
400	Average	1.24	49.6	62	68.2	74.4	80.6	86.8	93	99.2	112	124
	Heavy	1.96	78.4	98	108	118	128	137	147	157	176	196
500	Average	1.21	48.4	60.5	66.6	72.6	78.7	84.7	90.7	96.8	109	121
	Heavy	1.87	74.8	93.5	103	112	122	131	140	150	168	187
600	Average	1.17	46.8	58.5	64	70	76	82	88	94	105	117
	Heavy	1.85	74.0	92.5	102	111	120	130	139	148	167	185
800	Average	1.11	44.4	55.5	61.1	66.6	72.2	77.7	83.3	88.8	100	111
	Heavy	1.76	70.4	88.0	96.8	106	115	123	132	141	158	176
1,000	Average	1.10	44.0	55.0	60.5	66	71.5	77	82.5	88	99	110
	Heavy	1.67	66.8	83.5	91.9	100	108	117	125	134	150	167
1,200	Average	.995	39.8	49.8	54.7	59.7	64.7	69.7	74.7	79.6	89.6	99.5
	Heavy	1.58	63.2	79.0	86.9	94.8	103	111	119	126	142	158
1,500	Average	.920	36.8	46.0	50.6	55.2	59.8	64.4	69	73.6	82.8	92
	Heavy	1.50	60.0	75.0	82.5	90.0	97.5	105	113	120	135	150
2,000	Average	.835	33.4	41.8	45.9	50.1	54.3	58.5	62.7	66.8	75.2	83.5
	Long storage	.775	31.0	38.8	42.6	46.5	50.4	54.3	58.1	62	69.8	77.5
3,000	Average	.750	30.0	37.5	41.3	45.0	48.8	52.5	56.2	60.0	67.5	75.0
	Long storage	.576	23.0	28.8	31.7	34.6	37.3	40.3	43.2	46.1	51.8	57.6
5,000	Long storage	.403	16.1	20.2	22.2	24.2	26.2	28.2	30.2	32.2	36.3	40.3
7,500	Long storage	.305	12.2	15.3	16.8	18.3	19.8	21.4	22.9	24.4	27.5	30.5
10,000	Long storage	.240	9.6	12.0	13.2	14.4	15.6	16.8	18.0	19.2	21.6	24.0
20,000	Long storage	.187	7.48	9.35	10.3	11.2	12.2	13.1	14.0	15.0	16.8	18.7
50,000	Long storage	.178	7.12	8.90	9.79	10.7	11.6	12.5	13.4	14.2	16.0	17.8
75,000	Long storage	.176	7.04	8.80	9.68	10.6	11.5	12.3	13.2	14.1	15.8	17.6
100,000	Long storage	.173	6.92	8.65	9.52	10.4	11.2	12.1	13.0	13.8	15.6	17.3

* For average and heavy service, product load is based on product entering at 10 deg above the refrigerator temperature; for long storage the entering temperature is approximately equal to the refrigerator temperature.
Where the product load is unusual, do not use this table.

From *ASHRAE Data Book,* Fundamentals Volume, 1972 Edition, by permission of the American Society of Heating, Refrigerating, and Air-Conditioning Engineers.

and with the difference in temperature between the inside and outside of the cooler. Too, an allowance is made for normal and heavy usage. Normal and heavy usage have already been defined in Section 10-14. No safety factor is used when the cooling load is calculated by the short method. The total cooling load is divided by the desired operating time for the equipment to find the average hourly load used to select the equipment (see Example 10-13).

Example 10-12 A walk-in cooler 18 ft × 10 ft × 10 ft high has 4 in. of corkboard insulation and 1 in. of wood on each side (total wall thickness is 6 in.). The temperature outside the cooler is 85°F. Thirty-five hundred pounds of mixed vegetables are cooled 10°F to the storage temperature each day. Compute the cooling load in Btu per hour based on a 16-hr/day operating time for the equipment. The inside temperature is 40°F. Usage is heavy.

Solution

Outside surface area $= 4 \times 18 \text{ ft} \times 10 \text{ ft}$
$= 720 \text{ ft}^2$
$= 2 \times 10 \text{ ft} \times 10 \text{ ft}$
$= 200 \text{ ft}^2$
$\overline{920 \text{ ft}^2}$

Inside volume (since total wall thickness is 6 in., the inside dimensions are 1 ft less than the outside dimensions) $= 17 \text{ ft} \times 9 \text{ ft} \times 9 \text{ ft}$
$= 1377 \text{ ft}^3$

TABLE 10-17 Wall Heat Gain

(Btu per square foot per 24 hr)

Insulation: Cork or Equivalent, in.	Temp. Difference (Ambient Temp. Minus Refrigerator Temp)., F																	
	1	40	45	50	55	60	65	70	75	80	85	90	95	100	105	110	115	120
3	2.4	96	108	120	132	144	156	168	180	192	204	216	228	240	252	264	267	288
4	1.8	72	81	90	99	108	117	126	135	144	153	162	171	180	189	198	207	216
5	1.44	58	65	72	79	87	94	101	108	115	122	130	137	144	151	159	166	173
6	1.2	48	54	60	66	72	78	84	90	96	102	108	114	120	126	132	138	144
7	1.03	41	46	52	57	62	67	72	77	82	88	93	98	103	108	113	118	124
8	0.90	36	41	45	50	54	59	63	68	72	77	81	86	90	95	99	104	108
9	0.80	32	36	40	44	48	52	56	60	64	68	72	76	80	84	88	92	96
10	0.72	29	32	36	40	43	47	50	54	58	61	65	68	72	76	79	83	86
11	0.66	26	30	33	36	40	43	46	50	53	56	60	63	66	69	73	76	79
12	0.60	24	27	30	33	36	39	42	45	48	51	54	57	60	63	66	69	72
13	0.55	22	25	28	30	33	36	39	41	44	47	50	52	55	58	61	63	66
14	0.51	20	23	26	28	31	33	36	38	41	43	46	49	51	54	56	59	61
Single glass	27.0	1080	1220	1350	1490	1620	1760	1890	2030	2160	2290	2440	2560	2700	2840	2970	3100	3240
Double glass	11.0	440	500	550	610	660	715	770	825	880	936	990	1050	1100	1160	1210	1270	1320
Triple glass	7.0	280	320	350	390	420	454	490	525	560	595	630	665	700	740	770	810	840

Note: Where wood studs are used multiply the above values by 1.1.

From *ASRE Data Book*, Design Volume, 1955–56 Edition, by permission of the American Society of Heating, Refrigerating, and Air-Conditioning Engineers.

Wall gain factor (Table 10-17) (45°F TD and 4 in. insulation) $= 81$ Btu/(ft^2)(24 hr)

Air changes (Table 10-8A) (by interpolation) $= 14.9$ per 24 hr

Heat gain per cubic foot of air (Table 10-7A) (50% RH) $= 1.69$ Btu/ft^3

Specific heat of mixed vegetables (Table 10-10) $= 0.9$ Btu/lb °F

Average respiration heat of vegetables (Table 10-13) $= 0.09$ Btu/lb hr

Wall gain load:

Area × wall gain factor

= 920 ft^2 × 81 Btu/(ft^2)(24 hr)

= 74,500 Btu/24 hr

Air change load:

Inside volume × air changes × air change factor

= 1377 cu ft × 14.9 × 1.69 Btu/cu ft

= 34,700 Btu/24 hr

Product load:

Temperature reduction

$= m \times c \times \Delta T$

= 3500 lb × 0.9 Btu/lb °F × 10°F

= 31,500 Btu/24 hr

Respiration heat

= Mass × reaction heat × 24 hr

= 3500 lb × 0.09 Btu/(lb)(hr) × 24 hr

= 7600 Btu/24 hr

Summation: 148,300 Btu/24 hr

Safety factor (10%) = 14,800 Btu

Total cooling load = 163,100 Btu/24 hr

Required cooling capacity (Btu/hr) $= \dfrac{\text{Total cooling load}}{\text{Desired running time}}$

$= \dfrac{163{,}100\ \text{Btu/24 hr}}{16\ \text{hr}}$

= 10,200 Btu/hr

Example 10-13 Recalculate the refrigeration load for the cooler described in Example 10-12 using the short-form calculation.

Solution From Example 10-12, the outside surface area is 920 ft^2 and the inside volume is 1377 ft^3. From Table 10-17, the wall gain factor is 81 Btu/(ft^2)(24 hr), and from Table 10-16, the usage factor by interpolation and assuming heavy usage at 45°F TD is 69 Btu/(ft^3)(24 hr).

Wall gain load = area × wall gain factor

= 920 ft^2 × 81 Btu/(ft^2)(24 hr)

= 74,500 Btu/24 hr

Usage load = inside volume × usage factor

= 1377 ft^3 × 69 Btu/(ft^3)(24 hr)

= 95,000 Btu/24 hr

Summation = 169,500 Btu/24 hr

Required equipment capacity $= \dfrac{169{,}500\ \text{Btu/24 hr}}{16\ \text{hr}}$

= 10,600 Btu/hr

Example 10-14 A holding cooler 18 ft × 30 ft × 12 ft high is to be used for the short-term storage of fresh beef. Seventy-five hundred pounds of beef enter the cooler at 45°F and are cooled to the storage temperature of 35°F each day. All the walls are partitions adjacent to unconditioned spaces (90°F and 50% RH) except the east wall (18 ft × 12 ft), which is adjacent to a chilling room maintained at the same inside design temperature. Wall construction is 4-in. cinder block insulated with 4-in. corkboard equivalent. The floor, located over an unconditioned space, is a 5-in. concrete slab insulated with 4-in. corkboard equivalent and finished with 3 in. of concrete. The ceiling, situated under an unconditioned space, is a 4-in. concrete slab with wood sleepers and insulated with the equivalent of 4 in. of corkboard. Two people work in the space during the loading periods, usage is average, and the lighting load is 500 W. Determine the required equipment capacity in Btu per hour based on an 18-hr operating time.

Solution The U factor for the ceiling (Table 10-3) is 0.069 Btu/(hr)(ft^2)(°F); for the floor (Table 10-3), 0.066 Btu/(hr)(ft^2)(°F); and for the walls (Table 10-1), 0.066 Btu/(hr)(ft^2)(°F). Since the TD across all the walls except the east wall, including floor and ceiling, is the same, and since the difference in the wall U factors is slight, the walls may be lumped together for calculation. There is no TD across the east wall and consequently no heat gain through the wall. The outside surface area of the walls, except the east wall is 2000 ft^2. The inside volume is 5400 ft^3. The number of air changes per 24 hr (Table 10-8) by interpolation and assuming normal usage is 6.9, and the air change factor (Table 10-7) is 2.17 Btu/ft^3. The specific heat (Table 10-11) is 0.75 Btu/(lb)(°F).

Wall gain load $= (A)(U)(D)(24 \text{ hr})$

$= (2000)(0.067)(55)(24)$

$= 176{,}900$ Btu/24 hr

Air change load = (inside volume)(air changes)(air change factor)

$= (5400)(6.9)(2.17)$

$= 80{,}900$ Btu/24 hr

Product load $= (m)(c)(\Delta T)$

$= (7500)(0.75)(10)$

$= 56{,}200$ Btu/24 hr

Misc. load

People $= (2)(900)(24)$

$= 43{,}200$ Btu/24 hr

Lights $= (500 \text{ W})(3.4)(24)$

$= 40{,}800$ Btu/24 hr

Summation 398,000 Btu/24 hr
Safety factor (10%) 39,800 Btu/24 hr
Total cooling load = 437,800 Btu/24 hr

$$\text{Req. equip. capacity} = \frac{437{,}800 \text{ Btu/24 hr}}{18 \text{ hr}}$$

$= 24{,}300$ Btu/hr

Example 10-15 Three thousand lug boxes of apples are stored at 35°F in a storage cooler 50 ft × 40 ft × 10 ft. The apples enter the cooler at a temperature of 85°F and at the rate of 200 lugs per day each day for the 15 day harvesting period. The walls including floor and ceiling are constructed of 1 in. boards on both sides of 2 × 4 studs and are insulated with $3\frac{5}{8}$ in. of rock wool. All of the walls are shaded and the ambient temperature is 85°F. The average weight of apples per lug box is 59 lb. The lug boxes have an average weight of 4.5 lb and a specific heat value of 0.60 Btu/lb °F. Determine the average hourly cooling load based on 20 hr operating time for the equipment.

Solution The load calculation is made for the maximum loading that occurs on the fifteenth day. The outside surface area is 5800 ft^2 and the inside volume is 17,200 ft^3. The wall U factor from Table 10-2 is 0.072 Btu/(hr)(ft^2)(°F). From Table 10-8A by interpolation, the number of air changes per 24 hr for normal usage is 3.7, and the air change factor (Table 10-7A) is 1.86 Btu/ft^3. From Table 10-9, the specific heat of apples is 0.89 Btu/(lb)(°F), and the chilling rate factor is 0.67. The respiration heat from Table 10-13 by interpolation is 0.023 Btu/(lb)(hr).

Wall gain load $= (A)(U)(D)(24 \text{ hr})$

$= (5800)(0.072)(50)(24)$

$= 501{,}100$ Btu/24 hr

Air change load = (inside volume)(air changes)(air change factor)

$= (17{,}200)(3.7)(1.86)$

$= 118{,}400$ Btu/24 hr

$$\text{Product load} = \frac{(m)(c)(\Delta T)}{\text{chilling rate factor}}$$

$$\text{Apples} = \frac{(200 \times 59)(0.89)(50)}{0.67}$$

$= 783{,}700$ Btu/24 hr

$$\text{Lug boxes} = \frac{(200 \times 4.5)(0.6)(50)}{0.67}$$

$= 40{,}300$ Btu/24 hr

Respiration $= (m)(\text{reaction heat})(24 \text{ hr})$

$= (3000 \times 59)(0.023)(24 \text{ hr})$

$= 97{,}700$ Btu/24 hr

Summation	1,541,200 Btu/24 hr
Safety factor (10%)	154,000 Btu/24 hr
Total cooling load	1,695,200 Btu/24 hr

$$\text{Average hourly load} = \frac{\text{Total cooling load}}{\text{Operating time}}$$

$$= \frac{1{,}695{,}200 \text{ Btu/24 hr}}{20 \text{ hr}}$$

$$= 84{,}800 \text{ Btu/hr}$$

Example 10-16 Twenty-two thousand pounds of dressed poultry are blast-frozen on hand trucks each day (24 hr) in a freezing tunnel 14 ft × 9 ft × 10 ft high (see Fig. 10-4). The poultry is precooled to 45°F before entering the freezer where it is frozen and its temperature lowered to 0°F for storage. The lighting load is 200 watts. The hand trucks carrying the poultry total 1400 lb per day and have a specific heat of 0.25 Btu/lb/°F. The north and east partitions adjacent to the equipment room and vestibule are constructed of 6 in. clay tile insulated with 8 in. of corkboard. The south and west partitions adjacent to storage cooler are 4 in. clay tile with 2 in. corkboard insulation. The roof is a 6 in. concrete slab insulated with 8 in. of corkboard and covered with tar, felt, and gravel. The floor is a 6 in. concrete slab insulated with 8 in. of corkboard and finished with 4 in. of concrete. The floor is over a ventilated crawl space. Roof is exposed to the sun. The equipment room is well ventilated so that the temperature inside is approximately the outdoor design temperature for the region (92°F). The inside design temperature for both the storage room and the freezer is 0°F. The vestibule temperature and relative humidity are 50°F and 70%, respectively. Determine the average hourly refrigeration load based on 20 hr per day operating time for the equipment.

Solution

Outside surface area

Roof (9 ft × 14 ft) = 126 ft^2

Floor (9 ft × 14 ft) = 126 ft^2

N partition (9 ft × 10 ft) = 90 ft^2

E partition (14 ft × 10 ft) = 140 ft^2

Inside volume (8 ft × 9 ft × 13 ft) = 936 ft^3

U factors

Roof (Table 10-3) = 0.036 Btu/(hr)(ft^2)(°F)

Floor (Table 10-3) = 0.035 Btu/(hr)(ft^2)(°F)

N and *E* partitions (Table 10-1)

= 0.035 Btu/(hr)(ft^2)(°F)

Roof sun factor (Table 10-6) = 20°F

Air changes (Table 10-8B) = 14.0/2 per 24 hr

Air change factor (Table 10-7B) = 1.48 Btu/ft^3

Specific heat of poultry (Table 10-11)

Above freezing = 0.79 Btu/lb °F

Below freezing = 0.37 Btu/lb °F

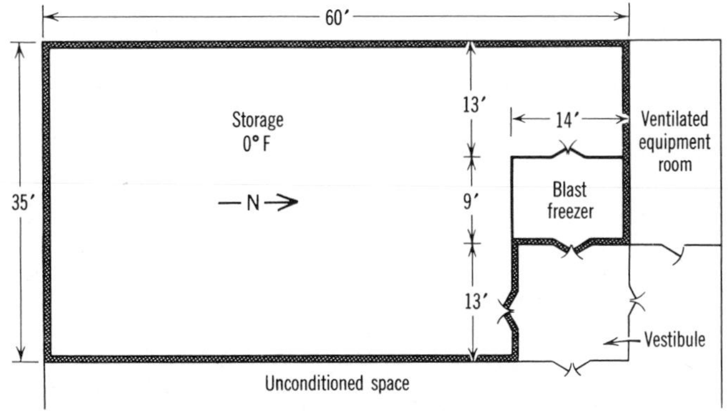

Fig. 10-4

Latent heat of poultry (Table 10-11) = 106 Btu/lb
Freezing temperature = 27°F
Wall gain load
= (A)(U)(D)(24 hr)
Roof
= (126)(0.036)
(92° + 20°)(24) = 12,200 Btu/24 hr
Floor
= (126)(0.035)
(92°)(24) = 9,700 Btu/24 hr
N partition
= (90)(0.035)(92°)(24) = 7,000 Btu/24 hr
E partition
= (140)(0.035)
(50°)(24) = 5,900 Btu/24 hr
Air change load = (inside volume)
(air changes)
(air change factor)
= (936)(7.0)(1.48)
= 9,700 Btu/24 hr
Product load
Temperature reduction
$= (m)(c)(\Delta T)$
Poultry
= (22,000)(0.79)
(45 − 27) = 312,800 Btu/24 hr
= (22,000)(0.37)
(27 − 0) = 219,800 Btu/24 hr
Trucks
= (1400)(0.25)
(45 − 0) = 15,700 Btu/24 hr
Freezing $= (m)(h_{if})$
= (22,000)
(106) = 2,332,000 Btu/24 hr
Misc. load (lights)
= (150 W)(3.4)(24) = 12,200 Btu/24 hr
Summation
= 2,927,300 Btu/24 hr
Safety factor (10%)
= 292,700 Btu/24 hr
Total cooling load
= 3,220,000 Btu/24 hr
Required equipment capacity

$$= \frac{3{,}220{,}000 \text{ Btu/24 hr}}{20 \text{ hr}} = 161{,}000 \text{ Btu/hr}$$

Example 10-17 Five hundred gallons of partially frozen ice cream at 25°F are entering a hardening room 10 ft × 15 ft × 9 ft each day. Hardening is completed and the temperature of the ice cream is lowered to −20°F in 10 hr. The walls, including floor and ceiling, are insulated with 8 in. of corkboard and the overall thickness of the walls is 12 in. The ambient temperature is 90°F and the lighting load is 300 watts. Assume the average weight of ice cream is 5 lb per gallon, the average specific heat below freezing is 0.5 Btu/lb °F, and the average latent heat per pound is 100 Btu. Determine the average hourly load based on 18 hr operation.

Solution

Outside surface area = 750 ft^3
Inside volume
(8 ft × 13 ft × 7 ft) = 728 ft^3
Wall gain factor
(Table 10-17) = 99 Btu/(ft^2)(24 hr)
Air changes
(Table 10-8B)
(interpolated) = 16.7 per 24 hr
Heat gain per cubic foot
(Table 10-7B)
(50% RH) = 3.88 Btu/ft^3

Wall gain load:
area × wall gain factor
= 750 × 99 = 74,250 Btu/24 hr
Air change load:
Inside volume × air changes × air change factor
= 728 × 16.7 × 3.88 = 47,170 Btu/24 hr
Product load:
Temperature reduction

$$= \frac{m \times C \times TD \times 24 \text{ hr}}{\text{Chilling time}}$$

$$= \frac{(500 \times 5) \times 0.5 \times (25 - -20) \times 24}{10}$$

= 135,000 Btu/24 hr

Freezing

$$= \frac{m \times \text{latent heat} \times 24 \text{ hr}}{\text{Freezing time}}$$

$$= \frac{(500 \times 5) \times 100 \times 24 \text{ hr}}{10}$$

= 600,000 Btu/24 hr

Miscellaneous load:

Lighting:

300 watts × 3.4 × 24 hr = 24,480 Btu/24 hr

Summation: 880,900 Btu/24 hr

Safety factor (10%) = 88,090 Btu

968,990 Btu/24 hr

Average hourly load (968,990/18 hr) = 53,800 Btu/hr

Example 10-18 A cooler 10 ft × 12 ft × 9 ft high equipped with four 2 ft × 3 ft triple-glass doors is used for general purpose storage in a grocery store. The cooler is maintained at 35°F and the service load is heavy. The walls are insulated with the equivalent of 4 in. of corkboard and the ambient temperature is 80°F. Determine the cooling load in Btu per hour based on a 16 hr operating time.

Solution The glass area is 24 ft^2 (4 × 2 × 3) and the net wall area is 612 ft^2 (636 − 24). From Table 10-17, the wall gain factor is 81 Btu/(ft^2)(24 hr), and the factor for triple glass is 320 Btu/(ft^2)(24 hr). The inside volume is approximately 790 ft^3, and from Table 10-16), the usage factor for a 45°F TD and heavy usage is 79 Btu/(ft^3)(24 hr).

Wall gain load

(wall area)(wall gain factor) = (612)(81) = 49,570 Btu/24 hr

(glass area)(glass factor) = (24)(320) = 7,680 Btu/24 hr

Usage load

(inside volume)(usage factor) = (790)(79) = 62,410 Btu/24 hr

Total cooling load = 119,660 Btu/24 hr

Average hourly load $= \dfrac{119{,}660 \text{ Btu/24 hr}}{16 \text{ hr}}$

= 7480 Btu/hr

10-23. METRIC CALCULATIONS

The following examples demonstrate the use of metric units in typical heat transfer calculations.

Example 10-19 A certain sample of polyurethane insulation has a thermal conductivity (k) of 0.17 (Btu)(in)/(hr)(ft^2)(°F). Determine the value of k in SI units.

Solution The thermal conductivity (k) has the units of (W)(m)/(m^2)(°K). By mathematical cancellation, this fraction reduces to W/(m)(°K). Applying the conversion factor from Table 1-1,

$$k = (0.17)(0.1442285) = 0.0245 \text{ W/(m)(°K)}$$

Example 10-20 Determine the rate of heat transfer in watts through an 8 m^2 section of the polyurethane insulation described in Example 10-19 if the thickness of the insulation is 10.16 cm (4 in.) and the temperature difference between the two sides is 14°C.

Solution From Example 10-19, $k =$ 0.0245 W/(m)(°K). Applying Equation 10-2,

$$Q = (A)\frac{k}{x}(T_2 - T_1) = (8 \text{ m}^2)\frac{0.0245}{0.1016}(14°)$$

$$= 27 \text{ W}$$

Example 10-21 From Table 10-4, the convection coefficient for an outside wall surface with a wind velocity of 7.5 mph (3.35 m/s) is 4 Btu/(hr)(ft^2)(°F). Determine the value of the convection coefficient in SI units.

Solution Applying the conversion factor from Table 1-1,

$$C(f_o) = (4)(5.678) = 22.7 \text{ W/(m}^2\text{)(°K)}$$

Example 10-22 The roof of a building 8 m by 10 m has a U value of 0.047 Btu/(hr)(ft^2)(°F). If the temperature difference is 52°C, compute the heat gain through the roof in watts.

Solution Using the conversion factor from Table 1-1,

$$U = (0.047)(5.678) = 0.267 \text{ W/(m}^2\text{)(°K)}$$

Applying Equation 10-2,

$$Q = (80 \text{ m}^2)(0.267)(52°) = 1.11 \text{ kW}$$

Customary Problems

Determine the value of C for a 4-in. thickness of an insulating material having a k value of 0.24 (Btu)(in.)/(hr)(ft^2)(°F).

10-2 Assuming a wind velocity of 7.5 mph, compute the value of U for a wall constructed of 8-in. hollow clay tile insulated with 6 in. of polyurethane foam and finished on the inside with 0.5 in. of cement plaster.

10-3 A cooler wall 10 ft by 18 ft is insulated with the equivalent of 3 in. of expanded polystyrene. Compute the heat gain through the wall in Btu per 24 hr if the inside temperature is 37°F and the outside temperature is 78°F.

10-4 A reach-in cooler is equipped with eight triple-pane glass doors, each measuring 2 ft by 2.5 ft. Compute the conduction heat gain through the doors in Btu per hour if the temperature difference between the inside and outside is 40°F.

10-5 The north wall of a cold storage warehouse is 12 ft by 60 ft and is constructed of 8-in. hollow clay tile insulated with 6 in. of corkboard. The location is Tulsa, Oklahoma, and the inside design temperature is −10°F. Determine the heat gain through the wall in Btu per 24 hr.

10-6 A cold storage warehouse in New Orleans, Louisiana, has a 30-ft by 50-ft flat roof constructed of 4 in. of concrete covered with tar and gravel and insulated with the equivalent of 4 in. of corkboard. If the roof is unshaded and the inside of the warehouse is maintained at 35°F, compute the heat gain through the roof in Btu per hour.

10-7 The floor of the cold storage room described in Problem 10-6 consists of a 2-in. concrete slab insulated with the equivalent of 3 in. of corkboard and finished with 2 in. of concrete. Determine the heat gain through the floor in Btu per hour.

10-8 A frozen storage room has an interior volume of 2000 ft^3 and is maintained at a temperature of −10°F. The usage is light, and the outside design conditions (anteroom) are 50°F and 70% RH. Compute the air change load in Btu per 24 hr.

10-9 Five thousand pounds of fresh lean beef enter a chilling cooler at 100°F and are chilled to 38°F in 24 hr. Compute the chilling load in Btu per 24 hr.

10-10 Four thousand pounds of green beans packed in bushel baskets (32 lb per bushel) enter a chilling cooler at a temperature of 80°F and are chilled to a temperature of 33°F in 20 hr. The empty baskets have a mass of 3 lb and a specific heat of 0.6 Btu/(lb)(°F). Compute the product load in Btu per 24 hr.

10-11 Six hundred pounds of prepared, packaged beef enter a freezer at a temperature of 36°F. The beef is to be frozen and its temperature reduced to 0°F in 5 hr. Compute the product load.

10-12 Fifty-five hundred crates of apples are in storage at 37°F. An additional 500 crates enter the storage cooler at a temperature of 85°F and are chilled to the storage temperature in 24 hr. The average mass of apples per crate is 60 lb. The crate has a mass of 10 lb and a specific heat of 0.6 Btu/(lb)(°F). Determine the product in Btu per 24 hr.

10-13 A walk-in cooler 8 ft by 16 ft by 9 ft high equipped with twelve 2 ft by 2 ft triple-glass doors is used for general purpose storage in a drive-in market (heavy usage). The walls are insulated with 2 in. of expanded polystyrene (equivalent to 3-in. corkboard), and the cooler is to be maintained at 35°F. Compute the cooling load in Btu per hour based on a 16-hr operating time if the ambient temperature is 75°F.

Metric Problems

10-14 A certain sample of polystyrene insulation has a thermal conductivity (k) of 0.2 (Btu)(in)/(hr)(ft^2)(°F). Determine the value of k in SI units.

10-15 Determine the rate of heat transfer in watts through a 32 m^2 section of the polystyrene insulation described in Problem 10-14 if the thickness of the insulation is 0.1525 m and the temperature difference between the two sides is 60°C.

10-16 From Table 10-4, the convection coefficient for an outside wall surface with a wind velocity of 15 mph (6.7 m/s) is 6 Btu/(hr)(ft^2)(°F). Determine the value of the convection coefficient in SI units.

10-17 The wall of a building 3.5 m by 8 m has a U value of 0.032 Btu/(hr)(ft^2)(°F). If the temperature difference is 60°C, calculate the heat gain through the wall in watts.

11

EVAPORATORS

11-1. Types of Evaporators

An evaporator is any heat transfer surface in which a volatile liquid is vaporized for the purpose of removing heat from a refrigerated space or product. Because of the many and diverse applications of mechanical refrigeration, evaporators are manufactured in a wide variety of types, shapes, sizes, and designs and may be classified in a number of different ways, such as type of construction, method of liquid feed, operating condition, method of air (or liquid) circulation, type of refrigerant control, and application.

11-2. Types of Construction

The three principal types of evaporator construction are (1) bare-tube, (2) plate-surface, and (3) finned. Bare-tube and plate-surface evaporators are sometimes classified together as prime-surface evaporators in that the entire surface of both these types is more or less in contact with the vaporizing refrigerant inside. With the finned evaporator, the refrigerant-carrying tubes are the only prime surface. The fins themselves are not filled with refrigerant and are, therefore, only secondary heat transfer surfaces whose function is to pick up heat from the surrounding air and conduct it to the refrigerant-carrying tubes.

Although prime-surface evaporators of both the bare-tube and plate-surface types give satisfactory service on a wide variety of applications operating in any temperature range, they are most frequently applied to liquid chilling applications and to air cooling applications where the space temperature is maintained below 34°F and frost accumulation on the evaporator surface cannot readily be prevented. Frost accumulation on prime-surface evaporators does not affect the evaporator capacity to the extent that it does on finned coils. Furthermore, most prime surface evaporators, particularly the plate-surface type, are easily cleaned and can readily be defrosted manually by either brushing or scraping off the frost accumulation. This can be accomplished without interrupting the refrigerating process and jeopardizing the quality of the refrigerated product.

11-3. Bare-Tube Evaporators

Bare-tube evaporators are usually constructed of either steel pipe or copper tubing. Steel pipe is used for large evaporators and for evaporators to be employed with ammonia, whereas copper tubing is utilized in the manufacture of smaller evaporators intended for use with refrigerants other than ammonia. Bare-tube coils are available in a number of sizes, shapes, and designs, and are usually custom made to the individual application. Common shapes for bare-tube coils are flat zigzag and oval trombone, as shown in Fig. 11-1. Spiral bare-tube coils are often employed for liquid chilling.

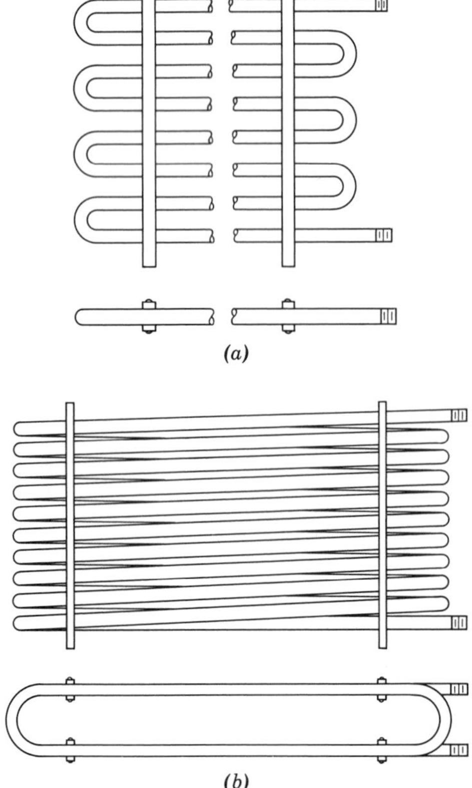

Fig. 11-1 Common designs for bare-tube coils. (*a*) Flat zigzag coil. (*b*) Oval trombone coil.

Large ceiling-hung bare-pipe coils employing natural convection air circulation are frequently used in frozen storage rooms and in storage coolers where the circulation of large quantities of low velocity air is desirable. They are used also, either "dry" or as "spray coils," in conjunction with centrifugal blowers to provide high-velocity chilled air for blast-cooling or freezing operations.

11-4. Plate-Surface Evaporators

Plate-surface evaporators are of several types. Some are constructed of two flat sheets of metal so embossed and welded together as to provide a path for refrigerant flow between the two sheets (Fig. 11-2). This particular type of plate-surface evaporator is widely used in household refrigerators and home freezers because it is easily cleaned, economical to manufacture, and can be readily formed into any one of the various shapes required (Fig. 11-3).

Another type of plate-surface evaporator consists of formed tubing installed between two metal plates which are welded together at the edges (Fig. 11-4). In order to provide good thermal contact between the welded plated and the tubing carrying the refrigerant, the space between the plates is either filled with a eutectic solution or evacuated so that the pressure of the atmosphere exerted on the outside surface of the plates holds the plates firmly against the tubing inside. Those containing the eutectic solution are especially useful where a holdover capacity is required. Many are used on refrigerated trucks. In such applications, the plates are mounted either vertically or horizontally from the ceiling or walls of the truck (Fig. 11-5) and are usually connected to a central plant refrigeration system while the trucks are parked at the terminal during the night. The refrigerating capacity thus stored in the eutectic solution is sufficient to refrigerate the product during the next day's operations. The temperature of the plates is controlled by the melting point of the eutectic solution.

Plate-type evaporators may be used singly or in banks. Figure 11-6 illustrates how the plates can be grouped together for ceiling mounting in holding rooms, locker plants, freezers, etc. The plates may be manifolded for parallel flow of the refrigerant (Fig. 11-7) or they may be connected for series flow.

Plate-surface evaporators provide excellent shelves in freezer rooms and similar applications (Fig. 11-8). They are also widely used as partitions in freezers, frozen food display cases, ice cream cabinets, soda fountains, etc. Plate evaporators are especially useful for liquid cooling installations where unusual peak load conditions are encountered periodically. By building up an ice bank on the surface of the plates during periods of light loads, a holdover refrigerating capacity is established which will help the refrigerating equipment carry the load through the heavy or peak condition (Fig. 11-9). Since this allows the use of smaller capacity equipment than would ordinarily be required by the peak load, a savings is affected in initial cost and, usually, also in operating expenses.

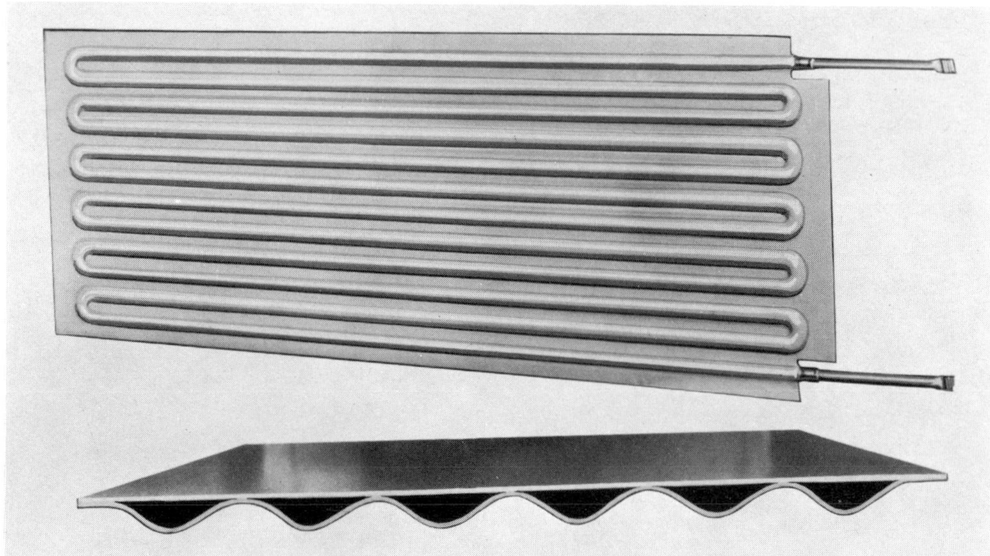

Fig. 11-2 Standard serpentine plate evaporator. (Courtesy Kold-Hold Division—Tranter Manufacturing, Inc.)

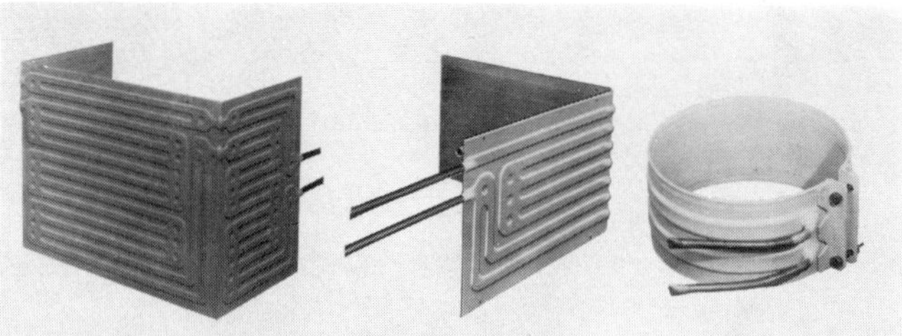

Fig. 11-3 Some typical shapes available in plate-type evaporators. (Courtesy Dean Products, Inc.)

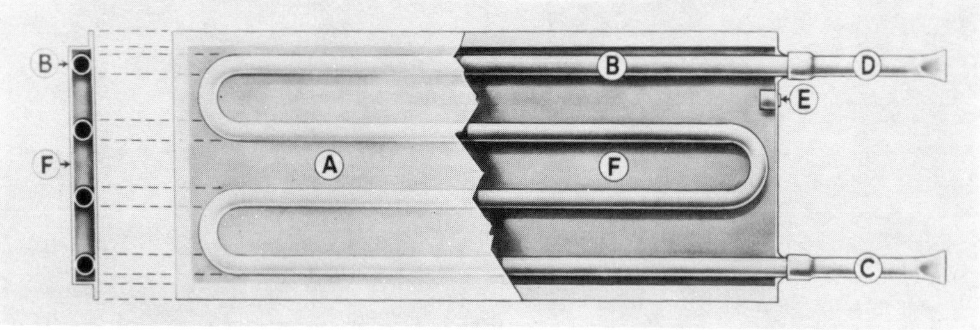

Fig. 11-4 Plate-type evaporator. (Courtesy Dole Refrigerating Company.)

(A) Outside jacket of plate. Heavy, electrically welded steel. Smooth surface.

(B) Continuous steel tubing through which refrigerant passes.

(C) Inlet from compressor.

(D) Outlet to compressor. Copper connections for all refrigerants except ammonia where steel connections are used.

(E) Fitting where vacuum is drawn and then permanently sealed.

(F) Vacuum space in dry plate. Space in hold-over plate contains eutectic solution under vacuum. No maintenance required due to sturdy, simple construction. No moving parts; nothing to wear or get out or order; no service necessary.

Fig. 11-5 Freezer plates installed in wholesale ice cream truck body. (Courtesy Dole Refrigerating Company.)

Fig. 11-6 Plate banks employed in low temperature storage rooms. (Courtesy Dole Refrigerating Company.)

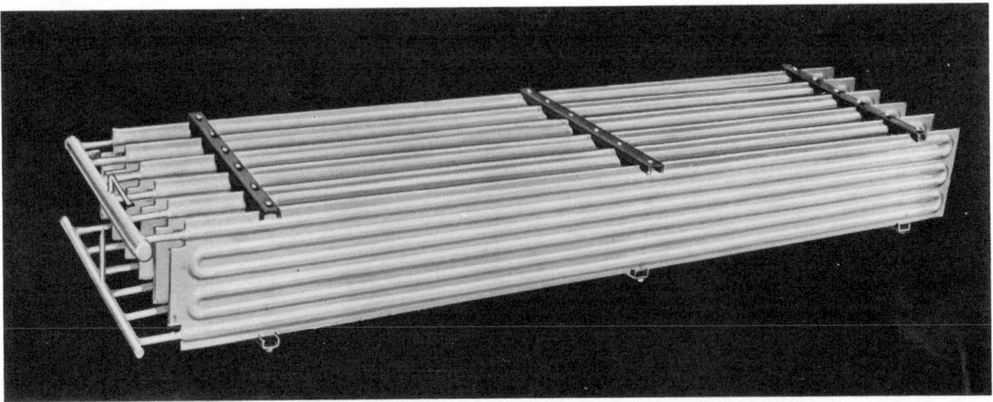

Fig. 11-7 Plate bank with plates manifolded for parallel refrigerant flow. Plates may also be connected for series flow. (Courtesy Kold-Hold Division—Tranter Manufacturing, Inc.)

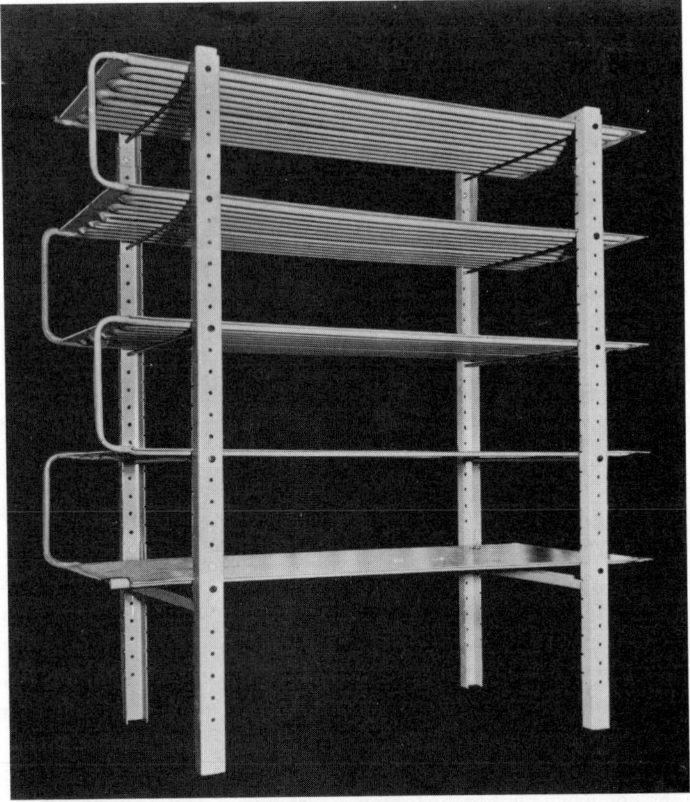

Fig. 11-8 Plate evaporators employed as freezer shelves. Note that plates are arranged for series refrigerant flow. (Courtesy Kold-Hold Division—Tranter Manufacturing, Inc.)

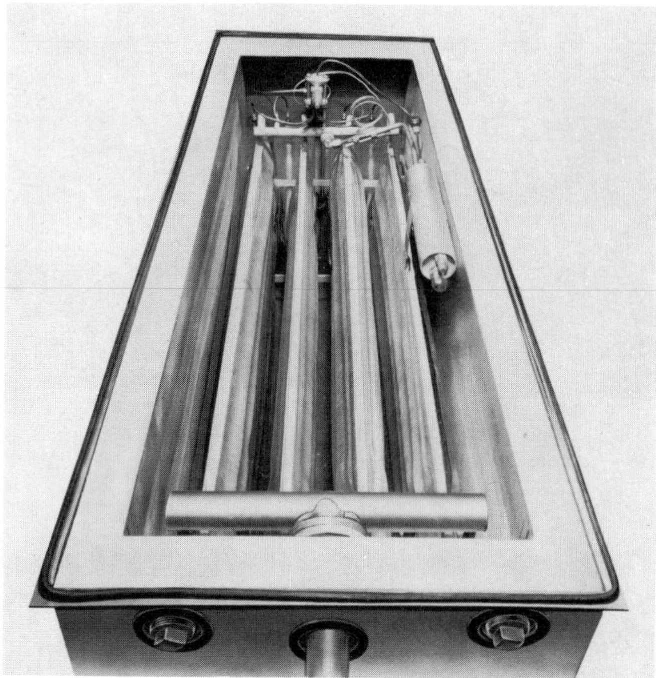

Fig. 11-9 Ice-Cel. Refrigeration holdover capacity is established by building up a bank of ice on plate evaporators. (Courtesy Dole Refrigerating Company.)

11-5. Finned Evaporators

Finned coils are bare-tube coils upon which metal plates or fins have been installed (Fig. 11-30). The fins, serving as secondary heat-absorbing surfaces, have the effect of increasing the outside surface area of the evaporator, thereby improving its efficiency for cooling air and other gases. With bare-tube evaporators, much of the air that circulates over the coil passes through the open spaces between the tubes and does not come in contact with the coil surface. When fins are added to a coil, the fins extend out into the open spaces between the tubes and act as heat collectors. They remove heat from that portion of the air which would not ordinarily come in contact with the prime surface and conduct it back to the tubing.

It is evident that to be effective the fins must be connected to the tubing in such a manner that good thermal contact between the fins and the tubing is assured. In some instances, the fins are soldered directly to the tubing. In others, the fins are slipped over the tubing and the tubing is expanded by pressure or some such means so that the fins bite into the tube surface and establish good thermal contact. A variation of the latter method is to flare the fin hole slightly to allow the fin to slip over the tube. After the fin is installed, the flare is straightened and the fin is securely locked to the tube.

Fin size and spacing depend in part on the particular type of application for which the coil is designed. The size of the tube determines the size of the fin. Small tubes require small fins. As the size of the tube increases, the size of the fin may be effectively increased. Fin spacing varies from 1 to 14 fins per inch, depending primarily on the operating temperature of the coil.

Frost accumulation on air-cooling coils operating at low temperatures is unavoidable, and since any frost accumulation on finned coils tends to restrict the air passages between the fins and to retard air circulation through the coil, evaporators designed for low temperature applications must have wide fin spacing (two or three fins per inch) in order to minimize the danger of restricting air circulation. On the other hand, coils designed for air conditioning and other installations where the coil operates at temperatures high enough so that no frost accumulates on the coil surface may have as many as 14 fins per inch.

When air circulation over finned coils is by gravity, it is important that the coil offer as little resistance to air flow as is possible; therefore, in general, fin spacing should be wider for natural convection coils than for coils employing fans.

It has been determined that a definite relationship exists between the inside and outside surfaces of an evaporator. Since external finning affects only the outside surface, the addition of fins beyond a certain limit will not materially increase the capacity of the evaporator. In fact, in some instances, excessive finning may actually reduce the evaporator capacity by restricting the air circulation over the coil unnecessarily.

Since their capacity is affected more by frost accumulation than any other type of evaporator, finned coils are best suited to air-cooling application where the temperature is maintained above 34°F. When finned coils are used for low temperature operation, some means of defrosting the coil at regular intervals must be provided. This may be accomplished automatically by several means which are discussed in another chapter.

Because of the fins, finned coils have more surface area per unit of length and width than prime-surface evaporators and can therefore be built more compactly. Generally, a finned coil will occupy less space than either a bare-tube or plate-surface evaporator of the same capacity. This provides for a considerable savings in space and makes finned coils ideally suited for use with fans as forced convection units.

11-6. Evaporator Capacity

The capacity of any evaporator or cooling coil is the rate at which heat will pass through the evaporator walls from the refrigerated space or product to the vaporizing liquid inside and is usually expressed in Btu per hour. An evaporator selected for any specific application should have sufficient heat transfer capacity to allow the vaporizing refrigerant to absorb heat at the rate necessary to produce the required cooling when operating at the design conditions.

Heat reaches the evaporator by all three methods of heat transfer. In air-cooling applications most of the heat is carried to the evaporator by convection currents set up in the refrigerated space either by action of a fan or by gravity circulation resulting from the difference in temperature between the evaporator and the space. Too, some heat is radiated directly to the evaporator from the product and from the walls of the space. Where the product is in thermal contact with the outer surface of the evaporator, heat is transferred from the product to the evaporator by direct conduction. This is always true for liquid cooling applications where the liquid being cooled is always in contact with the evaporator surface. However, some circulation of the cooled fluid either by gravity or by action of a pump is still necessary for good heat transfer.

Regardless of how the heat reaches the outside surface of the evaporator, it must pass through the walls of the evaporator to the refrigerant inside by conduction. Therefore, the capacity of the evaporator, that is, the rate at which heat passes through the walls, is determined by the same factors that govern the rate of heat flow by conduction through any heat transfer surface and is expressed by the equation

$$Q = A \times U \times D \tag{11-1}$$

where Q = the quantity of heat transferred in Btu per hour

A = the outside surface area of the evaporator (both prime and finned)

U = the overall conductance factor in Btu/(hr)(ft^2 outside surface)(°F D)

D = the logarithmic mean temperature difference in degrees Fahrenheit between the temperature outside the evaporator and the temperature of the refrigerant inside the evaporator

11-7. *U* or Overall Conductance Factor

The resistance to heat flow offered by the evaporator walls is the sum of three factors whose relationship is expressed by the following:

$$\frac{1}{U} = \frac{R}{f_i} + \frac{L}{K} + \frac{1}{f_0} \qquad (11\text{-}2)$$

where U = the overall conductance factor in Btu/(hr)(ft^2)(°F D)

f_i = the conductance factor of the inside surface film in Btu/(hr)(ft^2 inside surface)(°F D)

L/K = resistance to heat flow offered by metal of tubes and fins

f_0 = the conductance factor of the outside surface film in Btu/(hr)(ft^2 outside surface)(°F D)

R = ratio of outside surface to inside surface

Since a high rate of heat transfer through the evaporator walls is desirable, the U or conductance factor should be as high as possible. Metals, because of their high conductance factor, are usually employed in evaporator construction. However, a metal which will not react with the refrigerant must be selected. Iron, steel, brass, copper, and aluminum are the metals most commonly used. Iron and steel are not affected by any of the common refrigerants, but are apt to rust if any free moisture is present in the system. Brass and copper can be used with any refrigerant except ammonia, which dissolves copper. Aluminum may be used with any refrigerant except methyl chloride. Magnesium and magnesium alloys cannot be used with the fluorinated hydrocarbons or with methyl chloride.

Of the three factors involved in Equation 11-2, the metal of the evaporator walls is the least significant. The amount of resistance to heat flow offered by the metal is so small, especially where copper and aluminum are concerned, that it is usually of no consequence. Thus, the U factor of the evaporator is determined primarily by the coefficients of conductance of the inside and outside surface films.

In general, because of the effect they have on the inside and outside film coefficients, the value of U for an evaporator depends on the type of coil construction and the material used, the amount of interior wetted surface, the velocity and conductivity of the refrigerant inside the the coil, the amount of oil present in the evaporator, the material being cooled, the condition of the external surface, the fluid (either gaseous or liquid) velocity over the coil, and the ratio of inside to outside surface.

Heat transfer by conduction is greater through liquids than through gases and the rate at which the refrigerant absorbs heat from the evaporator walls increases as the amount of interior wetted surface increases. In this respect, flooded evaporators, since they are always completely filled with liquid, are more efficient than the dry-expansion type. This principle also applies to the external evaporator surface. When the outside surface of the evaporator is in direct contact with some liquid or solid medium, the heat transfer by conduction to the outside surface of the evaporator is greater than when air or some other gas is the medium in contact with the evaporator surface.

Any fouling of either the external or internal surfaces of the evaporator tends to act as thermal insulation and decreases the conductance factor of the evaporator walls and reduces the rate of heat transfer. Fouling of the external surface of air-cooling evaporators is usually caused by air borne dust, lint, grease, and other contaminants which adhere to the wet coil surfaces or by frost accumulation on the coil surface. In liquid-cooling applications, fouling of the external tube surface usually results from scale formation and corrosion. Fouling of the internal surface of the evaporator tubes is usually caused by excessive amounts of oil in the evaporator and/or low refrigerant velocities. At low velocities, vapor bubbles, formed by the boiling action of the refrigerant, tend to cling to the tube walls, thereby decreasing the amount of interior wetted surface. Increasing the refrigerant velocity produces a scrubbing action on the walls of the tube which carries away the oil and bubbles and improves the rate of heat flow. Thus, for a given tube size, the inside film coefficient increases as the refrigerant velocity increases. The refrigerant velocity is limited, however, by the maximum allowable pressure drop through the coil and, if increased beyond a certain point, will result in a decrease rather than an increase in coil capacity. This depends to some extent on the method of coil circuiting and is discussed later. It can be shown also that the conductance of the outside surface film is improved by increasing the fluid velocity over the outside surface of the coil. But, here again, in many cases the maximum velocity is limited, this time by consideration other than the capacity of the evaporator itself.

Any increase in the turbulence of flow either inside or outside the evaporator will materially increase the rate of heat transfer through the evaporator walls. In general, internal turbulence increases with the difference in temperature across the walls of the tube, closer spacing of the tubes, and the roughness of the internal tube surface. In some instances, heat transfer is improved by internal finning.

Outside flow turbulence is influenced by fluid velocity over the coil, tube spacing, and the shape of the fins.

11-8. The Advantage of Fins

The advantage of finning depends on the relative values of the coefficients of conductance of the inside and outside surface films and upon R, the ratio of the outside surface to the inside surface. In any instance where the rate of heat flow from the inside surface of the evaporator to the liquid refrigerant is such that it exceeds the rate at which heat passes to the outside surface from the cooled medium, the overall capacity of the evaporator is limited by the capacity of the outside surface. In such cases, the overall value of U can be increased by using fins to increase the outside surface area to a point such that the amount of heat absorbed by the outside surface is approximately equal to that which can pass from the inside surface to the liquid refrigerant.

Because heat transfer is greater to liquids than to vapors, this situation often exists in air-cooling applications where the rate of heat flow from the inside surface to the liquid refrigerant is much higher than that from the air to the outside surface. For this reason, the use of finned evaporators for air-cooling applications is becoming more and more prevalent. On the other hand, in liquid chilling applications where liquid is in contact with both sides of the tube and the heat transfer rate is approximately equal for both surfaces, bare-tube evaporators perform at high efficiency and finning is unnecessary. This is usually the case when ammonia is the refrigerant used. In some liquid chilling installations employing fluorocarbon refrigerants the heat transfer rate on the chilled fluid side may exceed the rate on the refrigerant side, in which case finning of the refrigerant side of the tube will improve evaporator performance. Several methods of inner finning are shown in Fig. 11-10. Where the refrigerant is on the outside of the tube, a small, low fin resembling a pipe thread has been used successfully.

Fig. 11-10 Some methods of inner finning.

11-9. Logarithmic Mean Temperature Difference (LMTD)

The temperature of air (or any fluid) decreases progressively as the air passes through a cooling coil. If it is assumed that the drop in temperature occurs at a constant rate as the air passes through the coil, the temperature reduction of the air can be represented by a straight line, as shown by the dashed line in Fig. 11-11, with the midpoint of that line representing the mean temperature of the air passing through the coil. However, in practice, the drop in air temperature is greatest across the first row of the coil and diminishes as the air passes across each succeeding row. This is accounted for by the fact that the temperature difference between the air and the refrigerant is greatest across the first row, becomes less and less as the temperature of the air is reduced in passing across each succeeding row, and is least across the last row, as shown in Fig. 11-12. Consequently, the actual drop in the temperature of the air is shown by the heavy curved line in Fig. 11-11, with the midpoint of the curved line representing the actual mean temperature of the air passing through the coil.

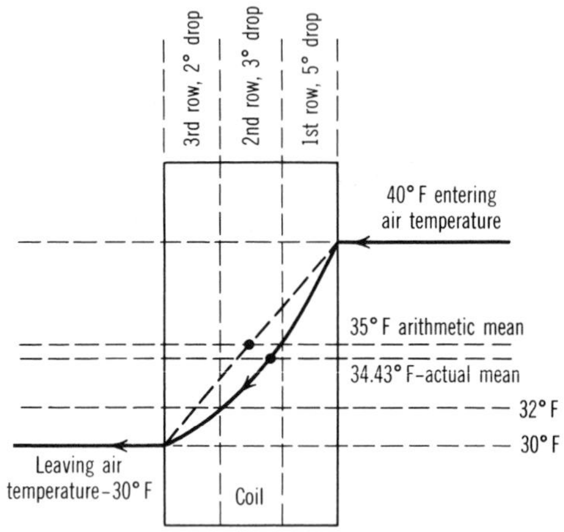

Fig. 11-11 Mean temperature of air passing through evaporator.

Although the value deviates slightly from the actual LMTD, an approximate mean temperature difference may be calculated by the following equation:

$$D = \frac{(T_e - T_r) + (T_L - T_r}{2} \tag{11-3}$$

where D = the arithmetic mean temperature
T_e = the temperature of the air entering the coil
T_L = the temperature of the air leaving the coil
T_r = the temperature of the refrigerant in the tubes

For the values given in Fig. 11-12, the arithmetic mean temperature difference is

$$D = \frac{(40 - 20) + (30 - 20)}{2} = 15°\text{F}$$

Since Equation 11-3 takes the mean temperature of the air to be the arithmetic average

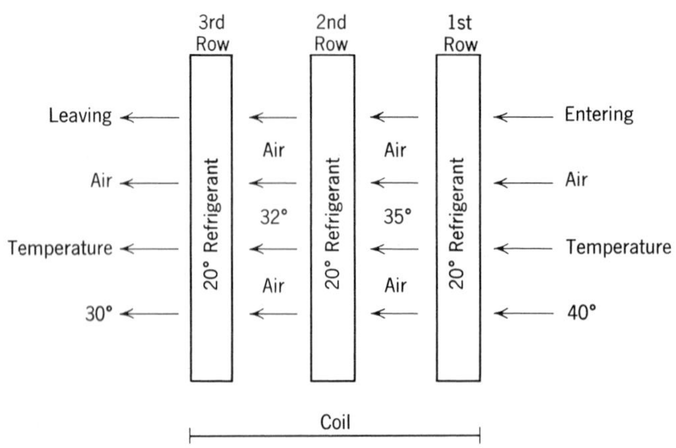

Fig. 11-12 Air temperature drop across typical three-row cooling coil.

TABLE 11-1 Mean Effective Temperature Differences

	1	2	3	4	5	6	7	8	9	10	11	12	13	14	15	16	17	18	19	20
1	1.00	1.44	1.82	2.16	2.48	2.79	3.08	3.37	3.64	3.91	4.17	4.43	4.68	4.93	5.17	5.41	5.65	5.88	6.11	6.34
2	1.44	2.00	2.47	2.89	3.28	3.64	3.99	4.33	4.65	4.97	5.28	5.58	5.88	6.17	6.45	6.73	7.01	7.28	7.55	7.82
3	1.82	2.47	3.00	3.51	3.95	4.33	4.73	5.11	5.40	5.82	6.17	6.49	6.82	7.15	7.46	7.77	8.08	8.37	8.67	8.97
4	2.16	2.89	3.51	4.00	4.48	4.93	5.36	5.77	6.17	6.55	6.92	7.28	7.64	8.00	8.32	8.66	8.98	9.31	9.63	9.94
5	2.48	3.28	3.95	4.48	5.00	5.49	5.94	6.38	6.81	7.21	7.61	8.00	8.37	8.74	9.10	9.46	9.81	10.15	10.49	10.82
6	2.79	3.64	4.33	4.93	5.49	6.00	6.37	7.01	7.40	7.85	8.27	8.70	9.08	9.47	9.98	10.22	10.61	10.96	11.30	11.67
7	3.08	3.99	4.73	5.36	5.94	6.37	7.00	7.63	7.86	8.39	8.87	9.32	9.67	10.10	10.52	10.86	11.26	11.65	12.04	12.37
8	3.37	4.33	5.10	5.77	6.38	7.01	7.63	8.00	8.49	8.96	9.42	9.86	10.30	10.72	11.13	11.54	11.94	12.33	12.72	13.10
9	3.64	4.65	5.40	6.17	6.81	7.40	7.86	8.49	9.00	9.58	10.06	10.52	10.97	11.24	11.70	12.14	12.57	12.99	13.39	13.92
10	3.91	4.97	5.82	6.55	7.21	7.85	8.39	8.96	9.58	10.00	10.49	10.97	11.43	11.89	12.33	12.77	13.19	13.61	14.02	14.43
11	4.17	5.28	6.17	6.92	7.61	8.27	8.87	9.42	10.06	10.49	11.00	11.49	11.96	12.42	12.94	13.33	13.79	14.22	14.65	15.06
12	4.43	5.58	6.49	7.28	8.00	8.70	9.32	9.86	10.52	10.97	11.49	12.00	12.50	12.99	13.45	13.90	14.45	14.80	15.23	15.66
13	4.68	5.88	6.82	7.64	8.37	9.08	9.67	10.30	10.97	11.43	11.96	12.50	13.00	13.48	13.91	14.44	14.90	15.35	15.80	16.26
14	4.93	6.17	7.15	8.00	8.74	9.47	10.10	10.72	11.24	11.89	12.42	12.99	13.48	14.00	14.58	14.93	15.46	15.90	16.38	16.81
15	5.17	6.45	7.46	8.32	9.10	9.98	10.52	11.13	11.70	12.33	12.94	13.45	13.91	14.58	15.00	15.87	16.00	16.46	16.90	17.39
16	5.41	6.73	7.77	8.66	9.46	10.22	10.86	11.54	12.14	12.77	13.33	13.90	14.44	14.93	15.87	16.00	16.29	16.98	17.31	17.93
17	5.65	7.01	8.08	8.98	9.81	10.61	11.26	11.94	12.57	13.19	13.79	14.45	14.90	15.46	16.00	16.29	17.00	17.51	18.07	18.51
18	5.88	7.28	8.37	9.31	10.15	10.96	11.65	12.33	12.99	13.61	14.22	14.80	15.35	15.90	16.46	16.98	17.51	18.00	18.35	18.99
19	6.11	7.55	8.67	9.63	10.49	11.30	12.04	12.72	13.39	14.02	14.65	15.23	15.80	16.38	16.90	17.31	18.07	18.35	19.00	19.23
20	6.34	7.82	8.95	9.94	10.82	11.67	12.37	13.10	13.92	14.43	15.06	15.66	16.26	16.81	17.39	17.93	18.51	18.99	19.23	20.00
21	6.57	8.08	9.25	10.25	11.15	12.00	12.74	13.47	14.19	14.83	15.47	16.08	16.69	17.26	17.83	18.35	18.96	19.43	20.24	20.49
22	6.79	8.34	9.54	10.56	11.47	12.35	13.11	13.84	14.57	15.22	15.87	16.50	17.11	17.71	18.28	18.84	19.40	19.96	20.45	20.99
23	7.02	8.60	9.82	10.86	11.79	12.68	13.44	14.20	14.89	15.61	16.27	16.92	17.53	18.12	18.72	19.27	19.90	20.38	20.90	21.46
24	7.24	8.85	10.01	11.16	12.11	13.02	13.79	14.56	15.27	15.99	16.64	17.31	17.95	18.55	19.15	19.73	20.33	20.86	21.48	21.94
25	7.46	9.11	10.38	11.46	12.43	13.34	14.14	14.92	15.65	16.37	17.05	17.74	18.35	18.95	19.58	20.14	20.76	21.30	21.86	22.41
26	7.67	9.36	10.65	11.75	12.74	13.67	14.46	15.26	16.02	16.75	17.43	18.11	18.76	19.38	20.01	20.60	21.20	21.77	22.34	22.87
27	7.89	9.61	10.92	12.05	13.05	13.99	14.81	15.62	16.38	17.11	17.82	18.50	19.20	19.79	20.42	21.01	21.63	22.19	22.76	23.33
28	8.10	9.85	11.19	12.33	13.35	14.31	15.15	15.96	16.75	17.48	18.20	18.89	19.55	20.20	20.83	21.44	22.04	22.62	23.20	23.77
29	8.32	10.01	11.46	12.62	13.65	14.63	15.49	16.31	17.10	17.85	18.57	19.27	19.94	20.60	21.24	21.85	22.49	23.07	23.66	24.22
30	8.53	10.34	11.73	12.90	13.95	14.94	15.79	16.64	17.46	18.20	18.94	19.64	20.33	20.99	21.64	22.27	22.90	23.48	24.08	24.66
31	8.74	10.58	11.98	13.19	14.25	15.25	16.12	16.98	17.81	18.56	19.31	20.02	20.71	21.27	22.09	22.67	23.31	23.92	24.50	25.10
32	8.94	10.82	12.26	13.47	14.55	15.57	16.45	17.31	18.11	18.91	19.66	20.39	21.09	21.77	22.45	23.08	23.72	24.33	24.94	25.53
33	9.15	11.06	12.51	13.74	14.84	15.87	16.75	17.64	18.46	19.26	20.03	20.76	21.47	22.18	22.83	23.47	24.13	24.75	25.35	25.96
34	9.36	11.29	12.76	14.02	15.13	16.17	17.08	17.97	18.80	19.61	20.37	21.12	21.85	22.53	23.22	23.88	24.53	25.15	25.79	26.30
35	9.56	11.53	13.03	14.29	15.47	16.48	17.40	18.29	19.14	19.96	20.72	21.48	22.22	22.92	23.60	24.27	24.94	25.58	26.19	26.95
36	9.77	11.76	13.28	14.56	15.70	16.77	17.71	18.62	19.48	20.30	21.08	21.85	22.58	23.30	23.99	24.66	25.33	25.97	26.62	27.28
37	9.97	12.00	13.53	14.83	15.99	17.07	18.01	18.94	19.81	20.64	21.43	22.20	22.95	23.66	24.37	25.04	25.72	26.36	27.01	27.64
38	10.17	12.23	13.78	15.10	16.27	17.36	18.32	19.25	20.14	20.97	21.78	22.55	23.30	24.05	24.73	25.43	26.11	26.77	27.41	28.06
39	10.37	12.45	14.04	15.37	16.55	17.67	18.63	19.57	20.47	21.31	22.13	22.91	23.67	24.41	25.12	25.81	26.50	27.16	27.80	28.48
40	10.57	12.68	14.29	15.63	16.83	17.95	18.92	19.88	20.80	21.64	22.46	23.26	24.02	24.77	25.49	26.19	26.89	27.56	28.21	28.23

of the entering and leaving air temperatures (midpoint of dashed line in Fig. 11-11), the resulting mean temperature difference will differ somewhat from the actual LMTD as determined by the following equation, which takes into account the actual mean temperature of the air (midpoint of curved line in Fig. 11-11):

$$D = \frac{(T_e - T_r) - (T_L - T_r)}{\ln \dfrac{(T_e - T_r)}{(T_L - T_r)}} \qquad (11\text{-}4)$$

Applying the values shown in Fig. 11-11 in Equation 11-4, the logarithmic mean temperature difference,

$$D = \frac{(40 - 20) - (30 - 20)}{\ln \dfrac{40 - 20}{30 - 20}} = \frac{10}{\ln 2} = 14.43°\text{F}$$

The preceding calculations were made on the assumption that the refrigerant temperature remains constant. When this is not the case, T_r will have two values (see Section 14-10). This condition is discussed in another chapter.

The log mean temperature difference, hereafter called mean effective temperature difference (METD), may also be determined from Table 11-1.

11-10. The Effect of Air Quantity on Evaporator Capacity

Although not a part of the basic heat transfer equation, there are other factors external to the coil itself which greatly affect coil performance. Principal among these are the circulation, velocity, and distribution of air in the refrigerated space and over the coil. These factors are closely related and in many cases are dependent one on the other.

Except in liquid cooling and in applications where the product is in direct contact with the evaporator, most of the heat from the product is carried to the evaporator by air circulation. If air circulation is inadequate, heat is not carried from the product to the evaporator at a rate sufficient to allow the evaporator to perform at peak efficiency. It is important also that the circulation of air is evenly distributed in all parts of the refrigerated space and over the coil. Poor distribution of the circulating air results in uneven temperatures and "dead spots" in the refrigerated space, whereas the uneven distribution of air over the coil surface causes some parts of the surface to function less efficiently than others and lowers evaporator capacity.

The velocity of the air passing over the coil has a considerable influence on both the value of U and the METD and is important in determining evaporator capacity. When air velocity is low, the air passing over the coil stays in contact with the coil surface longer and is cooled through a greater range. Thus, the METD and the rate of heat transfer is low. As air velocity increases, a greater quantity of air is brought in contact with the coil per unit of time, the METD increases, and the rate of heat transfer improves. In addition, high air velocities tend to break up the thin film of stagnant air which is adjacent to all surfaces. Since this film of air acts as a heat barrier and insulates the surface, its disturbance increases the conductance of the outside surface film and the over-all value of U improves.

11-11. Surface Area

Equation 11-1 indicates that the capacity of an evaporator varies directly with the outside surface area. This is true only if the U factor of the evaporator and the METD remain the same. In many cases, the value of U and the METD are affected when the surface area of the evaporator is changed. In such cases, the capacity of the evaporator does not increase or decrease in direct proportion to the change in surface area. To illustrate, in Fig. 11-13, coils B and C each have twice the surface area of coil A, yet the increase in capacity over the capacity of coil A will be greater for coil C than for coil B. Provided the air velocity is the same (the total quantity of air circulated over coil C must be twice that circulated over coil A), the METD across C will be exactly the same as that across A and the capacity of C will therefore be twice the capacity of coil A.

Figure 11-12 illustrates how the METD is affected when the surface area of the coil is increased by increasing the number of rows (depth). Notice again that the drop in air

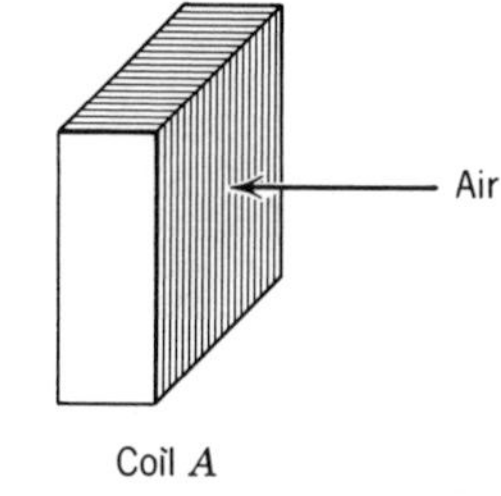

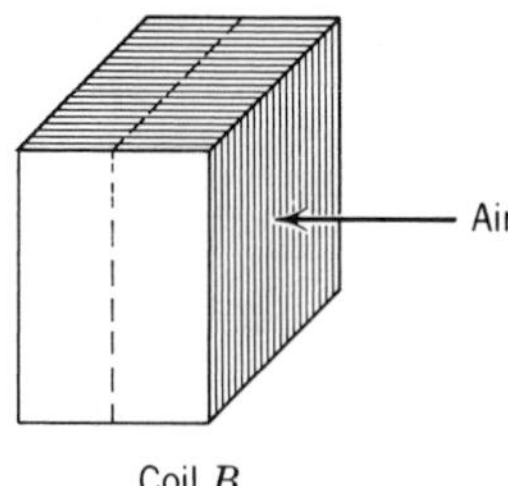

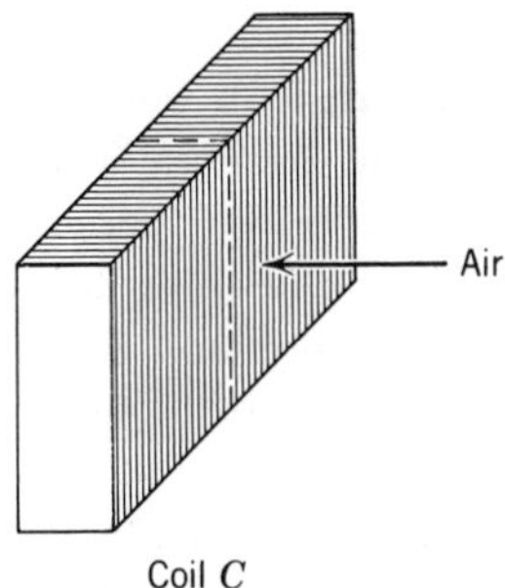

Fig. 11-13 Coils *B* and *C* both have twice the surface area of coil *A*. Coil *C* has twice the face area of coil *A* or coil *B*.

temperature is greatest across the first row and diminishes as the air passes across each succeeding row, with the result that each succeeding row has a lower heat transfer rate and performs less efficiently than the preceding row. For this reason, if the surface area of coil *A* in Fig. 11-13 is doubled by increasing the number of rows as in coil *B*, the METD will be decreased and the increase in capacity will not be nearly as great as when the surface area is increased as in coil *C*. For the same total surface area, a long, wide, flat coil will, in general, perform more efficiently than a short, narrow coil having more rows depth. However, in many instances, the physical space available is limited and compact coils arrangements must be used. In applications where it is permissible, the loss of capacity resulting from increasing the number of rows can be compensated for to some extent by increasing the air velocity over the coil. Too, in some applications, the use of deep coils is desirable for the purpose of dehumidification. As a general rule, the greater the coil depth, the longer the air stays in contact with the coil surface and the more closely the leaving temperature of the air will approach the surface temperature of the coil. Since the temperature of some part of the air passing through a cooling coil is usually reduced below the entering dew point temperature, dehumidification is accomplished. Obviously, the lower the leaving air temperature, the greater is the amount of dehumidification.

11-12. Evaporator Circuiting

It was demonstrated in Chapter 8 that excessive pressure drop in the evaporator results in the suction vapor arriving at the suction inlet of the compressor at a lower pressure than is actually necessary, thereby causing a loss of compressor capacity and efficiency.

To avoid unnecessary losses in compressor capacity and efficiency, it is desirable to design the evaporator so that the refrigerant experiences a minimum drop in pressure. On the other hand, a certain amount of pressure drop is required to flow the refrigerant through the evaporator, and since velocity is a function of pressure drop, the drop in pressure must be sufficient to assure refrigerant velocities high enough to sweep the tube surfaces free of vapor bubbles and oil and to carry the oil back to the compressor. Hence, good design requires that the method of evaporator circuiting be such that the drop in pressure through the evaporator is the minimum necessary to produce refrigerant velocities sufficient to provide a high rate of heat transfer and good oil return.

In general, the drop in pressure through any one evaporator circuit will depend upon the size of the tube, the length of the circuit, and the circuit load. By circuit load is meant the time rate of heat flow through the tube walls of

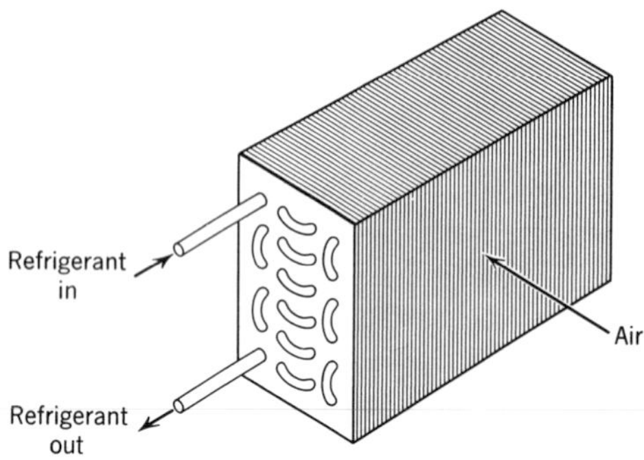

Fig. 11-14 Evaporator with one series refrigerant circuit.

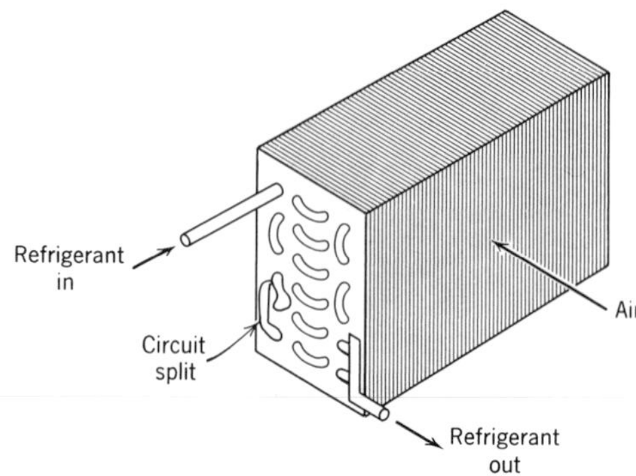

Fig. 11-15 Evaporator with split refrigerant circuit.

the circuit. The circuit load determines the quantity of refrigerant which must pass through the circuit per unit of time. The greater the circuit load, the greater must be the quantity of refrigerant flowing through the circuit and the greater will be the drop in pressure. Hence, for any given tube size, the greater the load on the circuit, the shorter the circuit must be in order to avoid excessive pressure drop.

Evaporators having only a single series refrigerant circuit, such as the one illustrated in Fig. 11-14, will perform satisfactorily within certain load limits. When the load limit is exceeded, the refrigerant velocity will be increased beyond the desired range and the pressure drop will be excessive.

Notice that the refrigerant enters at the top of the evaporator as a liquid and leaves at the bottom as a vapor. Since the volume of the refrigerant increases as the refrigerant vaporizes, the refrigerant velocity and the pressure drop per foot increase progressively as the refrigerant travels through the circuit, and are greatest at the end of the coil where the refrigerant is 100% vapor.

The excessive pressure drop occurring in the latter part of a single series circuit evaporator can be eliminated to a great extent by splitting the single circuit into two circuits in the lower portion of the evaporator (Fig. 11-15). When this is done, the refrigerant travels a single series path until the refrigerant velocity builds to the allowable maximum, at which time the circuit is split into two parallel paths for the balance of the travel through the evaporator. This has the effect of reducing the refrigerant velocity in the two paths to one-half the value it would have without the split, and the pressure drop per foot is reduced to one-eighth of the value it would have in the lower part of the evaporator with a single series circuit.* This, of course, will permit greater loading of the coil without exceeding the allowable pressure drop. At the same time, the velocity in all parts of the coil is maintained within the desirable limits so that the rate of heat transfer is not unduly affected.

Another method of reducing the pressure drop through the evaporator is to install refrigerant headers at the top and bottom of the evaporator so that the refrigerant is fed simultaneously through a multiple of parallel circuits (Fig. 11-16). The primary disadvantage of this type of circuiting is that the loading of the circuits is uneven, a circumstance that makes it difficult to distribute the refrigerant to the several circuits. Since the temperature difference between the air passing over the coil and the refrigerant in the tubes is much greater across the first circuit (first row) than across the last circuit (last row), the first circuit is more heavily loaded than the last

* Pressure drop increases as the square of the velocity. Reducing the velocity to one-half reduces the pressure drop to one-quarter of its original value. Then, since the length of each parallel branch is only one-half the length of a single circuit, the drop in pressure in the lower portion of the split coil is only one-eighth of the single circuit value.

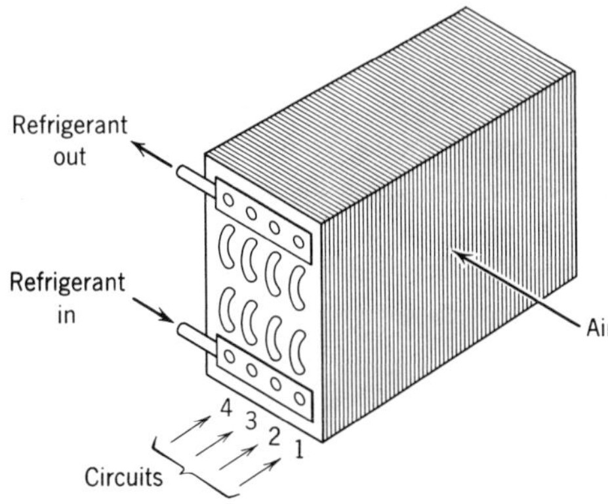

Fig. 11-16 Four-circuit evaporator with refrigerant headers on both inlet and outlet. Crossflow of air and refrigerant results in uneven circuit loading.

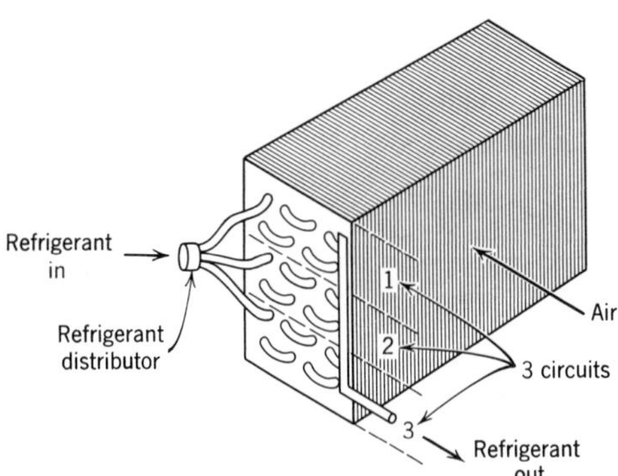

Fig. 11-17 Evaporator with refrigerant distributor and suction header. Notice counterflow arrangement for refrigerant and air.

circuit and consequently requires more refrigerant. Since these uneven circuit requirements make it impossible to adjust a single refrigerant feed to satisfy any one of the circuits without starving and/or overfeeding the other circuits, this type of coil circuiting is best used with "overfeed" or "flooded" systems.

The circuit arrangement shown in Fig. 11-17 is very effective and is widely used, particularly when circuit loading is heavy, as in the case of an air conditioning coil where the temperature differential between the refrigerant and the air is large and where external finning is heavy. Notice that the air passes in counterflow to the refrigerant so that the warmest air is in contact with the warmest part of the coil surface. This provides the greatest mean temperature differential and the highest rate of heat transfer. Notice also that loading of the circuits is even. The number and length of the circuits that such a coil should have are determined by the size of the tube and the load on the circuits.

For the multipass, headered evaporator, the arrangement shown in Fig. 11-18 is much more desirable than that shown in Fig. 11-16. Counterflowing of the air and the refrigerant increases the METD and permits more even loading of the circuits.

11-13. Methods of Refrigerant Feed

Evaporators may be classified according to the method of liquid feed as *dry-expansion*,

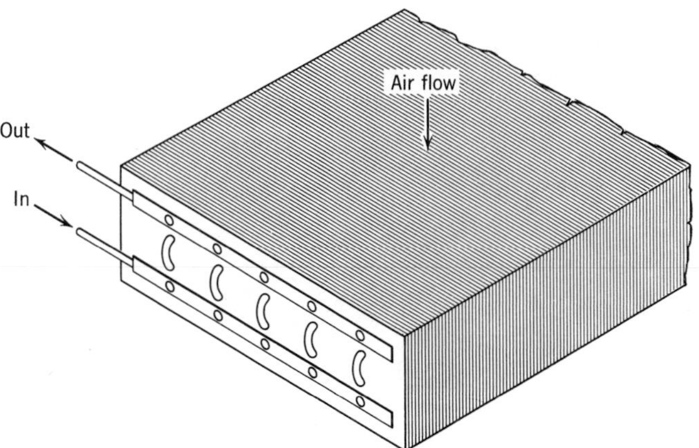

Fig. 11-18 Counterflow of refrigerant and air results in more even circuit loading and a higher mean temperature differential. Compare this arrangement with the crossflow arrangement in Fig. 11-16.

flooded, or *liquid overfeed*. With the dry-expansion method, the amount of liquid refrigerant fed into the evaporator is limited to that which can be completely vaporized by the time it reaches the end of the evaporator, so that refrigerant vapor only enters the suction line (Fig. 11-19*a*). The refrigerant flow control employed with this method of evaporator feed is usually either a thermostatic expansion valve or a capillary tube. To ensure the complete vaporization of the refrigerant in the evaporator and thereby prevent the carryover of liquid into the suction line and to the compressor, the refrigerant is permitted to become superheated approximately 10°F in the end of the evaporator, a practice that requires approximately 10% to 20% of the total evaporator surface.

In Chapter 6, it was shown that a portion of each unit mass of refrigerant circulated vaporizes in the refrigerant control as the pressure is reduced from the condensing pressure to the evaporator pressure. With dry-expansion feed, the resulting flash gas enters the evaporator along with the remaining liquid which vaporizes progressively as the refrigerant passes through the evaporator. It is evident from the foregoing that the refrigerant in the latter portion of a dry-expansion evaporator is nearly all in the vapor state and that this portion of the evaporator will not perform as effectively as the inlet portion of the evaporator where the refrigerant is largely in the liquid phase.

It is for this reason also that the surface temperature of a dry-expansion evaporator is always lowest near the refrigerant inlet and highest near the outlet, in spite of the fact that the saturation temperature of the refrigerant is lowest at the outlet because of the drop in pressure suffered by the refrigerant in flowing through the evaporator.

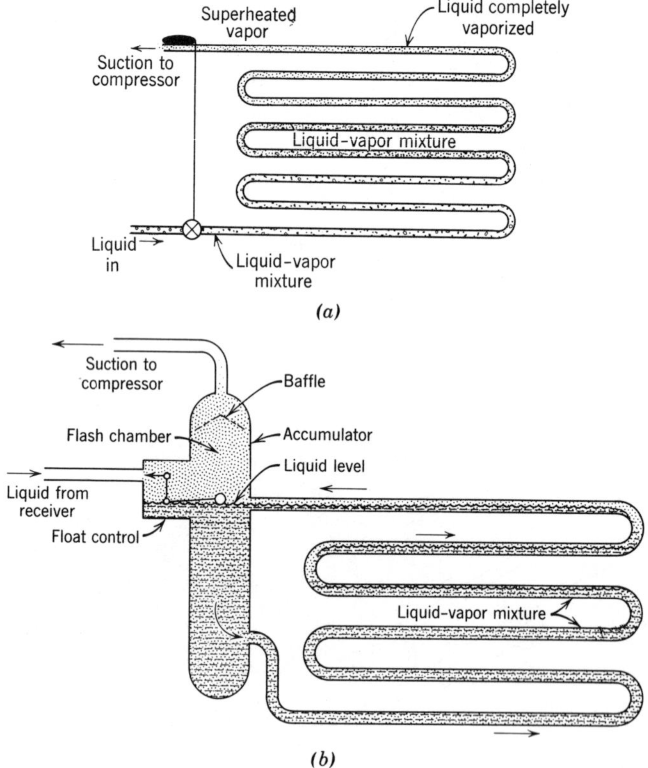

Fig. 11-19 (*a*) Dry-expansion evaporator. (*b*) Flooded evaporator. Notice accumulator and float control. Circulation of the refrigerant through the coil is by gravity. The vapor accumulated from the boiling action in the coil escapes to the top of the accumulator and is drawn off by the suction of the compressor.

While dry-expansion evaporators are somewhat less efficient than the flooded or liquid overfeed types, they are usually much simpler in design, are lower in initial cost, are much more compact, require a much smaller refrigerant charge, and have fewer oil return problems than the other types. For these reasons, the dry-expansion evaporator is the most popular type. This is particularly true for systems employing halocarbon refrigerants, since oil return from flooded halocarbon evaporators is sometimes difficult to accomplish.

The full-flooded evaporator is operated completely filled with liquid refrigerant, a condition that provides the greatest amount of interior wetted tube surface and consequently the highest possible heat transfer rate. As shown in Fig. 11-19*b*, the flooded evaporator is equipped with an accumulator or surge drum that serves as a liquid reservoir from which the liquid refrigerant is circulated by gravity through the evaporator circuits. The liquid level in the accumulator is maintained by a low-side or high-side float control, and the vapor generated by the boiling action of the refrigerant in the tubes is separated from the liquid in the upper part of the accumulator and is drawn directly into the suction line along with the flash gas that results when the pressure of the refrigerant is reduced from the condensing pressure to the evaporator pressure. Notice that the flash gas never enters the heat transfer portion of the evaporator.

A liquid overfeed evaporator is one wherein the amount of liquid refrigerant circulated through the evaporator is considerably in excess of that which can be vaporized. As shown in Fig. 17-29, the excess liquid is separated from the vapor in a low-pressure receiver or accumulator and recirculated to the evaporator, while the vapor is drawn off to the compressor suction. Recirculation rates range from a low of 2 to 1 up to a high of 6 or 7 to 1, with the higher rates being used with ammonia and the lower rates with Refrigerants 12, 22, and 502. A recirculation rate of 3 to 1 means that three times as much liquid is being circulated as can be vaporized, in which case the composition of the refrigerant in the return line to accumulator will be composed of two parts liquid and one part vapor by weight. With adequate liquid recirculation, the wetting of the internal tube surfaces and the performance of overfeed evaporators are similar to those obtained with full-flooded operation. The optimum recirculation rate for best evaporator performance varies with a number of factors and is often difficult to predict. In order to realize rated performance, it is important that the evaporator manufacturer's recommendations be followed as closely as possible. As in the case of dry-expansion evaporators, liquid flow into overfeed evaporators is controlled through some type of metering device, usually a hand expansion valve or an orifice that is designed or adjusted for the maximum flow rate required at peak loading.

Overfeed (liquid recirculation) evaporators are most commonly and most economically employed in multiple evaporator systems, as shown in Fig. 11-20. Whereas little difficulty is experienced in controlling the recirculation rate for a single evaporator, the balancing of a

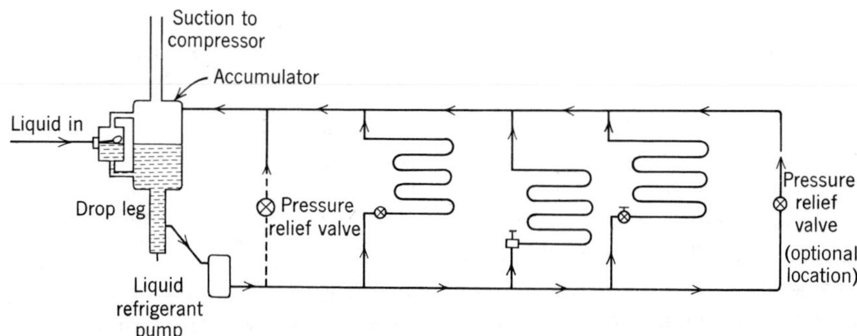

Fig. 11-20

multiple-evaporator system is somewhat more tedious but becomes easier as the recirculation rate increases. For this reason, recirculation rates are usually higher for a multiple-evaporator system than for a single evaporator. Also, in order to prevent excessive overfeeding of the remaining active evaporators, a bypass relief valve is installed on the discharge side of the pump to relieve the excess liquid back to the low-pressure receiver when one or more of the system evaporators are shut down. An alternate location for the bypass valve is at the end of the refrigerant circuit farthest from the pump (shown dotted in Fig. 11-20).

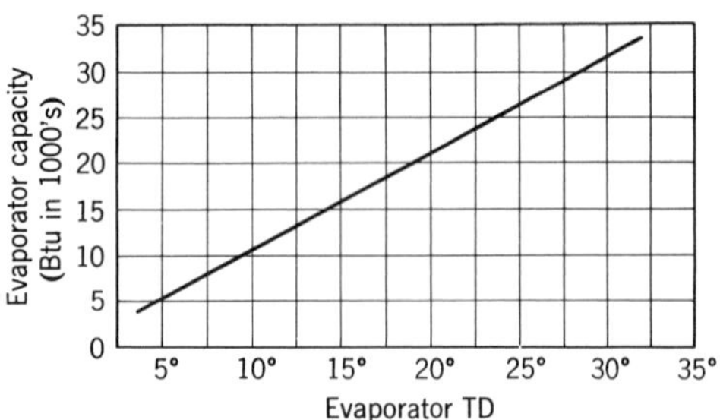

Fig. 11-21 Variation in evaporator capacity with evaporator TD.

11-14. Use of Manufacturers' Rating Data

Since the mathematical evaluation of all the variables that influence evaporator performance is not practical, evaporator performance is usually determined by actual testing of the evaporator. Because of the many variables encountered in refrigeration applications, there is no industrywide standard for rating low-temperature evaporators. Consequently, the method of rating varies somewhat with the type of evaporator and with the particular manufacturer involved. However, the rating methods do not differ greatly, and most manufacturers include, along with the evaporator rating data, instructions as to how to use the data.

The selection of evaporators from manufacturers' performance data is relatively simple once the conditions under which the evaporator is to operate have been determined. Typical evaporator rating tables, along with methods of evaporator selection, are discussed later in this chapter.

11-15. Evaporator TD

One of the most important factors to consider in selecting the proper evaporator for any given application is the evaporator TD. Evaporator TD is defined as the difference in temperature between the temperature of the air entering the evaporator, usually taken as the space design temperature, and the saturation temperature of the refrigerant corresponding to the pressure at the evaporator outlet.

Although more exact methods of rating evaporators are necessary in order to select evaporators for air conditioning applications and some product storage applications where space temperature and humidity are especially critical, ratings for most evaporators designed for product cooling applications are based on evaporator TD.

The relationship between evaporator capacity and evaporator TD is illustrated in Fig. 11-21. Notice that the capacity of the evaporator (Btu/hr) varies directly with the evaporator TD. That is, if an evaporator has a certain capacity at a 1°FTD, it will have exactly ten times that capacity if the TD is increased to 10°F, provided that all other conditions are the same.*

It is evident that an evaporator with a relatively small surface area operating at a relatively large TD can have the same capacity as another evaporator having a larger surface area but operating at a smaller TD. Thus, insofar as Btu per hour capacity alone is concerned, a small evaporator will have the

* Care should be taken not to confuse METD with evaporator TD. According to Equation 11-1, the Btu per hour capacity of any given evaporator (whose *U* factor and surface area are fixed at the time of manufacture) varies directly with the METD. However, assuming that the refrigerant temperature and all else remains constant, the METD between the air passing over the evaporator and the refrigerant in the evaporator will vary directly with the temperature of the air entering the evaporator. That is, if the temperature of the air entering the evaporator increases, the METD increases. Hence, the METD varies in proportion to the evaporator TD and, therefore, the capacity of the evaporator also varies in proportion to the evaporator TD.

same refrigerating effect as a larger one, provided that the TD at which the small evaporator operates is greater in proportion. However, it will be shown in the following sections that the temperature difference between the evaporator and the refrigerated space has considerable influence on the condition of the stored product and upon the operating efficiency of the entire system, and is usually the determining factor in coil selection. Before an evaporator can be selected, it is necessary to first determine the TD at which it is expected to function. Once the desired temperature difference is known, an evaporator having sufficient surface area to provide the required cooling capacity at the design TD can be selected.

11-16. The Effect of Coil TD on Space Humidity

The preservation of food and other products in optimum condition by refrigeration depends not only upon the temperature of the refrigerated space but also upon space humidity. When the humidity in the space is too low, excessive dehydration occurs in such products as cut meats, vegetables, dairy products, flowers, fruits, etc. On the other hand, when the humidity in the refrigerated space is too high, the growth of mold, fungus, and bacteria is encouraged and bad sliming conditions occur, particularly on meats and especially in the wintertime. Space humidity is of little importance, of course, when the refrigerated product is in bottles, cans, or other vapor-proof containers.

The most important factor governing the humidity in the refrigerated space is the evaporator TD. The smaller the difference in temperature between the evaporator and the space, the higher is the relative humidity in the space. Likewise, the greater the evaporator TD, the lower is the relative humidity in the space. Some of the other factors that influence the space humidity are air motion, system running time, type of system control, amount of exposed product surface, rate of infiltration, and outside air conditions.

When the product to be refrigerated is one that will be affected by the space humidity, an evaporator TD that will provide the optimum humidity conditions for the product should be selected. In such cases, the evaporator TD is the most important factor determining the evaporator selection. The design evaporator TD required for various space humidities is given in Table 11-2 for both natural-convection and forced-convection evaporators.

In applications where the space humidity is of no importance, the factors governing evaporator selection are: (1) system efficiency and economy of operation, (2) the physical space available for evaporator installation, and (3) initial cost.

TABLE 11-2 Evaporator Design TD

Relative Humidity, %	Design TD, °F	
	Natural Convection	Forced Convection
95–91	12–14	8–10
90–86	14–16	10–12
85–81	16–18	12–14
80–76	18–20	14–16
75–70	20–22	16–18

For temperatures 10° F and below, an evaporator TD of 10° F is generally used for forced convection evaporators.

11-17. The Effect of Air Circulation on Product Condition

As stated previously, circulation of air in the refrigerated space is essential to carry the heat from the product to the evaporator. When air circulation is inadequate, the capacity of the evaporator is decreased, the product is not cooled at a sufficient rate, the growth of mold and bacteria is encouraged, and sliming occurs on some products. On the other hand, too much air circulation can be as detrimental as too little. When the circulation of air is too great, the rate of moisture evaporation from the product surface increases and excessive dehydration of the product results. Excessive dehydration can be very costly in that it causes deterioration in product appearance and quality and shortens the life of the product. Furthermore, the loss of weight resulting from shrinkage and trimming is a considerable factor in dealer profits and in the price of perishable foods.

The desired rate of air circulation varies with the different applications and depends upon the space humidity, the type of product, and the length of the storage period.

With respect to product condition, air circulation and space humidity are closely associated. Poor air circulation has the same effect on the product as high humidity, whereas too much air circulation produces the same effect as low humidity. In many instances, it is difficult to determine whether product deterioration in a particular application is caused by faulty air circulation or poor humidity conditions. For the most part, product condition depends upon the combined effects of humidity and air circulation, rather than upon the effect of either one alone, and either of these two factors can be varied somewhat, provided that the other is varied in an off-setting direction. For example, higher than normal air velocities can be used without damage to the product when the space humidity is also maintained at a higher level.

The type of product and the amount of exposed surface should be given consideration when determining the desired rate of air circulation. Some products, such as flowers and vegetables, are more easily damaged by excessive air circulation than others and must be given special consideration. Cut meats, since they have more exposed surface, are more susceptible to loss of weight and deterioration than are beef quarters or sides, and air velocities should be lower. On the other hand, where the product is in vapor-proof containers, it will not be affected by high velocities and the rate of air circulation should be maintained at a high level to obtain the maximum cooling effect.

Recommended air velocities for product storage are given in Tables 10-9 through 10-12.

11-18. Natural Convection Evaporators

Natural convection evaporators are frequently used in applications where low air velocities and minimum dehydration of the product are desired. Typical installations are household refrigerators, display cases, walk-in coolers, reach-in refrigerators, and large storage rooms.

The circulation of air over the cooling coil by natural convection is a function of the temperature differential between the evaporator and the space. The greater the difference in temperature, the higher the rate of air circulation.

The circulation of air by natural convection is greatly influenced by the shape, size, and location of the evaporator, the use of baffles, and the placement of the stored product in the refrigerated space. Generally, shallow coils (one or two rows deep) extending the entire length of the cooler and covering the greater portion of the ceiling area are best. As the depth of the coil is increased, the coil offers greater resistance to the free circulation of air and the METD is thereby decreased with a resulting decrease in the coil capacity. Since cold air is more dense than warm air and tends to fall to the floor, evaporators should be located as high above the floor as possible, but care should be taken to leave sufficient room between the evaporator and the ceiling to permit the free circulation of air over the top of the coil.

For coolers less than 8 ft wide, single, ceiling-mounted evaporators are frequently used. When the width of the cooler exceeds 8 ft, two or more evaporators should be used. In coolers where there is not sufficient head room to permit the use of overhead coils, side-wall evaporators may be used. If properly installed, these will function with approximately the same efficiency as overhead coils. Typical overhead and side-wall installations for large storage rooms are shown in Figs. 11-22 and 11-23, respectively.

In small coolers, baffles are used with natural convection coils to assure good air circulation. The baffles are installed in such a manner that they aid and direct the free flow of air over the coil and throughout the refrigerated space. The cold and warm air flues should each have an area approximately equal to one-sixth or one-seventh of the floor area of the cooler. Assuming that the flues extend the full length of the fixture, the width of the flues will then be proportional to the width of the cooler.

Fig. 11-22 Overhead installation of natural convection evaporator. Evaporator has cast aluminum fins. (Courtesy Detroit Ice Machine Company.)

Fig. 11-23 Sidewall installation of natural convection evaporator. (Courtesy Detroit Ice Machine Company.)

Since warm air has a greater specific volume than cold air, some manufacturers recommend that the warm air flue be a little larger than the cold air flue. In Fig. 11-24, the width of the cold air flue is equal to $W/7$, whereas the width of the warm air flue is equal to $W/6$. The distance (A) from the coil to the ceiling should be approximately equal to the width of the warm air flue and never less than 3 in. Vertical side baffles should extend approximately 1 in. above and 3 to 4 in. below the coil. The horizontal baffles or coil decks should slope 1 to 2 in. per foot to give direction to the cold air flow and to drain off the condensate. Also, the coil decks must be insulated so that moisture does not condense on the undersurface of the deck and drip off on the product. The dimension (B) is 4 to 7 in., whereas (C) is usually 2 to 4 in.

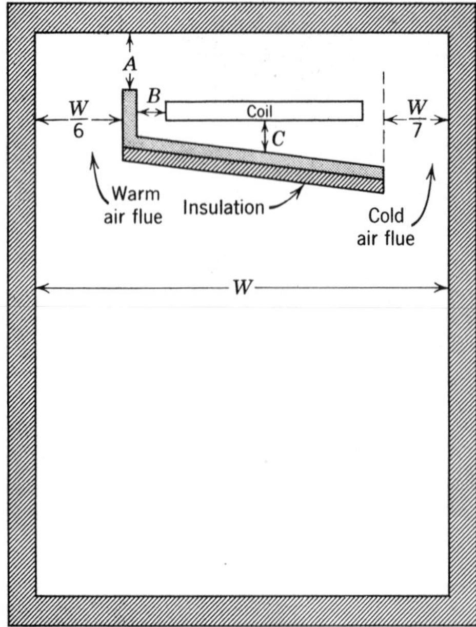

Fig. 11-24 Typical baffle arrangement for natural convection coil.

11-19. Coil-and-Baffle Assemblies

The availability of factory-built coil-and-baffle assemblies has practically eliminated the need for the custom building of baffles on the job. A typical ready built coil-and-baffle assembly is shown in Fig. 11-25. Since these assemblies are available in a wide variety of sizes and combinations (see Table R-1), they can be readily applied to almost any natural convection application.

11-20. Rating and Selection of Natural Convection Evaporators

Basic capacity ratings for natural convection evaporators, both prime surface and finned, are normally given in Btu per hour per degrees Fahrenheit TD. However, in some instances, as a matter of convenience, capacity ratings are given for TDs other than 1°F.

For the coil-and-baffle assemblies mentioned in the preceding section, the ratings given are per inch of finned length. For bare pipe evaporators, the ratings given are per square foot of external pipe surface, although in some instances bare pipe evaporators are rated per lineal foot of pipe. This is true also for the large-finned pipe shown in Fig. 11-22.

Ratings for plate evaporators are given per square foot of plate surface. Both sides of the plate are considered when computing the area of the plate. Frequently, ratings for plate evaporators apply to an entire plate or to a specific group or combination of plates.

Typical rating data for various types of natural convection evaporators are given in

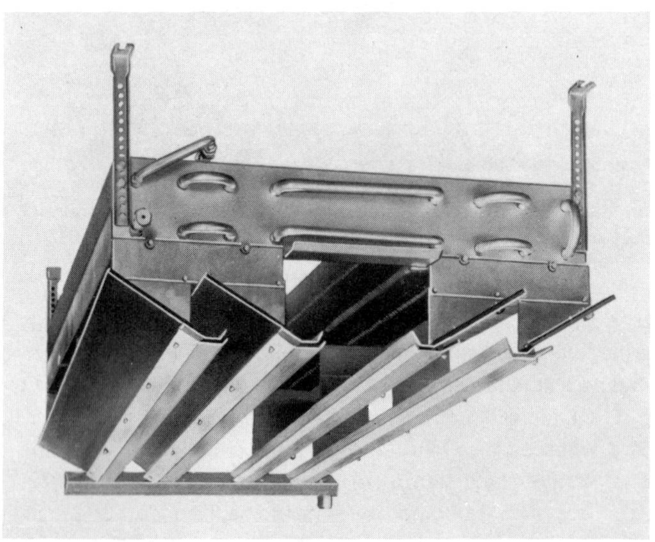

Fig. 11-25 Natural convection coil-and-baffle assembly. (Courtesy Dunham-Bush, Inc.)

Tables R-1 through R-7. The use of these rating data in the selection of the various types of evaporators is best illustrated through a series of examples.

Example 11-1 A vegetable storage cooler with inside dimensions of 18 ft by 9 ft by 9 ft has a calculated cooling load of 10,200 Btu/hr. Select a natural convection coil-and-baffle assembly (Fig. 11-25 and Table R-1) to meet these requirements.

Solution Since the capacity ratings for this type of evaporator are given in Btu/(hour)(°F TD)(inch of finned length), the required evaporator capacity must be reduced to this value before a selection can be made from the rating table. Also, recall that a natural convection evaporator should extend almost the full length of the cooler in order to assure adequate air circulation around the product.

From Table 10-10, the recommended design space humidity for mixed vegetable storage is approximately 85%, and from Table 11-2, the design evaporator TD required to provide an 85% RH with a natural convection evaporator is 16°F. The length of the cooler inside is 17 ft. Allowing 1 ft on each end of the evaporator for working space, the approximate overall length of the evaporator can be (17 ft − 2 ft) 15 ft. According to the manufacturer's specifications (Table R-1) the overall length of the evaporator is 7 in. longer than the finned length, so that the approximate finned length is (15 ft − 7 in.) 14 ft and 5 in. or 173 in.

The required capacity per degree Fahrenheit TD

$$= \frac{\text{Total evaporator capacity}}{\text{Design evaporator TD}}$$

$$= \frac{10{,}200 \text{ Btu/hr}}{16°\text{F TD}} = 637.5 \text{ Btu/(hr)(°F TD)}$$

Required capacity per inch of finned length

$$= \frac{\text{Required capacity per °F TD}}{\text{Approximate finned length}}$$

$$= \frac{637.5 \text{ Btu/(hr)(°F TD)}}{173 \text{ in.}}$$

$$= 3.68 \text{ Btu/(hr)(°F)(in.)}$$

Because of the width of the cooler, a two-section evaporator will give the best results. Reference to Table R-1 indicates that Model #PK-16 with two fins per inch ($\frac{1}{2}$-in. fin spacing) has a capacity of 3.65 Btu/(hr)(°F)(in.)

Using this model evaporator, the required finned length is

$$\frac{637.5 \text{ Btu/(hr)(°F TD)}}{3.65 \text{ Btu/(hr)(°F)(in.)}} = 175 \text{ in.}$$

The overall length of the evaporator is 182 in. (175 in. + 7 in.) and, since the overall length of the cooler inside is 204 in., the clearance between the evaporator and the cooler wall at each end is 11 in.

The width of the evaporator should always be checked against the width of the cooler to be sure that the evaporator can be installed in the space in accordance with the manufacturer's recommendations. For evaporators of this type, the manufacturer recommends that the side of the evaporator be not less than 6 in. or more than 12 in. from the cooler wall (installation dimension *A* of Table R-1) and that the distance between the two sections of the evaporator (dimension *C*) be not less than 6 in. or more than 8 ft.

The maximum allowable evaporator width can be determined by subtracting the minimum of dimensions *A* and *C* from the inside width of the cooler, viz:

Maximum width of evaporator = Inside width of cooler − ($A + C + C$). In this particular case, the maximum allowable width of the evaporator is

$$108 \text{ in.} - (6 \text{ in.} + 6 \text{ in.} + 6 \text{ in.}) = 90 \text{ in.}$$

Since the combined width of the two sections of evaporator Model #PK-16 is only 36 in. (18 in. × 2), the evaporator is suitable for the cooler. A logical arrangement of the two evaporator sections in the cooler is shown in Fig. 11-26.

To order the evaporator, specify the model number, fin spacing, and finned length, that is, Model #PK-16.2-175 in.

Note. There are other selections that would have been equally suitable for this installation. To avoid excessively long evaporators, which are inconvenient to ship and install, a multiple

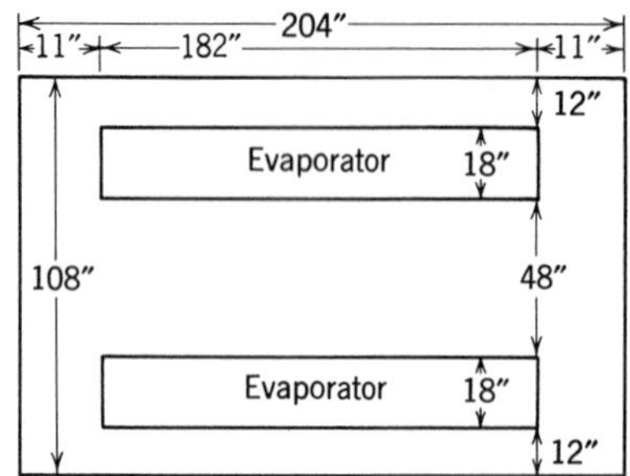

Fig. 11-26 Arrangement of natural convection evaporators in storage cooler (see Example 11-1).

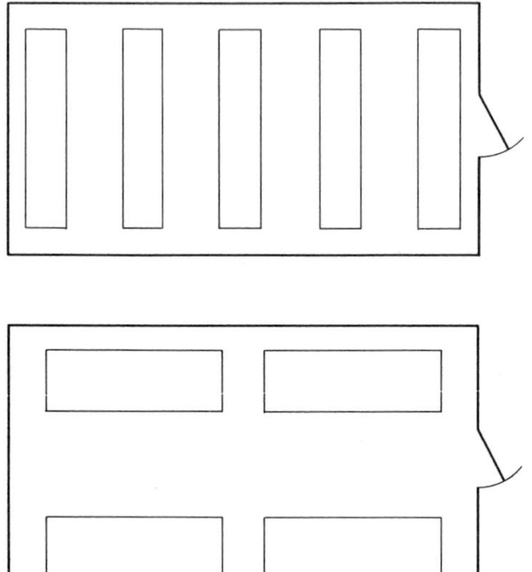

Fig. 11-27 Typical arrangements for natural convection evaporators in large coolers.

of evaporators should be used in large coolers. Typical arrangements for large coolers are shown in Fig. 11-27. These arrangements are also suitable for plate banks.

Example 11-2 A frozen storage room having inside dimensions of 18 ft by 32 ft by 10 ft has a calculated cooling load of 25,000 Btu/hr. The space temperature is to be maintained at 0°F. Using a 15°F evaporator TD, select an appropriate series of plate banks from Table R-3 for ceiling installation in the room.

Solution Analysis of room dimensions (32 ft × 18 ft) indicates that four to six evaporators (two or three banks installed end-to-end in two rows) will provide good ceiling coverage and adequate air distribution. Reference to Table R-3 will show that plate banks are available in stock lengths of 108 in. (9 ft) and 144 in. (12 ft). Three banks 108 in. long (a total of 27 ft) or two banks 144 in. long (a total of 24 ft) could be installed end-to-end in two rows and allow adequate working clearance at the ends.

Since the banks are already rated at the design TD of 15°F, the ratings can be used directly and the required capacity per bank can be determined by dividing the total hourly cooling load by the desired number of plate banks, that is,

$$\text{Required capacity per bank (Btu/hr)} = \frac{\text{Total cooling load (Btu/hr)}}{\text{Number of banks desired}}$$

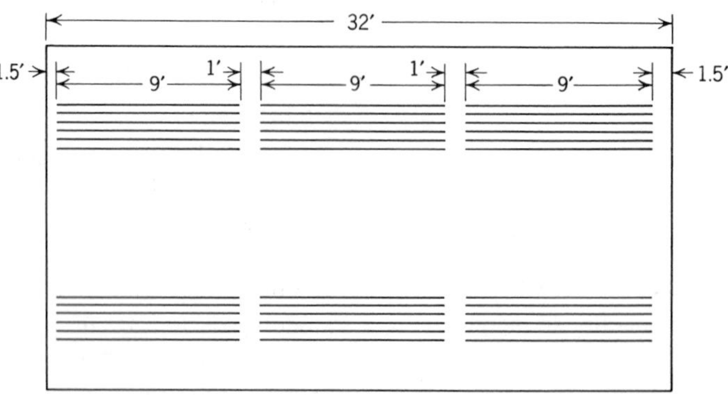

Fig. 11-28 Installation of plate banks in frozen storage room (see Example 11-2.)

In this instance, the capacity required per bank is

$$\frac{25{,}000 \text{ Btu/hr}}{6 \text{ banks}} = 4167 \text{ Btu/(hr)(15°F TD)}$$

or

$$\frac{25{,}000 \text{ Btu/hr}}{4 \text{ banks}} = 6250 \text{ Btu/(hr)(15°F TD)}$$

Referring to Table R-3, we see that plate bank Model #6-12108-B has a capacity of 4320 Btu/hr at a 15°F TD when operating below 32°F (frosted). This will permit good coverage of the ceiling and at the same time allow sufficient working space at the ends of the banks (see Fig. 11-28).

In ordering the evaporators, specify refrigerant and the type of connections desired (series or manifold), that is, Model #6-12108-B, series connected for Refrigerant-12. Notice (Table R-3) that the manufacturer specifies that two refrigerant flow controls should be used with each bank for Refrigerant-12, whereas only one is needed when ammonia is the refrigerant.

Example 11-3 For the conditions described in Example 11-2, determine the number of linear feet of $1\frac{1}{4}$ in. iron pipe that would be required for ceiling installation in the storage room.

Solution From Example 11-2, the design room temperature is 0°F and the evaporator TD is 15°F, so that the required refrigerant temperature is −15°F. From Table R-5, the *U* factor for bare-pipe coils in air with a 0°F room temperature and a 15°F refrigerant temperature is (by interpolation) 1.45 Btu/(hr)(ft^2)(°F TD). From Table R-6, 2.3 linear feet of $1\frac{1}{4}$ in. are required to provide 1 ft^2 of external pipe surface. In accordance with Equation 11-1, the pipe surface area required

$$= \frac{\text{Capacity required (Btu/hr)}}{\text{(Pipe } U \text{ factor)(Design TD)}}$$

$$= \frac{25{,}000 \text{ Btu/hr}}{1.45 \text{ Btu/(hr)(ft}^2\text{)(°F TD)} \times 15\text{°F}} = 1149.4 \text{ ft}^2$$

The linear feet of pipe required

$$= (1149.4 \text{ ft}^2)(2.3 \text{ ft/ft}^2) = 2644 \text{ ft}$$

Example 11-4 Assuming flooded operation and natural convection air circulation, determine the number of linear feet of Prestfin pipe coil (Fig. 11-29 and Table R-7) required to meet the conditions described in Example 11-2.

Solution From Example 11-2, the room design temperature is 0°F and the evaporator TD is 15°F, so the required refrigerant temperature

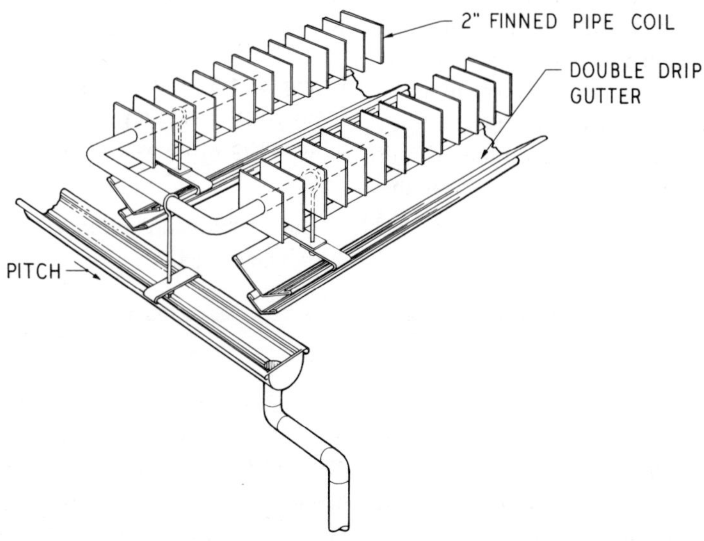

Fig. 11-29 Prestfin pipe coil.

is $-15°F$. From Table R-7, the U factor for Prestfin pipe coil operating flooded with natural convection air circulation and a refrigerant temperature below $-1°F$ is 1.2 Btu/(hr)(ft^2)(°F TD). From Equation 11-1, the square feet of surface required

$$= \frac{\text{Required capacity (Btu/hr)}}{(U \text{ factor})(\text{TD})}$$

$$= \frac{25{,}000 \text{ Btu/hr}}{(1.2)(15)} = 1389 \text{ ft}^2$$

The linear feet of Prestfin pipe required

$= 1389/8.1$ ft^2 per linear foot (1 in. fin spacing)

$= 172$ linear feet

11-21. Forced Convection Evaporators

Forced convection evaporators, commonly called "unit coolers," "fan coil units," or "blower coils" in commercial refrigeration, are essentially bare-tube or finned-tube coils encased in a metal housing and equipped with one or more fans to provide air circulation. Some typical unit coolers are shown in Fig. 11-30.

Larger units are available for floor mounting and may be operated either "wet" or "dry." "Wet" coils are coils that are continuously sprayed with brine or an antifreeze solution, as shown in Fig. 11-47.

The total cooling capacity of any evaporator is directly related to the air quantity (in cubic feet per minute) circulated over the evaporator. The air quantity required for a given evaporator capacity is basically a function of two factors: (1) the sensible heat ratio and (2) the drop in the temperature of the air passing over the evaporator, that is,

$$\text{cfm} = \frac{\text{total capacity (Btu/hr)} \times \text{sensible heat ratio}}{\text{temperature drop of the air} \times 1.08} \quad (11\text{-}5)$$

The sensible heat ratio is the ratio of the sensible cooling capacity of the evaporator to the total cooling capacity. When air is cooled below its dew point temperature, both the temperature and the moisture content of the air are reduced (Chapter 5). The temperature reduction is the result of sensible cooling, whereas the moisture removed is the result of latent cooling. Hence, for an evaporator having a total cooling capacity of one ton (12,000 Btu/hr) and a sensible heat ratio of 0.85, the sensible cooling capacity of the evaporator is 10,200 Btu/hr (85% of 12,000), whereas the latent cooling capacity is 1800 Btu/hr (12,000 − 10,200). Naturally, the sensible heat ratio of any evaporator will depend upon the conditions of the application, the design of the evaporator, and the air quantity. An average sensible heat ratio for unit coolers is approximately 0.85.

Design air quantities for unit coolers range from 1000 cfm/ton for low-velocity units up to 2500 cfm/ton for high-velocity blast coolers. Assuming a sensible heat ratio of 0.8 and applying Equation 11-5, the required temperature drop for the air passing through the high-velocity unit is

$$\frac{12{,}000 \text{ Btu/(hr)(ton)} \times (0.8)}{2500 \text{ cfm/ton} \times 1.08} = 3.56°\text{F}$$

The air velocity (feet per minute) over the coil is a function of the air quantity (cubic feet per minute) and the face area of the evaporator (square feet), that is,

$$\text{Velocity (fpm)} = \frac{\text{Air quantity (cfm)}}{\text{Face area (ft}^2)} \quad (11\text{-}6)$$

Example 11-5 Determine the coil face area required to maintain a face velocity of 400 fpm if the air flow rate over the coil is 2100 cfm.

Solution Applying Equation 11-6, the required face area

$$= \frac{2100 \text{ cfm}}{400 \text{ fpm}} = 5.25 \text{ ft}^2$$

Coil design face velocities range from 200 to 500 fpm for low-velocity units, from 500 to 800 fpm for medium-velocity units, and from 800 to 1200 fpm for high-velocity units. Where coils are to be operated wet, face velocities are limited to 600 fpm or less to avoid drawing

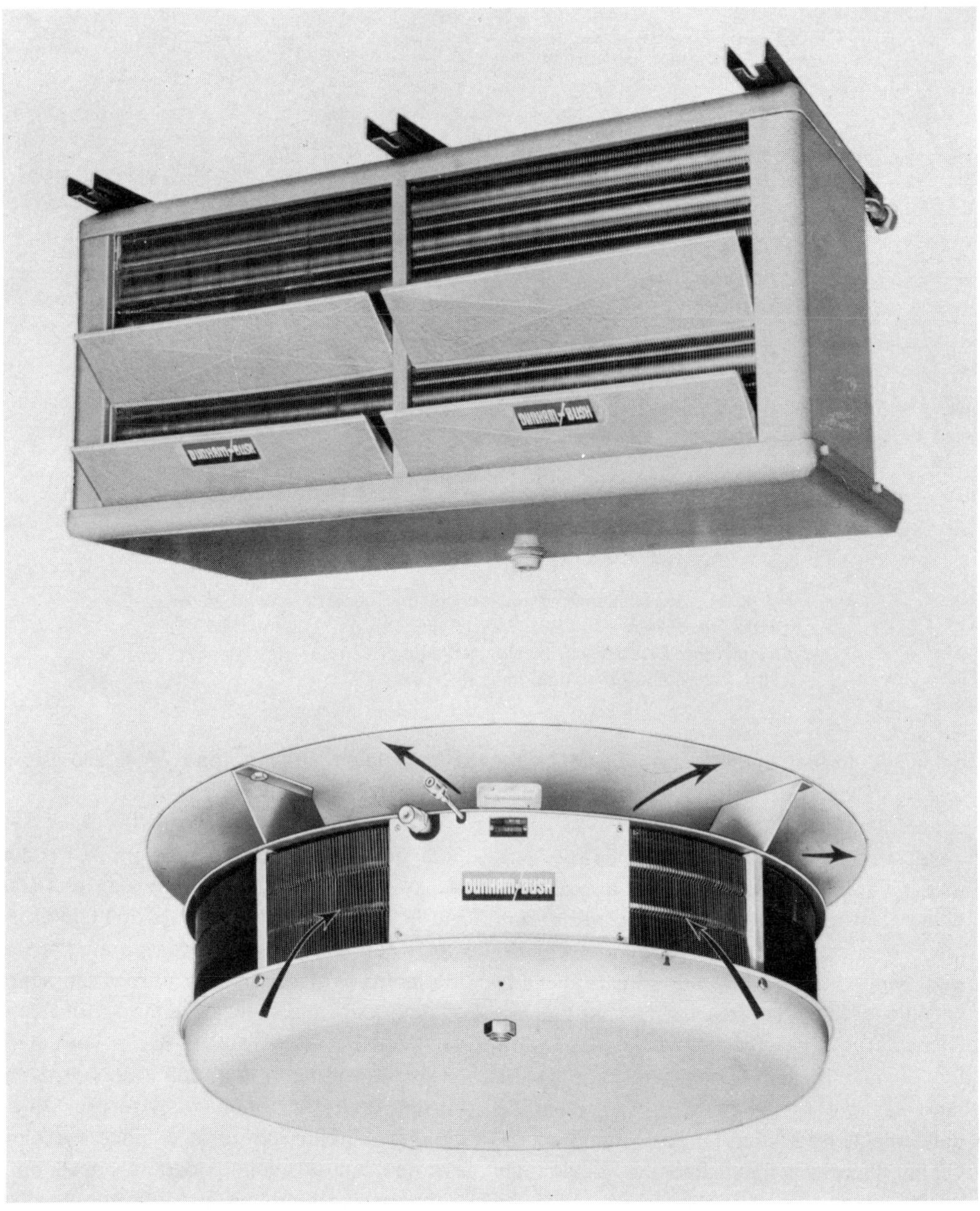

Fig. 11-30 Typical unit cooler designs. Notice that cooler designs are such that the air is not discharged directly on the stored produce. (Courtesy Dunham-Bush, Inc.)

the water or brine off the coil surface into the airstream.

Because of the tubes and fins, the actual free area of the coil is substantially less than the face area, and, consequently, the actual air velocity over the tubes and fins will always be considerably higher than the calculated face velocity.

11-22. Rating and Selection of Unit Coolers

Basic ratings for unit coolers are given in Btu/(hr)(°F TD). For convenience, sometimes ratings are given for 10°F and 15°F TDs. As in the case of natural convection evaporators, the design TD for unit coolers depends primarily on the space humidity requirements. In general, for any given space humidity, the

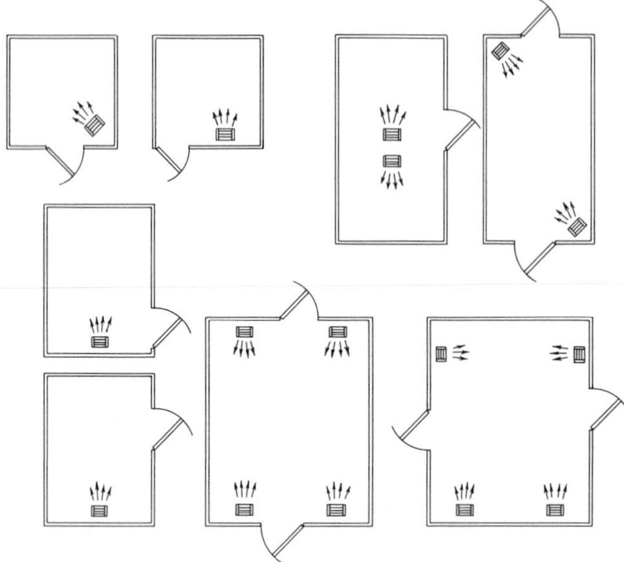

Fig. 11-31 Suggestions for location of unit coolers in walk-in refrigerators. (From the *ASRE Data Book*, Design Volume, 1957–58 edition, reproduced by permission of the American Society of Heating, Refrigerating and Air-Conditioning Engineers.)

design TD for unit coolers is about 4°F to 6°F less than those used for natural convection evaporators (see Table 11-2).

Because of the number of variables involved, unit coolers are usually rated for a fixed condition of air quantity and velocity, refrigerant feed, frost accumulation, operating temperature range, etc., with the selected rating conditions usually approximating those that are encountered in practice. In many cases where the air quantity and velocity are fixed by the fan selection at the time of manufacture, the only selection variable is the evaporator TD. Table R-8, which applies to the rectangular ceiling-hung units illustrated in Fig. 11-30, is representative of this simplified rating method. Realization of rated capacity will depend primarily on proper location of the evaporator in the space and on keeping the evaporator reasonably free of frost by adequate defrosting. Some suggested locations for ceiling-hung unit coolers are shown in Fig. 11-31.

Example 11-6 Using a design evaporator TD of 10°F, select a fan-coil unit from Table R-8 that will meet the requirements of a beer storage cooler maintained at 36°F and having a calculated cooling load of 17,200 Btu/hr based on a 16-hr operating time.

Solution From Table R-8, select unit cooler Model #UC-180 having a capacity of 18,000 Btu/hr at a 10°F TD. Since the unit cooler fan motor operates inside the refrigerated space, the motor heat becomes a part of the space cooling load and must be added to the load calculations. From Table R-8, the heat given off by the fan motor is 25,200 Btu/24 hr. Since the fan operates continuously to provide air circulation in the refrigerated space, while the average hourly cooling load is based on a 16-hr running time, the average Btu per hour load resulting from the fan motor heat is

$$\frac{25{,}200 \text{ Btu/24 hr}}{16 \text{ hr}} = 1575 \text{ Btu/hr}$$

When the fan motor heat is added, the average hourly cooling load for the storage cooler becomes 18,775 Btu/hr (17,200 + 1575). Although the calculated load is a little greater than the evaporator rating at a 10°F TD, the difference is insignificant and the unit cooler selected is well suited to the application. Assuming operation at design conditions, the evaporator operating TD will be

$$\frac{\text{Required capacity}}{\text{Rated capacity}} \times \text{design TD}$$

$$\frac{18{,}775 \text{ Btu/hr}}{18{,}000 \text{ Btu/hr}} \times 10°\text{F} = 10.4°\text{F}$$

Table R-9 is the manufacturer's rating data for a heavy-duty, floor-mounted product cooler and is representative of the more comprehensive rating and selection methods. Notice that ratings are given for both dry expansion and flooded operation and that correction factors are given for other air quantities and velocities and for dry expansion operation in the low-temperature range.

11-23. Liquid Chilling Evaporators

As with air-cooling evaporators, liquid chilling evaporators vary in type and design according to the type of duty for which they are intended. Five general types of liquid chillers are in common use: (1) the double-pipe cooler, (2) the Baudelot cooler, (3) the tank-type cooler, (4) the shell-and-coil cooler, and (5) the shell-and-tube cooler. In all cases, the factors which influence the performance of liquid chillers are the same as those which govern the performance of air-cooling evaporators and all other heat transfer surfaces. Heat transfer coefficients for average designs of some of the various chiller types are listed in the table in Fig. 11-32.

11-24. Double-Pipe Coolers

The double-pipe cooler consists of two tubes so arranged that one tube is inside the other. The chilled fluid flows in one direction through the inner tube while the refrigerant flows in the opposite direction through the annular space between the inner and outer tubes. One design of a double-pipe cooler is shown in Fig. 11-33. In this design the outer tubes are welded to vertical refrigerant headers while the inner tubes pass through the headers and are connected together by removable return bends. The advantages claimed for this unit are rigid construction, the elimination of refrigerant joints, and easy accessibility of the inner tubes for cleaning.

Double-pipe coolers may be operated either dry-expansion or flooded. In either case, counterflowing of the fluids in the tubes produces a relatively high heat transfer coefficient. However, this type of cooler has the disadvantage of requiring more space, particularly head room, than some of the other cooler designs. For this reason, the double-pipe cooler is used only in some few special applications. A number have been used in the wine-making and brewing industries to chill wine and wort, and in the petroleum industry for the chilling of oils.

Type of Evaporator	*Minimum*	*Maximum*	*Surface Side Basis for U*
Flooded shell-and-plain-tube (water to Refrigerants 12, 22, and 717)	130	190	Refrigerant
Flooded shell-and-finned-tube (water to Refrigerant 12, 22, or 500)	90	170	Refrigerant
Flooded shell-and-plain-tube (brine to Refrigerant 717)	45	100	Refrigerant
Flooded shell-and-plain-tube (brine to Refrigerant 12, 22, or 502)	30	90	Refrigerant
Direct expansion, shell-and-plain-tube (water to Refrigerant 12, 22, or 717) (Refrigerant in tubes)	80	220	Liquid
Direct expansion, shell-and-internal-finned-tubes (water to Refrigerant 12 or 22) (Refrigerant in tubes)	160	250	Liquid
Direct expansion, shell-and-plain-tube (brine to Refrigerant 12, 22, 717, or 502) (Refrigerant in tubes)	60	140	Liquid
Direct expansion, shell-and-internal-finned-tubes (non-salt brines to Refrigerant 12, 22, or 502)	100	170	Liquid
Shell-and-plain-tube coil (water in shell) (Refrigerant 12, 22, or 717 in coil)	10	25	Liquid
Baudelot cooler, flooded (Refrigerant 12 or 22 to water)	100	200	Liquid
Baudelot cooler, direct expansion (Refrigerant 717 to water)	60	150	Liquid
Baudelot cooler, direct expansion (Refrigerant 12 or 22 to water)	60	120	Liquid
Double-pipe cooler (Refrigerant 717 to water)	50	150	Liquid
Double-pipe cooler (Refrigerant 717 to brine)	50	125	Liquid
Tank-and-agitator, coil type water cooler (flooded, Refrigerant 717)	80	125	Liquid
Tank-and-agitator, coil type water cooler (flooded, Refrigerant 12, 22, or 500)	60	100	Liquid
Tank, ammonia (Refrigerant 717) to brine cooling, coils between cans in ice tank	15	40	Liquid
Tank, high velocity raceway type (Refrigerant 717 to brine)	80	110	Liquid

Fig. 11-32 From the *ASHRAE Data Book*, Equipment Volume, 1975 Edition, by permission of the American Society of Heating, Refrigerating, and Air-Conditioning Engineers.

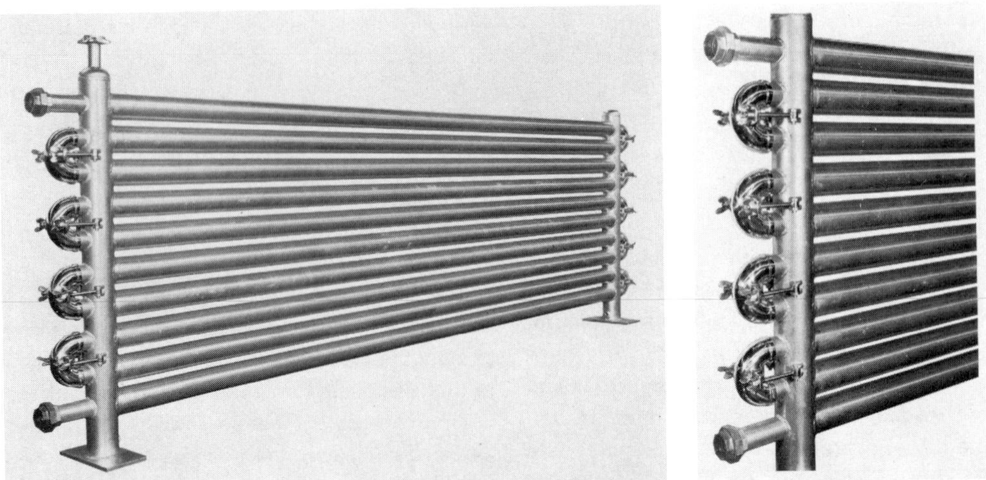

Fig. 11-33 Double pipe cooler. Removable return bends (right) are designed to make tube readily accessible for cleaning. (Courtesy Vilter Manufacturing Company.)

11-25. Baudelot Coolers

The Baudelot cooler shown in Fig. 11-34 consists of a series of horizontal pipes which are located one under the other and are connected together to form a refrigerant circuit or circuits. For either dry-expansion or flooded operation, the refrigerant is circulated through the inside of the tubes while the chilled liquid flows in a thin film over the outside. The liquid flows down over the tubes by gravity from a distributor located at the top of the cooler and is collected in a trough at the bottom. The fact that the chilled liquid is at atmospheric pressure and is open to the air makes the Baudelot cooler ideal for any liquid chilling application where aeration is a factor. The Baudelot chiller has been widely used for the cooling of milk, wine, and wort, and for the chilling of water for car-

Fig. 11-34 Baudelot cooler employed in milk-cooling application. (Courtesy Dole Refrigerating Company.)

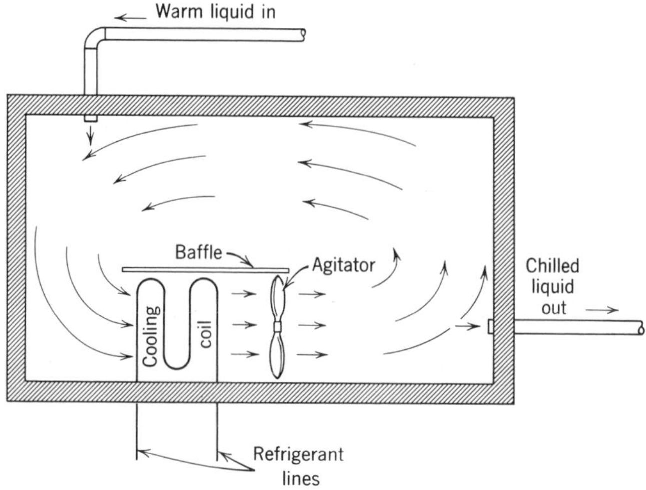

Fig. 11-35 Typical construction of tank-type liquid cooler.

bonation in bottling plants. With this particular type of chiller it is possible to chill liquid to a temperature very close to the freezing point without the danger of damaging the equipment if occasional freezing of the liquid occurs.

Another advantage of the Baudelot cooler, and one which is shared by the double-pipe cooler, is that the refrigerant circuit is readily split into several parts, a circumstance which permits precooling of the chilled liquid with cold water before the liquid enters the direct-expansion portion of the cooler (see Fig. 17-34).

11-26. Tank-Type Coolers

The tank-type liquid chiller consists essentially of a bare-tube refrigerant coil installed in the center or at one side of a large steel tank which contains the chilled liquid. Although completely submerged in the chilled liquid, the refrigerant coil is separated from the main body of the liquid by a baffle arrangement. As shown in Fig. 11-35, a motor driven agitator is utilized to circulate the chilled liquid over the cooling coil at relatively high velocity, usually between 100 and 150 fpm, the liquid being drawn in at one end of the coil compartment and discharged at the other end.

The spiral-shaped, bare-tube coils mentioned in Section 11-4 and the raceway-type coil illustrated in Fig. 11-36 are two coil designs frequently employed in tank-type chillers. With either design the coils are operated flooded. The Ice-Cel shown in Fig. 11-9 is another variation of the tank-type chiller.

Tank-type chillers can be applied to any liquid-chilling application where sanitation is not a primary factor, and are widely used for the chilling of water, brine, and other liquids to be used as secondary refrigerants. Because of their inherent holdover capacity, they are particularly suitable for applications subject to frequent and severe fluctuations in loading. In such cases, a comparatively large chilled-liquid storage tank is provided in order to minimize the rise in the temperature of the chilled liquid during periods of peak demand. The advantage gained by precooling is often considerable in cases where the liquid to be chilled enters the cooler at relatively high temperatures.

11-27. Shell-and-Coil Coolers

The shell-and-coil chiller is usually made up of one or more spiral-shaped, bare-tube coils enclosed in a welded steel shell (Fig. 11-37). As a general rule, the chiller is operated dry-expansion with the refrigerant in the coils and the chilled liquid in the shell. In a few cases, the chiller is operated flooded, in which case the refrigerant is in the shell and the chilled liquid passes through the tubes. The former arrangement has the advantage of providing a holdover capacity, thereby making this type of chiller ideal for small applications having

Fig. 11-36 Flooded raceway coil. (Courtesy Vilter Manufacturing Company.)

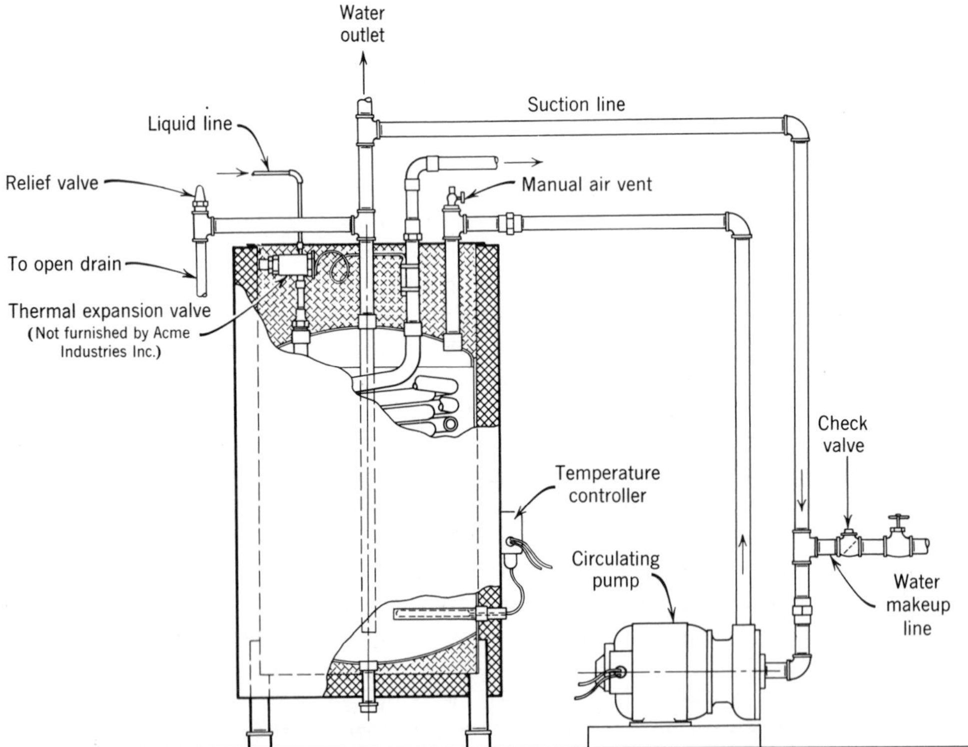

Fig. 11-37 Shell-and-coil cooler. (Courtesy Acme Industries.)

high but infrequent peak loads. It is used primarily for the chilling of water for drinking and for other purposes where sanitation is a prime factor, as in bakeries and photographic laboratories.

When operated flooded with the refrigerant in the shell, this type of chiller becomes what is commonly referred to as an "instantaneous" liquid chiller. One of the disadvantages of this arrangement is that there is no holdover capacity. Since the liquid is not recirculated, it must be chilled instantaneously as it passes through the coils. Another disadvantage is that the danger of damaging the chiller in the event of freeze-up is greatly increased in any chiller where the chilled liquid is circulated through the coils or tubes rather than over the outside of the tubes. For this reason, chillers employing this arrangement cannot be recommended for any application where it is required to chill the liquid below 38°F.

Instantaneous shell-and-coil chillers are used principally for the chilling of beer and other beverages in "draw-bars," in which case the beverage is usually precooled to some extent before entering the chiller.

11-28. Shell-and-Tube Chillers

Shell-and-tube chillers have a relatively high efficiency, require a minimum of floor space and head room, are easily maintained, and are readily adaptable to almost any type of liquid-chilling application. For these reasons, the shell-and-tube chiller is by far the most widely used type. Although individual designs differ somewhat, depending upon the refrigerant used and upon whether the chiller is operated dry-expansion or flooded, the shell-and-tube chiller consists essentially of a cylindrical steel shell in which a number of straight tubes are arranged in parallel and held in place at the ends by tube sheets. When the chiller is operated dry-expansion, the refrigerant is expanded into the tubes while the chilled liquid is circulated through the shell (Fig. 11-38*b*). When the chiller is operated flooded, the chilled

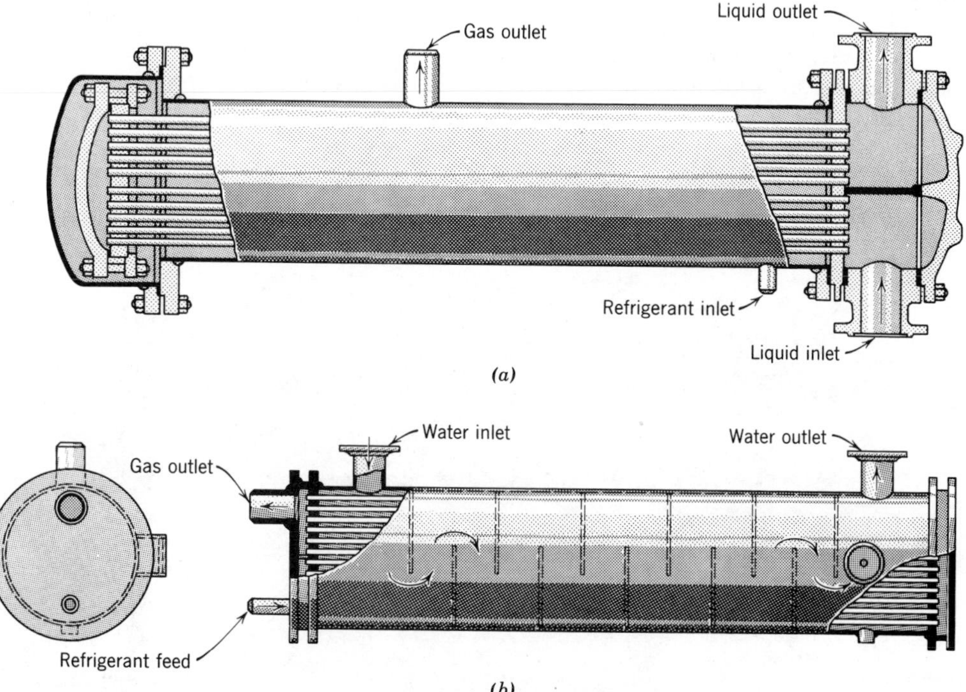

Fig. 11-38 Typical shell-and-tube chillers. (*a*) Hooded type. Tube bundle is removable. (*b*) Dry-expansion type (refrigerant in tubes). Note baffling of water circuit. Tube sheets are fixed. (Courtesy Worthington Corporation.)

liquid is circulated through the tubes and the refrigerant is contained in the shell, the level of the liquid refrigerant in the shell being maintained with some type of float control (Fig. 11-38*a*). In either case, the chilled liquid is circulated through the chiller and connecting piping by means of a liquid circulating pump, usually of the centrifugal type.

Shell diameters for shell-and-tube chillers range from approximately 6 to 60 in., and the number of tubes in the shell varies from fewer than 50 to several thousand. Tube diameters range from $\frac{5}{8}$ in. through 2 in., and tube lengths vary from 5 to 20 ft. Chillers designed for use with ammonia are equipped with steel tubes, whereas those intended for use with other refrigerants are usually equipped with copper tubes in order to obtain a higher heat transfer coefficient. Because of the relatively low film conductance of halocarbon refrigerants, chillers designed for use with these refrigerants are often equipped with tubes which are finned on the refrigerant side. In the case of dry-expansion chillers, the tubes are finned internally with longitudinal fins of the types shown in Fig. 11-10. For flooded operation, the tubes are finned externally using a very short fin which protrudes from the tube wall only approximately $\frac{1}{16}$th of an inch.

As a general rule, dry-expansion chillers are employed in small and medium tonnage installations requiring capacities ranging from 2 to approximately 250 tons, but are available in larger capacities. Flooded chillers, available in capacities ranging from approximately 10 through several thousand tons, are more frequently applied in the larger tonnage installations.

11-29. Dry-Expansion Chillers

The principal advantages of the dry-expansion chiller over the flooded type are the smaller refrigerant charge required and the assurance of positive oil return to the compressor. Too, as previously stated, the possibility of damage to the chiller in the event of freeze-up is always considerably less when the chilled liquid is circulated over the tubes rather than through them. The more important construction details of several designs of dry-expansion chillers are shown in Figs. 11-39 and 11-40.

In order to maintain the liquid velocity within the limits which will produce the most effective heat transfer-pressure drop ratio, the velocity of the chilled liquid circulated over the tubes is controlled by varying the length and spacing of the segmental baffles. When the flow rate and/or liquid viscosity is high, short, widely

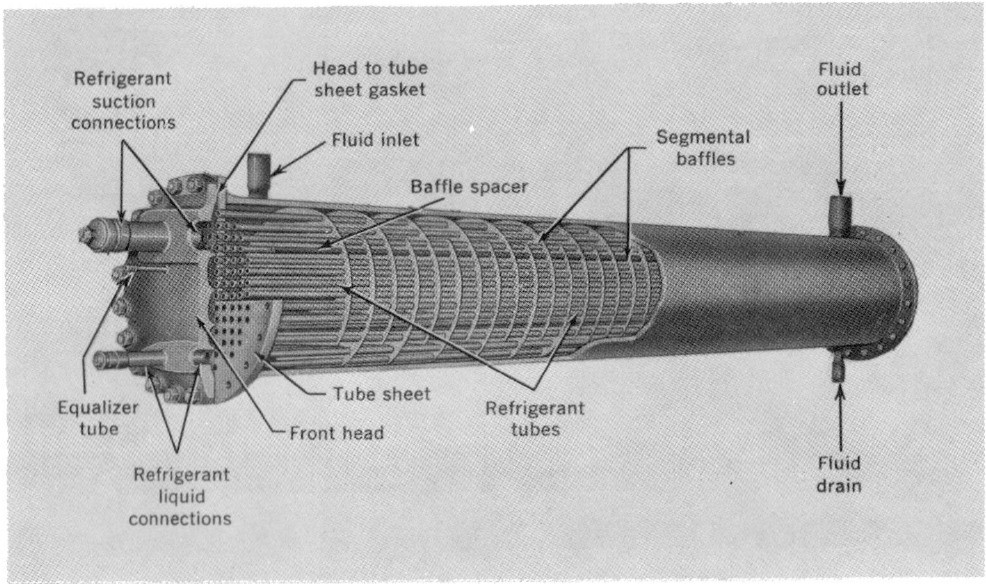

Fig. 11-39 Cutaway section illustrating construction details of dry-expansion chiller with fixed tube sheets. (Courtesy Acme Industries.)

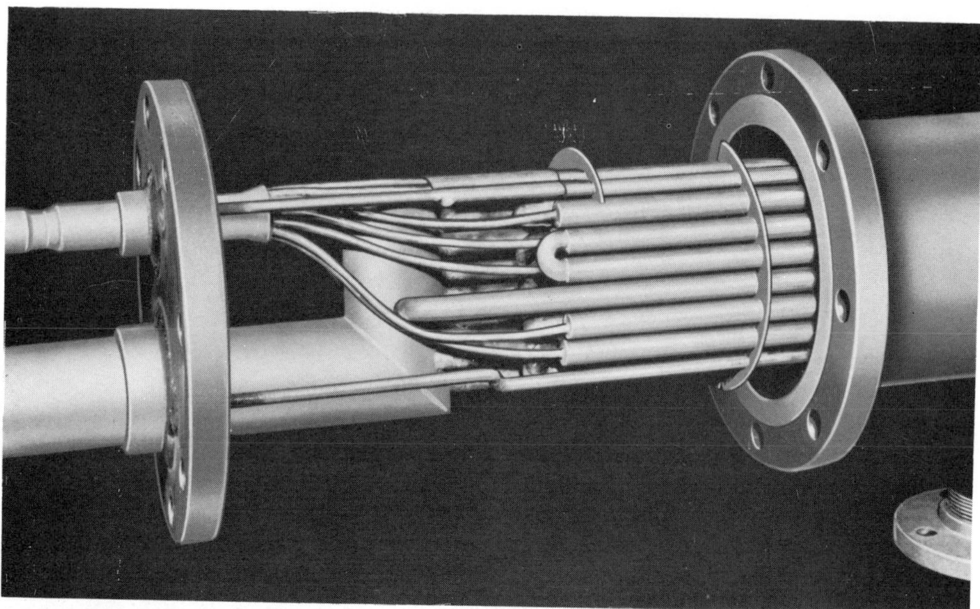

Fig. 11-40 Dry-expansion chiller with tube bundle partially removed to show tube arrangement and refrigerant distributors. Tube bundle can be removed as a unit. (Courtesy Kennard Division, American Air Filter Company, Inc.)

spaced baffles are used to reduce the velocity and minimize the pressure drop through the chiller. When the flow rate and/or liquid viscosity is low, longer, more closely spaced baffles are used in order to increase fluid velocity and improve the heat transfer coefficient (Fig. 11-41*a*).

The number and the length of the refrigerant circuits required to maintain the refrigerant velocity through the chiller tubes within reasonable limits depend on the total chiller load and on the relationship of the chilled liquid flow rate to the METD. Since these factors vary with the individual application, it follows that the

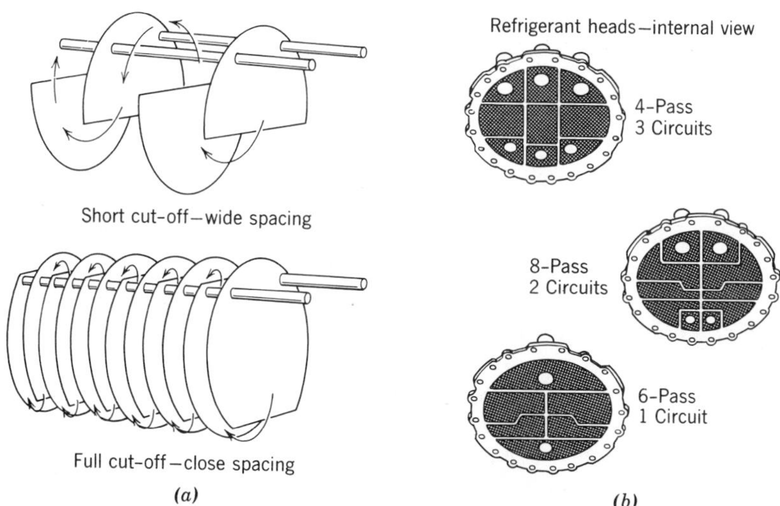

Fig. 11-41 (*a*) Baffle spacing in dry-expansion chillers. (*b*) Typical refrigerant heads for dry-expansion chiller. (Courtesy of Acme Industries.)

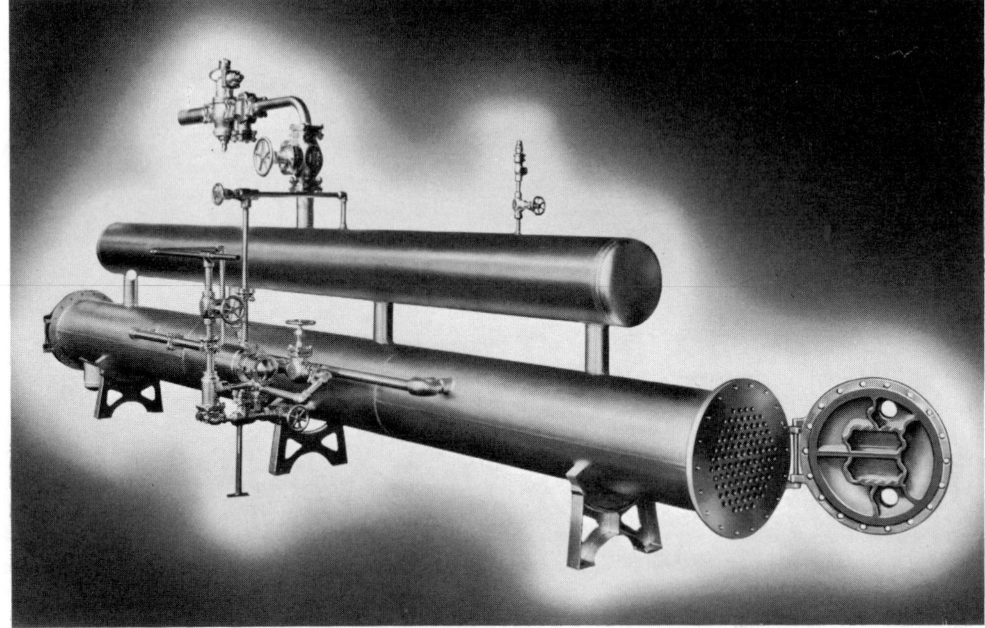

Fig. 11-42 Flooded chiller designed for multipass circulation of chilled liquid. Multipass circulation is accomplished by means of the baffled end-plates or water heads which are bolted to the ends of the chiller. (Courtesy Vilter Manufacturing Company.)

optimum refrigerant circuit design also varies with the individual application. For this reason, chillers are made available with either single or multiple refrigerant circuits of varying lengths. For the design shown in Fig. 11-39, the number and lengths of the refrigerant circuits depend on the tube length and on the arrangement of the baffling in the end-plates or refrigerant heads which are bolted to the tube sheets at the ends of the chiller. The refrigerant circuit arrangement for any one model chiller can be changed by changing the refrigerant heads (Fig. 11-41*b*).

11-30. Flooded Chillers

Standard flooded chiller designs include both single and multipass tube arrangements. For single pass flow, the tubes are so arranged that the chilled liquid passes through all the tubes simultaneously and in only one direction.

Multipass circulation of the chilled liquid through the chiller is accomplished through the use of baffled end-plates or heads which are bolted to the ends of the chiller (Fig. 11-42). The arrangement of the end-plate baffling determines the number of passes the chilled liquid makes from one end to the other before leaving the chiller. Although two-, four-, and six-pass arrangements are the most common, more passes are used in some instances.

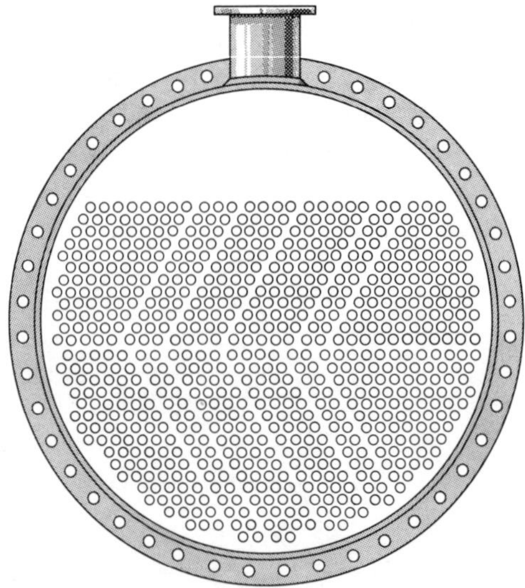

Fig. 11-43 Flooded chiller with shell only partially filled with tubes in order to provide a large vapor disengaging area above the tubes. (Courtesy Worthington Corporation.)

As in the case of the dry-expansion chiller, some flooded chillers are designed with removable tube bundles, whereas others have fixed tube sheets. In the fixed tube sheet design, the tube sheets are welded to the shell so that the tube bundle is not removable. However, when the end-plates are unbolted the tubes become readily accessible for cleaning and individual tubes can be removed and replaced if necessary. The chillers shown in Figs. 11-38*b* and 11-39 employ fixed tube sheets, whereas those in Figs. 11-38*a* and 11-40 have tube bundles.

In some flooded chiller designs, the shell is only partially filled with tubes in order to provide a large vapor-disengaging area and relatively low velocity in the space above the tubes (Fig. 11-43). This design eliminates the possibility of liquid carry-over into the suction line and therefore is particularly well suited to sudden heavy increases in loading.

In those chiller designs where the shell is completely filled with tubes, a surge drum or accumulator should be used to separate any entrained liquid from the vapor before the vapor enters the suction line. Some flooded chillers are equipped with built-in liquid-suction heat exchangers (Fig. 11-42). Although the primary function of the heat exchanger is to insure that only dry vapor enters the suction line, it has the additional effect of increasing the efficiency of the chiller in that it subcools the liquid approaching the chiller and thereby reduces the amount of flash gas that enters the cooler.

The vertical shell-and-tube chiller shown in Fig. 11-44 has the advantage of requiring a minimum amount of floor space. The chiller

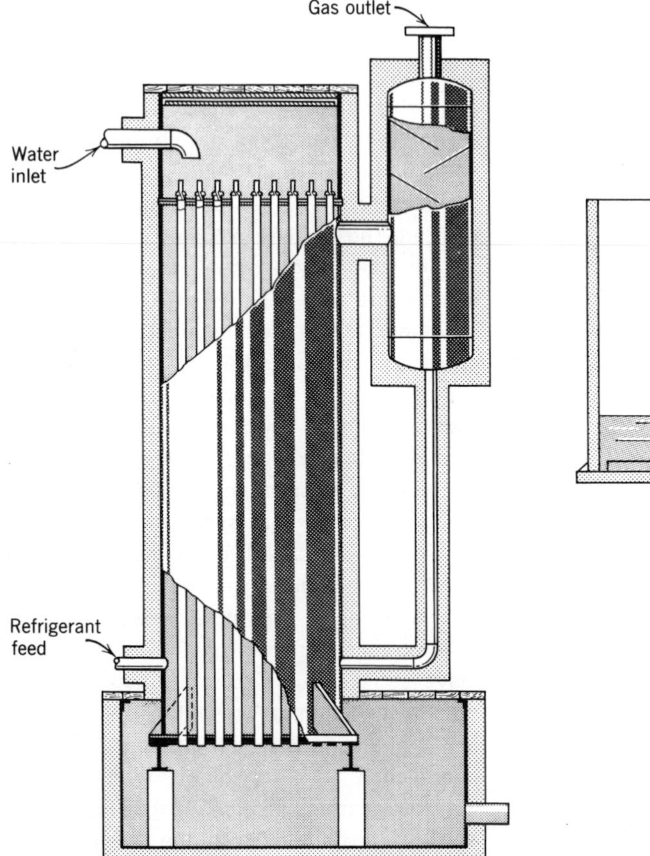

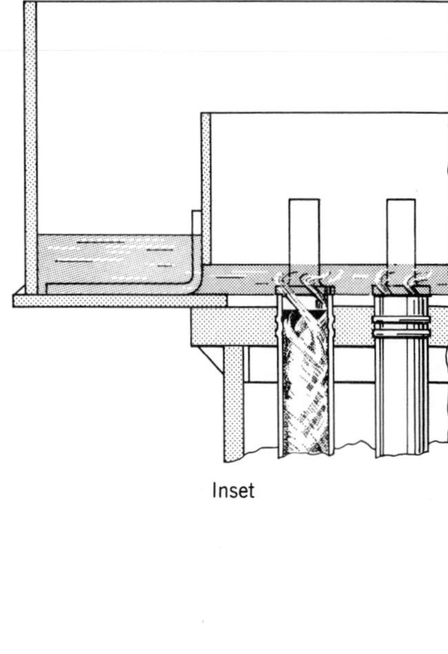

Fig. 11-44 Vertical shell-and-tube "Spira-Flo" chiller designed for flooded operation. The water flowing down through the tubes is given a swirling action by specially designed nozzles (inset). (Courtesy Worthington Corporation.)

is operated flooded. The chilled liquid enters the chiller at the top and flows by gravity down the inside of the tubes. A circulating pump draws chilled liquid from the storage tank at the bottom and delivers it through the connecting piping. The return liquid is piped to the distributor box at the top, from where it again flows through the tubes. A specially designed distributor installed at the top of each tube (inset) imparts a swirling motion to the chilled liquid, which causes the liquid to flow in a comparatively thin film down the inside tube surfaces.

11-31. Spray-Type Chillers

The spray-type chiller is similar in construction to the conventional flooded chiller except that the liquid refrigerant is sprayed over the outside of the water tubes from nozzles located in a spray header above the tube bundle (Fig. 20-19). The unevaporated liquid drains from the tube into a sump at the bottom of the chiller from which it is recirculated to the spray nozzles by a low head liquid pump. A high recirculation rate assures continuous wetting of the tube surfaces and results in a high rate of heat transfer.

The principal advantages of this type of chiller are its high efficiency and relatively small refrigerant charge. Disadvantages are the high installation cost and the need for a liquid recirculating pump.

11-32. Chiller Selection Procedure

Although selection methods differ somewhat depending upon the type of chiller and the particular manufacturer, all are based on the simple fundamentals of heat transfer and fluid flow which have already been described. Almost without exception, manufacturers include sample selection procedure along with the design and capacity data in their equipment catalogs. The following selection procedure follows very closely that given in the catalog of one manufacturer for the selection of dry-expansion chillers.*

* Acme Industries, Inc.

Example 11-7 It is desired to cool 50 gpm of water from 54°F to 46°F with a refrigerant temperature as measured at the cooler outlet of 40°F using Refrigerant-12.

Solution

Step 1. Determine the total chiller load in tons.

$$\frac{\text{Gpm} \times 500 \times \text{cooling range}}{12{,}000 \text{ Btu/hr ton}}$$

$$\frac{50 \times 500 \times (54 - 46)}{12{,}000} = 16.7 \text{ tons}$$

Step 2. Determine the mean effective temperature difference (METD).

Water in minus refrigerant temperature $54 - 40 = 14$°F LTD

Water out minus refrigerant temperature $46 - 40 = 6$°F STD

From Table 11-1, METD = 9.47°F

Step 3. Select trial chiller (shell diameter and baffles spacing) from Fig. 1 of Table R-10. Enter Fig. 1 at 50 gpm on the lower vertical scale and move horizontally across the chart to the diagonal line representing the type unit desired. The number indicates shell diameter and the letter indicates baffle spacing. Possible choices are 10M, 12L, 8M, 12K, 10K, and 8L. As a general rule, small diameter chillers are more economical, whereas large diameter chillers are more compact. Type M baffling produces the lowest pumping head, whereas type K baffling produces the highest pumping head. Hence, if space is not a problem, the most logical choice would seem to be type 8M. However, a check will show that neither 8M nor 8L is available with sufficient surface area in this instance. Therefore, select type 10M (8 to 30 tons). From the point of intersection move vertically upward to a diagonal line in the upper portion of Fig. 1 which represents a METD of 9.47°F as found in Step 2. From this intersection move horizontally to the scale at the left margin and read the loading of 1110 Btu/hr ft^2 (loading is the U value times the METD).

Step 4. Determine the surface area required.

$$\text{Surface area} = \frac{\text{Capacity (Btu/hr)}}{\text{Loading}}$$

$$\frac{200{,}000}{1110} = 180.2\ \text{ft}^2$$

Step 5. Select chiller length from Table R-10 to meet surface area requirements. A 10 in. × 14 ft DXH chiller has a surface area of 184 sq ft (Model No. DXH-1014).

Step 6. Determine the water pressure drop through the chiller. From the bottom of Fig. 1, Table R-10, the pressure drop per foot of length with type M baffling is 0.18 ft of water column. Pressure drop = length (feet) × pressure drop/foot

$$14\ \text{feet} \times 0.18 = 2.52\ \text{ft}\ H_2O$$

11-33. Direct and Indirect Systems

Any heat transfer surface into which a volatile liquid (refrigerant) is expanded and evaporated in order to produce a cooling effect is called a "direct-expansion" evaporator and the liquid so evaporated is called a "direct-expansion" refrigerant. A direct-expansion or "direct" refrigerating system is one wherein the system evaporator, employing a direct-expansion refrigerant, is in direct contact with the space or material being refrigerated, or is located in air ducts communicating with such spaces. Up to this point, only direct refrigerating systems have been considered.

Very often it is either inconvenient or uneconomical to circulate a direct-expansion refrigerant to the area or areas where the cooling is required. In such cases, an indirect refrigerating system is employed. Water or brine (or some other suitable liquid) is chilled by a direct-expansion refrigerant in a liquid chiller and then pumped through appropriate piping to the space or product being refrigerated. The chilled liquid, called a secondary refrigerant, may be circulated directly around the refrigerated product or vessel or it may be passed through an air-cooling coil or some other type of heat transfer surface (Fig. 11-45). In either case, the secondary refrigerant, warmed by the absorption of heat from the refrigerated space or product, is returned to the chiller to be chilled and recirculated.

Indirect refrigerating systems are usually employed to an advantage in any installation where the space or product to be cooled is located a considerable distance from the condensing equipment. The reason for this is that long direct-expansion refrigerant lines are seldom practical. In the first place, they are expensive to install and they necessitate a large refrigerant charge. Too, long refrigerant lines, particularly long risers, create oil return problems and cause excessive refrigerant pressure losses which tend to reduce the capacity and efficiency of the system. Furthermore, leaks are more serious and are much more likely to occur in refrigerant piping than in water or brine piping.

Indirect refrigeration is required also in many industrial process cooling applications where it is often impractical to maintain a vapor tight seal around the product or vessel being cooled. Too, indirect systems are used to an advantage in any application where the leakage of refrigerant and/or oil from the lines may cause contamination or other damage to a stored product. The latter applies particularly to meat packing plants and large cold storage applications when ammonia is used as a refrigerant.

11-34. Secondary Refrigerants

Some commonly used secondary refrigerants are water, calcium chloride and sodium chloride brines, ethylene and propylene glycols, Methanol (methyl alcohol), and glycerin.

Almost without exception, water is used as the secondary refrigerant in large air conditioning systems and also in industrial process cooling installations where the temperatures maintained are above the freezing point of water. Water, because of its fluidity, high specific heat value, and high film coefficient, is an excellent secondary refrigerant. It also has the advantage of being inexpensive and relatively noncorrosive. In air conditioning applications, the chilled water is circulated through an air cooling coil or through a water spray unit. In either case, the air is both cooled and dehumidified. In the water spray unit, the water is sprayed from nozzles and collected

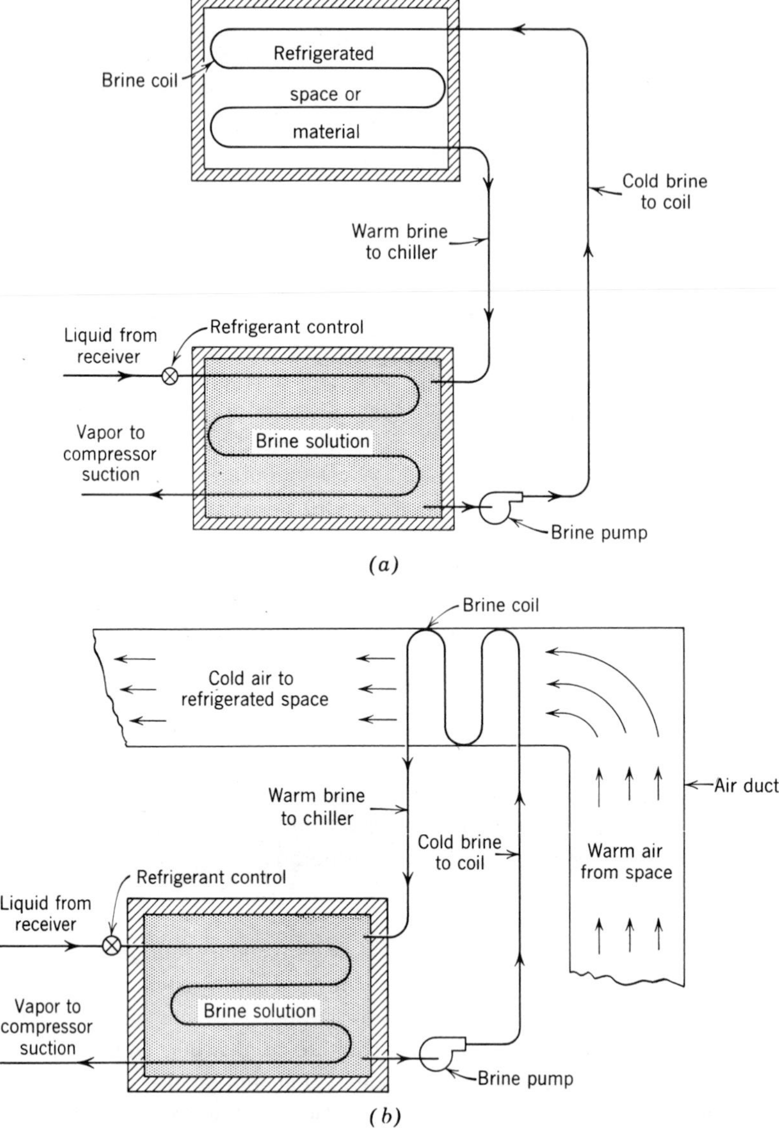

Fig. 11-45 (*a*) Indirect system. (*b*) Indirect system—brine coil in communicating duct.

in a pan or basin at the bottom of the spray unit, from which it is returned to the chiller. Since the air passing through the water spray is chilled below its dew point temperature, a certain amount of water vapor is condensed from the air and is carried to the basin with the spray water. With either the cooling coil or the spray unit, the amount of cooling and dehumidification can be controlled by varying the amount and temperature of the chilled water.

Water is also used frequently as a secondary refrigerant in small beverage coolers and in farm coolers designed for cooling milk cans. In such cases, the water, because of its high conductivity, permits more rapid chilling of the product than would be possible with air. Too, the water supplies a holdover capacity which tends to level out load fluctuations resulting from intermittent loading of the cooler.

11-35. Brines

Obviously, water cannot be employed as a secondary refrigerant in any application where the temperature to be maintained is below the freezing point of water. In such cases, a brine solution is often used.

Brine is the name given to the solution which results when various salts are dissolved in water. If a salt is dissolved in water, the freezing temperature of the resulting brine will be below the freezing temperature of pure water. Up to a certain point, the more salt dissolved in the solution, the lower will be the freezing temperature of the brine. However, if the salt concentration is increased beyond a certain point, the freezing temperature of the brine will be raised rather than lowered. Hence, a solution of any salt in water has a certain concentration at which the freezing point of the solution is lowest. A solution at the critical concentration is called a eutectic solution. At any concentration above or below this critical concentration, the freezing temperature of the solution will be higher, that is, above the eutectic temperature.* When the salt content of the brine is less than that which is required for a eutectic solution, the excess water will begin to precipitate from the solution in the form of ice crystals at some temperature above the eutectic temperature. The exact temperature at which the ice crystals will begin to form depends upon the degree of the salt concentration and upon the relative solubility of the salt in water, the latter factor decreasing as the temperature of the solution decreases. The continued precipitation of ice crystals from the solution as the temperature is reduced causes a progressive increase in the concentration of the remaining brine until, at the eutectic temperature, a slush consisting of ice and eutectic brine will exist. The further removal of heat from this mixture will result in solidification of the eutectic brine. Solidification of the eutectic brine will take place at a constant temperature.

On the other hand, when the salt content of the brine is in excess of the amount required for a eutectic solution, the excess salt will begin to precipitate from the solution in the form of salt crystals at some temperature above the eutectic temperature. Continued precipitation of salt from the mixture as the temperature is reduced will result in a mixture of salt and eutectic brine when the eutectic temperature is reached. The further removal of heat from the mixture will result in solidification of the eutectic brine at constant temperature.

Two types of brine are commonly used in refrigeration practice: (1) calcium chloride and (2) sodium chloride. The two brines are prepared from calcium chloride ($CaCl_2$) and sodium chloride (NaCl) salts, respectively, the latter salt being the common table variety.

Calcium chloride brine is used primarily in industrial process cooling, in product freezing and storage, and in other brine applications where temperatures below 0°F are required. The lowest freezing temperature which can be obtained with calcium chloride brine (the eutectic temperature) is approximately −67°F. The salt concentration in the eutectic solution is approximately 30% by weight. The freezing temperature of various concentrations of calcium chloride brine are given in Table 11-3, along with some of the other important properties of the brine.

The principal disadvantage of calcium chloride brine is its dehydrating effect and its tendency to impart a bitter taste to food products with which it comes in contact. For this reason, when calcium chloride brine is used in food freezing applications, the system must be designed so as to prevent the brine from coming into contact with the refrigerated product.

Sodium chloride brine is employed mainly in those applications where the possibility of product contamination prevents the use of calcium chloride brine. Sodium chloride brine is employed extensively in installations where the chilling and freezing of meat, fish, and other products are accomplished by means of a brine spray or fog.

The lowest temperature obtainable with sodium chloride brine is approximately −6°F. For this freezing temperature the salt concentration in the solution is approximately 23%. The thermal properties of sodium chloride

* At any temperature other than the eutectic temperature, the term "freezing temperature" is used to mean the temperature at which ice or salt crystals begin to precipitate from the solution.

TABLE 11-3 Properties of Pure Calcium Chloride Brine

Pure $CaCl_2$ % by wt	Specific gravity 60 F / 60 F	Baumé density 60 F	Specific heat 60 F Btu per lb F	Crystallization starts F	Weight per gallon			Weight per cubic foot		
					$CaCl_2$ lb/gal	Water lb/gal	Brine lb/gal	$CaCl_2$ lb/cu ft	Water lb/cu ft	Brine lb/cu ft
0	1.000	0.0	1.000	32.0	0.000	8.34	8.34	0.00	62.40	62.40
5	1.044	6.1	0.924	27.7	0.436	8.281	8.717	3.26	61.89	65.15
6	1.050	7.0	0.914	26.8	.526	8.234	8.760	3.93	61.59	65.52
7	1.060	8.2	0.898	25.9	.620	8.231	8.851	4.63	61.51	66.14
8	1.069	9.3	0.884	24.6	.714	8.212	8.926	5.34	61.36	66.70
9	1.078	10.4	0.869	23.5	.810	8.191	9.001	6.05	61.22	67.27
10	1.087	11.6	0.855	22.3	0.908	8.168	9.076	6.78	61.05	67.83
11	1.096	12.6	0.842	20.8	1.006	8.137	9.143	7.52	60.81	68.33
12	1.105	13.8	0.828	19.3	1.107	8.120	9.227	8.27	60.68	68.95
13	1.114	14.8	0.816	17.6	1.209	8.093	9.302	9.04	60.47	69.51
14	1.124	15.9	0.804	15.5	1.313	8.064	9.377	9.81	60.27	70.08
15	1.133	16.9	0.793	13.5	1.418	8.034	9.452	10.60	60.04	70.64
16	1.143	18.0	0.779	11.2	1.526	8.010	9.536	11.40	59.86	71.26
17	1.152	19.1	0.767	8.6	1.635	7.984	9.619	12.22	59.67	71.89
18	1.162	20.2	0.756	5.9	1.747	7.956	9.703	13.05	59.46	72.51
19	1.172	21.3	0.746	2.8	1.859	7.927	9.786	13.90	59.23	73.13
20	1.182	22.1	0.737	− 0.4	1.970	7.883	9.853	14.73	58.90	73.63
21	1.192	23.0	0.729	− 3.9	2.085	7.843	9.928	15.58	58.61	74.19
22	1.202	24.4	0.716	− 7.8	2.208	7.829	10.037	16.50	58.50	75.00
23	1.212	25.5	0.707	−11.9	2.328	7.792	10.120	17.40	58.23	75.63
24	1.223	26.4	0.697	−16.2	2.451	7.761	10.212	18.32	58.00	76.32
25	1.233	27.4	0.689	−21.0	2.574	7.721	10.295	19.24	57.70	76.94
26	1.244	28.3	0.682	−25.8	2.699	7.680	10.379	20.17	57.39	77.56
27	1.254	29.3	0.673	−31.2	2.827	7.644	10.471	21.13	57.12	78.25
28	1.265	30.4	0.665	−37.8	2.958	7.605	10.563	22.10	56.84	78.94
29	1.276	31.4	0.658	−49.4	3.090	7.565	10.655	23.09	56.53	79.62
29.87	1.290	32.6	0.655	−67.0	3.16	7.59	10.75	23.65	56.80	80.45
30	1.295	33.0	0.653	−50.8	3.22	7.58	10.80	24.06	56.70	80.76
32	1.317	34.9	0.640	−19.5	3.49	7.49	10.98	26.10	56.04	82.14
34	1.340	36.8	0.630	+ 4.3	3.77	7.40	11.17	28.22	55.35	83.57

From *ASRE Data Book*, Design Volume, 1957–58 Edition, by permission of The American Society of Heating, Refrigerating, and Air-Conditioning Engineers.

brine at various concentrations are given in Table 11-4.

It is of interest to notice that the thermal properties of both calcium chloride and sodium chloride brines are somewhat less satisfactory than those of water. As the salt content of the brines is increased, the fluidity, specific heat value, and thermal conductance of the brines all decrease. Hence, the stronger the brine solution, the greater the quantity of brine that must be circulated in order to produce a given refrigerating effect.

Since the specific gravity of the brine increases as the salt concentration increases, the degree of salt concentration and the thermal properties of the brine can be determined by measuring the specific gravity of the brine with a hydrometer.

11-36. Antifreeze Solutions

Certain water soluble compounds, generally described as antifreeze agents, are often used to depress the freezing point of water. The more widely known antifreeze agents are ethylene glycol, propylene glycol, Methanol (methyl alcohol), and glycerin. All these compounds are soluble in water in all proportions. The freezing temperature of water in solution with various percentages of each of these compounds is given in Table 11-5.

Propylene glycol is probably the most extensively used antifreeze agent in refrigeration service. In common with ethylene glycol, propylene glycol has a number of desirable properties. Unlike brine, glycol solutions are noncorrosive. They are also nonelectrolytic and therefore may be employed in systems

TABLE 11-4 Properties of Pure Sodium Chloride Brine

Pure NaCl % by wt	Specific gravity 59 F 39 F	Baumé density 60 F	Specific heat 59 F Btu/lb deg F	Crystallization starts F	Weight per gallon			Weight per cubic foot		
					NaCl lb/gal	Water lb/gal	Brine lb/gal	NaCl lb/cu ft	Water lb/cu ft	Brine lb/cu ft
0	1.000	0.0	1.000	32.0	0.000	8.34	8.34	0.000	62.40	62.4
5	1.035	5.1	0.938	27.0	0.432	8.22	8.65	3.230	61.37	64.6
6	1.043	6.1	0.927	25.5	0.523	8.19	8.71	3.906	61.19	65.1
7	1.050	7.0	0.917	24.0	0.613	8.15	8.76	4.585	60.91	65.5
8	1.057	8.0	0.907	23.2	0.706	8.11	8.82	5.280	60.72	66.0
9	1.065	9.0	0.897	21.8	0.800	8.09	8.89	5.985	60.51	66.5
10	1.072	10.1	0.888	20.4	0.895	8.05	8.95	6.690	60.21	66.9
11	1.080	10.8	0.879	18.5	0.992	8.03	9.02	7.414	59.99	67.4
12	1.087	11.8	0.870	17.2	1.090	7.99	9.08	8.136	59.66	67.8
13	1.095	12.7	0.862	15.5	1.188	7.95	9.14	8.879	59.42	68.3
14	1.103	13.6	0.854	13.9	1.291	7.93	9.22	9.632	59.17	68.8
15	1.111	14.5	0.847	12.0	1.392	7.89	9.28	10.395	58.90	69.3
16	1.118	15.4	0.840	10.2	1.493	7.84	9.33	11.168	58.63	69.8
17	1.126	16.3	0.833	8.2	1.598	7.80	9.40	11.951	58.36	70.3
18	1.134	17.2	0.826	6.1	1.705	7.76	9.47	12.744	58.06	70.8
19	1.142	18.1	0.819	4.0	1.813	7.73	9.54	13.547	57.75	71.3
20	1.150	19.0	0.813	+ 1.8	1.920	7.68	9.60	14.360	57.44	71.8
21	1.158	19.9	0.807	− 0.8	2.031	7.64	9.67	15.183	57.12	72.3
22	1.166	20.8	0.802	− 3.0	2.143	7.60	9.74	16.016	56.78	72.8
23	1.175	21.7	0.796	− 6.0	2.256	7.55	9.81	16.854	56.45	73.3
24	1.183	22.5	0.791	+ 3.8	2.371	7.51	9.88	17.712	56.09	73.8
25	1.191	23.4	0.786	+16.1	2.488	7.46	9.95	18.575	55.72	74.3
25.2	1.200			+32.0						

From *ASRE Data Book*, Design Volume, 1957–58 Edition, by permission of The American Society of Heating, Refrigerating, and Air-Conditioning Engineers.

TABLE 11-5 Freezing Points of Aqueous Solutions

Alcohol % by Wt	Alcohol Deg F	Glycerine % by Wt	Glycerine Deg F	Ethylene Glycol % by Vol	Ethylene Glycol Deg F	Propylene Glycol % by Vol	Propylene Glycol Deg F
5	28.0	10	29.1	15	22.4	5	29.0
10	23.6	20	23.4	20	16.2	10	26.0
15	19.7	30	14.9	25	10.0	15	22.5
20	13.2	40	4.3	30	3.5	20	19.0
25	5.5	50	−9.4	35	−4.0	25	14.5
30	−2.5	60	−30.5	40	−12.5	30	9.0
35	−13.2	70	−38.0	45	−22.0	35	2.5
40	−21.0	80	−5.5	50	−32.5	40	−5.5
45	−27.5	90	+29.1			45	−15.0
50	−34.0	100	+62.6			50	−25.5
55	−40.5					55	−39.5
						59	−57.0
						Above 60% fails to crystallize at −99.4 F	

From *ASRE Data Book*, 1957–58 Edition, by permission of the American Society of Heating, Refrigerating, and Air-Conditioning Engineers.

containing dissimilar metals. Being extremely stable compounds, glycols will not evaporate under normal operating conditions. Because of the many advantages of glycol solutions, they are being used to replace brines in a number of installations, particularly in the brewing and dairy industries. The change-over from brine to glycol can be accomplished with practically no change in the plant facilities.

11-37. Brine Spray Units

Like chilled water, the chilled brine (or antifreeze solution) may be circulated directly around the refrigerated product or container, or it may be used to cool the air in a refrigerated space. When used to cool air, the chilled brine is circulated through a serpentine coil or through a brine spray unit. Two types of brine spray units which have been used extensively are shown in Figs. 11-46 and 11-47. In the former unit chilled brine from a brine chiller

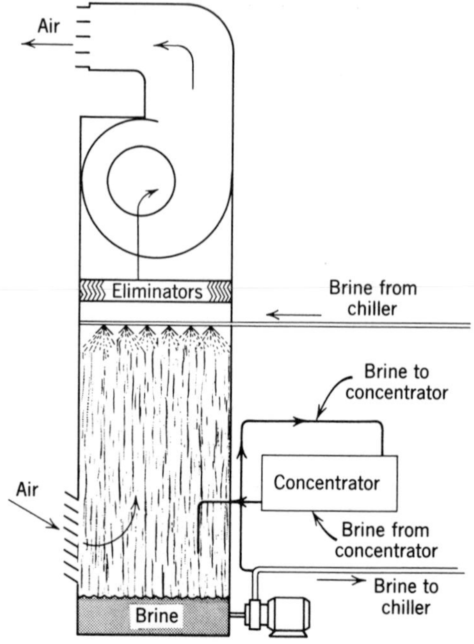

Fig. 11-46 Brine spray cooler.

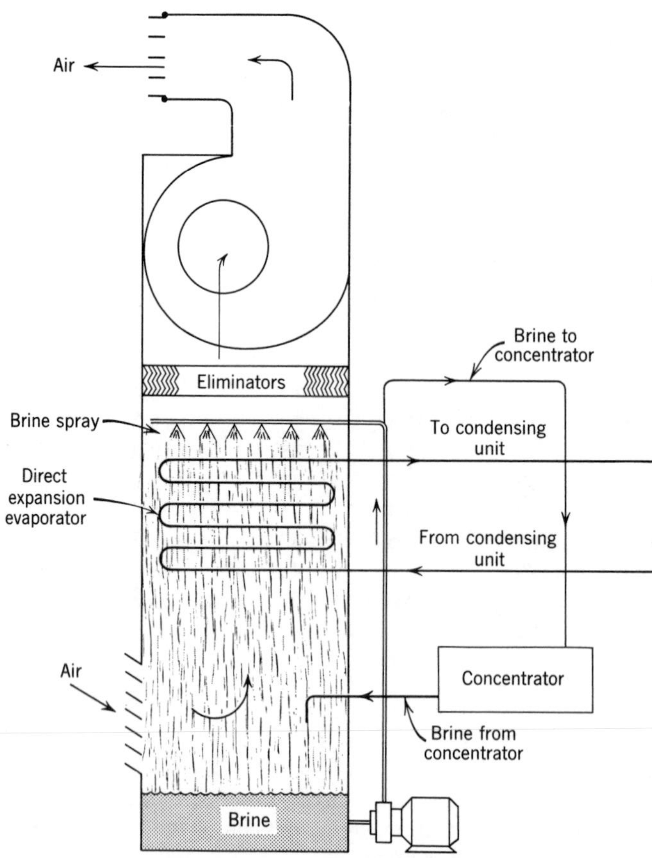

Fig. 11-47 Brine spray cooler.

located outside the refrigerated space is sprayed down from spray nozzles and collected in the basin of the unit, from where it is returned to the brine chiller. In the latter type the brine is chilled by a direct-expansion coil located within the brine spray unit itself.

Customary Problems

11-1 A flooded chiller is operating with a refrigerant temperature of 40°F. The chilled fluid enters the chiller at a temperature of 63°F and leaves at a temperature of 50°F. Calculate the mean effective temperature difference (LMTD) between the refrigerant and the chilled fluid.

11-2 Chilled water enters a cooling coil at 40°F and leaves at 50°F while the air being cooled enters at 80°F and leaves at 59°F. Compute the LMTD.

11-3 A flower storage cooler with inside dimensions of 9 ft by 11 ft by 10 ft high has a calculated cooling load of 6250 Btu/hr. Select a natural convection coil and baffle assembly (Fig. 11-25 and Table R-1) to meet the requirements of the cooler.

11-4 A frozen storage room having inside dimensions of 21 ft by 41 ft by 11 ft high has a calculated load of 37,500 Btu/hr. The space temperature is to be maintained at −20°F. Using an evaporator TD of 10°F, select an appropriate series of plate banks from Table R-3 for ceiling installation in the storage room.

11-5 For the conditions described in Problem 11-4, determine the number of linear feet of 2 in. iron pipe that would be required for ceiling installation in the storage room if the refrigerant temperature is maintained at −30°F.

11-6 Assuming flooded operation and natural convection air circulation, determine the number of linear feet of Prestfin pipe coil (Fig. 11-29 and Table R-7) required to meet the storage conditions described in Problem 11-4 using a coil TD of 10°F.

11-7 A holding cooler 18 ft by 30 ft by 14 ft high is used for the short term storage of fresh beef. Using a coil TD of 12°F, select four unit coolers to meet the requirements of the cooler if the calculated load is 27,500 Btu/hr.

11-8 With reference to Table R-9, a heavy-duty product cooler, Model #4256, with 4 fins per inch is operated direct expansion with R-502 at −50°F. What is the capacity of the product cooler at these conditions in Btu per hour per degree Fahrenheit?

11-9 Assuming all of the conditions of Problem 11-8, except that the coil face velocity is reduced from the nominal 600 fpm to 500 fpm, determine:

(a) the cubic feet per minute of air circulated over the coil,

(b) the capacity of the coil in Btu per hour per degree Fahrenheit.

11-10 The load on a heavy-duty product cooler is 132,000 Btu/hr. The cooler is to be operated flooded with ammonia with a 10°F TD and a refrigerant temperature of −50°F. If the desired coil face velocity is 400 fpm, select an appropriate product cooler from Table R-9.

11-11 Assume that the product cooler in Problem 11-10 is to be operated direct expansion with R-12 and select an appropriate cooling unit.

11-12 The load on a tank-type brine chiller is 6500 Btu/hr. The brine is to be maintained at a temperature of 0°F with a temperature of −9°F. Assuming little or no agitation of the brine, compute the linear feet of 1 in. iron pipe required for the evaporator.

11-13 Thirty-five gallons per minute of water are cooled from 58°F to 48°F with R-12 at 40°F. Select an appropriate chiller from Table R-10 and determine the water pressure drop through the chiller in pounds per square inch.

11-14 It is desired to cool 60 gpm of water from 56°F to 46°F with R-12 at a temperature of 38°F. Select an appropriate chiller from Table R-10 if the maximum permissible tube length is 14 ft.

12

PERFORMANCE OF RECIPROCATING COMPRESSORS

12-1. Refrigeration Compressors

Vapor compressors used in the refrigeration industry are of three principal types: (1) reciprocating, (2) rotary, and (3) centrifugal. Of the three, the reciprocating compressor is the one most frequently used.

Only the performance of reciprocating compressors is discussed in this chapter. The design of reciprocating compressors, along with the design and performance of rotory and centrifugal compressors, is discussed in Chapter 18. However, much of what is said in this chapter regarding the performance of reciprocating compressors, particularly as it relates to the overall performance of the refrigerating system, applies also to the performance of rotory and centrifugal compressors.

12-2. The Compression Cycle

Before attempting to analyze the performance of the compressor, it is necessary to become familiar with the series of processes which make up the compression cycle of a reciprocating compressor.

A compressor, with the piston shown at four points in its travel in the cylinder, is illustrated in Fig. 12-1. As the piston moves downward on the suction stroke, low-pressure vapor from the suction line is drawn into the cylinder through the suction valves. On the upstroke of the piston, the low-pressure vapor is first compressed and then discharged as a high-pressure vapor through the discharge valves into the head of the compressor.

To prevent the piston from striking the valve plate, all reciprocating compressors are designed with a small amount of clearance between the top of the piston and the valve plate when the piston is at the top of its stroke. The volume of this clearance space is called the clearance volume and is the volume of the cylinder when the piston is at top dead center.

Not all the high-pressure vapor will pass out through the discharge valves at the end of the compression stroke. A certain amount will remain in the cylinder in the clearance space between the piston and the valve plate. The vapor which remains in the clearance space at the end of each discharge stroke is called the clearance vapor.

Reference to Figs. 12-2 and 12-3 will help to clarify the operation of the compressor. Figure 12-2 is a time-pressure diagram in which cylinder pressure is plotted against crank position. Figure 12-3 is a theoretical pressure-volume

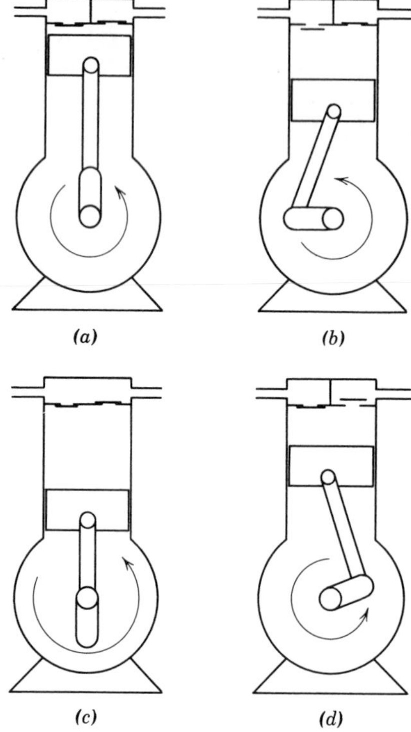

Fig. 12-1 (*a*) Piston at top dead center. (*b*) Suction valves open. (*c*) Piston at bottom dead center. (*d*) Discharge valves open.

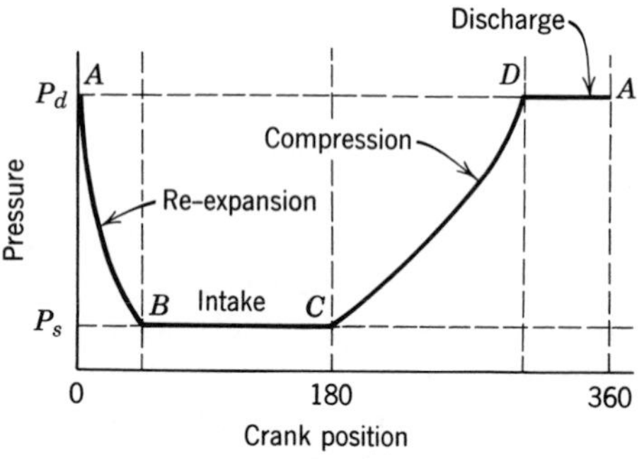

Fig. 12-2 Theoretical time-pressure diagram of compression cycle in which cylinder pressure is plotted against crank position.

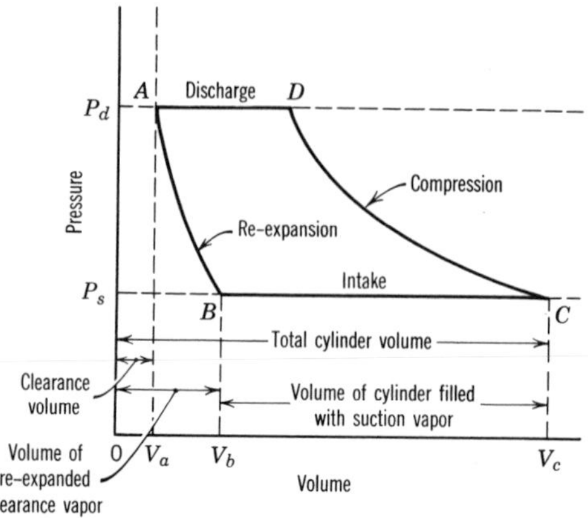

Fig. 12-3 Pressure-volume diagram of typical compression cycle.

diagram of a typical compression cycle. The lettered points on the Tp and pV diagrams correspond to the piston positions as shown in Fig. 12-1.

At point *A*, the piston is at the top of its stroke, which is known as top dead center. When the piston is at this position, both the suction and discharge valves are closed. The high pressure of the vapor trapped in the clearance space acts upward on the suction valves and holds them closed against the pressure of the suction vapor in the suction line. Because the pressure of the vapor in the head of the compressor is approximately the same as that of the vapor in the clearance volume, the discharge valves are held closed either by their own weight or by light spring loading.

As the piston moves downward on the suction stroke, the high-pressure vapor trapped in the clearance space is allowed to expand. The expansion takes place along line *A*–*B* so that the pressure in the cylinder decreases as the volume of the clearance vapor increases. When the piston reaches point *B*, the pressure of the re-expanded clearance vapor in the cylinder becomes slightly less than the pressure of the vapor in the suction line; whereupon the suction valves are forced open by the higher pressure in the suction line and vapor from the suction line flows into the cylinder. The flow of suction vapor into the cylinder begins when the suction valves open at point *B* and continues until the piston reaches the bottom of its stroke at point *C*. During the time that the piston is moving from *B* to *C*, the cylinder is filled with suction vapor and the pressure in the cylinder remains constant at the suction pressure. At point *C*, the suction valves close, usually by spring action, and the compression stroke begins.

The pressure of the vapor in the cylinder increases along line $C-D$ as the piston moves upward on the compression stroke. By the time the piston reaches point D, the pressure of the vapor in the cylinder has been increased until it is higher than the pressure of the vapor in the head of the compressor and the discharge valves are forced open; whereupon the high-pressure vapor passes from the cylinder into the hot gas line through the discharge valves. The flow of the vapor through the discharge valves continues as the piston moves from D to A while the pressure in the cylinder remains constant at the discharge pressure. When the piston returns to point A, the compression cycle is completed and the crankshaft of the compressor has rotated one complete revolution.

12-3. Piston Displacement

The piston displacement of a reciprocating compressor is the total cylinder volume swept through by the piston in any certain time interval and is usually expressed in cubic feet per minute (cfm) or cubic centimeters per second (cm^3/s). For any single-acting, reciprocating compressor, the piston displacement is computed as follows:

$$V_p = \frac{(0.7854D^2)(L)(N)(n)}{1728} \qquad (12\text{-}1)$$

where V_p = the piston displacement in cubic feet per minute
D = the diameter of the cylinder (bore) in inches
L = the length of stroke in inches
N = revolutions of the crankshaft per minute (rpm)
n = the number of cylinders

The volume of the cylinder which is swept through by the piston each stroke (each revolution of the crankshaft) is the difference between the volume of the cylinder when the piston is at the bottom of its stroke and the volume of the cylinder when the piston is at the top of its stroke. This part of the cylinder volume is found by multiplying the cross-sectional area of the bore by the length of stroke.

Cross-sectional area of the bore in square inches $= 0.7854D^2$

Volume of cylinder swept through by the piston each stroke in cubic inches $= (0.7854D^2)(L)$

Once the cylinder volume is known, the total cylinder volume swept through by the piston of a single cylinder compressor each minute in cubic inches can be determined by multiplying the cylinder volume by the rpm (N). When the compressor has more than one cylinder, the cylinder volume must also be multiplied by the number of cylinders (n). In either case, dividing the result by 1728 will give the piston displacement in cubic feet per minute.

Example 12-1 Calculate the piston displacement of a two-cylinder compressor rotating at 1450 rpm, if the diameter of the cylinder is 2.5 in. and the length of stroke is 2 in.

Solution Substituting in Equation 12-1,

$$V_p = \frac{(0.7854)(2.5^2)(2)(1450)(2)}{1728} = 16.48 \text{ cfm}$$

12-4. Theoretical Refrigerating Capacity

The refrigerating capacity of any compressor depends on the operating conditions of the system and, like system capacity, is determined by the mass of refrigerant circulated per unit time and by the refrigerating effect per unit mass circulated.

The mass flow rate produced by the compressor is equal to the mass of the suction vapor that the compressor takes in at the suction inlet per unit time. If it is assumed that the compressor is 100% efficient and that the cylinder of the compressor fills completely with suction vapor from the suction line with each downstroke of the piston, the volume of vapor drawn into the compressor cylinder and compressed per unit time will be exactly equal to the piston displacement of the compressor. The mass equivalent of this volume flow rate is the mass of refrigerant circulated by the compressor per unit time and can be calculated by multiplying the piston displacement of the compressor by the density of the suction vapor at the compressor inlet. Or, since specific volume is the reciprocal of density, the mass flow is also found by dividing the piston displacement by the specific volume of the suction vapor, viz:

$$m = (V_p)(\rho) \tag{12-2}$$

$$m = (V_p)/(v) \tag{12-3}$$

Once the mass flow rate is established, the theoretical refrigerating capacity of the compressor is determined by multiplying the mass flow rate by the refrigerating effect per unit mass.

Example 12-2 The compressor in Example 12-1 is operating on an R-12 system at an evaporator temperature of 20°F. If the suction vapor reaches the compressor inlet saturated and the liquid approaching the refrigerant control is at a temperature of 100°F, determine: (a) the mass flow rate of the refrigerant, and (b) the theoretical refrigerating capacity in Btu per minute, in tons and in watts.

Solution From Example 12-1, the piston displacement is 16.48 cfm, and from Table 16-3, the density of 20°F R-12 vapor is 0.8921 lb/ft^3. Applying Equation 12-2,

$$m = (16.48)(0.8921) = 14.70 \text{ lb/min}$$

From Table 16-3, the enthalpy of 20°F saturated vapor is 80.49 Btu/lb and the enthalpy of 100°F saturated liquid is 31.16 Btu/lb. Applying Equation 7-1, the refrigerating effect per unit mass, q_e

$$= 80.49 - 31.16 = 49.33 \text{ Btu/lb}$$

Applying Equation 6-1, the refrigerating capacity of the compressor (theoretical), Q_e

$$= (14.70 \text{ lb/min})(49.33 \text{ Btu/lb}) = 725.2 \text{ Btu/min}$$

or

$$\frac{725.2 \text{ Btu/min}}{200 \text{ Btu/min ton}} = 3.63 \text{ tons}$$

or

$$(3.63 \text{ tons})(3.517 \text{ kW/ton}) = 12.77 \text{ kW}$$

12-5. Actual Refrigerating Capacity

The actual refrigerating capacity of a compressor is always less than its theoretical capacity as calculated in the previous example. In the preceding example it is assumed: (1) that with each downstroke of the piston the cylinder of the compressor fills completely with suction vapor from the suction line and (2) that the density of the vapor filling the cylinder is the same as that in the suction line.

If these assumptions were correct, the actual refrigerating capacity would be equal to the theoretical capacity. Unfortunately, this is not the case. Because of the compressibility of the refrigerant vapor and the mechanical clearance between the piston and the valve plate of the compressor, the volume of suction vapor filling the cylinder during the suction stroke is always less than the cylinder volume swept through by the piston. Too, it will be shown later that the density of the vapor filling the cylinder is lower than the density of the vapor in the suction line. For these reasons, the actual volume of suction vapor at suction line conditions which is drawn into the cylinder of the compressor is always less than the piston displacement of the compressor and, therefore, the actual refrigerating capacity of the compressor is always less than its theoretical capacity.

12-6. Total Volumetric Efficiency

The actual volume of suction vapor removed from the suction line per unit time is the actual displacement of the compressor. The ratio of the actual displacement of the compressor to its piston displacement is known as the total or real volumetric efficiency of the compressor. Thus,

$$E_v = \frac{V_a}{V_p} \times 100 \tag{12-4}$$

where E_v = the total volumetric efficiency (%)
V_a = actual volume of suction vapor compressed per unit time
V_p = the piston displacement of the compressor

When the volumetric efficiency of the compressor is known, the actual displacement and refrigerating capacity can be found as follows:

$$V_a = V_p \times \frac{E_v}{100} \tag{12-5}$$

and

$$\begin{array}{c}\text{Actual}\\ \text{refrigerating}\\ \text{capacity}\end{array} = \begin{array}{c}\text{Theoretical}\\ \text{refrigerating}\\ \text{capacity}\end{array} \times \frac{E_v}{100} \quad (12\text{-}6)$$

Example 12-3 If the volumetric efficiency of the compressor in Example 12-2 is 76%, determine: (*a*) the actual displacement, (*b*) the actual refrigerating capacity.

Solution

(*a*) From Example 12-1, piston displacement $= 16.48\ \text{ft}^3/\text{min}$

Actual displacement $= 16.48 \times 0.76$

$= 12.52\ \text{ft}^3/\text{min}$

(*b*) From Example 12-2, theoretical refrigerating capacity $= 3.63$ tons

Actual refrigerating capacity $= 3.63 \times 0.76$

$= 2.76$ tons

The actual refrigerating capacity of the compressor may also be determined as illustrated in Example 12-2 if actual displacement is substituted for piston displacement.

12-7 Factors Influencing Total Volumetric Efficiency

The factors which tend to limit the volume of suction vapor compressed per working stroke, thereby determining the volumetric efficiency of the compressor, are the following:

1. Compressor clearance
2. Wiredrawing
3. Cylinder heating
4. Valve and piston leakage

12-8. The Effect of Clearance on Volumetric Efficiency

Because of compressor clearance and the compressibility of the refrigerant vapor, the volume of suction vapor flowing into the cylinder is less than the volume swept through by the piston. As previously shown, at the end of each compression stroke a certain amount of vapor remains in the cylinder in the clearance space after the discharge valves close. The vapor left in the clearance space has been compressed to the discharge pressure and, at the beginning of the suction stroke, this vapor must be reexpanded to the suction pressure before the suction valves can open and allow vapor from the suction line to flow into the cylinder. The piston will have completed a part of its suction stroke and the cylinder will already be partially filled with the reexpanded clearance vapor before the suction valves can open and admit suction vapor to the cylinder. Hence, suction vapor from the suction line will fill only that part of the cylinder volume which is not already filled with the reexpanded clearance vapor.

In Fig. 12-3, V_c is the total volume of the cylinder when the piston is at the bottom of its stroke. V_a, which represents the clearance volume, is the volume occupied by the clearance vapor at the end of the compression stroke. The difference between V_c and V_a then is the volume of the cylinder swept through by the piston each stroke. On the down stroke of the piston, the clearance vapor expands from V_a to V_b before the suction valves open. Therefore, the part of the cylinder volume which is filled with suction vapor during the balance of the suction stroke is the difference between V_b and V_c.

12-9. Theoretical Volumetric Efficiency

The volumetric efficiency of a compressor due to the clearance factor alone is known as the theoretical volumetric efficiency. It can be shown mathematically that the theoretical volumetric efficiency varies with the amount of clearance and with the suction and discharge pressures. The reason for this is easily explained.

12-10. Effect of Increasing the Clearance

If the clearance volume of the compressor is increased in respect to the piston displacement, the percentage of high-pressure vapor remaining in the cylinder at the end of the compression stroke will be increased. When reexpansion takes place during the suction stroke, a greater percentage of the total cylinder volume will be filled with the reexpanded clearance vapor and the volume of suction vapor taken in per stroke will be less than when the clearance volume is smaller. To obtain maximum volumetric efficiency, the clearance volume of a vapor compressor should be kept as small as possible.

It should be noted that this does not hold true for a reciprocating liquid pump. Since a liquid is not compressible, the liquid left in the clearance space at the end of the discharge stroke has the same specific volume as the liquid at the suction inlet. Therefore, there is no reexpansion of the liquid in the clearance during the suction stroke and the volume of liquid taken in each stroke is always equal to the volume swept by the piston, regardless of clearance.

12-11. Variation with Suction and Discharge Pressures

Increasing the discharge pressure or lowering the suction pressure will have the same effect on volumetric efficiency as increasing the clearance. If the discharge pressure is increased, the vapor in the clearance will be compressed to a higher pressure and a greater amount of reexpansion will be required to expand it to the suction pressure. Likewise, if the suction pressure is lowered, the clearance vapor must experience a greater reexpansion in expanding to the lower pressure before the suction valves will open.

On the other hand, for a constant discharge pressure, the amount of reexpansion that the clearance vapor experiences before the suction valves open diminishes as the suction pressure rises. It is evident, then, that the volumetric efficiency of the compressor increases as the suction pressure increases and decreases as the discharge pressure increases.

12-12. Compression Ratio

The ratio of the absolute suction pressure to the absolute discharge pressure is called the compression ratio. Thus,

$$R = \frac{\text{Absolute discharge pressure}}{\text{Absolute suction pressure}} \qquad (12\text{-}7)$$

where R = the compression ratio.

Example 12-4 Calculate the compression ratio of a R-12 compressor when the suction temperature is 20°F and the condensing temperature is 100°F.

Solution From Table 16-3,

Absolute pressure of R-12 saturated vapor at 20°F $= 35.75$ psi

Absolute pressure of R-12 saturated vapor at 100°F $= 131.6$ psi

Compression ratio $= \dfrac{131.6}{35.75}$

$= 3.68$

Examination of Equation 12-7 indicates that the compression ratio is increased by either increasing the discharge pressure or lowering the suction pressure, or both.

In the preceding section it was shown that increasing the discharge pressure or lowering the suction pressure decreases the volumetric efficiency. It follows, then, that when the suction and discharge pressures are varied in such a direction that the compression ratio is increased, the volumetric efficiency of the compressor decreases. For a compressor of any given clearance, the volumetric efficiency varies inversely with the compression ratio.

12-13. The Effects of Wiredrawing

Wiredrawing is defined as a "restriction of area for a flowing fluid, causing a loss in pressure by (internal and external) friction without the loss of heat or performance of work; throttling."*

In order to have a flow of vapor from the suction line through the suction valves into the compressor cylinder, there must be a pressure differential across the valves sufficient to overcome the spring tension of the valves and valve weight and inertia. This means that the suction vapor experiences a mild, throttling expansion or drop in pressure as it flows through the suction valves and passages of the compressor. Therefore, the pressure of the suction vapor filling the cylinder of the compressor is always less than the pressure of the vapor in the suction line. As a result of the expanded condition of the vapor filling the cylinder, the volume of suction vapor taken in from the

* *Asre Data Book*, 1957–58 (page 39–27).

suction line each stroke is less than if the vapor filling the cylinder was at the suction line pressure.

A similar pressure differential is required across the discharge valves in order to cause the discharge vapor to flow through the valves into the condenser. To provide the necessary pressure differential across the discharge valves, the vapor in the cylinder must be compressed to a pressure somewhat higher than the actual condensing pressure. The vapor left in the clearance space at the end of the discharge stroke will be at this higher pressure. To reexpand from this higher pressure during the suction stroke, the clearance vapor must suffer a greater amount of reexpansion than if it had been compressed only to the condensing pressure. As a result of the greater expansion of the clearance vapor, a larger portion of the cylinder volume is filled with the reexpanded clearance vapor during the down stroke of the piston and the amount of suction vapor drawn in from the suction line is reduced.

Unlike the other factors which determine volumetric efficiency, wiredrawing is not directly affected by the compression ratio. In general, wiredrawing is a function of the velocity of the refrigerant vapor flowing through the valves and passages of the compressor. As the velocity of the vapor through the valves is increased, the effect of wiredrawing increases.

The refrigerant velocity through the valves of a compressor depends upon the design of the valves, the refrigerant used, and the speed of the compressor.

Increasing the speed of the compressor increases the piston displacement. Hence, the velocity of the vapor through the valves and the effects of wiredrawing are increased as the rpm are increased.

12-14. The Effects of Cylinder Heating

Another factor which tends to reduce the volumetric efficiency of the compressor is the heating of the suction vapor in the compressor cylinder. The suction vapor entering the compressor cylinder is heated by heat conducted from the hot cylinder walls and by friction which results from the turbulence of the vapor in the cylinder and from the fact that the refrigerant vapor is not a perfect gas. The heating causes the vapor to expand after entering the cylinder so that a smaller mass of vapor will fill the cylinder and thereby still further reduce the volume of vapor taken in from the suction line.

Cylinder heating increases as the compression ratio increases. At high compression ratios, the work of compression is greater and the discharge temperature is higher. This causes a rise in the temperature of the cylinder walls and other compressor parts so that the transfer of heat to the suction vapor occurs at a higher rate.

12-15. The Effect of Piston and Valve Leakage

Any back leakage of gas through either the suction or discharge valves or around the piston will decrease the volume of vapor pumped by the compressor. Because of precision manufacturing processes, there is very little leakage of gas around the pistons of a compressor in good condition. However, since it is not possible to design valves that will close instantaneously, there is always a certain amount of back leakage of gas through the suction and discharge valves.

As the pressure in the cylinder is lowered at the beginning of the suction stroke, a small amount of high-pressure vapor in the head of the compressor will leak back into the cylinder before the discharge valves can close tightly. Similarly, at the start of the compression stroke, some of the vapor in the cylinder will flow back through the suction valves into the suction line before the suction valves can close.

To assure prompt closing of the valves, both the suction and discharge valves are usually constructed of lightweight materials and are slightly spring loaded. However, since the spring tension increases wiredrawing, the amount of spring loading is critical.

For any given compressor, the amount of back leakage through the valves is a function of the compression ratio and the speed of the compressor. The higher the compression ratio, the greater is the amount of valve leakage. The effect of compressor speed on valve leakage is discussed later.

12-16. Determining the Total Volumetric Efficiency

The combined effect of all of the foregoing factors on the volumetric efficiency of the compressor varies with the design of the compressor and with the refrigerant used. Furthermore, for any one compressor the volumetric efficiency is not a constant value; it changes with the operating conditions of the system. Therefore, the total volumetric efficiency of a compressor is difficult to predict mathematically and can be determined with accuracy only by actual testing of the compressor in a laboratory.

However, the results of such tests indicate that the volumetric efficiency of any one compressor is primarily a function of the compression ratio and, for any given compression ratio, remains practically constant, regardless of the operating range. It has been determined also that compressors having the same design characteristics will have approximately the same volumetric efficiencies, regardless of the size of the compressor.

The relationship between the compression ratio and the volumetric efficiency of a typical halocarbon compressor is illustrated by the curve in Fig. 12-4. The values given are for compressors ranging in size from 5 to 25 hp. Smaller compressors will have slightly lower efficiencies, whereas larger compressors will have slightly higher efficiencies. Efficiencies for ammonia compressors are usually 5% or 10% greater than those of halocarbon compressors.

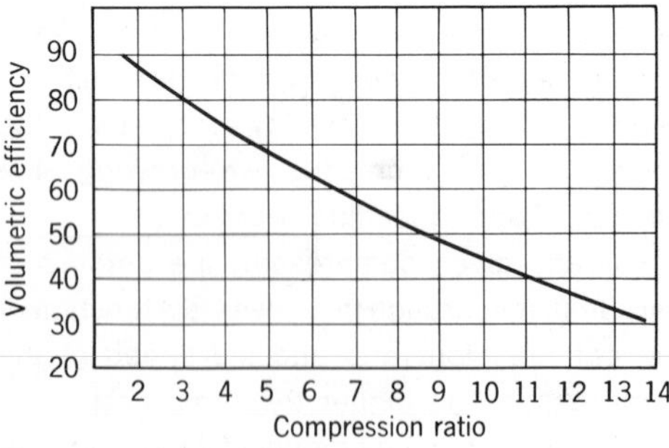

Fig. 12-4 Effect of compression ratio on volumetric efficiency of halocarbon compressor.

12-17. Variation in Compressor Capacity with Suction Temperature

Compressor performance and cycle efficiency vary considerably with the operating conditions of the system. The most important factor governing the capacity of the compressor is the vaporizing temperature of the liquid in the evaporator, that is, the suction temperature. The large variations in compressor capacity which accompany changes in the operating suction temperature are primarily a result of a difference in the density of the suction vapor entering the suction inlet of the compressor. The higher the vaporizing temperature of the liquid in the evaporator, the higher is the vaporizing pressure and the greater is the density of the suction vapor. Because of the difference in the density of the suction vapor, each unit volume of vapor compressed by the compressor will represent a greater mass of refrigerant when the suction temperature is high than when the suction temperature is low. This means that for any given piston displacement, the mass of refrigerant circulated by the compressor per unit of time increases as the suction temperature increases.

The effect of suction temperature on compressor capacity is best illustrated by an actual example.

Example 12-5 Assuming 100% efficiency, if the liquid reaches the refrigerant control at 100°F in each case, determine the mass of refrigerant circulated per minute and the theoretical refrigerating capacity of the compressor in Example 12-1 when operating at each of the following suction temperatures: (a) 10°F and (b) 40°F.

Solution

(a) From Table 16-3,
density of R-12
saturated vapor at 10°F $= 0.7402$ lb/ft^3

From Example 12-1,
piston displacement $= 16.48$ ft^3/min

Mass flow rate at 10°F
suction $= 16.48 \times 0.7402$
$= 12.20$ lb/min

From Table 16-3, enthalpy of R-12 saturated vapor at 10°F	= 79.36 Btu/lb
Enthalpy of R-12 liquid at 100°F	= 31.16 Btu/lb
Refrigerating effect	= 48.20 Btu/lb
Theoretical refrigerating capacity of compressor at 10°F suction, Btu/min	= 12.20 × 48.02 = 588.04 Btu/min
Theoretical refrigerating capacity in tons	$= \dfrac{588.04}{200}$ = 2.94 tons

(b)

From Table 16-3, density of R-12 saturated vapor at 40°F	= 1.263 lb/ft^3
From Example 12-1, piston displacement	= 16.48 ft^3/min
Mass flow rate at 40°F suction	= 16.48 × 1.263 = 20.81 lb/min
From Table 16-3, enthalpy of R-12 saturated vapor at 40°F	= 82.71 Btu/lb
Enthalpy of R-12 liquid at 100°F	= 31.16 Btu/lb
Refrigerating effect	= 51.55 Btu/lb
Theoretical refigerating capacity of compressor at 40°F suction, Btu/min	= 20.81 × 51.55 = 1072.76 Btu/min
Theoretical refrigerating capacity in tons	$= \dfrac{1072.76}{200}$ = 5.36 tons

In analyzing the results of Example 12-5, the following observations are of interest:

1. Although the piston displacement of the compressor is the same in each case, the mass flow rate increases from 12.20 lb/min to 20.81 lb/min when the operating suction temperature is raised from 10°F to 40°F. The increase in the mass flow rate results entirely from the greater density of the suction vapor entering the suction inlet of the compressor. In this instance, the percentage increase is

$$\frac{20.81 - 12.20}{12.20} \times 100 = 70.6\%$$

2. The theoretical refrigerating capacity of the compressor at the 10°F suction temperature is 2.94 tons, whereas at the 40°F suction temperature, the capacity increases to 5.36 tons. This represents an increase in refrigerating capacity of

$$\frac{5.36 - 2.94}{2.94} \times 100 = 82.3\%$$

Although the increased density of the suction vapor at the higher suction temperature accounts for the greater part of the increase in compressor capacity, it is not the only reason for it. As indicated, the increase in the mass flow rate is only 70.6%, whereas the total increase in compressor capacity is 82.3%. The additional 11.7% gain in capacity is brought about by an increase in the refrigerating effect per unit mass of refrigerant circulated. Although the actual gain in refrigerating effect per unit mass is only 6.95%, when this increase is applied to the higher mass flow rate occurring at the higher suction temperature, the net gain in capacity, which can be attributed to the greater refrigerating effect, increases to 11.7%.

The actual variation in compressor capacity with changes in suction temperature is more pronounced than that indicated by theoretical computations. The reason for this is that the compression ratio changes as the suction temperature changes. When the vaporizing temperature increases while the condensing temperature remains constant, the compression ratio decreases, and the volumetric efficiency of the compressor improves. Hence, at the higher suction temperature, in addition to pumping a greater mass of refrigerant per unit of volume, the volume of vapor pumped by the compressor is also larger because of the improved volumetric efficiency.

Example 12-6 Assuming that the saturated discharge temperature is 100°F, determine the actual refrigerating capacity of the compressor in Example 12-5 when operating at each of the suction temperatures in question.

Solution

(a) From Table 16-3, absolute pressure corresponding to 100°F saturation temperature = 131.6 psi

Absolute pressure corresponding to 10°F saturation temperature = 29.35 psi

Compression ratio $= \dfrac{131.6}{29.35}$ = 4.48

From Fig. 12-4, volumetric efficiency = 72%

From Example 12-5, theoretical refrigerating capacity at 10°F suction = 2.94 tons

Actual refrigerating capacity at 10°F suction = 2.94 × 0.72 = 2.12 tons

(b) From Table 16-3, absolute pressure corresponding to 100°F saturation temperature = 131.6 psi

Absolute pressure corresponding to 40°F saturation temperature = 51.68 psi

Compression ratio $= \dfrac{131.6}{51.68}$ = 2.55

From Fig. 12-4, volumetric efficiency = 84%

From Example 12-5, theoretical refrigerating capacity at 40°F suction = 5.36 tons

Actual refrigerating capacity at 40°F suction = 5.36 × 0.84 = 4.50 tons

Whereas the theoretical increase in compressor capacity is only 82.3%, the actual increase in refrigerating capacity

$$= \frac{4.50 - 2.12}{2.12} \times 100$$

$$= 112\%$$

12-18. Effect of Condensing Temperature on Compressor Capacity

In general, the refrigerating capacity of the compressor decreases as the condensing temperature increases. The effect that the condensing temperature has on compressor efficiency and capacity can be evaluated by comparing the results of the following example with those of Examples 12-5 and 12-6.

Example 12-7 Determine the theoretical and actual refrigerating capacities of the compressor in Example 12-1 for each of the two vaporizing temperatures given in Examples 12-5 and 12-6, if the condensing temperature in each case is 120°F rather than 100°F.

Solution

(a) For the 10°F vaporizing temperature.

From Example 12-1, piston displacement of compressor = 16.48 ft³/min

From Table 16-3, density of R-12 saturated vapor at 10°F = 0.7402 lb/ft³

Theoretical mass flow rate = 16.48 × 0.7402 = 12.20 lb/min

Refrigerating effect per pound at 10°F vaporizing and 120°F condensing = 43.20 Btu/lb

Theoretical refrigerating capacity of compressor = 12.20 × 43.20 = 527.04 Btu/min

Theoretical refrigerating capacity in tons $= \dfrac{527.04}{200}$ = 2.64 tons

From Table 16-3, absolute suction pressure = 29.35 psi

Absolute discharge pressure = 171.8 psi

Compression ratio $= \dfrac{171.8}{29.35}$ = 5.85

From Fig. 12-4,
volumetric efficiency = 64%
Actual refrigerating
capacity in tons = 2.64 × 0.64
= 1.69 tons

(b) For the 40°F suction temperature.
From Example 12-1,
piston displacement
of compressor = 16.48 ft^3/min

From Table 16-3, density
of R-12 saturated vapor
at 40°F = 1.263 lb/ft^3

Theoretical mass flow
rate = 16.48 × 1.263
= 20.81 lb/min

Refrigerating effect per
pound at 40°F
evaporating and
condensing 120°F = 46.55 Btu/lb

Theoretical refrigerating
capacity of
compressor = 20.81 × 46.55
= 968.7 Btu/min

Theoretical refrigerating
capacity in tons $= \frac{968.7}{200}$
= 4.84 tons

From Table 16-3,
absolute suction
pressure = 51.68 psi

Absolute discharge
pressure = 171.8 psi

Compression ratio $= \frac{171.8}{51.68}$
= 3.32

From Fig. 12-4,
volumetric efficiency = 78%
Actual refrigerating
capacity in tons = 4.84 × 0.78
= 3.78 tons

Examining first the 10°F cycle, notice that raising the condensing temperature from 100°F to 120°F reduces the theoretical refrigerating capacity of the compressor from 2.94 tons to 2.64 tons and the actual capacity from 2.12 tons to 1.69 tons.

Since a 100% efficient compressor is assumed to displace a theoretical volume of vapor equal to its piston displacement and since the density of the suction vapor entering the compressor for any one vaporizing temperature is always the same regardless of the condensing temperature, the theoretical weight of refrigerant displaced by the compressor is the same at all condensing temperatures, and therefore the theoretical refrigerating capacity of the compressor for any condensing temperature depends only upon the refrigerating effect per unit mass of refrigerant circulated. Hence, the difference in the theoretical refrigerating capacity of the compressor at the two condensing temperatures results entirely from the difference in the refrigerating effect per unit mass.

The reduction in actual compressor capacity may be attributed to several factors: (1) a reduction in the refrigerating effect per unit mass and (2) a reduction in the volumetric efficiency of the compressor.

Increasing the condensing temperature while the suction temperature remains constant increases the compression ratio and reduces the volumetric efficiency of the compressor so that the actual volume of vapor displaced by the compressor per unit of time decreases. Therefore, even though the density of the vapor entering the compressor remains the same at all condensing temperatures, the actual mass of refrigerant circulated by the compressor per unit of time decreases because of the reduction in compressor displacement.

Increasing the condensing temperature increases the isentropic discharge temperatre. In this instance, it is interesting to note (Fig. 7-7) that the increase in the isentropic discharge temperature is somewhat greater than that in the condensing temperature. Whereas the increase in condensing temperature is only 20°F (120° − 100°), the increase in the discharge temperature is 23.5°F (137.5° − 114°). This is accounted for by the greater work of compression at the higher compression ratio. Had the condensing temperature been increased in such a way that the compression ratio did not change (by increasing the suction temperature in proportion), the increase in the discharge temperature would have been approximately the same as that for the condensing temperature.

High discharge temperatures are undesirable and are to be avoided whenever possible. The higher the discharge temperature, the higher is the average temperature of the cylinder walls and the greater is the superheating of the suction vapor in the compressor cylinder. In addition to its adverse effect on compressor efficiency, high discharge temperatures tend to increase the rate of acid formation in the system, cause carbonization of the oil in the head of the compressor, and produce other effects detrimental to the equipment.

The loss of compressor efficiency and capacity resulting from an increase in the condensing temperature of the cycle is more serious when the suction temperature of the cycle is low than when the suction temperature is high.

When the cycle is operating at a 40°F vaporizing temperature, increasing the condensing temperature from 100°F to 120°F reduces the theoretical capacity of the compressor from 5.36 tons to 4.84 tons and the actual compressor capacity from 4.52 tons to 3.79 tons. The loss in theoretical capacity is

$$\frac{5.36 - 4.84}{5.36} \times 100 = 9.7\%$$

The loss in actual compressor capacity amounts to

$$\frac{4.50 - 3.78}{4.50} \times 100 = 16\%$$

For the 10°F cycle, the loss in theoretical compressor capacity is

$$\frac{2.94 - 2.64}{2.94} \times 100 = 10.2\%$$

and the loss in actual compressor capacity is

$$\frac{2.12 - 1.69}{2.12} \times 100 = 20\%$$

Note that the loss in theoretical capacity brought about by increasing the condensing temperature is approximately the same for both suction temperatures, whereas the loss in actual compressor capacity is much greater at the lower suction temperature. To a great extent, it is the loss in volumetric efficiency that causes the marked decrease in the actual capacity of the compressor at the higher condensing temperature. The change in volumetric efficiency for a given change in condensing temperature becomes greater as the suction temperature of the cycle decreases. This accounts for the greater effect that a change in condensing temperature has on compressor capacity when the suction temperature is low.

12-19. Compressor Power

The theoretical power required to drive the compressor may be found by multiplying the actual refrigerating capacity of the compressor in tons (or kilowatts) by the theoretical power required per unit capacity for the operating conditions in question.

Example 12-8 Find the theoretical power required to drive the compressor in Example 12-3.

Solution From Example 12-3, actual refrigerating capacity in tons = 2.76 tons

From Fig. 7-8, theoretical power required per ton = 0.965 hp

The theoretical power required to drive the compressor = 2.76 × 0.965 = 2.66 hp

Notice that it is the actual refrigerating capacity of the compressor, rather than the theoretical refrigerating capacity, which must be used in determining the theoretical power requirements of the compressor.

The theoretical power as calculated in the preceding example is only an indication of the power that would be required by a 100% efficient compressor operating on an ideal compression cycle and does not represent the total power that must be delivered to the shaft of the compressor. In practice, there are certain losses in power that accrue because of the mechanical friction in the compressor and because of the deviation of an actual compression cycle from the ideal compression cycle. Naturally, additional power must be supplied to the compressor to offset these losses. Therefore, the actual power required by a compressor will always be greater than the theoretical computations indicate.

12-20. Variation in Compressor Power with Suction Temperature

Although the power required per unit of refrigerating capacity diminishes as the suction temperature rises, the total power required by any one compressor may either increase or decrease, depending upon whether the total work done by the compressor increases or decreases.

The total amount of work done by the compressor per unit of time and, hence, the power required to drive the compressor, is the function of only two factors: (1) the work of compression per unit mass of vapor compressed and (2) the mass of vapor compressed per unit of time.

The amount of work which is done in compressing the vapor from the suction pressure to the discharge pressure varies with the compression ratio. The greater the compression ratio, the greater is the work of compression. Therefore, when the suction temperature is raised while the condensing temperature remains the same, the compression ratio becomes smaller and the work of compression per unit mass is reduced. However, at the same time, because of the greater density of the suction vapor, the mass of vapor compressed by the compressor per unit of time increases. Since the saving in work done resulting from the reduction in the work per unit mass is seldom sufficient to offset the increase in the work of the compressor because of the increase in the mass of vapor compressed, raising the suction temperature will usually increase the power requirements of the compressor.

Example 12-9 Compute the theoretical power required by the compressor in Example 12-7 at each of the suction temperatures listed.

Solution

(a) From Example 12-6, actual refrigerating capacity in tons at 10°F suction temperature = 2.12 tons

From Fig. 7-8, theoretical power per unit at 10° suction and 100°F condensing = 1.13 hp

Theoretical power of compressor at 10°F suction = 2.12 × 1.13
= 2.4 hp

(b) From Example 12-6, actual refrigerating capacity in tons at 40°F suction temperature = 4.50 tons

From Fig. 7-8, theoretical power per unit at 40°F suction and 100°F condensing = 0.683 hp

Theoretical power of compressor at 40°F suction = 4.50 × 0.683
= 3.07 hp

Although the power per ton decreases 39.5% as the suction temperature is raised from 10°F to 40°F, because of the increase in the refrigerating capacity of the compressor, the power required by the compressor increases from 2.51 hp to 3.15 hp. This represents an increased in the power required of

$$\frac{3.07 - 2.40}{2.40} \times 100 = 28\%$$

The increase in compressor power with the suction temperature is relatively small in comparison to the increase in compressor capacity. In this instance, for a 30°F rise in suction temperature, the capacity of the compressor increased 112%, whereas compressor power increased only 28%. The average increase in compressor capacity per degree of rise in suction temperature is 112%/30°F or 3.73%, whereas the increase in power amounts to only 1.0% per degree of rise.

The relationship between compressor capacity and compressor power requirements at

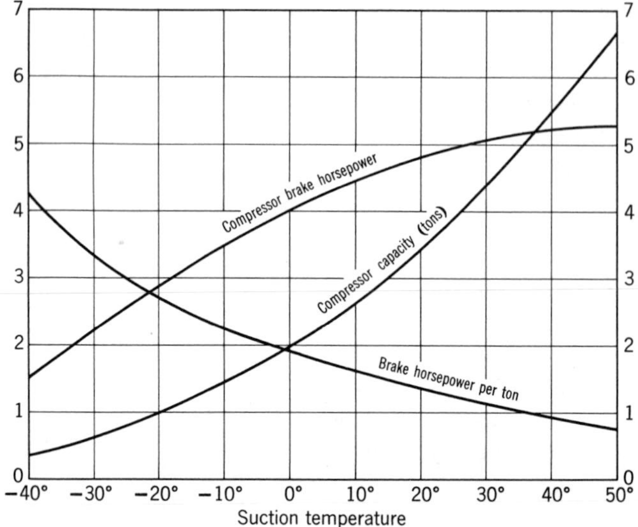

Fig. 12-5 Curves illustrate the effects of suction temperature on the capacity and horsepower of reciprocating compressors.

various suction temperatures is shown by the curves in Fig. 12-5. The curves are for a typical R-12 compressor operating at a constant condensing temperature of 100°F.

As shown by the curve in Fig. 12-5, the power required by a R-12 compressor increases as the suction temperature increases up to a certain point at which the power required by the compressor is at a maximum. On reaching this point, if the suction temperature is further increased, the power required by the compressor diminishes. This is not true, however, for compressors using ammonia as a refrigerant. For compressors using ammonia, the power does not reach a maximum value, but continues to increase indefinitely as the suction temperature increases.

The suction temperature at which the power required by a R-12 compressors reaches a maximum depends upon the condensing temperature and increases as the condensing temperature increases.

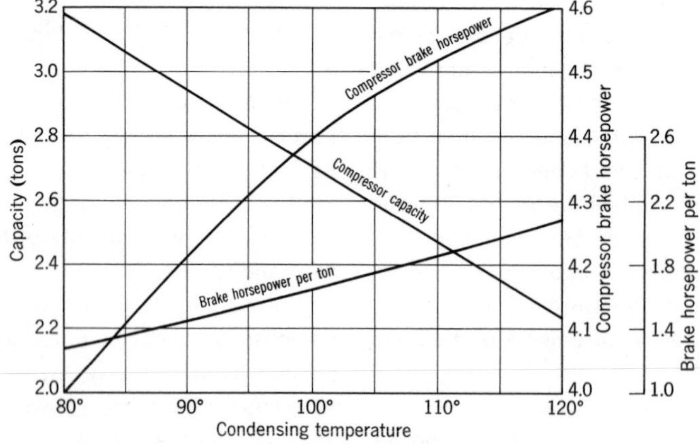

Fig. 12-6 Curves illustrate the effects of condensing temperature on capacity and horsepower of reciprocating compressors.

12-21. The Effect of Condensing Temperature on Compressor Power

The curves in Fig. 12-6 illustrate the relationship between the power required per ton of refrigerating capacity, the actual refrigerating capacity of the compressor, and the power required by the compressor at various condensing temperatures when the suction temperature is kept constant. Note that, although the theoretical power required per ton increases as the condensing temperature increases, the theoretical power required by any one compressor will not increase in the same proportion. This is true because the decrease in the refrigerating capacity of the compressor which is coincident with an increase in the condensing temperature will offset to some extent the increase in the power per ton.

For instance, according to Fig. 7-8, for a cycle operating at a 10°F vaporizing temperature, the theoretical power required per ton increases from 1.13 to 1.52 when the condensing temperature of the cycle is increased from 100°F to 120°F. At the same time, the actual refrigerating capacity of one particular compressor drops from 2.12 tons to 1.69 tons when the condensing temperature is raised from 100°F to 120°F. The theoretical power required by the compressor at the 100°F condensing temperature is

$$1.13 \times 2.12 = 2.4 \text{ hp}$$

For the 120°F condensing temperature, the theoretical power required by the compressor is

$$1.52 \times 1.69 = 2.57 \text{ hp}$$

12-22. Actual Power Requirements

The total power which must be supplied to the shaft of the compressor is called the shaft power and may be computed from the theoretical power by application of a factor called overall compressor efficiency. The overall efficiency is an expression of the relationship of the theoretical power to the shaft power in percent. Written as an equation, the relationship is

$$E_o = \frac{P_T}{P_S} \times 100 \qquad (12\text{-}8)$$

and

$$P_S = \frac{P_T}{E_o/100} \qquad (12\text{-}9)$$

where E_o = the overall efficiency in percent
P_T = the theoretical power
P_S = the shaft power

Example 12-10 Determine the shaft power required by the compressor in Example 12-8, if the overall efficiency of the compressor is 80%.

Solution From Example 12-8,

$$P_T = 2.66 \text{ hp}$$

Applying Equation 12-9, the

$$P_S = \frac{P_T}{E_o}$$

$$= \frac{2.66 \text{ hp}}{0.80}$$

$$= 3.33 \text{ hp}$$

The overall efficiency is sometimes broken down into two components: (1) the compression efficiency and (2) the mechanical efficiency. In such cases, the relationship is

$$E_o = E_c \times E_m \qquad (12\text{-}10)$$

where E_c = the compression efficiency in percent
E_m = the mechanical efficiency in percent

so that

$$P_S = \frac{P_T}{E_c \times E_m} \qquad (12\text{-}11)$$

The compression efficiency of a compressor is a measure of the losses resulting from the deviation of the actual compression cycle from the ideal compression cycle, whereas the mechanical efficiency of the compressor is a measure of the losses resulting from the mechanical friction in the compressor. The principal factors which bring about the deviation of an actual compression cycle from the ideal compression cycle are: (1) wiredrawing, (2) the exchange of heat between the vapor and the cylinder walls, and (3) fluid friction due to the turbulence of the vapor in the cylinder and to the fact that the refrigerant vapor is not an ideal gas. Notice that the factors which determine the compression efficiency of the compressor are the same as those which influence the volumetric efficiency. Consequently, for any one compressor, the volumetric and compression efficiencies are roughly the same and they vary with the compression ratio in about the same proportions. For this reason, the shaft power required per unit of refrigerating capacity can be approximated with reasonable accuracy by dividing the theoretical power per unit capacity by the volumetric efficiency of the compressor and then adding about 10% to offset the power loss due to the mechanical friction in the compressor. Written as an equation,

$$P_S = \frac{m(h_d - h_c) \times 1.1}{42.42 \times E_v} \qquad (12\text{-}12)$$

Since the relationship between the various factors which influence the compression efficiency are difficult to evaluate mathematically, the compression efficiency of a compressor can be determined accurately only by actual testing of the compressor.

12-23. Indicated Power

A device sometimes used in the past to determine the compression efficiency is the indicator diagram. An indicator diagram is a pressure-volume diagram of the compression cycle of a compressor which is produced during the actual testing of the compressor.

A theoretical indicator diagram for an ideal compression cycle is shown in Fig. 12-7. It has been illustrated previously that the area under a process diagram on a pressure-volume chart is a measure of the work of the process. In Fig. 12-7, notice that the area *dDCd* represents the work done by the piston in compressing the vapor during the isentropic process *CD*, and that the area *aADda* represents the work done by the piston in discharging the vapor from the cylinder during the constant pressure process *DA*, whereas the area *aABa* represents the work done back on the piston by the vapor during the isentropic reexpansion (of the clearance vapor) process *AB*. Since the work of process *AB* is work given back to the piston by the fluid, the net work input to the compression cycle is the sum of the work of processes *CD* and *DA*, less the work of process *AB*. Therefore, the net work of the compression cycle is represented by the area *BADCB*, the total area enclosed by the cycle diagram.

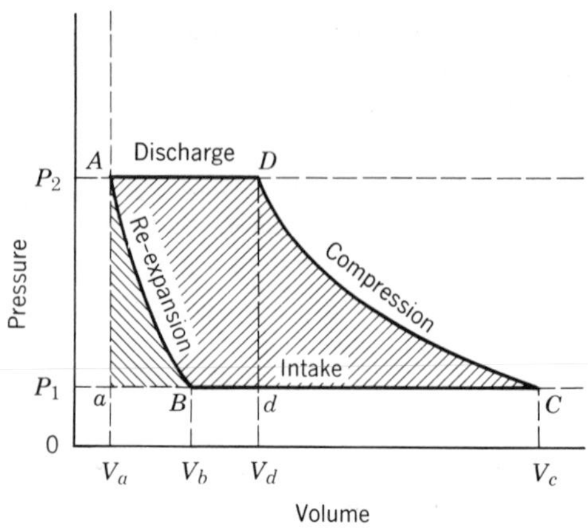

Fig. 12-7 Theoretical indicator diagram for an ideal compression cycle.

The work of the compression cycle as determined from the indicator diagram is called the indicated work and the power computed from the indicated work is called the indicated power.

Since the indicator diagram illustrated in Fig. 12-7 is of an ideal compression cycle, the indicated power computed from the indicated work would, of course, be equal to the theoretical power. However, in practice, since the indicator reproduces the true paths of the various processes which make up the compression cycle, the indicated work of the diagram is an accurate measure of the actual work of the compression cycle and, therefore, the indicated power is the actual power required to do the work of compressing the refrigerant vapor.

Care should be taken not to confuse indicated power with shaft power. Although the indicated power includes the power required to offset the losses resulting from the deviation of an actual compression cycle from the ideal cycle, it does not include the power required to overcome the losses resulting from the mechanical friction in the compressor. In other words, the indicated power takes into account the compression efficiency but not the mechan-

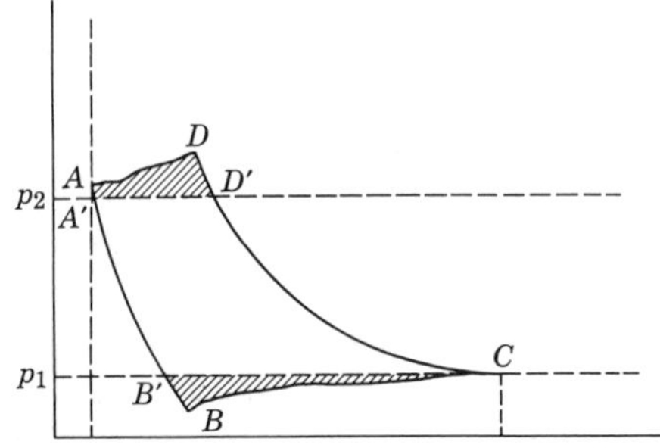

Fig. 12-8 Indicator diagram for an actual compression cycle.

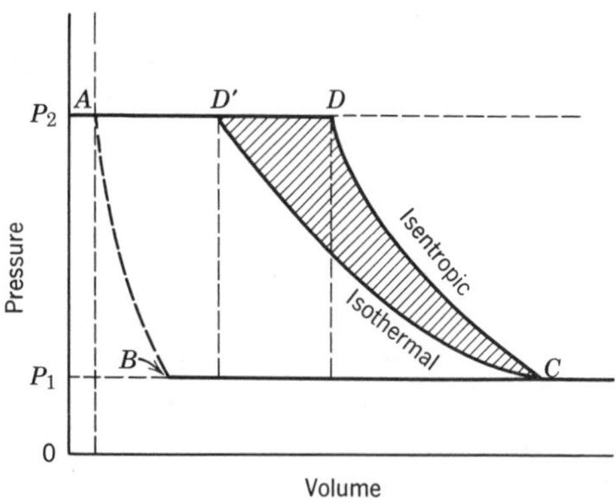

Fig. 12-9 Isentropic vs. isothermal compression.

ical efficiency. Hence, shaft power differs from indicated power in that the shaft power includes the power required to overcome the mechanical friction in the compressor, whereas the indicated power does not. The power necessary to overcome the mechanical friction in the compressor is sometimes referred to as the friction power (P_F), so that

$$P_S = P_I + P_F \tag{12-13}$$

The relationship of the indicated power to the theoretical power is

$$P_I = \frac{P_T}{E_c} \tag{12-14}$$

A simulated indicator diagram of an actual compression cycle is shown in Fig. 12-8. The area *ABCDA* enclosed by the cycle diagram is, of course, a measure of the work of the cycle. An ideal cycle, *A'B'C D'A'*, is drawn in for comparison. Pressures p_1 and p_2 represent the pressure of the vapor entering and leaving the compressor. The areas above line p_2 and below line p_1 represent the increased work of the cycle due to wiredrawing. Notice that at the end of the suction and discharge strokes (points *C* and *A*), the piston velocity diminishes to zero and the pressure of the vapor tends to return to p_1 and p_2, respectively.

12-24. Isothermal vs. Isentropic Compression

Reference to Fig. 12-9 will show that if the compression process in the compressor was isothermal rather than isentropic the net work of the compression cycle would be reduced even though the work of the compression process itself is greater for isothermal compression than for isentropic compression. The reduction in the work of the cycle which would be realized through isothermal compression is indicated by the crosshatched area in Fig. 12-9.

Isothermal compression is not practical for a refrigeration compressor since it would result in the discharge of saturated liquid from the compressor. Furthermore, if a cooling medium were available at a temperature low enough to cool the compressor sufficiently to produce isothermal compression, the cooling medium could be used directly as the refrigerant and there would be no need for the refrigeration cycle.

12-25. Water-Jacketing the Compressor Cylinder

Any heat which is given up by the compressor cylinder to some external cooling medium represents, in effect, heat given up by the vapor during the compression process. Cooling of the vapor during compression causes the path of the compression process to shift from the isentropic path toward an isothermal path. The greater the amount of cooling, the greater will be the shift toward the isothermal.

If the temperature of the air surrounding the compressor was exactly the same as the temperature of the compressor cylinder, there would be no transfer of heat from the cylinder to the air and any heat given up by the vapor to the cylinder would eventually be reabsorbed by the vapor and the compression process would be approximately adiabatic. However, since there is almost always some transfer of heat from the compressor to the surrounding air, compression is usually polytropic rather than isentropic. For an air-cooled compressor, the transfer of heat to the air will be small and, therefore, the value of the polytropic compression exponent, n, will very nearly approach the isentropic compression exponent, k. Hence, the assumption of isentropic compression for the ideal cycle is ordinarily not too much in error for an air-cooled compressor.

Water-jacketing of the compressor cylinder results in lowering the temperature of the cylinder walls, and cooling of the vapor during compression will be greater for the compressor having a water jacket. Too, cylinder heating is reduced and the vapor is discharged from the compressor at a lower temperature. All of this has the effect of reducing the work of the compression cycle. However, the gain is usually not sufficient to warrant the use of a water jacket on most compressors, particularly compressors designed for R-12. For the most part, water-jacketing of the compressor is limited to compressors designed for use with refrigerants which have unusually high discharge temperatures, such as ammonia. Even then, the purpose of the jacketing is not so much for increasing compressor efficiency as it is to reduce the rate of oil carbonization and the formation of acids, both of which increase rapidly as the discharge temperature increases.

12-26. Wet Compression

Wet compression occurs when small particles of unvaporized liquid are entrained in the suction vapor entering the compressor. Theoretical computations indicate that wet compression will bring about desirable gains in compression efficiency and reduce the work of compression. This would be true if the small particles of liquid vaporized during the actual compression of the vapor. However, in practice, this is not the case. Since heat transfer is a function of time and since compression of the vapor in a modern high-speed compressor takes place very rapidly, there is not sufficient time for the liquid to completely vaporize during the compression stroke. Hence, some of the liquid particles remain in the vapor in the clearance volume and vaporize during the early part of the suction stroke. This action reduces the volumetric efficiency of the compressor without benefit of the return of work to the piston by the expansion of the vaporizing particles.

A result similar to this is encountered when excessive cooling of the cylinder reduces the temperature of the vapor in the clearance below the saturation temperature corresponding to the discharge. Some of the clearance vapor will condense and the particles of liquid formed will vaporize during the early part of the suction stroke.

12-27. The Effect of Compressor Clearance on Horsepower

Theoretically, the clearance of the compressor has no effect on the horsepower, since the work done by the piston in compressing the clearance vapor is returned to the piston as the clearance vapor reexpands at the start of the suction stroke. However, since the refrigerant vapor is not an ideal gas, there is some loss of power in overcoming the internal friction of the fluid so that the power returned to the piston during the reexpansion of the clearance vapor will always be less than the power required to compress it. Hence, the clearance does have some, although probably slight, effect on the power requirements.

12-28. Compressor Speed

Since the speed of rotation is one of the factors determining piston displacement (Equation 12-1), the capacity of the compressor changes considerably when the speed of the compressor is changed. As the speed of the compressor is increased, the piston displacement is increased and the compressor displaces a greater volume of vapor per unit of time. Theoretically, based on the assumption that the volumetric efficiency of the compressor remains constant, the capacity of the compressor varies in direct proportion to the speed change. In practice, the volumetric efficiency may vary somewhat with a change in speed, but since the change is usually small, it is reasonable to assume that the change in capacity is proportional to the change in speed.

12-29. Mechanical Efficiency

The mechanical friction in the compressor varies with the speed of rotation, but for any one speed, the mechanical friction, and therefore the friction power, will remain practically the same at all operating conditions. Since the friction power remains the same, it follows that the mechanical efficiency of the compressor depends entirely upon the loading of the compressor. As the total shaft power of the compressor increases due to loading of the compressor, the friction power, being constant, will become a smaller and smaller percentage of the total power and the mechanical efficiency will increase. It is evident that the mechanical efficiency of the compressor will be greatest when the compressor is fully loaded. The mechanical efficiency of the compressor will vary with the design of the compressor and with compressor speed. An average compressor of good design operating fully loaded at a standard speed should have a mechanical efficiency somewhat above 90%.

12-30. The Effect of Suction Superheat on Compressor Performance

It has been shown that superheating of the suction vapor causes the vapor to reach the compressor in an expanded condition. Therefore, when the vapor reaches the compressor in a superheated condition, the mass of refrigerant circulated by the compressor per unit time is less than when the vapor reaches the compressor saturated. Whether or not the reduction in the mass flow rate of refrigerant reduces the refrigerating capacity of the compressor depends upon whether or not the superheating produces useful cooling. When the superheating produces useful cooling, the gain in refrigerating capacity resulting from the increase in the refrigerating effect is usually more than sufficient to offset the loss in refrigerating capacity resulting from the reduction in the mass of refrigerant circulated. On the other hand, when the superheating produces no useful cooling, there is no offsetting gain in capacity and the refrigerating capacity of the compressor is reduced accordingly. The additional heat energy so entering the system must be elevated to the discard temperature by the compressor and subsequently rejected to the condensing medium in the condenser. Not only does this place an additional and often unnecessary load on the compressor and condenser but also much of the beneficial suction vapor cooling of the compressor, which results from the superheating of the vapor in the compressor cylinder, is lost.

12-31. The Effect of Subcooling on Compressor Performance

When subcooling of the liquid refrigerant is accomplished in such a way that the heat given up by the liquid leaves the system, the specific volume of the suction vapor at the compressor inlet is unaffected by the subcooling and the mass of refrigerant circulated per unit time by the compressor is the same as when no subcooling takes place. Since the refrigerating effect per unit mass is increased by the subcooling, the capacity of the compressor is increased by an amount equal to the amount of subcooling. Notice that the increase in the refrigerating capacity of the compressor resulting from the subcooling is accomplished without increasing the power requirements of the compressor. Therefore, subcooling improves compressor efficiency, provided the heat given up during the subcooling leaves the system.

12-32. Compressor Rating and Selection

As previously stated, mathematical evaluation of all the factors which influence compressor performance is not practical. Hence, compressor capacity and power requirements are determined accurately only by actual testing of the compressor. Table R-11 is a typical compressor rating table supplied by the compressor manufacturer for use in compressor selection.

It has been shown in the foregoing sections that both the refrigerating capacity and the power requirements of a compressor vary with the condition of the refrigerant vapor entering and leaving the compressor. Notice in Table R-11 that compressor refrigerating capacities and power requirements are listed for various saturated suction and discharge temperatures. The saturated suction temperature is the saturation temperature corresponding to the pressure of the vapor at the suction inlet of the compressor, and the saturated discharge temperature is the saturation temperature corresponding to the pressure of the vapor at the discharge of the compressor.

Although compressor ratings are based on the saturated suction and discharge temperatures, test standards require a certain amount of suction superheat at the inlet of the compressor, and compressor ratings reflect this superheating. Notice that the ratings listed in Table R-11 are based on a suction line temperature of 65°F. When the temperature of the suction gas is less than 65°F, the tabulated ratings are corrected by multiplying by the appropriate correction factor.

The superheating is assumed to occur in the evaporator, in the suction line inside the refrigerated space, or in a liquid-suction heat exchanger, so that the superheating produces useful cooling. Any superheating that occurs outside the refrigerated space should be added to the compressor load.

The amount of superheating specified in the rating standards very nearly approaches the amount that would normally be expected in a well-designed application. Hence, the effect of the superheating requirement is to cause the compressor to be rated under conditions similar to those under which the compressor will be operating in the field. For this reason, except in unusual cases, no appreciable error will occur if the compressor ratings given in Table R-11 are used without correction of any kind. Furthermore, compressor capacity requirements are not usually critical within certain limits. There are several reasons for this. First of all, the methods of determining the required compressor capacity (cooling load calculations) are not in themselves exact. Too, it is seldom possible to select a compressor which has exactly the required capacity at the design conditions. Another reason that compressor capacity is not critical within reasonable limits is that the operating conditions of the system do not remain constant at all times, but vary from time to time with the loading of the system, the temperature of the condensing medium, etc. General practice is to select a compressor having a capacity equal to or somewhat in excess of the required capacity at the design operating conditions.

It was shown in Chapter 8 that subcooling of the liquid increases the refrigerating effect per pound and thereby increases compressor capacity. With regard to subcooling, the ratings given in Table R-11 are based on saturated liquid approaching the refrigerant control, that is, no subcooling. Where the liquid is subcooled by external means, the capacity of the compressor may be increased approximately 1% for each 2°F of subcooling. Here again, for reasons outlined in the preceding paragraph, the effect of subcooling is usually neglected in selecting the compressor.

To select a compressor for a given application, the following data are needed:

1. The required refrigerating capacity (tons)
2. The design saturated suction temperature
3. The design saturated discharge temperature

Naturally, the required refrigerating capacity is the average hourly load as determined by the cooling load calculations. However, if an evaporator selection is made prior to the compressor selection, the compressor should be selected to match the evaporator capacity rather than the calculated load. The reasons for this are discussed in Chapter 13.

The design saturated suction temperature depends upon the design conditions of the application. Specifically, it depends upon the evaporator temperature (the saturation temperature of the refrigerant at the evaporator outlet) and upon the pressure loss in the suction line. For instance, assume an evaporator temperature of 28°F and a suction line pressure loss of approximately 3 psi. From Table 16-3, the saturation pressure of Refrigerant-12 corresponding to a temperature of 28°F is 41.59 psia. Allowing for the 3 psi pressure loss in the suction line, the pressure of the vapor at the suction inlet of the compressor is 38.59 psia (41.59 − 3). From Table 16-3, the saturation temperature corresponding to a pressure of

38.59 psia, and, therefore, the saturated suction temperature, is approximately 24°F.

The design saturated discharge temperature depends primarily on the size of the condenser selected and upon the quantity and temperature of the available condensing medium. Methods of condenser selection are discussed in Chapter 14.

12-33. Condensing Unit Rating and Selection

Since condensing unit capacity depends upon the capacity of the compressor, methods of rating and selecting condensing units are practically the same as those for rating and selecting compressors. The only difference is that, whereas compressor capacities are based on the saturated suction and discharge temperatures, condensing unit capacities are based on the saturated suction temperature and on the quantity and temperature of the condensing medium. Since the size of the condenser is fixed at the time of manufacture, for any given condenser loading, the only variables determining the saturation temperature at the discharge of the compressor (and therefore the capacity of the compressor at any given suction temperature) is the quantity and temperature of the condensing medium. For air-cooled condensing units, when the quantity of the air passing over the condenser is fixed by the fan selection at the time of manufacture, the only variable determining the capacity of the condensing unit, other than the suction temperature, is the ambient air temperature (temperature of the air entering the condenser). Hence, ratings for air-cooled condensing units are based on the saturated suction temperature and the design ambient air temperature.

Ratings for water-cooled condensing units are based on the saturated suction temperature and on the entering and leaving water temperature.* Typical capacity ratings for air-cooled and water-cooled condensing units are given in Tables R-12 and R-13, respectively.

* For any given condenser loading and entering water temperature, the leaving water temperature depends only on the quantity of water (gallons per minute) flowing through the condenser.

Example 12-11 A certain refrigeration application has a calculated cooling load of 71,500 Btu/hr. If the design saturated suction and discharge temperatures are −15°F and 100°F, respectively, and the suction vapor is superheated to 45°, select an appropriate compressor from Table R-11. Disregard small amounts of liquid subcooling.

Solution The required cooling capacity is (71,500/12,000) 5.96 tons. From Table R-11, at a 100°F condensing temperature, compressor Model # EO36 has a rated capacity of 5.3 tons at a −20°F suction temperature and a capacity of 7.5 tons at a −10°F suction temperature. By interpolation, the rated capacity at −15°F suction temperature is 6.4 tons. From Table R-11, the correction factor for a superheated suction temperature of 45°F (20° difference) is 0.98, so that the corrected capacity is (6.4 × 0.98) 6.27 tons. This compressor will meet the stated requirements.

Example 12-12 From Table R-11, select a compressor to handle a cooling load of 18 tons if the design saturated suction and discharge temperatures are 20°F and 120°F, respectively. The suction vapor is superheated to 65°F and there is no appreciable subcooling.

Solution For a condensing temperature of 120°F, compressor Model # EO50 has a rated capacity of 22.2 tons at a saturated suction temperature of 20°F. The required capacity of 18 tons is 81% (18/22.2) of the rated capacity. The data in Table R-11 indicate that the compressor will have 85% of its rated capacity of 22.2 tons at 1740 rpm when the speed is reduced to 1450 rpm.

Example 12-13 A certain refrigeration application has a calculated cooling load of 8750 Btu/hr and the ambient temperature is 90°F. If the design saturated suction temperature is 20°F, select an air-cooled condensing unit which will satisfy the requirements of the application.

Solution From Table R-12, select a 1 hp condensing unit having a capacity of 9340 Btu/hr at the prescribed conditions.

Customary Problems

12-1 Compute the piston displacement of an eight-cylinder compressor having a 3-in. bore and a 3-in. stroke and rotating at 3500 rpm.

12-2 Assuming 100% efficiency, what is the mass flow rate of refrigerant through the compressor if R-12 reaches the suction inlet saturated at 0°F?

12-3 Determine the refrigerant mass flow rate for the compressor in Problem 2 if the refrigerant is R-22.

12-4 Calculate the refrigerant mass flow rate for the compressor in Problem 2 if the refrigerant is R-502.

12-5 Using the data in Fig. 4-9, determine the refrigerant mass flow rate for the compressor in Problem 2 if the R-12 vapor has a saturation temperature of 20.04°F and is superheated to 50°F before reaching the suction inlet of the compressor.

12-6 Assume that the compressor in Problem 12-1 is on an R-12 system having an evaporator temperature of −20°F and a condensing temperature of 110°F. Determine:

(a) the volumetric efficiency from Figure 12-4,

(b) the actual compressor displacement in cubic feet per minute,

(c) the refrigerating capacity of the compressor in tons.

12-7 Rework Problem 12-6 using R-502 as the refrigerant.

12-8 Assuming the compression efficiency to be approximately equal to the volumetric efficiency and adding 10% for mechanical losses, compute the shaft power required to drive the compressor in Problem 12-6.

12-9 From Table R-11, select an R-12 compressor to handle a load of 10 tons if the design saturated suction and discharge temperatures are 0°F and 100°F, respectively. The suction vapor is superheated to 35°F, and subcooling of the liquid is assumed to be negligible.

12-10 A refrigeration system has a calculated load of 1 ton. If the saturated temperature is 0°F and the ambient air temperature is 100°F, select an air-cooled condensing unit from Table R-12 that will satisfy the requirements of the system.

12-11 A refrigerating system with a calculated load of 9500 Btu/hr is designed to operate with a saturated suction temtemperature of −20°F. If the ambient air temperature is 90°F, select an appropriate air-cooled condensing unit from Table R-12.

12-12 Select a water-cooled condensing unit from Table R-13 for the system in Problem 12-10 if the entering water temperature is 75°F and compute the required condenser water quantity in gallons per minute.

12-13 From Table R-13, select a water-cooled condensing unit for the system in Problem 12-11 if the entering water temperature is 85°F.

13

SYSTEM EQUILIBRIUM AND CYCLING CONTROLS

13-1. System Balance

In the design of a refrigerating system, one of the most important considerations is that of establishing the proper relationship or "balance" between the vaporizing and condensing sections of the system. It is important to recognize that whenever an evaporator and a condensing unit are connected together in a common system, a condition of equilibrium or "balance" is automatically established between the two such that the rate of vaporization is always equal to the rate of condensation. That is, the rate at which the vapor is removed from the evaporator and condensed by the condensing unit is always equal to the rate at which the vapor is produced in the evaporator by the boiling action of the liquid refrigerant. Since all the components in a refrigerating system are connected together in series, the refrigerant flow rate through all components is the same. It follows, therefore, that the capacity of all the components must of necessity be the same.

When the system components are selected to have equal capacities at the system design conditions, the point of system equilibrium or balance will occur at the system design conditions. On the other hand, when the components selected do not have equal capacities at the system design conditions, system equilibrium will be established at operating conditions other than the system design conditions and the system will not perform satisfactorily. This concept is best illustrated through the use of a series of examples.

Example 13-1 A walk-in cooler, having a calculated cooling load of 11,000 Btu/hr, is to be maintained at 35°F. The desired evaporator TD is 12°F and the ambient temperature is 90°F. Allowing 3°F (equivalent to approximately 2 psi) for the pressure drop in the suction line (see Section 12-32), select an air-cooled condensing unit and a unit cooler to meet the system design conditions.

Solution Since the design space temperature is 35°F and the design evaporator TD is 12°F, the design evaporator temperature is 23°F (35°F − 12°F). Allowing for a 3°F loss in the suction line resulting from pressure drop, the saturation temperature at the compressor suction is 20°F (23°F − 3°F).

From Table R-12, select $1\frac{1}{2}$ hp condensing unit, which has a capacity of 12,630 Btu/hr at a 20°F saturated suction and 90°F ambient air temperature. Although the condensing unit capacity is somewhat in excess of the calculated load of 11,000 Btu/hr, it is sufficiently close to make the condensing unit acceptable. However, to assure proper system balance, the unit cooler selection must now be based on the condensing unit capacity of 12,630 Btu/hr rather than on the calculated load of 11,000 Btu/hr.* Hence, a unit cooler having a capacity of approximately 12,630 Btu/hr at a 12°F TD is required.

From Table R-8, unit cooler Model #105 has a capacity of 10,500 Btu/hr at a 10°F TD. Using the procedure outlined in Section 11-22, it can be determined that this unit cooler will have a capacity of 12,600 Btu/hr when operating at a 12°F TD. Since this is very close to the condensing unit capacity, the unit cooler is ideally suited for the application.

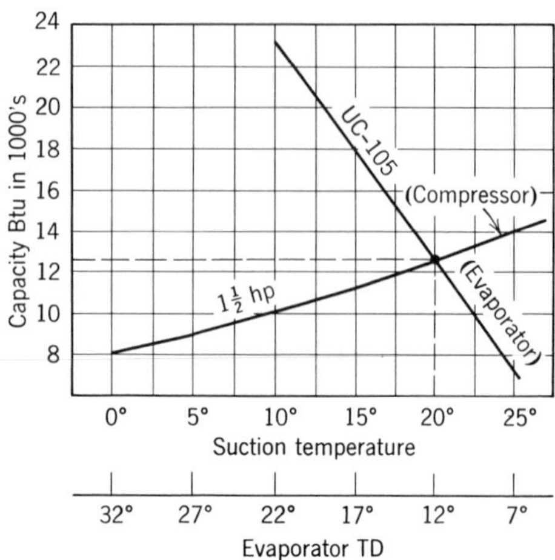

Fig. 13-1 Graphic analysis of system balance.

13-2. Graphical Analysis of System Equilibrium

For any particular evaporator and condensing unit connected together in a common system, the relationship established between the two, that is, the point of system balance, can be evaluated graphically by plotting evaporator capacity against condensing unit capacity on a common graph. Using data taken from the manufacturers' rating tables, condensing unit capacity is plotted against suction temperature, whereas evaporator capacity is plotted against evaporator TD. A graphical analysis of the system described in Example 13-1 is shown in Fig. 13-1.

In order to understand the graphical analysis of system equilibrium in Fig. 13-1, it is important to recognize that, for any given space temperature, there is a fixed relationship between the evaporator TD and the compressor suction temperature. That is, for any given space temperature, once the evaporator TD is selected, there is only one possible suction temperature which will satisfy the design conditions of the system.

In Example 13-1, notice that with a design space temperature of 35°F and assuming a 3°F loss in the suction line, the only possible suction temperature that can coexist with the design evaporator TD of 12°F is 20°F. In this instance, the evaporator TD will be 12°F when, and only when, the suction temperature is 20°F. Any suction temperature other than 20°F will result in an evaporator TD either greater or smaller than 12°F. For example, assume a suction temperature of 25°F. Adding 3°F to allow for the suction line loss, the evaporator temperature is found to be 28°F (25°F + 3°F). Then subtracting the evaporator temperature from the space temperature, it is determined that the evaporator TD will be 7°F (35°F − 28°F) when the suction temperature is 25°F. Using the same procedure, it can be shown that if the suction temperature is reduced to 15°F, the evaporator TD will increase to 17°F, and when the suction temperature is 10°F, the evaporator TD will be 22°F, and so on.

Apparently, then, raising or lowering the suction temperature always brings about a corresponding adjustment in the evaporator TD. Provided that the space temperature is kept constant, raising the suction temperature reduces the evaporator TD, whereas lowering the suction temperature increases the evaporator TD.

With regard to Fig. 13-1, the following procedure is used in making a graphical analysis of the system equilibrium conditions:

1. On graph paper, lay out suitable scales for capacity (Btu/hr), suction temperature (°F),

* Either the evaporator or the condensing unit may be selected first. However, once either one has been selected, the other must be selected for approximately the same capacity.

and evaporator TD (°F). The horizontal lines are used to represent capacity, whereas the vertical lines are given dual values, representing both suction temperature and evaporator TD. The latter is meaningful, however, only when the suction temperature and evaporator TD scales are so correlated that the two conditions which identify any one vertical line are conditions which, at the design space temperature, actually represent conditions that will occur simultaneously in the system. The procedure for correlating the suction temperature and evaporator TD scales was discussed in the preceding paragraphs.

2. Using manufacturer's catalog data, plot the capacity curve for the condensing unit. Since condensing unit capacity is not exactly proportional to suction temperature, the condensing unit capacity curve will ordinarily have a slight curvature. Hence, for accuracy, a capacity point is plotted for each of the suction temperatures listed in the table and these points are connected with the "best-fitting" curve.
3. From the evaporator manufacturer's catalog data, plot the evaporator capacity curve. Since evaporator capacity is assumed to be proportional to the evaporator TD, the evaporator capacity curve is a straight line, the position and direction of which is adequately established by plotting the evaporator capacity at any two selected TDs. The evaporator capacity at any other TD will fall somewhere along a straight line drawn through these two points. In Fig. 13-1, evaporator TDs of 7°F and 12°F are used in plotting the two capacity points required to establish the evaporator capacity curve.

Notice in Fig. 13-1, that as the suction temperature increases, the evaporator TD decreases. This means, in effect, that as the suction temperature increases the capacity of the evaporator decreases while the capacity of the condensing unit (compressor) increases. Likewise, as the suction temperature decreases, the capacity of the evaporator increases while the capacity of the condensing unit decreases. The intersection of the two capacity curves indicates the point of system equilibrium. In this instance, because the evaporator and condensing unit have been selected to have equal capacities at the system design conditions, the point of system equilibrium occurs at the system design conditions (12°F TD and 20°F suction temperature). Although the total system capacity is somewhat greater than the calculated load, the difference is not sufficiently great to be of any particular consequence, and means only that the system will operate fewer hours out of each 24 than was originally anticipated.* The relationship between system capacity and the calculated load is discussed more fully later in the chapter.

It has already been pointed out that where the evaporator and condensing unit selected do not have equal capacities at the system design conditions, the point of system equilibrium will occur at conditions other than the design conditions. For instance, assume that unit cooler Model #UC-180, rather than Model # UC-105, is selected in the foregoing example. At the design evaporator TD of 12°F, this unit cooler has capacity of 21,600 Btu/hr, whereas the condensing unit selected has a capacity of only 12,630 Btu/hr at the design suction temperature of 20°F. Consequently, at the design conditions, evaporator capacity will be greater than condensing unit capacity, that is, vapor will be produced in the evaporator at a greater rate than it is removed from the evaporator and condensed by the condensing unit. Therefore, the system will not be in equilibrium at these conditions. Rather, the excess vapor will accumulate in the evaporator and cause an increase in the evaporator temperature and pressure. Since raising the evaporator temperature increases the suction temperature and, at the same time, reduces the evaporator

* For simplicity, the heat given off by the evaporator fan motor has been neglected. If this heat is added to the cooling load, the total system capacity would be almost exactly equal to the calculated load. It is not often that equipment can be found which so nearly meets the requirements of an application as in this instance.

TD, the condensing unit capacity will increase and the evaporator capacity will decrease. System equilibrium will be established when the evaporator temperature rises to some point where the suction temperature and evaporator TD are such that the condensing unit capacity and the evaporator capacity are equal. In this instance, system equilibrium, as determined graphically (point *A* in Fig. 13-2), is established at a suction temperature of approximately 23°F. The evaporator TD is approximately 9°F, which is 3°F less than the design evaporator TD of 12°F and which will result in a space humidity somewhat higher than the design condition. The total system capacity is approximately 13,500 Btu/hr, which is about 23% greater than the calculated hourly load of 11,000 Btu/hr. This means that the system running time will be considerably shorter than originally calculated. For instance, if the original load calculation is based on a 16-hr running time, the system will operate only about 13 hr out of each 24.

The question immediately arises as to whether or not this system will perform satisfactorily. Although this would depend somewhat on the particular application, the answer is that it probably would not in the majority of cases. There are several reasons for this. First, the evaporator TD of 9°F is considerably less than the design TD of 12°F

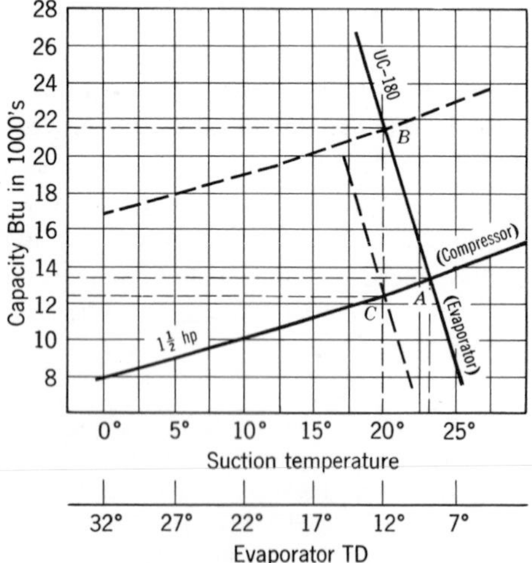

Fig. 13-2

and would probably result in a space humidity too high for the application. Ordinarily, the humidity in the refrigerated space must be maintained within certain fixed limits. Assuming that the design TD is selected to produce the median condition within these limits, a one degree deviation from the design TD in either direction is usually the maximum which can be allowed if the space humidity is to be maintained within the limits specified for the application.

Another consideration is the fact that the system capacity is some 23% greater than the calculated load so that the system operating time will be relatively short. Although the system capacity exceeds the calculated load by a larger margin than good practice prescribes, this in itself would not ordinarily cause any serious problem in a majority of applications. However, since the shorter running time will also tend to aggravate the already existing problem of high humidity, especially in the wintertime when the two conditions are taken together, it seems unlikely that the system would produce satisfactory results in any application where the space humidity is an important factor.

In the event that the equipment in question represents the best available selection, the question arises as to what can be done to bring the system into balance at conditions more in keeping with the design conditions. In this instance, since the problem is one of excessive evaporator capacity with relation to the condensing unit capacity at the design conditions, logical corrective measures prescribe either an increase in the condensing unit capacity or a reduction in the evaporator capacity in order to reestablish the point of system equilibrium at conditions nearer to the design conditions.

Which of these two measures will produce the most satisfactory results depends upon the relationship between the overall system capacity and the calculated load. Whereas increasing either the condensing unit capacity or the evaporator capacity will always bring about an increase in the overall system capacity, reducing either the condensing unit capacity or the evaporator capacity will always bring about a reduction in the overall system capacity. Referring to Fig. 13-2 for the system under consideration, if the condensing unit capacity is increased to the evaporator capacity at the design conditions, system equilibrium will shift from point *A* to point *B*. On the other hand, if the evaporator capacity is reduced to the condensing unit capacity at the design conditions, the point of system equilibrium will shift from *A* to *C*. Notice that, although the system is balanced at the design conditions at either points *B* or *C*, the overall system capacity at point *B* is considerably above the calculated load, whereas at point *C* the overall system capacity very nearly approaches the calculated load. Hence, in this instance, it is evident that increasing the condensing unit capacity as a means of bringing the system into balance at the design conditions cannot be recommended, since it would also increase the overall system capacity and therefore tend to aggravate the already existing problem of excessive system capacity with relation to the calculated load. On the other hand, in addition to bringing the system into balance at the design conditions, reducing the evaporator capacity will also have the beneficial effect of reducing the overall system capacity and thereby bringing it more into line with the calculated load.

13-3. Decreasing or Increasing Evaporator Capacity

Reducing the evaporator capacity can be accomplished in several ways. One is to "starve" the evaporator, that is, to reduce the amount of liquid refrigerant in the evaporator by adjusting the refrigerant flow control so that the evaporator is only partially flooded with liquid. This effectively reduces the size of the evaporator, since that part of the evaporator which is not filled with liquid becomes, in effect, a part of the suction line.

Another method of reducing the capacity of the evaporator is to reduce the air velocity over the evaporator by slowing the evaporator fan or blower. However, this method has its limitations in that the air velocity must be maintained at a level sufficient to assure adequate air circulation in the refrigerated space. Too, reducing the air quantity causes a change in the sensible heat ratio of the evaporator. Depending upon the particular application, this may or may not be desirable.

As a general rule, there is little, if anything, that can be done to increase the capacity of an undersized evaporator. Occasionally, the evaporator capacity can be increased by increasing the air quantity. However, the accompanying increases in air velocity and fan power requirements must also be taken into account.

In some cases, the evaporator surface area can be increased somewhat by using a length of either bare tubing or finned tubing as a "drier loop" or as additional evaporator surface. However, this too has its limitations because of the pressure drop accruing in the tubing.

13-4. Decreasing or Increasing Condensing Unit Capacity

Decreasing the condensing unit capacity can be accomplished in several ways, all of which involve decreasing the compressor displacement. Probably the simplest and most common method of reducing the condensing unit capacity is to reduce the speed of the compressor by reducing the size of the pulley on the compressor driver. The speed reduction required is approximately proportional to the desired capacity reduction.

The relationship between the speed of the compressor and the speed of the compressor driver is expressed in the following equation:

$$Rpm_1 \times D_1 = Rpm_2 \times D_2 \qquad (13\text{-}1)$$

where Rpm_1 = the speed of the compressor = (rpm)
D_1 = the diameter of the compressor flywheel (inches)
Rpm_2 = the speed of the compressor driver (rpm)
D_2 = the diameter of the driver pulley (inches)

Where the compressor driver is a four-pole, alternating-current motor operating on 60 cycle power, the approximate driver speed is 1750 rpm. For a two-pole, alternating-current motor, the approximate speed is 3500 rpm.

Example 13-2 A refrigeration compressor having a 10 in. flywheel is driven by a four-pole, alternating-current motor. If the diameter of the motor pulley is 4 in., determine the speed of the compressor.

Solution Rearranging and applying Equation 13-1,

$$Rpm_1 = \frac{Rpm_2 \times D_2}{D_1} = \frac{1750 \times 4}{10} = 700 \text{ rpm}$$

Example 13-3 Determine the diameter of the motor pulley required to reduce the speed of the compressor in Example 13-2 from 700 to 600 rpm.

Solution Rearranging and applying Equation 13-1,

$$D_2 = \frac{Rpm_1 \times D_1}{Rpm_2} = \frac{600 \times 10}{1750} = 3.5 \text{ in.}$$

Another method of reducing condensing unit capacity is to reduce the volumetric efficiency of the compressor by increasing the clearance volume. This increase is accomplished by installing a thicker gasket between the cylinder housing and the valve-plate.

In some cases, small increases in condensing unit capacity can be obtained by merely increasing the speed of the compressor. However, when the capacity increase needed is substantial, it is usually more practical and more economical to use a larger size condensing unit and reduce the capacity as necessary. The reasons for this are several.

First, since the increase in compressor capacity will be accompanied by an increase in the power requirements, any substantial increase in the compressor capacity will tend to overload the compressor driver and necessitate the use of a larger size. Too, some thought must be given to the condenser capacity. Here again, any increase in compressor capacity will tend to place a heavier load on the condenser. If the size of the condenser is not increased in proportion to the increase in the condenser load, excessive compressor discharge temperature and pressure will result. Not only will this materially reduce the life of the equipment and increase maintenance and operating costs, but it will also tend to nullify to some extent the gain in capacity originally accruing from the increase in compressor speed.

It is apparent from the foregoing that, in most cases, increasing the capacity of either the evaporator or the condensing unit is something which is not easily accomplished. Therefore, it is usually more practical and more economical to select oversized equipment rather than undersized equipment. When the evaporator or the condensing unit is oversized and capacity reduction is required to bring the system components into balance at the desired conditions, the capacity reduction can often be made with little, if any, loss in system efficiency.

13-5. System Capacity Versus Calculated Load

The relationship between system capacity and system load is one which warrants careful consideration and which can be best explained by comparing the refrigerating system to a water pumping system. For example, assume that it is desired to maintain a constant water level in the tank shown in Fig. 13-3. If the water flows into the tank at a fixed and constant rate which is readily computable, the water in the

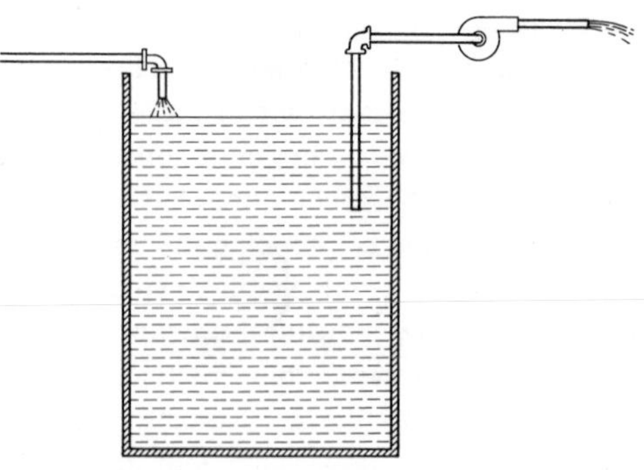

Fig. 13-3

tank can be maintained at a fixed level simply by installing a pumping system which has a capacity exactly equal to the flow rate of the water into the tank. Since the flow rate of the water entering the tank is constant and since the pumping rate is equal to the water flow rate, the pump will operate continuously and no other water level control of any kind will be needed.

On the other hand, if the flow rate of the water entering the tank varies from time to time, it is evident that if the level of the water is to be maintained within fixed limits, the pumping system must be selected to have a capacity equal to or somewhat in excess of the highest sustained flow rate of the water entering the tank. It is evident also that some means of cycling the pump "off" and "on" must be provided. Otherwise, during periods when the flow rate of the water entering the tank is less than maximum, the pumping rate will be excessive and the level of the water in the tank will be reduced below the desired level. One convenient and practical means of cycling the pump is to install a float control in the tank (Fig. 13-4). The float control is arranged to close the electrical contacts and start the pump when the water in the tank rises to a predetermined maximum level. When the water level in the tank falls to a predetermined lower limit, the float control acts to open the electrical contacts and stop the pump. In this way, intermittent operation of the pump will maintain a relatively constant water level in the tank.

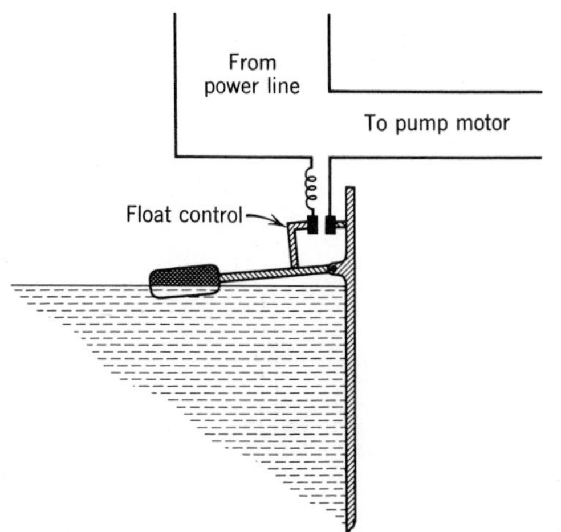

Fig. 13-4

The latter principle is readily applied to the refrigerating system. Since the cooling load on a refrigerating system varies from time to time, the system is usually designed to have a capacity equal to or somewhat in excess of the average maximum cooling load. This is done so that the temperature of the space or product can be maintained at the desired low level even under peak load conditions. As in the case of the water pumping system, since the system refrigerating capacity will always exceed the actual cooling load, some means of cycling the system "off" and "on" is needed in order to maintain the temperature of the space or product at a constant level within reasonable limits and to prevent the temperature of the space or product from being reduced below the desirable minimum.

For any refrigerating system, the relative length of the "off" and "on" cycles will vary with the loading of the system. During periods of peak loading, the "running" or "on" cycles will be long and the "off" cycles will be short, whereas during periods of minimum loading the "on" cycles will be short and the "off" cycles will be long.

Unlike the water pumping system, refrigerating systems are designed to have sufficient capacity to permit "off" cycles even during periods of peak loading. This is necessary in order to allow time for defrosting of the evaporator. However, allowances are made for defrosting time in the load calculations (the 24-hr load is divided by the desired running time to obtain the average hourly load) and need not be further considered when selecting the equipment.

13-6. Cycling Controls

The controls used to cycle a refrigerating system "on" and "off" are of two principal types: (1) temperature actuated (thermostatic) and (2) pressure actuated. Each of these types is discussed in the following sections.

13-7. Temperature Actuated Controls

Temperature actuated controls are called thermostats. Whereas float controls are sensitive to and are actuated by changes in liquid level, thermostats are sensitive to and are actuated by changes in temperature. Thermostats are used to control the temperature level of a refrigerated space or product by cycling the compressor (starting and stopping the compressor driving motor) in the same way that float controls are used to control liquid level by cycling the pump (starting and stopping the pump motor).

13-8. Temperature Sensing Elements

Two types of elements are commonly used in thermostats to sense and relay temperature changes to the electrical contacts or other actuating mechanisms. One is a fluid-filled tube or bulb which is connected to a bellows or diaphragm and filled with a gas, a liquid, or a saturated mixture of the two (Figs. 13-5*a* and 13-5*b*).* Increasing the temperature of the bulb or tube increases the pressure of the confined fluid which acts through the bellows or diaphragm and a system of levers to close electrical contacts or to actuate other compensating mechanisms (Fig. 13-6). Decreasing

* The thermostat described here is called a remote-bulb thermostat. Although there are a number of different types of thermostats, this is the type most frequently used in commercial refrigeration applications. Thermostats are used for many purposes other than controlling a compressor driving motor, as, for example, opening and closing valves, starting and stopping damper motors, etc.

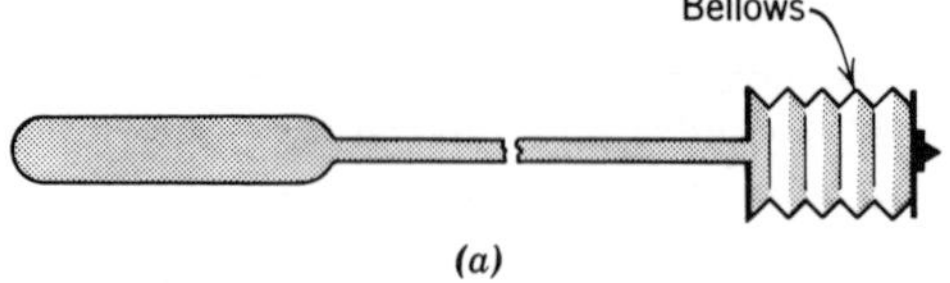

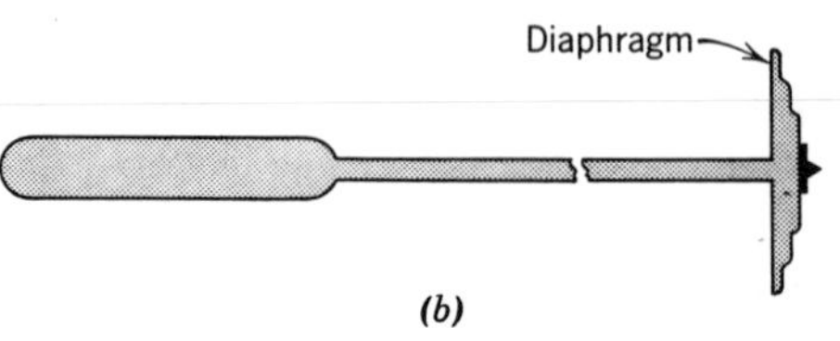

Fig. 13-5 Bulb-type temperature sensing element.

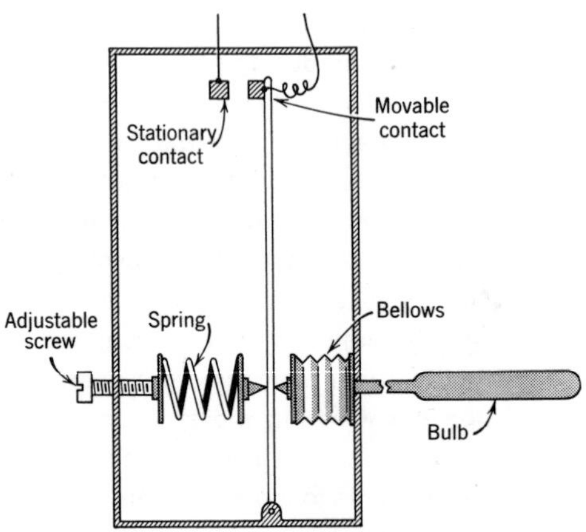

Fig. 13-6 Schematic diagram of simplified thermostatic control.

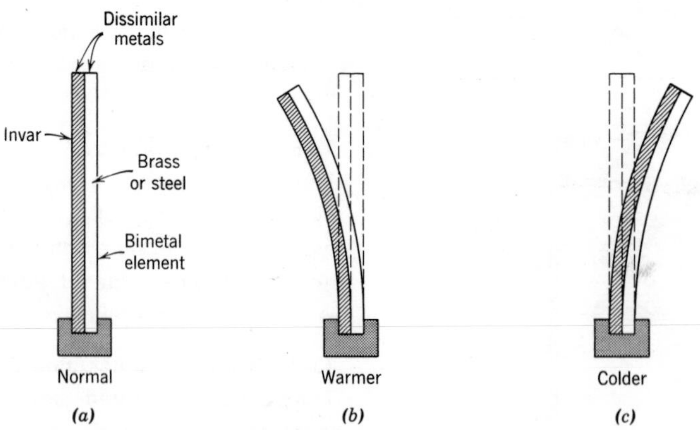

Fig. 13-7 Bimetal-type temperature sensing element.

the temperature of the tube or bulb will have the opposite effect.

Another and entirely different temperature sensing element is the compound bar, commonly called a bimetal element. The compound bar is made up of two dissimilar metals (usually Invar and brass or Invar and steel) bonded into a flat strip (Fig. 13-7*a*). Invar is an alloy which has a very low coefficient of expansion, whereas brass and steel have relatively high coefficients of expansions. Since the change in the length of the Invar per degree of temperature change will always be less than that of the brass or steel, increasing the temperature of the bimetal element causes the bimetal to warp in the direction of the Invar (the inactive metal) as shown in Fig. 13-7*b*, whereas decreasing the temperature of the bimetal element causes the bimetal to warp in the direction of the brass or steel (the active metal) as shown in Fig. 13-7*c*. This change in the configuration of the bimetal element with changes in temperature can be utilized directly or indirectly to open and close electrical contacts or to actuate other compensating mechanisms.

13-9. Differential Adjustment

Like float controls, thermostats have definite "cut-in" and "cut-out" points. That is, the thermostat is adjusted to start the compressor when the temperature of the space or product rises to some predetermined maximum (the cut-in temperature) and to stop the compressor when the temperature of the space or product is reduced to some predetermined minimum (the cut-out temperature).

The difference between the cut-in and cut-out temperatures is called the differential. In general, the size of the differential depends upon the particular application and upon the location of the temperature sensing element. Where the temperature sensing element of the thermostat is located in or on the product and controls the product temperature directly, the differential is usually small (2°F or 3°F). On the other hand, where the sensing element is located in the space and controls the space temperature, the differential is ordinarily about 6°F or 7°F. In many instances, the sensing element of the thermostat is clamped to the evaporator so that the space or product temperature is controlled indirectly by controlling the evaporator temperature, in which case the differential used must be larger (15°F to 20°F or more) in order to avoid short-cycling of the equipment.

When the thermostat controls the space or product temperature directly, the average space or product temperature is approximately the median of the cut-in and cut-out temperatures. Therefore, to maintain an average space temperature of 35°F, the thermostat can be adjusted for a cut-in temperature of approximately 38°F and a cut-out temperature of approximately 32°F.

On the other hand, when the space temperature is controlled indirectly by controlling the evaporator temperature, an allowance must be made in the cut-out setting to compensate for the evaporator TD. For example, for an average space temperature of 35°F and assuming an evaporator TD of 12°F, to compensate for the evaporator TD, the cut-out temperature would be set at 20°F (32°F − 12°) rather than at 32°F. Notice that the cut-in temperature is set at 38°F in either case. This is because the space temperature and the evaporator temperature are the same at the time that the system cycles on. After the compressor cycles off the evaporator continues to absorb heat from the space and warms up to the space temperature during the off cycle. Therefore, when the space temperature rises to the cut-in temperature of 38°F, the evaporator will also be at the cut-in temperature of 38°F. As soon as the compressor is started, the evaporator temperature is quickly reduced below the space temperature by an amount approximately equal to the design evaporator TD. Therefore, in this instance, when the space temperature is reduced to 32°F (the desired minimum), the evaporator temperature (which the thermostat is controlling) will be approximately 20°F (12°F less than the space temperature).

Regardless of whether the thermostat controls the space temperature directly or indirectly, proper adjustment of the cut-in and cut-out temperatures is essential to good operation. If the cut-in and cut-out temperatures are set too close together (differential too small) the system will have a tendency to short-cycle (start and stop too frequently). This will materially reduce the life of the equipment and may result in other unsatisfactory conditions. On the other hand, if the cut-in and cut-out temperatures are set too far apart (differential too large), the on and off cycles will be too long and unnecessarily large fluctuations in the average space temperature will result. Naturally, this too is undesirable.

Although approximate cut-in and cut-out temperature settings for various types of applications have been determined by field experience, in many cases it is necessary to use trial-and-error methods to determine the optimum settings for a specific installation.

13-10. Range Adjustment

In addition to the differential, cycling controls have another adjustment, called the "range," which is also associated with the cut-in and cut-out temperatures. Although, like the differential, the range can be defined as the difference between the cut-in and cut-out temperatures, the two are not the same. For example, assume that a thermostat is adjusted for a cut-in temperature of 30°F and a cut-out temperature of 20°F. Whereas the differential is said to be 10°F (30° − 20°), the range is said to be between 30°F and 20°F.

Although it is possible to change the range without changing the differential, it is not possible to change the differential without changing the range. For instance, suppose that the thermostat previously mentioned is readjusted so that the cut-in temperature is raised to 35°F and the cut-out temperature is raised to 25°F. Although the differential is still 10°F (35° − 25°), the operating range of the control is 5°F higher than it was originally, that is, the operating range is now between 35°F and 25°F, whereas previously it was between 30°F and 20°F. In this instance, the range of the control is changed, but the differential remains the same. Under the new control setting the average space temperature will be maintained approximately 5°F higher than under the old setting.

Suppose now that the differential is increased 5°F by raising the cut-in temperature from 30°F to 35°F while the cut-out temperature is left at the original setting of 20°F. Notice that both the differential and the range are changed. The differential, originally 10°F, is now 15°F and the range, originally between 30°F and 20°F, is now between 35°F and 20°F. With this control setting, the running cycle will be somewhat longer because the differential is larger. Too, the average space temperature will be 2 or 3 degrees higher because the cut-in temperature is higher. If the differential had been increased by lowering the cut-out temperature 5°F rather than by raising the cut-in temperature 5°F, the operating range of the control would have shifted to the opposite direction and the average space temperature would have been 2 or 3 degrees lower than the original space temperature.

Typical range and differential adjustments are shown in Fig. 13-8. Turning the range-adjusting screw clockwise increases the spring tension which the bellows pressure must overcome in order to close the contacts and, therefore, raises both the cut-in and cut-out temperatures. Turning the range-adjusting screw counterclockwise decreases the spring tension and lowers both the cut-in and cut-out temperatures.

Turning the differential-adjusting screw clockwise causes the limit bar *A* to move toward the screw head, thereby increasing the travel of the pin *B* in the slot. This has the effect of increasing the differential by lowering the cut-out temperature. Turning the differential adjusting screw counterclockwise raises the cut-out temperature and reduces the differential. By manipulating both range and differential adjustments, the thermostat can be adjusted for any desired cut-in and cut-out temperatures.

The arrangement shown in Fig. 13-8 represents only one of a number of methods which can be employed to adjust the cut-in and cut-out temperatures. The particular method used in any one control depends on the type of control and on the manufacturer. For example, for the control shown in Fig. 13-8, changing the range adjustment changes both the cut-in and

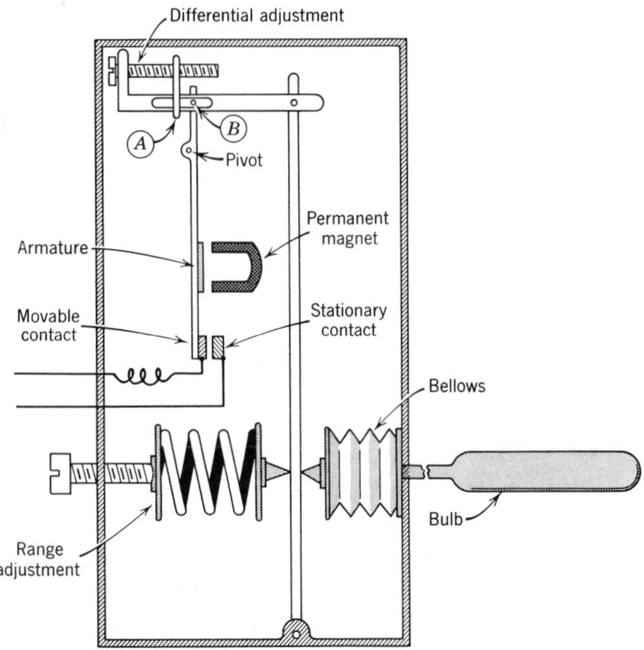

Fig. 13-8 Schematic diagram of thermostatic motor control illustrating range and differential adjustments.

cut-out temperatures simultaneously, whereas for another type of control changing the range adjustment changes only the cut-in temperature. For still another type of control, changing the range adjustment changes only the cut-out temperature. However, whatever the method of adjustment, the principles involved are similar and the exact method of adjustment is readily determined by examining the control. In many cases, instructions for adjusting the control are given on the control itself.

If electrical contacts are permitted to open or close slowly, arcing will occur between the contacts, and burning or welding together of the contacts will result. Therefore, cycling controls which employ electrical contacts must all be equipped with some means of causing the contacts to open and close rapidly in order to avoid arcing. In Fig. 13-8, the armature and permanent magnet serve this purpose. As the pressure in the bellows increases and the movable contact moves toward the stationary contact, the strength of the magnetic field between the armature and the horseshoe magnet increases rapidly. When the movable contact approaches to within a certain, predetermined, minimum distance of the stationary contact, the strength of the magnetic field becomes great enough to overcome the opposing spring tension so that the armature is pulled into the magnet and the contacts are closed rapidly with a snap action.

As the pressure in the bellows decreases, spring tension acts to open the contacts. However, since the force of the spring is opposed somewhat by the force of magnetic attraction, the contacts will not separate until a considerable force is developed in the spring. This causes the contacts to snap open quickly so that arcing is again avoided.

Toggle mechanisms are also frequently used as a means of causing the contacts to open and close with a snap action. Too, some controls employ a mercury switch as a means of overcoming the arcing problem. A typical mercury switch is illustrated in Fig. 13-9. As the glass tube is tilted to the right, the pool of mercury enclosed in the tube make contact between the two electrodes. As the bulb is tilted back to the left, contact is broken. The

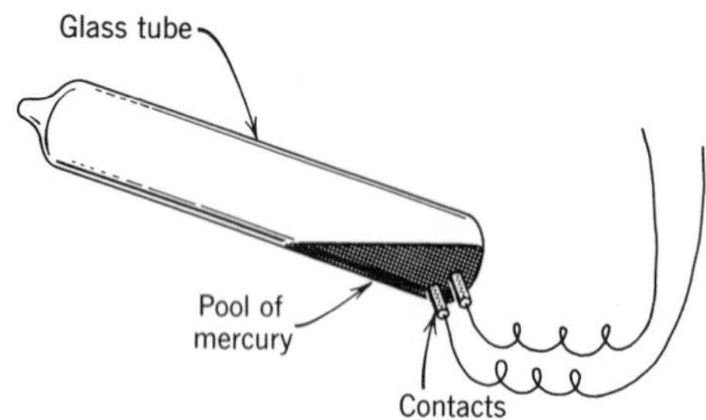

Fig. 13-9 Mercury contacts.

surface tension of the mercury provides the snap action necessary to prevent arcing.

13-11. Space Control vs. Evaporator Control

When the sensing element of the thermostat is located in the space or in the product, the thermostat controls the space temperature or product temperature directly. Likewise, when the sensing element is clamped to the evaporator, the thermostat controls the evaporator temperature directly. In such cases, control of the space or product temperature is accomplished indirectly through evaporator temperature control. Which of these two methods of control is the most suitable for any given application depends upon the requirements of the application itself.

For applications where close control of the space or product temperature is desired, a thermostat which controls the space or product temperature directly will ordinarily give the best results. On the other hand, for applications where off-cycle defrosting is required and where minor fluctuations in the space or product temperature are not objectionable, indirect control of the space temperature by evaporator temperature control is probably the better method.

In order to assure complete defrosting of the evaporator, the evaporator must be allowed to warm up to a temperature of approximately 37° or 38°F during each off cycle. When the thermostat controls the evaporator temperature, the cut-in temperature of the thermostat can be set for 38°F. Since the evaporator must warm up to this temperature before the compressor can be cycled on, complete defrosting of the evaporator during each off cycle is almost certain. On the other hand, when the thermostat controls the space temperature, there is no assurance that the evaporator will always warm up sufficiently during the off cycle to permit adequate defrosting.

If we assume that the thermostat is properly adjusted, if the load on the system is relatively constant and the capacity of the system is sufficient to handle the load, no defrosting problems are likely to arise with either method of temperature control. However, if the system is subject to considerable changes in load, defrosting problems are sometimes experienced in applications where the thermostat controls the space temperature. When the system is operating under peak load conditions, the temperature of the space tends to remain above the cut-out temperature of the thermostat for extended periods so that running cycles are long and frost accumulation on the evaporator is heavy. Too, under heavy load conditions, the space temperature warms up to the cut-in temperature of the thermostat very quickly during the off cycle so that the off cycles are usually short. Frequently, the off cycles are too short to allow adequate defrosting of the evaporator. In such cases, when the compressor cycles on again, the partially melted frost is caught on the evaporator and frozen into ice. Eventually the evaporator will be completely frozen over with ice, air flow over the evaporator will be severely restricted, and the system will become inoperative.

13-12. Pressure Actuated Cycling Controls

Pressure actuated cycling controls are of two types: (1) low-pressure actuated and (2) high-pressure actuated. Low-pressure controls are connected to the low-pressure side of the system (usually at the compressor suction) and are actuated by the low side pressure. High-pressure controls, on the other hand, are connected to the high-pressure side of the system (usually at the compressor discharge) and are actuated by the high side pressure.

The design of both the low-pressure and the high-pressure controls is similar to that of the remote-bulb thermostat. The principal difference between remote-bulb thermostat and the pressure controls is the source of the pressure which actuates the bellows or diaphragm. Whereas the pressure actuating the bellows of the thermostat is the pressure of the fluid confined in the bulb, the pressures actuating the bellows of the low- and high-pressure controls are the suction and discharge pressures of the compressor, respectively. Like the thermostat, both controls have cut-in and cut-out points which are usually adjustable in the field.

13-13. High-Pressure Controls

High-pressure controls are used only as safety controls. Connected to the discharge of the compressor, the purpose of the high-pressure control is to cycle the compressor off in the event that the pressure on the high-pressure side of the system becomes excessive. This is done in order to prevent possible damaging of the equipment. When the pressure on the high-pressure side of the system rises above a certain, predetermined pressure, the high-pressure control acts to break the circuit and stop the compressor. When the pressure on the high-pressure side of the system returns to normal, the high-pressure control acts to close the circuit and start the compressor. However, some high-pressure controls are equipped with "lock-out" devices which require that the control be reset manually before the compressor can be started again. Although high-pressure controls are desirable on all systems, because of the possibility of a water supply failure, they are essential on systems utilizing water-cooled condensers.

Since the condensing pressures of the various refrigerants are different, the cut-out and cut-in settings of the high-pressure control depend on the refrigerant used.

13-14. Low-Pressure Controls

Low-pressure controls are used both as safety controls and as temperature controls. When used as a safety control, the low-pressure control acts to break the circuit and stop the compressor when the low side pressure becomes excessively low and to close the circuit and start the compressor when the low side pressure returns to normal. Like high-pressure controls, some low-pressure controls are equipped with a lock-out device which must be manually reset before the compressor can be started.

Low-pressure controls are frequently used as temperature controls in commercial refrigeration applications. Since the pressure at the suction inlet of the compressor is governed by the saturation temperature of the refrigerant in the evaporator, changes in evaporator temperature are reflected by changes in the suction pressure. Therefore, a cycling control actuated by changes in the suction pressure can be utilized to control space temperature indirectly by controlling the evaporator temperature in the same way that the remote-bulb thermostat is used for this purpose. In such cases, the cut-in and cut-out pressures of the low-pressure control are the saturation pressures corresponding to the cut-in and cut-out temperatures of a remote-bulb thermostat employed in the same application. For example, assume that for a certain application the cut-in and cut-out temperature settings for a remote-bulb thermostat are 38°F and 20°F, respectively. If a low-pressure control is used in place of the thermostat, the cut-in pressure setting for the low-pressure control will be 50 psia (the saturation pressure of R-12 corresponding to a temperature of 38°F) and the cut-out pressure setting will be 36 psia (the saturation pressure of R-12 corresponding to a temperature of 20°F).*

As the evaporator warms up during the off cycle, the pressure in the evaporator increases accordingly. When the pressure in the evaporator rises to the cut-in pressure setting of the low-pressure control, the low-pressure control acts to close the circuit and start the compressor. Very soon after the compressor starts, the temperature and pressure of the evaporator are reduced to approximately the design evaporator temperature and pressure and they remain at this condition throughout most of the running cycle. Near the end of the running cycle the evaporator temperature and pressure are gradually reduced below the design conditions.

* When refrigerants other than R-12 are used in the system, the pressure settings will be the saturation pressures of those refrigerants corresponding to the desired cut-in and cut-out temperatures.

When the evaporator pressure is reduced to the cut-out pressure setting of the low-pressure control, the control acts to break the circuit and stop the compressor.

Since the refrigerant vapor undergoes a drop in pressure while flowing through the suction line, the pressure of the vapor at the suction inlet of the compressor is usually 2 or 3 lb less than the evaporator pressure. This is particularly true when the compressor is located some distance from the evaporator. Since the low-pressure control is actuated by the pressure at the suction inlet of the compressor, the pressure drop accruing in the suction line must be taken into account when the pressure control settings are made. To compensate for the pressure loss in the suction line, the cut-out pressure setting is lowered by an amount equal to the pressure loss in the line. For example, assuming a 3-lb pressure loss in the suction line, when the pressure in the evaporator is 36 psia, the pressure at the suction inlet of the compressor will be 33 psia. Hence, if it is desired to cycle the compressor off when the pressure in the evaporator is reduced to 36 psia, the cut-out pressure of the low-pressure control is set for 33 psia. In this instance, failure to make an allowance for the pressure loss in the suction line would cause the control to cycle the compressor when the pressure in the evaporator was reduced to only 39 psia rather than the desired 36 psia. The system would have a tendency to short cycle because the differential is too small and unsatisfactory operation would result.

Pressure loss in the suction line in no way affects the cut-in setting of the control. Since pressure drop is a function of velocity or flow, there is no pressure drop in the suction line when the system is idle. As soon as the compressor cycles off, the pressure at the suction of compressor rises to the evaporator pressure so that at the time the compressor cycles on the pressure at the compressor inlet is the same as the evaporator pressure. Hence, the cut-in pressure setting of the control is made without regard for the pressure drop in the suction line.

Since the low-pressure control controls evaporator temperature rather than space temperature, it is an ideal temperature control for applications requiring off-cycle defrosting. This is particularly true for "remote" installations where the compressor is located some distance from the evaporator. In such installations, low-pressure temperature control has a distinct advantage over thermostatic temperature control in that it ordinarily results in a considerable saving in electrical wiring. Because of the remote bulb, the thermostat must always be located near the evaporator or space whose temperature is being controlled. This requires that a pair of electrical conductors be installed between the fixture and the condensing unit. On the other hand, the low-pressure control is located at the compressor near the power source so that the amount of control wiring needed is much less.

13-15. Dual-Pressure Controls

A dual-pressure control is a combination of the low- and high-pressure controls in a single control. Ordinarily, only one set of electrical contact points are used in the control, although a separate bellows assembly is employed for each of the two pressures. A typical dual pressure control is shown in Fig. 13-10. This type of pressure control is frequently supplied as standard equipment on condensing units.

13-16. The Pump-Down Cycle

A commonly used method of cycling the condensing unit, known as a "pump-down cycle," employs both a thermostat and a low-pressure control. In a pump-down cycle, the space or evaporator temperature is controlled directly by the thermostat. However, instead of starting and stopping the compressor driver, the thermostat acts to open and close a solenoid valve installed in the liquid line, usually near the refrigerant flow control (Fig. 13-11). As the space or evaporator temperature is reduced to the cut-out temperature of the thermostat, the thermostat breaks the solenoid circuit, thereby deenergizing the solenoid and interrupting the flow of liquid refrigerant to the evaporator. Continued operation of the compressor causes evacuation of the refrigerant from that portion of the system beyond the point where the refrigerant flow is interrupted by the solenoid. When

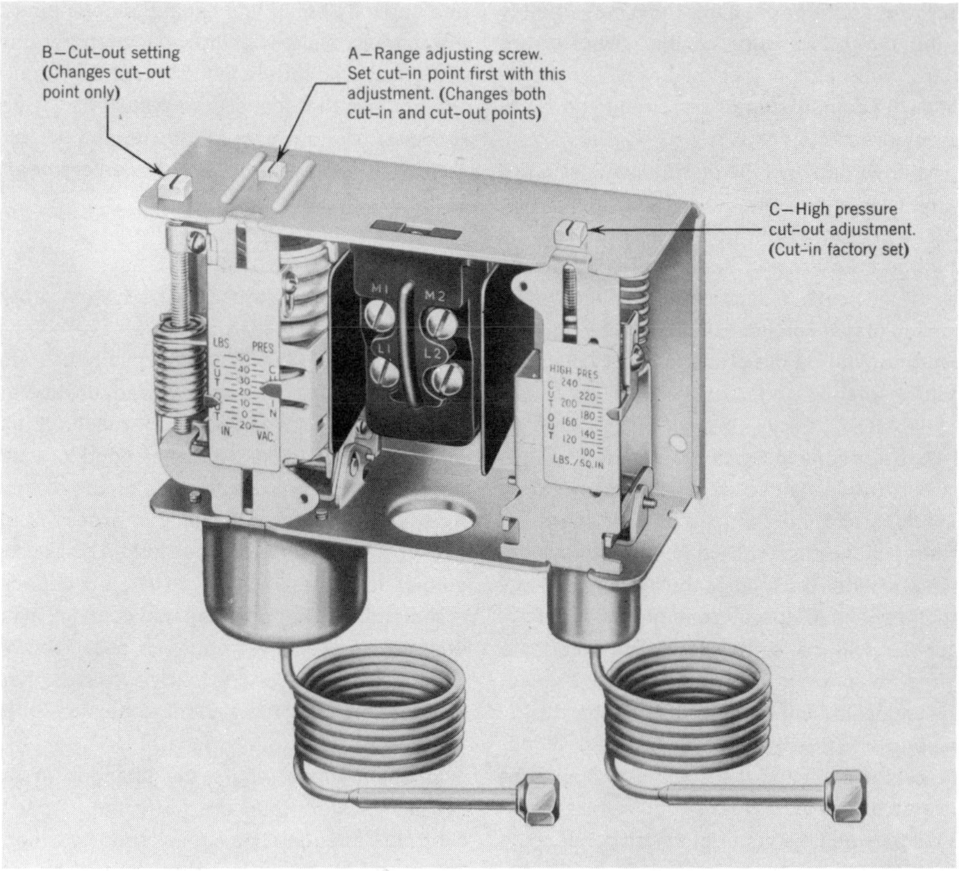

B—Cut-out setting
(Changes cut-out
point only)
A—Range adjusting screw.
Set cut-in point first with this
adjustment. (Changes both
cut-in and cut-out points)
C—High pressure
cut-out adjustment.
(Cut-in factory set)

Fig. 13-10 Dual pressure control. (Courtesy of Penn Controls, Inc.)

Power line
Refrigerant
flow control
Solenoid
S
Evaporator
T
Remote-bulb
Thermostat
Power line
Low-pressure
control
L.P.C.
Motor

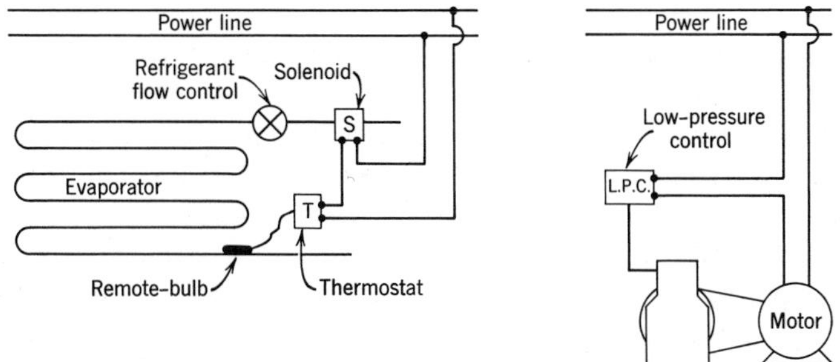

Fig. 13-11 Pump down cycle.

the pressure in the evacuated portion of the system is reduced to the cut-out pressure of the low pressure control, the low-pressure control breaks the compressor driver circuit and stops the compressor. When the temperature of the space or evaporator rises to the cut-in temperature of the thermostat, the thermostat closes the solenoid circuit and energizes the solenoid, thereby opening the liquid line and permitting liquid refrigerant to enter the evaporator. Since the evaporator is warm, the liquid entering the evaporator vaporizes rapidly so that the evaporator pressure rises immediately to the cut-in pressure of the low-pressure control, whereupon the low-pressure control closes the compressor driver circuit and starts the compressor.

The advantages of the pump-down cycle are many. One of the most important ones being that the amount of refrigerant absorbed by the oil in the crankcase of the compressor during the off cycle is substantially reduced. The problem of crankcase oil dilution by refrigerant absorption during the off cycle is fully discussed in Chapter 18.

13-17. Variations in System Capacity

It is worthwhile to notice that both the operating conditions and the capacity of a refrigerating system change as the load on the system changes. When the load on the system is heavy and the space temperature is high, the evaporator TD will be somewhat larger than the design evaporator TD and the capacity of the evaporator will be greater than the design evaporator capacity. Because of the higher evaporator capacity, the suction temperature will also be higher so that equilibrium is maintained between the vaporizing and condensing sections of the system. Hence, under heavy load conditions, the system operating conditions are somewhat higher than the average design conditions and the system capacity is somewhat greater than the average design capacity. Obviously, the power requirements of the compressor are greatest at this peak load condition and the compressor driver must be selected to have sufficient power to meet these requirements.

Conversely, when the load on the system is light, the space temperature will be lower than the average design space temperature, the evaporator TD will be less than the design TD, and the suction temperature will be lower than the design suction temperature. Therefore, the system operating conditions will be somewhat lower than the average design operating conditions and the system capacity will be somewhat less than the average design capacity.

During each running cycle the system passes through a complete series of operating conditions and capacities, the operating conditions and capacity being highest at the beginning of the running cycle when the space temperature is highest, and lowest at the end of the running cycle when the space temperature is lowest. However, during most of the running cycle, a well-designed system will operate very nearly at the design conditions.

13-18. Capacity Control

The importance of balancing the system capacity with the system load cannot be overemphasized. Any time the system capacity deviates considerably from the system load, unsatisfactory operating conditions will result. It has already been pointed out that good practice requires that the system be designed to have a capacity equal to or slightly in excess of the average maximum sustained load. This is done so that the system will have sufficient capacity to maintain the temperature and humidity at the desired level during periods of peak loading. Obviously, as the cooling load decreases, there is a tendency for the system to become oversized in relation to the load.

In applications where the changes in the average system load are not great, capacity control is adequately accomplished by cycling the system on and off as described in the preceding sections. In such cases, assuming that the cycling controls are properly adjusted, the relative length of the on and off cycles will vary with the load on the system. During periods when the load is heavy, on cycles will be long and off cycles will be short, whereas during periods when the load is light, on cycles will be short and off cycles will be long. Naturally, the degree of variation in the length of the on and off cycles will depend on the degree of load fluctuation. However, since the system must always be designed to have sufficient capacity to handle the maximum load, when the changes in the system load are substantial, it is evident that the system will be considerably oversized when the load is at a minimum. A system which is oversized for the load will usually prove to be as unsatisfactory as one that is undersized for the load. When the system is undersized for the load, the running time will be excessive, the space temperature will be high for extended periods, and the off cycles will be too short to permit adequate defrosting of the evaporator. On the other hand, where the system is oversized for the load, the off cycles will be too long and the equipment running time will be insufficient to remove the required amount of moisture from the space. This will result in unsatisfactory (higher than normal) humidity conditions in the refrigerated space.

For this reason, when changes in the system load are substantial, it is usually necessary to provide some means of automatically (or manually) varying the capacity of the system other than by cycling the system on and off. This is true also of large installations where the size of the equipment renders cycling the system on and off impractical.

There are many methods of bringing the refrigerating capacity into balance with the refrigerating load. Naturally, the most suitable method in any one case will depend upon the conditions and requirements of the installation itself. Some installations require only one or two steps of capacity control, whereas others require a number of steps. Frequently, several methods are employed simultaneously in order to obtain the desired flexibility. Too, in some cases, it is necessary to impose an artificial load on the equipment to achieve the proper balance between the sensible (temperature reduction) and latent (moisture reduction) loads. In applications where the latent load is too large a percentage of the total load, satisfaction of the latent load will result in overcooling of the space unless sensible heat is artificially introduced into the space. In such cases, the sensible heat is usually added to the space in the form of reheat. The air is first passed across a cooling coil and cooled to the temperature necessary to reduce the moisture content to the desired level and the air is then reheated to the required dry bulb temperature. The reheating is accomplished with steam or hot water coils, with electric strip-heaters, or with hot gas from the compressor discharge.

In some installations, the refrigerating capacity of the system is adequately controlled by controlling the capacity of the compressor only. Since the flow rate of the refrigerant must be the same in all components, any change in the capacity of any one component will automatically result in a similar adjustment in the capacity of all the other components. Therefore, increasing or decreasing the capacity of the compressor will, in effect, increase or decrease the capacity of the entire system. However, it is important to notice that with this method of capacity control the operating conditions of the system will change as the capacity of the system changes.

Where it is desired to vary the capacity of the system without allowing the operating conditions of the system to change, it is necessary to control both the evaporator capacity and the compressor capacity directly.*

Some of the more common methods of controlling evaporator and compressor capacities are considered in the following sections.

13-19. Evaporator Capacity Control

Probably the most effective method of providing evaporator capacity control is to divide the evaporator into several separate sections or circuits which are individually controlled so that one or more sections or circuits can be cycled out as the load decreases (Fig. 13-12). Using this method, any percentage of the evaporator capacity can be cycled out in any

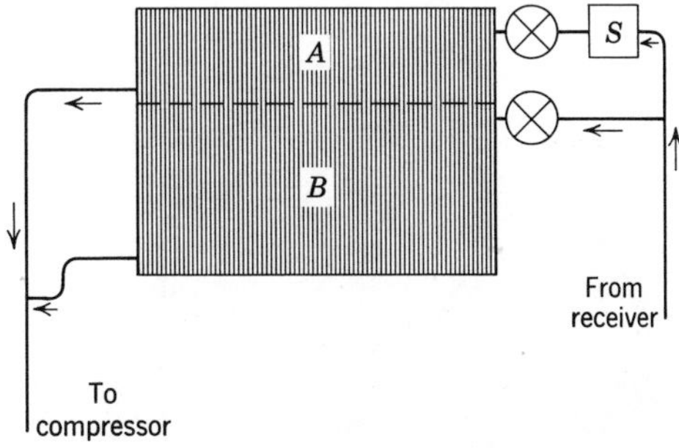

Fig. 13-12 Evaporator split into two segments for capacity control. Closing the solenoid valve in the liquid line of segment *A* renders this portion of the evaporator inoperative. The capacity reduction obtained is proportional to the surface area cycled out.

* Any reduction in system load and/or system capacity will also have some effect on the capacity of the condenser and on the size of the refrigerant lines. These topics are discussed in Chapters 14 and 17, respectively.

desired number of steps. The number and size of the individual evaporator sections depend on the number of steps of capacity desired and the percent change in capacity required per step, respectively. The arrangement of the evaporator sections or circuiting depends on the relationship of the sensible load to the total load at the various load conditions. Basically, two circuit arrangements are possible. Evaporator circuiting can be arranged to provide either "face" control of "depth" control, or both (Fig. 13-13 and 13-14). When "face" control is used, the "sensible heat ratio" of

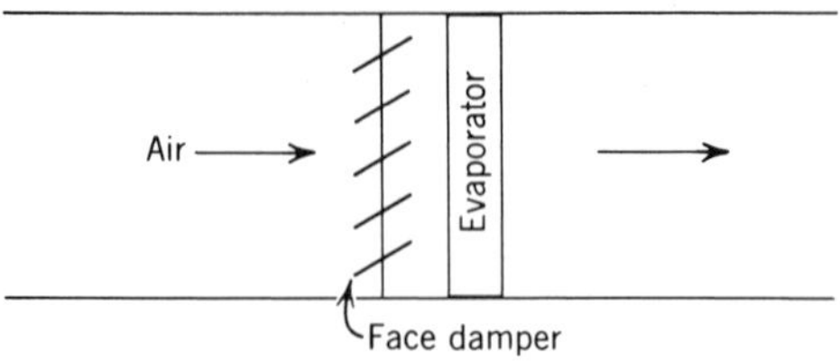

Fig. 13-15 Evaporator equipped with a face damper to vary the quantity of air passing over the evaporator. As the damper moves toward the closed position, the resistance against which the blower must work is increased so that the total quantity of air circulated decreases.

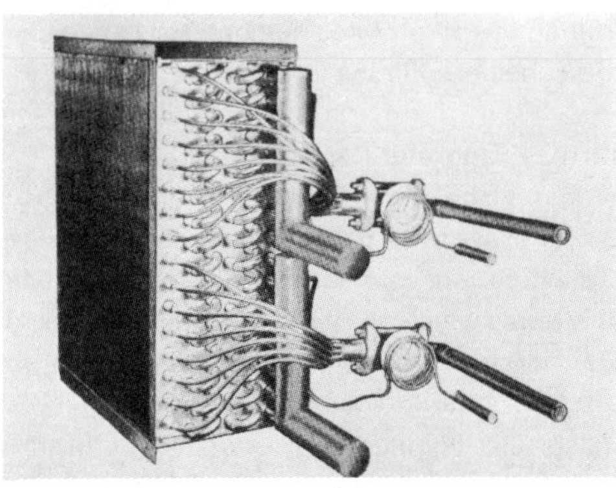

Fig. 13-13 Coil circuited for face control. (Courtesy Kennard Division, American Air Filter Company, Inc.)

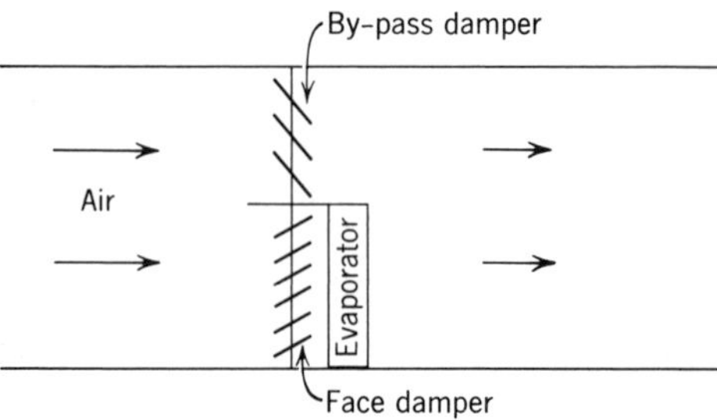

Fig. 13-16 Evaporator equipped with face and bypass dampers for capacity control. Dampers are interconnected so that bypass damper opens wider as face damper is closed off. With this arrangement the quantity of air passing over the evaporator can be regulated by allowing more or less air to bypass the evaporator. However, regardless of the position of the dampers, the total quantity of air circulated remains practically the same.

the evaporator is not affected.* On the other hand, "depth" control always changes the sensible heat ratio. As a general rule, the more depth the evaporator has the greater is its latent cooling (moisture removal) capacity. Hence, as one or more rows of the evaporator are cycled out, the sensible heat ratio increases.

Another common method of varying the evaporator capacity is to vary the amount of air circulated over the evaporator through the use of "face" or "face-and-by-pass" dampers. (Figs. 13-15 and 13-16). Multispeed blowers can also be used for this purpose. Also, in some instances, multispeed blowers and dampers

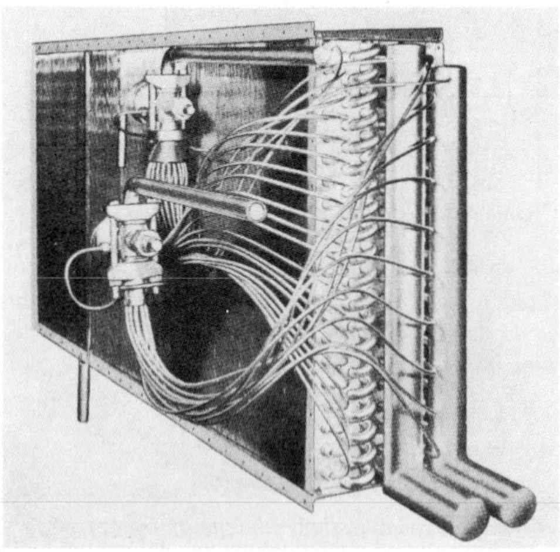

Fig. 13-14 Coil circuited for depth control. (Courtesy Kennard Division; American Air Filter Company, Inc.)

* Ratio of the sensible cooling capacity to the total cooling capacity.

are used together in order to provide the desired balance.

In nearly all cases, application of any of the foregoing methods of evaporator capacity control will necessitate simultaneous control of compressor capacity.

13-20. Compressor Capacity Control

There are a number of different methods of controlling the capacity of reciprocating compressors. One method, already mentioned, is to vary the speed of the compressor by varying the speed of the compressor driver. When an engine or turbine is employed to drive the compressor, the compressor capacity can be modulated over a wide range by governor control of the compressor driver.

When an electric motor drives the compressor, only two speeds are usually available so that the compressor operates either at full capacity or at 50% capacity. When more than two speeds are desired, it is necessary to use two separate windings in the motor, in which case four speeds will be possible.

Capacity control of multicylinder compressors is frequently obtained by "unloading" one or more cylinders so that they become ineffective. One method of accomplishing this is to bypass the discharge from one or more cylinders back into the suction line as shown in Fig. 13-17. When the suction pressure drops to a certain predetermined value a solenoid valve in the bypass line, actuated by a pressure control, opens and allows the discharge from one or more cylinders to flow through the bypass line back into the suction line where it

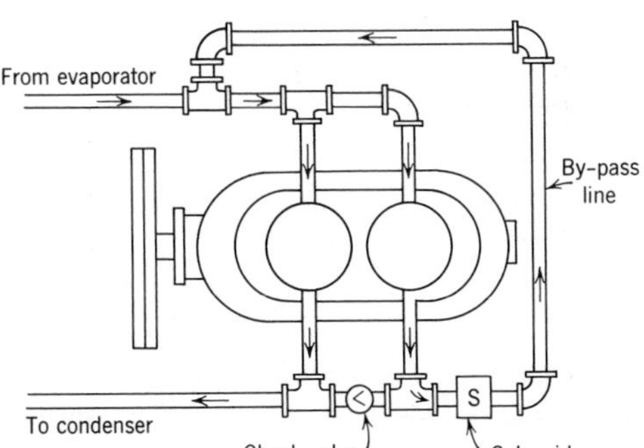

Fig. 13-17 Schematic diagram of cylinder bypass.

mixes with the incoming suction vapor. As long as the suction pressure remains below the cut-in setting of the control, the discharge from the unloaded cylinders continues to bypass to the suction line. When the suction pressure rises to the cut-out setting of the pressure-control, the solenoid valve is deenergized and the bypass line is closed so that the compressor is returned to full capacity operation.

Another method of unloading compressor cylinders is to depress the suction valves of the cylinder or cylinders to be unloaded so that they remain open during the compression (up) stroke. With the suction valves held open, the suction vapor drawn into the cylinder during the suction stroke is returned to the suction line during the compression stroke. A typical unloader of this type is shown in Fig. 13-18.

The operation of the unloader mechanism is as follows: when the suction pressure falls to the cut-in pressure of the pressure control, the control energizes the solenoid valve and admits high-pressure gas from the condenser to the unloader piston which acts to depress the suction valves and hold them open. When the suction pressure rises to the cut-out pressure of the pressure control, the solenoid valve is deenergized and the unloader piston is returned to the normal position.

In addition to providing capacity control, cylinder unloaders of all types are used to unload the compressor cylinders during compressor start-up so that the compressor starts in an unloaded condition, thereby reducing the inrush current demand.

With any of the capacity control methods described thus far the power required by the compressor decreases as the capacity decreases, although not exactly in the same proportion. Consequently, these control methods are the ones most frequently employed.

Another method of controlling compressor capacity is to throttle the suction of the compressor. However, this method is seldom employed with reciprocating compressors, since there is no concurrent reduction in the compressor power requirements.

Still another means of regulating compressor capacity is to artificially load the compressor as the system load drops off. This is

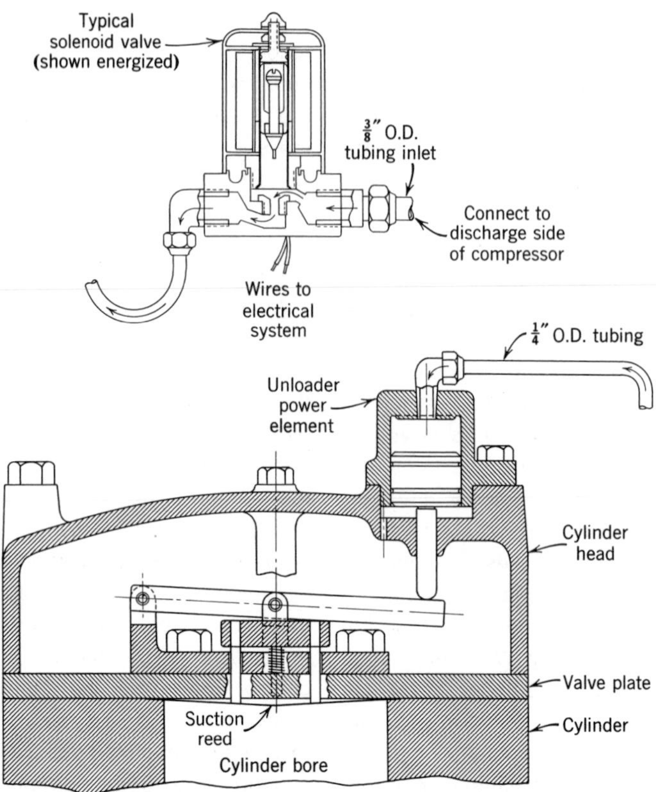

Fig. 13-18 Condenser pressure actuated cylinder unloader mechanism. (Courtesy Dunham-Bush, Inc.)

accomplished through the use of a hot gas bypass valve that is adjusted to open and bypass high-pressure hot gas from the discharge of the compressor back into the low-pressure side of the system when the low side pressure falls below some predetermined minimum (Fig. 13-19*a*). An automatic (constant pressure) expansion valve can be and has been employed as a hot gas bypass valve (Section 17-3), although valves especially designed for this purpose are now available.

There are several disadvantages of the hot gas bypass. First, since there is usually little or no reduction in compressor loading, the reduction in compressor power requirements is minimal. Also, unless the hot gas bypass is properly applied, excessive superheating of the suction vapor can occur, resulting in overheating of the compressor. This is especially true for the arrangement illustrated in Fig. 13-19*a*, which, for this reason, should be used only to provide for unloaded starting. Alternate arrangements, all of which provide some means of desuperheating the hot gas

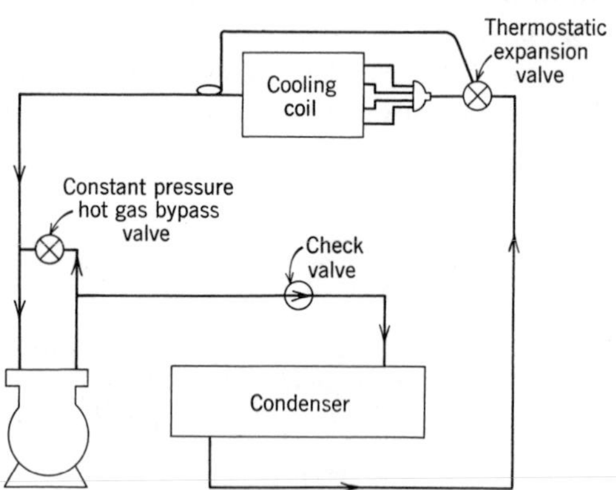

Fig. 13-19*a* For unloaded starting, the constant pressure bypass valve is replaced with a manual or solenoid stop valve. The check valve is required with unloaded starting to prevent blow-back of the high-pressure gas from the condenser.

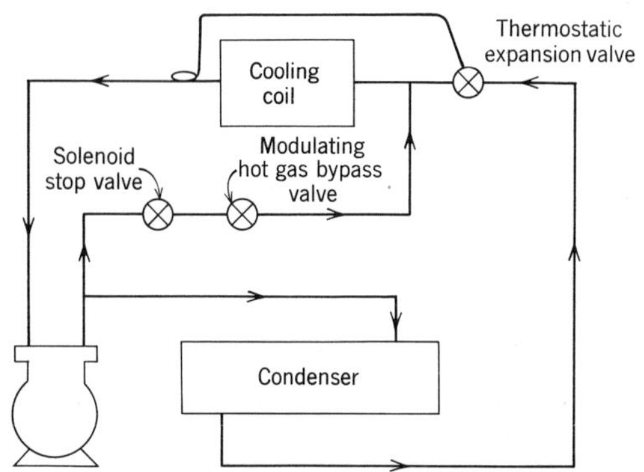

Fig. 13-19*b* Hot gas bypassed to the entrance of the evaporator mixes with, and is desuperheated by, the refrigerant in the evaporator, thereby eliminating the possibility of overheating the compressor as a result of excessive superheating of the suction vapor. In addition, bypassed hot gas keeps refrigerant velocity in the evaporator relatively high and improves oil return during periods of reduced loading. Where a refrigerant distributor is used, it is recommended that the bypass be connected directly.

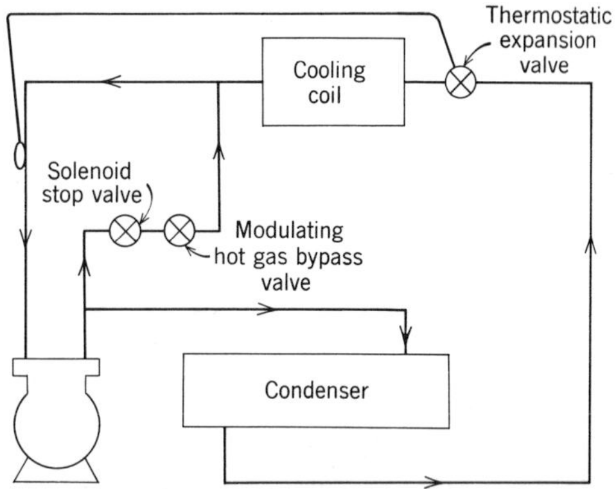

Fig. 13-19*c* Bypassing the hot gas to the end of the evaporator upstream of the expansion valve bulb makes it possible to reduce and control suction superheat by overfeeding the evaporator. The expansion valve bulb should be located a minimum of 3 ft downstream of the bypass connection. A solenoid stop valve is required in "pump-down" systems to provide positive shut-off of bypass line.

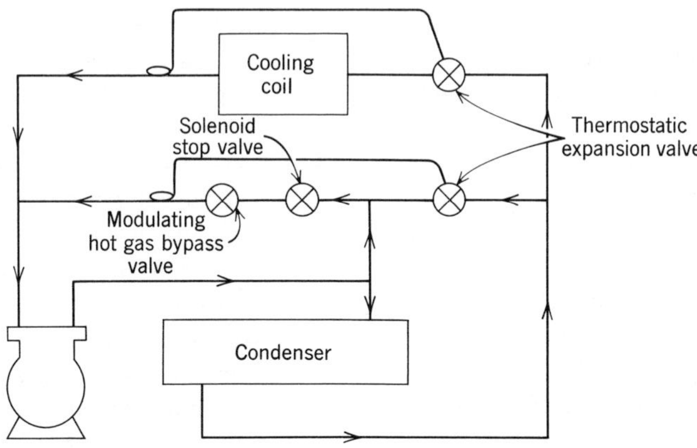

Fig. 13-19*d* Hot gas is desuperheated by direct expansion cooling from separate thermostatic expansion valve. This method is ideal for remote systems where the compressor is located some distance away from the evaporator section.

before it reaches the compressor suction, are illustrated in Figs. 13-19*b*, *c*, and *d*.

While the hot gas bypass is often employed as the sole means of controlling the capacity of small compressors (up to 10 hp), it is frequently used also to supplement other, more efficient methods of capacity control in situations where it is necessary to provide for capacity control down to 0% load or to provide for unloaded starting. To accomplish the latter, the bypass valve (Fig. 13-19*a*) is a manual or automatic (solenoid) valve that is full open during compressor startup. With the bypass valve full open between the discharge and suction of the compressor, the pressures are equalized, and the work of compression is only that which is required to offset friction, inertia, and the pressure losses in the compressor and bypass line. To prevent the blowback of high-pressure gas from the condenser,

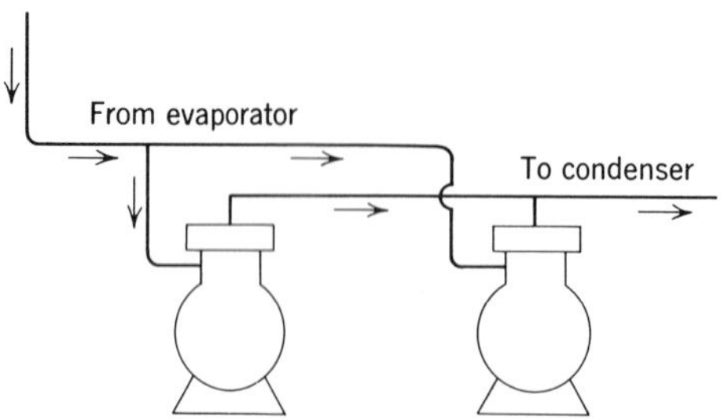

Fig. 13-20 Two compressors installed in parallel as a means of controlling compressor capacity. As the load diminishes, one compressor is cycled out to reduce the compressor capacity. Often, one compressor is equipped with cylinder unloaders to provide additional steps of control.

a check valve is installed between the condenser and the bypass line.

Still another means of controlling compressor capacity which is employed with good results is to operate two or more compressors in parallel (Fig. 13-20). Individual low-pressure controls are used to cycle the compressors. The cut-in and cut-out pressures of the individual controls are so adjusted that the compressors cycle off in sequence as the suction pressure decreases and cycle on in sequence when the suction pressure rises. Very often these compressors are equipped with cylinder unloaders to provide additional steps of control. Multiple compressor systems are discussed in detail in Chapter 20.

13-21. Multiple-System Capacity Control

Another method of controlling capacity is to employ two or more separate systems. The evaporators for the separate systems may be in the same housing and air stream or they may be in separate housings and air streams. In either case, separate compressors and condensers are used, although in some instances the condensers may be in a common housing.

This method of capacity control is well suited to installations where only two steps of capacity control are required, as in chilling or combination chilling and storage applications. The use of two or more separate systems has the added advantage of providing a certain amount of insurance against losses accruing from equipment failure. Should one system become inoperative, the other can usually hold the load until repairs can be made.

Customary Problems

13-1 A frozen storage room to be maintained at 0°F has a calculated refrigeration load of 12,000 Btu/hr. Four banks of plate evaporators, Model No. 4-12144 B, with a combined capacity of 12,096 Btu/hr at a 12°F TD, are installed in the room. From Table R-13, select a water-cooled condensing unit for system and plot capacity curves for the evaporators and condensing unit on a graph to determine the conditions at which the system will balance (operating suction temperature, evaporator TD, and system capacity) under the design load.

13-2 Using the conditions of Problem 13-1, plot an additional evaporator capacity curve to determine system operating conditions when two banks of plates are taken out of service.

13-3 Assuming that the system described in Problem 13-1 is charged with R-12, determine the cut-in and cut-out pressure settings for a low-pressure cycling control that will permit the system to cycle on and off with a space temperature differential of 4°F, i.e., cycle on when the space temperature is 2°F above 0°F and cycle off when the space temperature is 2°F below 0°F.

14

CONDENSERS AND COOLING TOWERS

14-1. Condensers

Like the evaporator, the condenser is a heat transfer surface. Heat from the hot refrigerant vapor passes through the walls of the condenser to the condensing medium. As the result of losing heat to the condensing medium, the refrigerant vapor is first cooled to saturation and then condensed into the liquid state.

Although brine or direct expansion refrigerants are sometimes used as condensing mediums in low temperature applications, in the great majority of cases the condensing medium employed is either air or water, or a combination of both.

Condensers are of three general types: (1) air-cooled, (2) water-cooled, and (3) evaporative. Air-cooled condensers employ air as the condensing medium, whereas water-cooled condensers utilize water to condense the refrigerant. In both the air-cooled and water-cooled condensers, the heat given off by the condensing refrigerant increases the temperature of the air or water used as the condensing medium.

Evaporative condensers employ both air and water. Although there is some increase in the temperature of the air passing through the condenser, condensation of the refrigerant in the condenser results primarily from the evaporation of the water sprayed over the condenser. The function of the air is to increase the rate of evaporation by carrying away the water vapor which results from the evaporating process.

14-2. The Condenser Load

The total heat rejection at the condenser includes both the heat absorbed in the evaporator and the energy equivalent of the work of compression. Any superheat absorbed by the suction vapor from the surrounding air also becomes a part of the load on the condenser.

Since the work of compression per unit of refrigerating capacity depends on the compression ratio, the quantity of heat rejected at the condenser per unit of refrigerating capacity varies with the operating conditions of the system. The compression heat also varies somewhat with the design of the compressor and will be greater for a suction-cooled hermetic compressor than for an open-type compressor because of the additional motor heat absorbed by the refrigerant gas.

Some compressor manufacturers publish total heat rejection data as a part of their compressor ratings (see column labled T.H.R in Table R-11), and when available, these data should be used as a basis for condenser selection. When such data are not available, the condenser load can be estimated by multiplying the compressor capacity by the appropriate factor from Table 14-1A or 14-1B, viz:

$$\text{Condenser load} = (\text{compressor capacity})(\text{heat rejection factor}) \tag{14-1}$$

Example 14-1 Determine the estimated condenser load for an open-type compressor

TABLE 14-1A Heat Rejection Factors: Open Compressors

Evaporator Temp. (°F)	Condensing Temperature (°F)					
	90	100	110	120	130	140
−30	1.37	1.42	1.47	·	·	·
−20	1.33	1.37	1.42	1.47	·	·
−10	1.28	1.32	1.37	1.42	1.47	·
0	1.24	1.28	1.32	1.37	1.41	1.47
10	1.21	1.24	1.28	1.32	1.36	1.42
20	1.17	1.20	1.24	1.28	1.32	1.37
30	1.14	1.17	1.20	1.24	1.27	1.32
40	1.12	1.15	1.17	1.20	1.23	1.28
50	1.09	1.12	1.14	1.17	1.20	1.24

* Outside of normal limits for single stage compressor application. Courtesy of Bohn Aluminum and Brass Company.

TABLE 14-1B Heat Rejection Factors: Suction-cooled Hermetic Compressors

Evaporator Temp. (°F)	Condensing Temperature (°F)					
	90	100	110	120	130	140
−40	1.66	1.73	1.80	2.00	·	·
−30	1.57	1.62	1.68	1.80	·	·
−20	1.49	1.53	1.58	1.65	·	·
−10	1.42	1.46	1.50	1.57	1.64	·
0	1.36	1.40	1.44	1.50	1.56	1.62
5	1.33	1.37	1.41	1.46	1.52	1.59
10	1.31	1.34	1.38	1.43	1.49	1.55
15	1.28	1.32	1.35	1.40	1.46	1.52
20	1.26	1.29	1.33	1.37	1.43	1.49
25	1.24	1.27	1.31	1.35	1.40	1.45
30	1.22	1.25	1.28	1.32	1.37	1.42
40	1.18	1.21	1.24	1.27	1.31	1.35
50	1.14	1.47	1.20	1.23	1.26	1.29

* Outside of normal limits for single stage compressor application. Courtesy of Bohn Aluminum and Brass Company.

having a cooling capacity of 16,300 Btu/hr when operating with a saturated suction temperature of 0°F and a saturated discharge temperature of 110°F.

Solution From Table 14-1A, the heat of compression factor for a saturated suction temperature of 0°F and a saturated condensing temperature of 110°F is 1.32. Applying Equation 14-1, the condenser load

$$= (16{,}300 \text{ Btu/hr})(1.32) = 21{,}500 \text{ Btu/hr}$$

Example 14-2 Determine the condenser load for the conditions given in Example 14-1 if a suction cooled hermetic compressor is employed.

Solution From Table 14-1B, the heat of compression factor is 1.44. Applying Equation 14-1, the condenser load

$$= (16{,}300 \text{ Btu/hr})(1.44) = 23{,}470 \text{ Btu/hr}$$

For operating conditions outside the normal range of single stage compression, the following equations can be used to estimate the condenser load:

For open compressors,

$$\text{condenser load} = \text{compressor capacity (Btu/hr)} + (2545 \times \text{Bhp}) \qquad (14\text{-}2)$$

For hermetic compressors,

$$\text{condenser load} = \text{compressor capacity (Btu/hr)} + (3413 \times \text{kW}) \qquad (14\text{-}3)$$

Example 14-3 A hermetic compressor operating at design conditions has a cooling capacity of 4.4 tons and a power input of 9 kW. Determine the total heat rejection at the condenser.

Solution Applying Equation 14-3, the condenser load

$$= (4.4 \text{ tons} \times 12{,}000) + (9 \text{ kW} \times 3413)$$
$$= 83{,}500 \text{ Btu/hr}$$

14-3. Condenser Capacity

Since heat transfer through the walls of the condenser is by conduction, condenser capacity is a function of the fundamental heat transfer equation:

$$Q_c = (A)(U)(D) \qquad (14\text{-}4)$$

where Q_c = the condenser capacity in Btu per hour

A = the surface area of the condenser in square feet

U = the overall heat transfer coefficient in Btu/(hr)(ft^2)(°F)

D = the log mean temperature difference between the condensing refrigerant and the condensing medium in °F

Examination of the factors in Equation 14-4 will show that for any fixed value of U, the capacity of the condenser is directly proportional to the surface area of the condenser and to the temperature difference between the condensing refrigerant and the condensing medium. It is evident also that for any one condenser of specific size and design, wherein both the surface area and the U factor are fixed at the time of manufacture, the capacity of the condenser is directly proportional to the temperature difference between the refrigerant and the condensing medium. Moreover, where the average temperature of the condensing medium is held constant, the capacity of the condenser is increased or decreased only by raising or lowering the condensing temperature.

14-4. Quantity and Temperature Rise of Condensing Medium

In both the air-cooled and water-cooled condensers, all the heat given off by the condensing refrigerant increases the temperature of the condensing medium. Therefore, in accordance with Equation 2-8, the temperature rise experienced by the condensing medium in passing through the condenser is directly proportional to the condenser load and inversely proportional to the quantity and specific heat of the condensing medium, that is,

$$\Delta T = \frac{Q_c}{(m)(c)} \tag{14-5}$$

where ΔT = the temperature rise of the condensing medium in the condenser

Q_c = the heat rejected at the condenser in Btu per hour

m = the mass flow rate of air or water in the condenser in pounds per hour

c = the specific heat of the condensing medium

Assuming that the value of c is constant, for any given condenser load (Q_c), the only variables in Equation 14-5 are the flow rate and temperature rise of the condensing medium, the value of each being inversely proportional to the value of the other. If either value is known, the other is easily calculated.

Average specific heat values for air and water are 0.24 Btu/lb and 1 Btu/lb, respectively. By substituting the appropriate value for c, Equation 14-5 can be written to apply specifically to either water or air, that is, for water

$$m = \frac{Q_c}{\Delta T} \tag{14-6}$$

$$\Delta T = \frac{Q_c}{m} \tag{14-7}$$

and for air

$$m = \frac{Q_c}{0.24 \times \Delta T} \tag{14-8}$$

$$\Delta T = \frac{Q_c}{0.24 \times m} \tag{14-9}$$

Since general practice is to express air and water quantities in cubic feet per minute (cfm) and in gallons per minute (gpm), respectively, it is usually desirable to compute condensing medium quantities in these units rather than in pounds per hour.

To convert pounds of water per hour into gallons per minute, divide by 60 min to reduce pounds per hour to pounds per minute, and then divide by 8.33 lb/gal to convert pounds per minute to gallons per minute, that is,

$$\text{gpm} = \frac{m\,(\text{lb/hr})}{60\text{ min} \times 8.33\text{ lb/gal}}$$

If these conversion constants are incorporated into Equation 14-6, the water quantity can be computed directly in gpm. The following equation results:

$$\text{gpm} = \frac{Q_c}{60 \times 8.33 \times \Delta T}$$

or, combining constants (60 × 8.33 = 500),

$$\text{gpm} = \frac{Q_c}{500 \times \Delta T} \tag{14-10}$$

To reduce pounds of air per hour to cubic feet per minute, divide pounds per hour by 60 min to determine pounds per minute and then multiply by the specific volume of the air to convert from pounds per minute to cubic feet per minute, that is,

$$\text{cfm} = \frac{m(\text{lb/hr}) \times v(\text{ft}^3/\text{lb})}{60 \text{ min}}$$

Assuming the specific volume of the air to be the specific volume of standard air (13.34 ft^3/lb), incorporation of these constants into Equation 14-7 results in the following:

$$\text{cfm} = \frac{Q_c \times 13.34 \text{ ft}^3/\text{lb}}{0.24 \times 60 \times \Delta T}$$

or, combining constants (13.34/0.24 × 60 = 1/1.08),

$$\text{cfm} = \frac{Q_c}{1.08 \times \Delta T} \qquad (14\text{-}11)$$

Example 14-4 If the load on a water-cooled condenser is 150,000 Btu/hr and the temperature rise of the water in the condenser is 10°F. What is the quantity of water circulated in gallons per minute?

Solution Applying Equation 14-10, the water quantity in gpm

$$= \frac{150{,}000}{500 \times 10} = 30 \text{ gpm}$$

Example 14-5 The load on a water-cooled condenser is 90,000 Btu/hr. If the quantity of water circulated through the condenser is 15 gpm, determine the temperature rise of the water in the condenser.

Solution Rearranging and applying Equation 14-10, ΔT

$$= \frac{90{,}000}{500 \times 15} = 12°\text{F}$$

Example 14-6 Thirty-six gallons of water per minute are circulated through a water-cooled condenser. If the temperature rise of the water in the condenser is 12°F, compute the load on the condenser in Btu per hour.

Solution Rearranging and applying Equation 14-10, the load on the condenser, Q_c

$$= 500 \times \Delta T \times \text{gpm} = 500 \times 12 \times 36 = 216{,}000 \text{ Btu/hr}$$

Example 14-7 The load on an air-cooled condenser is 121,500 Btu/hr. If the desired temperature rise of the air in the condenser is 25°F, determine the air quantity in cfm which must be circulated over the condenser.

Solution Applying Equation 14-11, the air quantity in cfm

$$= \frac{121{,}500}{1.08 \times 25} = 4500 \text{ cfm}$$

Example 14-8 Three thousand cubic feet per minute of air are circulated over an air-cooled condenser. If the load on the condenser is 64,800 Btu/hr, compute the temperature rise of the air passing over the condenser.

Solution Rearranging and applying Equation 14-11, ΔT

$$= \frac{64{,}800}{1.08 \times 3000} = 20°\text{F}$$

For any given condenser and condenser loading, the condensing temperature of the refrigerant in the condenser will depend only upon the average temperature of the condensing medium flowing through the condenser. The lower the average temperature of the condensing medium the lower is the condensing temperature. For example, assume that the size and loading of a condenser are such that the required mean temperature differential between the refrigerant and the condensing medium is 15°F. If the average temperature of the condensing medium is 90°F, the condensing temperature will be 105°F (90 + 15), whereas if the average temperature of the condensing medium is 85°F, the condensing temperature will be 100°F (85 + 15).

The average temperature of the condensing medium depends on the entering temperature and on the temperature rise experienced in the condenser, as shown in Fig. 14-1. Since the temperature rise of the condensing medium decreases as the flow rate increases, the higher the flow rate, the lower is the average temperature of the condensing medium and the lower is the condensing temperature. The flow rate of the condensing medium through the condenser is fixed within certain limits by the size and design of the condenser. If the flow rate through the condenser is too low, flow will be streamlined rather than turbulent and a low transfer coefficient will result. On the other hand, if the flow rate is too high, the pressure drop through the condenser becomes excessive, with the result that the power required to circulate the condensing medium also becomes excessive.

Since the design entering temperature of the condensing medium is usually fixed by conditions beyond the control of the system designer, it follows that the size and design of the condenser and the flow rate of the condensing medium are determined almost entirely by the design condensing temperature.

Although low condensing temperatures are desirable in that they result in high compressor efficiency and low power requirements for the compressor, this does not necessarily mean that the use of a large condensing surface and a high flow rate in order to provide a low condensing temperature will always result in the most practical and economical installation. Other factors that must be taken into account and that tend to limit the size of the condenser and/or the quantity of condensing medium circulated are initial cost, available space, and the power requirements of the fan, blower, or pump circulating the condensing medium. Too, where water is used as the condensing medium and the water leaving the condenser is wasted to the sewer (see Section 14-8), the availability and cost of the water must also be considered.

The limitations imposed on condenser size by the factors of initial cost and available space are self-evident. As for the power requirements of the fan, blower, or pump circulating the condensing medium, it has already been stated that the power required to circulate the condensing medium increases as the flow rate increases. If the flow rate is increased beyond a certain point, the increase in the power required to circulate the condensing medium will more than offset the reduction in the power requirements of the compressor accruing from the lower condensing temperature.

Obviously, the optimum flow rate for the condensing medium is the one which will result in the lowest over-all operating costs for the system. This will vary somewhat with the conditions of the individual installation, being influenced by the type of application, the size and type of condenser used, fouling rates, and the design conditions for the region, along with such practical considerations as the cost and availability of water, utility costs, local codes and restrictions, etc. For example, since good system efficiency prescribes lower condensing temperatures for low temperature applications than for high temperature applications, it follows that for the same condenser load the

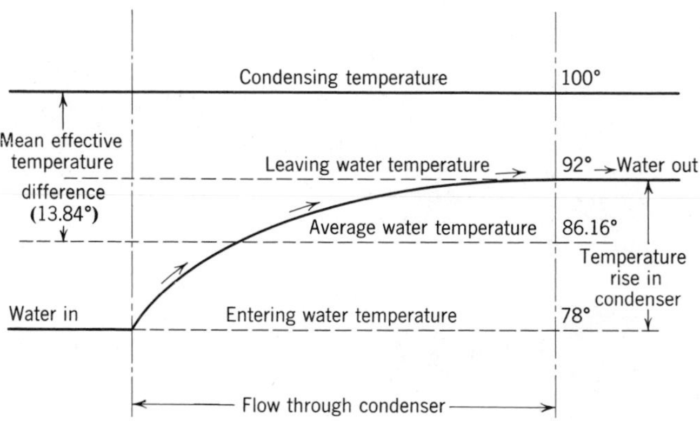

Fig. 14-1 Water temperature rise through condenser.

optimum condensing medium flow rate will usually be higher for a low temperature application than for a high temperature application. Too, where the entering temperature of the condensing medium is relatively high, larger condensing surfaces and higher flow rates are required to provide reasonable condensing temperatures than where the entering temperature of the condensing medium is lower.

14-5. Air-Cooled Condensers

The circulation of air over an air-cooled condenser may be either by natural convection or by action of a fan or blower. Where air circulation is by natural convection, the air quantity circulated over the condenser is low and a relatively large condensing surface is required. Because of their limited capacity natural convection condensers are used only on small applications, principally domestic refrigerators and freezers.

Natural convection condensers employed on domestic refrigerators are usually either plate surface or finned tubing. When finned tubing is used, the fins are widely spaced so that little or no resistance is offered to the free circulation of air. Too, wide fin spacing reduces the possibility of the condenser being fouled with dirt and lint.

The plate-type condenser is mounted on the back of the refrigerator in such a way that an air flue is formed to increase air circulation. Finned tube condensers are mounted either on the back of the refrigerator or at an angle underneath the refrigerator. Regardless of condenser type or location, it is essential that the refrigerator be so located that air is permitted to circulate freely through the condenser at all times. Too, warm locations, such as one adjacent to an oven, should be avoided whenever possible.

Air-cooled condensers employing fans or blowers to provide "forced-air" circulation can be divided into two groups according to the location of the condenser: (1) chassis-mounted and (2) remote.

A chassis-mounted air-cooled condenser is one that is mounted on a common chassis with the compressor and compressor driver so that it becomes an integral part of a packaged air-cooled "condensing unit" (Figs. 6-12 and 14-2). A "remote" air-cooled condenser is one that is located separately and usually some distance away from the compressor. Remote

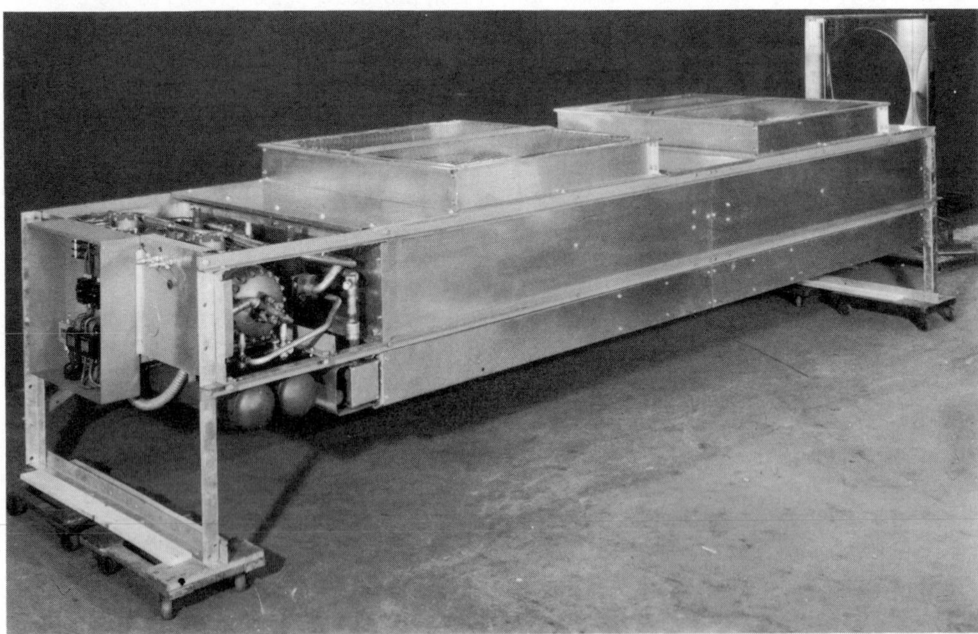

Fig. 14-2 Air-cooled condensing unit designed for remote installation. Notice generous size of condenser. (Courtesy Kramer Trenton Company.)

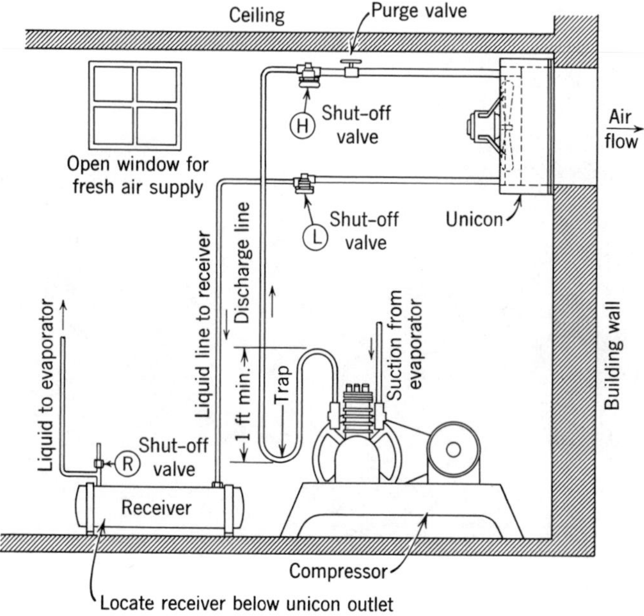

Fig. 14-3 Indoor installation of air-cooled condenser. (Courtesy Kramer Trenton Company.)

air-cooled condensers may be located either indoors or outdoors. When located indoors, provisions must be made for an adequate supply of outside air to the condenser (Fig. 14-3). If the condenser is installed in a warm location, such as in an attic or boiler room, ducts should be used to carry the air into the condenser and back to the outside. Because of the large quantity of air required, only the smaller sizes are mounted indoors.

When located outdoors, the air-cooled condenser may be mounted on the ground, on the roof, or on the side of a wall, with roof locations being the most popular. Typical wall and roof installations are shown in Figures 14-4 and 14-5, respectively. In all cases, the condenser should be so oriented that the prevailing winds for the area in the summertime will aid rather than retard the action of the fan. In the event that such orientation is not possible, wind deflectors should be installed on the discharge side of the condenser (Fig. 14-6).

Fig. 14-4 Remote air-cooled condensers installed on outside wall. (Courtesy Kramer Trenton Company.)

Air-cooled condensers are available in a variety of designs for either vertical or horizontal installation, and in sizes ranging from less than 1 ton up through 100 tons or more. Some are designed with two or more separate refrigerant circuits and can be used to serve several different refrigerating systems, including those employing different refrigerants.

Fig. 14-5 Remote air-cooled condensers mounted on roof. Notice that the condenser on the left is designed with multiple refrigerant circuits. (Courtesy Dunham-Bush, Inc.)

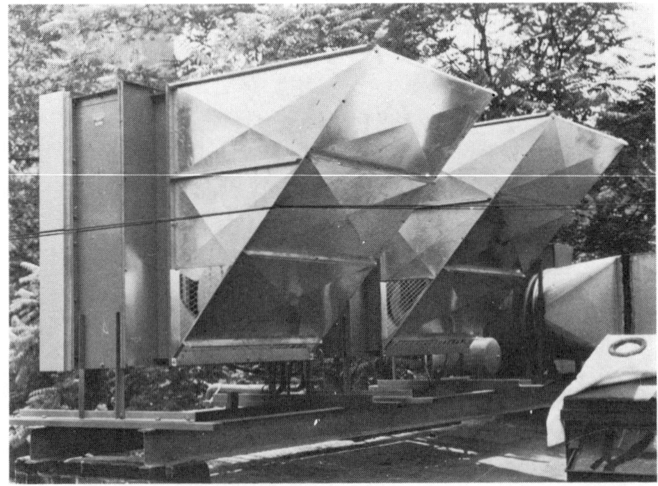

Fig. 14-6 Remote air-cooled condensers equipped with wind deflectors. (Courtesy Kramer Trenton Company.)

Other models are equipped with a liquid subcooling circuit. To ensure that the subcooling circuit performs effectively, these condensers should be employed without a liquid receiver tank in the system. Where a liquid receiver is used for pump-down, it should be installed upstream of the subcooling circuit.

14-6. Air Quantity and Velocity

For an air-cooled condenser there is a definite relationship between the size (face area) of the condenser and the quantity of air circulated in that the velocity of the air through the condenser is critical within certain limits. Good design prescribes the minimum air velocity that will produce turbulent flow and a high transfer coefficient. Increasing the air velocity beyond this point causes an excessive pressure drop through the condenser and results in an unnecessary increase in the power requirements of the fan or blower circulating the air.

The velocity of the air passing through an air-cooled condenser is a function of the face area of the condenser and the quantity of air circulated. The relationship is given in the following equation:

$$\text{Air velocity (fpm)} = \frac{\text{Air quantity (cfm)}}{\text{Face area (sq ft)}}$$

Normally, air velocities over air-cooled condensers are between 500 and 1000 fpm. However, because of the many variables involved, the optimum air velocity for a given condenser design is best determined by experiment. For this reason, most air-cooled condensers come from the factory already equipped with fans or blowers so that the air quantity and velocity over the condenser are fixed by the manufacturer. In all cases, to realize peak performance from an air-cooled condenser, the manufacturer's recommendations should be carefully followed.

14-7. Rating and Selection of Air-Cooled Condensers

Capacity ratings for air-cooled condensers are usually given in Btu per hour for various operating conditions. With the surface area and the value of U being fixed at the time of manufacture, the capacity of any one condenser depends only on the mean temperature difference between the air and the condensing refrigerant. Since most air-cooled condensers come equipped with fans or blowers, the quantity of air circulated over the condenser is also fixed so that the average temperature of the air passing over the condenser depends only on the dry bulb temperature of the entering air and the load on the condenser. In such cases, the capacity of the condenser is directly proportional to the temperature difference (TD) between the DB temperature of the entering air and the temperature of the condensing refrigerant.

In practice, the condenser TD ranges from 15°F to 35°F. Condenser TDs in the higher range should be limited to high-temperature applications and/or regions where the outdoor design DB temperature is relatively low. For refrigeration duty, a condenser design TD that will provide a condensing temperature of 110°F to 120°F will usually result in an economical condenser size. Using this as a basis for selecting the condenser, the design TD, and consequently the size of the condenser, will depend on the outdoor design DB temperature. The higher the DB temperature, the larger the condenser required. For example, for a condensing temperature of 110°F, if the DB temperature is 85°F, the condenser can be selected for a 25°F TD, whereas if the DB temperature is 90°F, the condenser should be selected for a 20°F TD, which will require a larger size.

Table R-14 is a typical manufacturer's rating table for air-cooled condensers. The capacity ratings listed in the table are based on a 30°F TD. For other design TDs, the condenser load is multiplied by the appropriate correction factor from Table R-14B before making a condenser selection from Table R-14A.

Since air-cooled condenser ratings are normally based on standard air density (sea level), if condensers are to be used at higher altitudes, the condenser total heat rejection (THR) should be corrected for the lower air densities by dividing the required condenser THR by the appropriate correction factor from Table 14-1C.

TABLE 14-1C Altitude Correction Factor

Altitude (Feet)	Correction Factor
0	1.0
500	0.989
1000	0.977
1500	0.966
2000	0.955
2500	0.945
3000	0.934
3500	0.924
4000	0.913
4500	0.903
5000	0.893
5500	0.883
6000	0.873
6500	0.863
7000	0.854

Courtesy of Bohn Aluminum and Brass Company.

Example 14-9 An open-type compressor employing R-22 has a cooling capacity of 75,000 Btu/hr operating at an evaporator temperature of 40°F and a condensing temperature of 120°F. If the ambient DB temperature is 95°F and the altitude is 5000 ft, select an air-cooled condenser to meet the design conditions.

Solution From Table 14-1A, the heat rejection factor is 1.20. From Table R-14B, the correction factor for a 25°F TD is 1.20, and the altitude correction factor from Table 14-1C is 0.893. Applying all these correction factors, the corrected condenser load (THR) is

$$Q_c = \frac{(75{,}000 \text{ Btu/hr})(1.20)(1.20)}{(0.893)}$$

$$= 120{,}940 \text{ Btu/hr}$$

Reference to Table R-14A indicates that Unit Size 8 has a capacity of 126,000 Btu/hr and will satisfy the design conditions.

Example 14-10 A suction-cooled hermetic compressor employing R-12 has a cooling capacity of 37,000 Btu/hr when operating with an evaporator temperature of 20°F and a condensing temperature of 110°F. If the ambient air DB temperature is 90°F, select an air-cooled condenser to satisfy design conditions.

Solution From Table 14-1B, the heat rejection factor is 1.33, and from Table R-14B, the correction factor for a 20°F TD is 1.50. Applying these two correction factors, the corrected THR is

$$Q_c = (34{,}000 \text{ Btu/hr})(1.33)(1.50)$$
$$= 67{,}830 \text{ Btu/hr}$$

Reference to Table R-14A indicates that Unit Size 5 has a capacity of 75,000 Btu/hr and will satisfy the design conditions. Using this size condenser, the actual TD will be

$$\frac{(\text{Design TD})(\text{Design THR})}{\text{Actual THR}} = \frac{(20°)(67{,}839 \text{ Btu/hr})}{75{,}000 \text{ Btu/hr}}$$

$$= 18.1°\text{F}$$

The actual condensing temperature at design conditions will be (90° = 18.1°) 108.1°F.

Example 14-11 (Selection procedure for multicircuited air-cooled condensers reprinted directly from manufacturer's engineering data).* Given six suction-cooled hermetic compressors with the type of refrigerant, design temperature difference (TD), evaporator temperature, and compressor capacities shown in the tabulation in Fig. 14-7. The design ambient temperature is 95°F.

Solution

1. Tabulate customer data in columns 2, 3, 4, and 5.
2. From Table 14-1B, select the heat rejection factor applicable to each evaporator temperature and condensing temperature (condensing temperature = ambient temperature F plus Design TD), and tabulate in column 6.
3. From Table R-14B, select the TD correction factor applicable to each design TD and tabulate in column 7. This converts capacities to the equivalent of 30° TD.
4. Multiply items in columns 5, 6, and 7, for each circuit and list in column 8.
5. Add all capacities in column 8 to obtain the total Btu per hour heat rejection required at 30° TD and use the total to select the proper size unit.

Based on the total heat rejection capacity for the six compressors of 280,139 Btu/hr at 30° TD, Table R-14A, shows the smallest unit that will meet the requirement is a Size 21, with a THR of 291,000 Btu/hr (R-12). Table R-14A, also lists the heat rejection capacity per tube circuit for Refrigerants-12, 22, and 502. The per tube circuit capacity applicable to the type refrigerant used is tabulated in column 9.

To determine the number of tube circuits required per circuit, divide column 8 by column 9, for each circuit, and list in column 10. If the decimal part of the tube circuits required is less than 10% of the whole number, then

* Bohn Aluminum and Brass Company.

1	2	3	4	5	×	6	×	7	=	8	÷	9	=	10	11
Circ. No.	Type Refr.	Design TD °F	Evap. Temp. °F	Comp. Capacity Btu/hr		Heat Reject. Factor	×	TD Corr. Factor	=	Total Heat Rejection Btu/hr	÷	Capacity per tube Circuit Table R-14A	=	No. Tube Circuits Req'd	No. Tube Circuits Assign.
1	R-502	25	+20	12,300	×	1.35	×	1.20	=	19,926	÷	18,750	=	1.06	1
2	R-12	15	−10	16,500	×	1.50	×	2.00	=	49,500	÷	18,187	=	2.72	3
3	R-22	30	+40	24,900	×	1.29	×	1.00	=	32,121	÷	19,125	=	1.68	2
4	R-12	15	−20	19,000	×	1.58	×	2.00	=	60,040	÷	18,187	=	3.30	3
5	R-12	20	−10	31,700	×	1.535	×	1.50	=	72,989	÷	18,187	=	4.01	4
6	R-502	20	+20	22,500	×	1.35	×	1.50	=	45,563	÷	18,750	=	2.43	3
										280,139					16

Fig. 14-7 Sample tabulation.

drop the decimal and enter the whole number in column 11. If the decimal is greater than 10% of the whole number, then round off to the next higher whole number and enter in column 11.

The sample tabulation indicates that 16 circuits are required, and Table R-14A shows that Unit Size 21 has 16 circuits available and therefore is the proper condenser selection.

If the total number of circuits required exceeds the number of circuits available, as listed in Table R-14A, it will be necessary to permit a slightly higher condensing temperature than specified for one or two of the circuits. A second alternative would be to select the next larger size unit.

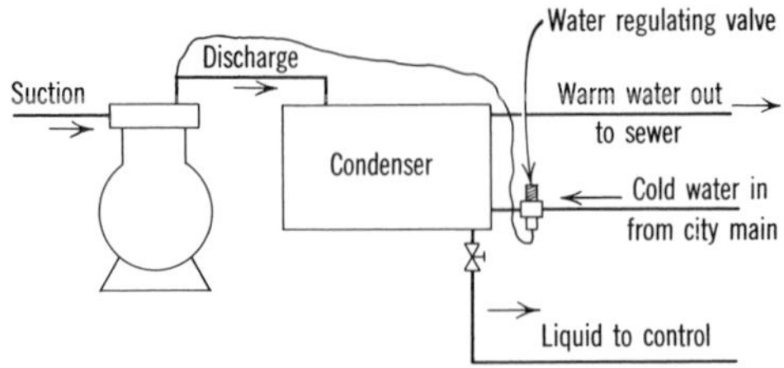

Fig. 14-8 Waste water system.

14-8. Water-Cooled Condenser Systems

Systems employing water-cooled condensers can be divided into two general categories: (1) waste-water systems and (2) recirculated water systems. In waste-water systems the water supply for the condenser is usually taken from the city main and wasted to the sewer after passing through the condenser (Fig. 14-8). In recirculated water systems the water leaving the condenser is piped to a water cooling tower where its temperature is reduced to the condenser entering temperature, after which the water is recirculated through the condenser (Fig. 14-9).

Naturally, where the condenser water is wasted to the sewer, the availability and cost of the water are important factors in determining the quantity of water circulated per unit of condenser load. As a general rule, an economical balance between water and power costs prescribes a water flow rate of approximately 1.5 gpm per ton of capacity.

The high cost of water, along with limited sewer facilities and recurring water shortages in many regions, has tended to limit waste-water systems to very small sizes. Too, many cities have placed severe restrictions on waste-water systems, particularly where the water supply is taken from the city main and wasted to the sewer.

When the condenser water is recirculated the power required to circulate the water

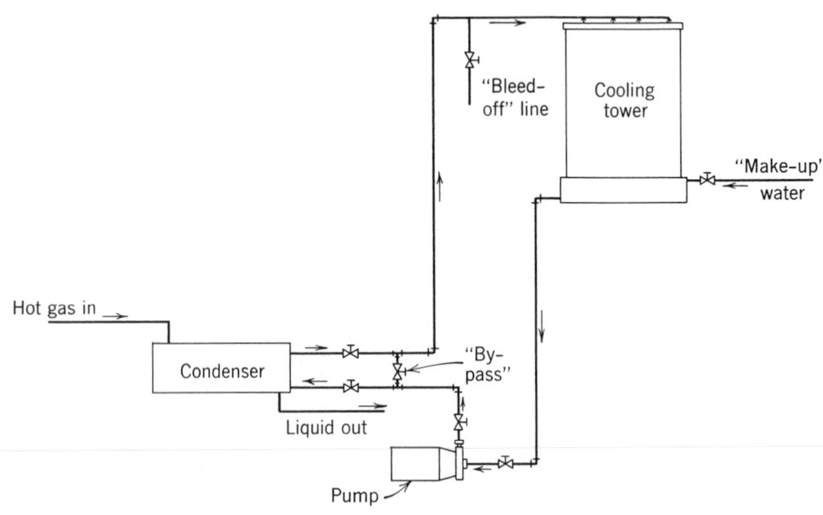

Fig. 14-9 Recirculating water system.

through the water system must be taken into account in determining the water flow rate. Experience has shown that, in general, a water flow rate of between 2.5 and 3 gpm per ton usually provides the most economical balance between the power required by the compressor and that required by the pump.

In some instances, the water supply for a waste-water system is taken from a well or from some nearby body of water, such as a river, lake, pond, etc., in which case both the cost of the water and the power requirements of the pump must be considered in determining the optimum water flow rate.

To a large extent, the quantity of water circulated through the condenser determines the design of the water circuit in the condenser. Since heat transfer is a function of time, it follows that where low water quantities necessitate a high temperature rise in the condenser, the water must remain in contact with the condensing refrigerant for a longer period than when the water flow rate is high and the temperature rise required is smaller. Hence, where the water flow rate is low, the number of water circuits through the condenser are few and the circuits are long so that the water will remain in the condenser for enough time to permit the required amount of heat to be absorbed. On the other hand, when the flow rate is high and the temperature rise low, more circuits are used and the circuits are shorter in order to reduce the pressure drop to a minimum. This is illustrated in Figs. 14-10*a* and 14-10*b*. In Fig. 14-10*a*, the two water circuits through the condenser are connected in series for a low flow rate and a high temperature rise. The water enters through opening *A* and leaves through opening *C*. Opening *B* is capped. In Fig. 14-10*b*, the two water circuits are connected in parallel for a high flow rate and a low temperature rise. The water enters through openings *A* and *C* and leaves through opening *B*.

It is of interest to notice that in Fig. 14-10 the pressure drop in circuit *b* is only one-eighth that in circuit *a*.* However, because of the higher velocity in circuit *a*, the heat transfer coefficient would be somewhat higher, and less condensing surface would be required for the same heat transfer capacity.

In designing the condenser water circuit particular attention must be given to the water velocity and pressure drop through the condenser. In all cases the minimum permissible velocity is that which will produce turbulent flow and a high transfer coefficient. Since pressure drop is a function of velocity, the pressure drop through the condenser increases as the water velocity increases. For this reason, the maximum permissible velocity in any one case is usually determined by the allowable pressure drop.† For waste-water systems, where the water is forced through the condenser by city main pressure, the pressure drop through the condenser is not critical as long as it is within the limits of turbulent flow

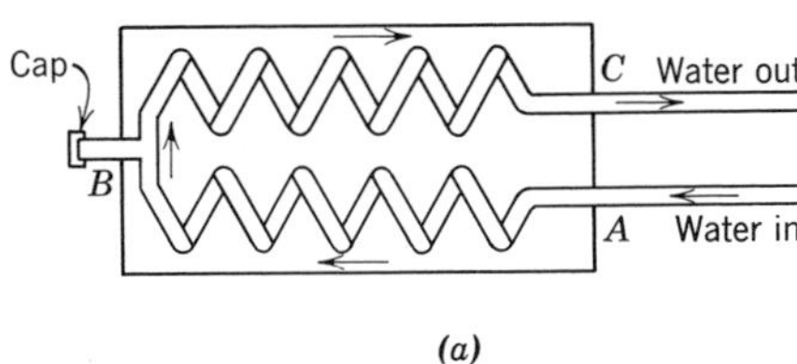

(a)

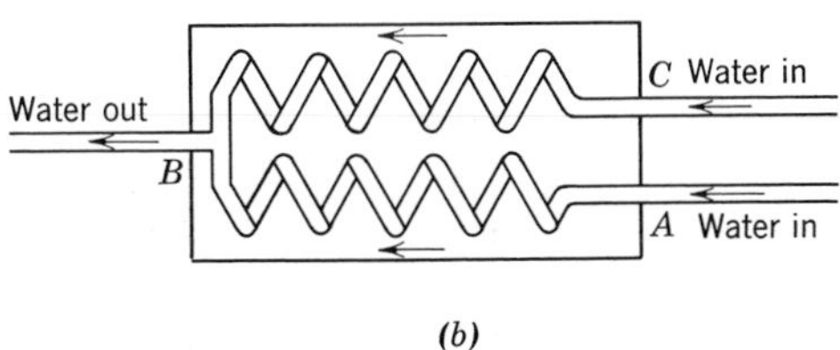

(b)

Fig. 14-10 (*a*) Water circuit connected for series flow. (*b*) Water circuit connected for parallel flow.

* Pressure drop increases as the square of the velocity. Reducing the velocity to one half reduces the pressure drop to one quarter of its original value. Then, since the length of each parallel circuit is only one-half the length of the single circuit, the drop in pressure through the parallel circuits is only one-eighth that of the single circuit, assuming that the total flow rate is the same in both cases.

† Excessive velocity will usually cause erosion of the water tubes, particularly at points where the water changes direction. For this reason, velocities should be maintained below 10 fps.

and the available city main pressure. In such cases, high velocities are recommended in order to take advantage of the higher transfer coefficient. On the other hand, when the water is circulated by action of a pump, a high pressure drop through the condenser will increase the pumping head and the power required to circulate the water. Therefore, for recirculating water systems, the optimum water velocity is one which will provide the most economical balance between a high transfer coefficient and a low pumping head.

14-9. Fouling Rates

Another factor which must be considered in selecting a water-cooled condenser is fouling of the tube surface on the water side. The fouling is caused primarily by mineral solids which precipitate out of the water and adhere to the tube surface. The scale thus formed on the tube not only reduces the water side transfer coefficient, but it also tends to restrict the water tube and reduce the quantity of water circulated, both of which will cause serious increases in the condensing pressure.

In general, the rate of tube fouling is influenced by (1) the quality of the water used with regard to the amount of impurities contained therein, (2) the condensing temperature, and (3) the frequency of tube cleaning with relation to the total operating time.

Most manufacturers of water-cooled condensers give condenser ratings for clean tubes and for several stages of tube fouling in accordance with the scale factors given in Table 14-2 for various types of water. These scale factors are an index of the reduction in the tube transfer coefficient resulting from the scale deposit. In selecting a water-cooled condenser, a minimum scale factor of 0.0005 should always be used. Under no circumstances should a condenser be selected on the basis of clean tubes. However, when the condensing temperature is low (leaving water temperature less than 100°F) and the condenser tubes are to be cleaned frequently, the fouling factor from Table 14-2 may be reduced to the next lowest value. The use of scale factors will be illustrated later in the chapter.

TABLE 14-2 Scale Factors—Water

Temperature of Water	125° F or Less	
Types of Water	Water Velocity Ft/Sec	
	3 Ft and Less*	Over 3 Ft†
Sea water	0.0005	0.0005
Brackish water	0.002	0.001
Cooling tower and artificial spray pond: Treated make-up	0.001	0.001
Untreated	0.003	0.003
City or well water (Such as Great Lakes)	0.001	0.001
Great Lakes	0.001	0.001
River water: Minimum	0.002	0.001
Mississippi	0.003	0.002
Delaware, Schuylkill	0.003	0.002
East River and New York Bay	0.003	0.002
Chicago Sanitary Canal	0.008	0.006
Muddy or silty	0.003	0.002
Hard (Over 15 grains) gal	0.003	0.003
Engine jacket	0.001	0.001
Distilled	0.0005	0.0005

* 2.16 gpm per tube is equivalent to a water velocity of 3 ft per second.

† This table is presented by permission of the Tubular Exchanger Manufacturers Association, Inc., New York.

14-10. Water-Cooled Condensers

Water-cooled condensers are of three basic types: (1) double-tube, (2) shell-and-coil, and (3) shell-and-tube.

As its name implies, the double-tube condenser consists of two tubes so arranged that one is inside of the other (Fig. 14-11). Water is piped through the inner tube while the refrigerant flows in the opposite direction in the space between the inner and outer tubes. With this arrangement, some air-cooling of the refrigerant is provided in addition to the water-cooling. Counterflowing of the fluids in any type of heat exchanger is always desirable since it results in the greatest mean temperature difference between the fluids and, therefore, the highest rate of heat transfer.

Several types of double-tube condensers are shown in Figs. 14-12 and 14-13. The type shown in Fig. 14-12 can be cleaned mechanically by removing the end-plates (inset). The type shown in Fig. 14-13 is cleaned by circulating approved chemicals through the water tubes (see Section 14-23).

Equipped with water-regulating valves (Section 14-19), double-tube condensers make excellent "booster" condensers for use with chassis-mounted air-cooled condensers during periods of peak loading. Since the water valve can be adjusted to open and allow water

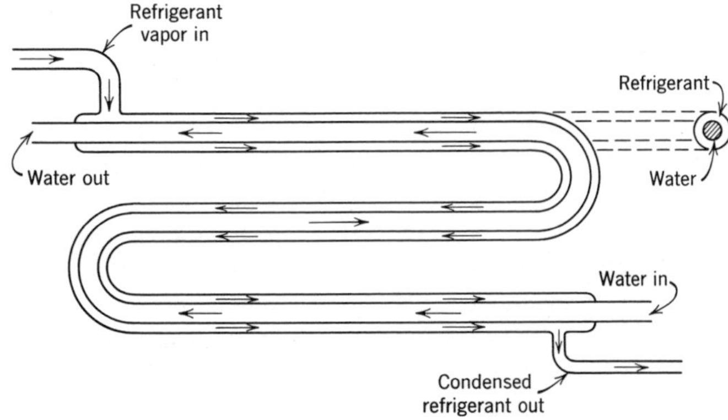

Fig. 14-11 Double-tube water-cooled condenser.

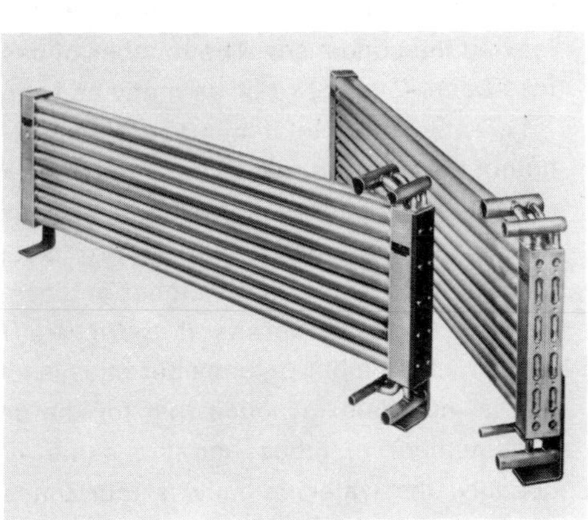

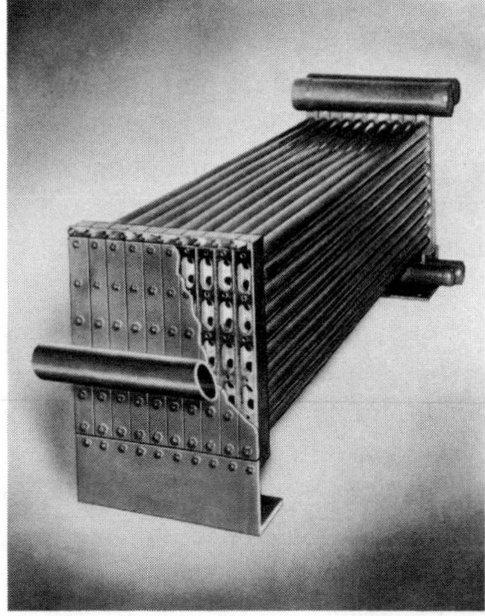

Fig. 14-12 Double-pipe condensers with mechanically cleanable tubes. (Courtesy Halstead and Mitchell.)

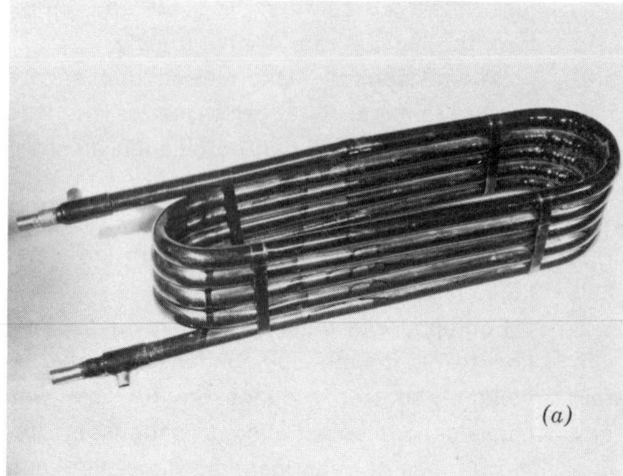

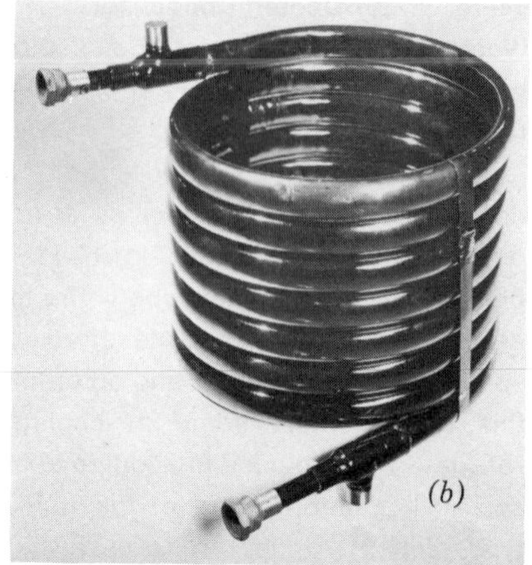

Fig. 14-13 Typical double-pipe condenser configurations. (*a*) Trombone configuration. (*b*) Helix configuration. (Courtesy Edwards Engineering Corporation.)

to flow through the condenser only when the condensing pressure rises to some predetermined level, the amount of water used is relatively small in comparison to the savings in power afforded by the increased compressor efficiency.

The shell-and-coil condenser is made up of one or more bare-tube or finned-tube coils enclosed in a welded steel shell (Fig. 14-10). The condensing water circulates through the coils while the refrigerant is contained in the shell surrounding the coils. Hot refrigerant vapor enters at the top of the shell and condenses as it comes in contact with the water coils. The condensed liquid drains off the coils into the bottom of the shell, which often serves also as the receiver tank. Care should be taken not to overcharge the system with refrigerant since an excessive accumulation of liquid in the condenser will tend to cover too much of the condensing surface and cause an increase in the discharge temperature and pressure.

Most shell-and-coil condensers are equipped with a split water circuit. The two parts of the circuit are connected in series for waste-water systems (Fig. 14-10*a*) and in parallel for recirculating systems (Fig. 14-10*b*). As a general rule, shell-and-coil condensers are used only for small installations up to approximately 10 tons capacity.

Shell-and-coil condensers are cleaned by circulating an approved chemical through the water coils.

The shell-and-tube condenser consists of a cylindrical steel shell in which a number of straight tubes are arranged in parallel and held in place at the ends by tube sheets. Construction is almost identical to that of the flooded-type shell-and-tube liquid chiller. The condensing water is circulated through the tubes, which may be either steel or copper, bare or extended surface. The refrigerant is contained in the steel shell between the tube sheets. Water circulates in the annular spaces between the tube sheets and the end-plates, the end-plates being baffled to act as manifolds to guide the water flow through the tubes. The arrangement of the end-plate baffling determines the number of passes the water makes through the condenser from one end to the other before leaving the condenser. The number of passes may be as few as two or as many as twenty.

For any given total number of tubes, the number of tubes per pass varies inversely with the number of passes. For example, assuming that a condenser has a total of forty tubes, if there are two passes, the number of tubes per pass is twenty, whereas if there are four passes, the number of tubes per pass is ten.

It is important to notice that for the same total number of tubes and the same water quantity, the water velocity is four times as great and the pressure drop through the condenser will be eight times as great for a four-pass condenser as for a two-pass condenser. Because of the higher velocity the transfer coefficient will be higher for the four-pass condenser and a smaller condensing surface will be required for a given heat transfer capacity. However, on the other hand, because of the high pressure drop, the power required to circulate the water will be greater. Hence, for a waste-water system, the four-pass condenser is probably the best selection, whereas for a recirculating system, the two-pass condenser is probably the better of the two. This example is intended only to illustrate the principles of design and should not be construed to mean that four-pass condensers are undesirable for recirculating systems.

Shell-and-tube condensers are available in capacities ranging from 2 tons up to several hundred tons or more. Shell diameters range from approximately 4 in. up to 60 in., whereas tube length varies from approximately 3 ft to 20 ft. The number and the diameter of the tubes depend on the diameter of the shell. Tube diameters of $\frac{5}{8}$ in. through 2 in. are common, whereas the number of tubes in the condenser varies from as few as six or eight to as many as a thousand or more. The end-plates of the condenser are removable to permit mechanical cleaning of the water tubes.

Single-pass, vertical shell-and-tube condensers are sometimes employed on large ammonia installations. The construction of the vertical shell-and-tube condenser is similar to that of the vertical shell-and-tube chiller illustrated in Fig. 11-44. The vertical condenser is equipped with a water box at the top to distribute the water to the tubes and a drain at the bottom to carry the water away. Each tube is equipped at the top with a distributor fitting which imparts a rotating motion to the water to assure adequate wetting of the tube. The hot refrigerant vapor usually enters at the side of the shell near the middle of the condenser and the liquid leaves the condenser at the side of the shell near the bottom. The height of vertical shell-and-tube condensers ranges from 12 ft to 18 ft. The tubes are mechanically cleanable and are readily cleaned while the system is still in operation, a circumstance that makes these condensers ideally suited for those installations where poor water quality and/or other operating conditions cause high scaling rates.

14-11. Rating and Selection of Water-Cooled Condensers*

The ratings shown in Table R-15 are based on condensing temperatures of 102° and 105°F, 20° and 10° water rise and 0.0005 scale factor which is the minimum recommended in ARI standards.

Where other conditions exist, the following procedure should be followed in selecting the proper condenser.

Condensers must not be selected for less than 0.5 gpm per tube below which streamline instead of turbulent water flow occurs. ARI standards indicate that the water velocity should not exceed 8 fps which is 5.75 gpm per tube for Acme STF and SRF condensers.

It is necessary to have the following information to select a proper condenser:

1. Total tons (low side),
2. Evaporator temperature,
3. Condensing temperature,
4. Water temperature "in,"
5. Water temperature "out," or gpm available,
6. Type of water or required scale factor.

Then proceed as follows:

1. Determine the corrected tons to be used in selecting the proper condenser by reference to Fig. 2, Table R-15. The factor obtained for the desired evaporator temperature and condensing temperature is multiplied by the actual tons to obtain corrected tons.
2. Determine the water temperature rise and gpm per ton. Knowing either factor, the other may be obtained by reference to Fig. 3, Table R-15. Use corrected tons to determine the total gallons per minute required.
3. Determine the temperature differences between the condensing temperature and the "water in" and "water out" temperatures and find the METD by referring to Table 11-1.
4. Make preliminary selection of condenser shell diameter by reference to Table R-15, basing the selection on the corrected tons found in step 1. Find the number of tubes

* The material in this section is reprinted directly from the manufacturer's catalog, the only alteration being the table designations. Courtesy of Acme Industries, Inc.

per pass and then by referring to step 2, find the gallons per minute per tube.

5. Select the desired scale factor by reference to Table 14-2, which suggests scale factors for various types of water. The most commonly used factor is 0.0005 and it should be borne in mind when selecting a factor that a determination is being made of the frequency of cleaning which will be required.
6. Referring to Fig. 1, Table R-15, determine the rate of heat transfer "U" for the gallons per minute per tube in step 4 and the scale factor in step 5.
7. Calculate the surface required by use of the following formula.

   Square feet of surface

$$= \frac{\text{Corrected tons} \times 14{,}400}{U \times \text{METD}}$$

8. Select a condenser having at least the required surface from Table R-15. Be sure to use the shell diameter determined in the preliminary selection of step 4.
9. Make final checks on selection.

   a. Using the gallons per minute per tube from step 4 and the nominal tube length shown in Table R-15 for the model selected in step 8, refer to Fig. 4 of Table R-15 to obtain water pressure drop through condenser.

   b. Obtain nominal operating charge from the last column of Table R-15. This is the maximum weight of liquid refrigerant which can be allowed in the shell during the operating period covering some of the lower tubes. Larger shell diameters or separate receivers may be used where greater storage capacity is needed during operation.

   c. Determine the pump down capacity from Table R-15. If less than the total weight of refrigerant to be used in the system and provision for complete pump-down are required, an additional receiver should be used.

Example 14-12 Select an R-12 condenser to meet the following conditions:

Refrigeration load	30 tons
Condensing temperature	100°F
Suction temperature	30°F
Water available 2 gpm/ton river water reasonably clean at	78°F
Maximum tube length	12 ft
Maximum water pressure drop	7.5 psi

Solution

1. From Fig. 2, the correction factor for 30°F suction temperature and 100°F condensing temperature is 1.013.
   Corrected tons 30 × 1.013 = 30.4 tons
2. From Fig. 3, for 2 gpm/ton the water temperature rise is found to be 14.4°.
   Total gpm 30.4 × 2 = 60.8
   Water "out" temperature 78 plus 14.4 = 92.4°F
3. GTD 100 − 78 = 22°
   LTD 100 − 92.4 = 7.6°
   From Table 11-1, METD = 13.55°F
4. Refer to Table R-15. Use of four passes will usually give an economical selection for 75°F water in and 95°F water out which approximates the required water conditions. Note that a $10\frac{3}{4}$ shell will probably be needed. This shell has 60 tubes.

   gpm per tube

$$\frac{\text{Total gpm} \times \text{number of passes}}{\text{Number of tubes in condenser}}$$

$$= \frac{60.8 \times 4}{60}$$

$$= 4.05 \text{ gpm per tube}$$

5. Referring to Table 14-2, for clean river water and over 3 fpm velocity, the suggested scale factor is 0.001.
6. From Fig. 1, the U factor for 4.05 gpm per tube and 0.001 scale factor is 121.5 Btu/hr per square foot of extended surface per °F METD.
7. Square feet required

$$\frac{\text{Corrected tons} \times 14{,}400}{U \text{ factor} \times \text{METD}}$$

$$= \frac{30.4 \times 14{,}400}{121.5 \times 13.55} = 266 \text{ sq ft}$$

8. Referring to Table R-15, a Model STF-1010 has 289 ft^2 external tube surface and should be selected. When installed the water con-

nection should be made for four-pass operation.

9. (a). For water pressure drop, refer to Fig. 4 and note that the pressure drop for 4.05 gpm per tube in an STF-1010 condenser connected for four passes is 7.1 psi.

   (b). Table R-15 shows a nominal operating charge of 38 lb of R-12, which will normally be sufficient for a 30-ton installation. However, if more operating storage is needed, a separate receiver may be chosen, or alternately a different condenser selection may be made if more economical.

   (c) Table R-15 also shows pump-down capacity which is 252 lb of R-12. Usually this will be sufficient, but if greater pump-down capacity is required, a separate receiver tank must be used.

14-12. Simplified Ratings

Simplified ratings, based on the horsepower of the compressor driver, are available for most air-cooled and water-cooled condensers, particularly in smaller sizes. Since the power required by the compressor varies with both the evaporator load and the compression ratio, it provides a reasonable index of the condenser load at all operating conditions. Table R-16, which applies to double-tube condensers of the type shown in Fig. 14-12, is a typical simplified condenser rating table.

14-13. Cooling Towers

Cooling towers are essentially water conservation or recovery devices. Warm water from the condenser is pumped over the top of the cooling tower from where it falls or is sprayed down to the tower basin. The temperature of the water is reduced as it gives up heat to the air circulating through the tower.

Although there is some sensible heat transfer from the water to the air, the cooling effect in a cooling tower results almost entirely from the evaporation of a portion of the water as the water falls through the tower. The heat to vaporize the portion of water that evaporates is drawn from the remaining mass of the water so that the temperature of the mass is reduced. The vapor resulting from the evaporating process is carried away by the air circulating through the tower. Since both the temperature and the moisture content of the air are increased as the air passes through the tower, it is evident that the effectiveness of the cooling tower depends to a large degree on the wet bulb temperature of the entering air. The lower the wet bulb temperature of the entering air, the more effective is the cooling tower.

Other factors that influence the performance of cooling towers are (1) the amount of exposed water surface and the length (time) of exposure, (2) the velocity of the air passing through the tower, and (3) the direction of the air flow with relation to the exposed water surface (parallel, transverse, or counter).

The exposed water surface includes (1) the surface of the water in the tower basin, (2) all wetted surfaces in the tower, and (3) the combined surface of the water droplets falling through the tower.

Theoretically, the lowest temperature to which the water can be cooled in a cooling tower is the wet bulb temperature of the entering air, in which case the water vapor in the leaving air will be saturated. In practice, it is not possible to cool the water to the wet bulb temperature of the air. In most cases, the temperature of the water leaving the tower will be 7° to 10°F above the wet bulb temperature of the entering air. Too, the air leaving the tower will always be somewhat less than saturated.

The temperature difference between the temperature of the water leaving the tower and the wet bulb temperature of the entering air is called the tower "approach." As a general rule, all other conditions being equal, the greater the quantity of water circulated over the tower the closer the leaving water temperature approaches the wet bulb temperature of the air. However, the quantity of water which can be economically circulated over the tower is somewhat limited by the power requirements of the pump.

The temperature reduction experienced by the water in passing through the tower (the difference between the entering and leaving water temperatures) is called the "range" of the tower. Naturally, to maintain equilibrium in the condenser water system, the tower "range" must always be equal to the temperature rise of the water in the condenser. An exception to this is where a condenser bypass is employed (see Section 14-16).

The load on a cooling tower can be approximated by measuring the water flow rate over the tower and the entering and leaving water temperatures. The following equation is applied:

Tower load (Btu/min) = flow rate (gpm) × 8.33 × (entering water temperature − leaving water temperature) (14-12)

Example 14-13 Determine the approximate load on a cooling tower if the entering and leaving water temperatures are 96°F and 88°F, respectively, and the flow rate of the water over the tower is 30 gpm.

Solution Applying 14-12, the tower load (Btu/min) $= 30 \times 8.33 \times (96 - 88)$
$= 2000$ Btu/min

Since the load on the tower is equal to the load on the condenser, the approximate refrigerating capacity of the system can be computed by dividing the tower load by the condenser load in Btu per minute per ton corresponding to the operating conditions of the system.

Example 14-14 Determine the approximate capacity of a refrigerating system connected to the cooling tower in Example 14-13 if the evaporating and condensing temperatures are 0°F and 100°F, respectively, and the system employs an open-type compressor.

Solution From Table 14-1A, the heat rejection factor is 1.28, so that the load on the condenser is (200 × 1.28) 256 Btu/(min)(ton). The approximate system capacity

$$= \frac{\text{Total tower load (Btu/min)}}{\text{Condenser Load Btu/(min)(ton)}} = \frac{2000}{256}$$

$= 7.8$ tons

Since the heat absorbed per pound of water evaporated is approximately 1000 Btu, assuming a condenser load of 250 Btu/(min)(ton), the quantity of water evaporated per ton of refrigeration (evaporator) is approximately 0.25 lb per minute or 2 gal per hour.

In addition to the water lost by evaporation, water is lost from the cooling tower by "drift" and by "bleed-off". A small amount of water in the form of small droplets is entrained and carried away by the air passing through the tower. Water lost in this manner is called the drift loss. The amount of drift loss from a tower depends on the design of the tower and the wind velocity.

"Bleed-off" is the continuous or intermittent wasting of a certain percentage of the circulated water in order to avoid a build-up in the concentration of dissolved mineral solids and other impurities in the condenser water. Without bleed-off the concentration of dissolved mineral solids in the condenser will build up quite rapidly as a result of the evaporation taking place in the cooling tower. Since the scaling rate is proportional to the quality of the water, as the concentration of mineral solids in the water increases the scaling rate also increases.

The amount of bleed-off required to maintain the concentration of dissolved mineral solids at a reasonable level depends upon the cooling range, the water flow rate, and the initial water conditions. Suggested bleed-off rates for various cooling ranges are given in Table 14-3. To determine the quantity of water loss by bleed-off, multiply the water flow rate over the tower by the factor obtained from Table 14-3.

Example 14-15 Determine the quantity of water lost by bleed-off if the water flow rate over the tower is 30 gpm and the range is 10°F.

Solution From Table 14-3, the percent bleed-off required $= 0.33\%$

The quantity of water lost by bleed-off $= 30 \text{ gpm} \times 0.0033$
$= 0.099$ gpm

The bleed-off line should be located in the hot water return line near the top of the tower so that water is wasted only when the pump is running (Fig. 14-9).

TABLE 14-3 Bleed-Off Rates

Cooling Range Deg. F	Percent Bleed-off
6	0.15
$7\frac{1}{2}$	0.22
10	0.33
15	0.54
20	0.75

Courtesy The Marley Company.

Makeup water, to replace that lost by evaporation, drift, and bleed-off, is piped to the tower basin through a float valve which tends to maintain a constant water level in the basin.

14-14. Cooling Tower Design

According to the method of air circulation, cooling towers are classified as either natural draft or mechanical draft. When air circulation through the tower is by natural convection, the tower is called a natural draft or atmospheric tower. When air circulation through the tower is by action of a fan or blower, the tower is called a mechanical draft tower. Mechanical draft towers may be further classified as either induced draft or forced draft, depending on whether the fan or blower draws the air through the tower or forces (blows) it through. A schematic diagram of a spray-type natural draft tower is shown in Fig. 14-14. Schematic diagrams of induced draft and forced draft towers are shown in Figs. 14-15 and 14-16, respectively.

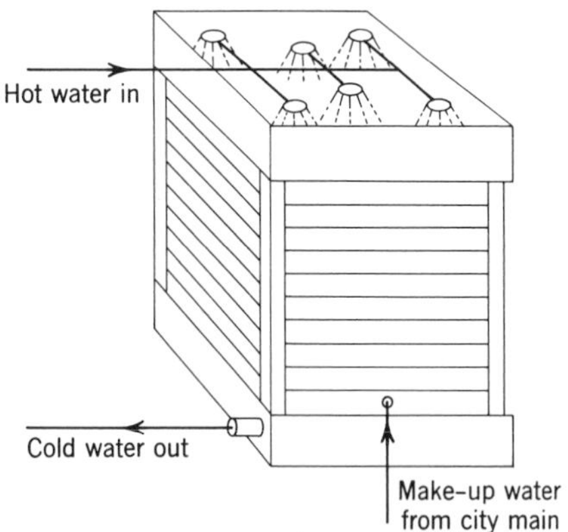

Fig. 14-14 Natural draft-cooling tower.

In the spray-type atmospheric tower, the warm water from the condenser is pumped to the top of the tower where it is sprayed down through the tower through a series of spray nozzles. Since the amount of exposed water surface depends primarily on the spray pattern, a good spray pattern is essential to high efficiency. The type of spray pattern obtained depends on the design of the nozzles. For most nozzle designs, a water pressure drop of 7 to 10 psi will produce a suitable spray pattern.

Some natural draft towers contain decking or filling (usually of redwood) to increase the amount of wetted surface in the tower and to break up the water into droplets and slow its fall to the bottom of the tower. Atmospheric

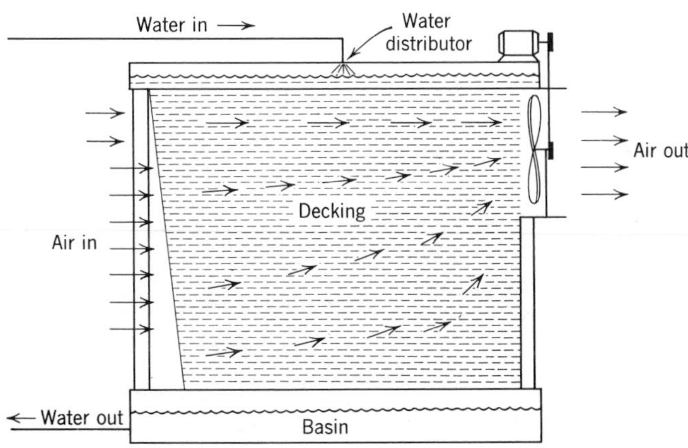

Fig. 14-15 Small induced draft-cooling tower.

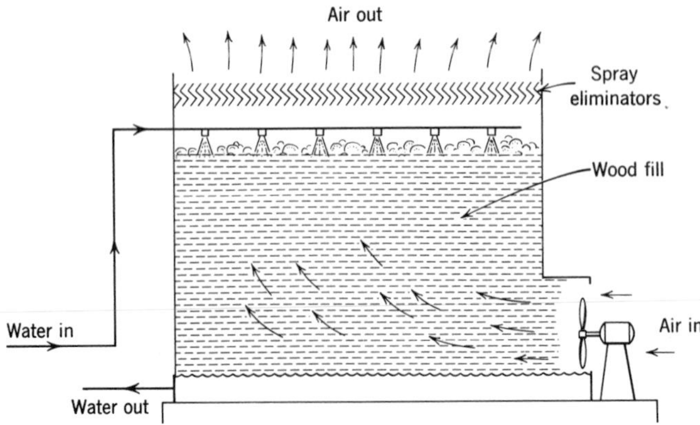

Fig. 14-16 Forced draft-cooling tower.

towers containing decking are called "splash-deck towers." Often, in splash-deck towers, no spray nozzles are used and the water is broken up into droplets by the "splash-impact" method.

The quantity and velocity of the air passing through a natural draft cooling tower depend on the wind velocity. Hence, the capacity of a natural draft tower varies with the wind velocity, as does the amount of "drift" experienced. Too, natural draft towers must always be located out-of-doors in places where the wind can blow freely through the tower. In commercial applications, roof installations are common.

Since air circulation through mechanical draft towers is by action of a fan or blower, small mechanical draft towers can be installed indoors as well as out-of-doors, provided that an adequate amount of outside air is ducted into and out of the indoor location. Too, since larger air quantities and higher velocities can be used, the capacity of a mechanical draft tower per unit of physical size is considerably greater than that of the natural draft tower. In addition, most mechanical draft towers contain some sort of decking or fill to improve further the efficiency. Spray eliminators must be used in mechanical draft towers to prevent excessive drift losses.

14-15. Cooling Tower Rating and Selection

Table R-17 contains rating data for the spray-type, natural draft cooling tower illustrated in Fig. 14-14 and is a typical cooling tower rating table. Notice that the tower ratings are given in tons, based on a heat transfer capacity of 250 Btu/(min)(ton). Nominal tower ratings are based on a 3 mph wind velocity, and 80°F design wet bulb temperature, and a water flow rate over the tower of 4 gpm per ton. Tower performance at conditions other than those listed in the table can be determined by using a rating correction chart.

To select the proper tower from the rating table, the following data must be known:

1. Desired tower capacity in tons (compressor capacity)
2. Design wet bulb temperature
3. Desired leaving water temperature (condenser entering water temperature or tower approach)

or

1. Desired flow rate over the tower (gallons per minute)
2. Design wet bulb temperature
3. Desired entering and leaving water temperatures (tower cooling range and tower approach)

Example 14-16 From Table R-17, select a cooling tower to meet the following conditions:

1. Required tower capacity = 20 tons
2. Design wet bulb temperature = 78°F
3. Desired leaving water temperature = 86°F

Solution From Table R-17, select tower, Model #CSA-66, which has a capacity of 20.7 tons at the desired conditions when the flow rate over the tower is 3 gpm per ton. Hence, for 20-tons capacity, a total of 60 gpm (20×3) must be circulated over the tower. As shown in the table, the entering water temperature will be approximately 96°F.

14-16. Condenser Bypass

For any given tower range and approach, the entering and leaving water temperatures will depend only on the wet bulb temperature of the air. Hence, in regions (particularly coastal areas) where the outdoor wet bulb temperature is relatively high, a closer approach to the wet bulb temperature is required in order to maintain a reasonable condensing temperature with an economical condenser size than in areas where the wet bulb temperature is lower. It has already been shown that, in general, the greater the quantity of water circulated over the tower per unit of capacity the closer the leaving water temperature will approach the wet bulb temperature. Therefore, in regions having a high wet bulb temperature, it is usually desirable to circulate a greater quantity of water over the tower than can be economically circulated through the condenser because of the excessive pumping head encountered. This can be accomplished by installing a condenser bypass line as shown in Fig. 14-9. Through the use of a condenser bypass, a certain, predetermined portion of the water circulated over the tower is permitted to bypass the condenser, thereby reducing the overall pumping head.

The advantage of the condenser bypass is that it makes possible the maintenance of reasonable condensing temperatures with moderate condenser and tower sizes without greatly increasing the pumping head. The quantity of water flowing through the bypass is regulated by the hand valve in the bypass line. Once the hand valve has been adjusted for the proper flow rate through the bypass, the handle should be removed from the valve so that the valve adjustment cannot be changed indiscriminately. An excessive amount of water flowing through the bypass will not only tend to starve the condenser and raise the condensing pressure, but it may also cause the pump motor to become overloaded, thereby rendering the entire system inoperative. The desired flow rate through the bypass is determined by subtracting the flow rate through the condenser from the flow rate over the tower. This will be illustrated presently.

Since the cooling tower capacity must of necessity be equal to the condenser capacity at the design conditions, it follows that

$$\text{Tower gpm} \times \text{tower range} \times 500 = \text{condenser gpm} \times \text{condenser rise} \times 500$$

Eliminating the constant,

$$\text{Tower gpm} \times \text{tower range} = \text{condenser gpm} \times \text{condenser rise} \qquad (14\text{-}13)$$

Example 14-17 A compressor on a refrigerating system has a capacity of 25 tons. The design wet bulb temperature is 80°F. The desired condenser water entering temperature is 87°F and the desired temperature rise through the condenser is 10°F. Select a cooling tower from Table R-17 and determine:

(a) The total gallons per minute circulated over the tower
(b) The temperature of the water entering the tower
(c) The tower cooling range
(d) The temperature of the water leaving the condenser
(e) The gallons per minute circulated through the condenser
(f) The gallons per minute circulated through the bypass

Solution From Table R-17, tower, Model #SA-58 has a capacity of 25 tons at an 80°F wet bulb temperature and a 7° approach. This capacity is based on a water flow rate of 4 gpm/ton and on a cooling range of 7.5° ($94.5 - 87$).

Total gpm over the tower for 25 tons

$$= 25 \text{ tons} \times 4 \text{ gpm/ton}$$
$$= 100 \text{ gpm}$$

From Table R-17, the tower entering water temperature

$$= 94.5°\text{F}$$

Tower range

$$= 94.5 - 87 = 7.5°$$

Water temperature leaving condenser

$$= 87 + 10 = 97°F$$

Rearranging and applying Equation 14-13, condenser gpm

$$= \frac{\text{Tower gpm} \times \text{tower range}}{\text{Condenser rise}}$$

$$= \frac{100 \times 7.5}{10}$$

$$= 75 \text{ gpm}$$

Gpm circulated through by-pass

$$= \text{Tower gpm} - \text{condenser gpm}$$
$$= 100 - 75$$
$$= 25 \text{ gpm}$$

14-17. Evaporative Condensers

An evaporative condenser is essentially a water conservation device and is, in effect, a condenser and a cooling tower combined into a single unit. A diagram of a typical evaporative condenser is shown in Fig. 14-17.

As previously stated, both air and water are employed in the evaporative condenser. The water, pumped from the sump up to the spray header, sprays down over the refrigerant coils and returns to the sump. The air is drawn in from the outside at the bottom of the condenser by action of the blower and is discharged back to the outside at the top of the condenser. In some cases, both pump and blower are driven by the same motor. In others, separate motors are used. The eliminators installed in the air stream above the spray header are to prevent entrained water from being carried over into the blower. An alternate arrangement, with the blower located on the entering air side of the condenser, is shown in Fig. 14-18.

Although the actual thermodynamic processes taking place in the evaporative condenser are somewhat complex, the fundamental process is that of evaporative cooling. Water is evaporated from the spray and from the wetted surface of the condenser into the air, the source of the vaporizing heat being the condensing refrigerant in the condenser coil.

The cooling produced is approximately 1000 Btu/lb of water evaporated. All the heat given up by the refrigerant in the condenser even-

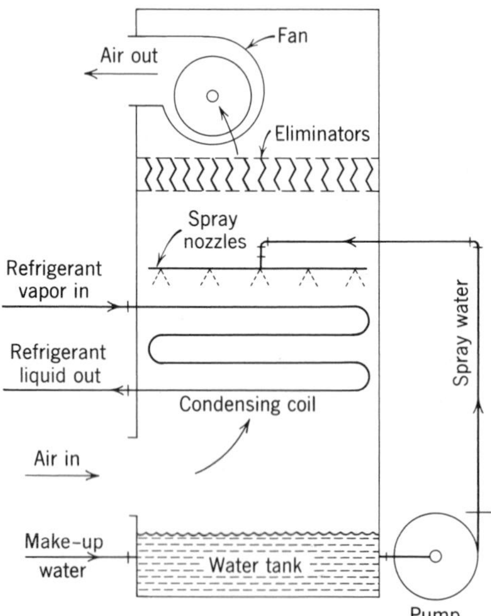

Fig. 14-17 Schematic diagram of evaporative condenser.

Fig. 14-18 Cutaway view of "Dri-Fan" evaporative condenser. Funnel-shaped overflow drain provides automatic bleed-off. (Courtesy Refrigeration engineering, Inc. A proprietary design of Refrigeration Engineering, Inc.)

tually leaves the condenser as either sensible heat or latent heat (moisture) in the discharge air. Since both the temperature and the moisture content of the air are increased as the air passes through the condenser, the effectiveness of the condenser depends, in part, on the wet bulb temperature of the entering air. The lower the wet bulb temperature of the entering air the more effective is the evaporative condenser.

To facilitate cleaning and scale removal, the condensing coil is usually made up of bare rather than finned tubing. The amount of coil surface used per ton of capacity varies with the manufacturer and depends to a large extent on the amount of air and water circulated.

Generally, the capacity of the evaporative condenser increases as the quantity of air circulated through the condenser increases. As a practical matter, the maximum quantity of air which can be circulated through the condenser is limited by the power requirements of the fan and by the maximum air velocity that can be permitted through the eliminators without the carry over of water particles.

The quantity of water circulated over the condenser should be sufficient to keep the tube surface thoroughly wetted in order to obtain maximum efficiency from the tube surface and to minimize the rate of scale formation. However, a water flow rate in excess of the amount required for adequate wetting of the tubes will only increase the power requirements of the pump without materially increasing the condenser capacity.

Assuming a condenser load of 15,000 Btu per hour per ton, the water lost by evaporation is approximately 15 lb (2 gal) per hour per ton (15,000/1000). In addition to the water lost by evaporation, a certain amount of water is lost by drift and by bleed-off. The amount of water lost by drift and by bleed-off is approximately 1.5 to 2.5 gal per hour per ton, depending upon the design of the condenser and the quality of water used. Hence, total water consumption for an evaporative condenser is between 3 and 4 gal per hour per ton.

Some evaporative condensers are available equipped with desuperheating coils, which are usually installed in the leaving air stream. The hot gas leaving the compressor passes first through the desuperheating coils where its temperature is reduced before it enters the condensing coils. The desuperheating coils tend to increase the overall capacity of the condenser and reduce the scaling rate by lowering the temperature of the wetted tubes. Too, often the receiver tank is located in the sump of the evaporative condenser in order to increase the amount of liquid subcooling.

14-18. Rating and Selection of Evaporative Condensers

Table R-18 is a typical evaporative condenser rating table. Notice that the ratings are based on the temperature difference between the condensing temperature and the design wet bulb temperature. The following sample selection is reprinted directly from the manufacturer's catalog data:*

Example 14-17 Select an evaporative condenser for the following conditions:

6-ton evaporator load (Refrigerant-12)
20° evaporator temperature
78° entering wet bulb temperature
105°F condensing temperature

Solution Since the rating table is in terms of evaporator load at 40°F, it is necessary to correct for other evaporator temperatures by using a correction factor from Table R-18B as follows:

Tons × evaporator correction factor
= Rating table tons

Therefore, 6 × 1.05 = 6.3 tons.

Referring to Table R-18A, the E-135F has a capacity of only 5.6 tons at 78°F entering wet bulb and 105°F condensing temperature. It does, however, have the required capacity of 6.3 tons at between 105°F and 110°F condensing temperature.

The compressor ratings should then be checked to see if the compressor originally selected has the required capacity at between 105°F and 110°F condensing temperature. If not, it will be necessary to select the next larger

* McQuay Products.

size evaporative condenser or compressor to do the job.

The next larger size evaporative condenser, the E-270F, has a capacity of 11.2 tons at the given conditions; however, the required capacity of 6.3 tons will be obtained at a condensing temperature between 90 and 95°F. The compressor selection should then be made for these conditions.

14-19. Water Regulating Valves

The water flow rate through a water-cooled condenser on a waste water system is automatically controlled by a water regulating valve (Fig. 14-19). The valve is installed on the water line at the inlet of the condenser and is actuated by the compressor discharge (Fig. 14-8). When the compressor is in operation, the valve acts to modulate the flow of water through the condenser in response to changes in the condensing pressure. An increase in the condensing pressure tends to collapse the bellows further and open the valve wider against the tension of the range spring, thereby increasing the water flow rate through the condenser. Likewise, as the condensing pressure decreases, the valve moves toward the closed position so that the flow rate through the condenser is reduced accordingly. Although the regulating valve tends to maintain the condensing pressure constant within reasonable limits, the condensing pressure will usually be considerably higher during periods of peak loading than during those of light loading.

When the compressor cycles off, the water valve remains open and water continues to flow through the condenser until the pressure in the condenser is reduced to a certain predetermined minimum, at which time the valve closes off completely and shuts off the water flow.

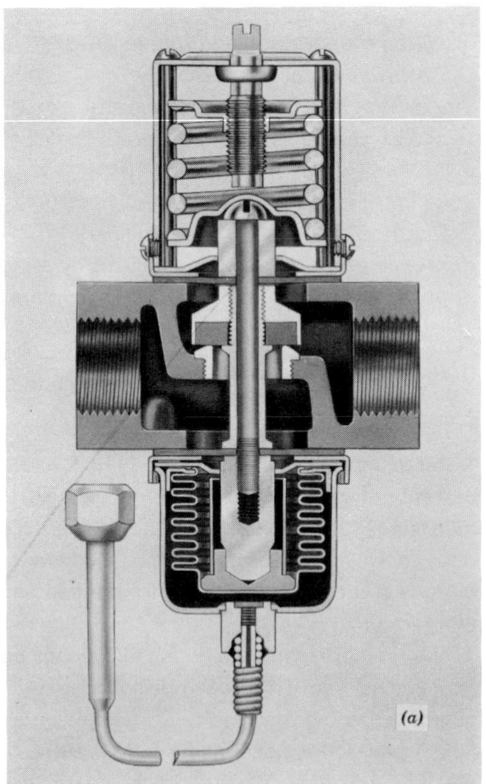

Fig. 14-19 Typical threaded-type water regulating valve. Larger sizes are available with flange connections. (*a*) Cross-sectional view showing principal parts. (*b*) Exterior view. (Courtesy Penn Controls, Inc.)

When the compressor cycles on again, the water valve remains closed until the pressure in the condenser builds up to the valve opening pressure, at which time the valve opens and permits water to flow through the condenser. The opening pressure of the valve is approximately 7 psi above the shut-off pressure.

The water valve is set for the desired shut-off pressure by adjusting the tension of the range spring. The minimum operating pressure for the valve, that is, the shut-off pressure, must be set high enough so that the valve will not remain open and permit water to flow through the condenser when the compressor is on the off cycle. Since the saturation temperature of the refrigerant in the condenser can never be lower than the ambient temperature at the condenser, the shut-off point of the water valve should be set at a saturation pressure corresponding to the maximum ambient temperature in the summertime at the condenser location. Too, the shut-off pressure of the valve must be high enough so that the minimum condensing temperature in the wintertime is sufficiently high to provide a pressure differential across the refrigerant control large enough to assure its proper operation.

The capacity of water regulating valves varies with the size of the valve and the pressure drop across the valve orifice. The available pressure drop across the valve orifice is determined by subtracting the pressure drop through the condenser and water piping from the total pressure drop available at the water main.

Water regulating valves are usually selected from flow charts (Table R-19). In order to select the proper valve from the flow chart, the following data must be known: (1) the desired water quantity in gpm; (2) the maximum ambient temperature in the summertime; (3) the desired condensing temperature; and (4) the available water pressure drop across the valve.

The following selection procedure and sample selection are reprinted directly from the literature of the manufacturer:*

* By permission of Penn Controls, Inc., Goshen, Indiana.

1. Draw horizontal line across upper half of flowchart (Table R-19) through the required flow rate.
2. Determine refrigerant condensing pressure rise above valve opening point.

   a. Valve closing point (to assure closure under all conditions) must be the refrigerant condensing pressure equivalent to the highest ambient air temperature expected at time of maximum load. Read this in psig from "Saturated Vapor Table" for refrigerant selected.

   b. Read from the same table the operating condensing pressure corresponding to selected condensing temperature.

   c. Valve opening point will be about 7 psi above closing point.

   d. Subtract opening pressure from operating pressure. This gives the condensing pressure rise.
3. Draw horizontal line across lower half of flowchart through this value.
4. Determine the water pressure drop through the valve—this is the pressure actually available to force the water through the valve.

   a. Determine the minimum water pressure available from city mains or other source.

   b. From condensing unit manufacturer's tables read pressure drop through condenser corresponding to required flow.

   c. Add to this estimated or calculated drop through piping, etc., between water valve and condenser, and from condenser to drain (or sump of cooling tower).

   d. Subtract total condenser and piping drop from available water pressure. This is the available pressure drop through the valve.

Example 14-19 The required flow for an R-12 system is found to be 27 gpm. Condensing pressure is 125 psig and the maximum ambient temperature estimated at 86°F. City water pressure is 40 psig and manufacturer's table gives drop through condenser and accompanying piping and valves as 15 psi. Drop through installed piping approximately 4 psi. Select proper size of water regulating valve from Table R-19.

Solution

1. Draw a line through 27 gpm—see dotted line, upper half of flowchart (Table R-19).
2. Closing point of valve is pressure of R-12 corresponding to 86°F ambient = 93 psig.
3. Opening point of valve is 93 + 7 = 100 psig.
4. Condensing pressure rise = 125 − 100 = 25 psi.
5. Draw line through 25 psi—see dotted line, lower half of flowchart.
6. Available water pressure drop through valve = 40 − 19 = 21 psi.
7. Interpolate just over the 20 psi curve—circle on lower half of flowchart.
8. Draw vertical line upward from this point to flow line—circle on flowchart marks this intersection.
9. This intersection falls between curves for 1 in. and $1\frac{1}{4}$ in. valves. The $1\frac{1}{4}$ in. valve is required.

14-20. Condenser Controls

For reasons of economy, the condensing medium is circulated through the condenser only when the compressor is operating. Hence, common practice is to cycle the condenser fan and/or pump on and off with the compressor. This is usually accomplished by electrically interlocking the fan and/or pump circuit with the compressor driver circuit. Method of interlocking electrical circuits are discussed in Chapter 21.

Whereas high pressure controls are always desirable as safety devices on any type of system, they are essential on all equipment employing water as the condensing medium in order to protect the equipment against damage from high condensing pressures and temperatures in the event that the water supply becomes restricted or is shut-off completely. The high pressure control has already been discussed in Section 13-13.

If a refrigerating system is to function properly and efficiently, the condensing temperature must be maintained within certain limits. As previously described, high condensing temperatures cause losses in compressor capacity and efficiency, overheating of the compressor, excessive power consumption, and in some cases, overloading of the compressor driver.

An abnormally low condensing temperature, on the other hand, will cause an insufficient pressure differential across the refrigerant control (condensing pressure to vaporizing pressure), which reduces the capacity of the control and results in starving of the evaporator and general unbalancing of the system.

As a general rule, low condensing temperatures result from either one or both of two principal causes: (1) low ambient temperatures and (2) light refrigerating loads. Naturally, the problem of low condensing temperatures is more acute in the wintertime when the ambient temperature and the refrigerating load are both apt to be low.

To maintain the condensing temperature at a sufficiently high level, it is necessary to make some provision for reducing or controlling the capacity of the condenser during periods when the ambient temperature is low and/or the refrigerating load is light. Although the methods employed to control the capacity of the condenser vary somewhat with the type of condenser used, all involve reducing either the quantity of condensing medium circulated or the amount of effective condensing surface. Condenser capacity control devices are usually actuated by pressure or temperature controls which respond to condensing pressure or temperature.

With regard to air-cooled condensers, the condensing temperature is maintained within the desired limits by varying the air quantity through the condenser or by causing a portion of the condenser to become filled with liquid so as to reduce the amount of effective condensing surface.

The quantity of air circulated over an air-cooled condenser may be controlled by the use of dampers placed in the airstream, by varying the speed of the fan, by cycling the condenser fan, or by some combination of these methods.

A volume damper installed in the fan discharge or used as a face damper and modulated by a damper motor responding either to the ambient temperature or to the condensing pressure or one modulated by a damper operator actuated directly by the refrigerant pressure in the condenser, will provide reasonably stable condenser capacity control. However, the dampers and controls are subject to mechanical and electrical problems caused by corrosion, dust accumulation, and freezing, and where propeller-type fans are used, the fan motor must be oversized to offset the increased static pressure imposed by partially or fully closed dampers.

A more satisfactory means of controlling the capacity of air-cooled condensers employs a recently developed solid-state modulating fan-speed control. This control utilizes a thermistor, which senses either the ambient air temperature or the temperature of the gas in the condenser and which changes electrical resistance as the temperature changes. An electronic circuit detects the change in resistance and regulates the voltage to the fan motor(s), which may be either a shaded-pole or a permanent-split-capacitor type. Depending on the temperature of the thermistor, the speed of the condenser fan can be modulated through the entire range from full speed to zero.

Since it tends to cause rather large fluctuations in the condensing temperature, along with short-cycling of the fan, cycling of the fan as a means of controlling the capacity of an air-cooled condenser usually is not practical for single-fan condensers. However, where a multiple of fans are employed with a single-coil condenser, cycling of the fans in sequence provides a satisfactory and convenient means of controlling the condensing pressure over a wide range of operating conditions, particularly when the cycling is used in conjunction with one condenser fan equipped with modulating damper or variable speed control.

Air side capacity control usually is not suitable for multicircuit condensers. Also, a strong wind blowing full into the face of a vertical air-cooled condenser can completely negate any capacity reduction achieved by cycling fans or reducing the fan speed. However, this difficulty is largely overcome by the use of wind deflectors (Fig. 14-6) or horizontal condensers.

Another method of controlling the capacity of air-cooled condensers is to vary the amount of effective condensing surface by causing the liquid refrigerant to back up into the lower portion of the condenser whenever the condensing pressure drops below the desired minimum. Most manufacturers of air-cooled condensers have their own proprietary method for accomplishing this type of refrigerant side capacity control, which usually includes also some means of alleviating the other difficulties associated with cold weather operation (Section 14-22).

One typical design employs a modulating valve installed in a bypass line between the inlet and outlet of the condenser (valve *B* in Fig. 14-20). As the receiver pressure drops below the set point of the valve, the valve modulates toward the open position and allows high-pressure vapor from the compressor discharge to flow through the bypass line, thereby restricting the flow of liquid refrigerant from the condenser and causing the liquid to back up into the lower portion of that unit. The amount of discharge vapor bypassed, and therefore the amount of liquid retained in the lower portion of the condenser, is automatically controlled by the modulating valve and depends on the receiver tank pressure. With an ambient temperature of 50°F, approximately 50% of the condenser will be filled with liquid. This increases to approximately 90% as the ambient

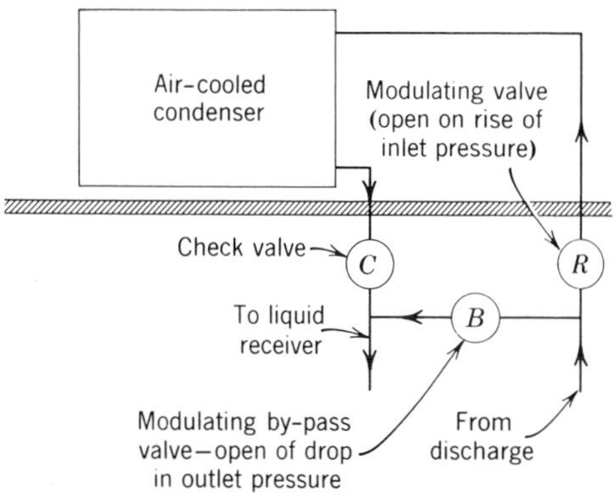

Fig. 14-20 Sure-start Winterstat provides normal head and receiver pressures when the compressor starts by allowing the compressor to impose its full discharge pressure on the liquid through the open (*B*) valve. When the receiver pressure is up to normal, the (*R*) valve opens and allows the discharge gas to flow to the condenser. (Courtesy Kramer Trenton Company.)

temperature falls to 0°F. An oversized receiver tank is required to store the excess liquid during normal operation.

When refrigerant side capacity control is employed with multiple circuit condensers, separate controls are required for each system.

With regard to evaporative condensers, capacity control is usually obtained by regulating the air circulated through the condenser. Methods for controlling the air quantity are similar to those used with air-cooled condensers. Another control method involves regulation of the entering air WB temperature. As shown in Fig. 14-21, this is accomplished by recirculating some part of or all the air passing through the condenser. As the condensing pressure falls below some predetermined minimum, the inlet and outlet dampers are moved toward the closed position while the bypass damper moves toward the open position, thereby causing more recirculation of the air and a higher resultant WB temperature.

Cycling of the pump as a means of controlling the capacity of an evaporative condenser cannot be recommended. Each time the pump cycles off a thin film of scale is formed on the condenser tubes. Consequently, frequent cycling of the condenser pump greatly increases the scaling rate, which reduces the efficiency of the condenser and increase maintenance costs. However, where some other means of capacity control is used, the pump can be shut down and the coil operated dry when the ambient air temperature drops below freezing. To prevent freezing, the water in the sump must be heated, usually electrically, or drained to an indoor tank (Fig. 14-22).

With reference to water-cooled condensers, recall that for a given load and condensing surface, the condensing temperature varies with the quantity and temperature of the water entering the condenser. Where waste water is used, the modulating action of the water-regulating valve controls the water flow rate through the condenser and maintains the condensing temperature above the desired minimum so that low condensing temperatures are not usually a problem with waste water systems. Where the water flow rate through the condenser on a recirculating water system is

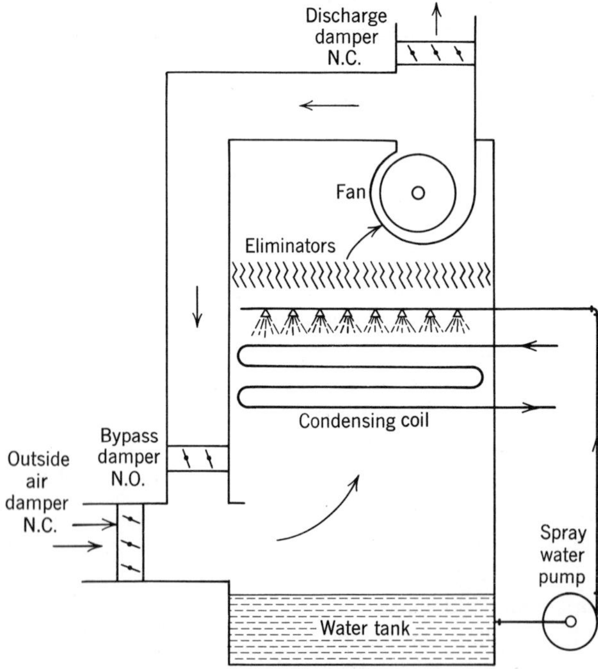

Fig. 14-21 Controlling evaporative condenser capacity by regulating entering air WB temperature.

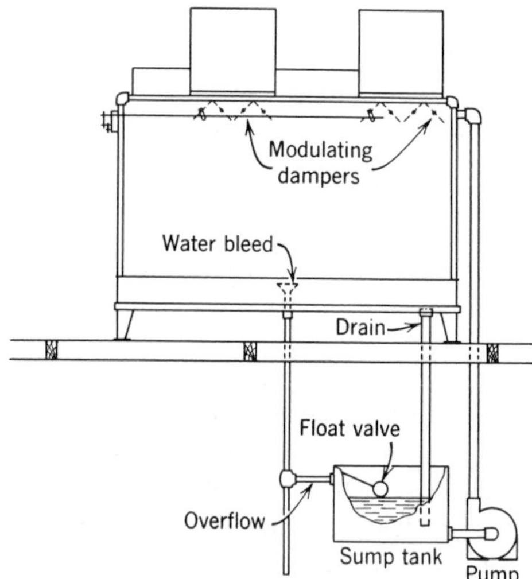

Fig. 14-22 Evaporative condenser equipped with modulating dampers for capacity control. Protected auxiliary pump is designed to prevent freezing during winter operation. (Courtesy Refrigeration Engineering Inc.)

maintained constant, the condensing temperature decreases as the temperature of the water leaving the tower decreases. Consequently, when the ambient air temperature is low, the condensing temperature will also be low unless some means is provided for regulating the water flow rate through the condenser or for increasing the temperature of the water leaving the tower.

One common method of controlling the water flow rate through the condenser utilizes a three-way modulating valve that varies the water flow rate through a condenser bypass to maintain the condenser leaving water temperature above some predetermined minimum, usually 75°F (see Fig. 14-23). As the condensing temperature tends to fall, the three-way valve moves to bypass more water and thereby reduce the condenser water flow rate and increase the temperature rise of the water in the condenser. With this method of control, the pump delivery rate remains relatively constant, which is a desirable feature when a number of condensers are served by a single tower and pump.

Where mechanical draft cooling towers are used, the condensing temperature can be maintained at the desired level through control of the tower leaving water temperature. As in the case of the evaporative condenser, this can be accomplished by regulating the air flow through the tower.

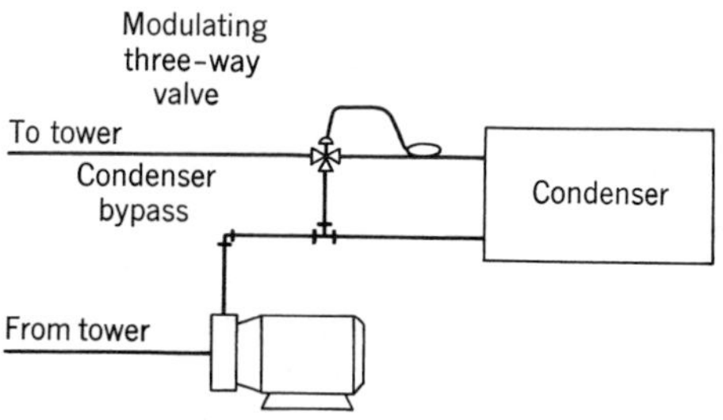

Fig. 14-23 Condenser bypass control.

Another method employs a three-way modulating or snap-action valve (shown dotted in Fig. 14-24) to divert water around the cooling tower as the tower leaving water temperature falls below some predetermined minimum. When the tower is to be operated in below-freezing weather, the diverting valve must be of the snap-action type.

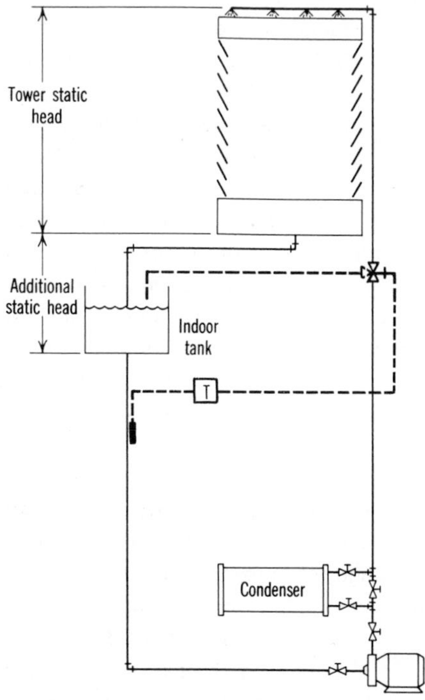

Fig. 14-24 Protected indoor tank.

14-21. Winter Operation

When the compressor and/or condenser are so located that they are exposed to low ambient temperatures, the pressure in these parts may fall considerably below that in the evaporator during the compressor off-cycle. In such cases, the liquid refrigerant, which otherwise would remain in the evaporator, very often tends to migrate to the area of lower pressure in the compressor and condenser. With no liquid refrigerant in the evaporator, an increase in evaporator temperature is not reflected by a corresponding increase in the evaporator pressure, and, where the system is controlled by a low pressure motor control, the rise in evaporator pressure may not be sufficient to actuate the control and cycle the system on in response to an increase in the evaporator temperature.

Corrective measures are several. One is to install a thermostatic motor control in series with the low pressure control. The thermostat is adjusted to cycle the system on and off, whereas the low pressure control serves only as a safety device. Another, and usually more practical, solution is to isolate the condenser during the off-cycle. One method of isolating the condenser during the off-cycle is illustrated in Fig. 14-20. The check valve (*C*) in the condenser liquid line prevents the refrigerant from boiling off in the receiver and backflowing to the condenser during the off-cycle. The (*R*) valve, which closes on drop of pressure at the valve inlet, closes when the compressor stops, preventing the flow of refrigerant from the evaporator, through the compressor valves and discharge line, into the condenser. With the condenser isolated, the evaporator pressure can build up and start the compressor regardless of the ambient temperature at the condenser.

Another and rather obvious problem concerning the operation of evaporative condensers and cooling towers in the wintertime is the danger of freezing when the equipment is exposed to freezing temperatures. In general, the measures employed to prevent freezing are similar to those used to prevent low condensing temperatures, that is, controlling the air quantity through the tower by the use of dampers or by cycling the fan. In addition, an auxiliary sump must be installed in a warm location and the piping arranged so that the water drains by gravity into the auxiliary sump and does not remain in the tower or condenser sump (Figs. 14-22 and 14-24).

An alternative to the auxiliary sump is to supply heat to the water in the tower basin with a steam coil or electric strip heaters.

14-22. Closed-Circuit Coolers

In some installations, the closed-circuit cooler is being used in place of the conventional cooling tower to cool the water from water-cooled condensers (Fig. 14-25). The closed-circuit cooler is very similar in operation to the evaporative condenser, except that the fluid circulated through the coil is the hot water from the refrigerant condenser rather than the hot refrigerant gas itself.

While closed-circuit coolers are used primarily for heat recovery, there is also the additional advantage of eliminating the problems of condenser water contamination and the scaling of the tubes in the condenser.

Capacity control methods for the closed-circuit cooler are the same as those for evaporative condensers and cooling towers, as are the protective measures necessary for operation in below-freezing weather. One simple method of protecting the condenser water circuit against freezing is to substitute glycol or some other antifreeze solution for the water in the condenser circuit.

14-23. Condenser and Tower Maintenance

As a general rule, air-cooled condensers require little maintenance other than regular lubrication of the fan and motor bearings. However, the fan blades and condensing surface should be inspected occasionally for the accumulation of dust and other foreign materials. These parts should be kept clean in order to obtain high efficiency from the condenser.

Any type of condenser employing water is subject to scaling of the condenser tubes, corrosion, and the growth of algae and bacterial slime on all wetted surfaces. The latter is controlled by frequent cleaning of the infected

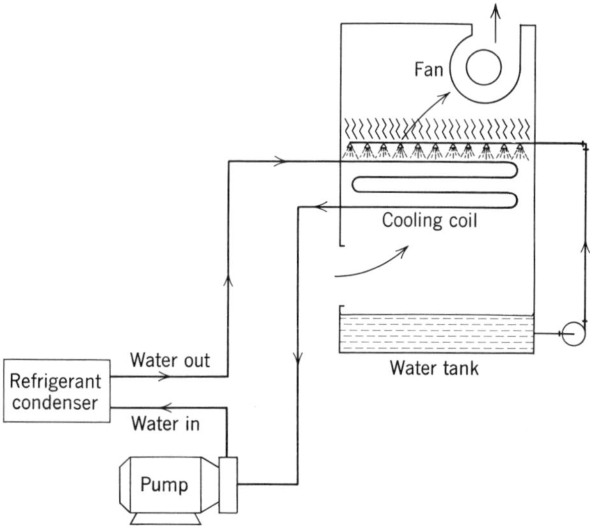

Fig. 14-25 Simplified closed circuit condenser water system.

parts and by the use of various algaecides which are available commercially.

As previously stated, the scaling rate depends primarily upon the condensing temperature and the quality of water used. The scaling rate will be relatively low where the condenser leaving water temperature is below 100°F. Too, the importance of providing for the recommended amount of bleed-off cannot be overemphasized with regard to keeping the scaling rate at a minimum. In addition, a number of chemical companies have products which when added to the sump water considerably reduce the scaling rate.

Scale can be removed from the condenser tubes by applying an approved inhibited acid compound, many of which are available in either liquid or powder form. After the tower or condenser sump has been drained, cleaned, and filled with fresh water, the cleaning compound can be added directly to the sump water. The pump is then started and the cleaner is circulated through the system until the system is clean, at which time the sump is again drained, flushed, and filled with clean water before the system is placed in normal operation.

It should be pointed out that descaling compounds have an acid base and should not be allowed to contact grass, shrubs, or painted surfaces. Therefore, it is usually advisable to remove the cooling tower spray nozzles, if any, in order to minimize the danger of damaging shrubs or painted surfaces with drift from the tower.

When rapid descaling of the condenser tubes is required, an inhibited solution (18%) of muriatic acid may be used. However, muriatic acid should be used only on the condenser tubes. The system pump should not be used to circulate the acid. A small pump having an acid resistant impeller (brass or nylon) may be used for this purpose (see Fig. 14-26). After the

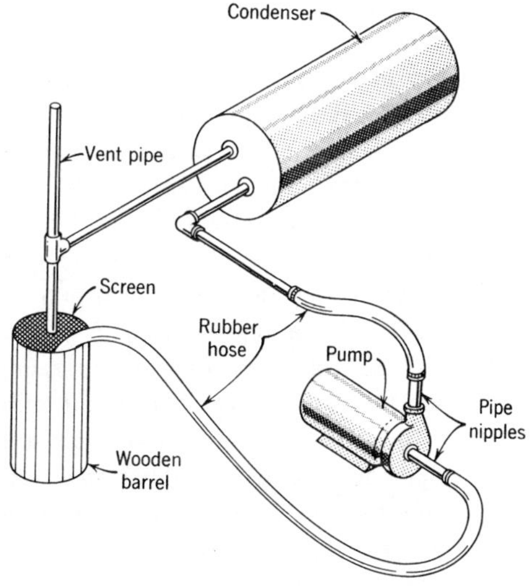

Fig. 14-26 Apparatus for descaling condenser.

condenser is clean, it should be flushed with clean water or with an acid neutralizer as recommended by the manufacturer.

Corrosion is usually greatest in areas near salt water or in industrial areas where relative large concentrations of sulfur and other industrial fumes are found in the atmosphere. Corrosion damage is minimized by regular cleaning and painting of the affected parts and by application of protective coatings of various types.

Customary Problems

14-1 Determine the estimated condenser load for an open-type compressor having a refrigerating capacity of 49,200 Btu/hr if the design saturated suction and discharge temperatures are −20°F and 100°F, respectively.

14-2 Determine the condenser load for the conditions given in Problem 14-1 if a suction-cooled hermetic compressor is employed.

14-3 Using Equation 14-2, compute the load on the condenser for the compressor in Problem 14-1 if the brake hp is 11.4 hp.

14-4 Using Equation 14-3, calculate the total heat rejection at the condenser for a suction-cooled hermetic compressor having a cooling capacity of 4.9 tons and a power input of 10.8 kW.

14-5 The heat load on the evaporator of a refrigerating system is 55,000 Btu/hr. If the coefficient of performance of the system is 4.3 to 1, what is the approximate heat rejection at the condenser?

14-6 The heat rejected to a water-cooled condenser is 130,000 Btu/hr. How many square feet of condenser surface is required if the U factor for the condenser is 110 Btu/(hr)(ft^2)(°F) and the METD is 7°F at the design gpm?

14-7 An R-502 waste water system operating with an open type compressor at a −20°F suction temperature and a 90°F condensing temperature has an evaporator load of 5 tons. If the condenser is selected for a 12°F water temperature rise, how many gallons per minute of water must be circulated through the condenser?

14-8 Seventy-five gallons per minute of water are circulated through a water-cooled condenser. If the temperature rise of the water in the condenser is 14°F, what is the rate of heat rejection in the condenser in Btu per hour?

14-9 An R-22 air conditioning system operating with an open type compressor at evaporator temperature of 40°F and a condensing temperature of 120°F has an evaporator load of 210,000 Btu/hr. Air is circulated over the condenser at the rate of 13,000 cfm. If the temperature of the air entering the condenser is 90°F, calculate (a) the temperature of the air leaving the condenser and (b) the METD.

14-10 An R-12 system having a cooling capacity of 60,000 Btu/hr is operating with a suction-gas cooled hermetic compressor at a saturated suction temperature of 20°F and a saturated discharge temperature of 110°F. If the temperature rise of the air through the condenser is 10°F, what is the approximate cubic feet per minute of air circulated over the condenser?

14-11 If the condenser in Problem 14-9 has a face area of 16.4 ft^2, what is the face velocity of the air?

14-12 An open-type compressor employing R-22 has a cooling capacity 120,000 Btu/hr operating at an evaporator temperature of 40°F and a condensing temperature of 110°F. If the ambient DB temperature is 90°F and the altitude is 3000 ft, select an air-cooled condenser to meet the design conditions.

14-13 Assume that a suction-cooled hermetic compressor is substituted for the open compressor in Problem 14-12 and select an air-cooled condenser to satisfy the design conditions.

14-14 Select a shell-and-tube water-cooled condenser for an R-12 system to meet the following conditions:

Evaporator load	60 tons
Evaporator temperature	40°F

Condensing temperature	110°F
Water quantity	2.5 gpm/ton

Untreated cooling tower water enters the condenser at 85°F.

14-15 Rework Problem 14-14 using treated cooling tower water.

14-16 A cooling tower and a water-cooled condenser (with bypass) are operating with a condenser load of 240,000 Btu/hr. Forty-eight gallons per minute are circulated through the condenser and 32 gpm are bypassed. The ambient WB temperature is 78°F, and the tower approach is 7°F. Determine (a) the temperature of the water entering the condenser, (b) the temperature of the water leaving the condenser, (c) the temperature of the water entering the cooling tower, and (d) the tower range.

14-17 A compressor on a Refrigerant-12 system has a capacity of 50 tons. The design WB temperature is 78°F. The desired condenser water entering temperature is 85°F and the desired temperature rise through the condenser is 12°F. Select a cooling tower from Table R-15 and determine (a) the total gallons per minute circulated over the tower, (b) the temperature of the water entering the tower, (c) the temperature of the water leaving the condenser, (d) the tower range, (e) the gallons per minute circulated through the condenser, and (f) the gallons per minute bypassed.

14-18 Select an evaporative condenser for the following conditions:

Refrigerant-12 system	
Evaporator load	10 tons
Evaporator temperature	40°F
Wet bulb temperature of entering air	78°F
Condensing temperature	105°F

14-19 Rework Problem 14-18 using an evaporator load of five tons and an evaporator temperature of −10°F.

14-20 For a Refrigerant-12 system, select a water regulating valve to meet the following conditions:

(a) Desired condensing temperature is 106°F.
(b) Maximum ambient temperature is 90°F.
(c) Desired water quantity through condenser at maximum loading—9 gpm.
(d) Pressure available at city main during period of peak loading—50 psi.
(e) Pressure loss through condenser and water piping—12 psi.

15

FLUID FLOW, CENTRIFUGAL LIQUID PUMPS, WATER AND BRINE PIPING

15-1. Fluid Pressure

The total pressure exerted by any fluid is the sum of the static and velocity pressures of the fluid, that is,

$$p_t = p_s + p_v \tag{15-1}$$

where p_t = the total pressure
p_s = the static pressure
p_v = the velocity pressure

All flowing fluids possess kinetic energy and therefore exert a force or pressure in the direction of flow. The pressure exerted by a fluid which is the direct result of fluid motion or velocity is called the velocity pressure of the fluid. Any pressure exerted by a fluid which is not the direct result of fluid motion or velocity, regardless of the force causing the pressure, is called the static pressure of the fluid. For fluids at rest (static), the velocity pressure is equal to zero and the total pressure is equal to the static pressure. Whereas velocity pressure acts only in the direction of flow, static pressure acts equally in all directions. This is easily demonstrated through the use of an example employing a gravitational column.

It was shown in Chapter 1 that the action of gravity on any body causes the body to exert a force which is commonly referred to as the weight of the body. For a solid material, because of the rigid molecular structure, the gravitational force or pressure is exerted in a downward direction only. However, because of the loose molecular structure of fluids, the gravitation force or pressure exerted at any point in a body of fluid acts equally in all directions—up, down, and sideways, and always at right angles to any containing surfaces. When no force other than the force of gravity is acting on the fluid, the pressure at any depth in a body of fluid is proportional to the weight of fluid above that depth. When an external force in addition to the force of gravity is applied to the liquid, the pressure at any depth in the fluid is proportional to the weight of the fluid above that depth, plus the pressure caused by the external force.

For example, assume that a flat-bottomed container 1 sq ft in cross section and 10 ft high is filled to the top with water (Fig. 15-1). Neglecting the pressure of the atmosphere on the surface of the water, and assuming a water density of 62.4 lb/ft^3, the total force acting on the bottom of the tank due to the weight (gravitational force) of the water alone is 624 lb (10 × 62.4). Since the base area of the tank is 1 sq ft, the pressure exerted on the bottom of the tank is 624 psf or 4.33 psi (624/144). Since this pressure

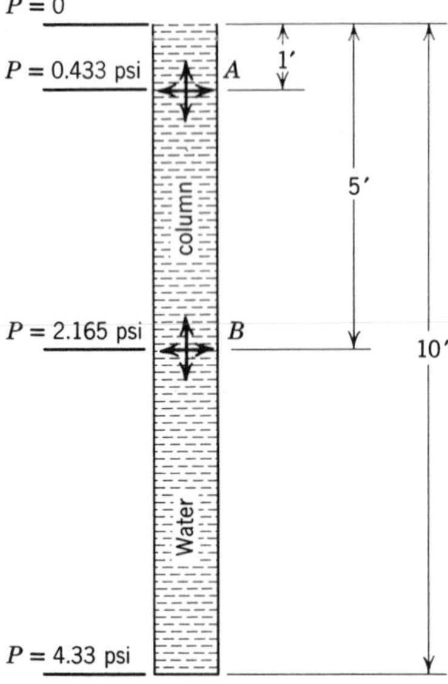

Fig. 15-1 Illustrating head-pressure relationship.

acts equally in all directions, it is exerted on the sides of the tank at the base as well as on the bottom of the tank.

Assume now that level *A* in the water column is exactly 1 ft below the surface of the water. The volume and weight of water above this level are 1 ft^3 and 62.4 lb, respectively. Since this weight of water is also evenly distributed over an area of 1 ft^2, the fluid pressure acting in all directions from any point at level *A* is 62.4 psf or 0.433 psi. Similarly, the volume and weight of water above level *B*, which is located 5 ft below the surface of the water, are 5 ft^3 and 312 lb (5 × 62.4), respectively, and the fluid pressure at this level is 312 psf or 2.165 psi.

If the force exerted on the top of the water by the pressure of the atmosphere is taken into account, the pressure of the water at any level in the tank will be increased by an amount equal to the pressure of the atmosphere. Assuming normal sea level pressure, the fluid pressures at levels *A* and *B* are 15.129 psi (0.433 + 14.696) and 16.861 (2.165 + 14.696), respectively, while the pressure at the base of the tank is 19.026 psi (4.33 + 14.696). However, it should be recognized that since the pressure of the atmosphere is exerted also on the outside of the tank the pressure tending to burst the tank is still only that resulting from the gravitational effect on the water alone.

For any noncompressible fluid (liquid) of uniform density, the pressure exerted by the fluid at any point in a column of fluid is directly proportional to the depth of the fluid at that point and can be determined by multiplying the depth by the density of the fluid,* that is,

$$\text{Pressure (psf)} = \text{depth (ft)} \times \text{density (lb/ft}^3) \tag{15-2}$$

$$\text{Pressure (psi)} = \frac{\text{depth (ft)} \times \text{density (lb/ft}^3)}{144} \tag{15-3}$$

15-2. Head-Pressure Relationship

The vertical distance between any two levels in a column of fluid is called the "head" of the fluid at the lower level with respect to the upper level. For example, with respect to level *B* in Fig. 15-1, the head of the water at the base of the column is 5 ft. With respect to the top of the column, the head of the water at the base of the column is 10 ft. Similarly, with respect to the top, the water heads at levels *A* and *B* are 1 ft and 5 ft, respectively.

Since the depth of the liquid at any level in a liquid column is equal to the head of the liquid at that level with respect to the top of the column, the head can be substituted for depth in Equation 15-3 and the following relationship between head and pressure is established:

$$\text{Pressure (psi)} = \frac{\text{Head (ft)} \times \text{density (lb/ft}^3)}{144} \tag{15-4}$$

$$\text{Head (ft)} = \frac{\text{Pressure (psi)} \times 144}{\text{Density (lb/ft}^3)} \tag{15-5}$$

It is evident from the foregoing that there is a definite and fixed relationship between the head and the pressure of any liquid, the head-pressure ratio for any given liquid being dependent upon the density of the liquid. For example, in the case of water, the head-pressure ratio is 2.31 ft to 1 psi. For mercury,

* This is not true of a compressible fluid because the density of a compressible fluid varies with the depth.

the head-pressure ratio is 2.04 in. to 1 psi. This means that a pressure of 1 psi is equivalent to head of 2.31 ft of water column or 2.04 in. of mercury column. Conversely, a 1 ft column of water (1 ft water head) is equivalent to 0.433 psi, whereas a 1 ft column of mercury (1 ft mercury head) is equivalent to 5.89 psi.

With respect to the head-pressure relationship, the following general statements can be made:

1. For any liquid of given and uniform density, the pressure exerted by the liquid is directly proportional to the head of the liquid.
2. At any given head, the pressure exerted by any liquid is directly proportional to the density of the liquid. Liquids having different densities will exert different pressures at the same head.

15-3. Static and Velocity Heads

The total head of any fluid is the sum of the static and velocity heads of the fluid, that is,

$$h_t = h_s + h_v \tag{15-6}$$

where h_t = the total head in feet
h_s = the static head in feet
h_v = the velocity head in feet

The static head of any liquid is expressed as the height in feet (or inches) of a gravitational column of that liquid which would be required to produce a base pressure equal to the static pressure of the liquid. That is, the head in feet of liquid column equivalent to the static pressure of the liquid is called the static head of the liquid. Likewise, the head in feet of liquid column equivalent to the velocity pressure of a liquid is called the velocity head of the liquid.

The fundamental relationship between velocity and velocity head is established by Galileo's law, which states in effect that all falling bodies, regardless of mass, accelerate at equal rates and that the final velocity of any falling body, neglecting friction, depends only upon the height from which the body falls. Hence, the height in feet from which a body must fall in order to attain a given velocity is the velocity head corresponding to that velocity. The velocity head corresponding to any given velocity can be determined by applying the following equation:

$$h_v = \frac{v^2}{2g} \tag{15-7}$$

where h_v = the velocity head in feet
v = the velocity in feet per second (fps)
g_c = the acceleration of gravity (32.2 ft/s^2)

By combining and/or rearranging Equations 15-7 and 15-4, the following relationships are established:

To convert velocity head to velocity pressure,

$$p_v = \frac{h_v \times \rho}{144} \tag{15-8}$$

To convert velocity to velocity pressure,

$$p_v = \frac{v^2 \times \rho}{2g \times 144} \tag{15-9}$$

To convert velocity head to velocity,

$$v = \sqrt{2g \times h_v} \tag{15-10}$$

To convert velocity pressure to velocity,

$$v = \sqrt{\frac{2g \times p_v \times 144}{\rho}} \tag{15-11}$$

15-4. Head-Energy Relationships

The basic relationship of head to energy or work is shown in the following equation:

$$\text{Energy or work (ft-lb)} = \text{weight (lb)} \times \text{head (ft)} \tag{15-12}$$

The fact that Equation 15-7 is identical to Equation 1-23 indicates that the velocity head of a fluid is an expression of the kinetic energy per pound of fluid. Similarly, it can be shown that the static head of a fluid is an expression of the potential energy per pound of fluid.

In any fluid column of uniform and constant density, the potential energy per pound of fluid is the same at all levels in the column. However, the potential energy at various levels is differently divided between the energy of position and the energy of pressure (head) depending upon the elevation. For example, in

Fig. 15-1, 1 lb of water at the uppermost level in the tank has a potential energy of position with relation to the base of 10 ft-lb (l lb × 10 ft) in accordance with Equation 1-8. Since the head at this level is zero, the potential energy of pressure (head) is also zero. On the other hand, 1 lb of water at the base of the tank has no potential energy of position, but has pressure or head energy of 10 ft-lb (1 lb × 10 ft), according to Equation 15-12. Likewise, 1 lb of water at a level midway in the water column also has potential energy in the amount 10 ft-lb, the energy being evenly divided between the energy of position and the energy of pressure.

15-5. Static Head-Velocity Head Relationship in Flowing Fluids

The fact that the static pressure of a fluid is exerted equally in all directions, whereas the velocity pressure of the fluid is exerted only in the direction of flow, makes it relatively simple to measure the static and velocity pressures (or heads) of a fluid flowing in a conduit. This is illustrated in Fig. 15-2. Notice that tube *A* is so connected to the conduit that the opening of the tube is exactly perpendicular to the line of flow. Since only the static pressure of the fluid will act in this direction, the height of the fluid column in tube *A* is a measure of the static pressure or static head of the fluid in the conduit. On the other hand, tube *B* is so arranged in the conduit that the opening of the tube is directly in the line of flow. Since both the static pressure and the velocity pressure of the flowing fluid act on the opening of tube *B*, the height of the liquid column in tube *B* is a measure of the total pressure or total head of the fluid. Since the total pressure or head of a fluid is the sum of the static and velocity pressures or heads, it follows that the difference in the heights of the two fluid columns is a measure of the velocity pressure or velocity head of the fluid in the conduit.

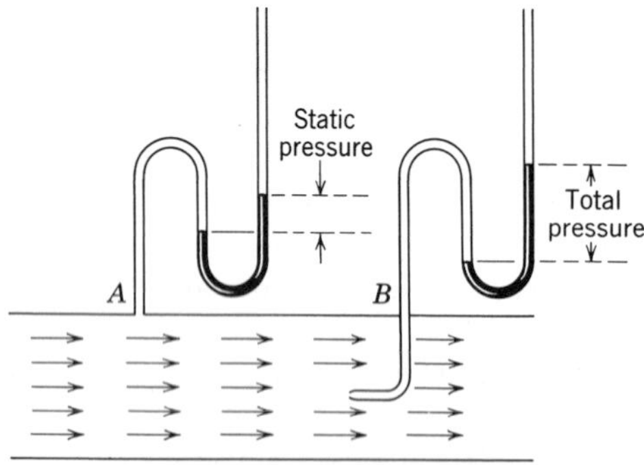

Fig. 15-2 Illustrating relationship between the static, velocity, and total pressures of a fluid flowing in a circuit.

If losses resulting from friction are neglected, the total pressure or head of a flowing fluid will be the same at all points along the conduit. However, the total head may be differently divided between static head and velocity head at the several points, depending upon the velocity of the fluid at these points.

For any given flow rate (quantity of flow), the velocity of the fluid flowing in a conduit varies inversely with the cross-sectional area of the conduit. This relationship is expressed by the basic equation

$$v = \frac{Q}{A} \tag{15-13}$$

where v = the velocity in feet per second
Q = the flow rate in cubic feet per second
A = the cross-sectional area of the conduit in square feet

Note. When Q is in cubic feet per minute, v will be in feet per minute.

In accordance with Equation 15-13, the fluid velocity (and velocity head) in section *B* of the conduit in Fig. 15-3 is greater than that in sections *A* and *C*, since the cross-sectional area of section *B* is less than that of sections *A* and *C*. Assuming that the total head of the fluid is the same at all points in the conduit, it follows then that the static head-velocity head ratio in section *B* is different from that in sections *A* and *C*. As the fluid flows through the reducer between sections *A* and *B*, static head is converted to velocity head (pressure is converted to velocity). Conversely, as the fluid flows through the increaser between sections *B* and *C*, velocity head is converted back into static head (velocity is converted to pressure).

In view of the head-energy relationship, it is evident that the conversion of static head to velocity head is in fact a conversion of potential

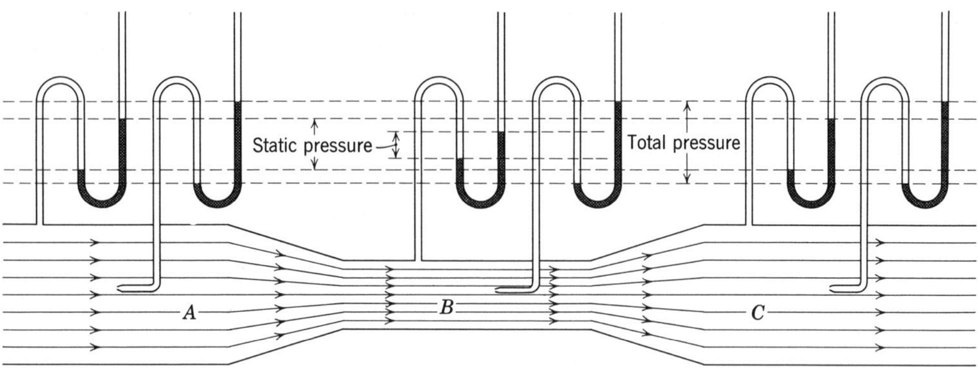

Fig. 15-3 Illustrating changes in static velocity pressure ratio resulting from changes in conduit area.

energy (pressure) into kinetic energy (velocity). Likewise, the conversion of velocity head to static head represents a conversion of kinetic energy (velocity) to potential energy (pressure).

15-6. Friction Head

It has already been established that a fluid flowing in a conduit will suffer losses in energy (converted into heat) as a result of the work of overcoming friction. These energy losses are

CHART 15-1 Resistance of Flow of Water Through Smooth Copper Tubing

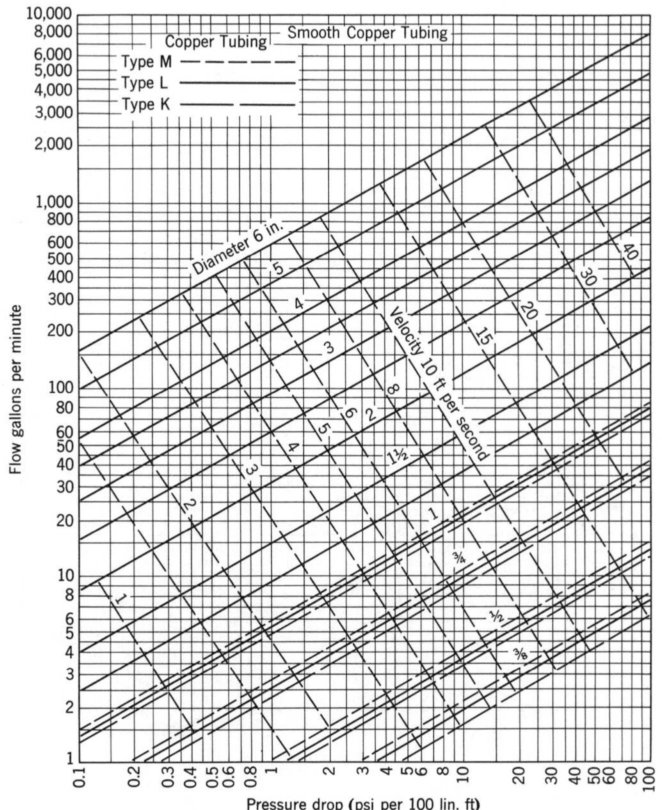

frequently expressed in terms of pressure drop or head loss. The pressure drop in psi or the head loss in feet experienced by a fluid flowing between any two points in a conduit is known as the friction head or friction loss between these two points.

The amount of pressure drop or head loss suffered by a fluid resulting from friction and turbulence in flowing through a conduit varies with a number of factors: (1) the viscosity and specific gravity of the fluid, (2) the velocity of the fluid, (3) the roughness of the internal surface of the conduit, and (4) the length of the conduit.

Obviously, the mathematical evaluation of all these factors is too laborious for most practical purposes. As a general rule, the friction loss in piping is determined from charts and tables.

The pressure (friction) loss in pounds per square inch per 100 ft of straight pipe is given in Charts 15-1 and 15-2 for various flow rates in various sizes of pipe. Chart 15-1 applies to smooth copper tube, whereas Chart 15-2 applies to fairly rough pipe. Since the pressure loss for a given pipe size and flow rate is proportional to the length of the pipe, the pressure loss through any given length of

CHART 15-2 Resistance to Flow of Water Through Fairly Rough Pipe

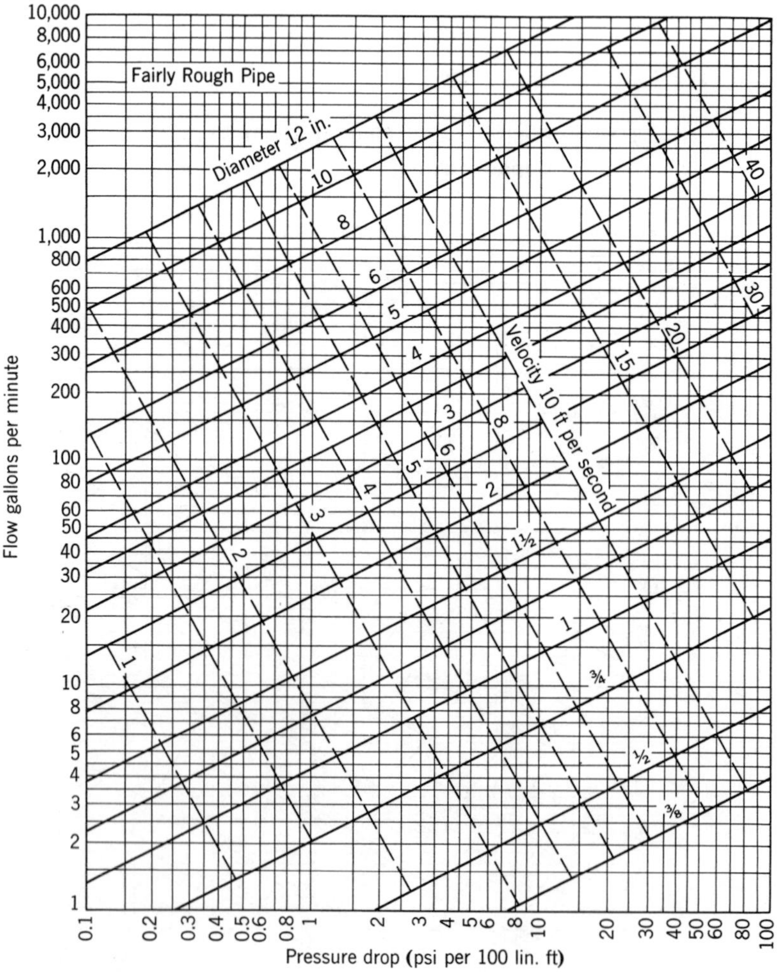

TABLE 15-1 Equivalent Length in Feet to Be Added to Run Owing to Valves and Fittings

Type of Fitting	Nominal Pipe Sizes—Inches																		
	$\frac{1}{2}$	$\frac{3}{4}$	1	$1\frac{1}{4}$	$1\frac{1}{2}$	2	$2\frac{1}{2}$	3	$3\frac{1}{2}$	4	$4\frac{1}{2}$	5	6	7	8	9	10	11	12
Gate valve—open	0.3	0.5	0.6	0.8	0.9	1.2	1.4	1.7	2.0	2.5	2.7	3.0	3.5	4.0	4.5	5.0	6.0	6.5	7.0
Globe valve—open	16	21	26	35	43	54	65	80	95	110	120	140	160	180	210	250	280	305	330
Angle valve—open	8	11	14	18	20	25	31	40	45	51	60	70	80	91	110	125	140	152	165
Standard 45 deg elbow	0.8	1.0	1.3	1.6	2.0	2.5	3.0	3.8	4.5	5.0	5.8	6	8	8.5	10	11	13	14	15
Standard 90 deg elbow	1.5	2.0	2.5	3.5	4.5	5.0	6.5	8.0	10	11	13	14	16	18	20	23	26	28	30
Medium sweep 90 deg elbow	1.4	1.8	2.3	3.0	3.5	4.5	5.2	6.8	8	9	10	11	14	15	17	19	21	23	25
Long sweep 90 deg elbow	1.0	1.5	2.0	2.5	3.0	3.5	4	5	6	7	8	9	10	12	14	16	18	19	20
Square elbow 90 deg	3.0	4.5	5.5	7.5	9	12	14	17	20	22	24	26	33	38	44	50	53	55	57
Close return bend	3.5	5	6	8	10	13	15	18	20	24	26	30	35	42	49	54	61	66	72
Stand tee—full-size branch*	3.0	4.5	5.5	7.5	9	12	14	17	20	22	24	26	33	38	44	50	53	55	57
Stand tee—through run	1.0	1.5	2.0	2.5	3.0	3.5	4	5	6	7	8	9	10	12	14	16	18	19	20
Sudden enlargement from d to D†																			
d/D = $\frac{1}{4}$	1.5	2.0	2.5	3.5	4.5	5.0	6.5	8.0	10	11	13	14	16	18	20	23	26	28	30
d/D = $\frac{1}{2}$	1.0	1.3	1.6	2.2	2.6	3.3	3.8	4.9	5.6	6.4	7.0	8.1	10	11	13	15	16	17	18
d/D = $\frac{3}{4}$	0.3	0.5	0.6	0.8	0.9	1.2	1.4	1.7	2.0	2.5	2.7	3.0	3.5	4.0	4.5	5.0	6.0	6.5	7.0
Sudden contraction from D to d†																			
d/D = $\frac{1}{4}$	0.8	1.0	1.3	1.6	2.0	2.5	3.0	3.8	4.5	5.0	5.8	6	8	8.5	10	11	13	14	15
d/D = $\frac{1}{2}$	0.6	0.8	1.0	1.3	1.5	1.8	2.3	2.8	3.4	3.6	4.3	4.8	5.6	6.4	7.5	8.5	9.5	11	12
d/D = $\frac{3}{4}$	0.3	0.5	0.6	0.8	0.9	1.3	1.4	1.7	2.0	2.5	2.7	3.0	3.5	4.0	4.5	5.0	6.0	6.5	7.0
Ordinary pipe entrance with upstream end of pipe flush with inside of tank	0.9	1.3	1.5	2.0	2.4	3.0	3.6	4.5	5.1	6.0	6.6	7.5	9.0	11	12	14	15	17	18
Entrance with pipe projecting into tank beyond inside face (borda entrance)	1.5	2.0	2.5	3.5	4.0	5.0	6.0	7.8	9.0	10	12	13	15	17	19	21	24	27	30

* Pressure drop through side outlet, or from side outlet through run.
† Equivalent feet of the smaller diameter pipe, "d."
Courtesy York Corporation.

straight pipe is determined by the following equation:

$$\text{Total pressure loss (ft)} = \frac{\text{Total length of pipe (ft)}}{100} \times \text{pressure loss/100 ft (psi)} \qquad (15\text{-}14)$$

Pipe fittings, such as elbows, tees, valves, etc., offer a greater resistance to flow than does straight pipe and therefore must be taken into account in determining the total friction loss through the piping. For convenience, this is frequently accomplished by considering the fittings as having a resistance equal to a certain length of straight pipe called the "equivalent length."* Table 15-1 lists the equivalent length of straight pipe for various types of fittings and valves. Notice that the equivalent length varies with size of the pipe.

When the equivalent length of the fittings is added to the actual length of straight pipe, the result is called the "total equivalent length." This value is then applied in Equation 15-14 to determine the total friction loss through the piping.

Example 15-1 A water piping system consists of 128 ft of 2 in. straight pipe, 6 standard

* Another method of determining the pressure loss through fittings, abrupt enlargements, and so forth, employs velocity heads.

TABLE 15-2 Pressure Drop Correction Factors*

		Specific Gravity at (° F)		Friction Correction Factor Temperature—° F								
Liquid	Freeze at ° F	−20	+20	−20	−10	0	10	20	30	40	50	60
Calcium brine												
Sp gr = 1.10	20.3	—	1.11	—	—	—	—	1.21	1.19	1.15	1.12	1.11
Sp gr = 1.20	−5.8	—	1.21	—	—	1.49	1.44	1.38	1.33	1.28	1.26	1.24
Sp gr = 1.25	−26.0	1.27	1.26	1.85	1.75	1.66	1.57	1.50	1.44	1.40	1.37	1.34
Sodium brine												
Sp gr = 1.10	14.9	—	1.11	—	—	—	1.27	1.21	1.19	1.15	1.12	1.11
Sp gr = 1.18	−6.0	—	1.19	—	1.58	1.50	1.44	1.39	1.33	1.28	1.25	1.22
Ammonia (liquid)	−107.8	0.68	0.65	0.65	0.65	0.65	0.65	0.65	0.65	0.65	0.65	0.65
Alcohol (ethyl) (100%)	−114.6	0.83	0.81	0.97	0.95	0.93	0.92	0.91	0.91	0.90	0.90	0.90
Alcohol (ethyl) (40%)	−22	0.93	0.91	1.45	1.39	1.33	1.29	1.23	1.19	1.15	1.12	1.10
Alcohol (methyl) (100%)	−97	0.84	0.82	0.85	0.85	0.85	0.84	0.84	0.84	0.83	0.83	0.83
Alcohol (methyl) (30%)	−5.8	—	0.96	1.32	1.26	1.22	1.19	1.16	1.12	1.09	1.07	1.05
Ethylene glycol (60%)	−59.0	1.10	1.09	1.87	1.83	1.78	1.72	1.62	1.57	1.46	1.40	1.36
Ethylene glycol (50%)	−38.0	1.09	1.08	1.83	1.74	1.64	1.54	1.48	1.42	1.37	1.31	1.26
Ethylene glycol (30%)	2.0	—	1.06	—	—	—	1.34	1.27	1.22	1.17	1.13	1.11
Refrigerant-11 (liquid)	−168	1.60	1.55	1.42	1.42	1.42	1.42	1.42	1.42	1.42	1.42	1.42
Refrigerant-12 (liquid)	−252	1.49	1.42	1.32	1.32	1.32	1.32	1.32	1.32	1.32	1.32	1.32
Methyl chloride	−144	1.02	0.97	1.13	1.13	1.13	1.12	1.12	1.12	1.11	1.11	1.11
Methylene chloride	−142.1	1.40	1.33	1.65	1.63	1.62	1.60	1.59	1.58	1.56	1.54	1.52

* To obtain pressure drop from flow of above liquids through pipes, multiply pressure drop for water flow (of equal quantity through same pipe) by factors from above table.

Courtesy York Corporation.

elbows, and 2 gate valves (full open). Using fairly rough pipe, if the flow rate through the system is 40 gpm, determine:

(a) The total equivalent length of straight pipe
(b) The total friction loss through the piping in pounds per square inch and in feet of water column.

Solution From Table 15-1, the equivalent lengths of 2 in. standard elbows and 2 in. gate valves (full open) are 5 ft and 1.2 ft, respectively. From Chart 15-2, for a flow rate of 40 gpm, the friction loss per hundred feet of 2 in. nominal pipe is 2.35 psi. From Table 1-1, a pressure of 1 psi is equivalent to 2.31 ft of water column.

(a) Total equivalent length

Straight pipe	= 128.0 ft
Six 2 in. elbows @ 5 ft	= 30.0
Two 2 in. gate valves @ 1.2 ft	= 2.4
	160.4 ft

(b) Applying Equation 15-14, the total friction loss through the piping

$$= \frac{160.4}{100} \times 2.35$$

$$= 3.77 \text{ psi}$$

Converting to ft H_2O

$$= 3.77 \times 2.31$$

$$= 8.71 \text{ ft } H_2O$$

Although the pressure loss determined from Charts 15-1 and 15-2 apply only to water, the charts can be used for other fluids by multiplying the water pressure loss obtained from these charts by the correction factors listed in Table 15-2.

15-7. Centrifugal Pumps

Liquid pumps used in the refrigerating industry to circulate chilled water or brine, and the condenser water are usually of the centrifugal type.

A centrifugal pump consists mainly of a rotating vane-type impeller that is enclosed in a stationary casing. The liquid being pumped is drawn in through the "eye" of the impeller and is thrown to the outer edge or periphery of the impeller by centrifugal force. Considerable velocity and pressure are imparted to the liquid in the process. The liquid leaving the periphery of the impeller is collected in the casing and directed through the discharge opening (Fig. 15-4).

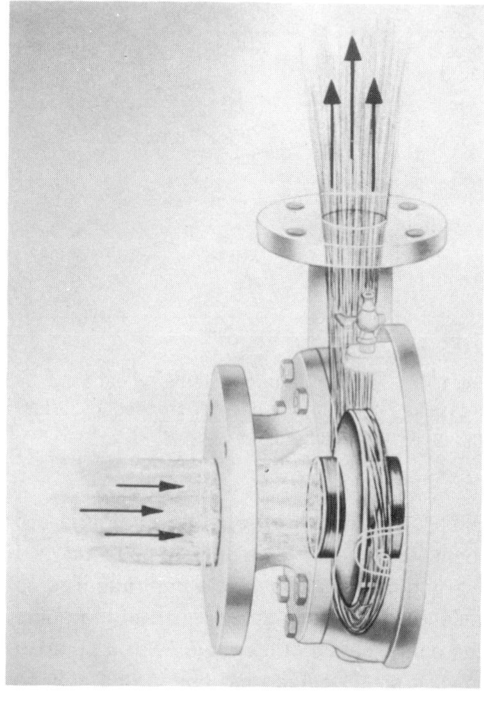

Fig. 15-4 Fluid flow through centrifugal pump. (Courtesy Ingersoll-Rand Company.)

Frequently, the impeller of the pump is mounted directly on the shaft of the pump-driving motor so that the pump and motor are an integral unit (Fig. 15-5). In other cases, the

Fig. 15-5 Typical centrifugal pump and motor assembly. (Courtesy Bell & Gossett Company.)

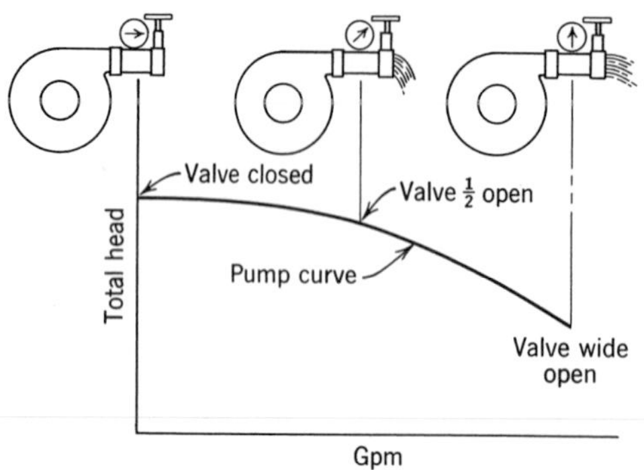

Fig. 15-6 Centrifugal pump delivery capacity increases as the pumping head decreases. (Courtesy Ingersoll-Rand Company.)

pump and motor are separate units and are connected together by a flexible coupling.

In general, the capacity of a centrifugal pump depends on the design and size of the pump and on the speed of the motor. For a pump of specific size, design, and speed, the volume of liquid handled varies with the pumping head against which the pump must work. A characteristic head-capacity curve for a typical centrifugal pump is shown in Fig. 15-6. Notice that the pumping head is maximum when the valve on the discharge of the pump is closed, at which time the pump delivery is zero. As the valve is opened, the pumping head decreases and the delivery rate increases.

Centrifugal pumps are rated in gpm of delivery at various pumping heads, that is, centrifugal pumps are rated to deliver a certain gpm against a certain pumping head. Although pump ratings are available in table form, more frequently they are taken from head-capacity curves (see Chart R-20). In either case, before the proper pump can be selected from the manufacturer's ratings, it is necessary to know the required gpm and the total pumping head against which the pump must operate.

15-8. Total Pumping Head

The total pumping head is the sum of the static head and the friction head.

The static head is the vertical distance between the "free liquid level" and the highest

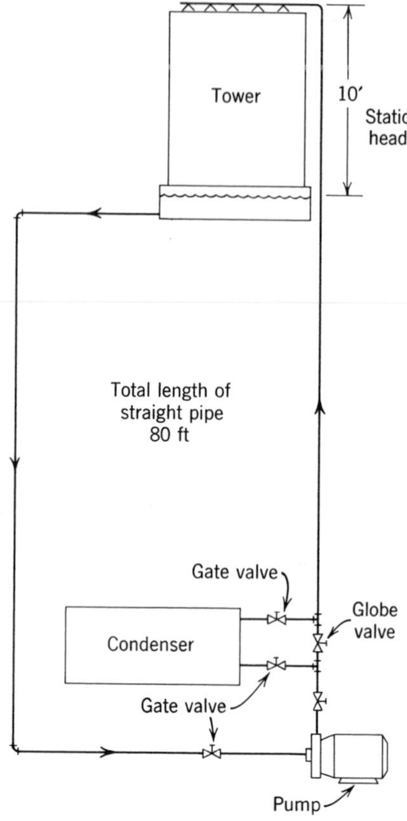

Fig. 15-7 Condenser-water circulating system.

point to which the liquid must be lifted by the pump. For the condenser-water circulating system in Fig. 15-7, the static head, measured in feet of water column, is the vertical distance in feet between the free water level in the tower basin and the tower spray header. Because of the water head in the tower basin, the water in the discharge pipe will stand to the level of the water in the tower basin of its own accord. Therefore, the distance the water is actually lifted by the pump is only the distance from the water level in the tower basin up to the spray header. Contrast this with the pumping system shown in Fig. 15-8.

When the piping system is a closed circuit, as in Fig. 15-9, there is no unbalanced static head on the pump, since the fluid on one side of the piping system will exactly balance the fluid on the other side.

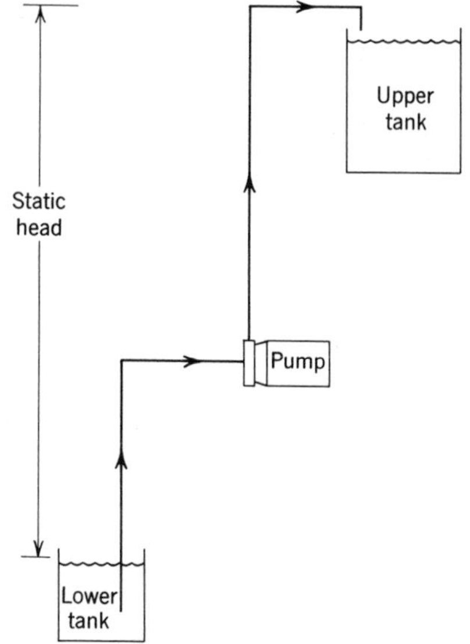

Fig. 15-8

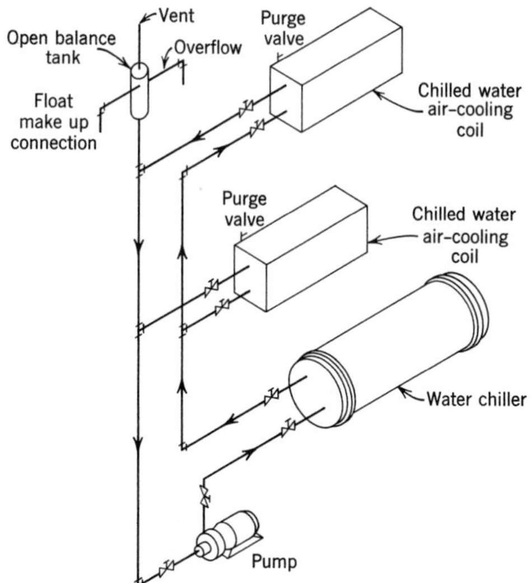

Fig. 15-9 Closed chilled water (or brine) circulating system. To compute pumping head use circuit having greatest friction loss. There is no unbalanced static head.

A typical piping system curve in which gpm is plotted against total head is shown in Fig. 15-11. Notice that the total head increases as the flow rate through the system increases and that the increase in the total head results

Gallons per Minute	Head
50	2.5 ft
100	10.0 ft
150	22.5 ft
200	40.0 ft
250	62.5 ft

Fig. 15-10

entirely from an increase in the friction head, the static head being constant.

It was shown previously that the pressure or head loss suffered by a flowing fluid as a result of friction varies as the square of the fluid velocity. Consequently, the pumping head for any given piping system resulting from friction varies as the square of the flow through the system, that is,

$$\left(\frac{Q_2}{Q_1}\right)^2 = \frac{h_2}{h_1} \tag{15-15}$$

$$h_2 = \left(\frac{Q_2}{Q_1}\right)^2 \times h_1 \tag{15-16}$$

where Q_1 = the initial flow rate in gallons per minute
Q_2 = the final flow rate in gallons per minute
h_1 = the initial head in feet
h_2 = the final head in feet

Employing Equation 15-16, data can be developed that can be used to plot a piping system curve.

Example 15-2 A piping system has a static head of 15 ft and a friction head of 10 ft when the flow rate is 100 gpm. Plot a system curve.

Solution Applying Equation 15-16 to find the system head at a flow rate of 50 gpm,

$$h_2 = \left(\frac{50\text{ gpm}}{100\text{ gpm}}\right)^2 \times 10\text{ ft} = 2.5\text{ ft}$$

Similar calculations produce the data listed in Fig. 15-10, from which the system curve in Figure 15-11 is plotted. Notice that the point at

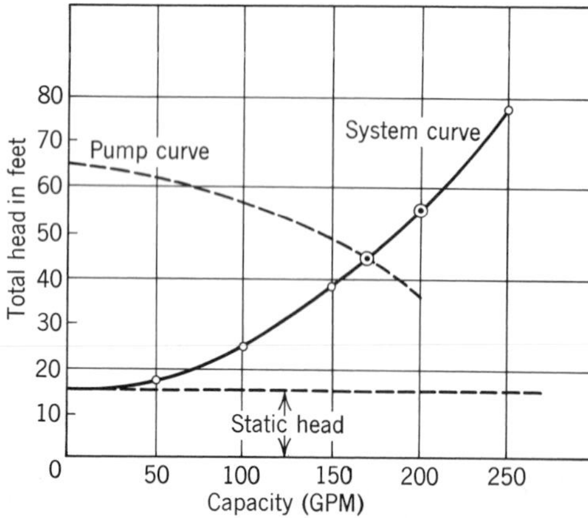

Fig. 15-11 Friction head of piping system increases as flow through system increases.

which the system curve intersects the pump curve is the balance point of the pumping system.

15-9. Determining the Total Pumping Head

The pressure loss through the various system components, such as condensers, chillers, and cooling towers, are found in the manufacturers' rating tables.

When more than one condenser (or chiller, etc.) is used in the system, the condensers are piped in parallel and only the condenser circuit with the largest pressure drop is considered in computing the pumping head.

The pressure loss through the cooling tower, as given by the tower manufacturer, is the total head and includes both the tower static and friction heads. Therefore, the static head of the tower should not be considered separately in determining the total pumping head. When the tower static head is the only static head in the system, the static head should be disregarded entirely. However, in the event that an auxiliary indoor storage tank is employed, as shown in Fig. 14-24, the vertical distance between the level of the water in the tank and the normal water level in the tower basin must be treated as a separate static head.

Since pump manufacturers usually express the pumping head in "feet of water column," it is necessary to compute the pumping head in these units. When the pressure loss through the several system components is given in psi or in other units of pressure, it must be converted to feet of water column before it can be used in computing the pumping head. The required conversion factors are found in Table 1-1.

Example 15-3 The recirculating water system shown in Fig. 15-7 is for a 15-ton refrigerating system. The flow rate over the tower is 60 gpm (4 gpm/ton). The flow rate through the condenser is 45 gpm (3 gpm/ton), with 15 gpm (1 gpm/ton) flowing through the condenser bypass. From the manufacturers' rating tables, the tower head based on 4 gpm/ton is 24 ft of water column, whereas the pressure drop through the condenser for 45 gpm is 6.1 psi or 14.1 ft of water column (6.1 × 2.31). If the size of the piping is 2 in. nominal, determine the total pumping head and select the proper pump from Chart R-19.

Solution Total equivalent length of pipe:

Straight pipe	= 80.0 ft
Three 2 in. standard elbows at 5 ft	= 15.0
Two 2 in. tees (side outlet) at 12 ft	= 24.0
Four 2 in. gate valves (open) at 1.2 ft	= 4.8
	= 123.8 ft

From Chart 15-2, the pressure loss per 100 ft of pipe (60 gpm and 2 in. pipe) = 5.25 psi

Applying Equation 15-14, the total pressure loss through the piping

$$= \frac{123.8}{100} \times 5.25$$

$$= 6.5 \text{ psi}$$

Converting to ft H_2O $= 6.5 \times 2.31$

$= 15.00$ ft H_2O

Total pumping head

Piping	= 15.00 ft
Condenser	= 14.10
Tower	= 24.00
	= 53.10 ft H_2O

From Chart R-20, select pump Model # 1531-28, which has a delivery capacity of 60 gpm at a 54 ft head.

Example 15-4 At the required flow rate of 100 gpm, a certain water system has a pumping head of 60 ft of water column. Select the proper pump from Chart R-20.

Solution Reference to Chart R-20 shows that pump Model #1531-30 is the smallest pump which can be used. However, since this pump will deliver 125 gpm at a 60-ft head, to obtain the desired flow rate of 100 gpm, the pumping head must be increased to 73 ft of water. This is accomplished by throttling the pump with a globe valve installed on the discharge side of the pump. (The pump should never be throttled on the suction side.)

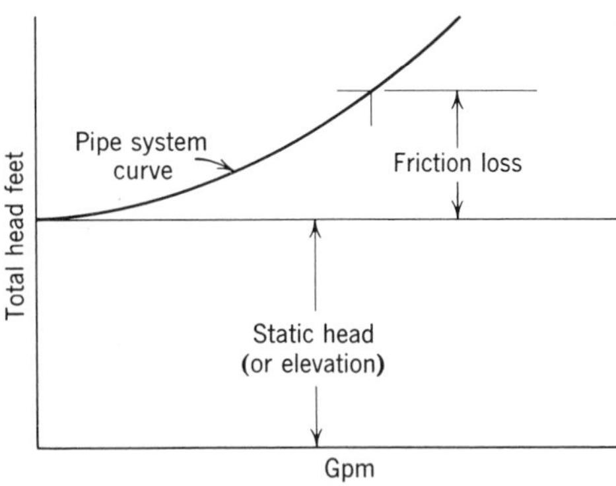

Fig. 15-12 Variations in pump, horsepower and efficiency with delivery rate. (Courtesy Ingersoll-Rand Company.)

15-10. Power Requirements

The power required to drive the pump depends upon the delivery rate in pounds per minute, the total pumping head, and the efficiency of the pump, that is,

$$\text{Bhp} = \frac{\text{Pounds per minute} \times \text{total head in feet}}{33{,}000 \times \text{pump efficiency}}$$

Since the flow rate is usually in gallons per minute, a more practical equation is

$$\text{Bhp} = \frac{\text{Gpm} \times \text{total head} \times 8.33 \text{ lb/gal}}{33{,}000 \times \text{efficiency}}$$

Combining constants,

$$\text{Bhp} = \frac{\text{Gpm} \times \text{total head in feet}}{3960 \times \text{efficiency}} \qquad (15\text{-}17)$$

Equation 15-17 applies to water. When a liquid other than water is handled, the specific gravity of the liquid must be taken into account, that is,

$$\text{Bhp} = \frac{\text{Gpm} \times \text{total head} \times \text{specific gravity}}{3960 \times \text{efficiency}} \qquad (15\text{-}18)$$

From Equation 15-18, it is evident that the power required by the pump increases as the delivery rate, total head, or specific gravity increases, and decreases as the pump efficiency increases.

Typical pump horsepower and efficiency curves are shown in Fig. 15-12. Notice that pump horsepower is lowest at no delivery and increases progressively as the delivery rate increases. Hence, for a given pump, any decrease in the pumping head will cause an increase in both the delivery rate and the power requirements of the pump.

Pump efficiency, also lowest at no delivery, increases to a maximum as the flow rate is increased and then decreases as the flow rate is further increased. The pump efficiency curve in Fig. 15-12 indicates that the highest efficiency is obtained when the pump is selected to deliver the desired gpm when operating at some point near the midpoint of its head-capacity curve.

15-11. Water Piping Design

In general, the water piping should be designed for the minimum friction loss consistent with reasonable initial costs so that the pumping requirements are maintained at a practical minimum. Water lines should be kept as short as possible and a minimum amount of fittings should be used.

Standard weight steel pipe, plastic pipe, or type "L" copper tubing are usually employed for condenser water piping. Pipe sizes which will provide water velocities in the neighborhood of 5 to 8 fps at the required flow rate will usually prove to be the most economical. For example, assume that 150 gpm of water are to be circulated through 100 equivalent feet of piping. The following approximate values of velocity and friction loss are shown in Chart 15-2 for a flow rate of 150 gpm through various sizes of pipe:

Pipe Size (inches)	Velocity (fps)	Friction Loss per 100 ft	
		(psi)	(ft H_2O)
2	15.5	31.5	72.8
$2\frac{1}{2}$	10.0	10.5	24.3
3	7.1	4.3	11.1
$3\frac{1}{2}$	5.2	2.0	4.6
4	3.9	1.1	2.5

Notice that whereas increasing the pipe size from 2 to 3 in. results in a considerable reduction in the friction loss (from 72.8 to 11.1 ft), a further increase in the pipe size from 3 to 4 in. reduces the friction loss by only an additional 8.6 ft of water column (11.1 to 2.5 ft), which will not ordinarily justify the increase in the cost of the pipe. Depending upon the characteristics of the available pump, either 3 in. or $3\frac{1}{2}$ in. pipe should be used. For instance, assume two separate systems having pumping heads, exclusive of the friction loss in the piping, of 55 ft of water column and 65 ft of water column, respectively. Reference to Chart R-20 indicates that the only suitable pump for either of the systems is Model 1531-32, which has a delivery rate of 150 gpm at a 70-ft head. Therefore, for the system having the 55-ft head, the permissible friction loss in the piping is 15 ft (70 − 55), whereas for the system having the 65-ft head, the permissible friction loss in the piping is only 5 ft (70 − 65). For the latter system, $3\frac{1}{2}$ in. pipe must be used, since the use of 3 in. pipe would result in a total pumping head in excess of the allowable 70 ft and necessitate the use of the next larger size pump. On the other hand, for the former system, 3 in. pipe is the most practical size. The use of $3\frac{1}{2}$ in. pipe in this instance would result in a total pumping head of only 61 ft and would necessitate throttling of the pump discharge in order to raise the pumping head to 70 ft and obtain the desired flow rate of 150 gpm.

In designing the piping system, care should be taken to include all valves and fittings necessary for the proper operation and maintenance of the water circulating system. It is good practice to install a globe valve on the discharge side of the pump to regulate the water flow rate when the latter is critical. Too, where the piping is long and/or the quantity of water in the system is large, shut-off valves installed on the inlet and outlet of both the pump and the condenser will permit repairs to these pieces of equipment without the necessity of draining the tower. A drain connection should be installed at the lowest point in the piping and the piping should be pitched downward so as to assure complete drainage during winter shut down.

The pump must always be located at some point below the level of the water in the tower basin in order to assure positive and continuous priming of the pump. When quiet operation is required, the pump may be isolated from the piping with short lengths of rubber hose. Automobile radiator hose is suitable for this purpose.

15-12. METRIC EQUIVALENTS

As shown in Chapter 1, in any location where g is equal to g_c, a 1-lb mass, by definition, exerts a 1-lb gravitational force, whereas a 1-kg mass exerts a force of 9.807 newtons (1 kg)(9.807 m/s^2). For this reason, the relationships among the head, density, and pressure of a fluid expressed in SI units differ somewhat from those expressed in British units and are given by the following equations:

$$p = (h)(\rho)(g) \tag{15-19}$$

$$h = \frac{p}{(\rho)(g)} \tag{15-20}$$

$$\rho = \frac{p}{(h)(g)} \tag{15-21}$$

where p = the pressure in newtons per square meter (N/m^2, or Pa)
ρ = the density of the fluid in kilograms per cubic meter (kg/m^3)
g = the acceleration of gravity 9.807 m/s^2).

Example 15-5 Determine the static pressure (p_s) exerted at the base of a column of water 25 m in height.

Solution Assuming a water density of 1000 kg/m^3 and applying Equation 15-19,

$$p_s = (25\text{ m})(1000\text{ kg/m}^3)(9.807\text{ m/s}^2)$$
$$= 245{,}175\text{ N/m}^2\text{ (Pa) or }2.45\text{ bar}$$

Example 15-6 Compute the height in meter of a column of water which is supported by a pressure of 6500 N/m^2.

Solution Applying Equation 15-20,

$$h_s = \frac{6500\text{ N/m}^2}{(1000\text{ kg/m}^3)(9.807\text{ m/s}^2)} = 0.663\text{ m}$$

The relationships among velocity, velocity head, and velocity pressure are given in the following equations:

To convert velocity to velocity head,

$$h_v = \frac{v^2}{(2)(g)} \tag{15-22}$$

To convert velocity to velocity pressure,

$$p_v = \frac{(v^2)(\rho)}{2} = \frac{v^2}{(2)(\nu)} \tag{15-23}$$

To convert velocity head to velocity pressure,

$$p_v = (h_v)(g)(\rho) = \frac{(h_v)(g)}{\nu} \tag{15-24}$$

To convert velocity head to velocity,

$$v = \sqrt{(h_v)(2)(g)} \tag{15-25}$$

$$v = \sqrt{\frac{(p_v)(2)}{(\rho)}} = \sqrt{(p_v)(2)(\nu)} \tag{15-26}$$

where v = the velocity in meters per second
h_v = the velocity head in meters
h_p = the velocity pressure in newtons per square meter (or Pa)
ν = the specific volume in cubic meters per kilogram (m^3/kg)

Example 15-7 Water flowing in a pipe has a velocity of 1.8 m/s. Compute:
(a) the velocity head,
(b) the velocity pressure.

Solution
(a) Applying Equation 15-22,

$$h_v = \frac{(1.8\text{ m/s})^2}{(2)(9.807\text{ m/s}^2)} = 0.165\text{ m}$$

(b) Applying Equation 15-23,

$$p_v = \frac{(1.8\text{ m/s})^2(1000\text{ kg/m}^3)}{2}$$
$$= 1620\text{ N/m}^2\text{ (Pa)}$$

or applying Equation 15-24,

$$p_v = (0.165\text{ m})(9.807\text{ m/s}^2)(1000\text{ kg/m}^3)$$
$$= 1618\text{ N/m}^2$$

Example 15-8 With reference to Fig. 15-3, assume that the pressure loss resulting from friction is negligible so that the total pressure of the fluid at point *B* is the same as that at point *A*. Calculate the static pressure (p_s) at point *B* if the static pressure at point *A* is 60,000 N/m^2 and the velocities at point *A* and *B* are 1.2 m/s and 3 m/s, respectively. The fluid flowing in the pipe is water with a density of 1000 kg/m^3.

Solution The total pressure at point *B*,

$$p_{tb} = p_{ta} = p_{s,a} + p_{v,a}$$
$$= (60{,}000\text{ N/m}^2)$$
$$+ (1.2\text{ m/s})^2(1000\text{ kg/m}^3)/2$$
$$= 60{,}720\text{ N/m}^2$$

The velocity pressure at point *B*,

$$p_{v,b} = (3\text{ m/s})^2(1000\text{ kg/m}^3)/2 = 4500\text{ N/m}^2$$

The static pressure at point *B*

$$p_s = p_t - p_v = 60{,}720\text{ N/m}^2 - 4500\text{ N/m}^2$$
$$= 56{,}220\text{ N/m}^2$$

The power required by a pump can be determined by employing the following equations:

$$P = \frac{(m)(g)(h)}{\text{pump efficiency}} \qquad (15\text{-}27)$$

$$P = \frac{(V)(\rho)(g)(h)}{\text{pump efficiency}} \qquad (15\text{-}28)$$

where P = the power in watts
m = the mass flow rate in kilograms per second
V = the volume flow rate in cubic meters per second
ρ = the density in kilograms per cubic meter
h = the head in meters
g = the acceleration of gravity

Example 15-9 A pump having an efficiency of 68% delivers 0.003 m^3/s of water against a head of 22 m. Compute the power requirements of the pump in watts.

Solution Assuming a water density of 1000 kg/m^3 and applying Equation 15-28,

$$P = \frac{(0.003\ m^3/s)(1000\ kg/m^3) \times (9.807\ m/s^2)(22\ m)}{0.68}$$

$$= 951.86\ W$$

Customary Problems

15-1 Assuming a water density of 62.4 lb/ft^3, determine the static pressure (p_s) in pounds per square inch exerted by a column of water 50 ft in height.

15-2 What is the head of water in feet equivalent to a pressure of 10 psi.

15-3 A column of fluid 30 ft high exerts a base pressure of 16.5 psi. What is the average density of the fluid?

15-4 Water flowing in a pipe has a velocity of 5 fps. Compute (a) the velocity head in feet, and (b) the velocity pressure in pounds per square inch. Assume a water density of 62.4 lb/ft^3.

15-5 What is the velocity of water flowing through a pipe with an internal area of 2.5 $in.^2$ if the water flow rate is 15 ft^3/min.

15-6 With reference to Fig. 15-3, assume that the pressure loss resulting from friction is negligible, so that the total pressure of the fluid at point *B* is the same as that at point *A*, and calculate the static pressure (p_s) at point *B* if the static pressure at point *A* is 10 psi and the velocities at points *A* and *B* are 6 fps and 8 fps, respectively. The fluid flowing in the conduit is water having a density of 62.4 ft^3/lb.

15-7 A water piping system consists of 90 ft of 3 in. nominal copper tube, nine standard elbows, and two gate valves. If the water flow rate through the pipe is 120 gpm, compute (a) the total equivalent length of straight pipe, and (b) the total friction loss in the pipe in feet of water column.

15-8 Rework Problem 15-7 using fairly rough pipe.

15-9 A piping system has a static head of 8 ft and a friction head of 12 ft when the flow rate through the system is 100 gpm. Plot a system curve from zero flow to 250 gpm using 50 gpm increments.

15-10 A recirculating condenser water system consists of 100 ft of 3 in. nominal copper tube and eight standard 90° elbows. At the desired flow rate of 150 gpm, the pressure loss in the condenser is 5 psi, and the pressure drop over the tower is 10 ft of water. An indoor storage tank is located 8 ft below the tower basin (8 ft static head). Determine (a) the total equivalent length of pipe, and (b) the total pumping head in feet of water column.

15-11 From Table R-20, select a pump for the condenser water system described in Problem 15-10.

15-12 Using Equation 15-17, compute the

brake horsepower required for the pump in Problem 15-11 if the pump has an efficiency of 65%.

Metric Problems

15-13 Determine the static pressure (p_s) in pascals exerted by a column of water 60 m in height.

15-14 Compute the height in meter of a column of fluid having a base pressure of 75,000 Pa and a density of 1200 kg/m^3.

15-15 Water flowing in a pipe has a velocity of 2.5 m/s. Compute (a) the velocity head in meters of water, and (b) the velocity pressure in pascals.

15-16 A pump having an efficiency of 58% is delivering 0.015 m^3/s of a fluid having a density of 1250 kg/m^3 against a pumping head of 30 m. Compute the power requirements of the pump in watts.

16

REFRIGERANTS

16-1. The Ideal Refrigerant

Generally speaking, a refrigerant is any body or substance which acts as a cooling agent by absorbing heat from another body or substance. With regard to the vapor-compression cycle, the refrigerant is the working fluid of the cycle which alternately vaporizes and condenses as it absorbs and gives off heat, respectively. To be suitable for use as a refrigerant in the vapor-compression cycle, a fluid should possess certain chemical, physical, and thermodynamic properties that make it both safe and economical to use.

It should be recognized at the onset that there is no "ideal" refrigerant and that, because of the wide differences in the conditions and requirements of the various applications, there is no one refrigerant that is universally suitable for all applications. Hence, a refrigerant approaches the "ideal" only to the extent that its properties meet the conditions and requirements of the application for which it is to be used.

Table 16-1 lists a number of fluids having properties which render them suitable for use as refrigerants. However, it will be shown presently that only a few of the more desirable ones are actually employed as such. Some, used extensively as refrigerants in the past, have been discarded as more suitable fluids were developed. Others, still in the development stage, show promise for the future. Tables 16-2 through 16-6 list the thermodynamic properties of some of the refrigerants in common use at the present time. The use of these tables has already been described in an earlier chapter.

16-2. Safe Properties

Ordinarily, the safe properties of the refrigerant are the prime consideration in the selection of a refrigerant. It is for this reason that some fluids, which otherwise are highly desirable as refrigerants, find only limited use as such. The more prominent of these are ammonia and some of the straight hydrocarbons.

To be suitable for use as a refrigerant, a fluid should be chemically inert to the extent that it is nonflammable, nonexplosive, and nontoxic both in the pure state and when mixed in any proportion with air. Too, the fluid should not react unfavorably with the lubricating oil or with any material normally used in the construction of refrigerating equipment. Nor should it react unfavorably with moisture which despite stringent precautions is usually present at least to some degree in all refrigerating systems. Furthermore, it is desirable that the fluid be of such a nature that it will not contaminate in any way foodstuff or other stored products in the event that a leak develops in the system.

16-3. Toxicity

Since all fluids other than air are toxic in the sense that they will cause suffocation when in

TABLE 16-1 ASRE Refrigerant Numbering System

ASRE Standard Refrigerant Designation	Chemical Name	Chemical Formula	Molecular Weight	Boiling Point, F	Status
Halocarbon Compounds					
10	Carbontetrachloride	CCl_4	153.8	170.2	
11	Trichloromonofluoromethane	CCl_3F	137.4	74.8	C
12	Dichlorodifluoromethane	CCl_2F_2	120.9	−21.6	C
13	Monochlorotrifluoromethane	$CClF_3$	104.5	−114.6	C
13B1	Monobromotrifluoromethane	$CBrF_3$	148.9	−72.0	S
14	Carbontetrafluoride	CF_4	88.0	−198.4	S
20	Chloroform	$CHCl_3$	119.4	142	
21	Dichloromonofluoromethane	$CHCl_2F$	102.9	48.1	D
22	Monochlorodifluoromethane	$CHClF_2$	86.5	−41.4	C
23	Trifluoromethane	CHF_3	70.0	−119.9	D
30	Methylene chloride	CH_2Cl_2	84.9	105.2	C
31	Monochloromonofluoromethane	CH_2ClF	68.5	48.0	
32	Methylene fluoride	CH_2F_2	52.0	−61.4	
40	Methyl chloride	CH_3Cl	50.5	−10.8	C
41	Methyl fluoride	CH_3F	34.0	−109	
(50	Methane	CH_4	16.0	−259	C)[1]
110	Hexachloroethane	CCl_3CCl_3	236.8	365	
111	Pentachloromonofluoroethane	CCl_3CCl_2F	220.3	279	
112	Tetrachlorodifluoroethane	CCl_2FCCl_2F	203.8	199.0	
112a	Tetrachlorodifluoroethane	CCl_3CClF_2	203.8	195.8	
113	Trichlorotrifluoroethane	CCl_2FCClF_2	187.4	117.6	C
113a	Trichlorotrifluoroethane	CCl_3CF_3	187.4	114.2	
114	Dichlorotetrafluoroethane	$CClF_2CClF_2$	170.9	38.4	C
114a	Dichlorotetrafluoroethane	CCl_2FCF_3	170.9	38.5	C
114B2	Dibromotetrafluoroethane	$CBrF_2CBrF_2$	259.9	117.5	D
115	Monochloropentafluoroethane	$CClF_2CF_3$	154.5	−37.7	D
116	Hexafluoroethane	CF_3CF_3	138.0	−108.8	
120	Pentachloroethane	$CHCl_2CCl_3$	202.3	324	
123	Dichlorotrifluoroethane	$CHCl_2CF_3$	153	83.7	
124	Monochlorotetrafluoroethane	$CHClFCF_3$	136.5	10.4	
124a	Monochlorotetrafluoroethane	CHF_2CClF_2	136.5	14	D
125	Pentafluoroethane	CHF_2CF_3	120	−55	
133a	Monochlorotrifluoroethane	CH_2ClCF_3	118.5	43.0	D
140a	Trichloroethane	CH_3CCl_3	133.4	165	
142b	Monochlorodifluoroethane	CH_3CClF_2	100.5	12.2	S
143a	Trifluoroethane	CH_3CF_3	84	−53.5	
150a	Dichloroethane	CH_3CHCl_2	98.9	140	
152a	Difluoroethane	CH_3CHF_2	66	−12.4	C
160	Ethyl chloride	CH_3CH_2Cl	64.5	54.0	
(170	Ethane	CH_3CH_3	30	−127.5	C)[1]
218	Octafluoropropane	$CF_3CF_2CF_3$	188	−36.4	
(290	Propane	$CH_3CH_2CH_3$	44	−44.2	C)[1]
Cyclic Organic Compounds					
C316	Dichlorohexafluorocyclobutane	$C_4Cl_2F_6$	233	140	
C317	Monochloroheptafluorocyclobutane	C_4ClF_7	216.5	77	
C318	Octafluorocyclobutane	C_4F_8	200	21.1	D
Azeotropes					
500	Refrigerants-12/152a 73.8/26.2 wt %*	CCl_2F_2/CH_3CHF_2	99.29	−28.0	C
501	Refrigerants-22/12 75/25 wt %	$CHClF_2/CCl_2F_2$	93.1	−42	
502	Refrigerants-11/115 48.8/51.2 wt %	$CHClF_2/CClF_2CF_3$	112	−50.1	

TABLE 16-1 *(Continued)*

ASRE Standard Refrigerant Designation	Chemical Name	Chemical Formula	Molecular Weight	Boiling Point, F	Status[1]
Miscellaneous Organic Compounds					
Hydrocarbons					
50	Methane	CH_4	16.0	−259	C
170	Ethane	CH_3CH_3	30	−127.5	C
290	Propane	$CH_3CH_2CH_3$	44	−44.2	C
600	Butane	$CH_3CH_2CH_2CH_3$	58.1	31.3	
601	Isobutane	$CH(CH_3)_3$	58.1	14	
(1150	Ethylene	$CH_2{=}CH_2$	28.0	−155.0	C)[3]
(1270	Propylene	$CH_3CH{=}CH_2$	42.1	−53.7	C)[3]
Oxygen Compounds					
610	Ethyl ether	$C_2H_5OC_2H_5$	74.1	94.3	
611	Methyl formate	$HCOOCH_3$	60.0	89.2	
Sulfur Compounds					
620					
Nitrogen Compounds					
630	Methyl amine	CH_3NH_2	31.1	20.3	
631	Ethyl amine	$C_2H_5NH_2$	45.1	61.8	
Inorganic Compounds					
717	Ammonia	NH_8	17	−28.0	C
718	Water	H_2O	18	212	
729	Air		29	−318	
744	Carbon dioxide	CO_2	44	−109 (subl.)	C
744A	Nitrous oxide	N_2O	44	−127	
764	Sulfur dioxide	SO_2	64	14.0	C
Unsaturated Organic Compounds					
1112a	Dichlorodifluoroethylene	$CCl_2{=}CF_2$	133	67	
1113	Monochlorotrifluoroethylene	$CClF{=}CF_2$	116.5	−18.2	
1114	Tetrafluoroethylene	$CF_2{=}CF_2$	100	−105	
1120	Trichloroethylene	$CHCl{=}CCl_2$	131.4	187	
1130	Dichloroethylene	$CHCl{=}CHCl$	96.9	118	
1132a	Vinylidene fluoride	$CH_2{=}CF_2$	64	−119	
1140	Vinyl chloride	$CH_2{=}CHCl$	62.5	7.0	
1141	Vinyl fluoride	$CH_2{=}CHF$	46	−98	
1150	Ethylene	$CH_2{=}CH_2$	28.0	−155.0	C
1270	Propylene	$CH_3CH{=}CH_2$	42.1	−53.7	C

* Carrier Corp. Document 2-D-127, p. 1.

1. The compounds methane, ethane, and propane appear in the halocarbon section in their proper numerical positions, but in parentheses since these products are not halocarbons.

2. The compounds ethylene and propylene appear in the hydrocarbon section as parenthetical items in order to indicate that these compounds are hydrocarbons. Ethylene and propylene are properly identified under Unsaturated Organic Compounds.

From the *ASRE Data Book*, Design Volume, 1957-58 Edition, by permission of the American Society of Heating, Refrigerating, and Air-Conditioning Engineers.

concentrations large enough to preclude sufficient oxygen to sustain life, toxicity is a relative term which becomes meaningful only when the degree of concentration and the time of exposure required to produce harmful effects are specified.

The toxicity of most commonly used refrigerants has been tested by National Fire Underwriters. As a result, the various refrigerants are separated into six groups according to their degree of toxicity, the groups being arranged in descending order (column 2 of Table 16-7). Those falling into group 1 are highly toxic and are capable of causing death or serious injury in relatively small concentrations and/or short exposure periods. On the

TABLE 16-2A Saturation Properties of Refrigerant 12 (SI Units)

TEMP. °C	PRES.	VOLUME m³/kg · 10³		DENSITY kg/m³		ENTHALPY kJ/kg			ENTROPY kJ/kg K		TEMP. °C
		LIQUID v_f	VAPOR v_g	LIQUID $1/v_f$	VAPOR $1/v_g$	LIQUID h_f	LATENT h_{fg}	VAPOR h_g	LIQUID s_f	VAPOR s_g	
-45	0.5044	0.65355	302.683	1.53010	0.00330	159.549	171.674	331.223	0.83901	1.59142	-45
-44	0.5298	0.65472	289.157	1.52736	0.00346	160.427	171.260	331.687	0.84285	1.59016	-44
-43	0.5562	0.65590	276.362	1.52462	0.00362	161.306	170.845	332.151	0.84667	1.58893	-43
-42	0.5836	0.65709	264.249	1.52186	0.00378	162.186	170.429	332.615	0.85047	1.58773	-42
-41	0.6121	0.65828	252.779	1.51910	0.00396	163.067	170.011	333.078	0.85427	1.58655	-41
-40	0.6417	0.65949	241.910	1.51633	0.00413	163.948	169.593	333.541	0.85805	1.58539	-40
-39	0.6724	0.66070	231.607	1.51355	0.00432	164.831	169.173	334.004	0.86181	1.58426	-39
-38	0.7043	0.66192	221.835	1.51076	0.00451	165.714	168.752	334.466	0.86557	1.58315	-38
-37	0.7373	0.66315	212.562	1.50796	0.00470	166.598	168.329	334.927	0.86931	1.58207	-37
-36	0.7716	0.66438	203.759	1.50515	0.00491	167.483	167.905	335.388	0.87304	1.58100	-36
-35	0.8071	0.66563	195.398	1.50233	0.00512	168.369	167.480	335.849	0.87676	1.57996	-35
-34	0.8438	0.66689	187.453	1.49951	0.00533	169.255	167.054	336.309	0.88046	1.57894	-34
-33	0.8819	0.66815	179.900	1.49667	0.00556	170.143	166.626	336.768	0.88415	1.57795	-33
-32	0.9213	0.66942	172.716	1.49382	0.00579	171.031	166.196	337.227	0.88783	1.57697	-32
-31	0.9620	0.67071	165.881	1.49097	0.00603	171.920	165.765	337.686	0.89150	1.57601	-31
-30	1.0041	0.67200	159.375	1.48810	0.00627	172.810	165.333	338.143	0.89516	1.57507	-30
-29	1.0477	0.67330	153.178	1.48522	0.00653	173.701	164.899	338.600	0.89880	1.57416	-29
-28	1.0927	0.67461	147.275	1.48234	0.00679	174.593	164.463	339.057	0.90244	1.57326	-28
-27	1.1392	0.67593	141.649	1.47944	0.00706	175.486	164.026	339.513	0.90606	1.57238	-27
-26	1.1872	0.67726	136.284	1.47653	0.00734	176.380	163.587	339.968	0.90967	1.57152	-26
-25	1.2368	0.67860	131.166	1.47361	0.00762	177.275	163.147	340.422	0.91327	1.57068	-25
-24	1.2880	0.67996	126.282	1.47068	0.00792	178.171	162.705	340.876	0.91686	1.56985	-24
-23	1.3408	0.68132	121.620	1.46774	0.00822	179.068	162.261	341.328	0.92043	1.56904	-23
-22	1.3953	0.68269	117.167	1.46479	0.00853	179.965	161.815	341.780	0.92400	1.56825	-22
-21	1.4515	0.68407	112.913	1.46183	0.00886	180.864	161.367	342.231	0.92756	1.56748	-21
-20	1.5093	0.68547	108.847	1.45886	0.00919	181.764	160.918	342.682	0.93110	1.56672	-20
-19	1.5690	0.68687	104.960	1.45587	0.00953	182.665	160.466	343.131	0.93464	1.56598	-19
-18	1.6304	0.68829	101.242	1.45288	0.00988	183.567	160.013	343.580	0.93816	1.56526	-18
-17	1.6937	0.68972	97.6841	1.44987	0.01024	184.470	159.558	344.028	0.94168	1.56454	-17
-16	1.7589	0.69115	94.2788	1.44685	0.01061	185.374	159.100	344.474	0.94518	1.56385	-16
-15	1.8260	0.69261	91.0182	1.44382	0.01099	186.279	158.641	344.920	0.94868	1.56317	-15
-14	1.8950	0.69407	87.8951	1.44078	0.01138	187.185	158.180	345.365	0.95216	1.56250	-14
-13	1.9660	0.69554	84.9027	1.43773	0.01178	188.093	157.716	345.809	0.95564	1.56185	-13
-12	2.0390	0.69703	82.0344	1.43466	0.01219	189.001	157.250	346.252	0.95910	1.56121	-12
-11	2.1140	0.69853	79.2842	1.43158	0.01261	189.911	156.783	346.693	0.96256	1.56059	-11
-10	2.1912	0.70004	76.6464	1.42849	0.01305	190.822	156.312	347.134	0.96601	1.55997	-10
-9	2.2704	0.70157	74.1155	1.42538	0.01349	191.734	155.840	347.574	0.96945	1.55938	-9
-8	2.3519	0.70310	71.6864	1.42227	0.01395	192.647	155.365	348.012	0.97287	1.55879	-8
-7	2.4355	0.70465	69.3543	1.41914	0.01442	193.562	154.888	348.450	0.97629	1.55822	-7
-6	2.5214	0.70622	67.1146	1.41599	0.01490	194.477	154.408	348.886	0.97971	1.55765	-6
-5	2.6096	0.70780	64.9629	1.41284	0.01539	195.395	153.926	349.321	0.98311	1.55710	-5
-4	2.7001	0.70939	62.8952	1.40967	0.01590	196.313	153.442	349.755	0.98650	1.55657	-4
-3	2.7930	0.71099	60.9075	1.40648	0.01642	197.233	152.955	350.187	0.98989	1.55604	-3
-2	2.8882	0.71261	58.9963	1.40328	0.01695	198.154	152.465	350.619	0.99327	1.55552	-2
-1	2.9859	0.71425	57.1579	1.40007	0.01750	199.076	151.972	351.049	0.99664	1.55502	-1
0	3.0861	0.71590	55.3892	1.39685	0.01805	200.000	151.477	351.477	1.00000	1.55452	0
1	3.1888	0.71756	53.6869	1.39361	0.01863	200.925	150.979	351.905	1.00335	1.55404	1
2	3.2940	0.71924	52.0481	1.39035	0.01921	201.852	150.479	352.331	1.00670	1.55356	2
3	3.4019	0.72094	50.4700	1.38708	0.01981	202.780	149.975	352.755	1.01004	1.55310	3
4	3.5124	0.72265	48.9499	1.38379	0.02043	203.710	149.468	353.179	1.01337	1.55264	4
5	3.6255	0.72438	47.4853	1.38049	0.02106	204.642	148.959	353.600	1.01670	1.55220	5
6	3.7414	0.72612	46.0737	1.37718	0.02170	205.575	148.446	354.020	1.02001	1.55176	6
7	3.8601	0.72788	44.7129	1.37384	0.02236	206.509	147.930	354.439	1.02333	1.55133	7
8	3.9815	0.72966	43.4006	1.37050	0.02304	207.445	147.411	354.856	1.02663	1.55091	8
9	4.1058	0.73146	42.1349	1.36713	0.02373	208.383	146.889	355.272	1.02993	1.55050	9

Courtesy E. I. du Pont & Co., Inc.

TABLE 16-2A (*Continued*)

TEMP. °C	PRES.	VOLUME m³/kg · 10³		DENSITY kg/m³		ENTHALPY kJ/kg			ENTROPY kJ/kg K		TEMP. °C
		LIQUID v_f	VAPOR v_g	LIQUID $1/v_f$	VAPOR $1/v_g$	LIQUID h_f	LATENT h_{fg}	VAPOR h_g	LIQUID s_f	VAPOR s_g	
10	4.2330	0.73327	40.9137	1.36375	0.02444	209.323	146.363	355.686	1.03322	1.55010	10
11	4.3631	0.73510	39.7352	1.36035	0.02517	210.264	145.834	356.098	1.03650	1.54970	11
12	4.4962	0.73695	38.5975	1.35694	0.02591	211.207	145.302	356.509	1.03978	1.54931	12
13	4.6323	0.73882	37.4991	1.35350	0.02667	212.152	144.766	356.918	1.04305	1.54893	13
14	4.7714	0.74071	36.4382	1.35006	0.02744	213.099	144.226	357.325	1.04632	1.54856	14
15	4.9137	0.74262	35.4133	1.34659	0.02824	214.048	143.683	357.730	1.04958	1.54819	15
16	5.0591	0.74455	34.4230	1.34310	0.02905	214.998	143.135	358.134	1.05284	1.54783	16
17	5.2076	0.74649	33.4658	1.33960	0.02988	215.951	142.584	358.535	1.05609	1.54748	17
18	5.3594	0.74846	32.5405	1.33608	0.03073	216.906	142.029	358.935	1.05933	1.54713	18
19	5.5145	0.75045	31.6457	1.33253	0.03160	217.863	141.470	359.333	1.06258	1.54679	19
20	5.6729	0.75246	30.7802	1.32897	0.03249	218.821	140.907	359.729	1.06581	1.54645	20
21	5.8347	0.75449	29.9429	1.32539	0.03340	219.783	140.340	360.122	1.06904	1.54612	21
22	5.9998	0.75655	29.1327	1.32179	0.03433	220.746	139.768	360.514	1.07227	1.54579	22
23	6.1684	0.75863	28.3485	1.31817	0.03528	221.712	139.192	360.904	1.07549	1.54547	23
24	6.3405	0.76073	27.5894	1.31453	0.03625	222.680	138.611	361.291	1.07871	1.54515	24
25	6.5162	0.76286	26.8542	1.31086	0.03724	223.650	138.026	361.676	1.08193	1.54484	25
26	6.6954	0.76501	26.1422	1.30718	0.03825	224.623	137.436	362.059	1.08514	1.54453	26
27	6.8782	0.76718	25.4524	1.30347	0.03929	225.598	136.841	362.439	1.08835	1.54423	27
28	7.0647	0.76938	24.7840	1.29974	0.04035	226.576	136.241	362.817	1.09155	1.54393	28
29	7.2550	0.77161	24.1362	1.29599	0.04143	227.557	135.636	363.193	1.09475	1.54363	29
30	7.4490	0.77386	23.5082	1.29222	0.04254	228.540	135.026	363.566	1.09795	1.54334	30
31	7.6468	0.77614	22.8993	1.28842	0.04367	229.526	134.411	363.937	1.10115	1.54305	31
32	7.8485	0.77845	22.3088	1.28460	0.04483	230.515	133.790	364.305	1.10434	1.54276	32
33	8.0541	0.78079	21.7359	1.28075	0.04601	231.506	133.164	364.670	1.10753	1.54247	33
34	8.2636	0.78316	21.1802	1.27688	0.04721	232.501	132.532	365.033	1.11072	1.54219	34
35	8.4772	0.78556	20.6408	1.27298	0.04845	233.498	131.894	365.392	1.11391	1.54191	35
36	8.6948	0.78799	20.1173	1.26906	0.04971	234.499	131.250	365.749	1.11710	1.54163	36
37	8.9164	0.79045	19.6091	1.26511	0.05100	235.503	130.600	366.103	1.12028	1.54135	37
38	9.1423	0.79294	19.1156	1.26113	0.05231	236.510	129.943	366.454	1.12347	1.54107	38
39	9.3723	0.79546	18.6362	1.25713	0.05366	237.521	129.281	366.802	1.12665	1.54079	39
40	9.6065	0.79802	18.1706	1.25309	0.05503	238.535	128.611	367.146	1.12984	1.54051	40
41	9.8451	0.80062	17.7182	1.24903	0.05644	239.552	127.935	367.487	1.13302	1.54024	41
42	10.088	0.80325	17.2785	1.24494	0.05788	240.574	127.252	367.825	1.13620	1.53996	42
43	10.335	0.80592	16.8511	1.24082	0.05934	241.598	126.561	368.160	1.13938	1.53968	43
44	10.587	0.80863	16.4356	1.23667	0.06084	242.627	125.864	368.491	1.14257	1.53941	44
45	10.843	0.81137	16.0316	1.23248	0.06238	243.659	125.158	368.818	1.14575	1.53913	45
46	11.104	0.81416	15.6386	1.22826	0.06394	244.696	124.445	369.141	1.14894	1.53885	46
47	11.369	0.81698	15.2563	1.22401	0.06555	245.736	123.725	369.461	1.15213	1.53856	47
48	11.639	0.81985	14.8844	1.21973	0.06718	246.781	122.996	369.777	1.15532	1.53828	48
49	11.914	0.82277	14.5224	1.21541	0.06886	247.830	122.258	370.088	1.15851	1.53799	49
50	12.193	0.82573	14.1701	1.21105	0.07057	248.884	121.512	370.396	1.16170	1.53770	50
51	12.477	0.82873	13.8271	1.20666	0.07232	249.942	120.757	370.699	1.16490	1.53741	51
52	12.766	0.83179	13.4931	1.20223	0.07411	251.004	119.993	370.997	1.16810	1.53712	52
53	13.060	0.83489	13.1678	1.19776	0.07594	252.072	119.220	371.292	1.17130	1.53682	53
54	13.359	0.83804	12.8509	1.19326	0.07782	253.144	118.437	371.581	1.17451	1.53651	54
55	13.663	0.84125	12.5421	1.18871	0.07973	254.222	117.644	371.865	1.17772	1.53620	55
56	13.972	0.84451	12.2412	1.18412	0.08169	255.304	116.841	372.145	1.18093	1.53589	56
57	14.286	0.84783	11.9479	1.17948	0.08370	256.392	116.027	372.419	1.18415	1.53557	57
58	14.605	0.85121	11.6620	1.17480	0.08575	257.486	115.202	372.688	1.18738	1.53524	58
59	14.929	0.85464	11.3832	1.17008	0.08785	258.585	114.367	372.952	1.19061	1.53491	59
60	15.259	0.85814	11.1113	1.16531	0.09000	259.690	113.519	373.210	1.19384	1.53457	60
61	15.594	0.86171	10.8460	1.16049	0.09220	260.801	112.660	373.461	1.19709	1.53422	61
62	15.935	0.86534	10.5872	1.15562	0.09445	261.918	111.789	373.707	1.20034	1.53387	62
63	16.280	0.86904	10.3346	1.15069	0.09676	263.042	110.905	373.947	1.20359	1.53351	63
64	16.632	0.87282	10.0881	1.14572	0.09913	264.172	110.008	374.180	1.20686	1.53313	64

TABLE 16-2B Refrigerant-12 Properties of Superheated Vapor (SI Units)

Temp. °C	31°C (7.6468 bar)			32°C (7.8485 bar)			33°C (8.0541 bar)			34°C (8.2636 bar)			Temp. °C
35	23.4435	366.921	1.5528	22.7108	366.557	1.5501	22.0000	366.182	1.5474	21.3103	365.793	1.5447	35
40	24.1047	370.616	1.5647	23.3634	370.277	1.5621	22.6446	369.928	1.5595	21.9475	369.568	1.5568	40
45	24.7474	374.278	1.5763	23.9970	373.961	1.5737	23.2697	373.635	1.5712	22.5644	373.299	1.5686	45
50	25.3741	377.914	1.5876	24.6142	377.617	1.5851	23.8778	377.311	1.5827	23.1640	376.997	1.5802	50
55	25.9868	381.529	1.5987	25.2170	381.250	1.5963	24.4712	380.962	1.5939	23.7485	380.667	1.5914	55
60	26.5871	385.129	1.6096	25.8072	384.865	1.6072	25.0517	384.594	1.6049	24.3197	384.316	1.6025	60
65	27.1764	388.717	1.6203	26.3862	388.467	1.6180	25.6208	388.210	1.6156	24.8793	387.947	1.6133	65
70	27.7560	392.295	1.6308	26.9552	392.058	1.6285	26.1798	391.815	1.6262	25.4286	391.566	1.6239	70
75	28.3268	395.867	1.6411	27.5154	395.642	1.6389	26.7297	395.411	1.6366	25.9686	395.174	1.6343	75
80	28.8897	399.436	1.6513	28.0675	399.221	1.6491	27.2714	399.001	1.6469	26.5004	398.776	1.6446	80
85	29.4455	403.002	1.6613	28.6124	402.797	1.6591	27.8059	402.587	1.6569	27.0247	402.372	1.6547	85
90	29.9948	406.568	1.6712	29.1508	406.372	1.6691	28.3337	406.171	1.6669	27.5424	405.966	1.6647	90
95	30.5383	410.134	1.6810	29.6833	409.947	1.6788	28.8555	409.755	1.6767	28.0539	409.558	1.6745	95
100	31.0764	413.704	1.6906	30.2103	413.524	1.6885	29.3718	413.340	1.6863	28.5599	413.151	1.6842	100
105	31.6097	417.276	1.7001	30.7324	417.103	1.6980	29.8832	416.926	1.6959	29.0609	416.746	1.6938	105
110	32.1385	420.853	1.7095	31.2500	420.686	1.7074	30.3900	420.516	1.7053	29.5572	420.343	1.7032	110
115	32.6631	424.434	1.7188	31.7635	424.274	1.7167	30.8926	424.111	1.7146	30.0494	423.943	1.7126	115
120	33.1840	428.022	1.7280	32.2731	427.867	1.7259	31.3913	427.710	1.7239	30.5376	427.548	1.7218	120
125	33.7015	431.616	1.7371	32.7792	431.467	1.7350	31.8865	431.314	1.7330	31.0222	431.159	1.7309	125
130	34.2157	435.216	1.7461	33.2821	435.072	1.7440	32.3785	434.925	1.7420	31.5036	434.775	1.7399	130
135	34.7269	438.824	1.7550	33.7820	438.685	1.7529	32.8674	438.543	1.7509	31.9819	438.398	1.7489	135
140	35.2353	442.440	1.7638	34.2791	442.305	1.7617	33.3535	442.168	1.7597	32.4574	442.027	1.7577	140
145	35.7412	446.064	1.7725	34.7736	445.933	1.7705	33.8370	445.800	1.7685	32.9303	445.663	1.7665	145
150	36.2447	449.696	1.7811	35.2656	449.569	1.7791	34.3181	449.440	1.7771	33.4007	449.308	1.7751	150
155	36.7460	453.336	1.7897	35.7555	453.213	1.7877	34.7969	453.088	1.7857	33.8688	452.959	1.7837	155

Temp. °C	35°C (8.4772 bar)			36°C (8.6948 bar)			37°C (8.9164 bar)			38°C (9.1423 bar)			Temp. °C
35	20.6408	365.392	1.5419	-------	-------	------	-------	-------	------	-------	-------	------	35
40	21.2711	369.196	1.5542	20.6145	368.812	1.5515	19.9770	368.416	1.5488	19.3577	368.007	1.5460	40
45	21.8804	372.953	1.5661	21.2168	372.597	1.5635	20.5727	372.229	1.5608	19.9473	371.849	1.5582	45
50	22.4719	376.673	1.5777	21.8006	376.340	1.5751	21.1492	375.997	1.5726	20.5171	375.644	1.5700	50
55	23.0478	380.363	1.5890	22.3684	380.051	1.5865	21.7094	379.729	1.5841	21.0700	379.399	1.5816	55
60	23.6102	384.030	1.6001	22.9224	383.736	1.5977	22.2554	383.434	1.5953	21.6083	383.123	1.5928	60
65	24.1607	387.677	1.6109	23.4642	387.400	1.6086	22.7889	387.115	1.6062	22.1339	386.822	1.6039	65
70	24.7007	391.310	1.6216	23.9952	391.047	1.6193	23.3113	390.778	1.6170	22.6482	390.501	1.6147	70
75	25.2313	394.931	1.6321	24.5167	394.682	1.6298	23.8241	394.427	1.6275	23.1525	394.165	1.6253	75
80	25.7534	398.545	1.6424	25.0296	398.308	1.6402	24.3281	398.065	1.6379	23.6480	397.816	1.6357	80
85	26.2681	402.152	1.6525	25.5349	401.926	1.6503	24.8244	401.695	1.6481	24.1356	401.458	1.6459	85
90	26.7759	405.756	1.6625	26.0332	405.540	1.6603	25.3136	405.319	1.6582	24.6160	405.093	1.6560	90
95	27.2775	409.357	1.6724	26.5253	409.151	1.6702	25.7965	408.940	1.6681	25.0900	408.724	1.6659	95
100	27.7736	412.958	1.6821	27.0118	412.761	1.6800	26.2736	412.559	1.6778	25.5582	412.352	1.6757	100
105	28.2645	416.560	1.6917	27.4930	416.371	1.6896	26.7455	416.177	1.6875	26.0211	415.979	1.6854	105
110	28.7508	420.165	1.7011	27.9695	419.983	1.6991	27.2126	419.797	1.6970	26.4791	419.606	1.6949	110
115	29.2328	423.772	1.7105	28.4418	423.597	1.7084	27.6754	423.418	1.7064	26.9327	423.235	1.7043	115
120	29.7108	427.384	1.7197	28.9100	427.215	1.7177	28.1341	427.043	1.7156	27.3823	426.866	1.7136	120
125	30.1852	431.000	1.7289	29.3746	430.837	1.7268	28.5892	430.671	1.7248	27.8281	430.501	1.7228	125
130	30.6564	434.621	1.7379	29.8359	434.464	1.7359	29.0409	434.304	1.7339	28.2706	434.140	1.7319	130
135	31.1245	438.249	1.7469	30.2940	438.097	1.7449	29.4895	437.942	1.7428	28.7099	437.784	1.7409	135
140	31.5897	441.883	1.7557	30.7493	441.736	1.7537	29.9351	441.586	1.7517	29.1462	441.433	1.7497	140
145	32.0523	445.524	1.7645	31.2019	445.382	1.7625	30.3781	445.237	1.7605	29.5799	445.088	1.7585	145
150	32.5124	449.173	1.7731	31.6520	449.035	1.7712	30.8186	448.894	1.7692	30.0110	448.750	1.7672	150
155	32.9702	452.828	1.7817	32.0999	452.695	1.7798	31.2567	452.558	1.7778	30.4398	452.419	1.7759	155

Courtesy E. I. du Pont de Nemours & Company, Inc.

TABLE 16-2B *(Continued)*

	V	H	S	V	H	S	V	H	S	V	H	S	
Temp. °C	39°C (9.3723 bar)			40°C (9.6065 bar)			41°C (9.8451 bar)			42°C (10.0880 bar)			Temp. °C
40	18.7558	367.584	1.5433	18.1706	367.146	1.5405	-------	-------	------	-------	-------	------	40
45	19.3398	371.458	1.5556	18.7495	371.054	1.5529	18.1757	370.636	1.5502	17.6177	370.205	1.5475	45
50	19.9033	375.279	1.5675	19.3072	374.904	1.5649	18.7281	374.517	1.5623	18.1652	374.117	1.5597	50
55	20.4494	379.058	1.5791	19.8469	378.708	1.5766	19.2618	378.347	1.5741	18.6933	377.975	1.5715	55
60	20.9805	382.804	1.5904	20.3711	382.475	1.5880	19.7795	382.137	1.5855	19.2049	381.790	1.5831	60
65	21.4985	386.521	1.6015	20.8819	386.212	1.5991	20.2834	385.895	1.5967	19.7023	385.569	1.5943	65
70	22.0050	390.217	1.6123	21.3809	389.926	1.6100	20.7753	389.626	1.6077	20.1874	389.319	1.6053	70
75	22.5013	393.896	1.6230	21.8695	393.620	1.6207	21.2565	393.337	1.6184	20.6615	393.046	1.6161	75
80	22.9885	397.561	1.6334	22.3489	397.299	1.6312	21.7283	397.031	1.6289	21.1261	396.755	1.6267	80
85	23.4678	401.215	1.6437	22.8201	400.966	1.6415	22.1918	400.711	1.6393	21.5821	400.450	1.6371	85
90	23.9397	404.862	1.6538	23.2839	404.625	1.6516	22.6477	404.382	1.6495	22.0305	404.133	1.6473	90
95	24.4052	408.503	1.6638	23.7410	408.276	1.6616	23.0969	408.045	1.6595	22.4721	407.807	1.6573	95
100	24.8647	412.140	1.6736	24.1922	411.924	1.6715	23.5401	411.702	1.6694	22.9074	411.475	1.6672	100
105	25.3188	415.776	1.6833	24.6379	415.569	1.6812	23.9777	415.356	1.6791	23.3372	415.139	1.6770	105
110	25.7681	419.412	1.6928	25.0787	419.212	1.6907	24.4102	419.009	1.6887	23.7619	418.800	1.6866	110
115	26.2129	423.048	1.7022	25.5150	422.856	1.7002	24.8383	422.661	1.6981	24.1819	422.461	1.6961	115
120	26.6536	426.686	1.7116	25.9471	426.502	1.7095	25.2621	426.314	1.7075	24.5977	426.121	1.7055	120
125	27.0905	430.327	1.7208	26.3754	430.150	1.7187	25.6821	429.968	1.7167	25.0096	429.783	1.7147	125
130	27.5240	433.972	1.7299	26.8003	433.801	1.7279	26.0985	433.626	1.7259	25.4180	433.447	1.7239	130
135	27.9543	437.622	1.7389	27.2219	437.456	1.7369	26.5118	437.287	1.7349	25.8231	437.115	1.7329	135
140	28.3817	441.276	1.7478	27.6406	441.116	1.7458	26.9220	440.953	1.7438	26.2252	440.786	1.7418	140
145	28.8063	444.937	1.7566	28.0565	444.782	1.7546	27.3294	444.624	1.7526	26.6244	444.462	1.7507	145
150	29.2284	448.603	1.7653	28.4698	448.453	1.7633	27.7343	448.300	1.7614	27.0211	448.144	1.7594	150
155	29.6482	452.276	1.7739	28.8808	452.131	1.7720	28.1368	451.982	1.7700	27.4154	451.831	1.7681	155
160	30.0657	455.956	1.7825	29.2896	455.815	1.7805	28.5371	455.671	1.7786	27.8074	455.524	1.7767	160
Temp. °C	43°C (10.3352 bar)			44°C (10.5870 bar)			45°C (10.8432 bar)			46°C (11.1039 bar)			Temp. °C
45	17.0748	369.758	1.5447	16.5463	369.296	1.5419	16.0316	368.818	1.5391	-------	-------	------	45
50	17.6178	373.705	1.5570	17.0854	373.279	1.5544	16.5673	372.839	1.5517	16.0629	372.384	1.5489	50
55	18.1408	377.592	1.5690	17.6037	377.197	1.5664	17.0813	376.790	1.5638	16.5730	376.369	1.5612	55
60	18.6467	381.432	1.5806	18.1043	381.063	1.5781	17.5769	380.684	1.5756	17.0641	380.293	1.5730	60
65	19.1380	385.233	1.5919	18.5897	384.888	1.5895	18.0569	384.533	1.5870	17.5390	384.168	1.5846	65
70	19.6165	389.003	1.6030	19.0621	388.679	1.6006	18.5235	388.345	1.5982	18.0001	388.003	1.5958	70
75	20.0840	392.748	1.6138	19.5231	392.442	1.6115	18.9784	392.128	1.6092	18.4491	391.805	1.6069	75
80	20.5416	396.473	1.6244	19.9741	396.183	1.6222	19.4229	395.886	1.6199	18.8876	395.581	1.6176	80
85	20.9905	400.182	1.6349	20.4161	399.907	1.6326	19.8584	399.625	1.6304	19.3168	399.336	1.6282	85
90	21.4316	403.878	1.6451	20.8502	403.616	1.6429	20.2858	403.349	1.6407	19.7377	403.074	1.6385	90
95	21.8657	407.564	1.6552	21.2773	407.315	1.6530	20.7060	407.060	1.6509	20.1513	406.799	1.6487	95
100	22.2936	411.243	1.6651	21.6979	411.005	1.6630	21.1197	410.762	1.6609	20.5583	410.512	1.6587	100
105	22.7158	414.917	1.6749	22.1128	414.689	1.6728	21.5275	414.456	1.6707	20.9593	414.218	1.6686	105
110	23.1328	418.587	1.6845	22.5225	418.369	1.6825	21.9301	418.146	1.6804	21.3550	417.918	1.6783	110
115	23.5452	422.256	1.6940	22.9274	422.047	1.6920	22.3278	421.833	1.6900	21.7457	421.614	1.6879	115
120	23.9532	425.924	1.7034	23.3279	425.723	1.7014	22.7210	425.517	1.6994	22.1320	425.307	1.6974	120
125	24.3574	429.594	1.7127	23.7245	429.400	1.7107	23.1104	429.202	1.7087	22.5143	429.000	1.7067	125
130	24.7579	433.265	1.7219	24.1175	433.078	1.7199	23.4960	432.887	1.7179	22.8928	432.692	1.7159	130
135	25.1551	436.938	1.7309	24.5071	436.758	1.7290	23.8782	436.574	1.7270	23.2679	436.386	1.7250	135
140	25.5493	440.616	1.7399	24.8936	440.442	1.7379	24.2574	440.264	1.7360	23.6399	440.083	1.7340	140
145	25.9406	444.297	1.7487	25.2773	444.129	1.7468	24.6336	443.957	1.7449	24.0089	443.782	1.7429	145
150	26.3294	447.984	1.7575	25.6583	447.821	1.7556	25.0072	447.655	1.7536	24.3753	447.485	1.7517	150
155	26.7157	451.676	1.7662	26.0369	451.518	1.7643	25.3783	451.357	1.7623	24.7391	451.192	1.7604	155
160	27.0997	455.373	1.7748	26.4132	455.220	1.7729	25.7471	455.064	1.7710	25.1007	454.904	1.7691	160
165	27.4817	459.077	1.7833	26.7874	458.929	1.7814	26.1138	458.777	1.7795	25.4601	458.622	1.7776	165
Temp. °C	47°C (11.3692 bar)			48°C (11.6392 bar)			49°C (11.9138 bar)			50°C (12.1932 bar)			Temp. °C

V = specific volume ($m^3/kg \times 10^3$) or (cm^3/g)

H = enthalpy (kJ/kg) or (cm^3/g)

S = entropy (kJ/kg °K) or (J/g °K)

TABLE 16-3 Refrigerant-12 (Dichlorodifluoromethane) Saturation Properties

Temp.	Pressure		Volume		Density		Heat content from − 40°			Entropy from − 40°	
°F t	Abs. lb/in.^{2} p	Gage lb/in.^{2} p_d	Liquid ft^3/lb v_f	Vapor ft^3/lb v_g	Liquid lb/ft^3 $1/v_f$	Vapor lb/ft^3 $1/v_g$	Liquid Btu/lb h_f	Latent Btu/lb h	Vapor Btu/lb h_g	Liquid Btu/lb °F s_f	Vapor Btu/lb °F s_g
155	0.1163	29.68*	0.00954	232.29	104.86	0.004305	− 24.61	84.61	60.00	− 0.0686	0.2092
− 105	0.1527	29.61*	0.00957	179.79	104.46	0.005562	− 23.50	84.07	60.57	− 0.0650	0.2065
− 145	0.1985	29.52*	0.00961	140.52	104.05	0.007117	− 22.39	83.53	61.14	− 0.0615	0.2040
− 140	0.2554	29.40*	0.00965	110.92	103.64	0.009016	− 21.29	83.01	61.72	− 0.0580	0.2017
− 135	0.3256	29.26*	0.00969	88.34	103.22	0.01132	− 20.19	82.49	62.30	− 0.0546	.1995
− 130	0.4116	29.08*	0.00973	70.94	102.80	0.01410	− 19.10	81.98	62.88	− 0.0512	.1975
− 125	0.5160	28.87*	0.00977	57.42	102.38	0.01742	− 18.02	81.48	63.46	− 0.0480	0.1955
− 120	0.6417	28.61*	0.00981	46.84	101.95	0.02125	− 16.94	80.98	64.04	− 0.0448	.1937
− 115	0.7921	28.31*	0.00985	38.49	101.52	0.02598	− 15.85	80.48	64.63	− 0.0416	0.1919
− 110	0.9709	27.94*	0.00989	31.84	101.08	0.03141	− 14.78	80.00	65.22	− 0.0385	0.1903
− 105	1.182	27.51*	0.00994	26.51	100.64	0.03773	− 13.71	79.52	65.81	− 0.0355	0.1888
− 100	1.430	27.01*	0.00998	22.20	100.20	0.04504	− 12.64	79.04	66.40	− 0.0325	0.1873
− 95	1.719	26.42*	0.01003	18.71	99.75	0.05344	− 11.58	78.57	66.99	− 0.0295	0.1860
− 90	2.054	25.74*	0.01007	15.86	99.30	0.06305	− 10.51	78.10	67.59	− 0.0266	0.1847
− 85	2.441	24.95*	0.01012	13.51	98.85	0.07400	− 9.46	77.64	68.18	− 0.0238	0.1835
− 80	2.885	24.05*	0.01016	11.57	98.39	0.08640	− 8.40	77.17	68.77	− 0.0210	0.1823
− 75	3.393	23.01*	0.01021	9.958	97.92	0.1004	− 7.35	76.71	69.36	− 0.0182	0.1813
− 70	3.971	21.84*	0.01026	8.608	97.46	0.1162	− 6.30	76.25	69.95	− 0.0155	0.1802
− 65	4.626	20.50*	0.01031	7.474	96.99	0.1338	− 5.25	75.79	70.54	− 0.0128	0.1793
− 60	5.365	19.00*	0.01036	6.516	96.51	0.1535	− 4.20	75.33	71.13	− 0.0102	0.1783
− 55	6.195	17.31*	0.01041	5.704	96.04	0.1753	− 3.15	74.87	71.72	− 0.0076	0.1774
− 50	7.125	15.42*	0.01047	5.012	95.55	0.1995	− 2.11	74.42	72.31	− 0.0050	0.1767
− 45	8.163	13.31*	0.01052	4.420	95.07	0.2263	− 1.06	73.97	72.91	− 0.0025	0.1759
− 40	9.32	10.92*	0.0106	3.911	94.58	0.2557	0	73.50	73.50	0	0.17517
− 38	9.82	9.91*	0.0106	3.727	94.39	0.2683	0.40	73.34	73.74	0.00094	0.17490
− 36	10.34	8.87*	0.0106	3.553	94.20	0.2815	0.81	73.17	73.98	0.00188	0.17463
− 34	10.87	7.80*	0.0106	3.389	93.99	0.2951	1.21	73.01	74.22	0.00282	0.17438
− 32	11.43	6.66*	0.0107	3.234	93.79	0.3092	1.62	72.84	74.46	0.00376	0.17412

− 30	12.02	5.45*	0.0107	3.088	93.59	0.3238	2.03	72.67	74.70	0.00471	0.17387
− 28	12.62	4.23*	0.0107	2.950	93.39	0.3390	2.44	72.50	74.94	0.00565	0.17364
− 26	13.26	2.93*	0.0107	2.820	93.18	0.3546	2.85	72.33	75.18	0.00659	0.17340
− 24	13.90	1.63*	0.0108	2.698	92.98	0.3706	3.25	72.16	75.41	0.00753	0.17317
− 22	14.58	0.24*	0.0108	2.583	92.78	0.3871	3.66	71.98	75.64	0.00846	0.17296
− 20	15.28	0.58	0.0108	2.474	92.58	0.4042	4.07	71.80	75.87	0.00940	0.17275
− 18	16.01	1.31	0.0108	2.370	92.38	0.4219	4.48	71.63	76.11	0.01033	0.17253
− 16	16.77	2.07	0.0108	2.271	92.18	0.4403	4.89	71.45	76.34	0.01126	0.17232
− 14	17.55	2.85	0.0109	2.177	91.97	0.4593	5.30	71.27	76.57	0.01218	0.17212
− 12	18.37	3.67	0.0109	2.088	91.77	0.4789	5.72	71.09	76.81	0.01310	0.17194
− 10	19.20	4.50	0.0109	2.003	91.57	0.4993	6.14	70.91	77.05	0.01403	0.17175
− 8	20.08	5.38	0.0109	1.922	91.35	0.5203	6.57	70.72	77.29	0.01496	0.17158
− 6	20.98	6.28	0.0110	1.845	91.14	0.5420	6.99	70.53	77.52	0.01589	0.17140
− 4	21.91	7.21	0.0110	1.772	90.93	0.5644	7.41	70.34	77.75	0.01682	0.17123
− 2	22.87	8.17	0.0110	1.703	90.72	0.5872	7.83	70.15	77.98	0.01775	0.17107
0	23.87	9.17	0.0110	1.637	90.52	0.6109	8.25	69.96	78.21	0.01869	0.17091
2	24.89	10.19	0.0110	1.574	90.31	0.6352	8.67	69.77	78.44	0.01961	0.17075
4	25.96	11.26	0.0111	1.514	90.11	0.6606	9.10	69.57	78.67	0.02052	0.17060
5†	26.51	11.81	0.0111	1.485	90.00	0.6735	9.32	69.47	78.79	0.02097	0.17052
6	27.05	12.35	0.0111	1.457	89.88	0.6864	9.53	69.37	78.90	0.02143	0.17045
8	28.18	13.48	0.0111	1.403	89.68	0.7129	9.96	69.17	79.13	0.02235	0.17030
10	29.35	14.65	0.0112	1.351	89.45	0.7402	10.39	68.97	79.36	0.02328	0.17015
12	30.56	15.86	0.0112	1.301	89.24	0.7687	10.82	68.77	79.59	0.02419	0.17001
14	31.80	17.10	0.0112	1.253	89.03	0.7981	11.26	68.56	79.82	0.02510	0.16987
16	33.08	18.38	0.0112	1.207	88.81	0.8288	11.70	68.05	80.05	0.02601	0.16974
18	34.40	19.70	0.0113	1.163	88.58	0.8598	12.12	68.15	80.27	0.02692	0.16961
20	35.75	21.05	0.0113	1.121	88.37	0.8921	12.55	67.94	80.49	0.02783	0.16949
22	37.15	22.45	0.0113	1.081	88.13	0.9251	13.00	67.72	80.72	0.02873	0.16938
24	38.58	23.88	0.0113	1.043	87.91	0.9588	13.44	67.51	80.95	0.02963	0.16926
26	40.07	25.37	0.0114	1.007	87.68	0.9930	13.88	67.29	81.17	0.03053	0.16913
28	41.59	26.89	0.0114	0.973	87.47	1.028	14.32	67.07	81.39	0.03143	0.16900

TABLE 16-3 *(Continued)*

Temp.	Pressure		Volume		Density		Heat content from − 40°			Entropy from − 40°	
°F t	Abs. lb/in.^{2} p	Gage lb/in.^{2} p_d	Liquid ft^3/lb v_f	Vapor ft^3/lb v_g	Liquid lb/ft^3 $1/v_f$	Vapor lb/ft^3 $1/v_g$	Liquid Btu/lb h_f	Latent Btu/lb h	Vapor Btu/lb h_g	Liquid Btu/lb °F s_f	Vapor Btu/lb °F s_g
30	43.16	28.46	0.0115	0.939	87.24	1.065	14.76	66.85	81.61	0.03233	0.16887
32	44.77	30.07	0.0115	0.908	87.02	1.102	15.21	66.62	81.83	0.03323	0.16876
34	46.42	31.72	0.0115	0.877	86.78	1.140	15.65	66.40	82.05	0.03413	0.16865
36	48.13	33.43	0.0116	0.848	86.55	1.180	16.10	66.17	82.27	0.03502	0.16854
38	49.88	35.18	0.0116	0.819	86.33	1.221	16.55	65.94	82.49	0.03591	0.16843
40	51.68	36.98	0.0116	0.792	86.10	1.263	17.00	65.71	82.71	0.03680	0.16833
42	53.51	38.81	0.0116	0.767	85.88	1.304	17.46	65.47	82.93	0.03770	0.16823
44	55.40	40.70	0.0117	0.742	85.66	1.349	17.91	65.24	83.15	0.03859	0.16813
46	57.35	42.65	0.0117	0.718	85.43	1.393	18.36	65.00	83.36	0.03948	0.16803
48	59.35	44.65	0.0117	0.695	85.19	1.438	18.82	64.74	83.57	0.04037	0.16794
50	61.39	46.69	0.0118	0.673	84.94	1.485	19.27	64.51	83.78	0.04126	0.16785
52	63.49	48.79	0.0118	0.652	84.71	1.534	19.72	64.27	83.99	0.04215	0.16776
54	65.63	50.93	0.0118	0.632	84.50	1.583	20.18	64.02	84.20	0.04304	0.16767
56	67.84	53.14	0.0119	0.612	84.28	1.633	20.64	63.77	84.41	0.04392	0.16758
58	70.10	55.40	0.0119	0.593	84.04	1.686	21.11	63.51	84.62	0.04480	0.16749
60	72.41	57.71	0.0119	0.575	83.78	1.740	21.57	63.25	84.82	0.04568	0.16741
62	74.77	60.07	0.0120	0.557	83.57	1.795	22.03	62.99	85.02	0.04657	0.16733
64	77.20	62.50	0.0120	0.540	83.34	1.851	22.49	62.73	85.22	0.04745	0.16725
66	79.67	64.97	0.0120	0.524	83.10	1.909	22.95	62.47	85.42	0.04833	0.16717
68	82.24	67.54	0.0121	0.508	82.86	1.968	23.42	62.20	85.62	0.04921	0.16709
70	84.82	70.12	0.0121	0.493	82.60	2.028	23.90	61.92	85.82	0.05009	0.16701
72	87.50	72.80	0.0121	0.479	82.37	2.090	24.37	61.65	86.02	0.05097	0.16693
74	90.20	75.50	0.0122	0.464	82.12	2.153	24.84	61.38	86.22	0.05185	0.16685
76	93.00	78.30	0.0122	0.451	81.87	2.218	25.32	61.10	86.42	0.05272	0.16677
78	95.85	81.15	0.0123	0.438	81.62	2.284	25.80	60.81	86.61	0.05359	0.16669

80	98.76	84.06	0.0123	0.425	81.39	2.353	26.28	60.52	86.80	0.05446	0.16662
82	101.7	87.00	0.0123	0.413	81.12	2.423	26.76	60.23	86.99	0.05534	0.16655
84	104.8	90.1	0.0124	0.401	80.87	2.495	27.24	59.94	87.18	0.05621	0.16648
86†	107.9	93.2	0.0124	0.389	80.63	2.569	27.72	59.65	87.37	0.05708	0.16640
88	111.1	96.4	0.0124	0.378	80.37	2.645	28.21	59.35	87.56	0.05795	0.16632
90	114.3	99.6	0.0125	0.368	80.11	2.721	28.70	59.04	87.74	0.05882	0.16624
92	117.7	103.0	0.0125	0.357	79.86	2.799	29.19	58.73	87.92	0.05969	0.16616
94	121.0	106.3	0.0126	0.347	79.60	2.880	29.68	58.42	88.10	0.06056	0.16608
96	124.5	109.8	0.0126	0.338	79.32	2.963	30.18	58.10	88.28	0.06143	0.16600
98	128.0	113.3	0.0126	0.328	79.06	3.048	30.67	57.78	88.45	0.06230	0.16592
100	131.6	116.9	0.0127	0.319	78.80	3.135	31.16	57.46	88.62	0.06316	0.16584
102	135.3	120.6	0.0127	0.310	78.54	3.224	31.65	57.14	88.79	0.06403	0.16576
104	139.0	124.3	0.0128	0.302	78.27	3.316	32.15	56.80	88.95	0.06490	0.16568
106	142.8	128.1	0.0128	0.293	78.00	3.411	32.65	56.46	89.11	0.06577	0.16560
108	146.8	132.1	0.0129	0.285	77.73	3.509	33.15	56.12	89.27	0.06663	0.16551
110	150.7	136.0	0.0129	0.277	77.46	3.610	33.65	55.78	89.43	0.06749	0.16542
112	154.8	140.1	0.0130	0.269	77.18	3.714	34.15	55.43	89.58	0.06836	0.16533
114	158.9	144.2	0.0130	0.262	76.89	3.823	34.65	55.08	89.73	0.06922	0.16524
116	163.1	148.4	0.0131	0.254	76.60	3.934	35.15	54.72	89.87	0.07008	0.16515
118	167.4	152.7	0.0131	0.247	76.32	4.049	35.65	54.36	90.01	0.07094	0.16505
120	171.8	157.1	0.0132	0.240	76.02	4.167	36.16	53.99	90.15	0.07180	0.16495
122	176.2	161.5	0.0132	0.233	75.72	4.288	36.66	53.62	90.28	0.07266	0.16484
124	180.8	166.1	0.0133	0.277	75.40	4.413	37.16	53.24	90.40	0.07352	0.16473
126	185.4	170.7	0.0133	0.220	75.10	4.541	37.67	52.85	90.52	0.07437	0.16462
128	190.1	175.4	0.0134	0.214	74.78	4.673	38.18	52.46	90.64	0.07522	0.16450
130	194.9	180.2	0.0134	0.208	74.46	4.808	38.69	52.07	90.76	0.07607	0.16438
132	199.8	185.1	0.0135	0.202	74.13	4.948	39.19	51.67	90.86	0.07691	0.16425
134	204.8	190.1	0.0135	0.196	73.81	5.094	39.70	51.26	90.96	0.07775	0.16411
136	209.9	195.2	0.0136	0.191	73.46	5.247	40.21	50.85	91.06	0.07858	0.16396
138	215.0	200.3	0.0137	0.185	73.10	5.405	40.72	50.43	91.15	0.07941	0.16380
140	220.2	205.5	0.0138	0.180	72.73	5.571	41.24	50.00	91.24	0.08024	0.16363

* Inches of mercury below one atmosphere.

† Standard ton temperatures.

TABLE 16-4 Refrigerant-22 (Monochlorodifluoromethane) Properties of Liquid and Saturated Vapor

Temp F	Pressure		Liquid, density	Vapor, sp vol	Enthalpy, datum −40 F Btu per lb		Entropy, datum −40 F Btu per lb F	
t	psia	psig	lb/cu ft $1/v_f$	cu ft/lb v_g	Liquid h_f	Vapor h_g	Liquid s_f	Vapor s_g
−155	0.19901	29.51*	97.67	188.1	−29.07	86.78	−0.0808	0.2996
−150	0.2605	29.39*	97.33	146.1	−27.79	87.36	−0.0767	0.2952
−145	0.3375	29.23*	96.99	114.5	−26.52	87.94	− .0727	.2912
−140	0.4332	29.04*	96.63	90.61	−25.25	88.53	−0.0687	0.2874
−135	0.5511	28.80*	96.27	72.33	−23.99	89.11	− .0647	.2837
−130	0.6949	28.51*	95.91	58.21	−22.73	89.70	−0.0609	0.2803
−125	0.8692	28.15*	95.53	47.23	−21.47	90.29	− .0571	.2770
−120	1.079	27.72*	95.15	38.60	−20.22	90.88	−0.0534	0.2738
−115	1.329	27.21*	94.76	31.77	−18.98	91.47	− .0497	.2708
−110	1.626	26.61*	94.37	26.33	−17.73	92.07	−0.0461	0.2680
−105	1.976	25.90*	93.97	21.96	−16.48	92.67	− .0425	.2653
−100	2.386	25.06*	93.56	18.43	−15.23	93.27	−0.0390	0.2627
− 95	2.865	24.09*	93.14	15.54	−13.98	93.87	− .0356	.2602
− 90	3.417	22.96*	92.72	13.20	−12.73	94.47	−0.0322	0.2579
− 85	4.055	21.67*	92.29	11.26	−11.47	95.08	− .0288	.2556
− 80	4.787	20.18*	91.85	9.650	−10.22	95.68	−0.0255	0.2535
− 78	5.100	19.55*	91.67	9.086	− 9.72	95.92	− .0242	.2526
− 76	5.430	18.87*	91.49	8.561	− 9.21	96.16	− .0229	.2518
− 74	5.79	18.14*	91.31	8.072	− 8.70	96.40	− .0216	.2510
− 72	6.17	17.37*	91.13	7.616	− 8.20	96.64	− .0203	.2502
− 70	6.57	16.55*	90.95	7.192	− 7.69	96.88	−0.0253	0.2494
− 68	6.99	15.70*	90.77	6.795	− 7.19	97.12	− .0177	.2487
− 66	7.40	14.86*	90.58	6.426	− 6.68	97.36	− .0164	.2479
− 64	7.86	13.93*	90.39	6.079	− 6.17	97.60	− .0151	.2472
− 62	8.35	12.93*	90.21	5.755	− 5.67	97.84	− .0138	.2465
− 60	8.86	11.89*	90.03	5.452	− 5.16	98.08	−0.0126	0.2458
− 58	9.39	10.81*	89.84	5.166	− 4.65	98.32	− .0113	.2451
− 56	9.94	9.69*	89.65	4.900	− 4.13	98.56	− .0100	.2444
− 54	10.51	8.53*	89.46	4.650	− 3.61	98.80	− .0087	.2438
− 52	11.11	7.31*	89.27	4.415	− 3.09	99.04	− .0075	.2431
− 50	11.74	6.03*	89.08	4.192	− 2.58	99.28	−0.0062	0.2425
− 48	12.40	4.68*	88.88	3.986	− 2.06	99.52	− .0050	.2418
− 46	13.09	3.28*	88.68	3.793	− 1.54	99.76	− .0037	.2412
− 44	13.80	1.83*	88.49	3.611	− 1.02	100.00	− .0025	.2406
− 42	14.54	0.326*	88.30	3.440	− 0.51	100.23	− .0012	.2400
− 40	15.31	0.610	88.10	3.279	0.00	100.46	0.0000	0.2394
− 38	16.12	1.42	87.90	3.126	0.53	100.70	.0013	.2389
− 36	16.97	2.27	87.70	2.981	1.05	100.93	.0025	.2383
− 34	17.85	3.15	87.50	2.844	1.58	101.17	.0037	.2377
− 32	18.77	4.07	87.29	2.713	2.10	101.40	.0050	.2372
− 30	19.72	5.02	87.09	2.590	2.62	101.63	0.0062	0.2367
− 28	20.71	6.01	86.89	2.474	3.15	101.86	.0074	.2361
− 26	21.73	7.03	86.69	2.365	3.69	102.10	.0086	.2356
− 24	22.79	8.09	86.48	2.262	4.22	102.33	.0099	.2351
− 22	23.88	9.18	86.27	2.165	4.75	102.56	.0111	.2346
− 20	25.01	10.31	86.06	2.074	5.28	102.79	0.0123	0.2341
− 18	26.18	11.48	85.85	1.987	5.82	103.02	.0135	.2336
− 16	27.39	12.69	85.64	1.905	6.40	103.25	.0147	.2331
− 14	28.64	13.94	85.43	1.827	6.90	103.48	.0159	.2326
− 12	29.94	15.24	85.21	1.752	7.43	103.70	.0170	.2321
− 10	31.29	16.59	84.99	1.681	7.96	103.92	0.0182	0.2316
− 8	32.69	17.99	84.78	1.613	8.49	104.14	.0194	.2312
− 6	34.14	19.44	84.56	1.549	9.02	104.36	.0205	.2307
− 4	35.64	20.94	84.34	1.488	9.55	104.58	.0217	.2302
− 2	37.19	22.49	84.12	1.429	10.09	104.80	.0228	.2298
0	38.79	24.09	83.90	1.373	10.63	105.02	0.0240	0.2293
2	40.43	25.73	83.68	1.320	11.17	105.24	0.0251	0.2289
4	42.14	27.44	83.45	1.270	11.70	105.45	.0262	.2285
5	43.02	28.33	83.34	1.246	11.97	105.56	.0268	.2283
6	43.91	29.21	83.23	1.221	12.23	105.66	.0274	.2280
8	45.74	31.04	83.01	1.175	12.76	105.87	.0285	.2276
10	47.63	32.93	82.78	1.130	13.29	106.08	0.0296	0.2272
12	49.58	34.88	82.55	1.088	13.82	106.29	.0307	.2268
14	51.59	36.89	82.32	1.048	14.36	106.50	.0319	.2264
16	53.66	38.96	82.09	1.009	14.90	106.71	.0330	.2260
18	55.79	41.09	81.86	0.9721	15.44	106.92	.0341	.2257

* Inches mercury below one atmosphere.

TABLE 16-4 *(Continued)*

Temp F t	Pressure		Liquid, density	Vapor, sp vol	Enthalpy, datum −40 F Btu per lb		Entropy, datum −40 F Btu per lb F	
	psia	psig	lb/cu ft $1/v_f$	cu ft/lb v_g	Liquid h_f	Vapor h_g	Liquid s_f	Vapor s_g
20	57.98	43.28	81.63	0.9369	15.98	107.13	0.0352	0.2253
22	60.23	45.53	81.39	.9032	16.52	107.33	.0364	.2249
24	62.55	47.85	81.16	.8707	17.06	107.53	.0375	.2246
26	64.94	50.24	80.92	.8398	17.61	107.73	.0379	.2242
28	67.40	52.70	80.69	.8100	18.17	107.93	.0398	.2239
30	69.93	55.23	80.45	0.7816	18.74	108.13	0.0409	0.2235
32	72.53	57.83	80.21	.7543	19.32	108.33	.0421	.2232
34	75.21	60.51	79.97	.7283	19.90	108.52	.0433	.2228
36	77.97	63.27	79.73	.7032	20.49	108.71	.0445	.2225
38	80.81	66.11	79.49	.6791	21.09	108.90	.0457	.2222
40	83.72	69.02	79.25	0.6559	21.70	109.09	0.0469	0.2218
42	86.69	71.99	79.00	.6339	22.29	109.27	.0481	.2215
44	89.74	75.04	78.76	.6126	22.90	109.45	.0493	.2211
46	92.88	78.18	78.51	.5922	23.50	109.63	.0505	.2208
48	96.10	81.40	78.26	.5726	24.11	109.80	.0516	.2205
50	99.40	84.70	78.02	0.5537	24.73	109.98	0.0528	0.2201
52	102.8	88.10	77.77	.5355	25.34	110.14	.0540	.2198
54	106.2	91.5	77.51	.5184	25.95	110.30	.0552	.2194
56	109.8	95.1	77.26	.5014	26.58	110.47	.0564	.2191
58	113.5	98.8	77.01	.4849	27.22	110.63	.0576	.2188
60	117.2	102.5	76.75	0.4695	27.83	110.78	0.0588	0.2185
62	121.0	106.3	76.50	.4546	28.46	110.93	.0600	.2181
64	124.9	110.2	76.24	.4403	29.09	111.08	.0612	.2178
66	128.9	114.2	75.98	.4264	29.72	111.22	.0624	.2175
68	133.0	118.3	75.72	.4129	30.35	111.35	.0636	.2172
70	137.2	122.5	75.46	0.4000	30.99	111.49	0.0648	0.2168
72	141.5	126.8	75.20	.3875	31.65	111.63	.0661	.2165
74	145.9	131.2	74.94	.3754	32.29	111.75	.0673	.2162
76	150.4	135.7	74.68	.3638	32.94	111.88	.0684	.2158
78	155.0	140.3	74.41	.3526	33.61	112.01	.0696	.2155
80	159.7	145.0	74.15	0.3417	34.27	112.13	0.0708	0.2151
82	164.5	149.8	73.89	.3313	34.92	112.24	.0720	.2148
84	169.4	154.7	73.63	.3212	35.60	112.36	.0732	.2144
86	174.5	159.8	73.36	.3113	36.28	112.47	.0744	.2140
88	179.6	164.9	73.09	.3019	36.94	112.57	.0756	.2137
90	184.8	170.1	72.81	0.2928	37.61	112.67	0.0768	0.2133
92	190.1	175.4	72.53	.2841	38.28	112.76	.0780	.2130
94	195.6	180.9	72.24	.2755	38.97	112.85	.0792	.2126
96	201.2	186.5	71.95	.2672	39.65	112.93	.0803	.2122
98	206.8	192.1	71.65	.2594	40.32	113.00	.0815	.2119
100	212.6	197.9	71.35	0.2517	40.98	113.06	0.0827	0.2115
102	218.5	203.8	71.05	.2443	41.65	113.12	.0839	.2111
104	224.6	209.9	70.74	.2370	42.32	113.16	.0851	.2107
106	230.7	216.0	70.42	.2301	42.98	113.20	.0862	.2104
108	237.0	222.3	70.11	.2233	43.66	113.24	.0874	.2100
110	243.4	228.7	69.78	0.2167	44.35	113.29	0.0886	0.2096
112	249.9	235.2	69.45	.2104	45.04	113.34	.0898	.2093
114	256.6	241.9	69.12	.2043	45.74	113.38	.0909	.2089
116	263.4	248.7	68.78	.1983	46.44	113.42	.0921	.2085
118	270.3	255.6	68.44	.1926	47.14	113.46	.0933	.2081
120	277.3	262.6	68.10	0.1871	47.85	113.52	0.0945	0.2078
122	384.4	269.7	67.75	.1825	48.6	113.57		
124	391.6	276.9	67.40	.1772	49.4	113.61		
126	398.8	284.1	67.05	.1724	50.2	113.65		
128	306.1	291.4	66.70	.1675	50.8	113.69		
130	313.5	298.8	66.35	0.1629	51.5	113.71		
132	321.0	306.3	66.00	.1585	52.3	113.74		
134	328.7	314.0	65.65	.1538	53.1	113.77		
136	336.6	321.9	65.25	.1492	53.8	113.79		
138	344.6	329.9	64.85	.1449	54.6	113.80		
140	352.7	338.0	64.45	0.1408	55.3	113.81		
142	361.0	346.3	64.05	.1368	56.1	113.80		
144	369.7	355.0	63.65	.1330	56.9	113.79		
146	379.0	364.3	63.25	.1292	57.7	113.78		
148	388.8	374.1	62.85	.1253	58.4	113.76		
150	399.0	384.3	62.45	0.1216	59.2	113.74		
152	407.0	392.3	62.02	.1179	60.0	113.71		
154	416.0	401.3	61.58	.1141	60.8	113.67		
156	426.0	411.3	61.13	.1105	61.6	113.62		
158	436.5	421.8	60.67	.1070	62.5	113.56		
160	448.0	433.3	60.20	0.1035	63.5	113.50		

Courtesy E. I. du Pont de Nemours & Company, Inc.

TABLE 16-5 Refrigerant 502-Saturation Properties

TEMP. °F	PRESSURE		VOLUME cu ft/lb		DENSITY lb/cu ft		ENTHALPY Btu/lb			ENTROPY Btu/(lb)(° R)		TEMP. °F
	PSIA	PSIG	LIQUID v_f	VAPOR v_g	LIQUID $1/v_f$	VAPOR $1/v_g$	LIQUID h_f	LATENT h_{fg}	VAPOR h_g	LIQUID s_f	VAPOR s_g	
− 51	14.36	0.69*	0.01069	2.7079	93.58	0.3693	− 2.82	76.57	73.75	−0.0068	0.1806	− 51
− 50	14.74	0.04	0.01070	2.6428	93.47	0.3784	− 2.57	76.44	73.87	−0.0062	0.1804	− 50
− 49	15.13	0.43	0.01071	2.5797	93.37	0.3876	− 2.32	76.32	74.00	−0.0056	0.1803	− 49
− 48	15.52	0.82	0.01072	2.5185	93.26	0.3971	− 2.06	76.19	74.13	−0.0049	0.1801	− 48
− 47	15.92	1.23	0.01073	2.4590	93.15	0.4067	− 1.80	76.06	74.26	−0.0043	0.1800	− 47
− 46	16.33	1.64	0.01075	2.4012	93.05	0.4165	− 1.55	75.94	74.39	−0.0037	0.1799	− 46
− 45	16.75	2.06	0.01076	2.3451	92.94	0.4264	− 1.29	75.81	74.52	−0.0031	0.1797	− 45
− 44	17.18	2.48	0.01077	2.2905	92.83	0.4366	− 1.03	75.68	74.65	−0.0025	0.1796	− 44
− 43	17.62	2.92	0.01078	2.2376	92.72	0.4469	− 0.77	75.55	74.78	−0.0018	0.1795	− 43
− 42	18.06	3.36	0.01080	2.1861	92.62	0.4574	− 0.52	75.42	74.90	−0.0012	0.1793	− 42
− 41	18.51	3.82	0.01081	2.1361	92.51	0.4682	− 0.26	75.29	75.03	−0.0006	0.1792	− 41
− 40	18.97	4.28	0.01082	2.0874	92.40	0.4791	0	75.16	75.16	0	0.1791	− 40
− 39	19.45	4.75	0.01084	2.0402	92.29	0.4902	0.26	75.03	75.29	0.0006	0.1790	− 39
− 38	19.92	5.23	0.01085	1.9942	92.18	0.5015	0.52	74.90	75.42	0.0012	0.1788	− 38
− 37	20.41	5.72	0.01086	1.9495	92.07	0.5130	0.78	74.77	75.55	0.0018	0.1787	− 37
− 36	20.91	6.21	0.01087	1.9060	91.96	0.5247	1.03	74.64	75.67	0.0024	0.1786	− 36
− 35	21.42	6.72	0.01089	1.8637	91.85	0.5366	1.30	74.50	75.80	0.0031	0.1785	− 35
− 34	21.93	7.24	0.01090	1.8226	91.74	0.5487	1.56	74.37	75.93	0.0037	0.1784	− 34
− 33	22.46	7.76	0.01091	1.7825	91.63	0.5610	1.82	74.24	76.06	0.0043	0.1783	− 33
− 32	23.00	8.30	0.01093	1.7436	91.52	0.5735	2.07	74.11	76.18	0.0049	0.1781	− 32
− 31	23.54	8.84	0.01094	1.7056	91.41	0.5863	2.34	73.97	76.31	0.0055	0.1780	− 31
− 30	24.10	9.40	0.01095	1.6687	91.30	0.5993	2.60	73.84	76.44	0.0061	0.1779	− 30
− 29	24.66	9.97	0.01097	1.6328	91.19	0.6124	2.85	73.71	76.56	0.0067	0.1778	− 29
− 28	25.24	10.54	0.01098	1.5978	91.08	0.6259	3.12	73.57	76.69	0.0073	0.1777	− 28
− 27	25.82	11.13	0.01099	1.5637	90.97	0.6395	3.38	73.44	76.82	0.0079	0.1776	− 27
− 26	26.42	11.72	0.01101	1.5305	90.85	0.6534	3.64	73.30	76.94	0.0085	0.1775	− 26
− 25	27.02	12.33	0.01102	1.4982	90.74	0.6675	3.90	73.17	77.07	0.0091	0.1774	− 25
− 24	27.64	12.95	0.01103	1.4667	90.63	0.6818	4.16	73.03	77.19	0.0097	0.1773	− 24
− 23	28.27	13.57	0.01105	1.4360	90.52	0.6964	4.43	72.89	77.32	0.0103	0.1772	− 23
− 22	28.91	14.21	0.01106	1.4061	90.40	0.7112	4.69	72.75	77.44	0.0109	0.1771	− 22
− 21	29.56	14.86	0.01108	1.3770	90.29	0.7262	4.95	72.62	77.57	0.0115	0.1770	− 21
− 20	30.22	15.52	0.01109	1.3486	90.18	0.7415	5.21	72.48	77.69	0.0121	0.1769	− 20
− 19	30.89	16.19	0.01110	1.3209	90.06	0.7571	5.48	72.34	77.82	0.0127	0.1768	− 19
− 18	31.57	16.88	0.01112	1.2939	89.95	0.7729	5.74	72.20	77.94	0.0133	0.1767	− 18
− 17	32.27	17.57	0.01113	1.2676	89.83	0.7889	6.01	72.06	78.07	0.0139	0.1766	− 17
− 16	32.97	18.28	0.01115	1.2419	89.72	0.8052	6.27	71.92	78.19	0.0145	0.1766	− 16
− 15	33.69	18.99	0.01116	1.2169	89.60	0.8218	6.54	71.78	78.32	0.0151	0.1765	− 15
− 14	34.42	19.72	0.01117	1.1925	89.49	0.8386	6.80	71.64	78.44	0.0156	0.1764	− 14
− 13	35.16	20.46	0.01119	1.1686	89.37	0.8557	7.06	71.50	78.56	0.0162	0.1763	− 13
− 12	35.91	21.22	0.01120	1.1454	89.26	0.8731	7.33	71.36	78.69	0.0168	0.1762	− 12
− 11	36.68	21.98	0.01122	1.1227	89.14	0.8907	7.60	71.21	78.81	0.0174	0.1761	− 11
− 10	37.46	22.76	0.01123	1.1006	89.02	0.9086	7.86	71.07	78.93	0.0180	0.1760	− 10
− 9	38.25	23.55	0.01125	1.0790	88.91	0.9268	8.13	70.93	79.06	0.0186	0.1760	− 9
− 8	39.05	24.35	0.01126	1.0579	88.79	0.9453	8.40	70.78	79.18	0.0192	0.1759	− 8
− 7	39.86	25.17	0.01128	1.0373	88.67	0.9640	8.66	70.64	79.30	0.0198	0.1758	− 7
− 6	40.69	26.00	0.01129	1.0172	88.55	0.9831	8.93	70.49	79.42	0.0204	0.1757	− 6
− 5	41.53	26.84	0.01131	0.9976	88.44	1.0024	9.20	70.34	79.54	0.0209	0.1756	− 5
− 4	42.39	27.69	0.01132	0.9784	88.32	1.0220	9.47	70.20	79.67	0.0215	0.1756	− 4

*Inches of mercury below one atmosphere.

Courtesy E. I. du Pont de Nemours & Company, Inc.

TABLE 16-5 *(Continued)*

TEMP. °F	PRESSURE PSIA	PRESSURE PSIG	VOLUME cu ft/lb LIQUID v_f	VOLUME VAPOR v_g	DENSITY lb/cu ft LIQUID $1/v_f$	DENSITY VAPOR $1/v_g$	ENTHALPY Btu/lb LIQUID h_f	ENTHALPY LATENT h_{fg}	ENTHALPY VAPOR h_g	ENTROPY Btu/(lb)(° R) LIQUID s_f	ENTROPY VAPOR s_g	TEMP. °F
− 3	43.26	28.56	0.01134	0.9597	88.20	1.0420	9.74	70.05	79.79	0.0221	0.1755	− 3
− 2	44.14	29.44	0.01135	0.9414	88.08	1.0622	10.00	69.91	79.91	0.0227	0.1754	− 2
− 1	45.03	30.33	0.01137	0.9236	87.96	1.0828	10.27	69.76	80.03	0.0233	0.1753	− 1
0	45.94	31.24	0.01138	0.9061	87.84	1.1036	10.54	69.61	80.15	0.0239	0.1753	0
1	46.86	32.16	0.01140	0.8891	87.72	1.1248	10.81	69.46	80.27	0.0244	0.1752	1
2	47.79	33.10	0.01142	0.8724	87.60	1.1463	11.08	69.31	80.39	0.0250	0.1751	2
3	48.74	34.05	0.01143	0.8561	87.48	1.1681	11.35	69.16	80.51	0.0256	0.1751	3
4	49.71	35.01	0.01145	0.8402	87.36	1.1902	11.62	69.01	80.63	0.0262	0.1750	4
5	50.68	35.99	0.01146	0.8247	87.24	1.2126	11.89	68.86	80.75	0.0268	0.1749	5
6	51.68	36.98	0.01148	0.8094	87.12	1.2354	12.16	68.70	80.86	0.0273	0.1749	6
7	52.68	37.99	0.01149	0.7946	87.00	1.2585	12.43	68.55	80.98	0.0279	0.1748	7
8	53.70	39.01	0.01151	0.7800	86.88	1.2820	12.70	68.40	81.10	0.0285	0.1747	8
9	54.74	40.04	0.01153	0.7658	86.76	1.3058	12.98	68.24	81.22	0.0291	0.1747	9
10	55.79	41.09	0.01154	0.7519	86.63	1.3300	13.25	68.08	81.33	0.0296	0.1746	10
11	56.86	42.16	0.01156	0.7383	86.51	1.3545	13.52	67.93	81.45	0.0302	0.1745	11
12	57.94	43.24	0.01158	0.7250	86.39	1.3793	13.80	67.77	81.57	0.0308	0.1745	12
13	59.03	44.34	0.01159	0.7120	86.26	1.4045	14.06	67.62	81.68	0.0314	0.1744	13
14	60.14	45.45	0.01161	0.6992	86.14	1.4301	14.34	67.46	81.80	0.0319	0.1743	14
15	61.27	46.57	0.01163	0.6868	86.02	1.4561	14.62	67.30	81.92	0.0325	0.1743	15
16	62.41	47.72	0.01164	0.6746	85.89	1.4824	14.89	67.14	82.03	0.0331	0.1742	16
17	63.57	48.88	0.01166	0.6626	85.77	1.5091	15.16	66.98	82.14	0.0336	0.1742	17
18	64.75	50.05	0.01168	0.6510	85.64	1.5362	15.44	66.82	82.26	0.0342	0.1741	18
19	65.94	51.24	0.01169	0.6395	85.52	1.5637	15.71	66.66	82.37	0.0348	0.1740	19
20	67.14	52.45	0.01171	0.6283	85.39	1.5915	15.99	66.50	82.49	0.0354	0.1740	20
21	68.37	53.67	0.01173	0.6174	85.26	1.6198	16.26	66.34	82.60	0.0359	0.1739	21
22	69.61	54.91	0.01175	0.6066	85.14	1.6485	16.54	66.17	82.71	0.0365	0.1739	22
23	70.86	56.17	0.01176	0.5961	85.01	1.6775	16.81	66.01	82.82	0.0371	0.1738	23
24	72.13	57.44	0.01178	0.5858	84.88	1.7070	17.10	65.84	82.94	0.0376	0.1738	24
25	73.42	58.73	0.01180	0.5757	84.76	1.7369	17.37	65.68	83.05	0.0382	0.1737	25
26	74.73	60.04	0.01182	0.5659	84.63	1.7672	17.65	65.51	83.16	0.0388	0.1736	26
27	76.06	61.36	0.01183	0.5562	84.50	1.7980	17.93	65.34	83.27	0.0393	0.1736	27
28	77.40	62.70	0.01185	0.5467	84.37	1.8292	18.21	65.17	83.38	0.0399	0.1735	28
29	78.76	64.06	0.01187	0.5374	84.24	1.8608	18.48	65.01	83.49	0.0405	0.1735	29
30	80.13	65.44	0.01189	0.5283	84.11	1.8928	18.76	64.84	83.60	0.0410	0.1734	30
31	81.53	66.83	0.01191	0.5194	83.98	1.9253	19.04	64.67	83.71	0.0416	0.1734	31
32	82.94	68.24	0.01193	0.5106	83.85	1.9583	19.32	64.49	83.81	0.0422	0.1733	32
33	84.37	69.67	0.01194	0.5021	83.72	1.9917	19.60	64.32	83.92	0.0427	0.1733	33
34	85.82	71.12	0.01196	0.4937	83.59	2.0256	19.88	64.15	84.03	0.0433	0.1732	34
35	87.28	72.59	0.01198	0.4854	83.46	2.0600	20.17	63.97	84.14	0.0438	0.1732	35
36	88.77	74.07	0.01200	0.4774	83.33	2.0948	20.44	63.80	84.24	0.0444	0.1731	36
37	90.27	75.58	0.01202	0.4695	83.20	2.1301	20.73	63.62	84.35	0.0450	0.1730	37
38	91.80	77.10	0.01204	0.4617	83.07	2.1659	21.01	63.44	84.45	0.0455	0.1730	38
39	93.34	78.64	0.01206	0.4541	82.93	2.2022	21.29	63.27	84.56	0.0461	0.1729	39
40	94.90	80.20	0.01208	0.4466	82.80	2.2390	21.57	63.09	84.66	0.0466	0.1729	40
41	96.48	81.78	0.01210	0.4393	82.67	2.2763	21.86	62.91	84.77	0.0472	0.1728	41
42	98.08	83.38	0.01212	0.4321	82.53	2.3142	22.14	62.73	84.87	0.0478	0.1728	42
43	99.70	85.00	0.01214	0.4251	82.40	2.3525	22.42	62.55	84.97	0.0483	0.1727	43
44	101.3	86.64	0.01216	0.4182	82.26	2.3914	22.71	62.36	85.07	0.0489	0.1727	44

other hand, those classified in group 6 are only mildly toxic, being capable of causing harmful effects only in relatively large concentrations. Since injury from the latter group is caused more by oxygen deficiency than by any harmful effects of the fluids themselves, for all practical purposes the fluids in group 6 are considered to be nontoxic. However, it should be pointed out that some refrigerants, although nontoxic when mixed with air in their normal state, are subject to decomposition when they come in contact with an open flame or an

TABLE 16-5 (*Continued*)

TEMP.	PRESSURE		VOLUME cu ft/lb		DENSITY lb/cu ft		ENTHALPY Btu/lb			ENTROPY Btu/(lb)(° R)		TEMP.
°F	PSIA	PSIG	LIQUID v_f	VAPOR v_g	LIQUID $1/v_f$	VAPOR $1/v_g$	LIQUID h_f	LATENT h_{fg}	VAPOR h_g	LIQUID s_f	VAPOR s_g	°F
45	103.0	88.30	0.01218	0.4114	82.13	2.4308	22.99	62.18	85.17	0.0494	0.1726	45
46	104.7	89.97	0.01220	0.4047	81.99	2.4708	23.28	61.99	85.27	0.0500	0.1726	46
47	106.4	91.67	0.01222	0.3982	81.86	2.5113	23.57	61.81	85.38	0.0505	0.1725	47
48	108.1	93.39	0.01224	0.3918	81.72	2.5524	23.85	61.62	85.47	0.0511	0.1725	48
49	109.8	95.13	0.01226	0.3855	81.58	2.5940	24.14	61.43	85.57	0.0517	0.1724	49
50	111.6	96.89	0.01228	0.3793	81.44	2.6362	24.42	61.25	85.67	0.0522	0.1724	50
51	113.4	98.66	0.01230	0.3733	81.31	2.6790	24.71	61.06	85.77	0.0528	0.1723	51
52	115.2	100.5	0.01232	0.3673	81.17	2.7224	25.00	60.87	85.87	0.0533	0.1723	52
53	117.0	102.3	0.01234	0.3615	81.03	2.7664	25.29	60.67	85.96	0.0539	0.1722	53
54	118.8	104.1	0.01236	0.3557	80.89	2.8110	25.58	60.48	86.06	0.0544	0.1722	54
55	120.7	106.0	0.01238	0.3501	80.75	2.8562	25.87	60.28	86.15	0.0550	0.1721	55
56	122.6	107.9	0.01241	0.3446	80.61	2.9020	26.16	60.09	86.25	0.0555	0.1721	56
57	124.5	109.8	0.01243	0.3392	80.47	2.9485	26.44	59.90	86.34	0.0561	0.1720	57
58	126.4	111.7	0.01245	0.3338	80.33	2.9956	26.73	59.70	86.43	0.0566	0.1720	58
59	128.4	113.7	0.01247	0.3286	80.18	3.0434	27.02	59.50	86.52	0.0572	0.1719	59
60	130.3	115.6	0.01249	0.3234	80.04	3.0918	27.32	59.30	86.62	0.0578	0.1719	60
61	132.3	117.6	0.01252	0.3184	79.90	3.1409	27.61	59.10	86.71	0.0583	0.1718	61
62	134.3	119.6	0.01254	0.3134	79.76	3.1907	27.91	58.89	86.80	0.0589	0.1717	62
63	136.4	121.7	0.01256	0.3085	79.61	3.2411	28.19	58.69	86.88	0.0594	0.1717	63
64	138.4	123.7	0.01258	0.3037	79.47	3.2923	28.48	58.49	86.97	0.0600	0.1716	64
65	140.5	125.8	0.01261	0.2990	79.32	3.3442	28.78	58.28	87.06	0.0605	0.1716	65
66	142.6	127.9	0.01263	0.2944	79.18	3.3968	29.08	58.07	87.15	0.0611	0.1715	66
67	144.8	130.1	0.01265	0.2898	79.03	3.4502	29.37	57.86	87.23	0.0616	0.1715	67
68	146.9	132.2	0.01268	0.2854	78.88	3.5043	29.67	57.65	87.32	0.0622	0.1714	68
69	149.1	134.4	0.01270	0.2810	78.74	3.5591	29.96	57.44	87.40	0.0627	0.1714	69
70	151.3	136.6	0.01272	0.2766	78.59	3.6147	30.25	57.23	87.48	0.0633	0.1713	70
71	153.5	138.8	0.01275	0.2724	78.44	3.6712	30.55	57.01	87.56	0.0638	0.1712	71
72	155.8	141.1	0.01277	0.2682	78.29	3.7284	30.85	56.80	87.65	0.0644	0.1712	72
73	158.0	143.3	0.01280	0.2641	78.14	3.7864	31.15	56.58	87.73	0.0649	0.1711	73
74	160.3	145.6	0.01282	0.2601	77.99	3.8452	31.45	56.36	87.81	0.0655	0.1711	74
75	162.7	148.0	0.01285	0.2561	77.84	3.9049	31.74	56.14	87.88	0.0660	0.1710	75
76	165.0	150.3	0.01287	0.2522	77.68	3.9654	32.04	55.92	87.96	0.0665	0.1709	76
77	167.4	152.7	0.01290	0.2483	77.53	4.0268	32.34	55.70	88.04	0.0671	0.1709	77
78	169.8	155.1	0.01292	0.2446	77.38	4.0890	32.64	55.47	88.11	0.0676	0.1708	78
79	172.2	157.5	0.01295	0.2408	77.22	4.1522	32.94	55.25	88.19	0.0682	0.1707	79
80	174.6	159.9	0.01298	0.2372	77.07	4.2162	33.24	55.02	88.26	0.0687	0.1707	80
81	177.1	162.4	0.01300	0.2336	76.91	4.2812	33.54	54.79	88.33	0.0693	0.1706	81
82	179.6	164.9	0.01303	0.2300	76.76	4.3471	33.84	54.56	88.40	0.0698	0.1706	82
83	182.1	167.4	0.01305	0.2266	76.60	4.4140	34.14	54.33	88.47	0.0704	0.1705	83
84	184.7	170.0	0.01308	0.2231	76.44	4.4819	34.45	54.09	88.54	0.0709	0.1704	84
85	187.2	172.5	0.01311	0.2197	76.29	4.5507	34.75	53.86	88.61	0.0715	0.1703	85
86	189.8	175.1	0.01314	0.2164	76.13	4.6206	35.06	53.62	88.68	0.0720	0.1703	86
87	192.5	177.8	0.01316	0.2132	75.97	4.6915	35.36	53.38	88.74	0.0726	0.1702	87
88	195.1	180.4	0.01319	0.2099	75.80	4.7634	35.67	53.14	88.81	0.0731	0.1701	88
89	197.8	183.1	0.01322	0.2068	75.64	4.8364	35.97	52.90	88.87	0.0737	0.1701	89
90	200.5	185.8	0.01325	0.2036	75.48	4.9105	36.28	52.65	88.93	0.0742	0.1700	90
91	203.2	188.5	0.01328	0.2006	75.32	4.9856	36.59	52.40	88.99	0.0747	0.1699	91
92	206.0	191.3	0.01331	0.1976	75.15	5.0619	36.89	52.16	89.05	0.0753	0.1698	92
93	208.8	194.1	0.01334	0.1946	74.99	5.1394	37.20	51.91	89.11	0.0758	0.1697	93

electrical heating element. The products of decomposition thus formed are highly toxic and capable of causing harmful effects in small concentrations and on short exposure. This is true of all the fluorocarbon refrigerants (see column 6 of Table 16-7).

16-4. Flammability and Explosiveness

With regard to flammability and explosiveness, most of the refrigerants in common use are entirely nonflammable and nonexplosive. Notable exceptions to this are ammonia and the straight hydrocarbons. Ammonia is slightly

TABLE 16-5 (*Continued*)

TEMP.	PRESSURE		VOLUME cu ft/lb		DENSITY lb/cu ft		ENTHALPY Btu/lb			ENTROPY Btu/(lb)(° R)		TEMP.
°F	PSIA	PSIG	LIQUID v_f	VAPOR v_g	LIQUID $1/v_f$	VAPOR $1/v_g$	LIQUID h_f	LATENT h_{fg}	VAPOR h_g	LIQUID s_f	VAPOR s_g	°F
94	211.6	196.9	0.01337	0.1916	74.82	5.2180	37.51	51.65	89.16	0.0764	0.1697	94
95	214.4	199.7	0.01340	0.1888	74.65	5.2979	37.82	51.40	89.22	0.0769	0.1696	95
96	217.3	202.6	0.01343	0.1859	74.48	5.3789	38.13	51.14	89.27	0.0775	0.1695	96
97	220.2	205.5	0.01346	0.1831	74.32	5.4612	38.44	50.88	89.32	0.0780	0.1694	97
98	223.1	208.4	0.01349	0.1804	74.15	5.5447	38.75	50.62	89.37	0.0786	0.1693	98
99	226.1	211.4	0.01352	0.1776	73.97	5.6296	39.06	50.36	89.42	0.0791	0.1692	99
100	229.1	214.4	0.01355	0.1750	73.80	5.7157	39.37	50.10	89.47	0.0796	0.1692	100
101	232.1	217.4	0.01358	0.1723	73.63	5.8033	39.68	49.83	89.51	0.0802	0.1691	101
102	235.1	220.4	0.01361	0.1697	73.45	5.8921	40.00	49.56	89.56	0.0807	0.1690	102
103	238.2	223.5	0.01365	0.1672	73.28	5.9824	40.31	49.29	89.60	0.0813	0.1689	103
104	241.3	226.6	0.01368	0.1646	73.10	6.0741	40.62	49.02	89.64	0.0818	0.1688	104
105	244.4	229.7	0.01371	0.1621	72.92	6.1673	40.94	48.74	89.68	0.0824	0.1687	105
106	247.6	232.9	0.01375	0.1597	72.74	6.2620	41.25	48.47	89.72	0.0829	0.1686	106
107	250.7	236.0	0.01378	0.1573	72.56	6.3582	41.57	48.18	89.75	0.0834	0.1685	107
108	254.0	239.3	0.01382	0.1549	72.38	6.4560	41.88	47.90	89.78	0.0840	0.1684	108
109	257.2	242.5	0.01385	0.1525	72.20	6.5554	42.20	47.62	89.82	0.0845	0.1683	109
110	260.5	245.8	0.01389	0.1502	72.01	6.6564	42.52	47.33	89.85	0.0851	0.1682	110
111	263.8	249.1	0.01392	0.1480	71.83	6.7590	42.83	47.04	89.87	0.0856	0.1680	111
112	267.1	252.4	0.01396	0.1457	71.64	6.8634	43.15	46.75	89.90	0.0862	0.1679	112
113	270.5	255.8	0.01400	0.1435	71.45	6.9695	43.47	46.45	89.92	0.0867	0.1678	113
114	273.9	259.2	0.01403	0.1413	71.26	7.0775	43.79	46.15	89.94	0.0872	0.1677	114
115	277.3	262.6	0.01407	0.1391	71.07	7.1872	44.11	45.85	89.96	0.0878	0.1676	115
116	280.8	266.1	0.01411	0.1370	70.87	7.2988	44.43	45.55	89.98	0.0883	0.1674	116
117	284.3	269.6	0.01415	0.1349	70.68	7.4124	55.75	45.24	89.99	0.0889	0.1673	117
118	287.8	273.1	0.01419	0.1328	70.48	7.5279	45.07	44.93	90.00	0.0894	0.1672	118
119	291.4	276.7	0.01423	0.1308	70.28	7.6454	45.39	44.62	90.01	0.0899	0.1671	119
120	295.0	280.3	0.01427	0.1288	70.08	7.7649	45.71	44.31	90.02	0.0905	0.1669	120
121	298.6	283.9	0.01431	0.1268	69.88	7.8866	46.04	43.99	90.03	0.0910	0.1668	121
122	302.2	287.5	0.01435	0.1248	69.68	8.0105	46.36	43.67	90.03	0.0916	0.1666	122
123	305.9	291.2	0.01439	0.1229	69.47	8.1365	46.68	43.35	90.03	0.0921	0.1665	123
124	309.7	295.0	0.01444	0.1210	69.26	8.2648	47.00	43.02	90.02	0.0926	0.1663	124
125	313.4	298.7	0.01448	0.1191	69.05	8.3955	47.33	42,69	90.02	0.0932	0.1662	125
126	317.2	302.5	0.01453	0.1173	68.84	8.5285	47.65	42.36	90.01	0.0937	0.1660	126
127	321.0	306.3	0.01457	0.1154	68.62	8.6639	47.97	42.03	90.00	0.0942	0.1659	127
128	324.9	310.2	0.01462	0.1136	68.41	8.8019	48.29	41.69	89.98	0.0948	0.1657	128
129	328.8	314.1	0.01467	0.1118	68.19	8.9424	48.62	41.35	89.97	0.0953	0.1655	129
130	332.7	318.0	0.01471	0.1101	67.96	9.0855	48.95	41.00	89.95	0.0958	0.1654	130
131	336.6	321.9	0.01476	0.1083	67.74	9.2313	49.27	40.65	89.92	0.0964	0.1652	131
132	340.6	325.9	0.01481	0.1066	67.51	9.3798	49.59	40.30	89.89	0.0969	0.1650	132
133	344.7	330.0	0.01486	0.1049	67.28	9.5312	49.91	39.95	89.86	0.0974	0.1648	133
134	348.7	334.0	0.01491	0.1032	67.05	9.6854	50.24	39.59	89.83	0.0979	0.1646	134
135	352.8	338.1	0.01497	0.1016	66.81	9.8425	50.56	39.23	89.79	0.0985	0.1644	135
136	357.0	342.3	0.01502	0.09997	66.58	10.003	50.88	38.87	89.75	0.0990	0.1642	136
137	361.1	346.4	0.01508	0.09837	66.33	10.166	51.21	38.50	89.71	0.0995	0.1640	137
138	365.3	350.6	0.01513	0.09679	66.09	10.332	51.53	38.13	89.66	0.1000	0.1638	138
139	369.6	354.9	0.01519	0.09522	65.84	10.502	51.86	37.75	89.61	0.1006	0.1636	139
140	373.8	359.1	0.01525	0.09368	65.59	10.674	52.17	37.38	89.55	0.1011	0.1634	140
141	378.2	363.5	0.01531	0.09216	65.33	10.850	52.49	37.00	89.49	0.1016	0.1632	141

flammable and explosive when mixed in rather exact proportions with air. However with reasonable precautions, the hazard involved in using ammonia as a refrigerant is negligible.

Straight hydrocarbons, on the other hand, are highly flammable and explosive, and their use as refrigerants except in special applications and under the surveillance of experienced operating personnel is not usually permissible. Because of their excellent thermal properties, the straight hydrocarbons are frequently employed in ultralow temperature applications.

TABLE 16-6 Refrigerant-717 (*Ammonia*) Properties of Liquid and Saturated Vapor

Temp F	Pressure		Liquid, density	Vapor, sp vol	Enthalpy, datum −40 F Btu per lb		Entropy, datum −40 F Btu per lb F	
t	psia	psig	lb/cu ft $1/v_f$	cu ft/lb v_g	Liquid h_f	Vapor h_g	Liquid s_f	Vapor s_g
−105	1.00	*27.9	45.71	223.14	−68.5	570.3	−0.1774	1.6243
−104	1.04	27.8	45.67	214.23	−67.5	570.7	− .1744	1.6205
−103	1.08	27.7	45.63	205.90	−66.4	571.2	− .1714	1.6167
−102	1.14	27.7	45.59	197.70	−65.4	571.6	− .1685	1.6129
−101	1.19	27.5	45.55	190.08	−64.3	572.1	− .1655	1.6092
−100	1.24	*27.4	45.51	182.90	−63.3	572.5	−0.1626	1.6055
− 99	1.29	27.3	45.47	175.42	−62.2	572.9	− .1597	1.6018
− 98	1.35	27.2	45.43	168.48	−61.2	573.4	− .1568	1.5982
− 97	1.41	27.0	45.40	161.98	−60.1	573.8	− .1539	1.5945
− 96	1.47	26.9	45.36	155.92	−59.1	574.3	− .1510	1.5910
− 95	1.52	*26.8	45.32	150.30	−58.0	574.7	−0.1481	1.5874
− 94	1.59	26.7	45.28	144.68	−57.0	575.1	− .1452	1.5838
− 93	1.66	26.5	45.24	139.27	−55.9	575.6	− .1423	1.5803
− 92	1.73	26.4	45.20	134.06	−54.9	576.0	− .1395	1.5768
− 91	1.79	26.2	45.16	129.06	−53.8	576.5	− .1366	1.5734
− 90	1.86	*26.1	45.12	124.28	−52.8	576.9	−0.1338	1.5699
− 89	1.94	26.0	45.08	119.75	−51.7	577.3	− .1309	1.5665
− 88	2.02	25.8	45.04	115.37	−50.7	577.8	− .1281	1.5631
− 87	2.11	25.6	45.00	111.31	−49.6	578.2	− .1253	1.5597
− 86	2.18	25.5	44.96	107.39	−48.6	578.6	− .1225	1.5504
− 85	2.27	*25.3	44.92	103.63	−47.5	579.1	−0.1197	1.5531
− 84	2.36	25.1	44.88	99.87	−46.5	579.5	− .1169	1.5498
− 83	2.46	24.9	44.84	96.28	−45.4	579.9	− .1141	1.5465
− 82	2.55	24.7	44.80	92.86	−44.4	580.4	− .1113	1.5432
− 81	2.65	24.5	44.76	89.65	−43.3	580.8	− .1085	1.5400
− 80	2.74	*24.3	44.73	86.54	−42.2	581.2	−0.1057	1.5368
− 79	2.85	24.1	44.68	83.50	−41.2	581.6	− .1030	1.5336
− 78	2.96	23.9	44.64	80.61	−40.1	582.1	− .1002	1.5304
− 77	3.07	23.6	44.60	77.90	−39.1	582.5	− .0975	1.5273
− 76	3.19	23.4	44.56	75.30	−38.0	582.9	− .0947	1.5242
− 75	3.30	*23.2	44.52	72.80	−37.0	583.3	−0.0920	1.5211
− 74	3.43	22.9	44.48	70.35	−35.9	583.8	− .0892	1.5180
− 73	3.56	22.7	44.44	68.01	−34.9	584.2	− .0865	1.5149
− 72	3.69	22.4	44.40	65.78	−33.8	584.6	− .0838	1.5119
− 71	3.82	22.2	44.36	63.70	−32.8	585.0	− .0811	1.5089
− 70	3.94	*21.9	44.32	61.65	−31.7	585.5	−0.0784	1.5059
− 69	4.09	21.6	44.28	59.60	−30.7	585.9	− .0757	1.5029
− 68	4.24	21.3	44.24	57.64	−29.6	586.3	− .0730	1.4999
− 67	4.39	21.0	44.19	55.78	−28.6	586.7	− .0703	1.4970
− 66	4.54	20.7	44.15	54.01	−27.5	587.1	− .0676	1.4940
− 65	4.69	*20.4	44.11	52.34	−26.5	587.5	−0.0650	1.4911
− 64	4.86	20.1	44.07	50.79	−25.4	588.0	− .0623	1.4883
− 63	5.03	19.6	44.03	49.26	−24.4	588.4	− .0596	1.4854
− 62	5.20	19.3	43.99	47.74	−23.3	588.8	− .0570	1.4826
− 61	5.38	18.9	43.95	46.23	−22.2	589.2	− .0543	1.4797
− 60	5.55	*18.6	43.91	44.73	−21.2	589.6	−0.0517	1.4769
− 59	5.74	18.2	43.87	43.37	−20.1	590.0	− .0490	1.4741
− 58	5.93	17.8	43.83	42.05	−19.1	590.4	− .0464	1.4713
− 57	6.13	17.4	43.78	40.79	−18.0	590.8	− .0438	1.4686
− 56	6.33	17.0	43.74	39.56	−17.0	591.2	− .0412	1.4658
− 55	6.54	*16.6	43.70	38.38	−15.9	591.6	−0.0386	1.4631
− 54	6.75	16.2	43.66	37.24	−14.8	592.1	− .0360	1.4604
− 53	6.97	15.7	43.62	36.15	−13.8	592.4	− .0334	1.4577
− 52	7.20	15.3	43.58	35.09	−12.7	592.9	.0307	1.4551
− 51	7.43	14.8	43.54	34.06	−11.7	593.2	.0281	1.4524

* Inches of mercury below one standard atmosphere (29.92 in.)
U. S. Dept. of Commerce, Bureau of Standards, Thermodynamic Properties of Ammonia, Circular No. 142 (1923) and Circular No. 472 (1948).

In such installations, the hazard incurred by their use is minimized by the fact that the equipment is constantly attended by operating personnel experienced in the use and handling of flammable and explosive materials.

The "American Standard Safety Code for Mechanical Refrigeration" sets forth in detail the conditions and circumstances under which the various refrigerants can be safely used. Most local codes and ordinances governing the use of refrigerating equipment are based on this code, which is sponsored jointly by the ASHRAE and ASA.

The degree of hazard incurred by the use of toxic refrigerants depends upon a number of factors, such as the quantity of refrigerant used with relation to the size of the space into which the refrigerant may leak, the type of occupancy,

TABLE 16-6 *(Continued)*

Temp F t	Pressure		Liquid, density	Vapor, sp vol	Enthalpy, datum −40 F Btu per lb		Entropy, datum −40 F Btu per lb F	
	psia	psig	lb/cu ft $1/v_f$	cu ft/lb v_g	Liquid h_f	Vapor h_g	Liquid s_f	Vapor s_g
− 50	7.67	*14.3	43.49	33.08	−10.6	593.7	−0.0256	1.4497
− 49	7.91	13.8	43.45	32.12	− 9.6	594.0	.0230	1.4471
− 48	8.16	13.3	43.41	31.20	− 8.5	594.4	.0204	1.4445
− 47	8.42	12.8	43.37	30.31	− 7.4	594.9	.0179	1.4419
− 46	8.68	12.2	43.33	29.45	− 6.4	595.2	.0178	1.4393
− 45	8.95	*11.7	43.28	28.62	− 5.3	595.6	.0127	1.4368
− 44	9.23	11.1	43.24	27.82	− 4.3	596.0		1.4342
− 43	9.51	10.6	43.20	27.04	− 3.2			1.4317
− 42	9.81	10.0	43.16	26.29	− 2.1			1.4292
− 41	10.10	9.3	43.12	25.56	− 1.1			1.4267
− 40	10.41	*8.7	43.08	24.86	0.0	597.6	0.0000	1.4242
− 39	10.72	8.1	43.04	24.18	1.1	598.0	.0025	1.4217
− 38	11.04	7.4	42.99	23.53	2.1	598.3	.0051	1.4193
− 37	11.37	6.8	42.95	22.89	3.2	598.7	.0076	1.4169
− 36	11.71	6.1	42.90	22.27	4.3	599.1	.0101	1.4144
− 35	12.05	*5.4	42.86	21.68	5.3	599.5	0.0126	1.4120
− 34	12.41	4.7	42.82	21.10	6.4	599.9	.0151	1.4096
− 33	12.77	3.9	42.78	20.54	7.4	600.2	.0176	1.4072
− 32	13.14	3.2	42.73	20.00	8.5	600.6	.0201	1.4048
− 31	13.52	2.4	42.69	19.48	9.6	601.0	.0226	1.4025
− 30	13.90	*1.6	42.65	18.97	10.7	601.4	0.0250	1.4001
− 29	14.30	0.8	42.61	18.48	11.7	601.7	.0275	1.3978
− 28	14.71	0.0	42.57	18.00	12.8	602.1	.0300	1.3955
− 27	15.12	0.4	42.54	17.54	13.9	602.5	.0325	1.3932
− 26	15.55	0.8	42.48	17.09	14.9	602.8	.0350	1.3909
− 25	15.98	1.3	42.44	16.66	16.0	603.2	0.0374	1.3886
− 24	16.42	1.7	42.40	16.24	17.1	603.6	.0399	1.3863
− 23	16.88	2.2	42.35	15.83	18.1	603.9	.0423	1.3840
− 22	17.34	2.6	42.31	15.43	19.2	604.3	.0488	1.3818
− 21	17.81	3.1	42.26	15.05	20.3	604.6	.0472	1.3796
− 20	18.30	3.6	42.22	14.68	21.4	605.0	0.0497	1.3774
− 19	18.79	4.1	42.18	14.32	22.4	605.3	.0521	1.3752
− 18	19.30	4.6	42.13	13.97	23.5	605.7	.0545	1.3729
− 17	19.81	5.1	42.09	13.62	24.6	606.1	.0570	1.3708
− 16	20.34	5.6	42.04	13.29	25.6	606.4	.0594	1.3686
− 15	20.88	6.2	42.00	12.97	26.7	606.7	0.0618	1.3664
− 14	21.43	6.7	41.96	12.66	27.8	607.1	.0642	1.3643
− 13	21.99	7.8	41.91	12.36	28.9	607.5	.0666	1.3621
− 12	22.56	7.9	41.87	12.06	30.0	607.8	.0690	1.3600
− 11	23.15	8.5	41.82	11.78	31.0	608.1	.0714	1.3579
− 10	23.74	9.0	41.78	11.50	32.1	608.5	0.0738	1.3558
− 9	24.35	9.7	41.74	11.23	33.2	608.8	.0762	1.3537
− 8	24.97	10.3	41.69	10.97	34.3	609.2	.0768	1.3516
− 7	25.61	10.9	41.65	10.71	35.4	609.5	.0809	1.3495
− 6	26.26	11.6	41.60	10.47	36.4	609.8	.0833	1.3474
− 5	26.92	12.2	41.56	10.23	37.5	610.1	0.0857	1.3454
− 4	27.59	12.9	41.52	9.991	38.6	610.5	.0880	1.3433
− 3	28.28	13.6	41.47	9.763	39.7	610.8	.0904	1.3413
− 2	28.98	14.3	41.43	9.541	40.7	611.1	.0928	1.3393
− 1	29.69	15.0	41.38	9.326	41.8	611.4	−.0951	1.3372
0	30.42	15.7	41.34	9.116	42.9	611.8	0.0975	1.3352
1	31.16	16.5	41.29	8.912	44.0	612.1	0.0998	1.3332
2	31.92	17.2	41.25	8.714	45.1	612.4	.1022	1.3312
3	32.69	18.0	41.20	8.521	46.2	612.7	.1045	1.3292
4	33.47	18.8	41.16	8.333	47.2	613.0	.1069	1.3273
5	34.27	19.6	41.11	8.150	48.3	613.3	.1092	1.3253
6	35.09	20.4	41.07	7.971	49.4	613.6	0.1115	1.3234
7	35.92	21.2	41.01	7.798	50.5	613.9	.1138	1.3214
8	36.77	22.1	40.98	8.629	51.6	614.3	.1162	1.3195
9	37.63	22.9	40.93	7.464	52.7	614.6	.1185	1.3176
10	38.51	23.8	40.89	7.304	53.8	614.9	.1208	1.3157
11	39.40	24.7	40.84	7.148	54.9	615.2	0.1231	1.3137
12	40.31	25.6	40.80	6.996	56.0	615.5	.1254	1.3118
13	41.24	26.5	40.75	6.847	57.1	615.8	.1277	1.3099
14	42.18	27.5	40.71	6.703	58.2	616.1	.1300	1.3081
15	43.14	28.4	40.66	6.562	59.2	616.3	.1323	1.3062

* Inches of mercury below one standard atmosphere (29.92 in.).

TABLE 16-6 *(Continued)*

Temp F t	Pressure		Liquid, density	Vapor, sp vol	Enthalpy, datum −40 F Btu per lb		Entropy, datum −40 F Btu per lb F	
	psia	psig	lb/cu ft $1/v_f$	cu ft/lb v_g	Liquid h_f	Vapor h_g	Liquid s_f	Vapor s_g
16	44.12	29.4	40.61	6.425	60.3	616.6	0.1346	1.3043
17	45.12	30.4	40.57	6.291	61.4	616.9	.1369	1.3025
18	46.13	31.4	40.52	6.161	62.5	617.2	.1392	1.3006
19	47.16	32.5	40.48	6.034	63.6	617.5	.1415	1.2988
20	48.21	33.5	40.43	5.910	64.7	617.8	.1437	1.2969
21	49.28	34.6	40.38	5.789	65.8	618.0	0.1460	1.2951
22	50.36	35.7	40.34	5.671	66.9	618.3	.1483	1.2933
23	51.47	36.8	40.29	5.556	68.0	618.6	.1505	1.2951
24	52.59	37.9	40.25	5.443	69.1	618.9	.1528	1.2897
25	53.78	39.0	40.20	5.334	70.2	619.1	.1551	1.2879
26	54.90	40.2	40.15	5.227	71.3	619.4	0.1573	1.2861
27	56.08	41.4	40.10	5.123	72.4	619.7	.1596	1.2843
28	57.28	42.6	40.06	5.021	73.5	619.9	.1618	1.2825
29	58.50	43.8	40.01	4.922	74.6	620.2	.1641	1.2808
30	59.74	45.0	39.96	4.825	75.7	620.5	.1663	1.2790
31	61.00	46.3	39.91	4.730	76.8	620.7	0.1686	1.2773
32	62.29	47.6	39.86	4.637	77.9	621.0	.1708	1.2755
33	63.59	48.9	39.82	4.547	79.0	621.2	.1730	1.2738
34	64.91	50.2	39.77	4.459	80.1	621.5	.1753	1.2731
35	66.26	52.6	39.72	4.373	81.2	621.7	.1775	1.2704
36	67.63	52.9	39.67	4.289	82.3	622.0	0.1797	1.2686
37	69.02	54.3	39.63	4.207	83.4	622.2	.1819	1.2669
38	70.43	55.7	39.58	4.126	84.6	622.5	.1841	1.2652
39	71.87	57.2	39.54	4.048	85.7	622.7	.1863	1.2635
40	73.32	58.6	39.49	3.971	86.8	623.0	.1885	1.2618
41	74.80	60.1	39.44	3.897	87.9	623.2	0.1908	1.2602
42	76.31	61.6	39.39	3.823	89.0	623.4	.1930	1.2585
43	77.83	63.1	39.34	3.752	90.1	623.7	.1952	1.2568
44	79.38	64.7	39.29	3.682	91.2	623.9	.1974	1.2552
45	80.96	66.3	39.24	3.614	92.3	624.1	.1996	1.2535
46	82.55	67.9	39.19	3.547	93.5	624.4	0.2018	1.2519
47	84.18	69.5	39.14	3.481	94.6	624.6	.2040	1.2502
48	85.82	71.1	39.10	3.418	95.7	624.8	.2062	1.2486
49	87.49	72.8	39.05	3.355	96.8	625.0	.2083	1.2469
50	89.19	74.5	39.00	3.294	97.9	625.2	.2105	1.2453
51	90.91	76.2	38.95	3.234	99.1	625.5	0.2127	1.2437
52	92.66	78.0	38.90	3.176	100.2	625.7	.2149	1.2421
53	94.43	79.7	38.85	3.119	101.3	625.9	.2171	1.2405
54	96.23	81.5	38.80	3.063	102.4	626.1	.2192	1.2389
55	98.06	83.4	38.75	3.008	103.5	626.3	.3214	1.2373
56	99.91	85.2	38.70	2.954	104.7	626.5	0.2236	1.2357
57	101.8	87.1	38.65	2.902	105.8	626.7	.2257	1.2341
58	103.7	89.0	38.60	2.851	106.9	626.9	.2279	1.2325
59	105.6	90.9	38.55	2.800	108.1	627.1	.2301	1.2310
60	107.6	92.9	38.50	2.751	109.2	627.3	.2322	1.2294
61	109.6	94.9	38.45	2.703	110.3	627.5	0.2344	1.2278
62	111.6	96.9	38.40	2.656	111.5	627.7	.2365	1.2262
63	113.6	98.9	38.35	2.610	112.6	627.9	.2387	1.2247
64	115.7	101.0	38.30	2.565	113.7	628.0	.2408	1.2231
65	117.8	103.1	38.25	2.520	114.8	628.2	.2430	1.2216
66	120.0	105.3	38.20	2.477	116.0	628.4	0.2451	1.2201
67	122.1	107.4	38.15	2.435	117.1	628.6	.2473	1.2186
68	124.3	109.6	38.10	2.393	118.3	628.8	.2494	1.2170
69	126.5	111.8	38.05	2.352	119.4	628.9	.2515	1.2155
70	128.8	114.1	38.00	2.312	120.5	629.1	.2537	1.2140
71	131.1	116.4	37.95	2.273	121.7	629.3	0.2558	1.2125
72	133.4	118.7	37.90	2.235	122.8	629.4	.2579	1.2110
73	135.7	121.0	37.84	2.197	124.0	629.6	.2601	1.2095
74	138.1	123.4	37.79	2.161	125.1	629.8	.2622	1.2080
75	140.5	125.8	37.74	2.125	126.2	629.9	.2643	1.2065
76	143.0	128.3	37.69	2.089	127.4	630.1	0.2664	1.2050
77	145.4	130.7	37.64	2.055	128.5	630.2	.2685	1.2035
78	147.9	133.2	37.58	2.021	129.7	630.4	.2706	1.2020
79	150.5	135.8	37.53	1.988	130.8	630.5	.2728	1.2006
80	153.0	138.3	37.48	1.955	132.0	630.7	.2749	1.1991

TABLE 16-6 *(Continued)*

Temp F t	Pressure		Liquid, density	Vapor, sp vol	Enthalpy, datum −40 F Btu per lb		Entropy, datum −40 F Btu per lb F	
	psia	psig	lb/cu ft $1/v_f$	cu ft/lb v_g	Liquid h_f	Vapor h_g	Liquid s_f	Vapor s_g
81	155.6	140.9	37.43	1.923	133.1	630.8	0.2769	1.1976
82	158.3	143.6	37.37	1.892	134.3	631.0	.2791	1.1962
83	161.0	146.3	37.32	1.861	135.4	631.1	.2812	1.1947
84	163.6	149.0	37.26	1.831	136.6	631.3	.2833	1.1933
85	166.4	151.7	37.21	1.801	137.8	631.4	.2854	1.1918
86	169.2	154.5	37.16	1.772	138.9	631.5	0.2875	1.1904
87	172.0	157.3	37.11	1.744	140.1	631.7	.2895	1.1889
88	174.8	160.1	37.05	1.716	141.2	631.8	.2917	1.1875
89	177.7	163.0	37.00	1.688	142.4	631.9	.2937	1.1860
90	180.6	165.9	36.95	1.661	143.5	632.0	.2958	1.1846
91	183.6	168.9	36.89	1.635	144.7	632.1	0.2979	1.1832
92	186.6	171.9	36.84	1.609	145.8	632.3	.3000	1.1818
93	189.6	174.9	36.78	1.584	147.0	632.3	.3021	1.1804
94	192.7	178.0	36.73	1.559	148.2	632.5	.3041	1.1789
95	195.8	181.1	36.67	1.534	149.4	632.6	.3062	1.1775
96	198.9	184.2	36.62	1.510	150.5	632.6	0.3083	1.1761
97	202.1	187.4	36.56	1.487	151.7	632.8	.3104	1.1747
98	205.3	190.6	36.51	1.464	152.9	632.9	.3125	1.1733
99	208.6	193.9	36.45	1.441	154.0	632.9	.3145	1.1719
100	211.9	197.2	36.40	1.419	155.2	633.0	.3166	1.1705
101	215.2	200.5	36.34	1.397	156.4	633.1	0.3187	1.1691
102	218.6	203.9	36.29	1.375	157.6	633.2	.3207	1.1677
103	222.0	207.3	36.23	1.354	158.7	633.3	.3228	1.1663
104	224.4	210.7	36.18	1.334	159.9	633.4	.3248	1.1649
105	228.9	214.2	36.12	1.313	161.1	633.4	.3269	1.1635
106	232.5	217.8	36.06	1.293	162.3	633.5	0.3289	1.1621
107	236.0	221.3	36.01	1.274	163.5	633.6	.3310	1.1607
108	239.7	225.0	35.95	1.254	164.6	633.6	.3330	1.1593
109	243.3	228.6	35.90	1.235	165.8	633.7	.3351	1.1580
110	247.0	232.3	35.84	1.217	167.0	633.7	.3372	1.1566
111	250.8	236.1	35.78	1.198	168.2	633.8	0.3392	1.1552
112	354.5	239.8	35.72	1.180	169.4	633.8	.3413	1.1538
113	258.4	243.7	35.67	1.163	170.6	633.9	.3433	1.1524
114	262.2	247.5	35.61	1.145	171.8	633.9	.3453	1.1510
115	266.2	251.5	35.55	1.128	173.0	633.9	.3474	1.1497
116	270.1	255.4	35.49	1.112	174.2	634.0	0.3495	1.1483
117	274.1	259.4	35.43	1.095	175.4	634.0	.3515	1.1469
118	278.2	263.5	35.38	1.079	176.6	634.0	.3515	1.1455
119	282.3	267.6	35.32	1.063	177.8	634.0	.3556	1.1441
120	286.4	271.7	35.26	1.047	179.0	634.0	.3576	1.1427
121	290.6	275.9	35.20	1.032	180.2	634.0	0.3597	1.1414
122	294.8	280.1	35.14	1.017	181.4	634.0	.3618	1.1400
123	299.1	284.4	35.08	1.002	182.6	634.0	.3638	1.1386
124	303.4	288.7	35.02	0.987	183.9	634.0	.3659	1.1372
125	307.8	293.1	34.96	0.973	185.1	634.0	.3679	1.1358

whether or not open flames are present, the odor of the refrigerant, and whether or not experienced personnel are on duty to attend the equipment. For example, a small quantity of even a highly toxic refrigerant presents little hazard when used in relatively large spaces in that it is not possible in the event of a leak for the concentration to reach a harmful level. Too, the danger inherent in the use of toxic refrigerants is somewhat tempered by the fact that toxic refrigerants (including decomposition products) have very noticeable odors which tend to serve as a warning of their presence. Hence, toxic refrigerants are usually a hazard only to infants and others who, by reason of infirmity or confinement, are unable to escape the fumes. At the present time, ammonia is the only toxic refrigerant that is used to any great extent, and its use is ordinarily limited to packing plants, ice plants, and large cold storage facilities where experienced personnel are usually on duty.

16-5. Economic and Other Considerations

Naturally, from the viewpoint of economical operation, it is desirable that the refrigerant have physical and thermal characteristics which will result in the minimum power requirements per unit of refrigerating capacity, that is, a high coefficient of performance. The more important properties of the refrigerant which influence the capacity and efficiency are

TABLE 16-7. Relative Safety of Refrigerants

			Toxicity Lethal or Serious Injury[3]					
			Refrigerant in Air			Products of Decomposition by Flame		Flammable or Explosive Limits of Concentration in Air
Refrigerant	ASAB9 Safety Code Group Classification	Nat'l Fire Underwriters Group Number	Duration of Exposure (hr)	% by Vol	lb/1000 cu ft	Duration of Exposure (min)	% by Vol[4]	% by Vol
Methane	3[1]	+5						4.9–15.0
R-14	1[1]	6[1]						Nonflam.
Ethylene	3[1]	+5						3.0–25.0
Nitrous oxide			8	0.0025				Nonflam.
R-13	1[1]	6[1]						Nonflam.
Ethane	3	5	2	37.4–51.7				3.3–10.6
Carbon dioxide	1	5	$\frac{1}{2}$ to 1	29–30	33.2–34.3			Nonflam.
Kulene-131	1[1]	6[1]						Nonflam.
Propane	3	5	2	37.5–51.7	42.4–58.5			2.3–7.3
R-22	1	5A				16	1.0	Nonflam.[2]
Ammonia	2	2	$\frac{1}{2}$	0.5–0.6	0.221–0.256			16.0–25.0
Carrene-7	1	5A	2	19.4–20.3	50.2–52.2	25	1.1	Nonflam.
R-12	1	6	2	28.5–30.4	89.6–95.7	20	1.0	Nonflam.
Methyl chloride	2	4	2	2–2.5	2.62–3.28	30	2.4	8.1–17.2
Isobutane	3	+5						1.8–8.4
Sulfur dioxide	2	1	$\frac{1}{12}$	0.7	1.165			Nonflam.
Butane	3	5	2	37.5–51.7				1.6–6.5
R-114	1	6	2	20.1–21.5	90.5–96.8	15	1.0	Nonflam.
R-21	1		$\frac{1}{2}$	10.2	27.1			Nonflam.
Ethyl chloride	2	4	1	4.0	6.72	18	2.0	3.7–12.0
R-11	1	5	2	10	35.7	5	1.0	Nonflam.
Methyl formate	2	3	1	2–2.5	3.12–3.9			4.5–20.0
Methylene chloride	1	4A	$\frac{1}{2}$	5.1–5.3	11.25–11.7	20	1.0	Nonflam.
R-113	1	4	1	4.8–5.2	23.3–25.2	16	1.2	Nonflam.
Dichlorethylene	2	4	1	2–2.5	5.04–6.3	5	2.1	5.6–11.4

[1] Unofficial.
[2] Very slightly flammable, but for practical purposes considered nonflammable.
[3] To guinea pigs.
[4] Initial concentration.
From the *ASRE Data Book*, Design Volume, 1957–58 Edition, by permission of the American Society of Heating Refrigerating, and Air-Conditioning Engineers.

(1) the latent heat of vaporization, (2) the specific volume of the vapor, (3) the compression ratio, and (4) the specific heat of the refrigerant in both the liquid and vapor states.

Except in very small systems, a high latent heat value is desirable in that the weight of refrigerant circulated per unit of capacity is less. When a high latent heat value is accompanied by a low specific volume in the vapor state, the efficiency and capacity of the compressor are greatly increased. This tends not only to decrease the power consumption but also to reduce the compressor displacement required, which permits the use of smaller, more compact equipment. However, in small systems, if the latent heat value of the refrigerant is too high, the amount of refrigerant circulated will be insufficient for accurate control of the liquid.

A low specific heat for the liquid and a high specific heat for the vapor are desirable in that both tend to increase the refrigerating effect per pound, the former by increasing the subcooling effect and the latter by decreasing the superheating effect. When both are found in a single fluid, the efficiency of a liquid-suction heat exchanger is much improved.

The effect of compression ratio on the work of compression and, consequently, on the coefficient of performance, has already been

TABLE 16-8 Comparative Refrigerant Characteristics Performance Based on 5 F Evaporation and 86 F Condensation

Based on 5 F Evaporation and 86 F Condensation[a]

Refrigerant No.	Refrigerant Name	Evaporator pressure, psig	Condensing pressure, psig	Compression Ratio	Net refrigerating effect, Btu/lb	Refrigerant circulated, lb/min	Liquid circulated, cu in./min	Specific volume of suction gas, cu ft/lb	Compressor displacement, cfm	Horsepower, hp	Coefficient of performance	Comp. discharge temp, F
170	Ethane	221.3	661.1	2.86	58.6	3.41	342.9	0.53	1.82	1.953	2.41	122
744A	Nitrous Oxide	294.3	922.3	3.03	85.2	2.35	71.2	0.28	0.66	1.310	3.60	
744	Carbon Dioxide	317.5	1031.0	3.15	55.5	3.62	167.1	0.27	0.96	1.840	2.56	151
13B1	Bromotrifluoromethane	63.2	247.1	3.36	29.3	6.86	123.8	0.38	2.63	1.030	4.25	124
1270	Propylene	37.0	167.0	3.51	173.0	1.1	61.5	2.61	3.03	1.046	4.51	108
290	Propane	27.2	140.5	3.70	121.0	1.65	94.0	2.48	4.09	1.030	4.58	97
502	22/115 Azeotrope[b]	36.0	175.1	3.75	45.7	4.38	99.4	0.82	3.61	1.079	4.37	99
22	Chlorodifluoromethane	28.2	158.2	4.03	70.0	2.86	67.4	1.24	3.55	1.011	4.66	128
115	Chloropentafluoroethane	24.0	135.8	3.89	29.1	6.88	151	0.77	5.30	1.17	4.02	86
717	Ammonia	19.6	154.5	4.94	474.4	0.422	19.6	8.15	3.44	0.989	4.76	210
500	12/152a Azeotrope[b]	16.4	112.9	4.12	60.6	3.30	80.3	1.50	4.95	1.01	4.65	105
12	Dichlorodifluoromethane	11.8	93.3	4.08	50.0	4.00	85.6	1.46	5.83	1.002	4.70	101
40	Methyl Chloride	6.5	80.0	4.48	150.2	1.33	40.9	4.47	5.95	0.962	4.90	172
600a	Isobutane	3.3*	44.8	4.54	111.5	1.79	91.0	6.41	11.50	1.083	4.36	80
764	Sulfur Dioxide	5.9*	51.8	5.63	141.4	1.41	26.6	6.42	9.09	0.968	4.87	191
630	Methylamine	9.9*	46.8	6.13	304.0	0.66	28.2	15.54	10.23	0.978	4.81	
600	Butane	13.2*	26.9	5.07	128.6	1.56	75.9	9.98	15.52	0.953	4.95	88
114	Dichlorotetrafluoroethane	16.1*	22.0	5.42	43.1	4.64	89.2	4.34	20.14	1.049	4.49	86
21	Dichlorofluoromethane	19.2*	16.5	5.96	89.4	2.24	45.7	9.13	20.43	0.941	5.01	142
160	Ethyl Chloride	20.5*	12.4	5.83	142.3	1.45	45.8	17.06	24.82	0.906	5.21	106
631	Ethylamine	23.1*	10.0	7.40	225.5	0.89	349.0	32.32	38.67	0.855	5.52	
11	Trichlorofluoromethane	23.9*	3.5	6.19	66.8	2.99	56.6	12.21	36.54	0.938	5.03	111
611	Methyl Formate	26.3*	1.6*	7.74	189.2	1.06	29.9	48.25	51.00			
610	Ethyl Ether	26.9*	4.9*	8.20	126.3	1.58	62.9	35.00	55.40	0.822	5.74	
30	Methylene Chloride	27.6*	9.5*	8.60	134.6	1.49	30.9	49.90	74.30	0.963	4.90	205
113	Trichlorotrifluoroethane	27.9*	13.9*	8.02	53.7	3.73	66.5	27.38	102.03	0.973	4.84	86
1130	Dichloroethylene	28.3*	15.8*	8.42	114.3	1.75	38.3	63.60	111.20	0.973	4.83	
1120	Trichloroethylene	29.6*	26.2*	11.65	91.7	2.18	41.6	229.40	502.00	0.980	4.82	

[a] Saturated suction vapor except for Refrigerants 113, 114, and 115. In these cases, enough suction superheat was assumed to give saturated discharge vapor.
[b] See Table 1 for composition.
* Inches of mercury vacuum.

From the ASHRAE *Data Book*, Fundamentals Volume, 1972 Edition, by permission of the American Society of Heating, Refrigerating, and Air-Conditioning Engineers.

discussed in a previous chapter. Naturally, all other factors being equal, the refrigerant giving the lowest compression ratio is the most desirable. Low compression ratios result in low power consumption and high volumetric efficiency, the latter being more important in smaller systems since it permits the use of small compressors.

A low adiabatic discharge temperature is highly desirable. When combined with a reasonable compression ratio, a low adiabatic discharge temperature greatly reduces the possibility of overheating of the compressor and contributes measurably to a maintenance-free long life for the compressor. Since the rate of chemical reactions approximately doubles with each 10°C rise in temperature, a low adiabatic discharge temperature is particularly important when hermetic motor-compressors are employed. Whereas the discharge temperature of any one refrigerant always decreases as the compression ratio decreases, it is important to recognize that for any given compression ratio, the discharge temperature of one refrigerant may be significantly higher than that of another refrigerant operating with the same compression ratio.

A high coefficient of conductance can often improve heat transfer rates, particularly in liquid chilling applications, and thereby reduce the size and cost of heat transfer equipment. Also, it is desirable that the pressure-temperature relationship of the refrigerant be such that the pressure in the evaporator is always above atmospheric. In the event of a leak on the low pressure side of the system, if the pressure in the low side is below atmospheric, considerable amounts of air and moisture may be drawn into the system, whereas if the vaporizing pressure is above atmospheric, the possibility of drawing in air and moisture in the event of a leak is minimized.

Reasonably low condensing pressures under normal atmospheric conditions are also desirable in that they allow the use of lightweight materials in the construction of the condensing equipment, thereby reducing the size, weight, and cost of the equipment.

Naturally, the critical temperature and pressure of the refrigerant must be above the maximum temperature and pressure which will be encountered in the system. Likewise, the freezing point of the refrigerant must be safely below the minimum temperature to be obtained in the cycle. These factors are particularly important in selecting a refrigerant for a low temperature application.

Since the power required per unit of refrigerating capacity is very nearly the same for all the refrigerants in common use, efficiency and economy of operation are not usually deciding factors in the selection of the refrigerant. More important are those properties which tend to reduce the size, weight, and initial cost of the refrigerating equipment and which permit automatic operation and a minimum of maintenance. The cost and the availability of the refrigerant itself are also important considerations in the selection of a refrigerant.

For the purpose of comparison, the theoretical performance at standard ton conditions of 5°F (−15°C) evaporating temperature and 86°F (30°C) condensing temperature is shown in Table 16-8 for a number of common refrigerants. Also, to illustrate the effect that changing the refrigerant has on the refrigerating capacity and power requirements of a given compressor, the refrigerating capacities and power requirements for one model of open-type reciprocating compressor at various operating conditions are compared for Refrigerants-12, 22, and 502 in Table 16-9.

16-6. Early Refrigerants

In earlier days, when mechanical refrigeration was limited to a few large applications, ammonia and carbon dioxide were practically the only refrigerants available. Later, with the development of small, automatic domestic and commercial units, refrigerants such as sulfur dioxide and methyl chloride came into use, along with methylene chloride, which was developed for use with centrifugal compressors. Methylene chloride and carbon dioxide, because of their safe properties, were extensively used in large air conditioning applications.

With the exception of ammonia, all these refrigerants have fallen into disuse and are found only in some of the older installations, having been discarded in favor of the more suitable fluorocarbon refrigerants as the latter were developed. The fluorocarbons are practically the only refrigerants in extensive use at the present time. Again, an exception to this is ammonia which, because of its excellent thermal properties, is still widely used in such installations as ice plants, skating rinks, etc. A few other refrigerants also find limited use in special applications.

16-7. Development of the Fluorocarbons

The search for a completely safe refrigerant with good thermal properties led to the development of the fluorocarbon refrigerants in the late 1920's. The fluorocarbons (fluoronated hydrocarbons) are one group of a family of compounds known as the halocarbons (halogenated hydrocarbons). The halocarbon family of compounds are synthesized by replacing one or more of the hydrogen atoms in methane (CH_4) or ethane (C_2H_6) molecules, both of which are pure hydrocarbons, with atoms of chlorine, fluorine, and/or bromine, the latter group being from the halogen family. Halocarbons developed from the methane molecule are known as "methane series halocarbons." Likewise, those developed from the ethane molecule are referred to as "ethane series halocarbons."

The composition of the methane series halocarbons is shown in Fig. 16-1. Notice that the basic methane molecule consists of one atom of carbon (C) and four atoms of hydrogen (H). If the hydrogen atoms are replaced progressively with chlorine (Cl) atoms, the resulting compounds are methyl chloride (CH_3Cl), methylene chloride (CH_2Cl_2), chloroform ($CHCl_3$), and carbontetrachloride (CCl_4), respectively, the last two being the base molecules for the more popular fluorocarbons of the methane series.

If the chlorine atoms in the carbontetrachloride molecule are now replaced progressively with fluorine atoms, the resulting compounds are trichloromonofluoromethane (CCl_3F), dichlorodifluoromethane (CCl_2F_2),

TABLE 16-9 Effect of Changing Refrigerants on the Refrigerating Capacity and Power Requirements of One Model Compressor

Evap. Temp. °F	Cond. Temp. °F	R-12			R-22			R-502		
		Cap (Tons)	Shaft BHP	THR (Tons)	Cap (Tons)	Shaft BHP	THR (Tons)	Cap (Tons)	Shaft BHP	THR (Tons)
	80	2.50	7.01	3.5				4.60	13.7	6.50
−40	100	2.00	8.2	3.0				3.60	14.1	5.60
	120	1.60	9.5	2.5				2.60	14.8	4.70
	80	3.76	9.0	4.8				6.30	16.4	8.60
−30	100	3.04	10.0	4.6				5.20	17.1	7.80
	120	2.40	11.4	4.4		Not Rated		4.10	18.3	6.90
	80	4.75	10.3	6.2				8.20	18.9	10.8
−20	100	4.10	11.4	5.6				7.10	20.2	10.0
	120	3.40	12.8	4.9				5.80	21.9	9.1
	80	6.40	11.4	8.1				10.8	21.1	13.7
−10	100	5.60	13.0	7.5				9.3	22.2	12.8
	120	4.70	14.7	6.7				7.9	24.8	11.8
	80	8.40	12.8	10.2	13.0	20.4	15.9	14.0	23.2	17.4
0	100	7.34	14.4	9.6	11.6	23.3	15.2	12.0	25.6	16.0
	120	6.30	16.7	9.6	9.8	26.6	13.6	10.0	28.7	14.4
	80	10.6	13.9	12.9	16.4	21.2	19.4	17.7	25.0	21.8
10	100	9.42	15.5	12.1	14.6	25.3	18.5	15.1	28.1	19.6
	120	8.16	17.9	11.3	12.7	29.6	17.1	12.7	31.6	17.8
	140	7.01	20.4	10.7	10.8	34.0	15.3	10.3	34.4	15.2
	80	13.0	14.5	14.8	20.8	22.0	24.2	21.7	27.3	26.2
20	100	11.8	17.2	14.3	18.5	27.2	23.0	18.7	30.5	24.0
	120	10.4	19.4	13.1	16.4	32.5	21.6	15.5	34.4	21.2
	120	9.3	22.2	12.2	14.2	37.5	19.6	12.7	37.7	18.3
	80	16.3	15.7	18.5	26.0	22.7	29.6	26.7	29.0	32.0
30	100	14.5	18.1	17.5	23.6	28.8	28.6	22.8	32.7	28.9
	120	12.8	20.4	16.0	21.0	34.8	27.0	18.8	37.2	25.4
	140	11.1	23.7	14.5	18.6	41.0	25.1	15.0	41.2	21.1
	80	19.8	16.3	22.4	32.0	23.3	36.0			
40	100	18.7	19.6	22.0	29.0	30.2	34.5			
	120	17.4	22.0	21.5	26.0	37.0	32.8			
	140	16.0	25.7	20.5	23.9	43.0	31.6			
	80	25.2	17.2	28.2	39.0	23.6	43.5			
50	100	22.3	20.1	26.3	35.2	31.0	41.5		Not Rated	
	120	19.8	23.2	24.5	31.5	38.3	39.1			
	140	17.4	27.3	22.3	28.5	44.0	37.0			
	80				47.5	23.8	52.5			
60	100		Not Rated		42.5	31.4	49.4			
	120				38.5	39.0	47.0			
	140				34.0	45.0	43.5			

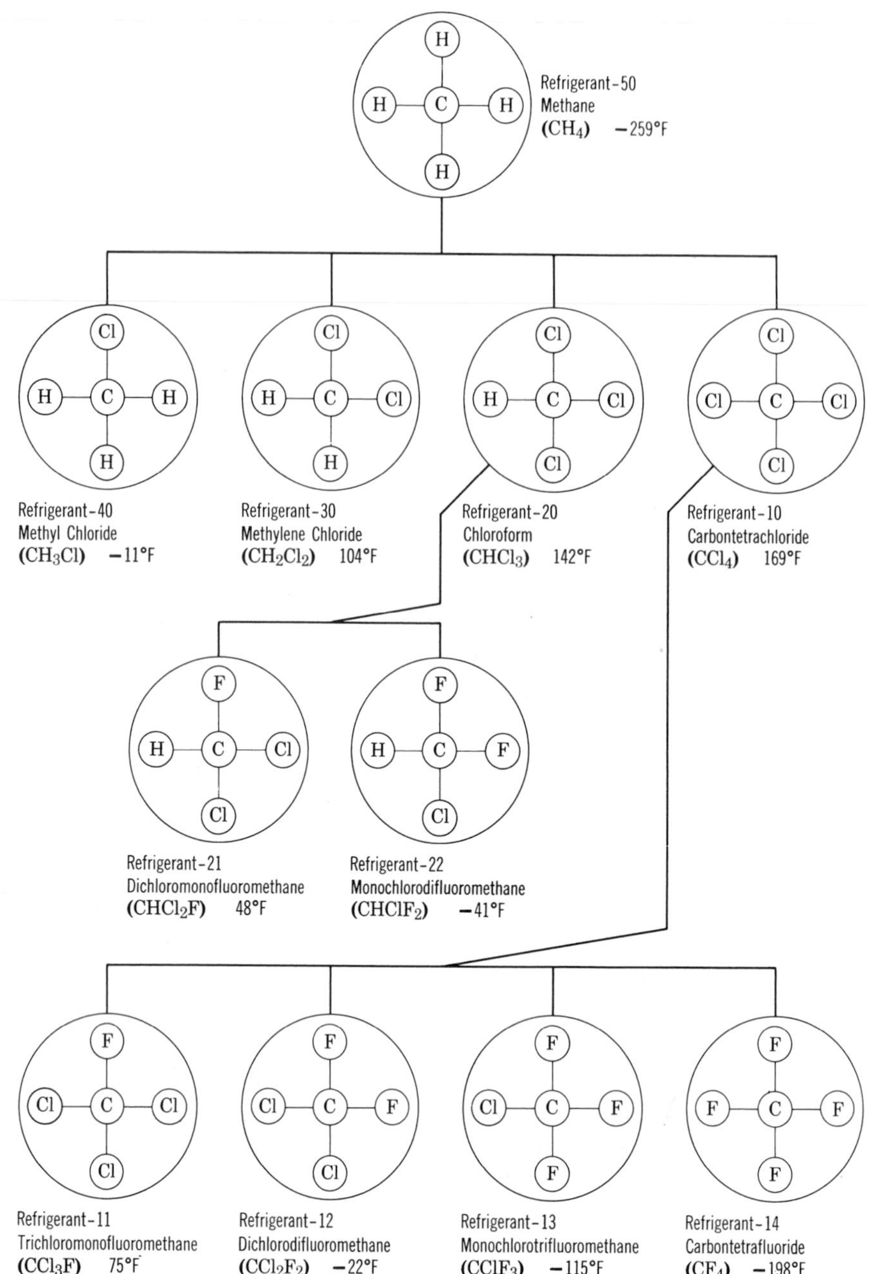

Fig. 16-1 Methane series refrigerants.

monochlorotrifluoromethane ($CClF_3$), and carbontetrafluoride (CF_4), respectively. In the same order, the ASHRAE refrigerant standard number designations for these compounds are Refrigerants-11, 12, 13, and 14, the last figure in the numbers being an indication of the number of fluorine atoms in the molecule.

The molecular structure of Refrigerants-21 and 22, which are also fluorocarbons of the methane series, is shown in Fig. 16-1. Notice the presence of the hydrogen atom in each of these two compounds, an indication that they are derivatives of the chloroform molecule rather than the carbontetrachloride molecule.

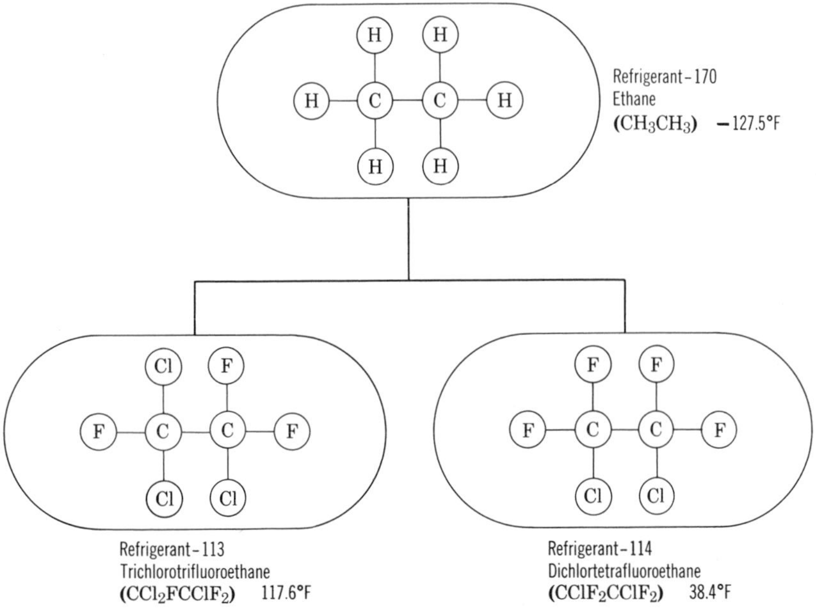

Fig. 16-2 Ethane series refrigerants.

Figure 16-2 shows the molecular structure of Refrigerants-113 and 114, the only two fluorocarbons of the ethane series which have been used in any substantial quantity. The presence of the two carbon atoms identifies the basic molecule as ethane, rather than methane, which has only one carbon atom.

The individual characteristics of these and other refrigerants are discussed in the following sections.

16-8. The Effect of Moisture

It is a well-established fact that moisture will combine in varying degrees with most of the commonly used refrigerants, causing the formation of highly corrosive compounds (usually acids) which will react with the lubricating oil and with other materials in the system, including metals. This chemical action often results in pitting and other damage to valves, seals, bearing journals, cylinder walls, and other polished surfaces. It may also cause deterioration of the lubricating oil and the formation of metallic and other sludges which tend to clog valves and oil passages, score bearing surfaces, and otherwise reduce the life of the equipment. Moisture corrosion also contributes to compressor valve failure and, in hermetic motor-compressors, often causes breakdown of the motor winding insulation, which results in shorting or grounding of the motor.

Although a completely moisture-free refrigerating system is not possible, good refrigerating practice demands that the moisture content of the system be maintained below the level which will produce harmful effects in the system. The minimum moisture level which will produce harmful effects in a refrigerating system is not clearly defined and will vary considerably, depending upon the nature of the refrigerant, the quality of the lubricating oil, and the operating temperatures of the system, particularly the compressor discharge temperature.

Moisture in a refrigerating system may exist as "free water" or it may be in solution with the refrigerant. When moisture is present in the system in the form of free water, it will freeze into ice in the refrigerant control and/or in the evaporator, provided that the temperature of the evaporator is maintained below the freezing point of the water. Naturally, the formation of ice in the refrigerant control orifice will prevent the flow of liquid refrigerant through that part and render the system inoperative until such time that the ice melts and flow through the control is restored. In such cases, refrigeration is usually intermittent as the flow of liquid is started and stopped by alternate melting and freezing of the ice in the control orifice.

Since free water exists in the system only when the amount of moisture in the system exceeds the amount that the refrigerant can hold in solution, freeze-ups are nearly always an indication that the moisture content of the system is above the minimum level that will produce corrosion. On the other hand, the mere absence of freeze-ups cannot be taken to mean that the moisture content of the system is necessarily below the level which will cause corrosion, since corrosion can occur with some refrigerants at levels well below those which will result in free water. Too, it must be recognized that freeze-ups do not occur in air conditioning systems or in any other system where the evaporator temperature is above the freezing point of water. For this reason, high temperature systems are often more subject to moisture corrosion than are systems operating at lower evaporator temperatures, since relatively large quantities of moisture can go unnoticed in such systems for relatively long periods of time.

Since the ability of an individual refrigerant to hold moisture in solution decreases as the temperature decreases, it follows that the moisture content in low temperature systems must be maintained at a very low level in order to avoid freeze-ups. Hence, moisture corrosion in low temperature systems is usually at a minimum.

The various refrigerants differ greatly both as to the amount of moisture they will hold in solution and as to the effect that the moisture has upon them. For example, the straight hydrocarbons, such as propane, butane, ethane, etc., absorb little if any moisture. Therefore, any moisture contained in such systems will be in the form of free water and will make its presence known by freezing out in the refrigerant control. Since this moisture must be removed immediately in order to keep the system operative, moisture corrosion will not usually be a problem when these refrigerants are used.

Ammonia, on the other hand, has an affinity for water and therefore is capable of absorbing moisture in such large quantities that free water is seldom found in systems employing this refrigerant.

The combination of water and ammonia produces aqua ammonia, a strong alkali, which attacks nonferrous metals, such as copper and brass, but has little if any effect on iron or steel or any other materials in the system. For this reason, ammonia systems can be operated successfully even when relatively large amounts of moisture are present in the system.

The halocarbon refrigerants hydrolyze only slightly and therefore form only small amounts of acids or other corrosive compounds. As a general rule, corrosion will not occur in systems employing halocarbon refrigerants when the moisture content is maintained below the level which will cause freeze-ups, provided that high quality lubricating oils are used and that discharge temperatures are reasonably low.

16-9. Refrigerant-Oil Relationship

With a few exceptions, the oil required for lubrication of the compressor is contained in the crankcase of the compressor where it is subject to contact with the refrigerant. Hence, as already stated, the refrigerant must be chemically and physically stable in the presence of oil, so that neither the refrigerant nor the oil is adversely affected by the relationship.

Although some refrigerants, particularly sulfur dioxide and the halocarbons, react with the lubricating oil to some extent, under normal operating conditions the reaction is usually slight and therefore of little consequence, provided that a high quality lubricating oil is used and that the system is relatively clean and dry. However, when contaminants, such as air and moisture, are present in the system in any appreciable amount, chemical reactions involving the contaminants, the refrigerant, and the lubricating oil often occur which can result in decomposition of the oil, the formation of corrosive acids and sludges, copper plating, and/or serious corrosion of polished metal surfaces. High discharge temperatures greatly accelerate these processes, particularly oil decomposition, and often result in the formation of carbonaceous deposits on discharge valves and pistons and in the compressor head and discharge line. This condition is aggravated by the use of poorly refined lubricating oils containing a high percentage of unsaturated hydrocarbons, the latter being very unstable chemically.

Because of the naturally high discharge temperature of Refrigerant-22 (see Table 16-8), breakdown of the lubricating oil, accompanied by motor burnouts, is a common problem with hermetic motor-compressor units employing this refrigerant, particularly when used in conjunction with air-cooled condensers and long suction lines.

Copper plating of various compressor parts is often found in systems employing halocarbon refrigerants. The parts usually affected are the highly polished metal surfaces which generate heat, such as seals, pistons, cylinder walls, bearing surfaces, and valves. The exact cause of copper plating has not been definitely determined, but considerable evidence does exist that moisture and poor quality lubricating oils are contributing factors.

Because copper is never used with ammonia, copper plating is not found in ammonia systems.

In any event, regardless of the nature of and/or the cause of unfavorable reactions between the refrigerant and the lubricating oil, these disadvantages can be greatly minimized or eliminated by the use of high quality lubricating oils, having low "pour" and/or "floc" points (see Section 18-16), by maintaining the system relatively free of contaminants, such as air and moisture, and by designing the system so that discharge temperatures are reasonably low.

16-10. Oil Miscibility

With regard to the refrigerant-oil relationship, one important characteristic which differs for the various refrigerants is oil miscibility, that is, the ability of the refrigerant to be dissolved into the oil and vice versa.

With reference to oil miscibility, refrigerants may be divided into three groups: (1) those which are miscible with oil in all proportions under conditions found in the refrigerating system, (2) those which are miscible under conditions normally found in the condensing section, but separate from the oil under the conditions normally found in the evaporator section, and (3) those which are not miscible with oil at all (or only very slightly so) under conditions found in the system.

As to whether or not oil miscibility is a desirable property in a refrigerant there is some disagreement. In any event, the fact of oil miscibility, or the lack of it, is not usually a major factor in the selection of the refrigerant. However, since it greatly influences the design of the compressor and other system components, including the refrigerant piping, the degree of oil miscibility is an important refrigerant characteristic and therefore should be considered in some detail.

With regard to the oil, one of the principal effects of an oil miscible refrigerant is to dilute the oil in the crankcase of the compressor, thereby lowering the viscosity (thinning) of the oil and reducing its lubricating qualities. To compensate for refrigerant dilution, the compressor lubricating oil used in conjunction with oil-miscible refrigerants should have a higher initial viscosity than that used for similar duty with nonmiscible refrigerants.

Viscosity may be defined as a measure of fluid friction or as a measure of the resistance that a fluid offers to flow. Thin, low viscosity fluids will flow more readily than thicker, more viscous fluids. To provide adequate lubrication for the compressor, the viscosity of the lubricating oil must be maintained within certain limits. If the viscosity of the oil is too low, the oil will not have sufficient body to form a protective film between the various rubbing surfaces and keep them separated. On the other hand, if the viscosity of the oil is too high, the oil will not have sufficient fluidity to penetrate between the rubbing surfaces, particularly where tolerances are close. In either case, lubrication of the compressor will not be adequate.

Any oil circulating through the system with the refrigerant will have an adverse affect on the efficiency and capacity of the system, the principal reason being that the oil tends to adhere to and to form a film on the surface of the condenser and evaporator tubes, thereby lowering the heat transfer capacity of these two units. Since the oil becomes more viscous and tends to congeal as the temperature is reduced, the problem with oil is greatest in the evaporator and becomes more acute as the temperature of the evaporator is lowered.

Since the only reason for the presence of oil in the refrigerating system is to lubricate the compressor, it is evident that the oil will best serve its function when confined to the compressor and not allowed to circulate with the refrigerant through other parts of the system. However, since, with few exceptions, the system refrigerant unavoidably comes into contact with the oil in the compressor, a certain amount of oil in the form of small particles will be entrained in the refrigerant vapor and carried over through the discharge valves into the discharge line. If the oil is not removed from the vapor at this point, it will pass into the condenser and liquid receiver from where it will be carried to the evaporator by the liquid refrigerant. Obviously, in the interest of system efficiency and in order to maintain the oil in the crankcase at a constant level, some provision must be made for removing this oil from the system and returning it to the crankcase where it can perform its lubricating function.

The degree of difficulty experienced in bringing about the return of oil to the crankcase depends primarily on three factors: (1) the oil miscibility of the refrigerant, (2) the type of evaporator used, and (3) the evaporator temperature.

When an oil-miscible refrigerant is employed, the problem of oil return is greatly simplified by the fact that the oil remains in solution with the refrigerant. This permits the oil to be carried along through the system by the refrigerant and, subsequently, to be returned to the crankcase through the suction line, provided that the evaporator and the refrigerant piping are properly designed.

Unfortunately, when nonmiscible refrigerants are used, once the oil passes into the condenser, the return of the oil to the crankcase is not so easily accomplished. The reason for this is that, except for a small amount of mechanical mixing, the refrigerant and the oil will remain separate, so that only a small portion of the oil is actually carried along with the refrigerant. For example, in the case of ammonia, which is lighter than oil, a large percentage of the oil will separate from the liquid ammonia and settle out at various low points in the system. For this reason, oil drains should be provided at the bottom of all receivers, evaporators, accumulators, and other vessels containing liquid ammonia, and provisions should be made for draining the oil from these points, either continuously or periodically, and returning it to the crankcase. This may be accomplished manually or automatically.

When flooded-type evaporators are used, the refrigerant velocity will not usually be sufficient to permit the refrigerant vapor to entrain the oil and carry it over into the suction line and back to the crankcase. Hence, even with oil miscible refrigerants, where flooded-type evaporators are employed, it is often necessary to make special provisions for oil return. The methods used to ensure the continuous return of the oil from the evaporator to the crankcase in such cases are described in Chapter 19.

Since the oil acts to lubricate the refrigerant flow control and other valves which may be in the system, the circulation of a small amount of oil with the refrigerant is not ordinarily objectionable. However, because of the adverse effect on system capacity, the amount of oil should be kept to a practical minimum. Too, since the oil in circulation comes initially from the compressor crankcase, an excessive amount in circulation may cause the oil level in the crankcase to fall below the minimum level required for adequate lubrication of the compressor parts.

In order to minimize the circulation of oil, an oil separator or trap is sometimes installed in the discharge line between the compressor and the condenser (see Section 19-12).

As a general rule, discharge line oil separators should be employed in any system where oil return is likely to be inadequate and/or where the amount of oil in circulation is apt to be excessive or to cause an undue loss in system capacity and efficiency. Specifically, discharge line oil separators are recommended for all systems employing nonmiscible refrigerants (or refrigerants which are not oil miscible at the evaporator conditions), not only because of the difficulty experienced in returning the oil from the evaporator to the crankcase but also because the presence of even small amounts of oil in the evaporators of such systems will usually cause considerable loss of evaporator efficiency and capacity.

The same thing is usually true for systems employing miscible refrigerants when the evaporator temperature is below 0°F. Oil separators are recommended also for all systems using flooded evaporators, since oil return from this type of evaporator is apt to be inadequate because of low refrigerant velocities.

Although oil separators are very effective in removing oil from the refrigerant vapor, they are not 100% efficient. Therefore, even though an oil separator is used, some means must still be provided for returning to the crankcase the small amount of oil which will always pass through the separator and find its way into other parts of the system. Too, since oil separators can often cause serious problems in the system if they are not properly installed, the use of oil separators should ordinarily be limited to those systems where the nature of the refrigerant or the particular design of the system requires their use. Oil separators are discussed in more detail in Chapter 19.

16-11. Leak Detection

Leaks in a refrigerating system may be either inward or outward, depending on whether the pressure in the system at the point of leakage is above or below atmospheric pressure. When the pressure in the system is above atmospheric at the point of leakage, the refrigerant will leak from the system to the outside. On the other hand, when the pressure in the system is below atmospheric, there is no leakage of refrigerant to the outside, but air and moisture will be drawn into the system. In either case, the system will usually become inoperative in a very short time. However, as a general rule, outward leaks are less serious than inward ones, usually requiring only that the leak be found and repaired and that the system be recharged with the proper amount of refrigerant. In the case of inward leaks, the air and moisture drawn into the system increase the discharge pressure and temperature and accelerates the rate of corrosion. The presence of moisture in the system may also cause freeze-up of the refrigerant control. Furthermore, after the leak has been located and repaired, the system must be completely evacuated and dehydrated before it can be placed in operation. A refrigerant drier should also be installed in the system.

The necessity of maintaining the system free of leaks demands some convenient means for checking a new system for leaks and for detecting leaks if and when they occur in systems already in operation. New systems should be checked for leaks under both vacuum and pressure.

One method of leak detection universally used with all refrigerants employs a relatively viscous soap solution which is relatively free of bubbles. The soap solution is first applied to the pipe joint or other suspected area and then examined with the help of a strong light. The formation of bubbles in the soap solution indicates the presence of a leak. For adequate testing with a soap solution, the pressure in the system should be 50 psig or higher.

The fact that sulfur and ammonia vapors produce a dense white smoke (ammonia sulfite) when they come into contact with one another provides a convenient means of checking for leaks in ammonia systems. To check for leaks in an ammonia system, a sulfur candle is held near, but not in contact with, all pipe joints and other suspected areas. A leak is indicated when the sulfur candle gives off a white smoke. Dampened phenophthalein paper, which turns red on contact with ammonia vapor, may also be used to detect ammonia leaks.

A halide torch is often used to detect leaks in systems employing any of the halocarbon refrigerants. The halide torch consists of a copper element which is heated by a flame. Air to support combustion is drawn in through a rubber tube, one end of which is attached to the torch. The free end of the tube is passed around all suspected areas. The presence of a halocarbon vapor is indicated when the flame changes from its normal color to a bright green or purple. The halide torch should be used only in well-ventilated spaces.

For carbon dioxide and the straight hydrocarbons, the only method of leak detection is the soap solution previously mentioned.

16-12. Ammonia

Ammonia is the only refrigerant outside of the fluorocarbon group that is being used to any great extent at the present time. Although ammonia is toxic and also somewhat flammable and explosive under certain conditions, its excellent thermal properties make it an ideal refrigerant for ice plants, packing plants, skating rinks, large cold storage facilities, etc., where experienced operating personnel are usually on duty and where its toxic nature is of little consequence.

Ammonia has the highest refrigerating effect per pound of any refrigerant, which, despite a rather high specific volume in the vapor state, makes possible a high refrigerating capacity with a relatively small piston displacement.

The boiling point of ammonia at standard atmospheric pressure is −28°F (−2.22°C). The evaporator and condenser pressures at standard ton conditions of 5°F (−15°C) and 86°F (30°C) are 34.27 psia (2.37 bar) and 169.2 psia (11.67 bar), respectively, which are moderate, so that lightweight materials can be used in the construction of the refrigerating equipment. However, the adiabatic discharge temperature is relatively high, being 210°F (98.89°C) at standard ton conditions, which makes water cooling of the compressor head and cylinders desirable. Too, high suction superheats should be avoided in ammonia systems.

Although pure anhydrous ammonia is noncorrosive to all metals normally used in refrigerating systems, in the presence of moisture, ammonia becomes corrosive to nonferrous metals, such as copper and brass. Obviously, these metals should never be used in ammonia systems.

Ammonia is not oil miscible and therefore will not dilute the oil in the compressor crankcase. However, provisions must be made for the removal of oil from the evaporator and an oil separator should be used in the discharge line of all ammonia systems.

Ammonia systems may be tested for leaks with sulfur candles, which give off a dense white smoke in the presence of ammonia vapor, or by applying a thick soap solution around the pipe joints, in which case a leak is indicated by the appearance of bubbles in the solution.

Ammonia is readily available almost anywhere and is by far the least expensive of any of the commonly used refrigerants. These two considerations, along with its chemical stability, affinity for water, and nonmiscibility with oil, make ammonia an ideal refrigerant for use in large systems where toxicity is not a factor. Because of its relatively high heat transfer coefficient and the consequent improvement in the refrigerant side heat transfer rate, ammonia is particularly suitable for large liquid chilling installations. Ammonia is used with open-type reciprocating, rotary, and centrifugal compressors.

16-13. Refrigerant-11

Refrigerant-11 is a fluorocarbon of the methane series and has a boiling point at atmospheric pressure of 74.7°F (23.7°C). Operating pressures at standard ton conditions are 2.94 psia (0.2 bar) and 18.19 psia (1.25 bar). Because of these low operating pressures and the relatively large compressor displacement required, R-11 is employed with centrifugal compressors, mainly in air conditioning systems for small office buildings, factories, department stores, theaters, etc. R-11 is widely used as a secondary refrigerant and as a solvent.

Like other fluorocarbon refrigerants, R-11 is noncorrosive, nontoxic, and nonflammable but dissolves natural rubber. A halide torch may be used for leak detection.

16-14. Refrigerant-12

Refrigerant-12 (CCl_2F_2) is probably the most widely used refrigerant at the present time. It is a completely safe refrigerant in that it is nontoxic, nonflammable, and nonexplosive. Furthermore, it is a highly stable compound which is difficult to break down even under extreme operating conditions. However, if brought into contact with an open flame or with an electrical heating element, Refrigerant-12 will decompose into products which are highly toxic (see Section 16-3).

Along with its safe properties, the fact that Refrigerant-12 condenses at moderate pressures under normal atmospheric conditions and has a boiling temperature of −21.6°F (−29.8°C) at atmospheric pressure makes it a suitable refrigerant for use in high, medium, and low temperature applications and with all three types of compressors. When employed in conjunction with multistage centrifugal type compressors, Refrigerant-12 has been used to cool brine to temperatures as low as −110°F (−80°C).

The fact that Refrigerant-12 is oil miscible under all operating conditions not only simplifies the problem of oil return but also tends to increase the efficiency and capacity of the system in that the solvant action of the refrigerant maintains the evaporator and condenser tubes relatively free of oil films which otherwise would tend to reduce the heat transfer capacity of these two units.

Although the refrigerating effect per pound for Refrigerant-12 is relatively small as compared to that of some of the other popular refrigerants, this is not necessarily a serious disadvantage. In fact, in small systems, the greater weight of Refrigerant-12 which must be circulated is a decided advantage in that it permits closer control of the liquid. In larger systems, the disadvantage of the low latent heat value is offset somewhat by a high vapor density, so that the compressor displacement required per ton of refrigeration is not much greater than that required for the other common refrigerants. The power required per ton of capacity also compares favorably with that required for other commonly used refrigerants.

A halide torch is used for leak detection.

16-15. Refrigerant-13

Refrigerant-13 ($CClF_3$) was developed for and is being used in ultra-low temperature applications, usually in the low stage of a two or three stage cascade system.

The boiling temperature of Refrigerant-13 is −144.5°F (−98°C) at atmospheric pressure. Evaporator temperatures down to −150°F (−100°C) are practical. The critical temperature is 83.9°F (28.9°C). Since condensing pressure and the compressor displacement required are both moderate, Refrigerant-13 is suitable for use with all three types of compressors.

Refrigerant-13 is a safe refrigerant. It is not miscible with oil. A halide torch may be used for leak detection.

16-16. Refrigerant-22

Refrigerant-22 ($CHClF_2$) has a boiling point at atmospheric pressure of −41.4°F (−40.8°C). Developed originally as a low temperature refrigerant, it has been used extensively in domestic and farm freezers and in commercial and industrial low temperature systems down to evaporator temperatures as low as −125°F (−87°C). Its primary use today is in packaged air conditioners, where, because of space limitations, the relatively small compressor displacement required is a decided advantage.

Both the operating pressures and the adiabatic discharge temperature are higher for Refrigerant-22 than for Refrigerant-12. Power requirements are approximately the same.

Because of the high discharge temperatures experienced with Refrigerant-22, suction superheat should be kept to a minimum, particularly where hermetic motor-compressors are employed. In low temperature applications, where compression ratios are likely to be high, water cooling of the compressor head and cylinders is recommended in order to avoid overheating of the compressor. Air-cooled condensers used with Refrigerant-22 should be generously sized.

Although miscible with oil at temperatures found in the condensing section, Refrigerant-22 will often separate from the oil in the evaporator. The exact temperature at which separation occurs varies considerably with the type of oil and the amount of oil mixed with the refrigerant. However, no difficulty is usually experienced with oil return from the evaporator when a properly designed serpentine evaporator is used and when the suction piping is properly designed. When flooded evaporators are employed, oil separators should be used and special provisions should be made to insure the return of oil from the evaporator. Oil separators should always be used on low temperature applications.

The principal advantage of Refrigerant-22 over Refrigerant-12 is the smaller compressor displacement required, being approximately 60% of that required for Refrigerant-12. Hence, for a given compressor displacement, the refrigerating capacity is approximately 60% greater with Refrigerant-22 than with Refrigerant-12. Too, refrigerant pipe sizes are usually smaller for Refrigerant-22 than for Refrigerant-12.

The ability of Refrigerant-22 to absorb moisture is considerably greater than that of Refrigerant-12 and therefore less trouble is experienced with freeze-ups in Refrigerant-22 systems. Although some consider this to be an advantage, the advantage gained is questionable, since any amount of moisture in a refrigerating system is undesirable.

Being a fluorocarbon, Refrigerant-22 is a safe refrigerant. A halide torch may be used for leak detection.

16-17. Refrigerant-113

Refrigerant-113 (CCl_2FCClF_2) boils at 117.6°F (47.5°C) under atmospheric pressure. Operating pressures at standard ton conditions are 0.9802 psia (0.068 Bar) and 7.86 psia (0.204 Bar), respectively. Although the compressor displacement per ton is somewhat high (100.76 ft^3/min/ton at standard ton conditions), the power required per ton compares favorably with other common refrigerants. The low operating pressures and the large displacement required necessitate the use of a centrifugal type compressor.

Although used mainly in comfort air conditioning applications, it is also employed in industrial process water and brine chilling down to 0°F (−18°C).

Refrigerant-113 is a safe refrigerant. A halide torch may be used for leak detection.

16-18. Refrigerant-114

Refrigerant-114 (CCl_2CClF_2) has a boiling point of 38.4°F (3.6°C) under atmospheric pressure. Evaporating and condensing pressures at standard ton conditions are 6.75 psia (0.89 bar) and 36.27 psia (2.5 bar), respectively. The compressor displacement required is relatively low for a low pressure refrigerant (19.59 ft^3/(min)(ton) at standard conditions) and the power required compares favorably with that required by other common refrigerants.

Refrigerant-114 is used with centrifugal compressors in large commercial and industrial air conditioning installations and for industrial process water chilling down to −70°F (−57°C). It is also used with vane-type rotary compressors in domestic refrigerators and in small drinking water coolers.

Like Refrigerant-22, Refrigerant-114 is oil miscible under conditions found in the condensing section, but separates from oil in the evaporator. However, because of the type of equipment used with Refrigerant-114 and the conditions under which it is used, oil return is not usually a problem.

Refrigerant-114 is a safe refrigerant. A halide torch may be used for leak detection.

16-19. Refrigerant-500

Refrigerant-500 is an azeotropic mixture* of Refrigerant-12 (73.8% by weight) and Refrigerant-152a (26.2%). It has a boiling point at atmospheric pressure of −28°F (−33°C). Evaporator and condenser pressures at standard ton conditions are 16.4 psig and 113.4 psig, respectively. Although the power requirements of Refrigerant-500 are approximately the same as those for Refrigerants-12 and 22, the compressor displacement required is greater than that required for Refrigerant-22, but somewhat less than that required for Refrigerant-12.

The principal advantage of Refrigerant-500 lies in the fact that its substitution for Refrigerant-12 results in an increase in compressor capacity of approximately 18%. This makes it possible to use the same direct connected compressor (as in a hermetic motor-compres-

* An azeotropic mixture is a mixture of two or more liquids, which, when mixed in precise proportions, form a compound having a boiling temperature which is independent of the boiling temperatures of the individual liquids.

sor unit) on either 50 or 60 hertz power with little or no change in the refrigerating capacity or in the power requirements.

It will be shown in Chapter 21 that the speed of an alternating current motor varies in direct proportion to the cycle frequency. Therefore, an electric motor operating on 50 hertz power will have only five-sixths of the speed it has when operating on 60 hertz power. For this reason, the displacement of a direct connected compressor is reduced approximately 18% when a change is made from 60 to 50 hertz power. Since the increase in capacity per unit of displacement accruing from the substitution of Refrigerant-500 for Refrigerant-12 is almost exactly equal to the loss of displacement suffered when changing for 60 to 50 hertz power, the same motor-compressor assembly is made suitable for use with both frequencies by the simple expedient of changing refrigerants.

16-20. Refrigerant-502

Refrigerant-502 is an azeotropic mixture of 48.8% by mass of R-22 and 51-2% of R-115. Initially developed as a low-temperature refrigerant to replace R-22 in some low-temperature, high-compression ratio applications, R-502 has been widely employed in the frozen and cold storage temperature ranges and in some comfort cooling applications, particularly where heat pumps are employed. The particular advantage of R-502 over R-22 is its lower adiabatic discharge temperature of 99°F (37.2°C), as compared to 128°F (53.3°C) at standard ton conditions. However, both the compressor displacement and the power required per unit of refrigerating capacity are somewhat higher for R-502, as are the operating pressures, although the latter are still in the moderate range.

R-502 has a boiling temperature at standard barometric pressure of −49.8°F (−45.4°C) and a critical temperature of 179.9°F (91.78°C). It is nonflammable and nontoxic (classification 5a) and has a rather low oil miscibility. A halide torch may be used for leak detection.

16-21. Refrigerant-503

Refrigerant-503 is an azeotropic mixture of 40.1% by mass of R-23 and 59.9% of R-13. It has a boiling temperature at standard barometric pressure of −127.6°F (−88.7°C) and a critical temperature of 67.1°F (19.5°C).

R-503 is a relative new refrigerant that has found widespread use as a replacement for R-13 in the temperature range of −100°F (−73.3°C) to −150°F (−101°C). At an evaporator temperature of −120°F (−84.4°C) and a condensing temperature of 20°F (6.67°C), the compressor displacement required for R-503 is only approximately 64% of that required for R-13 for the same refrigerating capacity. However, condensing pressures are somewhat higher for R-503, being 330.5 psia at 20°F as compared to 240.4 psia for R-13.

R-503 is used in reciprocating compressors in the low stage of cascade systems, with R-12, R-22, or R-502 being employed in the high stage.

16-22. Straight Hydrocarbons

The straight hydrocarbons are a group of fluids composed in various proportions of the two elements hydrogen and carbon. Those having significance as refrigerants are methane, ethane, butane, propane, ethylene, and isobutane. All are extremely flammable and explosive. Too, since all act as anesthetics in varying degrees, they are considered mildly toxic. Although none of these compounds will absorb moisture to any appreciable extent, all are extremely miscible with oil under all conditions.

Although a few of the straight hydrocarbons (butane, propane, and isobutane) have been used in small quantities for domestic refrigeration, their use is ordinarily limited to special applications where an experienced attendant is on duty. Ethane, methane, and ethylene are employed to some extent in ultra-low temperature applications, usually in the lower stage of two and three stage cascade systems. However, even in these applications, it is likely that they will be replaced in the future by Refrigerants-13, 14, and 503.

Leak detection is by soap solution only.

16-23. Refrigerant Drying Agents

Refrigerant drying agents, called dessicants are frequently employed in refrigerating systems to remove moisture from the refrigerant. Some of the most commonly used dessicants are silica gel (silicon dioxide), activated alumina (aluminum oxide), and Drierite (anhydrous calcium sulfate). Silica gel and activated alumina are adsorption-type dessicants and are available in granular form. Drierite is an absorption type dessicant and is available in granular form and in cast sticks.

17

REFRIGERANT FLOW CONTROLS

17-1. Types and Function

There are six basic types of refrigerant flow controls: (1) the hand expansion valve, (2) the automatic expansion valve, (3) the thermostatic expansion valve, (4) the capillary tube, (5) the low pressure float, and (6) the high pressure float.

Regardless of type, the function of any refrigerant flow control is twofold: (1) to meter the liquid refrigerant from the liquid line into the evaporator at a rate commensurate with the rate at which vaporization of the liquid is occurring in the latter unit, and (2) to maintain a pressure differential between the high and low pressure sides of the system in order to permit the refrigerant to vaporize under the desired low pressure in the evaporator while at the same time condensing at a high pressure in the condenser.

17-2. Hand Expansion Valves

Hand expansion valves are hand-operated needle valves (Fig. 17-1). The rate of liquid flow through the valve depends on the pressure differential across the valve orifice and on the degree of valve opening, the latter being manually adjustable. Assuming that the pressure differential across the valve remains the same, the flow rate through a hand expansion valve will remain constant at all times without regard for either the evaporator pressure or the evaporator loading.

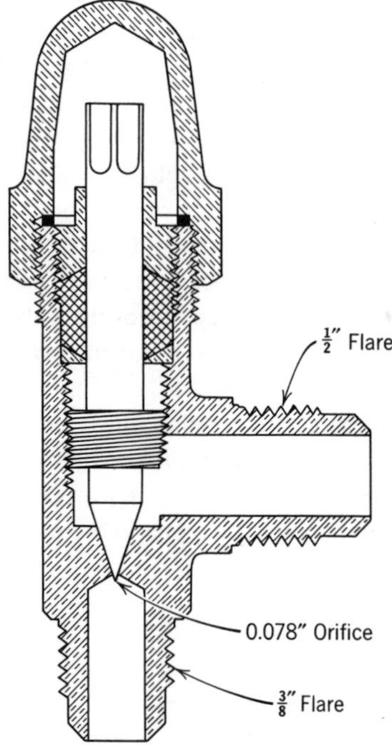

Fig. 17-1 Small capacity hand-expansion valve. (Courtesy Mueller Brass Company.)

The principal disadvantage of the hand expansion valve is that it is unresponsive to changes in the system load and therefore must be manually readjusted each time the load on the system changes in order to prevent either starving or overfeeding of the evaporator, depending upon the direction of the load shift. Too, the valve must be opened and closed manually each time the compressor is cycled on and off.

Obviously the hand expansion valve is suitable for use only on large systems where an operator is on duty and where the load on the system is relatively constant. When automatic control is desired and/or when the system is subject to frequent load fluctuations, some other type of refrigerant flow control is required.

At the present time, the principal use of the hand expansion valve is as an auxiliary refrigerant control installed in a by-pass line (Fig. 17-29). It is also frequently used to control the flow rate through oil bleeder lines (Fig. 19-12).

17-3. Automatic Expansion Valves

A schematic diagram of an automatic expansion valve is shown in Fig. 17-2. The valve consists mainly of a needle and seat, a pressure bellows or diaphragm, and a spring, the tension of the latter being variable by means of an adjusting screw. A screen or strainer is usually installed at the liquid inlet of the valve in order to prevent the entrance of foreign materials which may cause stoppage of the valve. The construction of a typical automatic expansion valve is shown in Fig. 17-3

The automatic expansion valve functions to maintain a constant pressure in the evaporator by flooding more or less of the evaporator surface in response to changes in the evaporator

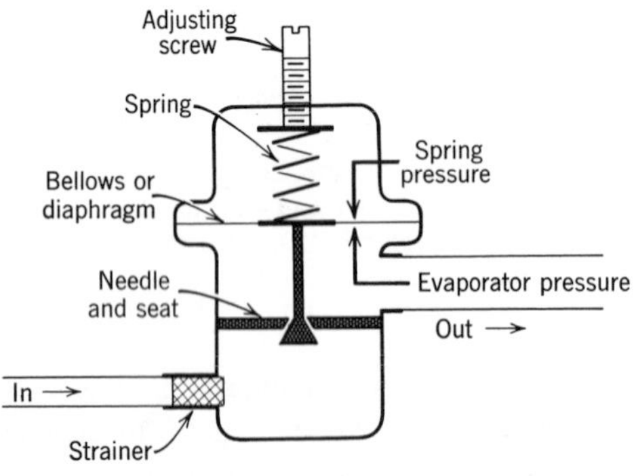

Fig. 17-2 Schematic diagram of automatic expansion valve.

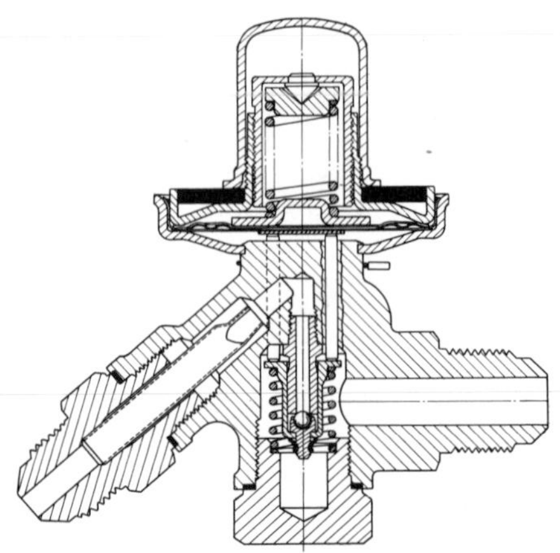

Fig. 17-3 Typical automatic expansion valve. (Courtesy Controls Company of America.)

load. The constant pressure characteristic of the valve results from the interaction of two opposing forces: (1) the evaporator pressure and (2) the spring pressure. The evaporator pressure, exerted on one side of the bellows or diaphragm, acts to move the valve in a closing direction, whereas the spring pressure, acting on the opposite side of the bellows or diaphragm, acts to move the valve in an opening direction. When the compressor is running, the valve functions to maintain the evaporator pressure in equilibrium with the spring pressure.

As the name implies, the operation of the valve is automatic and, once the tension of the spring is adjusted for the desired evaporator pressure, the valve will operate automatically to regulate the flow of liquid refrigerant into the evaporator so that the desired evaporator pressure is maintained, regardless of evaporator loading. For example, assume that the tension of the spring is adjusted to maintain a constant pressure in the evaporator of 10 psig. Thereafter, any time the evaporator pressure tends to fall below 10 psig, the spring pressure will exceed the evaporator pressure causing the valve to move in the opening direction, thereby increasing the flow of liquid to the evaporator and flooding more of the evaporator surface. As more of the evaporator surface becomes effective, the rate of vaporization increases and the evaporator pressure rises until equilibrium is established with the spring pressure. Should the evaporator pressure tend to rise above the desired 10 psig, it will immediately override the pressure of the spring and cause the valve to move in the closing direction, thereby throttling the flow of liquid into the evaporator and reducing the amount of effective evaporator surface. Naturally, this decreases the rate of vaporization and lowers the evaporator pressure until equilibrium is again established with the spring pressure.

It is important to notice that the operating characteristics of the automatic expansion valve are such that the valve will close off tightly when the compressor cycles off and remain closed until the compressor cycles on again. As previously described, vaporization continues in the evaporator for a short time after the compressor cycles off and, since the resulting vapor is not removed by the compressor, the pressure in the evaporator rises. Hence, during the off cycle, the evaporator pressure will always exceed the spring pressure and the valve will be tightly closed. When the compressor cycles on, the evaporator pressure will be immediately reduced below the spring pressure, at which time the valve will open and admit sufficient liquid to the evaporator to establish operating equilibrium between the evaporator and spring pressures.

The chief disadvantage of the automatic expansion valve is its relatively poor efficiency as compared to that of other refrigerant flow controls. In view of the evaporator-compressor relationship, it is evident that maintaining a constant pressure in the evaporator requires that the rate of vaporization in the evaporator be kept constant. To accomplish this necessitates severe throttling of the liquid in order to limit the amount of effective evaporator surface when the load on the evaporator is heavy and the heat transfer capacity per unit of evaporator surface is relatively high (Fig. 17-4*a*). As the load on the evaporator decreases and the heat transfer capacity per unit of evaporator surface is reduced, more and more of the evaporator surface must be flooded with liquid if a constant rate of vaporization is to be maintained (Fig. 17-4*b*). In fact, if the load on the evaporator is permitted to fall below a certain level, the

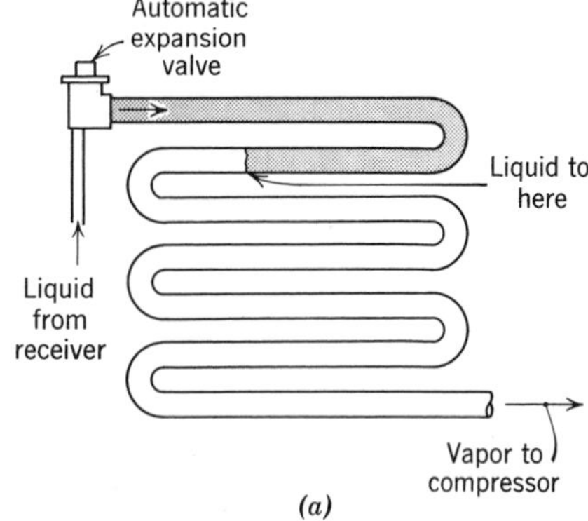

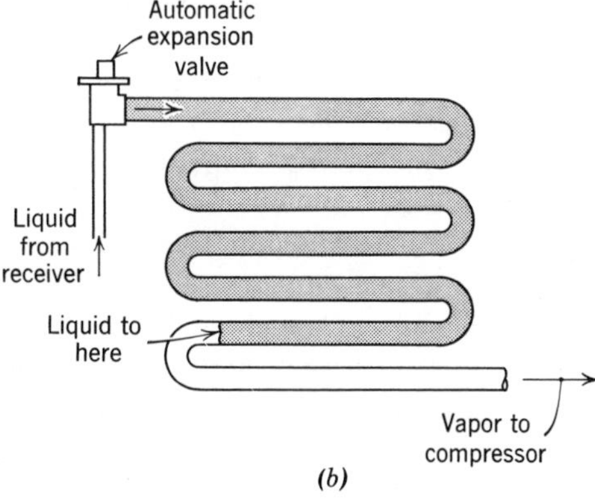

Fig. 17-4 Operating characteristics of the automatic expansion valve under varying load conditions. matic expansion valve under varying load conditions. (*a*) Heavy load conditions. (*b*) Minimum load conditions.

automatic expansion valve, in an attempt to keep the evaporator pressure up, can overfeed the evaporator to the extent that liquid will enter the suction line and be carried to the compressor where it may cause serious damage. However, in a properly designed system, overfeeding is not likely to occur, since the thermostat will usually cycle the compressor off before the space or product temperature is reduced to a level such that the load on the evaporator will fall below the critical point.

Since it permits only a small portion of the evaporator to be filled with liquid during periods when the load on the system is heavy, the constant pressure characteristic of the automatic expansion valve limits the capacity and efficiency of the refrigerating system at a

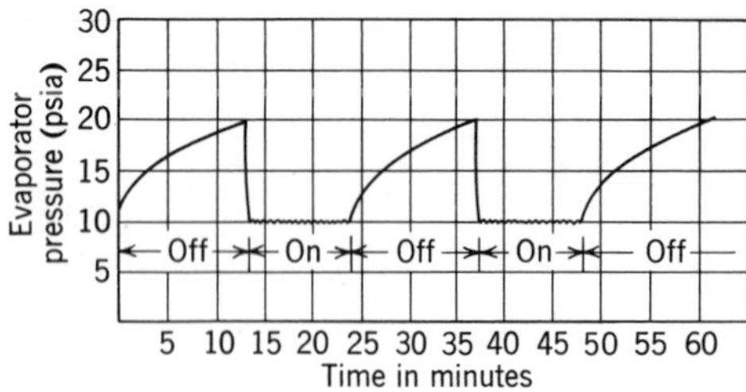

Fig. 17-5 Operating characteristics of the automatic expansion valve.

time when high capacity and high efficiency may be most desired. Too, because the evaporator pressure is maintained constant throughout the entire running cycle of the compressor, the valve must be adjusted for a pressure corresponding to the lowest evaporator temperature required during the entire running cycle (see Fig. 17-5). This results in some loss in compressor capacity and efficiency, since advantage cannot be taken of the higher suction temperatures which would ordinarily exist with a full-flooded evaporator during the early part of the running cycle.

Another disadvantage of the automatic expansion valve, which can also be attributed to its constant pressure characteristic, is that it cannot be used in conjunction with a low pressure motor control, since proper operation of the latter part depends on a rather substantial change in the evaporator pressure during the running cycle, a condition which obviously cannot be met when an automatic expansion valve is used as the refrigerant flow control.

In view of its poor efficiency under heavy load conditions, the automatic expansion valve is best applied only to small equipment having relatively constant loads, such as domestic refrigerators and freezers and small, retail ice-cream storage cabinets. However, even in these applications the automatic expansion valve is seldom used at the present time, having given way to other types of refrigerant flow controls which are more efficient and sometimes lower in cost.

17-4. Thermostatic Expansion Valve

Because of its high efficiency and its ready adaptability to any type of refrigeration application, the thermostatic expansion valve is probably the most widely used refrigerant control at the present time. Whereas the operation of the automatic expansion valve is based on maintaining a constant pressure in the evaporator, the operation of the thermostatic expansion valve is based on maintaining a constant degree of suction superheat at the evaporator outlet, a circumstance which permits the latter control to keep the evaporator completely filled with refrigerant under all conditions of system loading, without the danger of liquid slopover into the suction line. Because of its ability to provide full and effective use of all the evaporator surface under all load conditions, the thermostatic expansion valve is a particularly suitable refrigerant control for systems which are subject to wide and frequent variations in loading.

Figure 17-6 is a schematic diagram of a thermostatic expansion valve showing the principal parts of the valve, which are (1) a needle and seat, (2) a pressure bellows or diaphragm, (3) a fluid-charged remote bulb which is open to one side of the bellows or diaphragm through a capillary tube, and (4) a spring, the tension of which is usually adjustable by an adjusting screw. As in the case of the automatic expansion valve and all other refrigerant controls, a screen or strainer is usually installed at the liquid inlet of the valve to prevent the entrance of foreign material which may cause stoppage of the valve.

The characteristic operation of the thermostatic expansion valve results from the interaction of three independent forces, viz: (1) the evaporator pressure, (2) the spring pressure, and (3) the pressure exerted by the saturated liquid-vapor mixture in the remote bulb.*

As shown in Fig. 17-6, the remote bulb of the expansion valve is clamped firmly to the suction line at the outlet of the evaporator, where it is responsive to changes in the temperature of the refrigerant vapor at this point. Although

* With some exceptions which are discussed later, the fluid in the remote bulb is the refrigerant used in the system. Hence, the remote bulb of a thermostatic expansion valve employed on a Refrigerant-12 system would ordinarily be charged with Refrigerant-12.

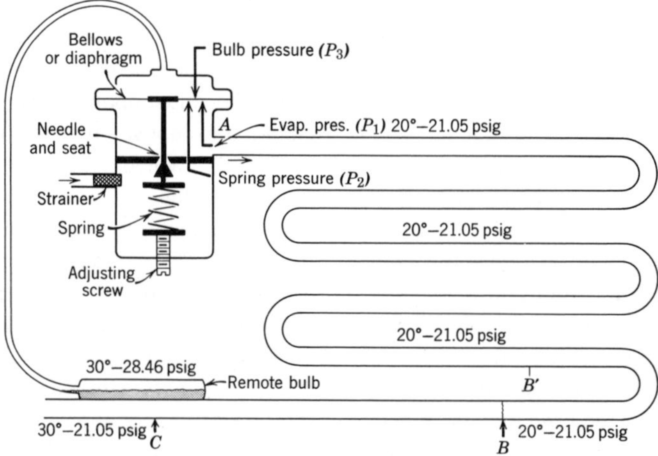

Fig. 17-6 Illustrating operating principle of conventional liquid-charged thermostatic expansion valve.

there is a slight temperature differential between the temperature of the refrigerant vapor in the suction line and the temperature of the saturated liquid-vapor mixture in the remote bulb, for all practical purposes the temperature of the two are the same and therefore it may be assumed that the pressure exerted by the fluid in the bulb is always the saturation pressure of the liquid-vapor mixture in the bulb corresponding to the temperature of the vapor in the suction line at the point of bulb contact.

Notice that the pressure of the fluid in the remote bulb acts on one side of the bellows or diaphragm through the capillary tube and tends to move the valve in the opening direction, whereas the evaporator pressure and the spring pressure act together on the other side of the bellows or diaphragm and tend to move the valve in a closing direction. The operating principles of the thermostatic expansion valve are best described through the use of an example.

With reference to Fig. 17-6, assume that Refrigerant-12 liquid is vaporizing in the evaporator at a temperature of 20°F so that the evaporator pressure (p_1) is 21.05 psig, the saturation pressure of Refrigerant-12 corresponding to a temperature of 20°F. Assume further that the tension of the spring is adjusted to exert a pressure (p_2) of 7.41 psi, so that the total pressure tending to move the valve in the closing direction is 28.46 psi, the sum of p_1 and p_2 (21.05 + 7.41). If the pressure drop in the evaporator is ignored, it can be assumed that the temperature and pressure of the refrigerant are the same throughout all parts of the evaporator where a liquid-vapor mixture of the refrigerant is present. However, at some point *B* near the evaporator outlet all the liquid will have vaporized from the mixture and the refrigerant at this point will be in the form of a saturated vapor at the vaporizing temperature and pressure. As the refrigerant vapor travels from point *B* through the remaining portion of the evaporator, it will continue to absorb heat from the surroundings, thereby becoming superheated so that its temperature is increased while its pressure remains constant. In this instance, assume that the refrigerant vapor is superheated 10° from 20 to 30°F during its travel from point *B* to the remote bulb location at point *C*. The saturated liquid-vapor mixture in the remote bulb, being at the same temperature as the superheated vapor in the line, will then have a pressure (p_3) of 28.46 psig, the saturation pressure of Refrigerant-12 at 30°F, which is exerted on the diaphragm through the capillary tube and which constitutes the total force tending to move the valve in the opening direction.

Under the conditions just described, the force tending to open the valve is exactly equal to the force tending to close the valve ($p_1 + p_2 = p_3$) and the valve will be in equilibrium.

The valve will remain in equilibrium until such time that a change in the degree of suction superheat unbalances the forces and causes the valve to move in one direction or the other.

By careful analysis of the foregoing example it can be seen that for the conditions described the valve will be in equilibrium when and only when the degree of superheat of the suction vapor at the remote bulb location is 10°F, which is exactly the amount required to offset the pressure exerted by the spring. Any change in the degree of suction superheat will cause the valve to move in a compensating direction in order to restore the required amount of superheat and reestablish equilibrium. For instance, if the degree of suction superheat becomes less than 10°F, the pressure in the remote bulb will be less than the combined evaporator and spring pressures and the valve will move toward the closed position, thereby throttling the flow of liquid into the evaporator until the superheat is increased to the required 10°F. On the other hand, if the superheat becomes greater than 10°F, the pressure in the remote bulb will exceed the combined evaporator and spring pressures and the valve will move toward the open position, thereby increasing the flow of liquid into the evaporator until the superheat is reduced to the required 10°F.

In all cases, the amount of superheat required to bring a thermostatic expansion valve into equilibrium depends upon the pressure setting of the spring. It is for this reason that the spring adjustment is called the "superheat adjustment." Increasing the tension of the spring increases the amount of superheat required to offset the spring pressure and bring the valve into equilibrium. A high degree of superheat is usually undesirable in that it tends to reduce the amount of effective evaporator surface. On the other hand, decreasing the spring tension reduces the amount of superheat required to maintain the valve in a condition of equilibrium and therefore tends to increase the amount of effective surface. However, if the valve superheat is set too low, the valve will lose control of the refrigerant to the extent that it will alternately "starve" and "overfeed" the evaporator, a condition often called "hunting." As a general rule, thermostatic expansion valves are adjusted for a superheat of 7° to 10° by the manufacturer. Since this superheat setting is ordinarily satisfactory for most applications, it should not be changed except when absolutely necessary.

Once the valve is adjusted for a certain superheat, the valve will maintain approximately that superheat under all load conditions, regardless of the evaporator temperature and pressure, provided that the capacity and operating range of the valve are not exceeded. For instance, in the preceding example, assume that because of an increase in system loading the rate of vaporization in the evaporator increases to the extent that all the liquid is vaporized by the time the refrigerant leaves point *B′*, rather than point *B*, in Fig. 17-6. The greater travel of the vapor before reaching point *C* will cause the superheat to exceed 10°F, in which case the increased bulb pressure resulting from the higher vapor temperature at point *C* will cause the valve to open wider and increase the flow of liquid to the evaporator, whereupon more of the evaporator surface will be filled with liquid so that the superheat is again reduced to the required 10°F. However, it is important to notice that when equilibrium is again established, the evaporator temperature and pressure will be higher than before because of the increased rate of vaporization. Furthermore, since the valve will maintain a constant superheat of approximately 10°F, the temperature of the vapor at point *C* will also be higher because of the increase in the evaporator temperature, as will the temperature and pressure of the fluid in the remote bulb.

Obviously, then, unlike the automatic expansion valve, the thermostatic expansion valve cannot be set to maintain a certain evaporator temperature and pressure, only a constant superheat. When a thermostatic expansion valve is used as a refrigerant control, the evaporator temperature and pressure will vary with the loading of the system, as described in Chapter 13. A typical internally equalized thermostatic expansion valve is shown in Fig. 17-7.

17-5. Externally Equalized Valves

Since the refrigerant undergoes a drop in pressure because of friction as it flows through the

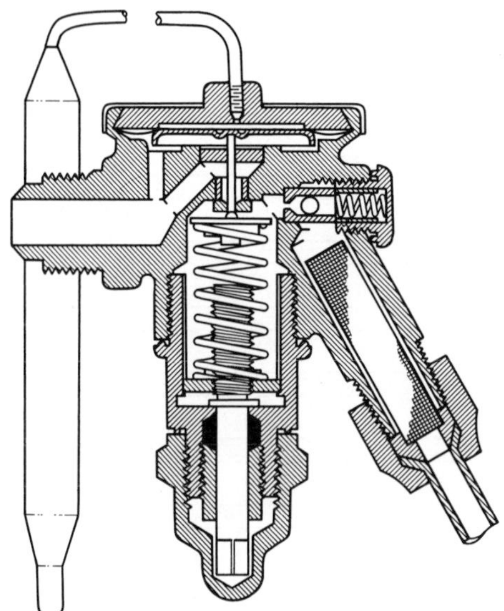

Fig. 17-7 Conventional liquid-charged, internally equalized thermostatic expansion valve. (Courtesy General Controls.)

evaporator, the saturation temperature of the refrigerant is always lower at the evaporator outlet than at the evaporator inlet. When the refrigerant pressure drop through the evaporator is relatively small, the drop in saturation temperature is also small and therefore of little consequence. However, when the pressure drop experienced by the refrigerant in the evaporator is of appreciable size, the saturation temperature of the refrigerant at the evaporator outlet will be considerably lower than that at the evaporator inlet, a circumstance which adversely affects the operation of the expansion valve in that it necessitates a higher degree of suction superheat in order to bring the valve into equilibrium. Since more of the evaporator surface will be needed to satisfy the higher superheat requirement, the net effect of the evaporator pressure drop, unless compensated for through the use of an external equalizer, will be to reduce seriously the amount of evaporator surface which can be used for effective cooling.

For example, assume that a Refrigerant-12 evaporator is fed by a standard, internally equalized thermostatic expansion valve and that the saturation pressure and temperature of the refrigerant at the evaporator inlet (point *A*) is 21.05 psig and 20°F, respectively, the former being the evaporator pressure (p_1) exerted on the diaphragm of the valve. If the valve spring is adjusted for a pressure (p_2) of 7.41 psi, a bulb pressure (p_3) of 28.46 psig (21.05 + 7.41) will be required for valve equilibrium.

If it is assumed that the refrigerant pressure drop in the evaporator is negligible, as in the previous example, the saturation pressure and temperature of the refrigerant at the evaporator outlet will be approximately the same as those at the evaporator inlet, 21.05 psig and 20°F, and the amount of suction superheat required for operation of the valve will be only 10° (30° − 20°), as shown in Fig. 17-6. On the other hand, assume now that in flowing through the evaporator, the refrigerant experiences a drop in pressure of 10 psi, in which case the saturation pressure at the evaporator outlet will be approximately 11 psig (21 − 10), 10 psi less than the inlet pressure. Since the saturation temperature corresponding to 11 psig is approximately 4°F, it is evident that a suction superheat of approximately 26°F will be required to provide the 30°F suction vapor temperature which is necessary at the point of bulb contact in order to bring the valve into equilibrium.

In order to satisfy the greater superheat requirement, vaporization of the liquid must be completed prematurely in the evaporator (point *B′* in Fig. 17-6) so that a considerable portion of the evaporator surface becomes relatively ineffective. Naturally, the loss of effective evaporator surface will materially reduce the overall capacity and efficiency of the system.

Although an external equalizer does not reduce the evaporator pressure drop in any way, it does permit the full and effective use of all of the evaporator surface. Notice in Fig. 17-8 that the externally equalized valve is so constructed that the evaporator pressure (p_1) which acts on the valve diaphragm is the evaporator outlet pressure rather than the evaporator inlet pressure. This is accomplished by completely isolating the valve diaphragm from the evaporator inlet pressure, while at the same time permitting the evaporator outlet pressure to be exerted on the diaphragm

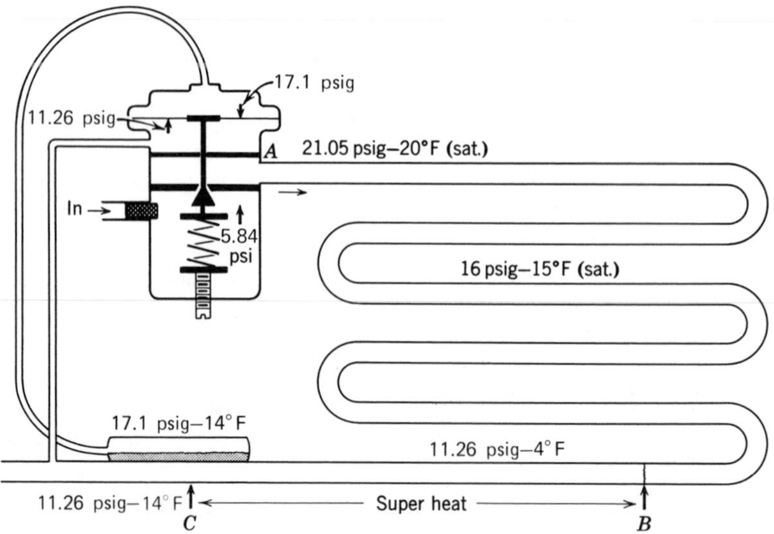

Fig. 17-8 Schematic diagram of externally equalized thermostatic expansion valve.

through a small diameter tube which is connected to the evaporator outlet or to the suction line 6 to 8 in. beyond the remote bulb location on the compressor side, as shown in Fig. 17-8. An expansion valve equipped with an external equalizer connection is shown in Fig. 17-9.

Since the evaporator pressure (p_1) exerted on the diaphragm of the externally equalized valve is the evaporator outlet pressure rather than the evaporator inlet pressure, the effect of the evaporator pressure drop is nullified to the extent that the degree of suction superheat required to operate the valve is approximately the same as when the evaporator pressure drop is negligible. Notice in Fig. 17-8 that the evaporator (outlet) pressure (p_1) exerted on the diaphragm is 11.26 psig, which, when added to the spring pressure (p_2) of 5.84 psi, constitutes a total pressure of 17.1 psi, tending to move the valve in the closing direction. Hence, a bulb pressure of 17.1 psig, corresponding to a saturation temperature of approximately 14°F, is required for equilibrium. Since the saturation temperature corresponding to the suction vapor pressure of 11.26 psig is 4°F, a suction superheat of only 10°F (14° − 4°) is necessary to provide valve equilibrium.

17-6. Pressure Limiting Valves

The propensity of a conventional liquid charged thermostatic expansion valve for keeping the evaporator completely filled with refrigerant, without regard for the evaporator temperature and pressure, has some disadvantages as well as advantages. Although this characteristic is desirable in that it insures full and efficient use of all the evaporator surface under all conditions of loading, it is, at the same time, undesirable in that it also permits overloading of the compressor driver because of excessive evaporator pressures and temperatures during periods of heavy loading.

Another disadvantage of the conventional thermostatic expansion valve is its tendency to open wide and overfeed the evaporator when the compressor cycles on, which in many cases permits liquid to enter the suction line with possible damage to the compressor. Overfeeding at start-up is caused by the fact that the evaporator pressure drops rapidly when the compressor is started and the bulb pressure remains high until the temperature of the bulb is cooled to the normal operating temperature by the suction vapor. Naturally, because of the high bulb pressure, the valve will be unbalanced in the open direction during this period and overfeeding of the evaporator will occur until the bulb pressure is reduced.

Fortunately, these operating difficulties can be overcome through the use of thermostatic expansion valves which have built-in pressure limiting devices. The pressure limiting devices

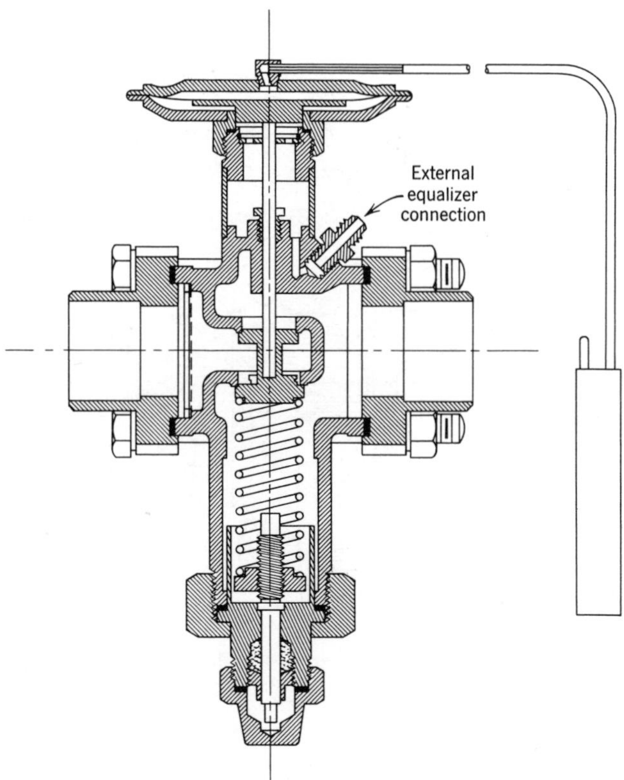

Fig. 17-9 Externally equalized thermostatic expansion valve. (Courtesy Sporian Valve Company.)

act to throttle the flow of liquid to the evaporator by taking control of the valve away from the remote bulb when the evaporator pressure rises to some predetermined maximum. Not only does this protect the compressor driver from overload during periods of heavy loading, it also tends to eliminate liquid flood-back to the compressor because of overfeeding at start-up.

The maximum operating pressure (MOP) of the expansion valve can be limited either by mechanical means or by the use of a gas charged remote bulb. The former is accomplished by placing a spring or a collapsible cartridge between the diaphragm and the valve stem or push-rods which actuate the valve pin. In the collapsible cartridge-type (Fig. 17-10), the cartridge, which is filled with a noncondensible gas, acts as a solid link between the diaphragm and the valve stem as long as the evaporator pressure is less than the pressure of the gas in the cartridge. Hence, control of the valve is vested in the remote bulb and the valve operates as a conventional thermostatic expansion valve as long as the evaporator pressure is less than the pressure of the gas in the cartridge. However, when the evaporator pressure exceeds the cartridge pressure, the cartridge collapses, thereby taking control of the valve away from the bulb and allowing the superheat spring to throttle the valve until the evaporator pressure is reduced below the cartridge pressure, at which time the cartridge will again act as a solid link, thereby returning control of the valve to the remote bulb.

The maximum evaporator pressure, called the maximum operating pressure (MOP) of the valve, depends on the pressure of the gas in the cartridge and can be changed simply by changing the valve cartridge. Cartridges are available for almost any desired maximum operating pressure.

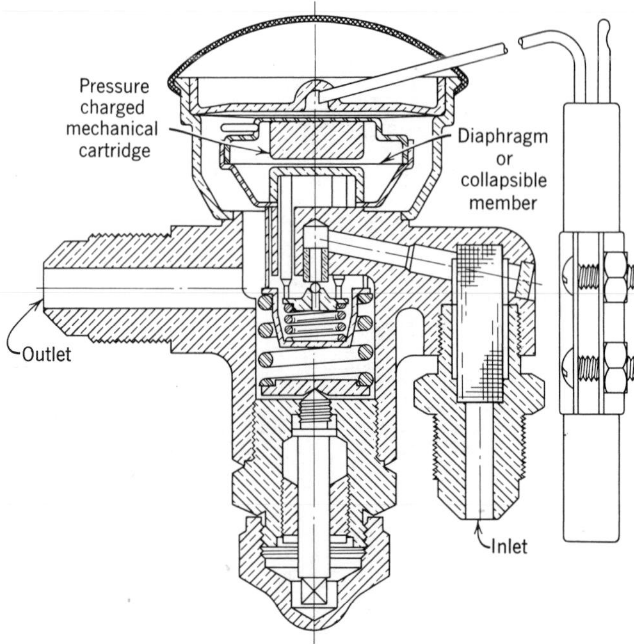

Fig. 17-10 Cartridge-type pressure limiting valve. (Courtesy Alco Valve Company.)

The operation of the spring-type pressure limiting valve (Fig. 17-11) is similar to that of the collapsible cartridge type in that the spring acts as a solid link between the valve diaphragm and the valve stem or push-rods whenever the pressure in the evaporator is less than the spring tension. When the evaporator pressure rises to a point where it exceeds the tension for which the spring is adjusted, the spring collapses and the flow of refrigerant to the evaporator is throttled until the evaporator pressure is again reduced below the spring tension. Obviously, the maximum operating pressure of the valve depends on the degree of spring tension, which in some cases is adjustable in the field.

In addition to the overload protection afforded by pressure limiting valves, they also tend to eliminate the possibility of liquid flood-back to the compressor at start-up. The fact that the evaporator pressure must be reduced below the MOP of the valve before the valve can open delays the valve opening sufficiently to permit the suction vapor to cool the remote bulb and reduce the bulb pressure before the valve opens. Hence, the valve does not open wide and overfeed the evaporator when the compressor is started.

The "pull-down" characteristics of a pressure limiting expansion valve as compared to those of a conventional expansion valve is shown in Fig. 17-12.

17-7. Gas-Charged Expansion Valves

The gas-charged thermostatic expansion valve is essentially a pressure limiting valve, the pressure limiting characteristic of the valve being a function of its limited bulb charge.

The remote bulb of the gas-charged expansion valve, like that of the liquid charged valve, is charged with the system refrigerant. However, whereas in the liquid charged valve, the remote bulb charge is sufficiently large to assure that a certain amount of liquid is always present in the remote bulb, in the gas charged valve the bulb charge is limited so that at some predetermined bulb temperature all the liquid will have vaporized and the bulb

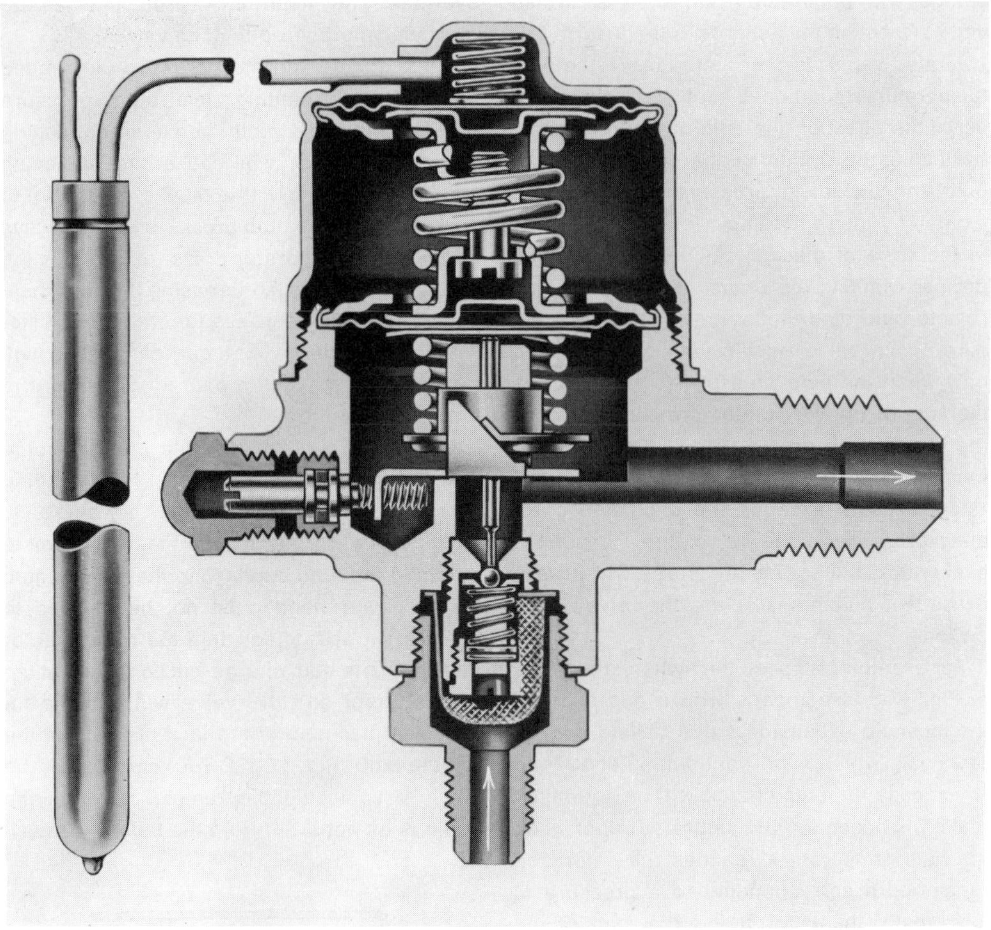

Fig. 17-11 Spring-type pressure limiting valve. (Courtesy Detroit Controls Division, American Radiator and Standard Sanitary Corporation.)

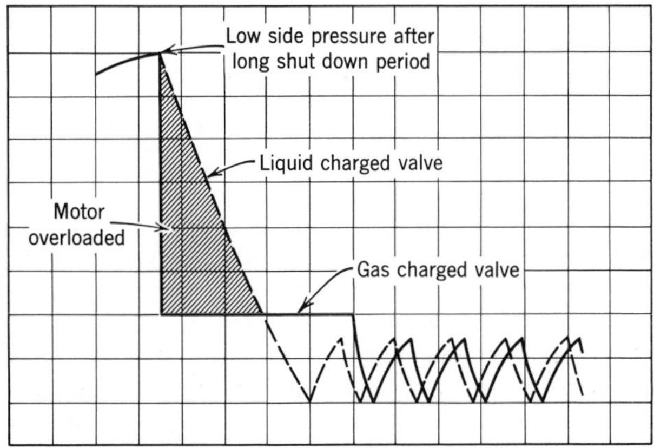

Fig. 17-12 Chart showing comparative performance of the gas charged and liquid charged valves during "pull-down." (Courtesy Detroit Controls Division, American Radiator and Standard Sanitary Corporation.)

charge will be in the form of a saturated vapor. Once the bulb charge is in the form of a saturated vapor, further increases in the bulb temperature (additional superheat) will have very little effect on the bulb pressure. Hence, by limiting the amount of charge in the remote bulb, the maximum pressure which can be exerted by the remote bulb is also limited. As in the case of mechanical pressure limiting devices, limiting the pressure exerted by the remote bulb also limits the evaporator pressure, since valve equilibrium is established only when the bulb pressure (p_3) is equal to the sum of the evaporator pressure (p_1) and the spring pressure (p_2), the latter pressure being constant. Therefore, any time the evaporator pressure exceeds the maximum operating pressure of the valve, the sum of the evaporator and spring pressures will always exceed the bulb pressure and the valve will be closed.

For example, suppose the system illustrated in Fig. 17-6 is equipped with a gas charged thermostatic expansion valve having a MOP of 25 psig with a superheat setting of 10°F, in which case the bulb charge will be so limited that it will become 100% saturated vapor when the bulb temperature reaches the saturation temperature corresponding to a pressure of 32.41 psig, the latter being the sum of the maximum evaporator pressure (25 psig) and the spring pressure equivalent to 10°F of superheat (7.41 psi). Once the bulb reaches this temperature, additional superheating of the suction vapor will have very little effect on the bulb pressure and therefore will not cause the valve to open wider.

In this instance, any time the evaporator pressure exceeds 25 psig, the sum of the evaporator and spring pressures will exceed the maximum bulb pressure of 32.41 psig and the valve will be closed. On the other hand, any time the evaporator pressure is below 25 psig, the sum of the evaporator and spring pressures will be less than the maximum bulb pressure and the bulb will be in control of the valve. The valve will then respond normally to any changes in the suction superheat.

Because of its pressure limiting characteristics, the gas charged thermostatic expansion valve provides the same compressor overload and flood-back protection as do mechanical pressure limiting valves.

Since the evaporator pressure is limited indirectly by limiting the bulb pressure (charge), any change in the superheat setting (spring pressure) will cause the maximum evaporator pressure (the MOP of the valve) to change. Since the bulb pressure (p_3) is always equal to the evaporator pressure (p_1) plus the spring pressure (p_2), increasing the superheat setting (p_2) will decrease the MOP (p_1). Likewise, decreasing the superheat setting will increase the maximum operating pressure of the valve.

In view of the limited bulb charge, some precautions must be observed when installing a gas charged expansion valve. The valve body must be in a warmer location than the remote bulb and the tube connecting the remote bulb to the power head must not be allowed to touch a surface colder than the remote bulb; otherwise the bulb charge will condense at the coldest point and the valve will become inoperative because of the lack of liquid in the remote bulb (Fig. 17-13). Too, care should be taken to so locate the remote bulb that the liquid does not drain from the bulb by gravity.

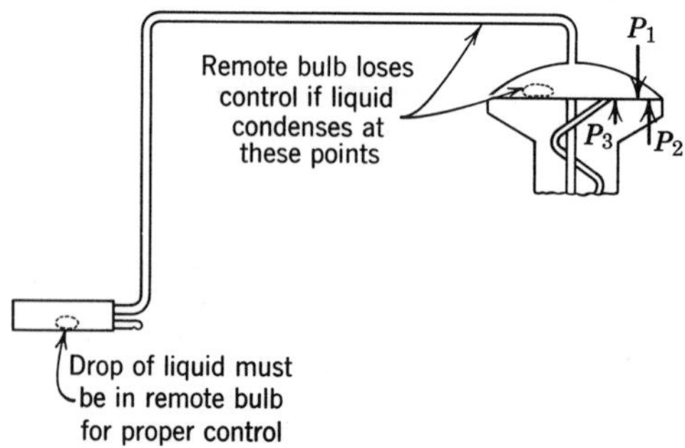

Fig. 17-13 Gas-charged thermo expansion valve. (Courtesy Alco Valve Company.)

17-8. Importance of Pressure Limiting Valves

The importance of pressure limiting valves is readily understood when it is recognized that many refrigeration systems are subject to occasional "pull-down" loads which are substantially greater than the average system load under normal operation. Since evaporator pressures and temperatures are abnormally high during these pull-down periods, the capacity and power requirements of the compressor are greatly increased, which often results in overloading of the compressor driver.

Obviously, two solutions to the problem are possible. One is to increase the size of the compressor driver so that it has sufficient power to carry the load during the overload period. The other is to limit the maximum evaporator pressure in order to avoid excessive compressor loading. Which is the better solution depends upon the requirements of the particular installation. In applications where rapid reduction of the space or product temperature is desirable, the former solution is the one recommended. However, since motoring the compressor for an occasional peak load condition unnecessarily increases both the initial and operating costs of the system, in applications where rapid reduction of the load is not required, it is usually more practical to limit the evaporator pressure to some maximum reasonably near the average evaporator pressure under normal operating conditions and then motor the compressor accordingly. This will ordinarily result in the use of a smaller size motor, thereby effecting a saving in both the initial and operating costs of the system.

For this reason, pressure limiting expansion valves of all types are widely used at the present time, particularly in air conditioning applications. As a general rule, a pressure limiting expansion valve is selected to have a MOP approximately 5 to 10 psi above the average evaporator pressure encountered at normal loading of the system. In ordering such a valve, the desired MOP must be stated.

17-9. Cross-Charged Expansion Valves

Although expansion valves having a bulb charged with the system refrigerant are suitable for most medium and high temperature applications, they are not ordinarily satisfactory for low temperature applications. The reason for this becomes evident upon examination of the pressure-temperature relationship of any common refrigerant.

A pressure-temperature curve for Refrigerant-12 is shown in Fig. 17-14. Notice that the change in pressure per degree of temperature change decreases considerably as the temperature of the refrigerant decreases. Therefore, the amount of superheat required to cause a given increase in remote bulb pressure is much greater at low temperatures than at high temperatures. For example, notice in Fig. 17-14, that at a temperature of 45°F, a temperature change of only approximately 5.5°F will cause a pressure change of 5 psi, whereas at −20°F, a temperature change of approximately 13°F is required for a 5 psi change in pressure. Obviously, then, when the expansion valve bulb is charged with the

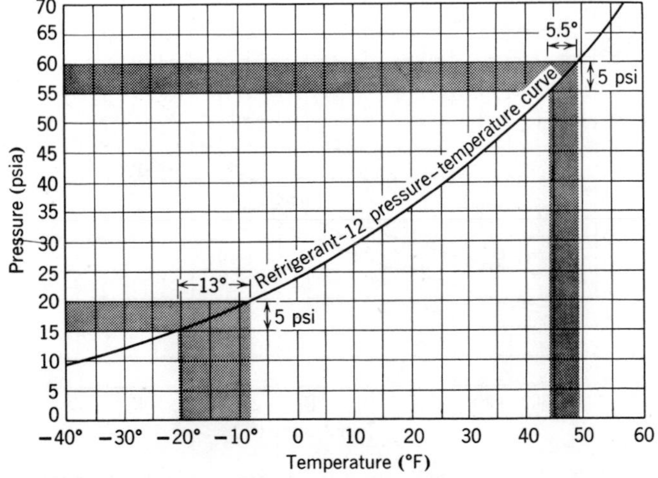

Fig. 17-14 The effect of temperature range on valve superheat.

system refrigerant, the amount of suction superheat necessary to actuate the valve becomes excessive at low temperatures, with the result that much of the evaporator surface becomes ineffective. Hence, for low temperature applications, good practice prescribes the use of an expansion valve having a bulb charged with some fluid other than the system refrigerant, usually one which has a boiling point somewhat below that of the system refrigerant, so that the pressure change in the bulb per degree of suction superheat is more substantial at the desired operating temperature of the valve. This will permit operation of the valve with a normal amount of suction superheat.

Expansion valves whose bulbs are charged with fluids other than the system refrigerant are called "cross-charged" valves because the pressure-temperature curve of the fluid crosses the pressure-temperature curve of the system refrigerant. Notice in Fig. 17-15 that the pressure-temperature curve of the bulb fluid (curve *B*) is somewhat flatter than that of the system refrigerant (curve *A*), so that as the evaporator pressure increases, a greater amount of suction superheat is required to bring the valve into equilibrium. Curve *C* in Fig. 17-15 indicates the bulb temperature necessary to bring the valve into equilibrium if the superheat spring is adjusted for a pressure of 5 psi. Because of the higher superheat requirement at high evaporator temperatures, the cross-charged valve has a pressure limiting effect which affords a certain amount of protection against motor overload and compressor flood-back at start-up.

Cross-charged expansion valves are by no means limited to low temperature applications. They are also used extensively in commercial applications where pressure limiting is desired and where their varying superheat characteristic is not objectionable. In such applications, they are often preferable to the gas-charged valve because they have less tendency to "hunt" and because the remote bulb location is not so critical.

Since a cross-charged valve will perform satisfactorily only within a given temperature range, several different types of cross charges are required for the various temperature ranges. Naturally, in ordering a cross-charged valve, the desired operating temperature range must be stated so that a valve with the proper cross charge can be selected.

17-10. Multioutlet Valves and Refrigerant Distributors

When an evaporator has more than one refrigerant circuit, the refrigerant from the expansion valve is delivered to the various evaporator circuits through a refrigerant dis-

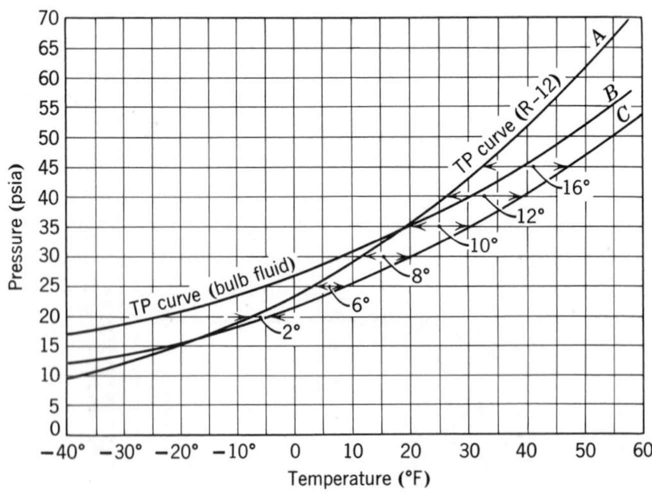

Fig. 17-15 Operating characteristics of cross-charged thermostatic expansion valve.

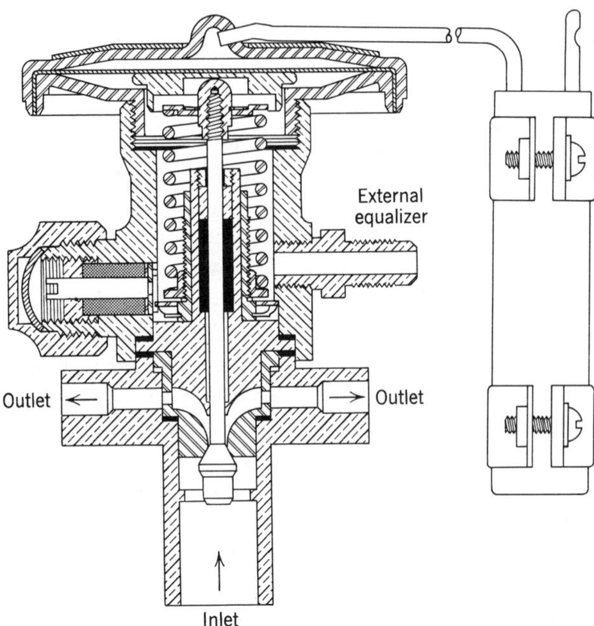

Fig. 17-16 Multioutlet thermostatic expansion valve. (Courtesy Alco Valve Company.)

tributor. In some instances the refrigerant distributor is an integral part of the valve itself. In others, it is a completely separate unit. In either case, it is important that the design of the distributor is such that the liquid-vapor mixture leaving the valve be evenly distributed to all of the evaporator circuits, if peak evaporator performance is to be expected.

A multioutlet expansion valve incorporating a refrigerant distributor is shown in Fig. 17-16. Since expansion and distribution of the refrigerant occur simultaneously within the valve itself, this, along with the carefully proportioned passages through the radial distributor, assures even distribution of a homogeneous mixture of liquid and vapor to each of the several evaporator circuits.

Four different types of refrigerant distributors are in common use at the present time: (1) the venturi type, (2) the pressure drop type, (3) the centrifugal type, and (4) the manifold type. Any of these distributors can be used with any standard, single-outlet expansion valve.

A venturi-type distributor, along with its flow characteristics, is shown in Fig. 17-17. This type of distributor utilizes the venturi principle of a large percentage of pressure

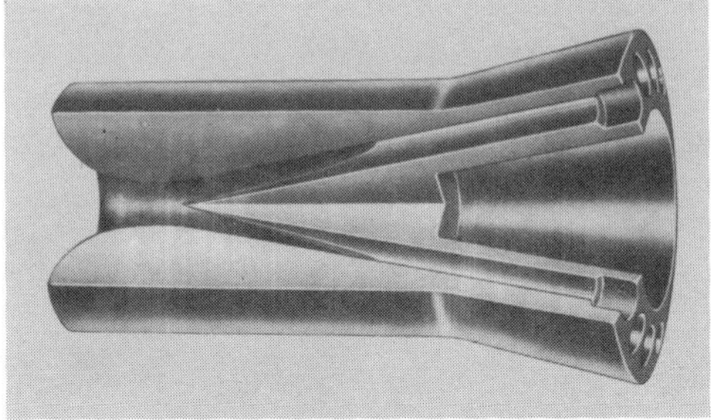

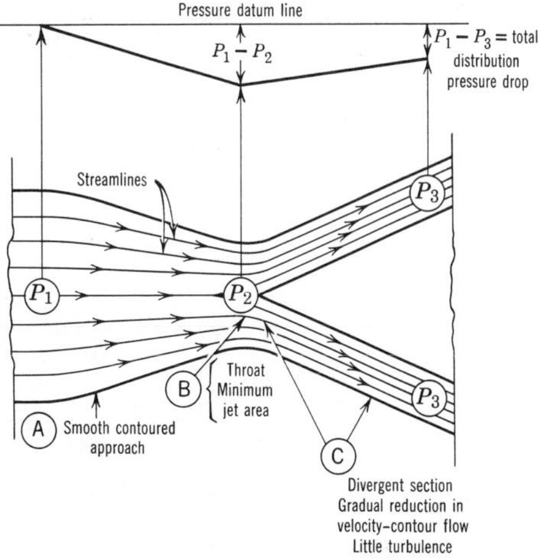

Fig. 17-17 Flow through a venturi-type distributor. (Courtesy Alco Valve Company.)

recovery, and depends on contour flow for equal distribution of the liquid-vapor mixture to each of the evaporator circuits. This distributor provides a minimum of turbulence and a minimum over-all pressure loss, the pressure loss being confined only to wall frictional losses. It may be mounted in any position.

A pressure drop-type distributor and its flow pattern is shown in Fig. 17-18. The following description is condensed from the manufacturer's catalog data.*

The distributor consists of a body or housing, the outlet end of which is drilled to receive the tubes connecting the distributor to the evaporator. The inlet end is recessed to receive an interchangeable nozzle which is held in place by a snap ring. The refrigerant, after leaving the expansion valve, enters the distributor inlet and passes through the nozzle. The nozzle orifice is sized to produce a pressure drop which increases the velocity of the liquid, thereby homogeneously mixing the liquid and vapor and eliminating the effect of gravity. The nozzle orifice centers the flow of refrigerant so that it impinges squarely on the center of the conical button inside the distributor body. The outlet passage holes are accurately spaced around the base of the conical button so that the mixture coming off the button divides evenly as it enters these holes. The orifice size of the nozzle determines the capacity of the distributor, and the pressure drop prevents separation of flash gas from the liquid, causing a homogeneous mixture of liquid and vapor to pass through the distributor.

A centrifugal-type distributor is illustrated in Fig. 17-19. This type of distributor depends upon a high entrance velocity to create a swirling effect which maintains a homogeneous mixture of the liquid and flash gas and which distributes the mixture evenly to each of the evaporator tubes.

A manifold or Weir type distributor is illustrated in Fig. 17-20. This type of distributor depends upon level mounting and low entrance velocities to insure even distribution of the refrigerant to the evaporator circuits. A baffle is often installed in the header in order to minimize the tendency to overfeed the evaporator circuits directly in front of the header

* Sporlan Valve Company.

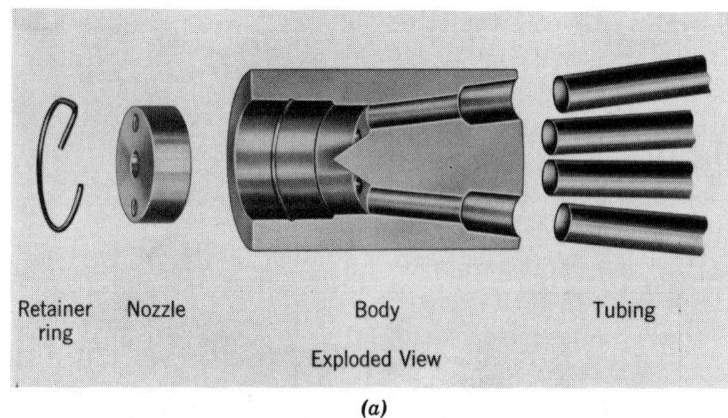

(a)

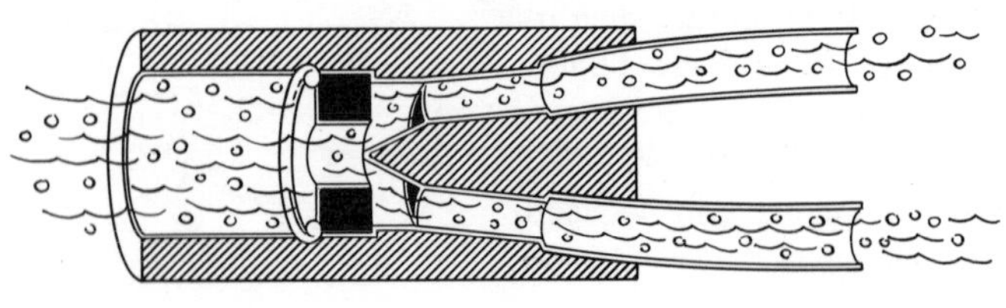

(b)

Fig. 17-18 Pressure drop-type distributor. (Courtesy Sporlan Valve Company.)

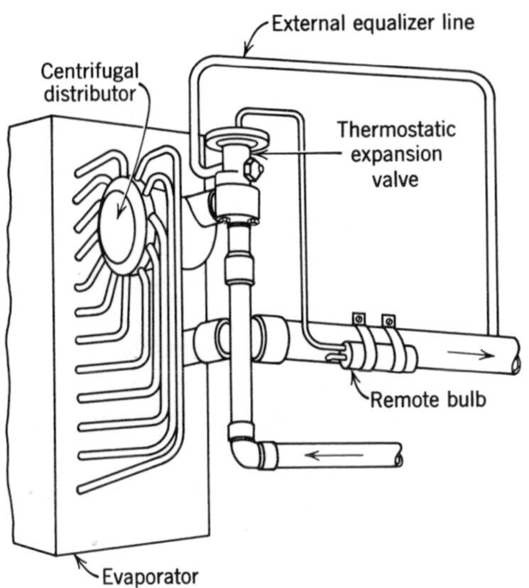

Fig. 17-19 Single outlet thermostatic expansion valve and centrifugal-type distributor. (Courtesy Alco Valve Company.)

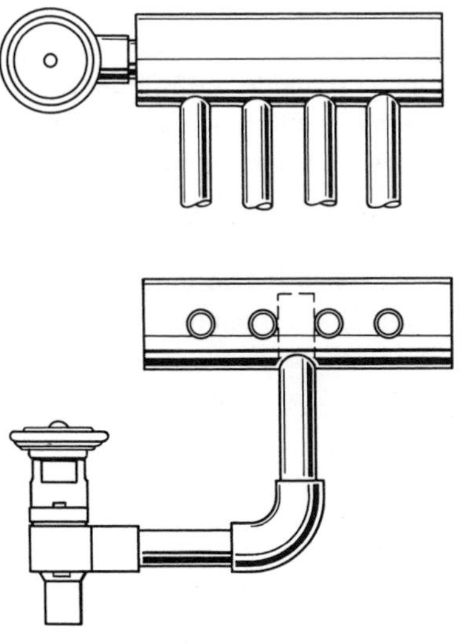

Fig. 17-20 Single outlet thermostatic expansion valve and manifold-type distributor. (Courtesy Alco Valve Company.)

inlet connection. Too, an elbow installed between the expansion valve and the header inlet will usually reduce the refrigerant velocity and help prevent unequal distribution of the refrigerant to the evaporator circuits.

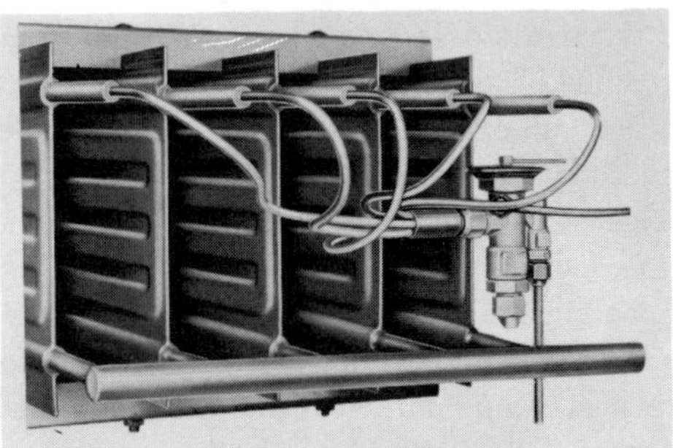

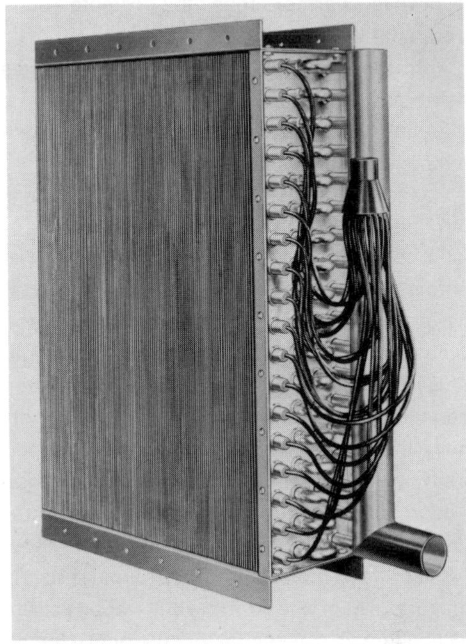

Fig. 17-21 Illustrating application of refrigerant distributors. (Courtesy Sporlan Valve Company.)

Some typical distributor applications are shown in Fig. 17-21.

17-11. Expansion Valve Location

For best performance, the thermostatic expansion valve should be installed as close to the evaporator as possible. With the exception of a refrigerant distributor, where one is used, there should be no restrictions of any kind between the evaporator and the expension valve. When it is necessary to locate a hand valve on the outlet side of the valve, the hand valve should have a full sized port.

Since there is enough liquid in a liquid charged expansion valve to insure that control of the valve will remain with the bulb under all conditions, a liquid charged thermostatic expansion valve can be installed in any position (power head up, down, or sideways), either inside or outside of the refrigerated space, without particular concern for the relative temperatures of the valve body and remote bulb. On the other hand, gas charged valves must be installed so that the valve body is always warmer than the remote bulb, preferably with the power head up.

With the exception of the manifold-type distributor, when a refrigerant distributor is used, the valve should be installed as close to the distributor as possible.

17-12. Remote Bulb Location

To a large extent, the performance of the thermostatic expansion valve depends upon the proper location and installation of the remote bulb. When an external remote bulb is used (mounted on the outside, rather than the inside of the refrigerant piping) as is normally the case, the bulb should be clamped firmly (with metal clamps) to a horizontal section of the suction line near the evaporator outlet, preferably inside the refrigerated space.

Since the remote bulb must respond to the temperature of the refrigerant vapor in the suction line, it is essential that the entire length of the remote bulb be in good thermal contact with the suction line. When an iron pipe or steel suction line is used, the suction line should be cleaned thoroughly at the point of bulb location and painted with aluminum paint in order to minimize corrosion. On suction lines under $\frac{7}{8}$ in. OD, the remote bulb is usually installed on top of the line. For suction lines $\frac{7}{8}$ in. OD and above, a remote bulb located in a position of 4 or 8 o'clock (Fig. 17-22) will normally give satisfactory control of the valve. However, since this is not true in all cases, the optimum bulb location is often best determined by trial and error.

It is important also that the remote bulb be so located that it is not unduly influenced by temperatures other than the suction line temperature, particularly during the compressor off cycle. If the temperature of the bulb is permitted to rise substantially above that of the evaporator during the off cycle, the valve will open allowing the evaporator to become filled with liquid refrigerant, with the result that liquid will flood back to the compressor when the compressor starts. When the remote bulb is located inside the refrigerated space, the temperature difference between the fixture temperature and the evaporator temperature is not usually large enough to affect adversely expansion valve operation. However, when it is necessary to locate the bulb outside the refrigerated space, both the bulb and the suction line must be well insulated from the surroundings. The insulation must be non-

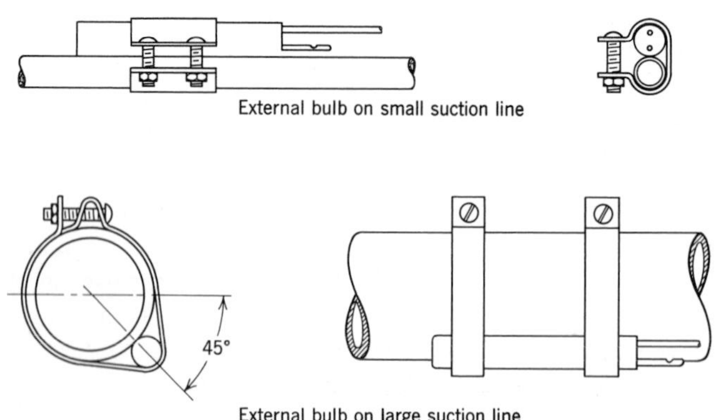

Fig. 17-22 Usual location of expansion valve remote bulb. (Courtesy Alco Valve Company.)

hydroscapoic and must extend at least 1 ft or more beyond the bulb location on both sides of the bulb.

Care must be taken also to locate the thermal bulb at least $1\frac{1}{2}$ ft from the point where an uninsulated suction line leaves a refrigerated fixture. When the bulb is located on the suction line too close to the point where the line leaves the refrigerated space, heat conducted along the suction line from the outside may cause the bulb pressure to increase to the extent that the valve will open and permit liquid to fill the evaporator during the off cycle.

On air conditioning applications, when suitable pressure limiting valves are employed, the remote bulb may be located either outside or inside the air duct, but always out of the direct air stream. On brine tanks or water coolers, the bulb should always be located below the liquid level at the coldest point.

Whenever the bulb location is such that there is a possibility that the valve may open on the off cycle, a solenoid valve should be installed in the liquid line directly in front of the expansion valve so that positive shut-off of the liquid during the off-cycle is assured. The system then operates on a pump-down cycle.

Under no circumstances should a remote bulb ever be located where the suction line is trapped. Any accumulation of liquid in the suction line at the point where the remote bulb is located will cause irregular operation (hunting) of the expansion valve. Except in a few special cases, the remote bulb must be located on the evaporator side of a liquid-suction heat exchanger.

Several of the more common incorrect remote bulb applications are shown in Figs. 17-23 through 17-25, along with recommended corrections for piping and remote bulb locations to avoid these conditions. In Fig. 17-23, liquid can trap in the suction line at the evaporator outlet, causing the loss of operating superheat and resulting in irregular operation of the valve due to alternate drying and filling of the trap. If valve operation becomes too irregular, liquid may be blown back to the compressor by the gas which forms in the evaporator behind the trap.

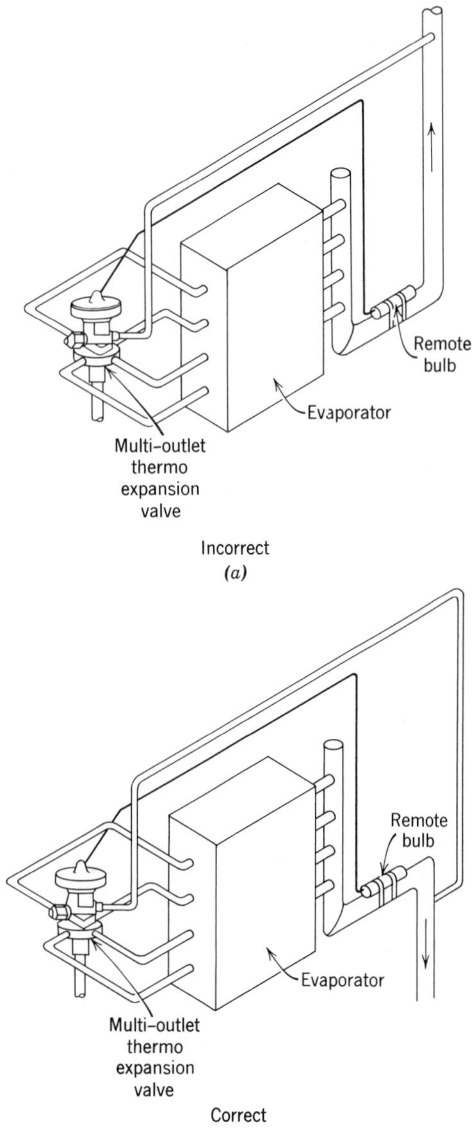

Fig. 17-23 (*a*) Remote bulb location shown trapped. (*b*) Remote bulb location shown free draining. (Courtesy Alco Valve Company.)

Figure 17-24 illustrates the proper remote bulb location to avoid trapped oil or liquid from affecting the operation of the expansion valve when the suction line must rise at the evaporator outlet. Liquid or oil accumulating in the trap during the off cycle will not affect the remote bulb and can evaporate without "slugging" to the compressor when the compressor is started. This piping arrangement

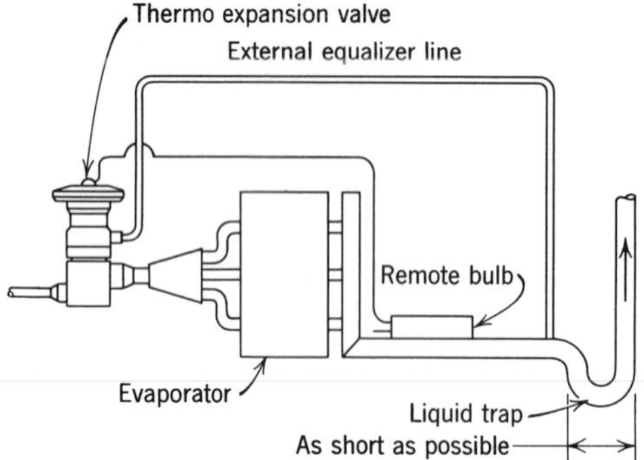

Fig. 17-24 Recommended remote bulb location and schematic piping for rising suction line. (Courtesy Alco Valve Company.)

is often used deliberately on large installations to avoid the possibility of liquid slugging to the compressor.

Figure 17-25 illustrates the incorrect application of the remote bulb on the suction header of an evaporator. With poor air circulation through the evaporator, liquid refrigerant can pass through some of the evaporator circuits without being evaporated and without affecting the remote bulb, a condition which can cause flood back to the compressor. The correct remote bulb location is shown by the dotted lines. However, correcting the remote bulb location in this instance will do nothing to improve the poor air distribution, but will only prevent flood back to the compressor. The air distribution must be approached as a separate problem.

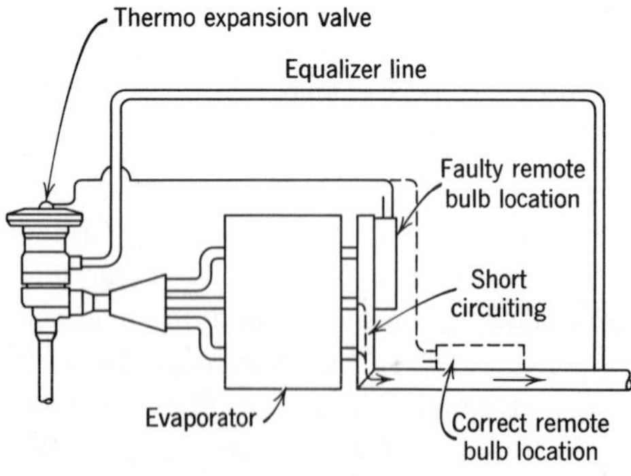

Fig. 17-25 Correct remote bulb location on "short circuiting" evaporator to prevent "flood back." (Courtesy Alco Valve Company.)

Since a trapped or partially trapped suction line at the remote bulb location will cause poor expansion valve performance, care should always be taken to arrange the suction piping from the evaporator so that oil and liquid will be drained away from the remote bulb location by gravity.

Location of the remote bulb on a vertical section of suction line should be avoided whenever possible. However, in the event that no other location is possible, the bulb should be installed well above the liquid trap on a suction riser.

In some instances, it is necessary or desirable to employ a remote bulb well, so that the remote bulb is in effect installed inside of the suction line (see Fig. 17-26). As a general rule, a remote bulb well should be used when low superheats are required or when the remote bulb is likely to be influenced by heat conducted down the suction line from a warm space. It is desirable also in installations where the suction line is very short or where the size of the suction line exceeds $2\frac{1}{8}$ in. OD.

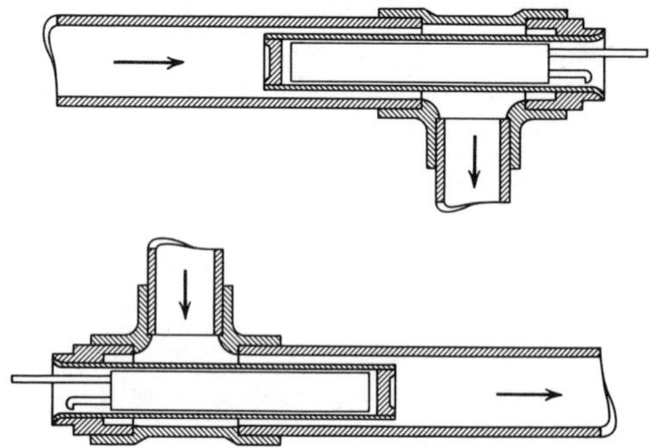

Fig. 17-26 Remote bulb well and suggested piping. (Courtesy Alco Valve Company.)

17-13. External Equalizer Location

In general, an external equalizer should be used in any case where the pressure drop through the evaporator is sufficient to cause a drop in the saturation temperature of the refrigerant in excess of 2°F at evaporator temperatures above 0°F or in excess of approximately 1°F at evaporator temperatures below 0°F. Naturally, because of the pressure-temperature relationship of the refrigerant, the maximum permissible pressure drop will vary with the individual refrigerant and with the operating temperature range. For example, with regard to Refrigerant-12, the permissible pressure drop is approximately 2.5 psi when the evaporator temperature is about 40°F, whereas for evaporator temperatures below 0°F, the permissible evaporator pressure drop is only approximately 0.5 psi.

External equalizers are required also whenever multioutlet expansion valves or refrigerant distributors are employed, regardless of whether or not the pressure drop through the evaporator is excessive. In such installations, the external equalizer is required in order to compensate for the pressure drop suffered by the refrigerant in passing through the distributor.

It should be emphasized at this point that the refrigerant pressure loss in a refrigerant distributor in no way affects the capacity or efficiency of the system, provided that the expansion valve has been properly selected (see Section 17-15). In any system, with or without a distributor, the refrigerant must undergo a drop in pressure between the high and low pressure sides of the system. On systems without distributors, all of this pressure drop is taken across the valve. When a distributor is employed, a portion of the pressure drop occurs in the distributor and the remainder across the valve. The total pressure drop and refrigerating effect are the same in either case.

This is not true, however, when the pressure loss is in the evaporator itself. As described in Section 8-9, any refrigerant pressure drop in the evaporator tends to reduce the capacity and efficiency of the system. Hence, evaporators having excessive pressure drops should be avoided whenever possible.

As a general rule, the external equalizer connection is made on the suction line 6 to 8 in. beyond the expansion valve bulb on the compressor side. However, in applications where the external equalizer is used to offset pressure drop through a refrigerant distributor and the pressure drop through the evaporator is not excessive, the external equalizer may be connected either to one of the feeder tubes or to one of the evaporator return bends at approximately the midpoint of the evaporator. When the external equalizer is connected to a horizontal line, it should be installed on top of the line in order to avoid the drainage of oil or liquid into the equalizer tube.

17-14. Thermal-Electric Expansion Valve

The electric expansion valve is a simple heat-motor-operated needle valve that is a part of a valve and sensor package and that can be positioned in response to a change in input voltage (Fig. 17-27). An increase in voltage opens the valve and increases refrigerant flow, whereas a decrease in voltage diminishes the flow or closes the valve completely.

The changes in voltage necessary to modulate the valve can be initiated by any one of a number of temperature, pressure, or liquid refrigerant sensors, and the function of the valve can be changed by switching from one sensing device to another. Since the valve responds only to input voltage and does not depend on refrigerant temperature or pressure to create modulation, the same valve will work equally well with any common refrigerant except ammonia. Moreover, operation of the valve is unaffected by evaporator pressure drop (does not require external equalizer) and by the operating temperature range.

When the valve is used as a superheat control, a small negative coefficient thermistor is installed in the suction line so that it is directly exposed to the refrigerant in the line. The

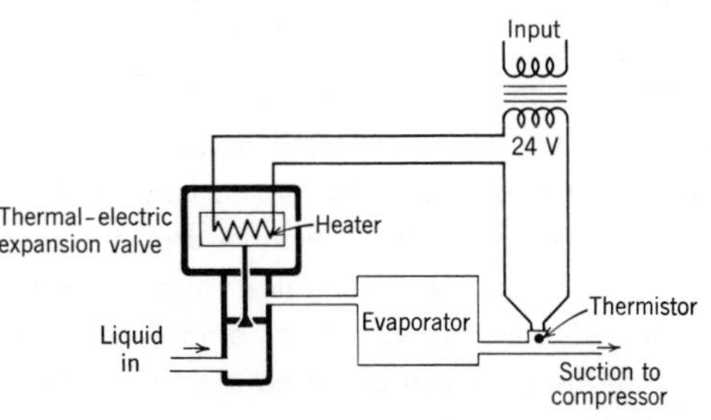

Fig. 17-27 Thermal-electric expansion valve.

thermistor is placed in series (electrically), with the expansion valve heater making the current input to the heater a function of thermistor resistance, which in turn is a function of the refrigerant conditions.

Exposure to gaseous or superheated refrigerant permits the thermistor to self-heat, thereby lowering its resistance and allowing the current input to the expansion valve heater to increase. The expansion valve responds by modulating in an opening direction to increase the refrigerant flow to the evaporator. This process continues until saturated suction conditions occur, permitting wet refrigerant to contact the thermistor. Liquid refrigerant or wet refrigerant gas will immediately cool the thermistor, thereby increasing its resistance and reducing the current input to the heater, which causes the expansion valve to modulate in a closing direction.

The termination point of the refrigerant saturation is controlled by the location of the thermistor and can be shifted from one point to another by using more than one thermistor, which can be switched in and out on demand.

The thermistor can be used also to control the liquid level in a suction accumulator or riser to provide flooded or semiflooded evaporator control with dry suction gas returning to the compressor.

17-15. Expansion Valve Rating and Selection Before the proper size valve can be selected, a decision must be made as to the exact type of valve desired with respect to bulb charge, pressure limiting, the possible need for an external equalizer, and the size of the valve inlet and outlet connections. Obviously, the nature and conditions of the application will determine the type of bulb charge and also whether or not a pressure limiting valve is needed. An externally equalized valve should be used whenever the pressure drop through the evaporator is substantial and/or when a refrigerant distributor is employed. The size of the inlet and outlet connections of the valve should be equal to those of the liquid line and evaporator, respectively. A slight reduction in size at the evaporator inlet is permissible.

Once a decision has been made on all of the foregoing, the proper size valve can be selected from the manufacturer's catalog ratings. Table R-21 is a typical thermostatic expansion valve rating table. The expansion valves are rated in tons of refrigerating capacity (or Btu/hr) at various operating conditions. Normally, valve ratings are based on a condensing temperature of 100°F with zero degrees of subcooling, but with solid liquid approaching the valve.

In order to select the proper size valve from the rating table, the following data should be known: (1) the evaporator temperature, (2) the system capacity in tons, and (3) the available pressure difference across the valve. In general, the first two factors determine the required liquid flow rate through the valve, whereas the latter determines the size orifice required to deliver the desired flow rate, the flow rate through the orifice being proportional to the pressure differential across the valve.

The pressure difference across the valve can never be taken as the difference between the suction and discharge pressures as measured at the compressor. When these two pressures are used as a basis for determining the pressure difference across the expansion valve, an allowance must always be made for the pressure losses which accrue between the expansion valve and the compressor on both the low and high pressure sides of the system. This includes the refrigerant distributor when one is used.

When the available pressure difference across the expansion valve has been determined, a valve should be selected from the manufacturer's rating table which has a capacity equal to or slightly in excess of the system capacity at the system design operating conditions.

Example 17-1. From Table R-21, select the proper size expansion valve for a Refrigerant-12 system, if the desired capacity is 8 tons at a 20°F evaporator temperature and the available pressure drop across the expansion valve is approximately 65 psi.

Solution. From Table R-21, select expansion valve Model # TJL1100F, which has a capacity of 8.1 tons at a 20°F evaporator temperature when the pressure drop across the valve is 60 psi. The letters in the model number indicate valve type and refrigerant.

Thermostatic expansion valves will not operate satisfactorily at less than 50% of their rated capacity. Therefore, when the load on the system is likely to fall below 50% of the design load, the evaporator should be split into two or more separate circuits, with each circuit being fed by an individual expansion valve. With this arrangement it is possible to cycle out portions of the evaporator as the load on the system fall off so that the load on any one expansion valve never drops below 50% of the design capacity of the valve.

17-16. Capilliary Tubes

The capilliary tube is the simplest of the refrigerant flow controls, consisting merely of a fixed length of small diameter tubing installed between the condenser and the evaporator, usually in place of the conventional liquid line (Fig. 17-28). Because of the high frictional resistance resulting from its length and small bore and because of the throttling effect resulting from the gradual formation of flash gas in the tube as the pressure of the liquid is reduced below its saturation pressure, the capillary tube acts to restrict or meter the flow of liquid from the condenser to the evaporator and also to maintain the required operating pressure differential between those two units.

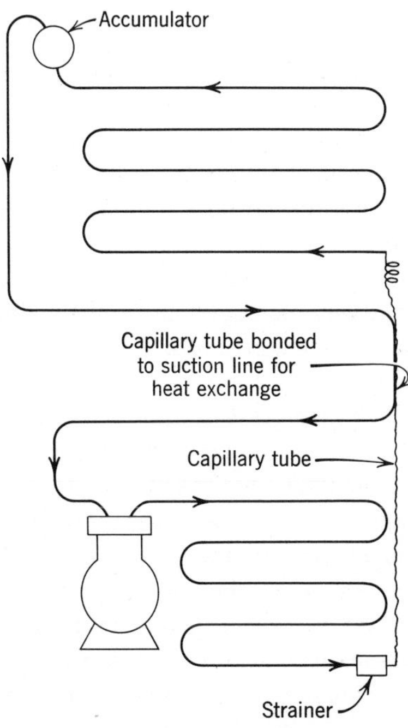

Fig. 17-28 Capillary tube system.

For any given tube length and bore, the resistance of the tube is fixed or constant so that the liquid flow rate through the tube at any one time is proportional to the pressure differential across the tube, said pressure differential being the difference between the vaporizing and condensing pressures of the system.

Since the capillary tube and the compressor are connected in series in the system, it is evident that the flow capacity of the tube must of necessity be equal to the pumping capacity of the compressor when the latter is in operation. Consequently, if the system is to perform efficiently and balance out at the design operating conditions, the length and bore of the tube must be such that the flow capacity of the tube at the design vaprizing and condensing pressures is exactly equal to the pumping capacity of the compressor at these same conditions.

In the event that the resistance of the tube is such that the flow capacity of the tube is either greater than or less than the pumping capacity of the compressor at the design conditions, a balance will be established between these two components at some operating conditions other than the system design conditions. For example, if the resistance of the tube is too great (tube too long and/or bore too small), the capacity of the tube to pass liquid refrigerant from the condenser to the evaporator will be less than the pumping capacity of the compressor at the design conditions, in which case the evaporator will become starved while the excess liquid will back-up in the lower portion of the condenser at the entrance to the capillary tube. Naturally, starving of the evaporator will result in lowering the suction pressure, whereas the build-up of liquid in the condenser will result in a reduction of the effective condensing surface and, consequently, an increase in the condensing temperature. Hence, the net effect of too much restriction in the capillary tube is to lower the suction pressure and raise the condensing pressure. Since both these conditions tend to increase the flow capacity of the tube and, at the same time, decrease the pumping capacity of the compressor, it is evident that the system will eventually establish equilibrium at some operating conditions where the capacity of the tube and the capacity of the compressor are exactly the same. In this instance, the point of balance will be at a lower suction pressure and a higher condensing pressure than the system design pressures. Too, since the capacity of the compressor is reduced at these conditions, the overall system capacity will be less than the design capacity.

On the other hand, when the tube does not have enough resistance (tube too short and/or bore too large), the flow capacity of the tube will be greater than the pumping capacity of the compressor at the design conditions, in which case overfeeding of the evaporator will result with the danger of possible liquid flood-back to the compressor. Also, there will be no liquid seal in the condenser at the entrance to the tube and, therefore, uncondensed gas will be allowed to enter the tube along with the liquid. Obviously, the introduction of latent heat into the evaporator in the form of uncondensed gas will have the effect of reducing the system capacity. Furthermore, because of the excessive flow rate through the tube, the compressor will not be able to reduce the evaporator pressure to the desired low level.

A system employing a capillary tube will operate at maximum efficiency only at one set of operating conditions. At all other operating conditions, the efficiency of the system will be somewhat less than maximum. However, it should be pointed out that the capillary tube is self-compensating to some extent and, if properly designed and applied, will give satisfactory service over a reasonable range of operating conditions. Normally, as the load on the system increases or decreases, the flow capacity of the capillary tube increases or decreases, respectively, partially because of the change in condensing pressure which ordinarily accompanies these changes in system loading and partially because of the change in the amount of liquid subcooling taking place in the condenser. For example, as the load on the evaporator is increased, liquid refrigerant will be vaporized in the evaporator and condensed in the condenser at a rate momentarily higher than that at which the capillary tube is passing liquid to the evaporator, with the result that the excess liquid will accumulate in the end of the condenser. With the condenser partially filled with liquid, the condensing pressure will be increased. At the same time, the liquid in the end of the condenser will be subject to a greater degree of subcooling, so that there will be less flash gas formed in the capillary tube. Both of these conditions tend to increase the flow capacity of the tube and thereby bring the system capacity more in line with the increased system load. Conversely, as the system load diminishes, the condensing pressure and the degree of subcooling decrease, so that the flow capacity of the tube decreases.

The capillary tube differs from other types of refrigerant flow controls in that it does not close off and stop the flow of liquid to the evaporator during the off cycle. When the compressor cycles off, the high and low side pressures equalize through the open capillary tube and any residual liquid in the condenser passes to the low pressure evaporator where it remains until the compressor cycles on again. For this reason, the refrigerant charge in a capillary tube system is critical and no receiver tank is employed between the condenser and the capillary tube. In all cases, the refrigerant charge should be the minimum which will satisfy the requirements of the evaporator and at the same time maintain a liquid seal in the condenser at the entrance to the capillary tube during the latter part of the operating cycle. Any refrigerant in excess of this amount will only back up in the condenser, thereby increasing the condensing pressure which, in turn, reduces the system efficiency and tends to unbalance the system by increasing the flow capacity of the tube. If the overcharge is sufficiently large, overloading of the compressor driver may also result. However, of more importance is the fact that all of the excess liquid in the condenser will pass to the evaporator during the off cycle. Being at the condensing temperature, a substantial amount of such liquid will cause the evaporator to warm up rapidly, thereby causing defrosting of the evaporator and/or short cycling of the compressor. Moreover, where a considerable amount of liquid enters the evaporator during the off cycle, flood-back to the compressor is likely to occur when the compressor cycles on.

Other than its simple construction and low cost, the capillary tube has the additional advantage of permitting certain simplifications in the refrigerating system which further reduce manufacturing costs. Because the high and low pressure equalize through the capillary tube during the off cycle, the compressor starts in an "unloaded" condition. This allows the use of a low starting torque motor to drive the compressor; otherwise a more expensive type of motor would be required. Furthermore, the small and critical refrigerant charge required by the capillary tube system results not only in reducing the cost of the refrigerant but also in eliminating the need for a receiver tank. Naturally, all these things represent a substantial savings in the manufacturing costs. Thus, capillary tubes are employed almost universally on all types of domestic refrigeration units, such as refrigerators, freezers, and room coolers. Many are used also on small commercial packaged units, particularly packaged air conditioners.

Capillary tubes should be employed only on those systems which are especially designed for their use. They are best applied to close-coupled, packaged systems having relatively constant loads and employing hermetic motor-compressors. Specifically, a capillary tube should not be used in conjunction with an open type compressor. Because of the critical refrigerant charge, an open type compressor may lose sufficient refrigerant by seepage around the shaft seal to make the system inoperative in only a very short time.

The use of capillary tubes on remote systems (compressor located some distance from the evaporator) should also be avoided as a general rule. Such systems are very difficult to charge accurately. Furthermore, because of the long liquid and suction lines, a large charge of refrigerant is required, all of which concentrates in the evaporator during the off cycle. Serious flood-back to the compressor is likely to occur at start-up unless an adequate accumulator is installed in the suction line.

Condensers designed for use with capillary tubes should be so constructed that liquid drains freely from the condenser into the capillary tube in order to prevent the trapping of liquid in the condenser during the off cycle. Any liquid trapped in the condenser during the off cycle will evaporate and pass through the tube to the evaporator in the vapor state rather than in the liquid state. The subsequent condensation of this vapor in the evaporator will unnecessarily add latent heat to the evaporator, thereby reducing the capacity of the system.

Too, the diameter of the condenser tubes should be kept as small as is practical so that a minimum amount of liquid backed-up in the condenser at the tube inlet will cause a maximum increase in the condensing pressure and therefore a maximum increase in the flow capacity of the tube.

Evaporators intended for use with capillary tubes should provide for liquid accumulation at the evaporator outlet in order to prevent liquid flood-back to the compressor at start-up (Fig. 17-28). The function of the accumulator is to absorb the initial surge of liquid from the evaporator as the compressor starts. The liquid then vaporizes in the accumulator and returns to the compressor as a vapor. To expedite the return of oil to the compressor crankcase, liquid from the evaporator usually enters at the bottom of the accumulator, whereas the suction to the compressor is taken from the top.

In most cases, best performance is obtained when the capillary tube is connected directly between the condenser and the evaporator without an intervening liquid line. When the condenser and evaporator are too far apart to make direct connection practical, some other type of refrigerant control should ordinarily be used.

Bonding (soldering) the capillary tube to the suction line for some distance in order to provide a heat transfer relationship between the two is usually desirable in that it tends to minimize the formation of flash gas in the tube. Flash gas, formed in the tube because of the gradual expansion of the liquid as its pressure is reduced, seriously reduces the flow capacity of the tube. When the tube is not bonded to the suction line, the tube must be shortened sufficiently to offset the throttling action of the vapor in the tube.

17-17. Flooded Evaporator Control

Refrigerant flow controls employed with flooded evaporators are usually of the float type. The float control consists of a buoyant member (hollow metal ball, cylinder or pan) which is responsive to refrigerant liquid level and which acts to open and close a valve assembly to admit more or less refrigerant into the evaporator in accordance with changes in the liquid level in the float chamber. The float chamber may be located on either the low pressure side or high pressure side of the system. When the float is located on the low pressure side of the system, the float control is called a low pressure float control. When the float is located on the high pressure side of the system, the float control is known as a high pressure float control.

The principal advantage of the flooded evaporator lies in the higher evaporator capacity and efficiency which is obtained therefrom. With flooded operation, the refrigerant in all parts of the evaporator is predominately in the liquid state and a high refrigerant side tube coefficient is produced, as compared to that obtained with the dry-expansion type evaporator wherein the refrigerant in the evaporator is predominately in the vapor state, especially in the latter part of the evaporator. For this reason, float controls (flooded evaporators) are used extensively in large liquid chilling installations where advantage can be taken of the high refrigerant side conductance coefficient. On the other hand, because of their bulk and the relatively large refrigerant charge required, float controls are seldom employed on small applications, having been discarded in this area in favor of the smaller, more versatile thermostatic expansion valve or the simpler, more economical capillary tube.

17-18. Low Pressure Float Control

The low pressure float control (low side float) acts to maintain a constant level of liquid in the evaporator by regulating the flow of liquid refrigerant into that unit in accordance with the rate at which the supply of liquid is being depleted by vaporization. It is responsive only to the level of liquid in the evaporator and will maintain the evaporator filled with liquid refrigerant to the desired level under all conditions of loading without regard for the evaporator temperature and pressure.

Operation of the low pressure float valve may be either continuous or intermittent. With continuous operation, the low pressure float valve has a throttling action in that it modulates toward the open or closed position to feed more or less liquid into the evaporator in direct response to minor changes in the evaporator liquid level. For intermittent operation, the valve is so designed that it responds only to minimum and maximum liquid levels, at which points the valve is either fully open or fully closed as the result of a toggle arrangement built into the valve mechanism.

The low pressure float may be installed directly in the evaporator or accumulator in which it is controlling the liquid level (Fig. 11-19*b*), or it may be installed external to these units in a separate float chamber (see Fig. 17-29).

On large capacity systems, a by-pass line equipped with a hand expansion valve is usually installed around the float valve in order to provide refrigeration in the event of float valve failure. Too, hand stop valves are usually installed on both sides of the float valve so that the latter can be isolated for servicing without the necessity for evacuating the large refrigerant charge from the evaporator (Fig. 17-29). Notice also in Fig. 17-29 the liquid pump em-

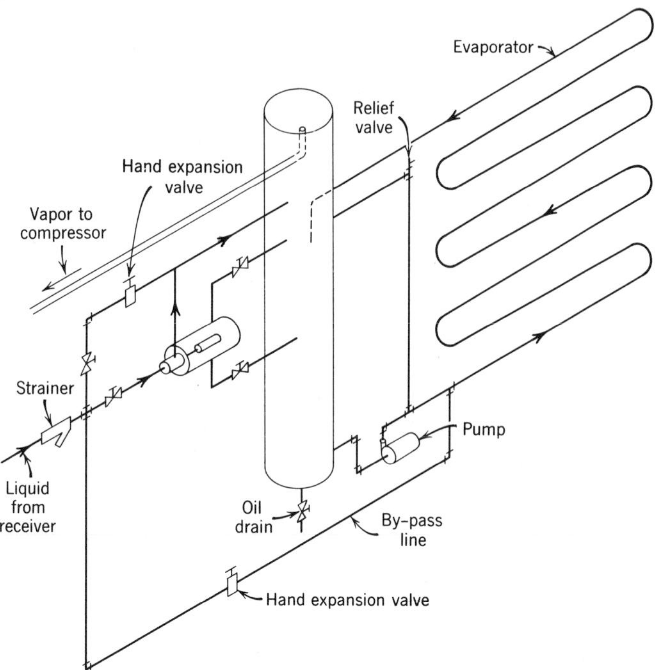

Fig. 17-29 Low side float valve controlling liquid level in accumulator. Note liquid pump used to recirculate refrigerant through evaporator.

ployed to provide forced circulation of the refrigerant through the evaporator tubes. It is of interest to compare this method of recirculation to the injection method of recirculation shown in Fig. 17-30, and with the gravity recirculation method shown in Fig. 11-19*b*. The hand expansion valve in the by-pass line around the liquid pump in Fig. 17-29 is to provide refrigeration in the event of pump failure.

Low pressure float valves may be used in multiple or in parallel with thermostatic expansion valves. In many instances, a single low pressure float valve can be used to control the liquid flow into several different evaporators.

17-19. High Pressure Float Valves

Like the low pressure float valve, the high pressure float valve is a liquid level actuated refrigerant flow control which regulates the flow of liquid to the evaporator in accordance with the rate at which the liquid is being vaporized. However, whereas the low pressure float valve controls the evaporator liquid level directly, the high pressure float valve is located on the high pressure side of the system and controls the amount of liquid in the evaporator indirectly by maintaining a constant liquid level in the high pressure float chamber (Fig. 17-31).

The operating principle of the high pressure float valve is relatively simple. The refrigerant vapor from the evaporator condenses into the liquid state in the condenser and passes into the float chamber and raises the liquid level in that component, thereby causing the float ball to rise and open the valve port so that a proportional amount of liquid is released from the float chamber to replenish the supply of liquid in the evaporator. Since vapor is always condensed in the condenser at the same rate that the liquid is vaporized in the evaporator, the high pressure float valve will continuously and automatically feed the liquid back to the evaporator at a rate commensurate with the rate of vaporization, regardless of the system load. When the compressor stops, the liquid level in the float chamber drops, causing the float valve to close and remain closed until the compressor is started again.

Since the high pressure float valve permits only a small and fixed amount of refrigerant to

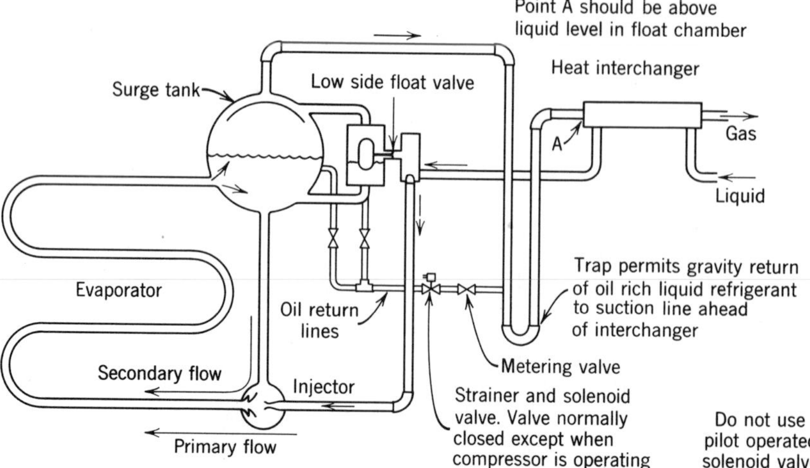

Fig. 17-30 Flooded evaporator (recirculating injector circulation). (Courtesy General Electric.)

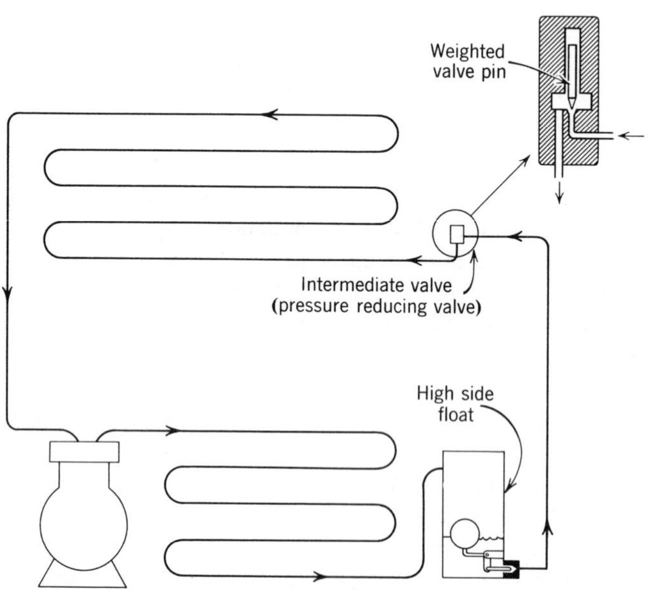

Fig. 17-31 High-pressure float valve.

remain in the high pressure side of the system, it follows that the bulk of the refrigerant charge is always in the evaporator, and that the refrigerant charge is critical. An overcharge of refrigerant will cause the float valve to overfeed the evaporator with the result that liquid refrigerant will flood back to the compressor. Moreover, if the system is seriously overcharged, the float valve will not throttle the liquid flow sufficiently to allow the compressor to reduce the evaporator pressure to the desired low level. On the other hand, if the system is undercharged, operation of the float will be erratic and the evaporator will be starved.

The high pressure float valve may be used with a dry-expansion type evaporator as shown in Fig. 17-31, or with a flooded-type evaporator as shown in Fig. 17-32. With the latter, liquid refrigerant is expanded into the surge drum (low pressure receiver) from where it flows into the evaporator through the drop leg pipe attached to the bottom of the surge drum. The suction vapor is drawn off at the top of the surge drum, as is the flash gas resulting from the expansion of the liquid as it passes through the float valve. To prevent liquid flood-back during changes in loading, the surge drum should have a volume equal to at least 25% of the evaporator volume.

The construction of a typical high pressure float valve is illustrated in Fig. 17-33. Notice that the float valve opens on a rising liquid level and that the construction of the valve pin and float arm pivot are such that the weight of the float ball will move the valve pin in a closing direction as the liquid refrigerant recedes in the float chamber. Notice also that the float ball is so positioned that the valve seat is always submerged in the liquid refrigerant in order to eliminate the possibility of wire-drawing by high velocity gas passing through the valve pin and seat. Too, the high pressure float assembly contains a vent tube to prevent the float chamber from becoming gas bound by non-condensible gases which may otherwise collect in the chamber and build up a pressure,

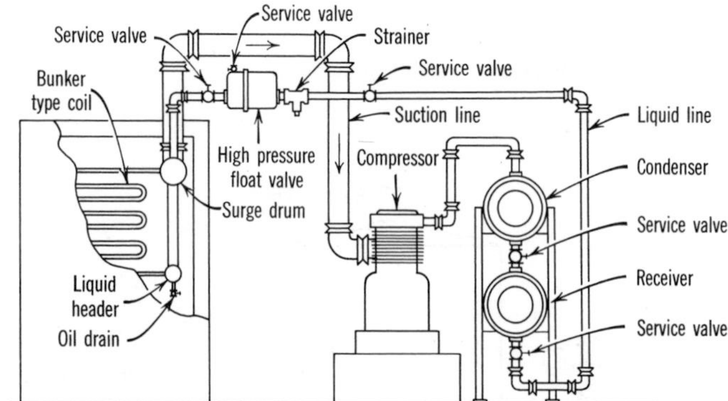

Fig. 17-32 Typical high-pressure float valve application. (Courtesy Alco Valve Company.)

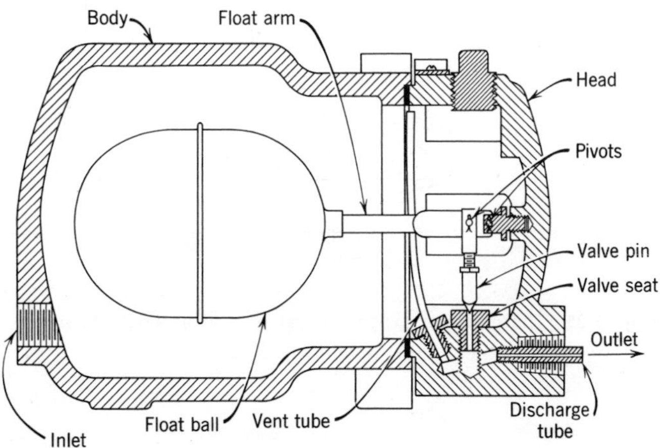

Fig. 17-33 High-pressure float valve. (Courtesy Alco Valve Company.)

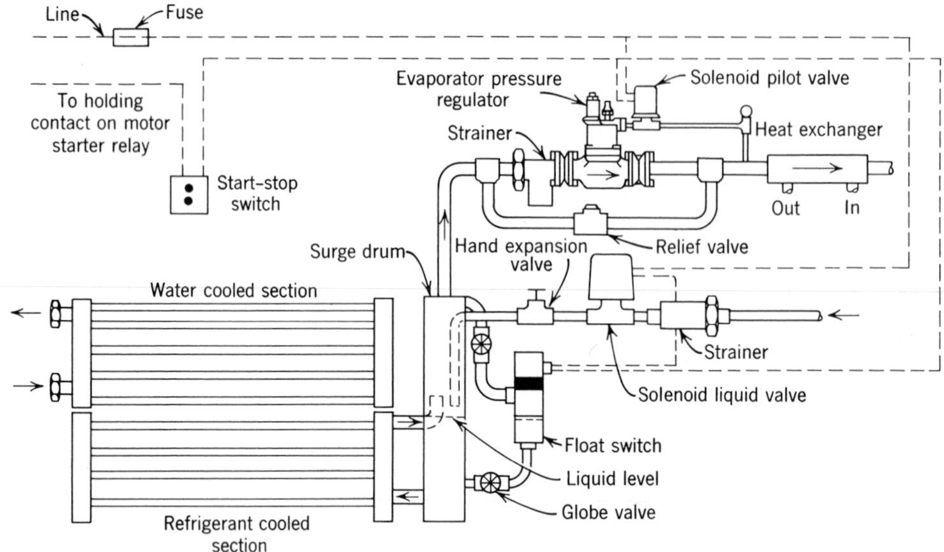

Fig. 17-34 Baudelot cooler with float switch, solenoid liquid valve, evaporator pressure regulator, and solenoid pilot valve. (Courtesy Alco Valve Company.)

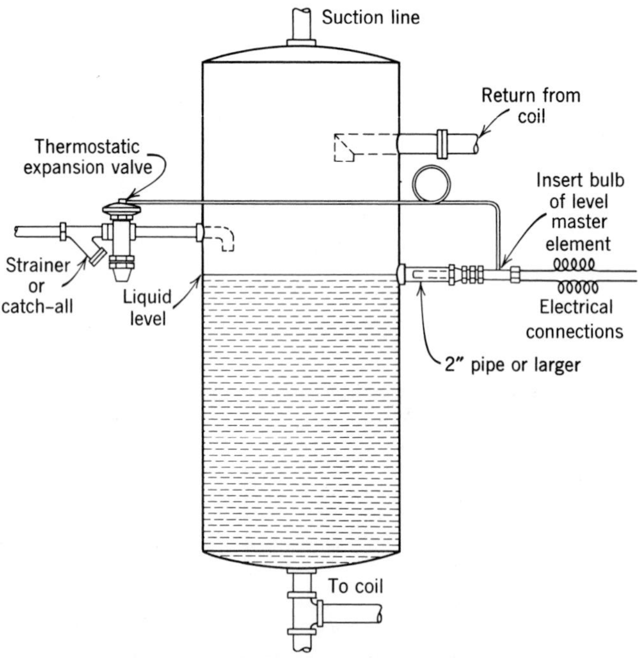

Fig. 17-35 Thermostatic expansion valve used as liquid levelcontrol. (Courtesy Sporlan Valve Company.)

thereby preventing liquid refrigerant from entering the chamber. The use of the vent tube makes possible the installation of the high pressure float valve at a point above or below the condenser without the danger of gas binding.

Unlike the low pressure float control, the high pressure float control, being independent of the evaporator liquid level, may be installed either above or below that unit. However, the float valve should be located as close to the evaporator as possible and always in a horizontal line in order to insure free action of the float ball and valve assembly. When the float is located some distance from the evaporator, it is usually necessary to provide some means of maintaining a high liquid pressure in the line between the float valve and the evaporator in order to prevent premature expansion of the liquid before it reaches the evaporator. In small systems, this is accomplished by installing an "intermediate" valve in the liquid line at the entrance to the evaporator (see inset of Fig. 17-31). In larger systems, a pilot valve is ordinarily employed for this purpose (see Section 17-22).

Because of their operating characteristics, high pressure float controls cannot be employed in multiple or in parallel with other types of refrigerant flow controls.

17-20. Float Switch

As shown in Fig. 17-34 a float switch can be employed to control the level of liquid in the evaporator. The float switch consists of two principal parts: (1) a float chamber equipped with a ball float which rises and falls with the liquid level in the evaporator and float chamber and (2) a mercury switch which is actuated by the ball float to open and close a liquid line solenoid valve when the level of liquid in the evaporator falls and rises, respectively. A hand expansion valve is installed in the liquid line between the solenoid valve and the evaporator to throttle the liquid refrigerant and prevent surging in the evaporator from the sudden inrush of liquid when the liquid line solenoid opens and to eliminate short cycling of the solenoid valve.

Float switches have many applications for operating electrical devices associated with the refrigerating system. They may be arranged for reverse action (close on rise) by employing a reverse acting switch.

17-21. Liquid Level Control with Thermostatic Expansion Valve

A thermostatic expansion valve with a specially designed thermal element can also be used to control the liquid level in flooded-type evaporators. Figure 17-35 illustrates a typical installation on a vertical surge drum.

The specially designed thermal element is an insert bulb consisting of a low wattage electric heating element (approximately 15 watts) and a reservoir for the thermostatic charge. The heating element of the thermal bulb is a means of providing an artificial superheat to the thermostatic charge, which increases the bulb pressure and results in opening the port of the expansion valve allowing more refrigerant to be fed to the evaporator. As the liquid level in the evaporator rises and more liquid comes in contact with the bulb the effect of the heater element is overcome, thereby decreasing the superheat and allowing the thermostatic expansion valve to throttle to the point of equilibrium or eventual shut-off.

The thermostatic expansion valve is installed in the liquid line and may be arranged to feed liquid directly into the evaporator or accumulator (surge tank), into an accumulator drop leg, or into a coil header.

17-22. Pilot Control Valves

Pilot-operated liquid control valves are employed on large tonnage installations. The pilot valve actuating the liquid control valve is usually a thermostatic expansion valve, a low pressure float valve, or a high pressure float valve. A liquid control valve designed for use with a thermostatic expansion valve pilot is illustrated in Fig. 17-36. The liquid control valve opens when pressure is supplied to the top of piston "A" from the pilot line. The small bleeder port "B" in the top of the piston vents this pressure to the outlet (evaporator) side of

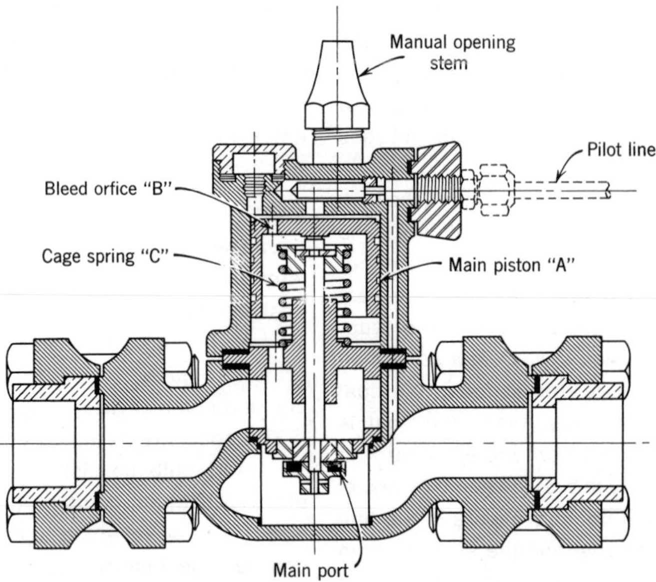

Fig. 17-36 Pilot-operated expansion valve—main regulator. (Courtesy Alco Valve Company.)

the liquid control valve. When the pressure supply to the top of the piston is cut off, the cage spring "C" closes the liquid control valve.

Figure 17-37 illustrates a pilot operated expansion valve installed on a direct expansion shell and tube chiller. The externally equalized thermostatic pilot valve supplies pressure to the top of the piston in response to changes in the temperature and pressure of the suction vapor. When the superheat in the suction vapor increases, indicating the need of greater refrigerant flow, the pilot thermo valve moves in an opening direction and supplies a greater pressure to the top of the piston thereby moving the piston in an opening direction and providing a greater flow of refrigerant. Conversely, when

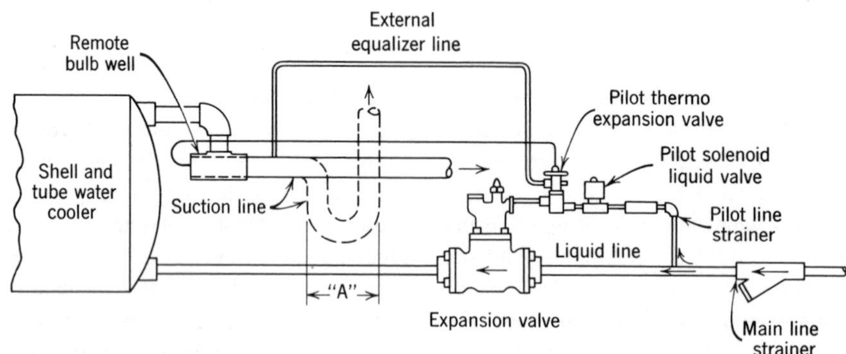

Fig. 17-37 Pilot-operated thermo expansion valve on shell-and-tube-water cooler with refrigerant in the tubes. (Courtesy Alco Valve Company.)

the suction vapor superheat decreases, indicating the need of a reduction in refrigerant flow, the pilot valve moves in a closing direction. This provides less pressure on top of the piston, permitting the piston to move in a closing direction and provide a smaller flow. In operation, the pilot valve and the main piston assume intermediate or throttling positions depending on the load.

A liquid control valve designed for use with a high pressure float valve as a pilot is illustrated in Fig. 17-38. A system employing this type of refrigerant control is illustrated in Fig. 17-39. Operation of the high pressure float pilot is similar to that of the thermo expansion valve pilot. On a rise in the liquid level in the pilot receiver, the float opens and admits high pressure liquid through the pilot line to the liquid control valve piston. This pressure acts against a spring and opens the valve stem to admit liquid to the evaporator. As the level in the pilot receiver descends, the pilot valve closes and the high pressure in the pilot line is bled off to the low side through an adjustable internal bleeder port in the liquid control valve. A pressure gage should be installed in the pilot line to facilitate adjusting the control valve during the initial start-up.

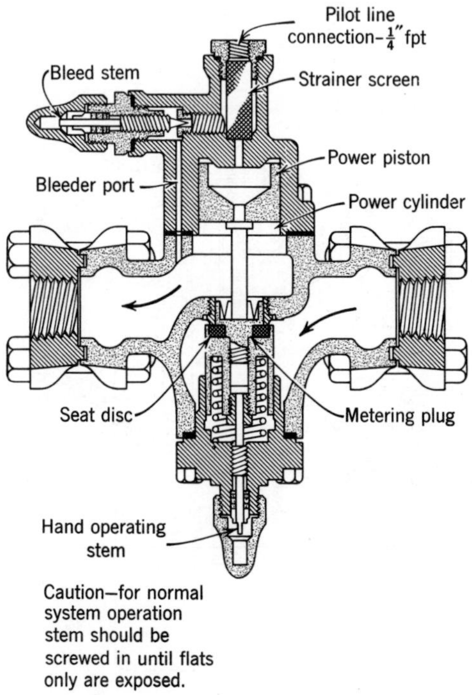

Fig. 17-38 Liquid control valve—high pressure. (Courtesy York Corporation.)

A liquid control valve designed for use with a low pressure float pilot is shown in Fig. 17-40. A typical application of a low pressure float pilot is shown in Fig. 17-41. As the liquid level in the cooler drops, the low pressure float pilot opens and allows the pressure in the pilot line to be relieved to the cooler so that the high pressure liquid acting on the bottom of the liquid control valve piston can lift the piston and admit more refrigerant to the cooler. As the level in the cooler builds up, the pilot float closes and high pressure liquid is bled into the area above the control piston through an internal bleeder in the bottom of the piston. The resulting high pressure on top of the piston causes the piston to drop down and close the main valve port. The latter bleeder is not adjustable, but modulation can be obtained by adjusting the pressure on the spring above the control piston. A gage is installed in the pilot line to aid in the adjustment of the valve at initial start-up.

17-23. Solenoid Valves

Solenoid valves are widely used in refrigerant, water, and brine lines in place of manual stop valves in order to provide automatic operation. A few of their many functions in the refrigerating system are described at appropriate places in this book.

A solenoid valve is simply an electrically operated valve which consists essentially of a coil of insulated copper wire and an iron core or armature (sometimes called a plunger) which is drawn into the center of the coil magnetic field when the coil is energized. By attaching a valve stem and pin to the coil armature, a valve port can be opened and closed as the coil is energized and deenergized, respectively.

Although there are a number of mechanical variations, solenoid valves are of two principal types: (1) direct acting and (2) pilot operated. Small solenoid valves are usually direct acting

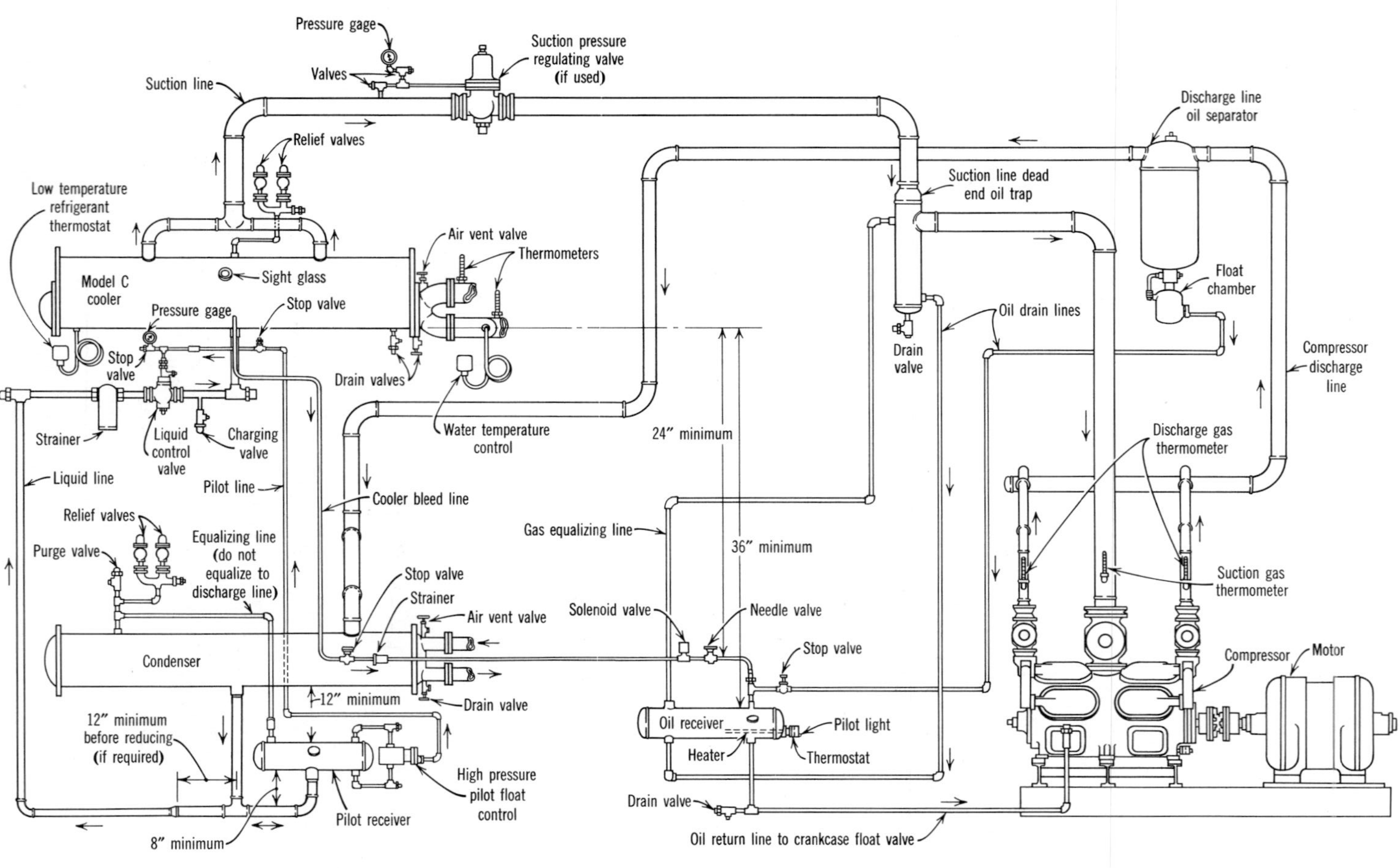

Fig. 17-39 Application of high pressure float pilot control valve. (Courtesy York Corporation.)

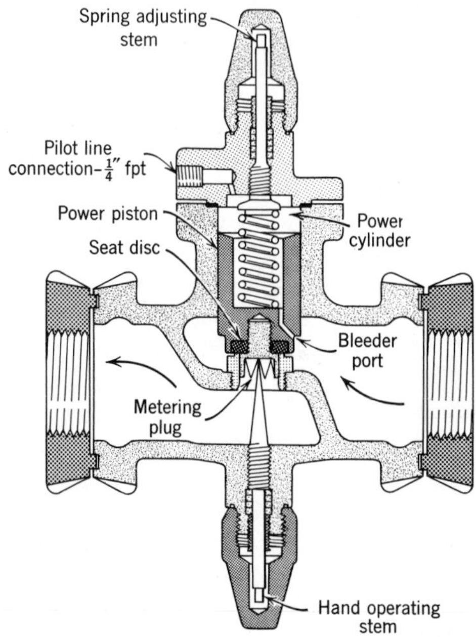

Fig. 17-40 Liquid control valve—low pressure. (Courtesy York Corporation.)

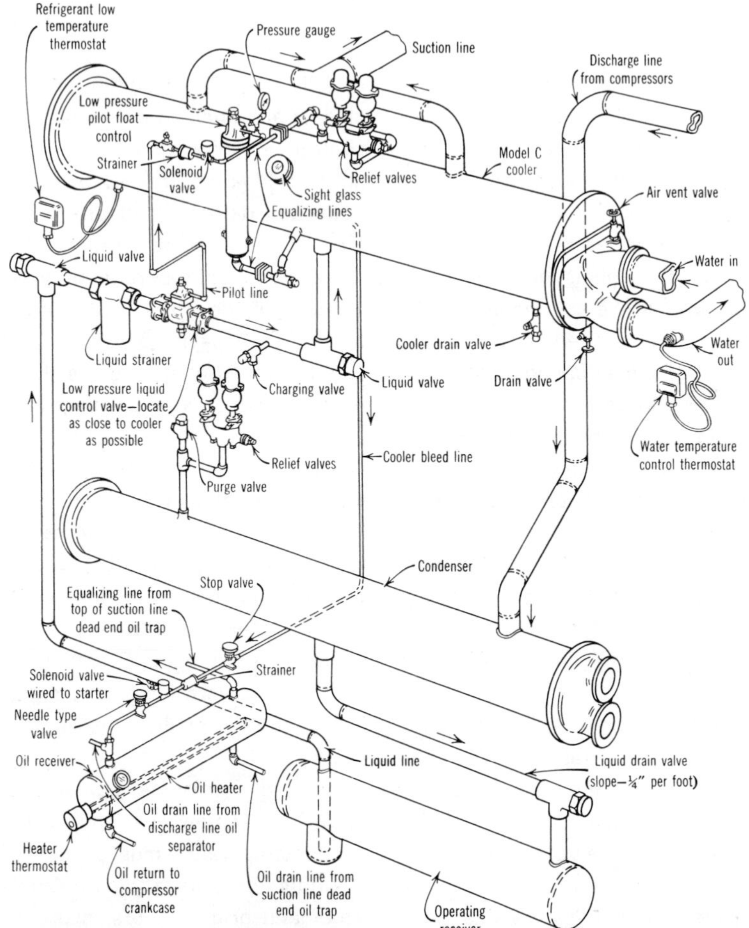

Fig. 17-41 Typical system—low-pressure float control. (Courtesy York Corporation.)

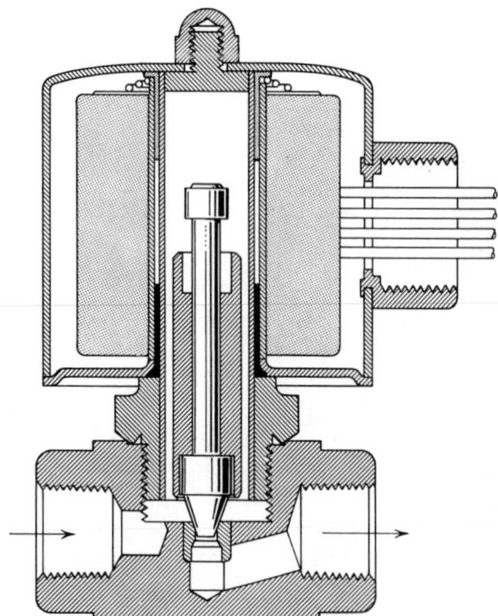

Fig. 17-42 Small, direct-acting solenoid valve. (Courtesy Sporlan Valve Company.)

(Fig. 17-42), whereas the larger valves are pilot operated (Fig. 17-43). In the direct acting valve, the valve stem attached to the coil armature controls the main valve port directly. In the pilot operated type, the coil armature controls only the pilot port rather than the main valve port. When the coil is energized, the armature is drawn into the coil magnetic field and the pilot port *A* is opened. This releases the pressure on top of the floating main piston *B* through the open pilot port, thereby causing a pressure unbalance across the piston. The higher pressure under the piston forces the piston to move upward, opening the main valve port *C*. When the coil is de-energized, the armature drops out of the coil magnetic field and closes the pilot port. The pressure immediately builds up on top of the main piston, causing the piston to drop and close off the main valve port.

Except where the solenoid valve is specially designed for horizontal installation, the solenoid valve must always be mounted in a vertical position with the coil on top.

In selecting a solenoid valve, the size of the valve is determined by the desired flow rate through the valve and never by the size of the line in which the valve is to be installed. Consideration must also be given to the maximum allowable pressure difference across the valve and to the pressure drop through the valve.

17-24. Suction Line Controls

Suction line controls are of two general types: (1) evaporator pressure regulators and (2) suction pressure regulators.

The function of the evaporator pressure regulator is to prevent the evaporator pressure, and therefore the evaporator temperature, from dropping below a certain predetermined minimum, regardless of how low the pressure in the suction line may drop because of the action of the compressor. It is important to recognize that the evaporator pressure regulator does not maintain a constant pressure in the evaporator but merely limits the minimum evaporator pressure. Evaporator pressure regulators are available with either throttling action (modulating) or snap-action (fully open or fully closed). The differential between the closing and opening points of the snap-action control not only gives close control of the product temperature but also provides for automatic defrosting of air-cooling evaporators when the temperatures in the space are sufficiently high to permit off-cycle defrosting.

The throttling-type evaporator pressure regulator (Fig. 17-44) is never fully closed while the compressor is operating. As the load on the evaporator decreases and the evaporator pressure tends to fall below the preset minimum regulator pressure, the regulator modulates toward the closed position to throttle the suction vapor to the compressor, thereby maintaining the evaporator pressure above the desired minimum. As the load on the evaporator increases and the evaporator pressure rises above the regulator setting, the regulator modulates toward the open position so that at full load the regulator is in the full open position.

Evaporator pressure regulators can be used in any installation when the evaporator pressure or temperature must be maintained above a certain minimum. They are widely used with water and brine chillers in order to prevent

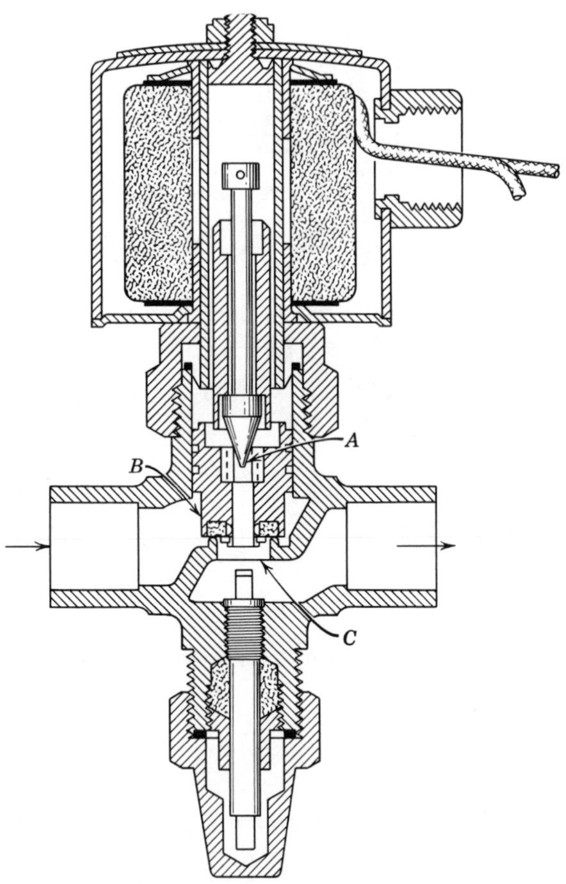

Fig. 17-43 Pilot-operated solenoid valve of the floating piston type. (Courtesy Sporlan Valve Company.)

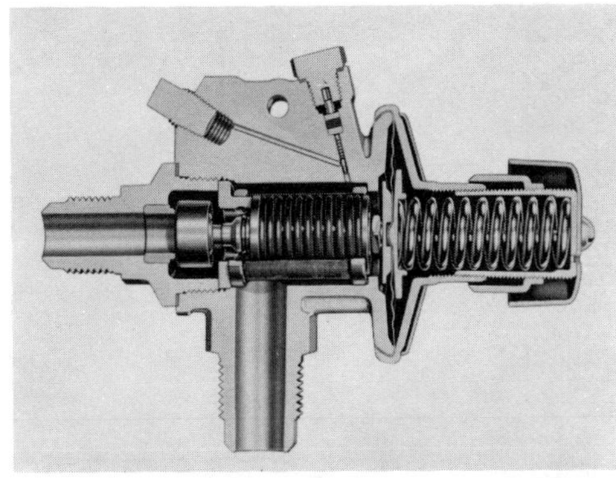

Fig. 17-44 Throttling-type evaporator pressure regulator. (Courtesy Controls Company of America.)

freeze-ups during periods of minimum loading. They are also used frequently in air-cooling applications where proper humidity control prescribes a minimum evaporator temperature. In multiple evaporator systems where the evaporators are all operated at approximately the same temperature, a single evaporator pressure regulator can be installed in the suction main to control the pressure in all the evaporators. On the other hand, where a multiple of evaporators connected to a single compressor are operated at different temperatures, a separate evaporator pressure regulator must be installed in the suction line of each of the higher temperature evaporators (see Section 20-16). This arrangement prevents the pressures in the warmer evaporators from dropping below the desired minimum while the compressor continues to operate to satisfy the coldest evaporator.

In large sizes, evaporator pressure regulators are pilot operated. The valve shown in Fig.

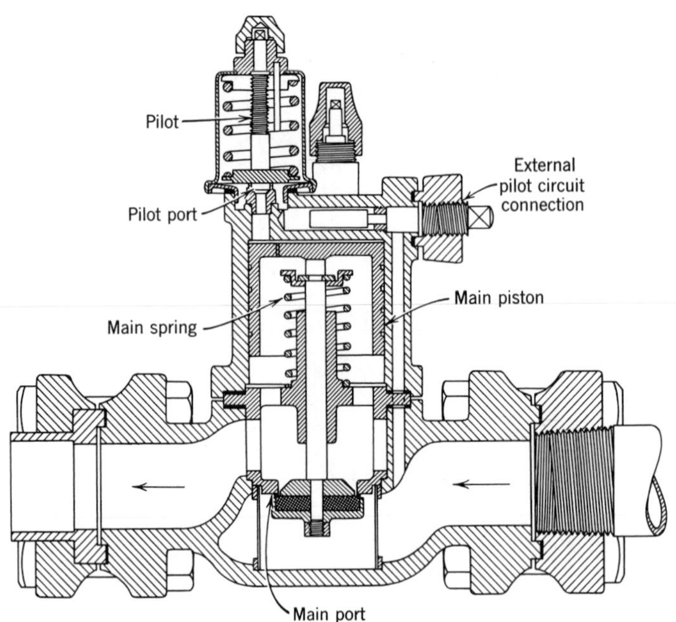

Fig. 17-45 Pilot-operated evaporator pressure regulator. (Courtesy Alco Valve Company.)

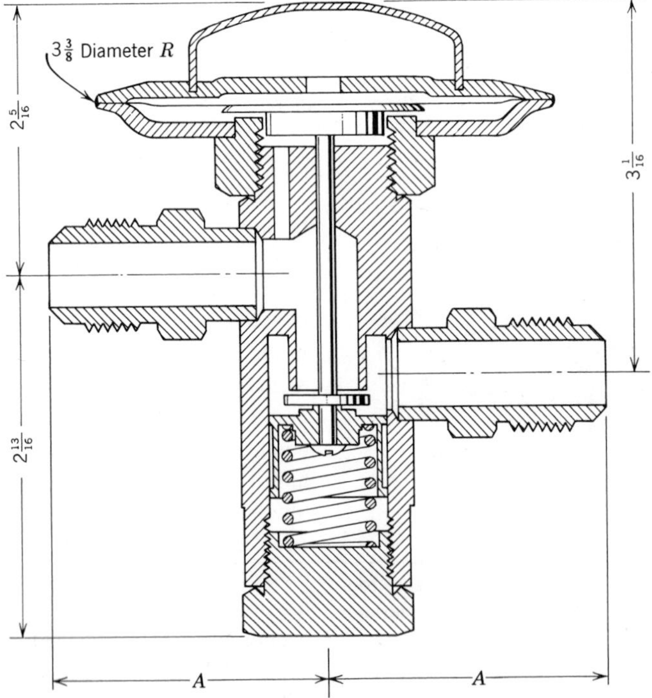

Fig. 17-46 Crankcase pressure regulator. (Courtesy Sporlan Valve Company.)

17-45 is designed for either internal or external pilot control. With internal pilot control, operation of the regulator is similar to that of the pilot solenoid described in the previous section. With external pilot control, operation of the regulator is similar to that of the pilot operated liquid control valve described in Section 17-22.

It is of interest to notice that the solenoid pilot shown in Fig. 17-34 has nothing to do with the pressure regulating function of the evaporator pressure regulator. The use of the solenoid pilot permits the evaporator pressure regulator to serve also as a suction stop valve.

The function of the suction pressure regulator (Fig. 17-46), sometimes called a "crankcase pressure regulator" or a "suction pressure hold-back valve," is to limit the suction pressure at the compressor inlet to a predetermined maximum, regardless of how high the pressure in the evaporator rises because of an increase in the evaporator load. The purpose of the suction pressure regulator is to protect the compressor driver from overload during periods when the evaporator pressure is above the normal operating pressure for which the compressor driver was selected. Suction pressure regulators are recommended for use on any installation where motor protection is desired because the system is subject to:

1. High starting loads.
2. Surges in suction pressure.
3. High suction pressure caused by hot gas defrosting or reverse cycle (heat pump) operation.
4. Prolonged operation at excessive suction pressures.

Like evaporator pressure regulators, suction pressure regulators in large sizes are pilot operated.

Customary Problem

17-1 An R-12 refrigerating system operating with saturated suction and discharge temperatures of 40°F and 110°F respectively, has the following pressure losses: 20 psi in the liquid line including the static head, 2 psi in the evaporator, 2 psi in the suction line, 3 psi in the discharge line, and 2 psi in the condenser. Determine the available pressure drop across the refrigerant flow control and select an appropriate thermostatic expansion valve from Table R-21 for a 10-ton load.

18

COMPRESSOR CONSTRUCTION AND LUBRICATION

18-1. Types of Compressors

Three types of compressors are commonly used for refrigeration duty: (1) reciprocating, (2) rotary, and (3) centrifugal. The reciprocating and rotary types are positive displacement compressors, compression of the vapor being accomplished mechanically by means of a compressing member. In the reciprocating compressor, the compressing member is a reciprocating piston, whereas in the rotary compressor, the compressing member takes the form of a roller, vane, or lobe. The centrifugal compressor, on the other hand, has no compressing member, compression of the vapor being accomplished primarily by action of the centrifugal force which is developed as the vapor is rotated by a high speed impeller.

All three compressor types have certain advantages in their own field of use. For the most part, the type of compressor employed in any individual application depends on the size and nature of the installation and on the refrigerant used.

18-2. Reciprocating Compressors

The reciprocating compressor is the most widely used type, being employed in all fields of refrigeration. It is especially adaptable for use with refrigerants requiring relatively small displacement and condensing at relatively high pressures. Among the refrigerants used extensively with reciprocating compressors are Refrigerants-12, 22, 500, 502, and 717 (ammonia).

As a general rule, because of limited valve areas, reciprocating compressors cannot be employed economically with low pressure refrigerants which require a large volumetric displacement per unit capacity. Although best applied to systems having evaporator pressures above one atmosphere, reciprocating compressors have also been used very successfully in both low temperature and ultra-low temperature installations.

Reciprocating compressors are available in sizes ranging from $\frac{1}{8}$ hp (approximately 90 W input) in small domestic units up through 250 tons or more in large industrial installations. The fact that reciprocating compressors can be manufactured economically in a wide range of sizes and designs, when considered along with its durability and efficiency under a wide variety of operating conditions, accounts for its widespread popularity in the refrigeration field.

Reciprocating compressors may be either single-acting or double-acting. In single-acting compressors, compression of the vapor occurs only on one side of the piston and only once during each revolution of the crankshaft, whereas in double-acting compressors, compression of the vapor occurs alternately on both sides of the piston so that compression occurs twice during each revolution of the crankshaft.

Single-acting compressors are usually of the enclosed type wherein the piston is driven directly by a connecting rod working off the crankshaft, both connecting rod and crankshaft being enclosed in a crankcase which is pressure tight to the outside, but open to contact with the system refrigerant (Fig. 18-1). Double-acting compressors, on the other hand, usually employ crankcases which are open (vented) to the outside, but isolated from the system refrigerant, in which case the piston is driven by a piston rod connected to a crosshead, which in turn, is actuated by a connecting rod working off the crankshaft (Fig. 18-2).

Because of its design, the double-acting compressor is obviously not practical in small sizes and therefore is limited to the larger industrial applications. Although the double-acting compressor is more expensive than the single-acting type, it is also more accessible for maintenance since the crankcase is not exposed to the system refrigerant. The principal disadvantage of the double-acting compressor is that the packing or seal around the piston rod is subject to both the suction and discharge pressures, whereas in the single-acting type compressor, the packing or seal around the crankshaft is subject only to the suction pressure. This disadvantage is made more serious because it is usually more difficult to maintain a pressure tight seal around the reciprocating piston rod of the double-acting compressor than it is around the rotating shaft of the single-acting type.

While a few double-acting compressors are still in service in some of the older refrigeration applications, they are no longer selected for refrigeration duty.

Single-acting reciprocating compressors differ considerably in design according to the type of duty for which they are intended. As previously stated, they can be further classified according to type as open, hermetic, or semi-hermetic (Section 6-15). While all reciprocating compressors used with ammonia are of the open type, those intended for use with the halo-

Fig. 18-1 Single-acting, vertical compressor with enclosed crankcase. (Courtesy Vilter Manufacturing Company.)

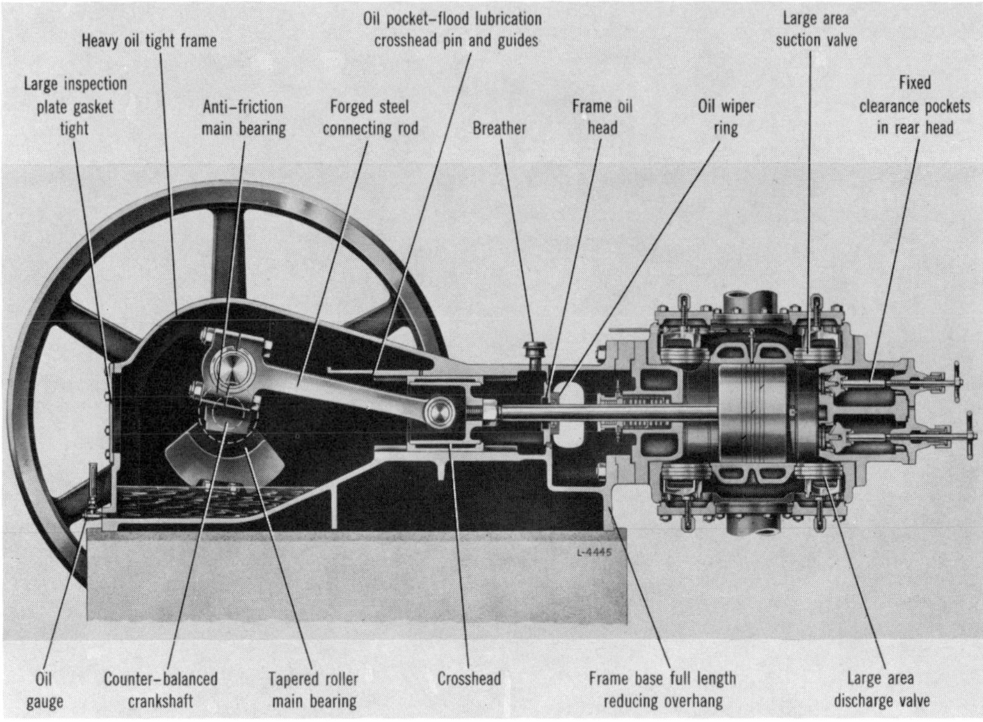

Fig. 18-2 Double-acting, horizontal compressor. Cylinder clearance can be adjusted manually to obtain capacity control. (Courtesy Worthington Corporation.)

carbon refrigerants are frequently of the hermetic and semihermetic types, particularly in the small and medium sizes.

Numerous combinations of the following design features are also used in order to obtain the desired flexibility: (1) the number and arrangement of the cylinders, (2) type of pistons, (3) type and arrangement of valves, (4) crank and piston speeds, (5) bore and stroke, (6) type of crankshaft, (7) method of lubrication, etc.

18-3. Cylinders

The number of cylinders varies from as few as one to as many as sixteen. In multicylinder compressors, the cylinders may be arranged in line, radially, or at an angle to each other to form a *V* or *W* pattern. For two and three cylinder compressors the cylinders are usually arranged in line. Where four or more cylinders are employed, *V*, *W*, or radial arrangements are ordinarily used. In-line arrangements have the advantage of requiring only a single valve plate, whereas *V*, *W*, and radial arrangements provide better running balance and permit the cylinders to be staggered so that the over-all compressor length is less. A modern direct-drive, open-type *V*/*W* compressor is shown in Fig. 18-3.

Compressor cylinders are usually constructed of close-grained cast iron which is easily machined and not subject to warping. For small compressors, the cylinders and crank-case housing are often cast in one piece, a practice which permits very close alignment of the working parts. For larger compressors, the cylinders and crankcase housing are usually cast separately and flanged and bolted together. As a general rule, the cylinders of the larger compressors are usually equipped with replaceable liners or sleeves.

Small compressors often have fins cast integral with the cylinders and cylinder head to increase cylinder cooling, whereas cylinder

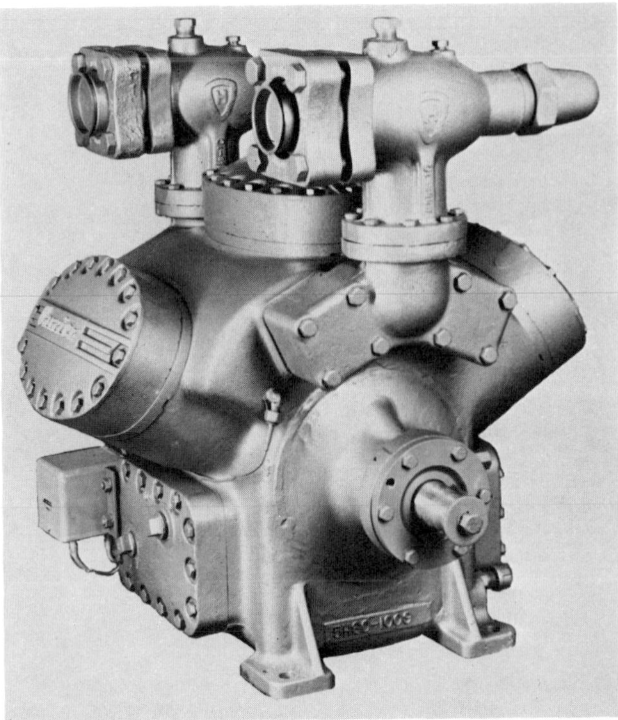

Fig. 18-3 Modern open-type, direct-drive *V/W* multicylinder compressor.

castings for larger compressors frequently contain water jackets for this purpose.

18-4. Pistons

Pistons employed in refrigeration compressors are of two common types: (1) automotive and (2) double-trunk. For the most part, the type of piston used depends on the method of suction gas intake and on the location of the suction valves. Automotive-type pistons are used when the suction gas enters the cylinder through suction valves located in the cylinder head (valve plate) as shown in Fig. 18-4. Double-trunk pistons are ordinarily used in medium and large compressors, in which case the suction gas enters through ports in the cylinder wall and in the side of the piston and passes into the cylinder through suction valves located in the top of the piston (Fig. 18-1). Notice that the bottom of the piston contains a bulkhead that seals off the hollow portion of the piston from the crankcase.

Because of small piston clearances (approximately 0.003 in. per inch of cylinder diameter), the oil film on the cylinder wall is usually sufficient to prevent gas blow-by around the pistons in small compressors. For this reason, rings are seldom used on pistons less than 2 in. (5 cm) in diameter. However, these pistons are provided with oil grooves to facilitate lubrication of the cylinder walls. Automotive-type pistons having diameters above 2 in. are usually equipped with two compression rings and one oil ring, the latter sometimes being located at the bottom of the piston. Double-trunk pistons are equipped with from one to three compression rings at the top and one or two oil rings at the bottom.

As a general rule, pistons are manufactured from close-grained cast iron, as are the rings. However, a number of aluminum pistons are in use. The use of cast iron permits closer tolerances. When aluminum pistons are used, they are usually equipped with at least one compression ring.

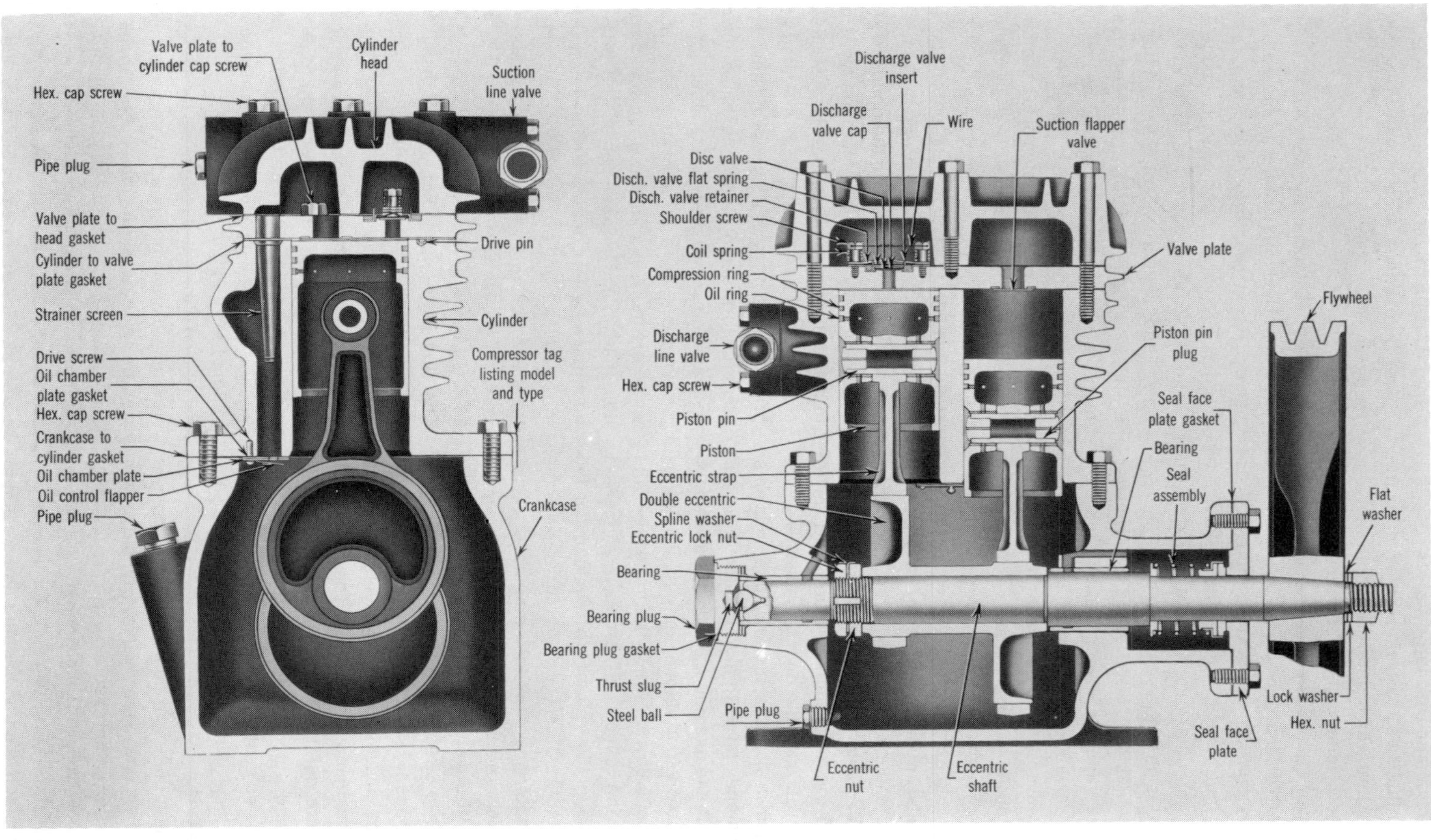

Fig. 18-4 Typical cross section of two cylinder, eccentric-type refrigeration compressor. (Courtesy Dunham-Bush, Inc.)

18-5. Suction and Discharge Valves

Since it influences to a greater or lesser degree all the factors which determine both the volumetric and compression efficiencies of the compressor, the design of the compressor suction and discharge valves is one of the most important considerations in compressor design. Furthermore, it will be shown later that valve design determines to a large extent the over-all design of the compressor.

The friction loss (wiredrawing effect) suffered by the vapor in flowing through the compressor valves and passages is primarily a function of vapor velocity and increases as the velocity of the vapor increases. Therefore, in order to minimize the wiredrawing losses, the valve should be designed to provide the largest possible restricted area (opening) and to open with the least possible effort. Too, whenever practical, the valves should be so located as to provide for straight-line flow (uniflow) of the vapor through the compressor valves and passages. In all cases, the valve openings must be sufficiently large to maintain vapor velocities within the maximum limits. The maximum permissible vapor velocity may be defined as that velocity beyond which the increase in the wiredrawing effect will produce a marked reduction in the volumetric efficiency of the compressor and/or a material increase in the power requirements of the compressor.

To minimize back leakage of the vapor through the valves, the valves should be designed to close quickly and tightly. In order to open easily and close quickly, the valves should be constructed of lightweight material and be designed for a low lift. They should be strong and durable and they should operate quietly and automatically. Furthermore, they should be so designed and placed that they do not increase the clearance volume of the compressor.

Although there are numerous modifications within any one type, valves employed in refrigeration compressors can be grouped into three basic types: (1) the poppet, (2) the ring plate, (3) the flexing or reed. All three types operate automatically, opening and closing in response to pressure differentials caused by changes in the cylinder pressure. To facilitate rapid closing of the valve, most discharge valves and some suction valves are spring loaded.

18-6. Poppet Valves

The poppet valve is similar to the automotive valve, except that the valve stem is much shorter. The valve is enclosed in a cage which serves both as a valve seat and valve stem guide and also as a retainer for the valve spring. A spring, dashpot, or bleeder arrangement is also included in the assembly to cushion and limit the valve travel. Except for minor differences, the design of the suction and discharge poppet valves is essentially the same, the principal difference being that the suction poppet valve is beveled on the stem side of the valve face, whereas the discharge poppet valve is beveled on the opposite side.

The poppet valve, one of the first types used in refrigeration compressors, is essentially a slow speed valve and as such is limited at the present time to a few types of slow speed compressors. In high speed machines, the poppet valve has been discarded in favor of either the ring plate valve or the flexing valve, both of which are more adaptable to high speed operation than is the poppet valve. The principal advantage of the poppet valve is that it can be mounted flush and therefore does not increase the clearance volume of the compressor.

18-7. Ring Plate Valves

The ring plate valve (Fig. 18-5) consists of a valve seat, one or more ring plates, one or more valve springs, and a retainer. The ring plates are held firmly against the valve seat by the valve springs, which also help to provide rapid closure of the valves. The function of the retainer is to hold the valve springs in place and to limit the valve lift.

The ring plate valve is suitable for use in both slow speed and high speed compressors and it may be used as either the suction or the discharge valve. In fact, when both suction and discharge valves are located in the head, they are usually contained in the same ring plate assembly. For example, in Fig. 18-5, the outer ring serves as the suction valve, whereas the two smaller rings serve as the discharge valve.

Fig. 18-5 Ring plate valve assembly. Outer ring plate is the suction valve. The two inner rings constitute the discharge valve. (Courtesy Frick Company.)

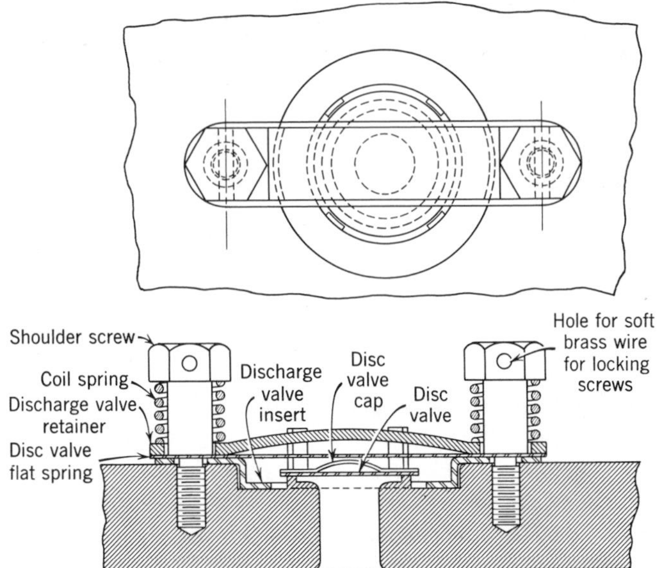

Fig. 18-6 Discharge valve assembly (disc valve). (From the *ASRE Data Book*, Design Volume, 1957–58 Edition. Reproduced by permission of the American Society of Heating. Refrigerating and Air-Conditiong Engineers)

One modification of the ring plate valve is the disc valve (Fig. 18-6), which is simply a thin disc held in place on the valve seat by a retainer.

18-8. Flexing Valves

Flexing valves vary in individual design to a a much greater extent than do either the poppet or ring plates-types. One popular type of flexing valve suitable for use in medium and large compressors is the Feather valve* (Fig. 18-7), which consists of a valve seat, a series of ribbon steel strips, and a valve guard or

* A proprietary design of the Worthington Pump and Machinery Company.

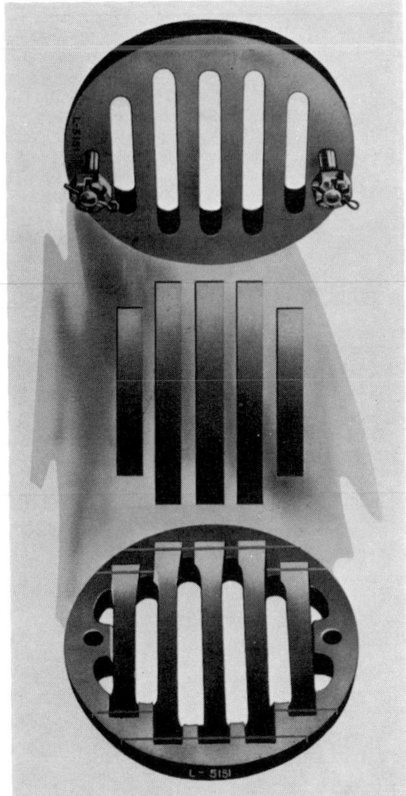

Fig. 18-7 One popular design of flexing valve. (Courtesy Worthington Corporation.)

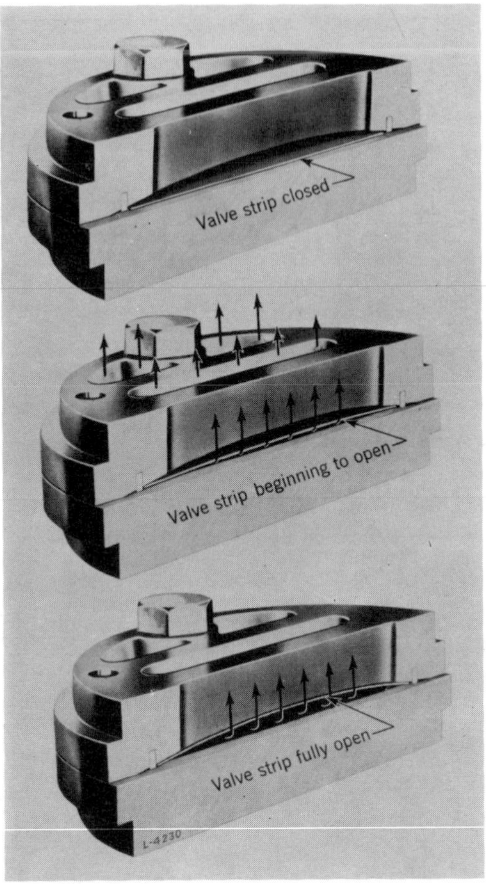

Fig. 18-8 Illustrating the operation of the Worthington Feather valve. (Courtesy Worthington Corporation.)

retainer. The flexible metal strips fit over slots in the valve seat and are held in place by the valve guard. The operation of the Feather valve is illustrated in Fig. 18-8. It is important to notice that in order to allow the valve reeds to flex under pressure, they are not tightly secured at either end. The principal advantage of the Feather valve is that the reeds are lightweight and easily opened and are so designed that they provide a large restricted area, all of which tend to reduce the wire-drawing effect to a minimum.

One disadvantage of all flexing valves, and one that is shared by the ring plate type, is that they cannot be mounted flush as can the poppet valve. Because of the presence of the valve port spaces, the clearance volume is necessarily increased in all compressors employing either ring plate or flexing valves of any design.

A flexing valve design widely used in smaller compressors is the flapper valve, of which there are innumerable variations. The flapper valve is a thin steel reed, which is usually fastened securely at one end while the opposite unfastened end rests on the valve seat over the valve port. The free end of the reed flexes or "flaps" to cover and uncover the valve port (Fig. 18-9).

A flapper valve design frequently employed in discharge valves, called a "beam" valve, is shown in Fig. 18-10. The valve reed is held in place over the valve port by a spring-loaded beam which is arched in the center to permit the reed to flex upward at this point. The ends of the reed are slotted and are held down by only the tension of the coil springs in order to allow the ends of the reed to move as the reed flexes up and down at the center. The spring-loaded beam also acts as a safety device to

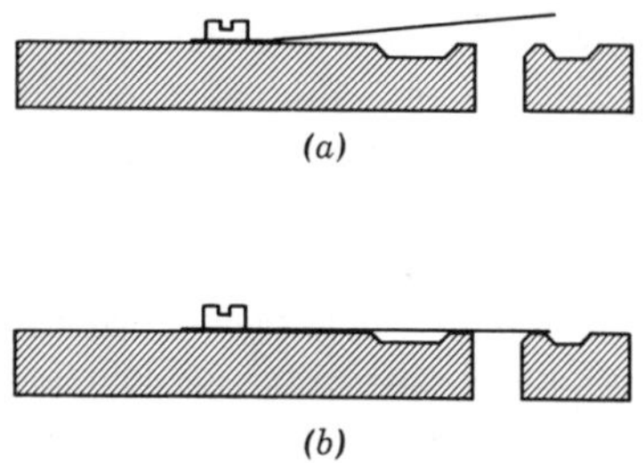

Fig. 18-9 Flapper-type flexing valve. (*a*) Port open. (*b*) Port closed.

Fig. 18-10 Compressor valve plate assembly. (Courtesy Tecumseh Products Company.)

protect the compressor against damage in the event that a slug of liquid refrigerant or oil enters the valve port. Since the valves are designed to handle vapor, there is not usually sufficient clearance to pass a slug of liquid of any kind. However, with the arrangement in Fig. 18-10, the whole valve assembly will lift to pass liquid slugs. Under ordinary discharge pressures, the tension of the springs is ample to hold the beam down firmly on the ends of the reed.

Another type of flexing valve in common use is the diaphragm valve. The diaphragm valve consists of a flexible metal disc which is held down on the valve seat by a screw or bolt through the center of the disc. The disc flexes up and down to uncover and cover the valve port. A diaphragm valve used as a suction valve mounted in the crown of a piston is illustrated in Fig. 18-11.

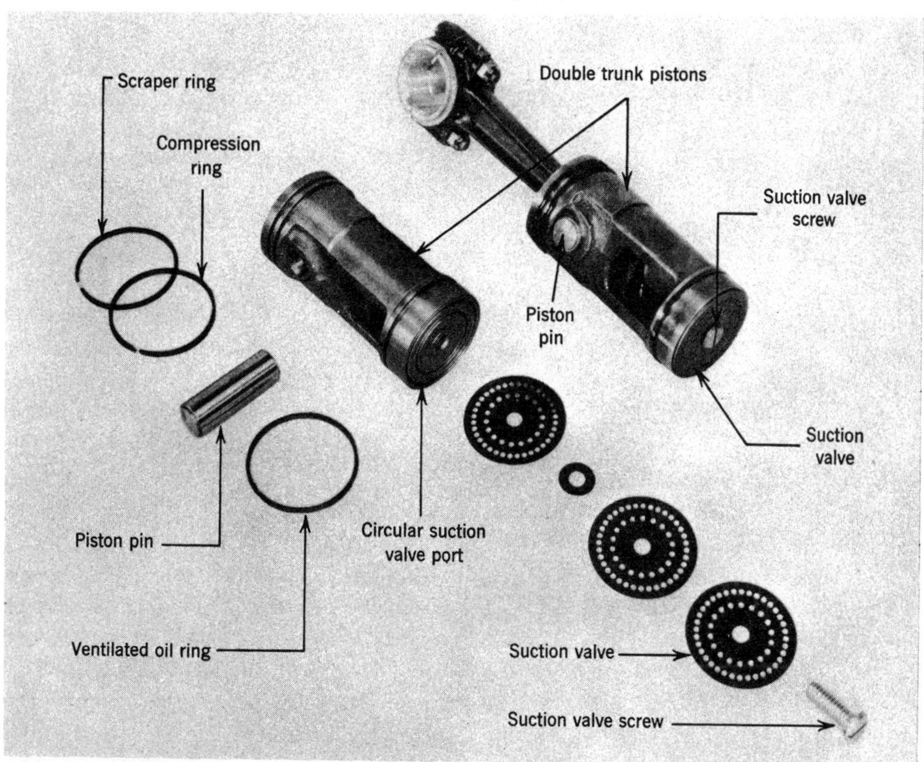

Fig. 18-11 Diaphragm-type suction valve. This same type of valve is often mounted in a valve plate and used as a discharge valve. (Courtesy York Corporation.)

18-9. Valve Location

As previously described, the discharge valves are usually located in the cylinder head, whereas the suction valves may be located either in the head, in which case the suction vapor enters the cylinder through the cylinder head, or in the crown of the piston, in which case the suction vapor enters through the side of the cylinder. As a general rule, with larger compressors, the suction valves are located in the piston and the suction vapor enters through the cylinder wall. With small and medium compressors, the suction valves are usually located in the cylinder head. When both valves are placed in the cylinder head, the head must be partitioned to permit separation of the suction and discharge vapors.

Most large compressors are equipped with secondary safety heads which are located at the end of the cylinder and held in place by heavy coil springs (Fig. 18-1). Under normal discharge pressures, the safety head is held firmly in place by the springs. However, in the event that a slug of liquid or some other noncompressible material enters the cylinder, the safety head will rise under the increased pressure and permit the material to pass into the cylinder head, thereby preventing damage to the compressor. In small compressors, the discharge valve is usually designed to provide this protection.

In large compressors, the valves and seats are removable for replacement. In small compressors, the suction and discharge valves are usually incorporated into a valve plate assembly, which is removed and replaced as a unit (see Fig. 18-10).

18-10. Crank and Piston Speeds

In an effort to reduce the size and weight of the compressor, the trend in modern compressor design is toward higher rotational speeds. Since single-cylinder piston displacement is a function of bore, stroke, and rpm, it follows that as the rpm is increased, the bore and stroke can be decreased proportionally without loss of displacement, provided that the volumetric efficiency of the compressor remains the same.

Rotative speeds between 500 and 1750 rpm are quite common, whereas some compressors are being operated successfully at speeds up to 3500 rpm. The maximum rotational speed of the compressor is more or less limited by the maximum allowable piston velocity.

Theoretically, there is no limit to piston speeds. However, as a practical matter, piston speeds are limited to a maximum of approximately 800 fpm, the limiting factor being the available valve area.

Since considerable difficulty is experienced in finding sufficient space in the compressor for valve arrangements, valve areas tend to be somewhat limited. Hence, when piston velocities are increased beyond 800 fpm, the velocity of the vapor through the valves will usually become excessive, with the result that the volumetric efficiency of the compressor is decreased while the power required by the compressor is increased.

Piston velocity is a function of compressor rpm and the length of the piston stroke. The following relationship exists:

$$\text{Piston velocity (fpm)} = \text{rpm} \times \text{stroke (ft)} \times 2 \text{ strokes per revolution}$$

For example, a compressor having a 4 in. stroke and rotating at 1200 rpm will have a piston speed of

$$\frac{1200 \times 4 \text{ in.} \times 2}{12} = 800 \text{ fpm}$$

If the rotational speed of the compressor is increased to 3600 rpm, in order to maintain a piston velocity of 800 fpm, the length of stroke will have to be reduced to

$$\frac{12 \times 800 \text{ fpm}}{3600 \text{ rpm} \times 2} = 1.33 \text{ in.}$$

Obviously, then, the maximum speed at which an individual compressor can be rotated without exceeding allowable piston velocities depends upon the length of stroke. The shorter the stroke, the higher is the maximum permissible rpm. This accounts for the fact that the volumetric efficiency of a compressor will usually remain constant or increase slightly as the speed of the compressor is increased up to a certain point beyond which, if the speed is further increased, the efficiency of the compressor will decrease.

18-11. Bore and Stroke

The relationship of the bore to the stroke differs somewhat with the individual compressor. Although the bore dimension may be either less than or greater than that of the stroke, the general trend in high speed compressors is toward a large bore and a short stroke. When the suction and discharge valves are both located in the head, a large bore is usually required in order to provide sufficient valve area. Too, since the piston stroke and the compressor rpm are both limited somewhat by the maximum allowable piston speed, it follows that the only practical means of increasing single-cylinder piston displacement is to increase the size of the bore. The increase in piston displacement accruing from an increase in the size of the bore need not increase the vapor velocity through the compressor valves since the increase in the bore also increases the available valve area.

However, since the amount of blow-by around the piston increases as the size of the bore is increased with relation to the stroke, good design practice limits the bore dimension to approximately 125% of that of the stroke. If the bore is increased beyond this point, the blow-by around the piston becomes excessive.

Cylinder bores range from approximately 1 in. in small domestic compressors up to approximately 18 in. in some of the large industrial types.

18-12. Cranks, Rods, and Bearings

Crank-shafts employed in large compressors are of the crank-throw type and are usually constructed of forged steel or alloy cast iron. All bearing-journals are highly polished and are usually case-hardened, particularly where brass or aluminum bearings are used. As a general rule, the crankshaft has a standard taper on the flywheel end, the flywheel being fastened to the crankshaft with one or more woodruff keys and a locknut arrangement. Crankshaft bearings are usually of the sleeve type, although antifriction (roller or ball) bearings are sometimes used for the mains. Common bearing materials are bronze, aluminum, and babbit.

The eccentric-type shaft, which consists of a cast iron eccentric mounted on a straight steel shaft (see Fig. 18-4) is often used in smaller compressors. The eccentric is counterbalanced and is fastened to the shaft by a key-and-lock screw arrangement. Since the bearing of the connecting rod completely encircles the eccentric, the entire eccentric acts as a bearing surface.

Connecting rods are constructed of bronze, aluminum, forged steel, or cast iron. Wrist pins are usually case-hardened steel. Wrist-pins bearings are generally of the sleeve type made of bronze and pressed into the rod. Bronze, aluminum, and cast iron are often used without bearings, in which case the shaft is usually case-hardened.

18-13. Crankshaft Seals

In order to prevent the leakage of refrigerant and oil from the crankcase (or the leakage of air into the crankcase in the event that the pressure in the crankcase is below atmospheric), a seal or packing must be provided at the point where the crankshaft passes through the crankcase. One of the oldest methods of sealing the crankshaft, still employed on some large ammonia compressors, is through the use of a stuffing box (Fig. 18-2). The stuffing box is a cylindrical housing which is cast as an integral part of the crankcase where the shaft emerges, and which is bored to an inside diameter somewhat larger than the diameter of the crankshaft. A series of packing rings, placed over the shaft and inserted into the stuffing box, fills the space in the stuffing box between the shaft and the stuffing box housing. The packing is held in place by a threaded gland nut which, when tightened, causes the packing rings to swell and press tightly against the shaft and housing, thereby affecting a vapor tight seal between the two. Because of the pressure of the rings against the rotating shaft, the rings will eventually wear and permit refrigerant to seep around the shaft, whereupon the packing gland

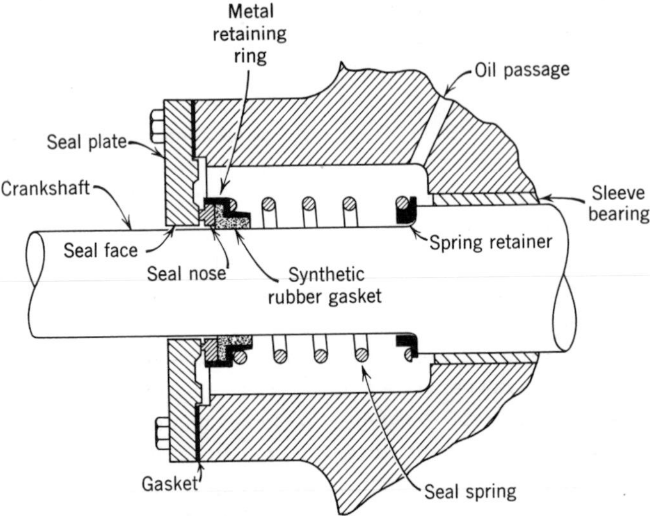

Fig. 18-12 Crankshaft seal.

nut must be tightened again to reestablish a tight seal. Although the stuffing box seal is satisfactory for large ammonia installations where an operator is on duty to tighten the packing gland as the occasion requires, they are not suitable for small compressors or for large compressors designed for automatic operation.

A crankshaft seal suitable for use on automatic equipment must be self-adjusting to compensate for wear and for varying crankcase pressures. It must not leak under pressure or vacuum when the shaft is rotating or idle. It must be self-lubricating, have a reasonably long life, and be easily replaceable in the field. Although there are a number of different seal designs which meet these qualifications and which are in use at the present time, one relatively simple design of crankshaft seal, which is rapidly gaining in popularity, is shown in Fig. 18-12. The seal consists essentially of a spring-loaded bronze or hard carbon seal nose which is sealed to the crankshaft with a synthetic rubber gasket. The spring holds the seal nose firmly against a highly polished steel seal face which is a part of the seal plate. An oil film between these two smooth surfaces form an effective vapor-tight seal. Notice that sealing occurs in three places: (1) at the rubber gasket between the seal nose and the crankshaft, (2) between the seal nose and seal face, and (3) at the gasket between the seal plate and the crankcase housing.

The compressor shown in Fig. 18-1 employs a double shaft seal. Notice that the seal remains completely submerged in oil during both the running and off cycles.

18-14. Compressor Lubricating Oils

The fact that the compressor lubricating oil usually comes into contact with, and often mixes with, the system refrigerant makes it necessary that the oil used to lubricate refrigeration compressors be specially prepared for that purpose. Some of the more important properties of the oil which must be considered when selecting the compressor lubricating oil are: (1) chemical stability, (2) pour and/or floc point, (3) dielectric strength, and (4) viscosity. In evaluating these oil properties with relation to an individual compressor, all the following factors should be taken into account: (1) the type and design of the compressor, (2) the nature of the refrigerant to be used, (3) the evaporator temperature, and (4) the compressor discharge temperature.

18-15. Chemical Stability

The importance of chemical stability is emphasized by the fact that it is necessary for the compressor lubricating oil to perform its lubricating function continuously and effectively without undergoing change for long periods of time. Since changing the oil in a hermetic motor-compressor is not usually practical, the same oil frequently remains in these units throughout the life of the unit, which is often ten years or more. Because of the high discharge temperatures encountered in hermetic motor-compressor units, particularly where air-cooled condensers are used, the ability of the oil to remain stable and resist decomposition under high temperature is especially important when selecting a lubricating oil for these units.

For the most part, the chemical stability of an oil is closely related to the amount of unsaturated hydrocarbons present in the oil. The smaller the percentage of unsaturated hydrocarbons contained in the oil, the more stable is the oil. For refrigeration service, a high quality oil with a very low percentage of unsaturated hydrocarbons is desired. These oils are usually light in color, being just off from a water-white.

18-16. Pour, Cloud, and Floc Points

The pour point of an oil is the lowest temperature at which the oil will flow, or "pour," when tested under certain specified conditions. Of two oils having the same viscosity, one may have a higher pour point than the other because of a greater wax content. Pour point is an important consideration in selecting an oil for low temperature systems. Naturally, the pour point of the oil should be well below the lowest temperature to be obtained in the evaporator. If the pour point of the oil is too high, the oil tends to congeal on the surface of the evaporator tubes, causing a loss in evaporator efficiency. Since this oil is not returned to the compressor, inadequate lubrication of the compressor may also result.

Since all lubricating oils contain a certain amount of paraffin, wax will precipitate from any oil if the temperature of the oil is reduced to a sufficiently low level. Because the oil becomes cloudy at this point, the temperature at which the wax begins to precipitate from the oil is called the cloud point of the oil. If the cloud point of the oil is too high, wax will precipitate from the oil in the evaporator and in the refrigerant control. Although a small amount of wax in the evaporator does little harm, a small amount of wax in the refrigerant control will cause stoppage of that part, with the result that the system will become inoperative.

The floc point of the oil is the temperature at which wax will start to precipitate from a mixture of 90% Refrigerant-12 and 10% oil by volume. Since the use of an oil soluble refrigerant lowers the viscosity of the oil and affects both the pour and cloud points, where oil miscible refrigerants are employed, the floc point of the oil is a more important property than the pour or cloud points. The use of 10% oil in the oil-refrigerant mixture to determine the floc point seems quite realistic, since the tendency of an oil-refrigerant mixture to separate wax increases as the amount of oil in the mixture increases and since the amount of oil circulating with the refrigerant seldom exceeds 10% and is usually much less.

Because the floc point of the oil is a measure of the relative tendency of the oil to separate wax when mixed with an oil soluble refrigerant, it is an important consideration when selecting an oil for use with an oil miscible refrigerant at evaporator temperatures below 0°F (−20°C). However, floc point has no significance when a nonmiscible refrigerant is used.

18-17. Dielectric Strength

The dielectric strength of an oil is a measure of the resistance that the oil offers to the flow of electric current. It is expressed in terms of the voltage required to cause an electric current to arc across a gap one-tenth of an inch wide between two poles immersed in the oil. Since any moisture, dissolved metals, or other impurities contained in the oil will lower its dielectric strength, a high dielectric strength is an indication that the oil is relatively free of contaminants. This is especially important in oils used with hermetic motor-compressor units, since an oil of low dielectric strength may contribute to grounding or shorting of the motor windings.

18-18. Viscosity

Viscosity has already been defined in Section 16-10 as the resistance that a fluid offers to flow. With regard to the lubricating oil, viscosity may also be defined as a measure of the "body" of the oil or of the ability of the oil to perform its lubricating function by forming a protective film or coating between the various moving parts of the compressor, thus keeping the parts separated and preventing wear. In order to provide adequate lubrication for the compressor, the viscosity of the oil must be maintained within reasonable limits. If the viscosity of the oil is too low, the oil will not have sufficient body to keep the moving parts separated and thin film lubrication will result, accompanied by excessive wearing of the rubbing surfaces. Too, since, in addition to its lubricating function, the oil frequently must serve as a sealing agent between the low and high pressures in the compressor, excessive blow-by of vapor around the pistons (in a reciprocating compressor) or vanes (in a rotary compressor) may occur when the viscosity of the oil is low. On the other hand, when the viscosity of the oil is too high, fluid friction will be excessive and the power consumption of the compressor will be increased. Furthermore, in extreme cases, a high viscosity oil may not have sufficient fluidity to penetrate between the various rubbing surfaces, particularly when tolerances are close, with the result that the lubrication of the compressor parts will be inadequate.

The viscosity of a lubricating oil is usually measured in Saybolt Seconds Universal (SSU), which is an index of the time in seconds required for a given quantity of oil at a controlled temperature, usually 100°F (37.78°C), to flow by gravity from a reservoir into a flask through a capillary tube of specified internal diameter and length. An oil having a temperature of 100°F and requiring 300 sec to pass through the tube is said to have a viscosity of 300 SSU at 100°F.

The viscosity of the lubricating oil changes considerably with the temperature, increasing as the temperature decreases. The effect of temperature on the viscosity of a typical lubricating oil is shown graphically in Fig. 18-13 (see top line—0% refrigerant dilution). Notice that the oil has a viscosity at 100°F (37.78°C) of approximately 175 SSU, but increases to approximately 1700 SSU when the temperature of the oil is reduced to 40°F (4.4°C).

Shown also in Fig. 18-13 is the effect of refrigerant dilution on the viscosity of the

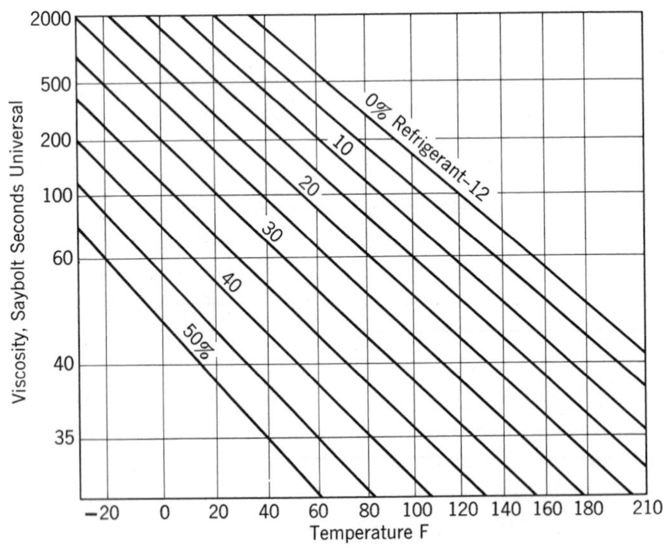

Fig. 18-13 Viscosity temperature curves of solution of Refrigerant-12 in oil. (From *ASRE Data Book,* Design Volume, 1957–58 Edition. Reproduced by permission of the American Society of Heating, Refrigerating and Air-Conditioning Engineers.)

lubricating oil. Notice, for example, that pure oil having a viscosity of 175 SSU at 100°F has a viscosity of about 60 SSU at this same temperature when diluted with 15% Refrigerant-12.

It is evident from the foregoing that both the operating temperature range and the effect of refrigerant dilution must be taken into account in selecting the proper viscosity oil. In all cases, the compressor manufacturer's recommendations should be followed when they are available.

18-19. Methods of Lubrication

Methods of lubricating the compressor vary somewhat depending upon the type and size of the compressor and upon the individual manufacturer. However, for the most part, lubrication methods can be grouped into two general types: (1) splash and (2) forced feed. Although forced feed lubrication can be found even in very small compressors, as a general rule, small, vertical, enclosed compressors up through approximately 15 hp (10 kW inpot) are splash lubricated. Above this size, most compressors employ some type of forced feed lubrication. Often, a combination of the splash and forced feed methods is found in a single compressor.

In the splash method of lubrication, the compressor crankcase acts as an oil sump and is filled with oil to a level approximately even with the bottom of the main crank bearings. With each revolution of the crankshaft, the connecting rod and crankshaft (or eccentric) dip into the oil causing the oil to be splashed up on the cylinder walls, bearings, and other rubbing surfaces. Usually, small cavities or oil reservoirs are located at each end of the crankcase housing immediately over the main bearings. These cavities collect oil which feed by gravity down into the main bearings and shaft seal (Fig. 18-4). In some instances, connecting rods are rifle-drilled to carry oil to the wrist-pin bearings. Too, oil scoops or dippers are sometimes installed on the end of the connecting rods to increase splashing and/or to aid in forcing oil through rifle-drilled oil passages.

A modified type of splash lubrication, sometimes called flooded lubrication, employs slinger rings, discs, screws, or similar devices to raise the oil to a level above the crankshaft or main bearings, from where it is allowed to flood over the bearings and/or feed through oil channels to the various rubbing surfaces (Fig. 6-14). This method is particularly suitable for small, high speed compressors where the conventional splash system may result in excessive oil carryover because of violent splashing of the oil in the crankcase.

In the forced feed method of lubrication, the oil is forced under pressure through oil tubes and/or rifle-drilled passages in the crankshaft and connecting rods to the various rubbing surfaces. After performing its lubricating function, the oil drains by gravity back into a sump located in the crankcase of the compressor. The oil is circulated under pressure developed by a small oil pump located in the crankcase of the compressor, usually at the end of the crankshaft (Fig. 18-1). Since most oil pumps are automatically reversible, the direction of crank rotation is not usually critical with regard to compressor lubrication. However, this is not true of all compressors, particularly those employing oil dippers with splash lubrication. When rotation is critical, an arrow denoting the proper direction of rotation is usually embossed on the flywheel or crankcase housing.

Oil strainers are always placed at the suction inlet of the oil pump to prevent the entrance of foreign material into the pump or bearings. Although not required, oil filters are worthwhile in all forced feed lubrication systems to eliminate the possibility of plugged oil line resulting from the accumulation of sludges or other residue. An oil pressure failure safety switch (Section 21-20) should be employed in conjunction with all forced feed lubrication systems.

In some large compressors, cylinders are lubricated by mechanical forced feed lubricators which are located external to the compressor crankshaft. In such cases, the cylinder lubrication system is entirely separate from the internal pressure lubricating system.

The bearings and cross-heads of horizontal, open crankcase compressors are usually splash lubricated (Fig. 18-2). Oil from the crankcase is carried to the cross-head by splash-fed troughs. Cylinders and piston rod packing glands are lubricated by mechanical forced feed lubricators similar to those used to lubricate the cylinders of large vertical compressors.

18-20. Liquid Refrigerant in the Compressor Crankcase

The presence of liquid refrigerant in the compressor crankcase is always undesirable for a number of reasons. In the first place, excessive dilution of the crankcase oil by liquid refrigerant can result in inadequate lubrication of the compressor parts. More important, however, is that fact that the liquid refrigerant will vaporize in the crankcase and cause foaming of the oil, with the result that the amount of oil carried over into the discharge line is materially increased. Under certain conditions, oil foaming may become so severe that all the oil is pumped out of the crankcase. Not only will this leave the compressor without lubrication but there is also the possibility that noncompressible slugs of liquid refrigerant and oil will enter the cylinder and cause serious damage to the compressor in the form of broken valves and pistons and bent or broken rods and shafts. Too, where considerable oil foaming occurs in compressors employing forced feed lubrication, lubrication will often be inadequate because the oil pump is unable to develop sufficient pressure to deliver the oil to the various rubbing surfaces.

Furthermore, vaporization of liquid refrigerant in the crankcase tends to reduce the capacity and efficiency of the compressor in that the resulting vapor is drawn into the cylinder and displaces vapor which would otherwise be taken from the suction line.

Liquid refrigerant may gain entrance into the crankcase in a number of ways:

1. Improper application or adjustment of the refrigerant flow control will often cause continuous or intermittent overfeeding of the evaporator, in which case liquid refrigerant will slop over from the evaporator into the suction line and be carried to the compressor crankcase. As described in Chapter 17, this condition is more likely to occur during start-up than at any other time. In any event, it is easily prevented or corrected by proper application and adjustment of the refrigerant control and/or by properly designed suction piping.
2. Liquid refrigerant may drain by gravity into the compressor crankcase from the evaporator and/or suction piping during the off cycle. This condition is also caused by faulty system design, particularly with reference to the evaporator and suction piping. A leaking refrigerant control may also be a contributing factor. Here, again, the condition is readily corrected or prevented by proper design.
3. Any time the temperature at the compressor crankcase falls below that of the evaporator, liquid refrigerant will boil off in the evaporator and condense in the compressor crankcase. Naturally, this can occur only during the off cycle and only when the compressor is so located that the ambient temperature at the crankcase can fall below that of the evaporator. It is prevalent in the wintertime in installations where the compressor is located outside or in a basement or some other cold location. The solution, of course, is to maintain the temperature of the crankcase above the saturation temperature of the refrigerant vapor. This can be accomplished by installing an electrical heating element in the crankcase or by moving the compressor to a warmer location. Another corrective measure is to operate the system on a pump-down cycle.

Because of the tendency of the lubrication oil to absorb oil miscible refrigerant vapors, a certain amount of liquid refrigerant will always be dissolved into the lubricating oil in any system employing an oil miscible refrigerant, assuming that the refrigerant vapor and the oil are permitted to come in contact with one another, as is usually the case. For the most part the percentage refrigerant that can be dissolved into the crankcase oil depends on three factors: (1) the degree of miscibility of the refrigerant, (2) the pressure of the refrigerant vapor, and (3) the temperature of the lubricating oil. For any one refrigerant, the percentage refrigerant that will be dissolved into the oil depends only on the pressure of the refrigerant vapor, the temperature of the oil, and the length of time that the two are in contact under steady conditions.

The solubility of Refrigerant-12 in oil under various conditions of temperature and pressure is shown graphically in Fig. 18-14. Notice that the percentage of Refrigerant-12 which can be dissolved in the oil increases considerably as the temperature of the oil decreases and as the pressure of the refrigerant vapor increases. For example, when the temperature of the oil is 100°F (38°C) and the refrigerant vapor pressure is 20 psi (1.38 bar), the maximum percentage or Refrigerant-12 which can be present in the oil-refrigerant mixture in the crankcase is only approximately 8% by weight. However, if the oil is cooled to 80°F (27°C) while the refrigerant vapor pressure is increased to 60 psi (4.14 bar), the percentage refrigerant in the oil-refrigerant mixture could be as high as 42%. In other words, under the latter conditions, the so-called lubricating oil in the crankcase could actually be 42% Refrigerant-12.

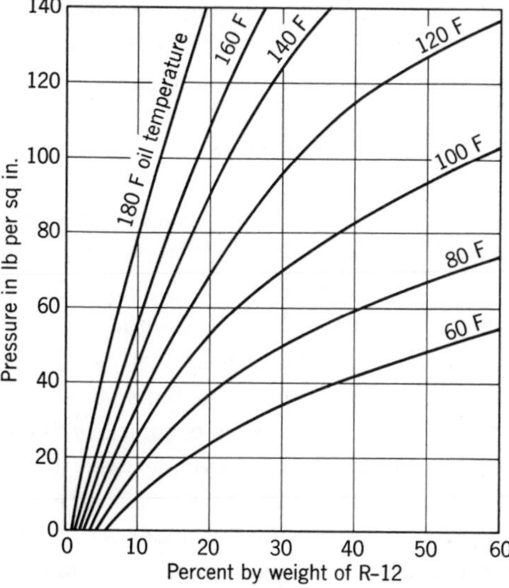

Fig. 18-14 Temperature-pressure relationship of Refrigerant-12 oil mixtures (pressure in psig). (From *ASRE Data Book,* Design Volume, 1957–58 Edition. Reproduced by permission of the American Society of Heating. Refrigerating and Air-Conditioning Engineers.)

The practical significance of the foregoing can be illustrated by the use of an example. Suppose that during the compressor off cycle, the pressure on the low side of a Refrigerant-12 system rises to 38 psig (3.6 bar), whereas the crankcase cools to a temperature of 80°F (27°C). Assuming that the off cycle is of sufficient length to permit equilibrium to be established, the percentage of liquid refrigerant in the oil-refrigerant mixture in the crankcase will be approximately 20% by weight, as determined from Fig. 18-14. Suppose now that the compressor cycles on and that the pressure in the evaporator and in the crankcase is immediately reduced to 25 psig (2.74 bar). At this lower pressure, the maximum percentage of refrigerant that can be present in the oil-refrigerant mixture is only approximately 13%. Therefore, in a matter of only a few seconds approximately one-third of the refrigerant (7% by weight of the total mixture) must of necessity vaporize out of the mixture in order to establish the new percentage. Naturally, the vaporization of this much refrigerant out of the mixture in such a short time will cause severe foaming of the oil with the result that a considerable amount of oil is drawn into the cylinder out of the crankcase.

There are several ways to reduce oil foaming and the loss of oil from the crankcase during compressor start-up. One common method is to equip the compressor with an oil check valve, which is installed in the oil passage between the suction inlet of the compressor and the crankcase (Fig. 18-15). With oil miscible refrigerants, the oil pumped over into the system ordinarily returns to the compressor with the suction vapor. On entering the suction inlet, the oil is separated from the vapor by impingement before the vapor enters the cylinder. The separated oil drains from the inlet chamber to the crankcase through an oil passage provided for this purpose. Since this oil passage also serves to equalize the crankcase pressure to the compressor suction, a check valve installed in the oil passage will

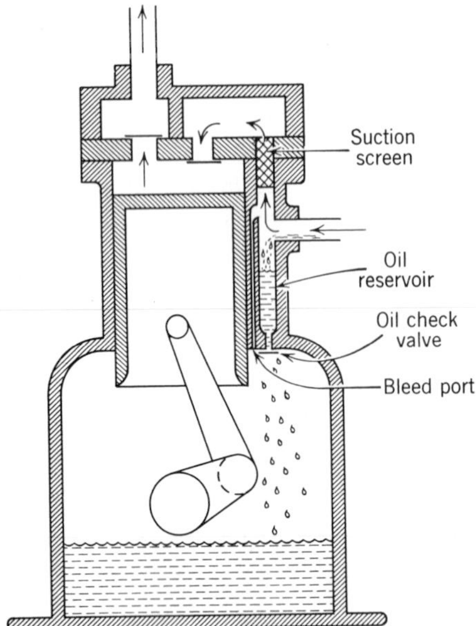

Fig. 18-15 Illustrating oil check value and bleed port.

prevent the crankcase pressure from venting to the suction, thereby eliminating the sudden reduction in crankcase pressure which produces oil foaming at start-up.

However, since no oil can drain through the passage to the crankcase until the crankcase pressure is reduced to the suction pressure, a small bleed port must be provided around the check (or through it) to permit the crankcase pressure to bleed off slowly into the compressor suction after the compressor cycles on. The bleed port is required also to relieve cylinder blow-by gases back to the suction of the compressor. If blow-by gases are not vented to the suction, the crankcase pressure will build up to the discharge pressure. Not only would this prevent the oil from returning to the crankcase, it would also cause a material increase in the power requirements of the compressor.

During the normal running cycle, the crankcase pressure is approximately the same as the suction, and minor fluctuation in the suction pressure produced by the throttling action of the refrigerant control will supply the pressure differential necessary to cause the oil to flow through the check valve into the crankcase. However, the suction inlet chamber of the compressor must be large enough to serve as a reservoir for all the oil that returns to the compressor during the time that the crankcase pressure is too high to permit oil drainage into the crankcase.

Another method of reducing the amount of oil foaming at start-up, and one which is rapidly growing in popularity, is to install a small wattage heating element in the compressor crankcase. The crankcase heater is wired to come on when the compressor cycles off and serves to keep the oil in the crankcase warm during the offcycle so that the amount of refrigerant which can be dissolved into the oil is relatively small. However, care should be taken to wire the heater to the secondary of the main disconnect so that it cannot be turned off unless the main disconnect is pulled.

Still another method of reducing oil foaming at start-up is to operate the system on a pump-down cycle, in which case the evaporator is completely evacuated and the crankcase pressure reduced to a low level before the compressor cycles off. The resulting low pressure in the crankcase limits the amount of refrigerant absorbed by the oil. The pump-down cycle used alone or in conjunction with either the oil check valve or the crankcase heater is very effective in reducing oil foaming.

18-21. Rotary Compressors

Rotary compressors in common use are of three general designs: (1) rolling piston, (2) rotating vane, and (3) helical lobe (screw). The rolling piston type employs a cylindrical steel roller which revolves on an eccentric shaft, the latter being mounted concentrically in a cylinder (Fig. 18-16). Because of the shaft eccentric, the cylindrical roller is eccentric with the cylinder and touches the cylinder wall at the point of minimum clearance. As the shaft turns, the roller rolls around the cylinder wall in the direction of shaft rotation, always maintaining contact with the cylinder wall. With relation to the camshaft, the inside surface of the cylinder roller moves counter to the direction of shaft rotation in the manner of a crankpin bearing. A spring-loaded blade, mounted in a slot in the cylinder wall, bears firmly against

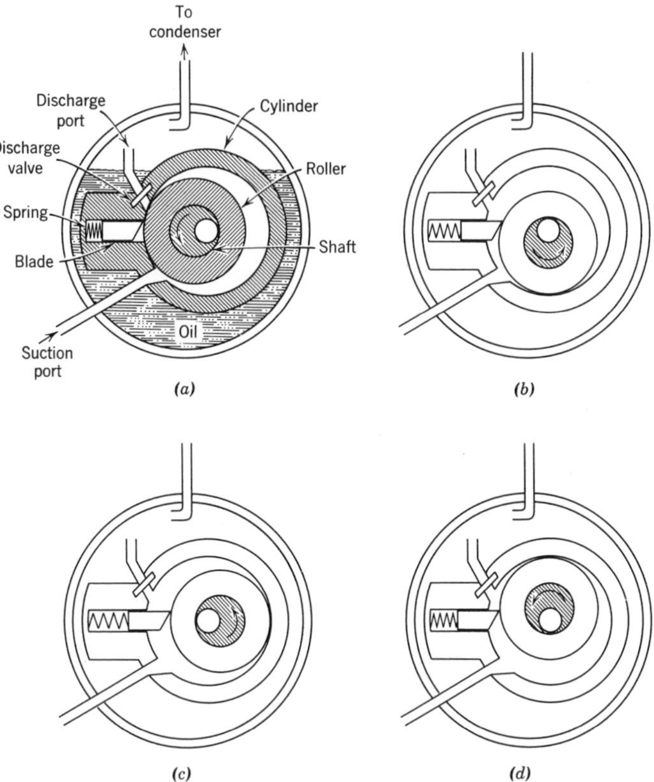

Fig. 18-16 Blade-type rotary compressor.

the roller at all times. The blade moves in and out of the cylinder slot to follow the roller as the latter rolls around the cylinder wall.

Cylinder heads or end-plates are used to close the cylinder at each end and to serve as supports for the camshaft. Both the roller and blade extend the full length of the cylinder with only working clearance being allowed between these parts and the end-plates. Suction and discharge ports are located in the cylinder wall near the blade slot, but on opposite sides. The flow of vapor through both the suction and discharge ports is continuous, except for the instant that the roller covers one or the other of the ports. The suction and discharge vapors are separated in the cylinder at the point of contact between the blade and roller on one side and between the roller and cylinder wall on the other side.

The point on the cylinder wall in contact with the roller changes continuously as the roller travels around the cylinder. At one point during each compression cycle the roller will cover the discharge ports, at which time only low pressure vapor will be in the cylinder. The manner in which the vapor is compressed by the roller is illustrated by the sequence of drawings in Fig. 18-16.

The whole cylinder assembly is enclosed in a housing and operates submerged in a bath of oil. Notice that the high pressure vapor is discharged into the space above the oil level in the housing from where it passes into the discharge line. All rubbing surfaces in the compressor including the end-plates are highly polished and closely fitted. Although no suction valves are needed, a check or flapper valve is installed in the discharge passage to eliminate back-feeding of the discharge vapor into the cylinder. When the compressor is operating, an oil film forms a seal between the high and low pressure areas. However, when the compressor stops, the oil seal is lost and the high and low pressures equalize in the compressor. A check must be placed in the suction line (or discharge line) to prevent the high pressure discharge gas from backing up through the compressor and suction line into the evaporator when the compressor cycles off.

The rotating vane type of rotary compressor employs a series of rotating vanes or blades which are installed equidistant around the periphery of a slotted rotor (Fig. 18-17). The rotor shaft is mounted eccentrically in a steel cylinder so that the rotor nearly touches the cylinder wall on one side, the two being separated only by an oil film at this point. Directly opposite this point the clearance between the rotor and the cylinder wall is maximum. Heads or end-plates are installed on the ends of the cylinder to seal the cylinder and to hold the rotor shaft. The vanes move back and forth radially in the rotor slots as they follow the contour of the cylinder wall when the rotor is turning. The vanes are held firmly against the cylinder wall by action of the centrifugal force developed by the rotating rotor. In some instances, the blades are springloaded to obtain a more positive seal against the cylinder wall.

The suction vapor drawn into the cylinder through suction ports in the cylinder wall is entrapped between adjacent rotating vanes. The vapor is compressed by the reduction in volume which results as the vanes rotate from the point of maximum rotor clearance to the point of minimum rotor clearance. The compressed vapor is discharged from the cylinder

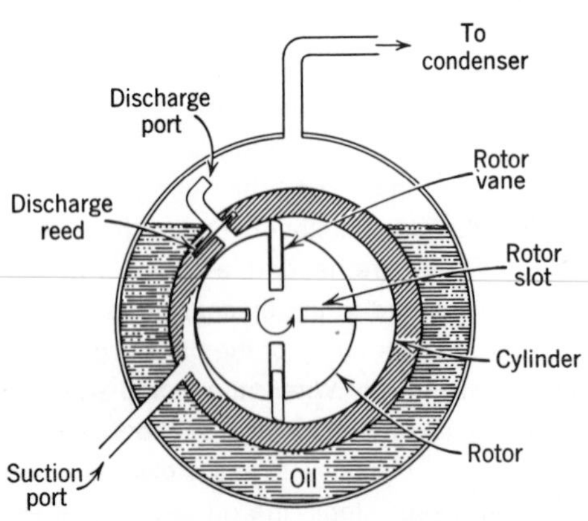

Fig. 18-17 Vane-type rotary compressor.

through ports located in the cylinder wall near the point of minimum rotor clearance. The discharge ports are so located as to allow discharge of the compressed vapor at the desired point during the compression process, that point being the design point of the compressor. Operation of the compressor at compression ratios either above or below the design point results in compression losses and in increased power requirements. Current practice limits compression ratios to a maximum of 7 to 1.

Like the rolling piston type, the rotating vane type of rotary compressor also requires the use of a check valve in the suction or discharge line to prevent the discharge gas from leaking back through the compressor and suction line to the evaporator when the compressor cycles off.

Although rotary compressors are positive displacement machines, because of their rotary motion and the smoother, more constant flow of the suction and discharge gases, they are much less subject to mechanical vibration and to the pronounced discharge pulsations associated with the reciprocating compressor. As in the case of reciprocating compressors, rotary compressors experience volumetric and compression losses resulting from backleakage and blow-by around the compressing element, cylinder heating, clearance, and wiredrawing. However, since clearance volumes and the associated reexpansion of the clearance vapor are small, the volumetric efficiency of rotary compressors is relatively high, being about 65% to 80%, depending on the individual design and the operating conditions.

Small rotary compressors of both the rolling piston and rotating vane types have been employed with Refrigerants-12 and 114 in domestic refrigerators and freezers for many years. Recently, the rolling piston type has been applied in sizes ranging up to 5 hp (approximately 4 kW input) in some small packaged air-conditioning units.

Large rotary compressors of the rotating vane type (Fig. 18-18) are widely used with Refrigerants-12, 22, and ammonia as the low stage or "booster" compressor in multistage compression systems with saturated suction temperatures ranging down to −125°F

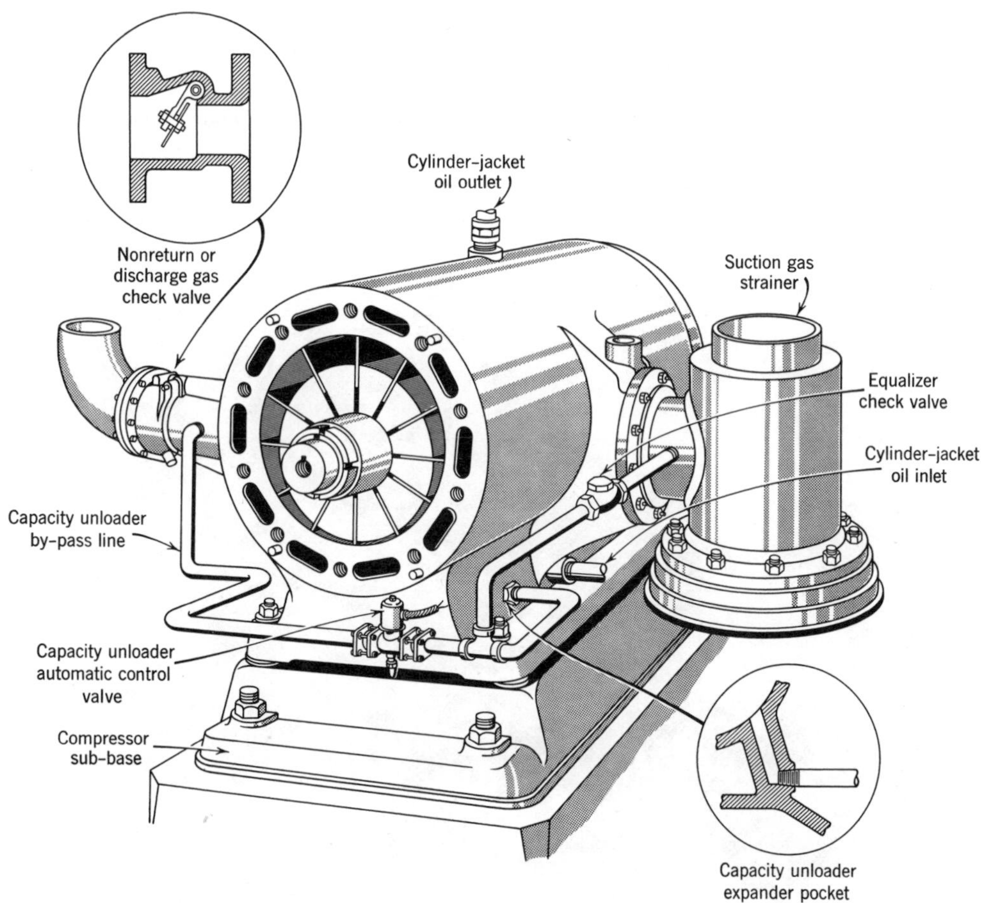

Fig. 18-18 Large capacity, rotating vane-type rotary compressor. (Courtesy Freezing Equipment Sales, Inc.)

(−87°C). These compressors are equipped with jacket and/or oil cooling to prevent overheating and to improve compressor efficiency. Jacket cooling is usually accomplished by circulating water or oil through the jacket, although one manufacturer has an alternate method involving the direct expansion of R-11.

Large compressors are usually pressure lubricated by oil from a mechanical lubricator or rotary gear pump driven directly from the compressor crankshaft. In some cases the oil is chilled and introduced into the rotary cylinder at key points to provide lubrication and cooling for the blades as well as to provide an oil seal for all running surfaces. In such cases, jacket cooling is not required. Although water/glycol oil-coolers are sometimes used in conjunction with evaporative condensers, the oil is usually chilled in a direct expansion oil chiller. Another method of providing the required cooling is to inject liquid refrigerant into the cylinder or housing at or immediately following the discharge ports.

Although the capacity of a rotary compressor varies directly with speed, capacity control is most frequently accomplished by relieving low-pressure refrigerant gas at the lowest degree of compression by bypassing the blade pocket to the suction line and thereby providing only partial compression of the total gas flow.

18-22. Helical Rotary (Screw) Compressors

The helical rotary or screw compressor is a positive displacement compressor in which compression is accomplished by the enmeshing of two mating helically groved rotors suitably housed in a cylinder equipped with appropriate inlet and discharge ports (Fig. 18-19). The male rotor is normally the driving rotor and consists of a series of lobes (usually four) along the length of the rotor that mesh with similarly formed corresponding helical flutes (usually six) on the female (driven) rotor (Fig. 18-20). As the rotors turn, gas is drawn through the inlet opening to fill the space between the male lobe and the female flute. As the rotors continue to rotate, the gas is moved past the suction port and sealed in the interlobe space. The gas so trapped in the interlobe space is moved both axially and radially and is

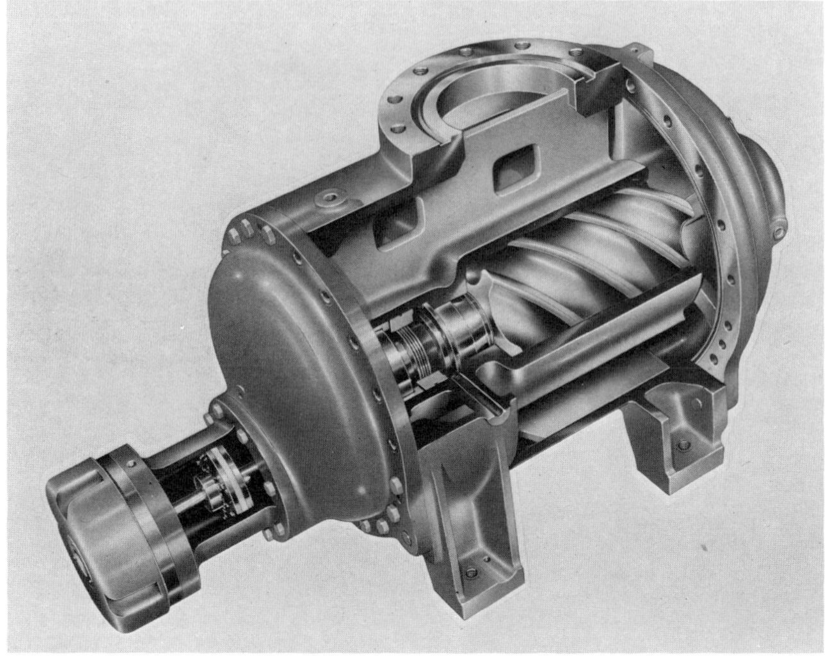

Fig. 18-19 (Courtesy York Corporation.)

Fig. 18-20 (Courtesy Frich Company.)

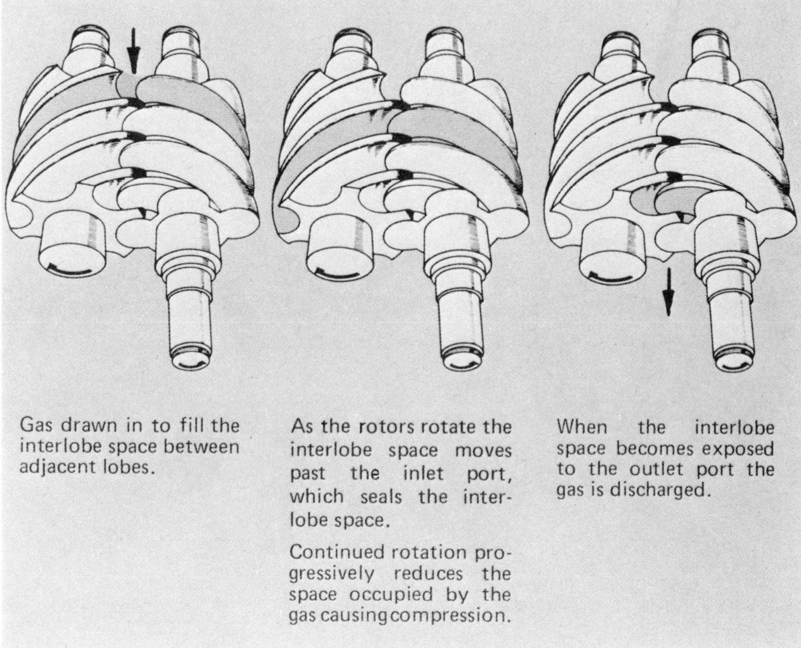

Fig. 18-21 (Courtesy York Corporation.)

compressed by direct volume reduction as the enmeshing of the lobes of the compressor progressively reduces the space occupied by the gas. Compression of the gas continues until the interlobe space communicates with the discharge ports in the cylinder and the compressed gas leaves the cylinder through these ports (Fig. 18-21).

From the foregoing it is evident that the screw compressor is a fixed volume ratio compressor, the volume ratio being a function of internal compressor design. As previously indicated, fixed value ratio machines operate most efficiently (without overcompression or undercompression) when the system compression ratio is the same as the internal compression ratio. Since it is not possible to manufacture screw compressors with a perfect internal ratio for every system, several internal ratios have been standardized. Manufacturers' rating data, which reflect deviations from optimum compression ratios, permit selection of an internal volume ratio that will result in the minimum power input.

Compressor capacity control is accomplished through the use of a unique sliding valve that is located inside the compressor housing underneath the rotors and operated by a piston in a hydraulic cylinder mounted on the compressor (Fig. 18-22). The piston is actuated by lubricating oil, which is fed from the oil pump to either side of the piston, thereby moving the slide valve and altering the point in the rotor travel at which compression begins. This permits internal gas recirculation and provides smooth, stepless capacity regulation to match system requirements down to 10% of design capacity with a roughly proportional reduction in power requirements. The slide valve also provides for unloaded starting, and its initial selection determines the internal volume ratio, thereby providing a means for the optimization of power requirements for a given application.

Lubricating systems are rather elaborate, consisting of an external oil pump, an oil separator and receiver or sump, and some means of cooling the oil, along with associated filters and safety devices. Screw compressors were not used in the United States for refrigeration duty until the late 1960s, when the principle of injecting oil into the compressor

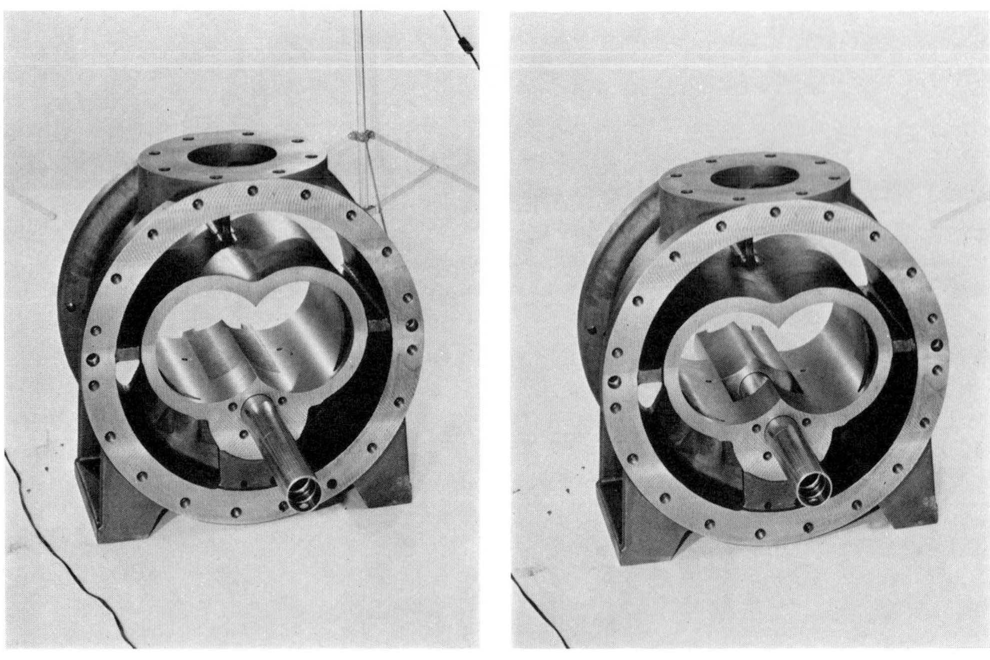

Fig. 18-22 (Courtesy Frich Company.)

Fig. 18-23 (Courtesy Frich Company.)

to absorb much of the heat of compression was applied. As in the case of some vane-type rotary compressors, oil injection provides internal cooling and prevents overheating of the compressor, and at the same time, oil seals the running clearances between the rotors. Since internal oil cooling maintains relatively uniform discharge temperatures below 100°C regardless of the compression ratio, screw compressors can be operated at compression ratios as high as 25 to 1, which makes single-stage compression practical in some low-temperature applications where multistage compression has heretofore been required.

Because the relatively large quantities of oil that are injected into the compressor are carried out of the compressor in the discharge gas stream, an efficient oil separator is an absolute necessity. Since the oil draining from the separator into the receiver is at the discharge temperature, the oil must be cooled before being reinjected into the compressor if it is to perform its cooling function adequately.

Oil cooling is accomplished in a water, glycol, or refrigerant-cooled shell-and-tube heat exchanger or by the direct injection of liquid refrigerant into the discharge of the compressor.

The screw compressor is suitable for use with all common refrigerants and has a high efficiency over a wide range of compression ratios. Because of its simplicity, versatility, durability, and reliability, the screw compressor has experienced widespread acceptance and growing usage in industrial refrigeration and air-conditioning applications in the capacity range of 50 tons (37 kW input) and above.

A complete screw compressor unit, including driver, oil separator and receiver, oil pump, and oil cooler, is illustrated in Fig. 18-23.

18-23. Centrifugal Compressors

The centrifugal compressor consists essentially of a series of impeller wheels mounted on a steel shaft and enclosed in a cast iron casing (Fig. 18-24). The number of impeller

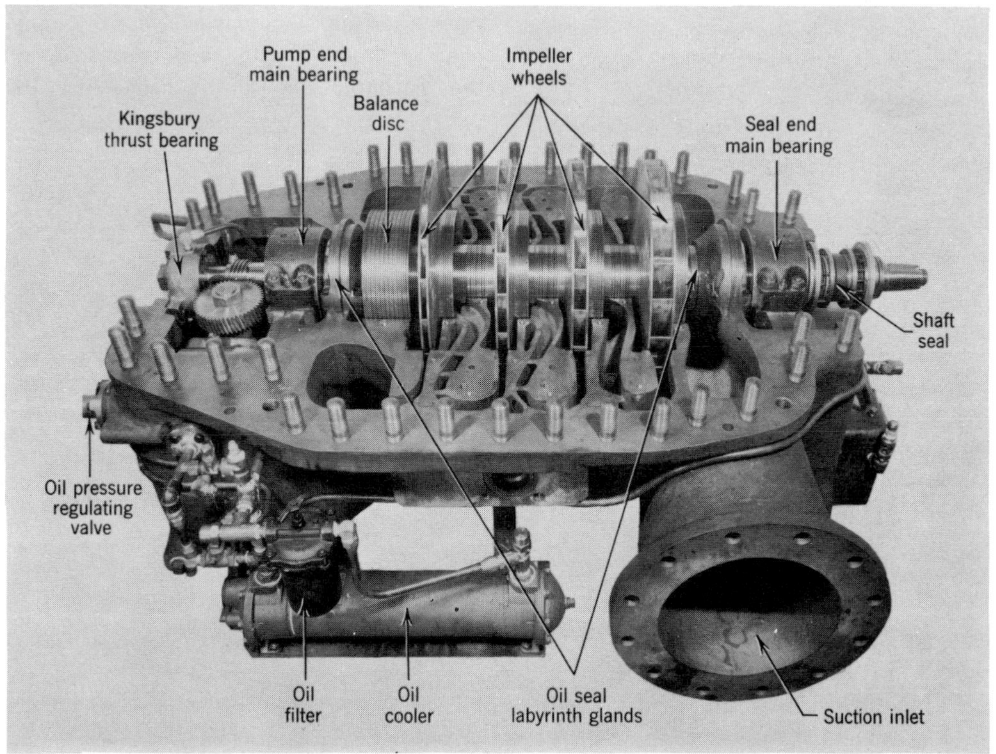

Fig. 18-24 Four-stage centrifugal compressor with upper half of housing removed to show construction of details. (Courtesy York Corporation.)

wheels employed depends primarily on the magnitude of the thermodynamic head which the compressor must develop during the compression process. Compressors employing two, three, and four wheels (stages of compression) are common. More wheels may be used when the required increase in head is sufficiently large to demand it. As many as twelve wheels have been used in some individual cases.

As shown in Fig. 18-25, the impeller wheel of a centrifugal compressor consists of two discs, a hub disc and a cover disc, with a number of blades or vanes mounted radially between them. To resist corrosion and erosion, the impeller blades are usually constructed either of stainless steel or of high carbon steel with a lead coating. A typical two-stage rotor is shown in Fig. 18-26.

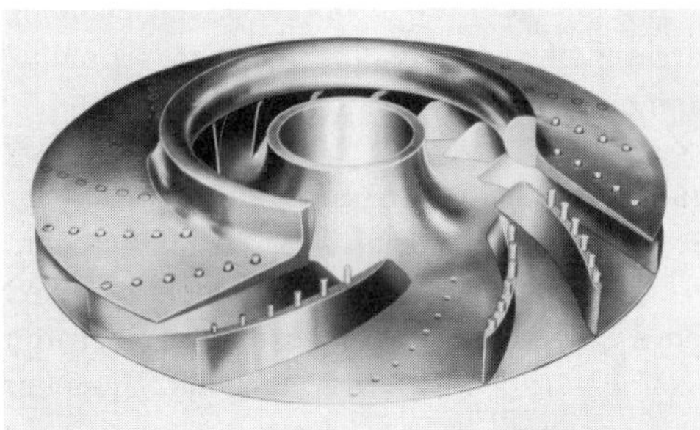

Fig. 18-25 Cutaway view of centrifugal compressor impeller wheel. (Courtesy York Corporation.)

The operating principles of the centrifugal compressor are similar to those of the centrifugal fan or pump. Low-pressure, low-velocity vapor from the suction line is drawn in the inlet cavity or "eye" of the impeller wheel along the axis of the rotor shaft. On entering the impeller wheel, the vapor is forced radially outward between the impeller blades by action of the centrifugal force developed by the rotating wheel, and is discharged from the blade tips into the compressor housing at high velocity and at increased temperature and pressure. The high-pressure, high-velocity vapor discharged from the periphery of the wheel is collected in specially designed passages in the casing which reduce the velocity of the vapor and direct the vapor to the inlet of the next stage impeller or, in the case of the last stage impeller, to a discharge chamber, from where the vapor passes through the discharge line to the condenser. The

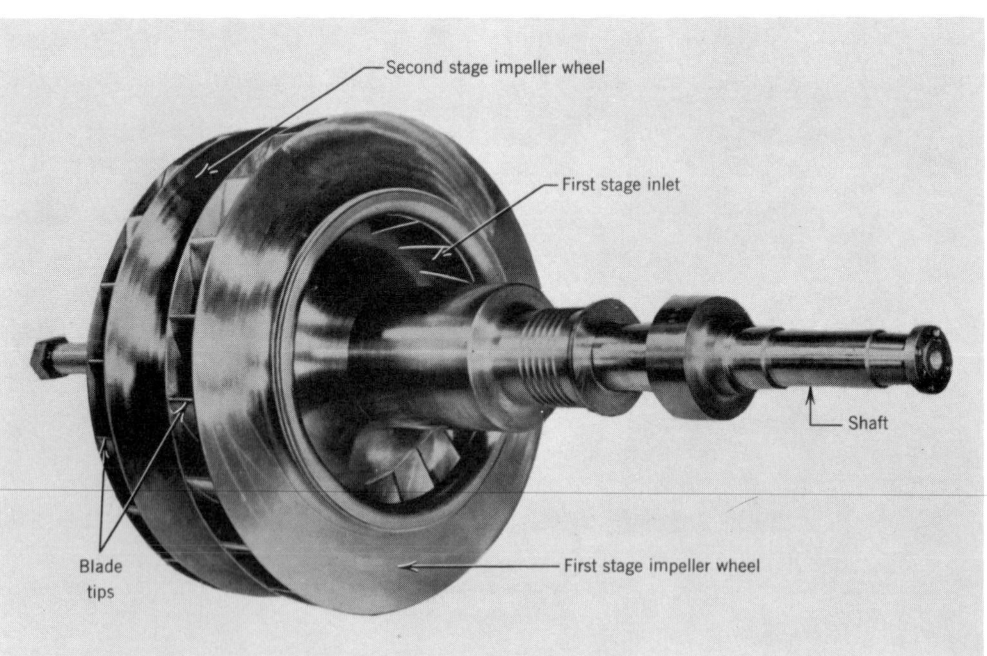

Fig. 18-26 Two-stage rotor centrifugal compressor rotor assembly. (Courtesy York Corporation.)

Fig. 18-27 Diagram of gas flow through centrifugal compressor. (Courtesy York Corporation.)

refrigerant flow path through a two-stage centrifugal compressor is shown diagrammatically in Fig. 18-27.

The rotating impeller wheels are essentially the only moving parts of the centrifugal compressor and as such are the source of all the energy imparted to the vapor during the compression process. The action of the impeller is such that both the static and velocity heads of the vapor are increased by the energy so imparted to the vapor. The centrifugal force exerted on the vapor confined between, and rotated with, the blades of the impeller wheels causes self-compression of the vapor in much the same manner that the force of gravity causes the upper layers of a gas column to compress the lower layers of the column. Hence, the static head produced centrifugally within the impeller wheels is equal to the static head which would be produced by an equivalent gravitational column (Section 15-3).

In addition to the static head which is produced centrifugally, a velocity head is also developed within the impeller wheel because of the increase in the velocity of the vapor as the vapor passes from the eye to the periphery of the wheel. As the mass of refrigerant vapor passes through, and is rotated by the impeller wheel, it attains a rotational velocity approaching that of the wheel. Since the greater portion of this velocity head is subsequently converted to static head within the casing surrounding the wheels, the total increase in pressure developed by a single wheel is the sum of the increases in both the static and velocity pressures of the vapor.

By assuming radial blades, the total head developed by a single impeller wheel is directly proportional to the square of the peripheral velocity of the wheel, that is,

$$h = \frac{v^2}{g_c}$$

where h = the total head in feet

v = the peripheral velocity of the wheel in feet per second

g_c = the gravitational constant

The total increase in pressure produced per wheel is

$$p = \frac{h \times \rho}{144} = \frac{v^2 \times \rho}{144 \times g_c}$$

where p = the pressure in pounds per square inch

ρ = the mean density of the vapor in pounds per cubic foot

From the foregoing it is evident that for a refrigerant of given density, the total increase in pressure developed by a single wheel depends only on the tip velocity of the impeller blades, this tip velocity in turn being proportional to the rotational speed of the rotor shaft and to the diameter of the impeller wheel. However, since the maximum tip velocity is limited by the strength of materials and by the sonic speed of the refrigerant, it follows that the maximum increase in pressure which can be obtained with a single impeller wheel is also limited. For this reason, single-stage centrifugal compressors, such as the one shown in Fig. 18-28, can be used only in those few applications where, because of a small temperature head (difference between vaporizing and condensing temperatures), the increase in pressure (head) required is relatively small. As a general rule, two or more impeller wheels must be used in order to obtain the necessary pressure increase, in which case compression of the vapor occurs in stages as the vapor passes from one wheel to the next. Assuming equal vapor velocities at the inlet and outlet of the compressor, the total increase in pressure in the compressor is the sum of the pressure increases produced by the individual wheels. Notice that in any series of wheels the wheels are made progressively smaller in size (width) in the direction of vapor flow in order to compensate for the reduced volume of the vapor resulting from prior compression in the preceding wheel or wheels (Fig. 18-24).

Since the head of a fluid is an expression of the energy per pound of fluid, it follows that the head developed by the compressor during the compression process is numerically equal to the work done in foot-pounds per pound of vapor compressed, and that the magnitude of the head which must be produced by the compressor depends on the refrigerant used and on the difference between the saturated suction and discharge temperatures. Therefore, it is evident that, for any given set of operating conditions, the head which must be produced by the compressor, that is, the

Fig. 18-28 Single-stage centrifugal compressor. (Courtesy York Corporation.)

diameter, speed, and number of wheels required, will be the same for small capacity compressors as for large capacity compressors. For this reason, centrifugal compressors are not practical in small sizes.

For good wheel performance, the diameter, width, and eye dimensions of the impeller wheel must be maintained within certain ratio limits. Since the width of the impeller must be reduced as the volume of vapor handled is reduced in order to insure stable operation at low gas volumes, the wheel width could become very narrow, resulting in high friction losses and poor wheel performance. Therefore, to keep the wheels in proportion, it becomes necessary to reduce the diameter of the wheels as the width of the wheels is reduced. At the same time, the speed of rotation and/or the number of wheels must be increased in order to maintain the required head. Since this tends to increase manufacturing and other costs, most manufacturers agree that the smallest practical size of centrifugal compressor is one where the volume flow rate of vapor discharged from the compressor is approximately 200 cfm. The exact tonnage this represents depends upon the refrigerant used and on the operating conditions. At the present time, centrifugal compressors are available in sizes ranging from approximately 35 tons up to 10,000 tons.

Centrifugal compressors are essentially high speed machines. Rotative speeds of between 3000 and 18,000 rpm are quite common, with much higher speeds being used in some individual cases. Because of their high rotative speeds, centrifugal compressors are capable of handling large volumes of vapor in relatively small sizes. Although especially suited for use with low pressure refrigerants requiring a large compressor displacement at moderate compression ratios, they have been applied successfully in all temperature ranges with both low and high pressure refrigerants.

Some of the more common refrigerants employed with centrifugal compressors are Refrigerants-11, 12, 113, 500, and ammonia. The high displacement required per ton of refrigeration with Refrigerants-11 and 113 make these refrigerants ideal for use with centrifugal compressors in high temperature applications where the displacement required per ton of capacity is relatively low. Their use in such applications permits small refrigerating capacities without requiring small compressor frames and wheel sizes. On the other hand, when the required refrigerating capacity is large and/or the evaporator design temperature is low, refrigerants which require a relatively small displacement per ton capacity, such as Refrigerants-12, 500, and ammonia, will ordinarily allow the use of smaller compressors to produce the same tonnage. In any event, because of the difference in the operating pressures, head requirements, and other characteristics of the several refrigerants, the compressor must be designed to fit the refrigerant as well as the application.

Centrifugal compressor efficiencies are relatively high in all sizes and over a wide range of operating conditions, being about 70 to 80% as a general rule, although values well over 80% are obtained in many instances. Efficiency losses in a centrifugal compressor are due primarily to irreversible changes resulting from turbulence and fluid friction.

18-24. Centrifugal Compressor Construction and Lubrication

For maximum-pressor efficiency, the conversion of velocity pressure into static pressure in the casing must occur gradually and smoothly and without an appreciable loss in the total pressure head. To accomplish this, a series of diffuser vanes is often installed in the casing passages which convey the vapor from the discharge of one wheel to the inlet of the next. The diffuser vanes are curved in a direction opposite to that of vapor discharge from the impeller wheels and are so designed (area increasing in the direction of vapor flow) that velocity reduction and the accompanying increase in static pressure take place gradually and smoothly and with a minimum loss of energy. When diffuser vanes are not employed, gradual velocity reduction is obtained by discharging the vapor directly into scroll- or volute-shaped passages which guide the vapor from one wheel to the next. A compressor of volute design is shown in Fig. 18-29. In some

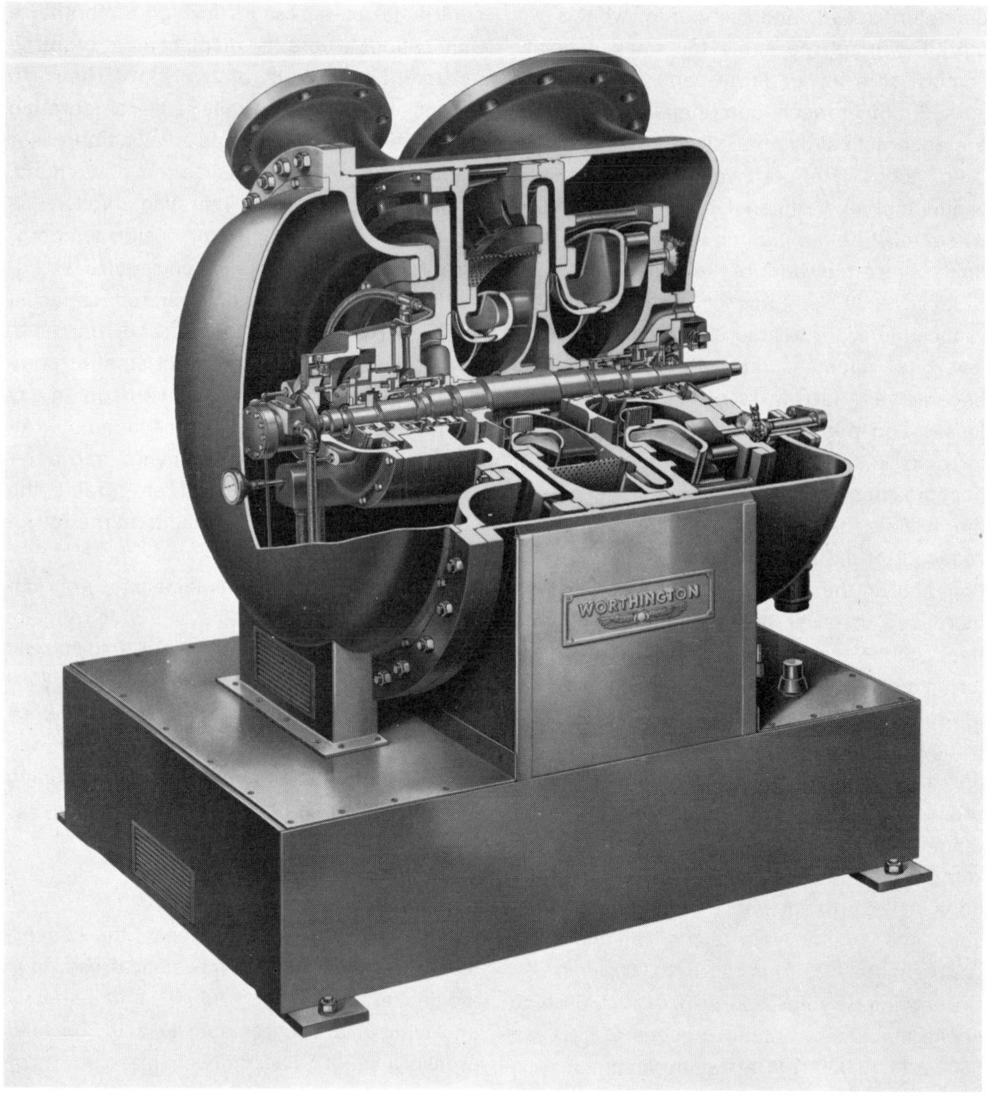

Fig. 18-29*a* Volute-type compressor. (Courtesy Worthington Corporation.)

instances, diffuser vanes and volutes are used together in a single compressor.

The back leakage of refrigerant between the several wheels or stages is limited to a practical minimum by the use of labyrinth-type seals which are arranged between the rotor and the stationary partitions. The labyrinth seal consists essentially of a series of thin steel strips which are fastened to the rotor and which match lands and grooves in the stationary partitions (Fig. 18-30). The labyrinth of passages provided by this type of seal causes a drop in the pressure of the refrigerant gas as it passes through each restricted area formed by the shaft sealing strips and housing. As the pressure drops, the velocity of the gas increases. However, on entering the next pocket, the gas encounters a large quantity of gas at rest and the increased velocity acquired during passage through the restriction is dissipated by the production of turbulence in the pockets. The leakage through the labyrinth seal is proportional to the clearance between the shaft sealing strips and the compressor housing and

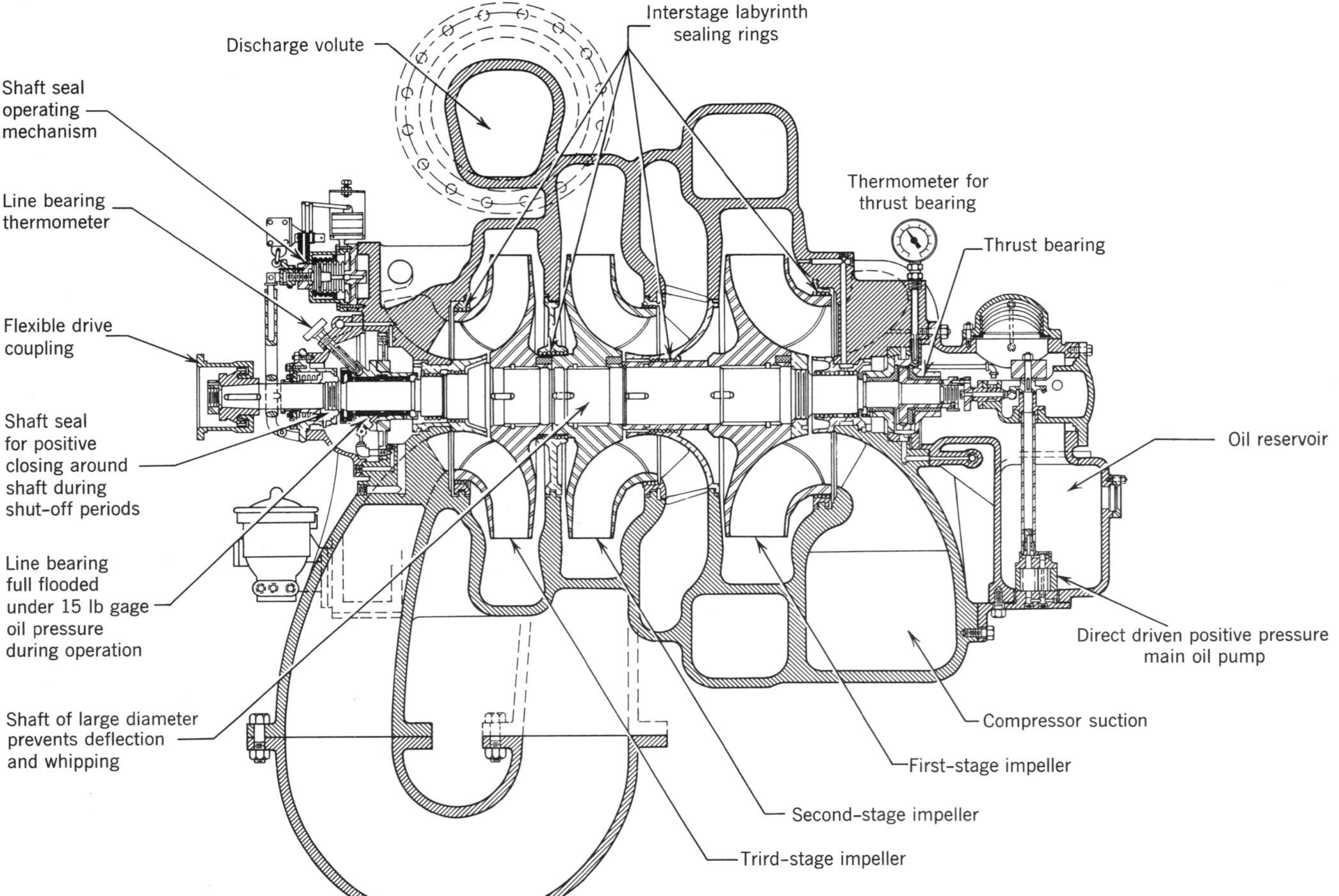

Fig. 18-29*b* Sectional elevation of three-stage volute centrifugal compressor. (Courtesy Worthington Corporation.)

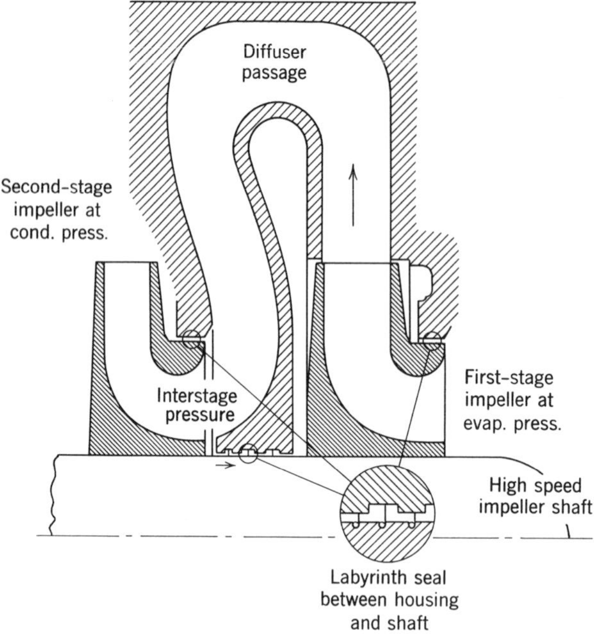

Fig. 18-30 Labyrinth seal between impellers. (Courtesy York Corporation.)

is also a function of the number of restrictions or pockets provided.

The fact that the vapor pressure on the discharge side of the impeller wheels is always greater than the pressure on the suction or inlet side of the wheels causes the rotor assembly to develop an axial thrust toward the suction inlet of the compressor. To offset this thrust, a balance disc is usually installed on the rotor shaft on the discharge side of the high stage impeller wheel (Fig. 18-24). This disc, equipped with a labyrinth seal, acts as a floating partition at the end of the discharge space. The pressure on the outboard side of the balance disc is equalized to the suction inlet of the low stage impeller through an equalizer line (Fig. 18-27), whereas the inboard side of the disc is subject to the discharge pressure of the high stage wheel. When the balance disc is properly sized, the pressure differential across the disc will exactly balance the natural thrust of the rotor assembly (Fig. 18-31).

In one design of a three-stage compressor (Fig. 18-29*b*), the impellers are so positioned on the shaft that the axial thrust developed by the third-stage impeller opposes the thrust developed by the first- and second-stage impellers. Since the third-stage impeller is the highest pressure impeller, the thrust produced by it is sufficient to counter substantially the combined thrust of the other two impellers.

The rotor assembly is supported radially in the housing by two main bearings, one located at each end of the rotor shaft (Fig. 18-24). A Kingsbury-type thrust bearing mounted on the discharge end of the shaft positions the rotor assembly axially in the casing. Since the axial

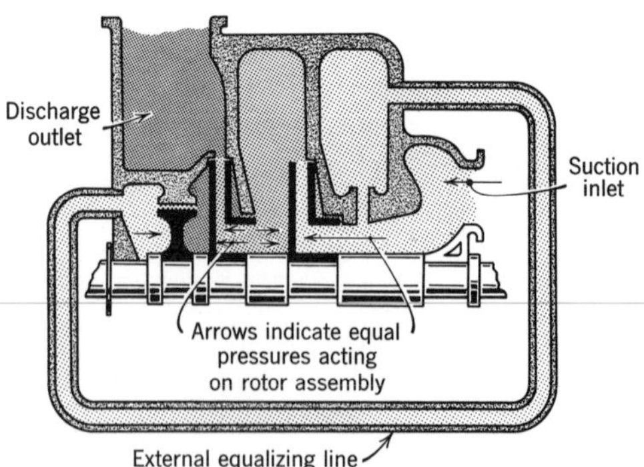

Fig. 18-31 Diagrammatic sketch of thrust balance. (Courtesy York Corporation.)

thrust of the rotor is usually neutralized by one means or another, the load on the thrust bearing is ordinarily very light. As in the case of the open-type reciprocating and rotary compressors, a shaft seal is employed between the compressor housing and the rotor shaft in order to prevent inward or outward leakage at the point where the shaft protrudes from the compressor housing.

Centrifugal compressors are pressure lubricating either by a submerged type oil pump driven directly from the rotor shaft or by a separate, externally mounted, motor-driven oil pump with an external oil reservoir. The principal parts of the compressor requiring lubrication are the two main bearings, the Kingsbury thrust bearing, and the shaft seal. Since these parts are so located that they do not come into direct contact with the system refrigerant during normal operation, lubrication is simplified in that there is little or no contamination of the refrigerant by the compressor lubricating oil. The leakage of oil along the rotor shaft from the main bearings into the refrigerant spaces is minimized by the use of oil seal labyrinth glands which are installed on the shaft on the inboard side of each of the main bearings (Fig. 18-24). Oil coolers are employed to maintain oil temperature during normal operation. Oil heaters are usually installed in the oil reservoir to prevent excessive refrigerant dilution of the oil during periods of shut-down. Oil filters are standard equipment on all centrifugal compressors. Compressors employing a shaft-driven oil pump must also be equipped with an auxiliary oil pump to supply oil pressure during start-ups and at other times when the shaft driven pump cannot supply adequate lubrication for the compressor parts.

18-25. Performance of Centrifugal Compressors

In addition to its ability to maintain a relatively high efficiency over a wide range of load conditions, and its high volumetric displacement per unit of size, there are certain other desirable performance characteristics inherent in the design of a centrifugal compressor. Principal among these is its relatively flat head-capacity characteristic as compared to that of positive displacement compressors. This, along with an extreme sensitivity to changes in speed, greatly simplifies the problem of capacity control and tends to give the centrifugal compressor a decided advantage over the reciprocating type in any large tonnage installation where the evaporator temperature must be maintained relatively constant despite wide variations in evaporator loading.

Like the centrifugal pump or blower, the delivery capacity (in cfm or in tons refrigeration) of an individual centrifugal compressor will decrease as the thermodynamic head produced by the compressor increases. Conversely, it is true also that as the delivery rate of the compressor is reduced the head produced by the compressor must increase. Therefore, since the maximum head which the compressor is capable of developing is limited by the peripheral speed of the impeller wheels, it follows that the minimum delivery capacity of the compressor is also limited. If the load on the evaporator becomes too small, the thermodynamic head necessary to handle the reduced volume of vapor will exceed the maximum head which the compressor can produce at a given speed. When this point, known as the "surging point" or "pumping limit," is reached, compressor operation becomes unstable and the compressor begins to "surge" or "hunt."* However, with proper capacity control methods, the load on a centrifugal compressor can be reduced to as little as 10% of the design load without exceeding the pumping limit of the compressor.

As in the case of the reciprocating compressor, the capacity and the power requirements of the centrifugal vary with the vaporizing and condensing temperatures of the cycle and with the speed of the compressor. With reference to these variables, the performance of the centrifugal compressor is compared to that of the reciprocating type in Figs. 18-32 through 18-35.

* Surging is an operating characteristic of all centrifugal compressors and is recognizable by the noise created as the flow of gas momentarily stops and reverses when the compressor attempts to remove vapor from the evaporator at a higher rate than the load is capable of generating the vapor.

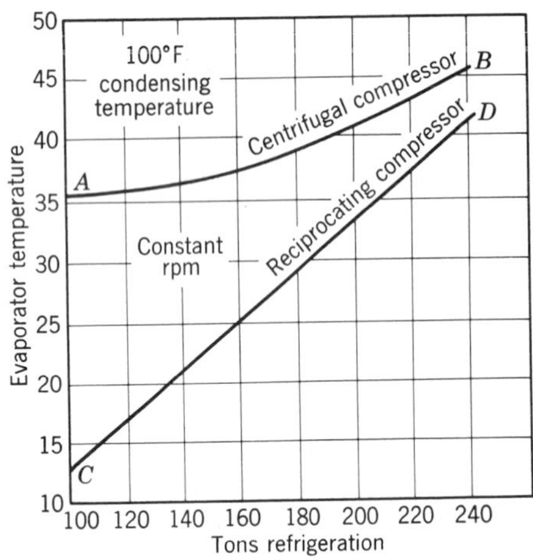

Fig. 18-32 Centrifugal versus reciprocating performance. (Courtesy York Corporation.)

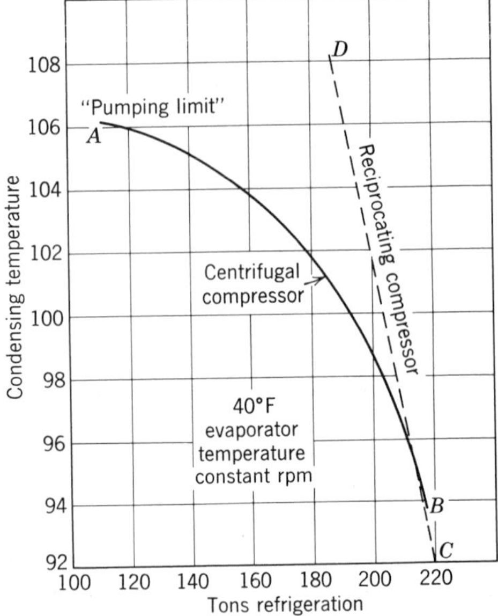

Fig. 18-33 Centrifugal versus reciprocating performance. (Courtesy York Corporation.)

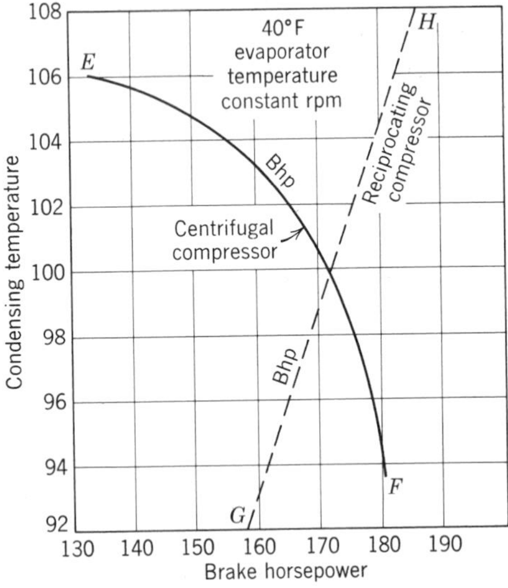

Fig. 18-34 Centrifugal versus reciprocating performance. (Courtesy York Corporation.)

Some of the more important differences in the performance of the two compressors become apparent on careful examination of these data.

Notice in Fig. 18-32 that a reduction in refrigerating capacity from 240 to 100 tons is accomplished by the centrifugal compressor with a corresponding change in evaporator temperature of only 10°F (5.5°C) as compared to a 29°F (16°C) change required by the reciprocating compressor to effect the same capacity reduction. This means in effect that a centrifugal compressor will maintain a more constant evaporator temperature over a much wider range of loading than will the reciprocating type. Naturally, this is an important advantage in any

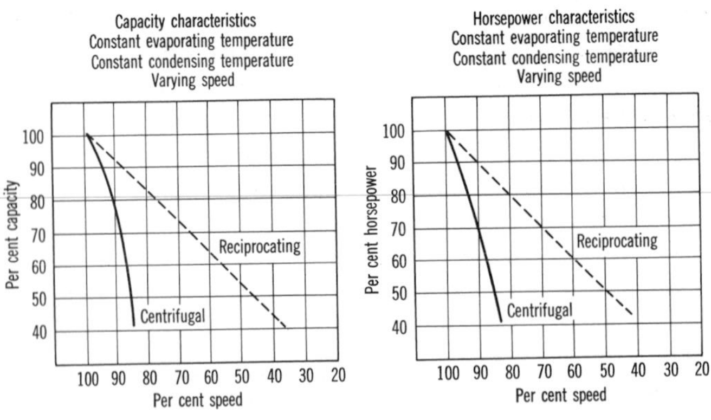

Fig. 18-35 (Courtesy Worthington Corporation.)

installation requiring the maintenance of a constant evaporator temperature under varying load conditions.

Too, the fact that a rather substantial change in capacity is brought about by only a small change in the suction temperature makes practical the use of suction throttling devices as a means of controlling the capacity of a centrifugal compressor, a practice which cannot be recommended for reciprocating compressors.

It is of interest to notice also that the operating range of the centrifugal compressor is definitely limited by the "surging" or "hunting" characteristic of the compressor. In Fig. 18-32, the centrifugal evaporator temperature cannot fall below 35°F (1.67°C) regardless of the reduction of the evaporator loading. A further decrease in evaporator load would cause the compressor to reach its "pumping limit" and a rise in the evaporator temperature would occur.* By comparison, the positive displacement reciprocating compressor will continue to reduce the evaporator temperature and pressure as the evaporator load is reduced until a capacity balance is obtained between the evaporator loading and the compressor capacity.

Figure 18-33 shows a performance comparison between centrifugal compressors operating at constant speed and evaporator temperature but with varying condensing temperature. Curve *A–B* illustrates that the centrifugal compressor experiences a rapid reduction in capacity as the condensing temperature increases. This characteristic of a centrifugal compressor makes it possible to control compressor capacity by varying the quantity and temperature of the condenser water. The capacity of the compressor can be reduced by this means until point *A* is reached, beyond which a further increase in the condensing temperature will cause the required thermodynamic head to exceed the developed head of the compressor for the given speed and tons capacity, with the result that hunting will occur.

The reduction in the capacity of the reciprocating compressor with a rise in the condensing temperature is relatively small as compared to that experienced by the centrifugal compressor. Regardless of the increase in condensing temperature, the reciprocating compressor will continue to have a positive displacement and produce a refrigerating effect.

Figure 18-34 compares the power requirements of the centrifugal and reciprocating compressors under conditions of varying condensing temperature. Whereas the centrifugal shows a reduction in power requirements with an increase in the condensing temperature to correspond with the rapid fall off in capacity shown in Fig. 18-33, the reciprocating compressor shows a small increase in power requirements to correspond with the small reduction in refrigeration tonnage shown in Fig. 18-33 for that machine.

Figure 18-34 also illustrates the nonoverloading characteristic of the centrifugal compressor. Notice that an increase in condensing temperature causes a reduction in both the refrigerating capacity and the power requirements of the compressor, although the horsepower required per ton increases.

With regard to compressor speed, the centrifugal compressor is much more sensitive to speed changes than is the reciprocating type. Whereas the change in the capacity of the reciprocating compressor is approximately proportional to the speed change, according to the performance curves in Fig. 18-35, a speed change of only 12% will cause a 50% reduction in the capacity of the centrifugal compressor.

18-26. Capacity Control

Capacity control of centrifugal compressors is usually accomplished either by varying the speed of the compressor or through the use of

* Capacity reductions somewhat below the surging point are possible through the use of a hot gas bypass, which maintains the compressor volume flow rate above the design limit by partial recirculation of the gas through the compressor, or through the use of a suction damper, which throttles the gas at the suction inlet of the compressor, thereby reducing the mass flow rate of refrigerant, and consequently the refrigerating capacity, while at the same time maintaining the compressor volume flow rate above the design minimum.

As always, where a hot gas bypass is employed, the bypass gas must be desuperheated to avoid overheating of the compressor.

Fig. 18-36 Centrifugal refrigerating machine. (Courtesy Worthington Corporation.)

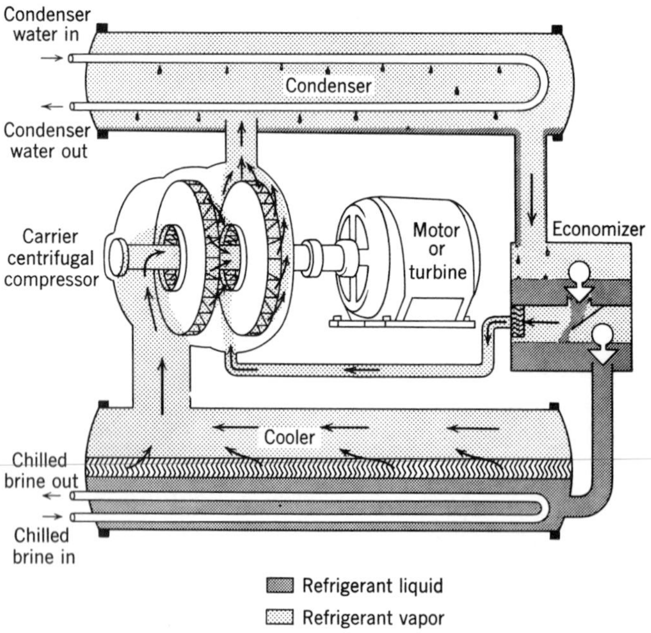

Fig. 18-37 Flow diagram for typical centrifugal refrigerating machine. (Courtesy Carrier Corporation.)

variable inlet guide vanes, called prerotation vanes.

Because of its extreme sensitivity to changes in speed, the centrifugal compressor is ideally suited for capacity regulation by means of variable speed drives, such as steam turbines and wound-rotor induction motors. When constant speed drives, such as synchronous or squirrel cage motors, are employed, speed control can be obtained through the use of a hydraulic or magnetic clutch installed between the drive and the step-up gear.

Prerotation vanes are installed in the compressor housing directly ahead of the impeller inlet and accomplish capacity reduction by changing the direction of flow of the refrigerant vapor immediately prior to its entering the impeller. A swirling action is imparted to the vapor as it passes through the vanes, that increases, and has a more pronounced effect on compressor capacity, as the vanes approach the closed position.

The use of suction throttling dampers and condenser temperature control as means of capacity regulation has already been mentioned, although these methods are not widely used. Bypassing of the discharge gas, also previously mentioned, is employed primarily to extend the pumping limit at minimum loading and is supplemental to other methods of capacity control.

18-27. Centrifugal Refrigeration Machines

Centrifugal compressors are available for refrigeration duty only as an integral part of a centrifugal refrigerating machine. Because of the relatively flat head-capacity characteristic of the centrifugal compressor, and the resulting limitation in the operating range, the component parts of a centrifugal refrigerating system must be very carefully balanced. Too, since very marked changes in capacity accrue with only minor changes in the suction or condensing temperatures, the centrifugal refrigerating system must be close-coupled, as shown in Fig. 18-36, in order to reduce the refrigerant line pressure losses to an absolute minimum.

A schematic diagram of a centrifugal refrigerating system is shown in Fig. 18-37. Except for the introduction of a flash intercooler (Section 20-12) between the condenser and evaporator, the centrifugal refrigerating system operates on a conventional vapor-compression cycle. High pressure liquid drains from the bottom of the condenser into the high pressure chamber of the intercooler, from where it passes through a high pressure float valve into the intermediate chamber of the intercooler. In passing through the float valve, a portion of the liquid flashes into the vapor state, thereby cooling the balance of the liquid to the temperature corresponding to the pressure in the intermediate chamber. From the intermediate chamber the cool liquid passes through the intermediate float valve into the evaporator, at which time the temperature of the liquid is reduced to the evaporator temperature by additional flashing. Hence, the effect of the intercooler is to increase the refrigerating effect per pound and to reduce the amount of flash gas in the evaporator. Since the flash vapor from the intermediate chamber is taken into the suction of the second-stage impeller, the pressure of this vapor will be above the evaporator pressure and therefore the power required to compress it to the condensing pressure will be less. Too, the cool vapor from the intercooler reduces the temperature of the discharge vapor from the first-stage impeller with the result that the capacity and efficiency of the system are increased.

19

REFRIGERANT PIPING AND ACCESSORIES

19-1. Piping Materials

In general, the type of piping material employed for refrigeration piping depends upon the size and nature of the installation, the refrigerant used, and the cost of materials and labor. Specific minimum requirements for refrigerant piping, with regard to type and weight of piping materials, methods of joining, etc., are set forth in the American Standard Safety Code for Mechanical Refrigeration (ASA Standard B9.1). Since the specifications in this standard represent good, safe, piping practice, they should be closely followed. Too, in all cases, local codes and ordinances must be taken into account.

The materials most frequently used for refrigerant piping are black steel, wrought iron, copper, and brass. All these are suitable for use with all the common refrigerants, except that copper and brass may not be used with ammonia, since, in the presence of moisture, ammonia attacks nonferrous metals.

Copper tubing has the advantage of being lighter in weight, more resistant to corrosion, and easier to install than either wrought iron or black steel. With all refrigerants except ammonia, refrigerant lines up to $4\frac{1}{8}$ in. OD (100 mm) may be either copper or steel. All lines above this size should be steel. However, general practice is to use all steel pipe in any installation where a considerable amount of piping exceeds 2 in. (50 mm) in size. Wrought iron pipe, although more expensive than black steel, is sometimes used in place of the latter because of its greater resistance to corrosion.

Steel pipe should be of either the seamless or lap-welded types, except that butt-welded pipe may be used in sizes up to 2 in (50 mm). All steel pipe 1 in. (25 mm) or smaller should be Schedule-80 (extra heavy). Above this size, Schedule-40 (standard weight) pipe may be used, except that liquid lines up to $1\frac{1}{2}$ in. (40 mm) should be Schedule-80.

Copper tubing is available in either hard or soft temper. The hard drawn tubing comes in 20 ft straight lengths, whereas the soft temper is usually packaged in 25 and 50 ft coils. Only types K and L are suitable for refrigerant lines.

Soft temper copper tubing may be used for refrigerant lines up to $\frac{7}{8}$ in. OD (20 mm), and is recommended for use where bending is required, where the tubing is hidden, and/or where flare connections are used. Hard temper tubing should be used for all sizes above $\frac{7}{8}$ in. OD (20 mm) and for smaller sizes when rigidity is desired.

19-2. Pipe Joints

Depending on the type and size of the piping, joints for refrigerant piping may be screwed, flanged, flared, welded, brazed, or soldered. When refrigerant pressures are below 250 psi (17 bar screwed joints may be used on pipe sizes up to 3 in. (80 mm). For higher pressures, screwed joints are limited to pipe sizes $1\frac{1}{4}$ in. (35 mm) and smaller. Above these sizes, flanged joints of the tongue and groove type should be used. Screwed-on flanges are limited to the pipe sizes listed above. For larger sizes, welded-neck flanges are required. A joint compound, suitable for refrigerant piping and applied to the male threads only, should be used with all screw connections.

Welding is probably the most commonly used method of joining iron and steel piping. Pipes 2 in. (50 mm) and over are usually butt-welded, whereas those $1\frac{1}{2}$ in. (40 mm) and smaller are generally socket-welded. Branch connections should be reinforced.

Flared compression fittings may be used for connecting soft temper copper tubing up to size $\frac{3}{4}$ in. OD (20 mm). Above this size and for hard temper copper tubing, joints should be made with sweat fittings using a hard solder. Hard solders are silver brazing alloys with melting temperatures above 1000°F (550°C). Soft solder (95% tin and 5% antimony), having a melting point below 500°F (260°C), may be used for tubing $\frac{1}{2}$ in. OD (15 mm) and smaller. A suitable noncorrosive soldering flux should be used with both types of solder.

Flare fittings should be forged brass, and sweat fittings may be either wrought copper or forged brass. Cast sweat fittings are not suitable for refrigeration duty. As a general rule, better fitting joints will result when tubing and fittings are obtained from the same manufacturer.

19-3. Location

In general, refrigerant piping should be located so that it does not present a safety hazard, obstruct the normal operation and maintenance of the equipment, or restrict the use of adjoining spaces. When the requirements of refrigerant flow will permit, piping would be at least $7\frac{1}{2}$ ft (2.3 m) above the floor, unless installed against the wall or ceiling. The piping code prohibits refrigerant piping in public hallways, lobbies, stairways, elevator shafts, etc., except that it may be placed across a hallway provided that there are no joints in the hallway and that nonferrous pipe 1 in. (25 mm) and smaller is encased in rigid metal conduit.

The arrangement of the piping should be such that it is easily installed and readily accessible for inspection and maintenance. In all cases, the piping should present a neat appearance. All lines should be run plumb and straight, and parallel to walls, except that horizontal suction lines, discharge lines, and condenser to receiver lines should be pitched in the direction of flow.

All piping should be supported by suitable ceiling hangers or wall brackets. The supports should be close enough together to prevent the pipe from sagging between the supports. As a general rule, supports should not be more than 10 ft (3 m) apart. A support should be placed not more than 2 ft (0.6 m) away from each change in direction, preferably on the side of the longest run. All valves in horizontal piping should be installed with the valve stems in a horizontal position whenever possible. All valves in copper tubing smaller than 1 in. OD (25 mm) should be supported independently of the piping. Risers may be supported either from the floor or from the ceiling.

When piping must pass through floors, walls, or ceilings, sleeves made of pipe or formed galvanized steel should be placed in the openings. The pipe sleeves should extend 1 in. (25 mm) beyond each side of the openings and curbs should be used around pipe sleeves installed in floors.

Provisions must be made also for the thermal expansion and contraction of the piping which usually amounts to approximately $\frac{3}{4}$ in. (20 mm) per hundred feet of piping. This is not ordinarily a serious problem, since refrigerant piping is usually three dimensional and therefore sufficiently flexible to absorb the small changes in length. However, care should be taken not to anchor rigidly both ends of a long straight length of pipe.

19-4. Vibration and Noise

In most cases, the vibration and noise in refrigerant piping originates not in the piping itself but in the connected equipment. However, regardless of the source, vibration, and the objectional noise associated with it, is greatly reduced by proper piping design. Often, relatively small vibrations transmitted to the piping from the connected equipment are amplified by improperly designed piping to the extent that serious damage to the piping and/or the connected equipment results.

For the most part, vibration in refrigerant piping is caused by the rigid connection of the piping to a reciprocating compressor, by gas pulsations resulting from the opening and closing of the valves in a reciprocating compressor, and by turbulence in the refrigerant gas due to high velocity. When centrifugal and rotary compressors are used, vibration and noise in the refrigerant piping is not usually a serious problem, being caused only by the latter of the above three factors. The reason lies in the rotary motion of the centrifugal and rotary compressors and in the smooth flow of the gas into and out of these units, as compared to the pulsating flow through the reciprocating-type compressor.

Since a small amount of vibration is inherent in the design of certain types of equipment, such as reciprocating compressors, it is not possible to eliminate vibration completely. However, if the piping immediately adjacent to such equipment is designed with sufficient flexibility, the vibration will be absorbed and dampened by the piping rather than transmitted and amplified by it. On small units piped with soft temper copper tubing, the desired flexibility is obtained by forming vibration loops in the suction and discharge lines near the point where these lines are connected to the compressor. If properly designed and placed, these loops will act as springs to absorb and dampen compressor vibration and prevent its transmission through the piping to other parts of the system. Where the compressor is piped with rigid piping, vibration eliminators (Fig. 19-1) installed in the suction and discharge lines near the compressor are usually effective in dampening compressor vibration. Vibration eliminators should be placed in a vertical line for best results.

On larger systems, adequate flexibility is ordinarily obtained by running the suction and discharge piping approximately 30 pipe diameters in each of two or three directions before anchoring the pipe. In all cases, isolation type hangers and brackets should be used when piping is supported by or anchored to building construction which may act as a sounding board to amplify and transmit vibrations and noise in the piping.

Although vibration and noise resulting from gas pulsations can occur in both the suction and discharge lines of reciprocating compressors, it is much more frequent and more intense in the discharge line. As a general rule, these gas pulsations do not cause sufficient vibration and noise to be of any consequence. Occasionally, however, the frequency of the pulsations and the design of the piping are such that resonance is established, with the result that the pulsations are amplified and sympathetic vibration (as with a tuning fork) is set up in the piping. In some instances, vibration can become so severe that the piping is torn loose from its supports. Fortunately, the condition can be remedied by changing the speed of the compressor, by installing a discharge muffler, and/or by changing the size of length of the discharge line. Since changing the speed of the compressor is not usually practical, the latter two methods are better solutions, particularly when used together.

Fig. 19-1 Vibration eliminator. (Courtesy Anaconda Metal Hose Division. The American Brass Company.)

When vibration and noise are caused by gas turbulence resulting from high velocity, the usual remedy is to reduce the gas velocity by increasing the size of the pipe. Sometimes this can be accomplished by installing a supplementary pipe.

19-5. General Design Considerations

Since many of the operational problems encountered in refrigeration applications can be traced directly to improper design and/or installation of the refrigerant piping and accessories, the importance of proper design and installation procedures cannot be overemphasized. In general, refrigerant piping should be so designed and installed as to:

1. Assure an adequate supply of refrigerant to all evaporators
2. Assure positive and continuous return of oil to the compressor crankcase
3. Avoid excessive refrigerant pressure losses which unnecessarily reduce the capacity and efficiency of the system
4. Prevent liquid refrigerant from entering the compressor during either the running or off cycles, or during compressor start-up
5. Avoid the trapping of oil in the evaporator or suction line which may subsequently return to the compressor in the form of a large "slug" with possible damage to the compressor.

19-6. Suction Line Size

Because of its relative location in the system, the size of the suction piping is usually more critical than that of the other refrigerant lines. Undersizing of the suction piping will cause an excessive refrigerant pressure drop in the suction line and result in a considerable loss in system capacity and efficiency. On the other hand, oversizing of the suction piping will often result in refrigerant velocities which are too low to permit adequate oil return from the evaporator to the compressor crankcase. Therefore, the optimum size for the suction piping is one that will provide the minimum practical refrigerant pressure drop commensurate with maintaining sufficient vapor velocity to insure adequate oil return.

Most systems employing oil miscible refrigerants are so designed that oil return from the evaporator to the compressor is through the suction line, either by gravity flow or by entrainment in the suction vapor. When the evaporator is located above the compressor and the suction line can be installed without risers or traps, the oil will drain by gravity from the evaporator to the compressor crankcase, provided that all horizontal piping is pitched downward in the direction of the compressor. In such cases, the minimum vapor velocity in the suction line is of little importance and the suction piping can be sized to provide the minimum practical pressure drop without regard for the velocity of the vapor. This holds true also for any system employing a nonmiscible refrigerant and for any other system where special provisons are made for oil return.

On the other hand, when the location of the evaporator and/or other conditions are such that a riser is required in the suction line, the riser must be sized small enough so that the resulting vapor velocity in the riser under minimum load conditions will be sufficiently high to entrain the oil and carry it up the riser and back to the compressor.

Since oil return up a riser results primarily from the oil being "dragged" or "pulled" up the wall of the riser by the gas flow, it is the vapor velocity at the pipe surface, rather than the overall average gas velocity, that is important. The relationship of the surface velocity to the average velocity depends on the density of the vapor and the inside diameter of the pipe. The lower the pressure and density of the vapor and the larger the inside diameter of the pipe, the higher is the average velocity required to produce a given surface velocity.

Some of the other factors that determine the minimum vapor velocity that will carry the oil up the riser are the viscosity and density of the oil and the amount of refrigerant dilution. The more viscous the oil, the higher is the surface velocity required to drag the oil up the riser.

Table 19-1A gives minimum capacities in tons of refrigeration that will result in gas velocities sufficiently high to ensure oil return up various sizes of suction risers for various saturated suction temperatures. Table 19-1B

TABLE 19-1A Minimum Tonnage for Oil Entrainment up Suction Risers (Type L Copper Tubing)

Refrigerant	Sat. Suction Temp, F	Pipe OD ½	⅝	¾	⅞	1⅛	1⅜	1⅝	2⅛	2⅝	3⅛	3⅝	4⅛
		Area, Sq. In.											
		0.146	0.233	0.348	0.484	0.825	1.256	1.78	3.094	4.77	6.812	9.213	11.97
R-12*	−40	0.061	0.110	0.182	0.27	0.54	0.91	1.4	2.79	4.78	7.49	10.9	15.1
	−20	.077	.138	.228	.34	.67	1.13	1.75	3.49	5.99	9.36	13.7	19.0
	0	.093	.167	.278	.42	.82	1.38	2.14	4.26	7.32	11.4	16.6	23.2
	20	.112	.201	.332	.50	.97	1.65	2.55	5.1	8.73	13.6	19.9	27.6
	40	.132	.238	.390	.59	1.15	1.94	3.0	6.0	10.3	16.1	23.4	32.6
R-22*	−40	0.09	0.16	0.27	0.41	0.79	1.34	2.1	4.1	7.1	11.1	16.1	22.4
	−20	.11	.20	.33	.50	.96	1.60	2.5	5.0	8.7	13.5	19.6	27.4
	0	.13	.24	.39	.59	1.2	1.96	3.0	6.1	10.4	16.2	23.6	32.8
	20	.16	.28	.46	.70	1.4	2.30	3.5	7.1	12.1	18.9	27.6	38.1
	40	.18	.33	.54	.81	1.6	2.70	4.1	8.2	14.1	22.0	32.1	44.6
R-500*	−40	0.068	0.12	0.20	0.31	0.60	1.0	1.6	3.1	5.4	8.4	12.2	16.9
	−20	.086	.16	.26	.39	.75	1.3	2.0	3.9	6.8	10.5	15.3	21.4
	0	.110	.19	.31	.47	.92	1.6	2.4	4.8	8.2	12.8	18.7	26.0
	20	.130	.23	.37	.56	1.1	1.9	2.9	5.7	9.9	15.3	22.4	31.2
	40	.150	.27	.44	.67	1.3	2.2	3.4	6.8	11.6	18.2	26.6	36.8
R-502†	−60	0.053	0.10	0.16	0.24	0.46	0.78	1.2	2.4	4.1	6.4	9.4	13.0
	−40	.070	.12	.20	.30	.59	1.0	1.5	3.1	5.3	8.3	12.0	16.8
	−20	.084	.15	.25	.38	.74	1.3	1.9	3.8	6.6	10.3	15.0	20.9
	0	.104	.19	.31	.47	.91	1.5	2.4	4.7	8.1	12.7	18.4	25.7
	20	.120	.22	.37	.56	1.1	1.8	2.9	5.7	9.8	15.2	22.2	30.8
	40	.146	.26	.43	.65	1.3	2.2	3.3	6.7	11.4	17.8	26.0	36.1

[a] Minimum tonnage values are based on the indicated saturation temperatures (SST) with 15 F deg of superheat and 90 F liquid temperature.
* R-12, R-22, and R-500, reduce or increase table values 1% for 10 F deg less or more superheat.
† For R-502, reduce or increase table values 2% for 10 F deg less or more superheat.
For liquid temperatures other than 90 F, multiply the table values by the corresponding factor listed in the following table:

Liquid Temperature, F		50	60	70	80	90	100	110	120	130	140
Correction	R-12, R-22, R-500	1.20	1.15	1.10	1.05	1.00	0.95	0.90	0.85	0.80	0.75
Factors	R-502	1.26	1.20	1.13	1.07	1.00	0.94	0.88	0.82	0.76	0.70

TABLE 19-1B Minimum Tonnage for Oil Entrainment up Hot Gas Risers (Type L Copper Tubing)

Refrigerant	Sat. Discharge Temp, F	Pipe OD ½	⅝	¾	⅞	1⅛	1⅜	1⅝	2⅛	2⅝	3⅛	3⅝	4⅛
		Area, Sq. In.											
		.146	.233	.348	.484	.825	1.256	1.78	3.094	4.77	6.812	9.213	11.97
R-12*	80	.17	.31	.50	.77	1.51	2.54	3.93	7.84	13.5	21.0	30.7	42.6
	90	.17	.31	.51	.77	1.51	2.54	3.92	7.84	13.5	21.0	30.7	42.6
	100	.17	.31	.51	.77	1.51	2.54	3.92	7.84	13.5	21.0	30.7	42.6
	110	.17	.31	.51	.77	1.50	2.53	3.90	7.81	13.4	20.9	30.5	42.2
	120	.17	.30	.50	.75	1.47	2.49	3.84	7.66	13.2	20.6	30.0	41.6
	130	.17	.30	.49	.72	1.45	2.44	3.77	7.54	12.9	20.3	29.4	40.8
	140	.16	.28	.47	.71	1.38	2.33	3.61	7.20	12.4	19.4	28.2	39.9
R-22*	80	.23	.42	.69	1.04	2.0	3.4	5.3	10.6	18.2	28.3	41.5	57.5
	90	.23	.42	.69	1.04	2.0	3.4	5.3	10.6	18.2	28.2	41.3	57.3
	100	.23	.42	.69	1.03	2.0	3.4	5.3	10.5	18.0	28.1	41.0	56.7
	110	.23	.41	.67	1.02	2.0	3.4	5.2	10.4	17.9	27.9	40.8	56.5
	120	.22	.40	.66	1.00	2.0	3.3	5.1	10.2	17.5	27.4	39.9	55.4
	130	.22	.39	.64	.98	1.9	3.2	5.0	10.0	17.2	26.8	39.0	54.0
	140	.21	.38	.63	.96	1.9	3.2	4.9	9.7	16.7	26.1	38.0	52.6
R-500*	80	.20	.36	.59	.89	1.73	2.92	4.51	9.0	15.5	24.2	35.4	49.0
	90	.20	.35	.58	.88	1.73	2.86	4.49	8.9	15.4	24.0	35.0	48.5
	100	.20	.35	.58	.88	1.73	2.86	4.47	8.8	15.3	23.8	34.9	48.2
	110	.20	.35	.57	.87	1.70	2.86	4.45	8.7	15.2	23.7	34.7	48.0
	120	.19	.34	.56	.86	1.66	2.82	4.44	8.7	15.0	23.3	34.1	47.3
	130	.19	.34	.56	.85	1.64	2.78	4.29	8.6	14.7	23.0	33.6	46.5
	140	.18	.33	.54	.83	1.61	2.71	4.20	8.4	14.4	22.5	32.8	45.5
R-502†	80	.18	.32	.53	.80	1.55	2.7	4.1	8.2	14.1	21.9	32.5	44.3
	90	.17	.31	.51	.77	1.49	2.52	3.92	7.8	13.4	20.9	30.5	42.3
	100	.165	.30	.50	.74	1.44	2.45	3.8	7.55	13.0	20.2	29.5	40.9
	110	.160	.29	.48	.72	1.41	2.38	3.71	7.35	12.7	19.7	28.7	39.8
	120	.154	.28	.46	.69	1.33	2.26	3.52	7.0	12.4	18.7	27.3	37.9
	130	.145	.26	.43	.65	1.27	2.14	3.34	6.62	11.4	17.8	25.9	35.9
	140	.135	.24	.40	.61	1.18	1.98	3.08	6.15	10.6	16.4	24.0	33.3

* Minimum tonnages are based on a saturated suction temperature of +20 F with 15 F deg of superheat at the indicated saturated condensing temperatures with 15 F deg subcooling and actual discharge temperature based on 70% compressor efficiency. For suction temperatures other than 20 F, multiply the table values by the following factors:

Sat. Suct. Temperature	−40	−20	0	+20	+40
Correction Factor	0.85	0.90	0.95	1.0	1.06

† Minimum tonnages are based on a saturated temperature of −20 F. All other conditions are the same as above. For suction temperatures other than −20 F, multiply the table values by the following factors:

Sat. Suct. Temperature	−60	−40	−20	0	+20	+40
Correction Factor	0.87	0.94	1.0	1.08	1.15	1.21

TABLE 19-2 Refrigerant Line Capacities in Tons for Refrigerant-12 (Single or High Stage Applications) Tons of Refrigeration Resulting in a Line Friction Drop per 100 Ft Equivalent Pipe Length Corresponding to 2°F (ΔT) Change in Saturation Temp

Line Size Type L Copper OD		*Suction Lines** *Suction Temp F* −40 $\Delta P=0.49$	−20 $\Delta P=0.72$	0 $\Delta P=1.01$	20 $\Delta P=1.38$	40 $\Delta P=1.82$	*Discharge Lines** $\Delta P=3.66$ Sat. −40	Suct. 0	Temp F 40	*Liquid Lines Line Size Type L Copper OD*		*Condenser to Receiver Velocity = 100 fpm*	*Receiver* to System* $\Delta T=1$ F $\Delta P=1.8$ psi
1/2					0.21	0.31	0.46	0.54	0.67	1/2		1.16	2.03
5/8			0.17	0.26	0.40	0.58	0.85	0.98	1.23	5/8		2.65	3.81
7/8		0.25	0.42	0.68	1.04	1.50	2.23	2.58	3.22	7/8		6.94	10.10
1 1/8		0.51	0.87	1.39	2.10	3.10	4.60	5.30	6.65	1 1/8		11.85	20.5
1 3/8		0.87	1.52	2.40	3.70	5.36	7.8	9.0	11.3	1 3/8		18.10	35.1
1 5/8		1.41	2.44	3.86	5.82	8.50	12.4	14.4	18.0	1 5/8		25.5	57.5
2 1/8		2.94	5.03	8.00	12.1	17.6	25.8	30.0	37.4	2 1/8		44.4	117.8
2 5/8		5.20	8.94	14.2	21.3	31.4	45.5	52.5	66.0	2 5/8		68.4	207.8
3 1/8		8.35	14.3	22.7	34.0	49.5	73.0	85.0	106.0	3 1/8		97.5	344.0
3 5/8		12.4	21.2	33.8	50.6	73.5	107.0	124.0	155.0	3 5/8		132.0	508.0
4 1/8		17.4	29.9	47.7	71.0	103.0	152.0	176.0	220.0	4 1/8		173.0	704.0
5 1/8		31.7	54.0	85.3	128.0	187.0	270.0	314.0	392.0				
6 1/8		50.8	86.0	137.0	206.0	299.0	428.0	494.0	620.0				
Steel IPS	*SCH*									*Steel IPS*	*SCH*		
1/2	40			0.30	0.45	0.64	0.92	1.07	1.34	1/2	80	3.43	3.23
3/4	40	0.24	0.41	0.64	0.96	1.39	1.96	2.26	2.83	3/4	80	6.25	7.27
1	40	0.46	0.78	1.22	1.82	2.68	3.75	4.35	5.42	1	80	10.4	14.3
1 1/4	40	0.97	1.60	2.52	3.78	5.41	7.8	9.0	11.3	1 1/4	80	18.6	30.1
1 1/2	40	1.50	2.41	3.76	5.62	8.12	11.4	13.2	16.5	1 1/2	80	25.5	47.3
2	40	2.81	4.69	7.40	10.9	15.7	21.6	25.1	31.4	2	40	48.0	111.9
2 1/2	40	4.44	7.42	11.6	17.3	24.7	34.7	40.2	50.2	2 1/2	40	68.3	173.0
3	40	8.04	13.2	20.6	30.6	43.8	61.0	70.8	88.4	3	40	104.0	311.8
4	40	16.03	27.0	42.8	62.9	90.2	125.0	146.0	182.0	4	40	179.0	634.0
5	40	30.0	49.1	78.7	114.0	165.0	228.0	264.0	330.0				
6	40	48.2	78.6	124.0	182.0	268.0	365.0	421.0	528.0				
8	48	98.4	161.0	254.0	376.0	541.0	745.0	865.0	1080.0				
10	40	180.0	297.0	458.0	678.0	972.0	1350.0	1570.0	1960.0				
12	ID	286.0	475.0	729.0	1080.0	1520.0	2130.0	2460.0	3090.0				

NOTES:

* (1) *Basis of Table: 100 F Condensing Temp, 2 F ΔT per 100 ft Equivalent Length (except liquid lines).*

(2) *For other ΔT's and Equivalent Lengths,*

$$\text{Line Capacity (Tons)} = \text{Table Tons} \times \left(50 \times \frac{\text{Actual } \Delta T \text{ loss desired, F}}{\text{Actual equiv. length, ft}}\right)^{0.55}$$

(3) *For other Tons and Equivalent Lengths–*

$$\Delta T \text{ for a given pipe size} = \frac{\text{Actual Equiv. Length, ft}}{50} \times \left(\frac{\text{Actual Tons}}{\text{Table Tons}}\right)^{1.8}$$

(4) *Values based on 100 F Condensing Temp. For capacities at other condensing temp. multiply table value by line capacity multiplier below:*

Line	*Condensing Temp, F*					
	80	*90*	*100*	*105*	*110*	*120*
Suction Lines	1.11	1.06	1.00	0.97	0.94	0.88
Discharge Lines	0.88	0.94	1.00	1.04	1.07	1.16

(5) *Tabulated data taken from chapter 9 of the ASRE 1957–58 Design Data Book. Initially developed from ARI preliminary data.*

(6) *For the Equivalent Change in saturation temp for various line pressure drops in psi refer to Fig.*

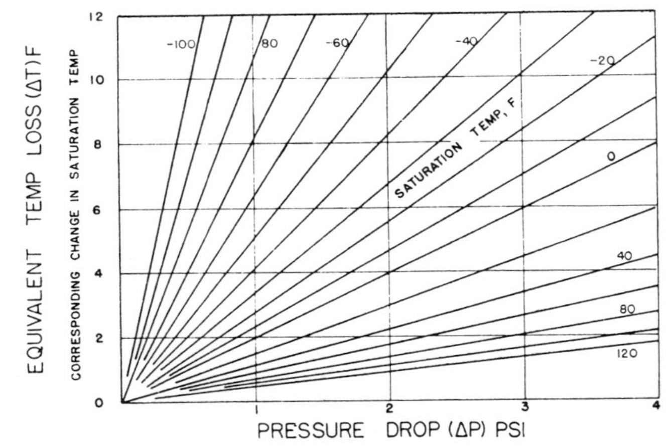

TEMPERATURE EQUIVALENT (ΔT) OF PRESSURE DROP, F (REFRIGERANT 12)

TABLE 19-3 Refrigerant Line Capacities for Refrigerant-22 (Single or High Stage Applications) Tons of Refrigeration Resulting in a Line Friction Drop per 100 Ft Equivalent Pipe Length Corresponding to 2°F (ΔT) Change in Saturation Temp

Line Size Type L Copper OD		*Suction Lines**					*Discharge Lines* $\Delta P = 6.1$*			*Liquid Lines*			
		Suction Temp, F								*Line Size Type L Copper OD*		*Condenser to Receiver Velocity = 100 fpm*	*Receiver to System $\Delta T = 1$ F*
		−40 $\Delta P = 0.79$	*−20* $\Delta P = 1.15$	*0* $\Delta P = 1.6$	*20* $\Delta P = 2.22$	*40* $\Delta P = 2.93$	*Sat. −40*	*Suct. 0*	*Temp 40*				
1/2					0.40	0.59	1.0	1.1	1.2	1/2		2.24	3.5
5/8			0.32	0.49	0.75	1.10	2.1	2.3	2.5	5/8		3.57	6.4
7/8		0.35	0.87	1.31	2.00	1.89	4.9	5.4	5.2	7/8		7.41	17.0
1 1/8		1.08	1.74	2.65	4.03	5.82	9.8	10.7	11.8	1 1/8		12.7	34.4
1 3/8		1.88	3.01	4.61	7.03	9.98	17.0	18.6	20.5	1 3/8		19.2	60.0
1 5/8		2.90	4.78	7.23	10.26	15.95	26.4	29.0	31.9	1 5/8		27.2	95.0
2 1/8		6.21	9.97	15.2	23.2	33.2	55.0	60.3	66.3	2 1/8		47.3	200.0
2 5/8		10.8	17.5	26.5	40.3	58.1	96.0	105.0	115.6	2 5/8		73.2	354.0
3 1/8		17.3	27.1	43.3	64.5	23.1	155.0	170.0	187.3	3 1/8		104.1	572.0
3 5/8		25.9	41.9	63.9	96.8	139.5	233.0	255.0	281.0	3 5/8		141.1	860.0
4 1/8		36.9	59.2	89.7	136.0	196.0	327.0	358.0	394.0	4 1/8		183.0	1200.0
5 1/8		66.1	106.7	162.0	245.0	355.0	588.0	644.0	709.0				
Steel IPS	*SCH*									*Steel IPS*	*SCH*		
1/2	40		0.38	0.56	0.84	1.20	2.0	2.2	2.4	1/2	80	4.66	5.5
3/4	40	0.49	0.78	1.18	1.77	2.52	4.1	4.5	5.0	3/4	80	6.17	12.2
1	40	0.94	1.49	2.24	3.24	4.72	7.8	8.5	9.4	1	80	13.2	24.4
1 1/4	40	1.95	3.01	4.67	6.83	9.73	16.1	17.6	19.4	1 1/4	80	22.9	51.5
1 1/2	40	2.89	4.63	7.04	10.4	14.7	24.1	26.5	29.1	1 1/2	80	37.1	78.0
2	40	5.60	8.90	13.0	19.9	28.2	46.6	51.0	56.2	2	40	51.5	185.0
2 1/2	40	8.90	14.2	21.5	31.9	45.9	74.7	82.0	90.0	2 1/2	40	73.3	297.0
3	40	15.9	25.2	38.0	56.5	80.1	132.0	144.0	159.0	3	40	113.0	510.0
3 1/2	40	23.1	36.1	55.1	81.0	116.0	189.0	207.0	228.0	3 1/2	40	151.5	704.0
4	40	32.1	50.8	76.7	112.7	159.5	260.0	285.0	314.0	4	40	195.0	1060.0
5	40	57.8	91.0	138.6	204.0	292.0	477.0	520.0	575.0				
6	40	94.1	148.6	224.0	329.0	472.0	775.0	850.0	937.0				
8	40	199.0	316.0	474.0	704.0	996.0	1650.0	1810.0	1992.0				
10	40	294.2	550.0	840.0	1226.0	1760.0	2880.0	3150.0	3470.0				
12	ID	555.0	877.0	1340.0	1935.0	2795.0	4640.0	5080.0	5590.0				

NOTES:

* (1) *Basis of Table: 105 F Condensing Temperature, 2 F ΔT per 100 ft Equivalent Length (except liquid lines).*

(2) *For Other ΔT's and Equivalent Lengths*

Line Capacity (Tons) = *Table Tons* ×

$$\left(50\times\frac{\text{Actual }\Delta T\text{ loss desired, F}}{\text{Actual Equiv. length, ft}}\right)^{0.55}$$

(3) *For other Tons and Equivalent Lengths*

$$\Delta T\text{ for a given pipe size}=\frac{\text{Actual Equiv. Length, ft}}{50}\times\left(\frac{\text{Actual Tons}}{\text{Table Tons}}\right)^{1.8}$$

(4) *For the equivalent change in saturation temperature for various line pressure drops in psi, refer to Fig.*

(5) *Data developed from preliminary ARI information. Subject to correction.*

(6) *For other condensing temperatures, multiply table tons by the following factors:*

Condensing Temp F	*Suction Lines*	*Hot Gas Lines*
80	1.13	0.77
90	1.08	0.86
100	1.03	0.95
110	0.97	1.04
120	0.91	1.13

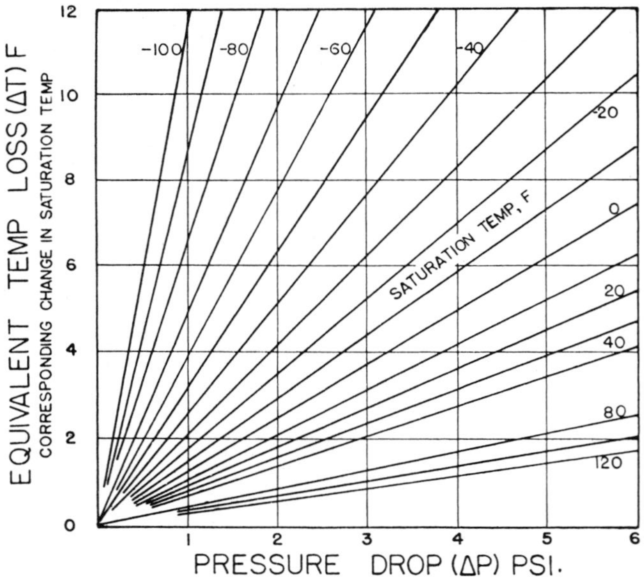

EQUIVALENT TEMP LOSS (ΔT) F
CORRESPONDING CHANGE IN SATURATION TEMP
SATURATION TEMP, F
PRESSURE DROP (ΔP) PSI.

TEMP EQUIV. (ΔT) OF PRESSURE DROP, F
(REFRIGERANT 22)

TABLE 19-4 Refrigerant Line Capacities for Refrigerant 502 (Single or High Stage Applications)

(Tons of Refrigeration Resulting in a Line Friction Drop (ΔP in psi) per 100 Ft Equivalent Pipe Length as Shown, with Corresponding (ΔT) Change in Saturation Temp.)

Line Size Type L Copper, OD	Suction Lines ΔT=2 F — Suction Temp, F						Discharge Lines ΔT=1.0 F ΔP=3.15 — Saturated Suction Temp			Liquid Lines[a]		
	−60 ΔP=0.31	−40 ΔP=0.94	−20 ΔP=1.33	0 ΔP=1.83	20 ΔP=2.43	40 ΔP=3.14	−40	0	40	Line Size Type L Copper, OD	Velocity =100 fpm	ΔT=1 F ΔP=3.15
1/2	0.10	0.11	0.15	0.22	0.34	0.49	0.61	0.62	0.78	1/2	1.61	2.40
5/8	0.11	0.15	0.26	0.42	0.63	0.91	1.14	1.27	1.45	5/8	2.58	4.52
7/8	0.23	0.41	0.68	1.09	1.64	2.39	2.98	3.34	3.80	7/8	5.35	12.01
1 1/8	0.46	0.82	1.38	2.20	3.33	4.83	6.02	6.74	7.66	1 1/8	9.13	24.43
1 3/8	0.80	1.44	2.42	3.84	5.80	8.41	10.49	11.74	13.34	1 3/8	13.90	42.71
1 5/8	1.27	2.28	3.83	6.07	9.16	13.29	16.51	18.49	21.01	1 5/8	19.68	67.69
2 1/8	2.65	4.76	7.97	12.63	18.98	27.45	34.03	38.14	43.36	2 1/8	34.23	140.87
2 5/8	4.71	8.44	14.12	22.29	33.50	48.38	59.93	67.18	76.35	2 5/8	52.79	249.43
3 1/8	7.56	13.54	22.58	35.56	53.38	77.02	95.34	107.2	121.5	3 1/8	75.35	398.62
3 5/8	11.30	20.15	33.58	52.83	79.25	114.56	141.4	158.6	180.1	3 5/8	101.9	593.10
4 1/8	15.98	28.47	47.39	74.49	111.78	160.90	199.0	223.1	253.5	4 1/8	132.5	837.24
5 1/8	28.71	51.07	84.85	133.32	199.37	286.92	354.3	397.2	451.2	—	—	—
6 1/8	46.35	82.31	136.77	214.07	319.89	459.97	567.6	636.5	723.1	—	—	—

NOTES:

(1) For Other ΔT's and Equivalent Lengths, L_e Line Capacity (Tons)

$$= \text{Table Tons} \times \left(\frac{100}{L_e} \times \frac{\text{Actual } \Delta T \text{ Loss Desired}}{\text{Table } \Delta T \text{ Loss}}\right)^{0.55}$$

(2) For other Tons and Equivalent Lengths in a given pipe size

$$\Delta T = \text{Table } \Delta T \times \frac{L_e}{100} \times \left(\frac{\text{Actual Tons}}{\text{Table Tons}}\right)^{1.8}$$

(3) Data developed from Reference 13.

(4) Values are based on 105 F condensing temperature. For other condensing temperatures, multiply table tons by the following factors:

Condensing Temp F	Suction Lines	Hot Gas Lines
80	1.20	.83
90	1.12	.91
100	1.04	.97
110	.96	1.02
120	.88	1.08
130	.80	1.16

* From *ASHRAE Data Book*, Fundamentals Volume, 1972 Edition, by permission of the American Society of Heating, Refrigerating, and Air-Conditioning Engineers.

provides similar information for various sizes of discharge risers at various saturated discharge temperatures.

Example 19-1 Determine the minimum size suction riser that will ensure oil return a minimum loading for a 75-ton R-502 system that is equipped with a reciprocating compressor having capacity steps of 25%, 50%, 75%, and 100% if the design saturated suction temperature at minimum loading is −20°F and the liquid refrigerant approaching the refrigerant flow control is 70°F.

Solution Since the minimum system capacity will occur when the compressor is operating at the lowest step of capacity, which is 25% of full load, the minimum capacity is (75 tons × 0.25) 18.75 tons. For R-502 at a saturated suction temperature of −20°F, Table 19-1A indicates a minimum capacity of 15 tons for a 3 5/8 in. OD copper pipe. The correction factor listed for 70°F liquid is 1.13, so that the corrected minimum capacity for the 3 5/8 in. OD pipe is (15 tons × 1.13) 16.95 tons. Since the minimum pipe capacity is less than the system minimum load, this size suction riser will ensure oil return during periods of minimum loading.

In the interest of high system efficiency, good design requires that the suction piping be sized so that the over-all refrigerant pressure drop in the line does not cause a drop in the saturated suction temperature of more than one or two degrees for Refrigerants-12, 22, and 502, or more than one degree for ammonia. Since the pressure-temperature relationship of all refrigerants changes with the temperature range, the maximum permissible pressure drop in the suction piping varies with the evaporator temperature, decreasing as the evaporator temperature decreases. For instance, for Refrigerant-12 vapor at 40°F, the maximum permissible pressure drop in the suction piping (equivalent to a 2°F drop in saturation temperature) is 1.8 psi, whereas for Refrigerant-12 vapor at −40°F, the maximum permissible pressure drop in the suction line is only 0.5 psi.

Tonnage capacities of various sizes of iron pipe and type L copper tubing at various suction temperatures are listed in Tables 19-2, 19-3, 19-4, and 19-5 for Refrigerants-12, 22, 502, and ammonia, respectively. The values listed in the tables are based on a suction line pressure loss equivalent to 2°F per 100 ft of pipe for Refrigerants-12, 22 and 502, and 1°F for ammonia. The condensing temperature is taken as 100°F for Refrigerant-12 and as 105°F for Refrigerants-22, 502 and ammonia. In all cases, tonnage capacities at other condensing temperatures can be determined by applying the correction factors given at the bottom of each table. Equations are also given at the bottom of the tables for correcting tonnages for other pressure loses and equivalent lengths. The following example will serve to illustrate the use of the tables.

Example 19-2 A 40-ton, Refrigerant-12 system has an evaporator temperature of 20°F and a condensing temperature of 110°F. If a suction pipe 30 ft long containing three standard elbows is required, determine:

(a) the size of type L copper tubing required,
(b) the overall pressure drop in the suction line in pounds per square inch.

Solution From Table 19-2, $3\frac{1}{8}$ in. OD copper tubing has a capacity of 34 tons based on a condensing temperature of 100°F and a suction line pressure loss equivalent to 2°F per 100 ft of pipe. Since the pressure loss is proportional to the length of pipe and since the length of pipe is relatively short in this instance, this pipe size may be sufficient and a trail calculation should be made. From Table 15-1, $3\frac{1}{8}$ in. OD (3 in. nominal) standard elbows have an equivalent length of 8 ft.

Actual equivalent length of suction piping:

Straight pipe length	= 30 ft
3 ells at 8 ft	= 24 ft
Total equivalent length	= 54 ft

Correction factor from Table 19-2 to correct tonnage for 110°F condensing temperature is 0.94.

$$= 34 \times 0.94$$

Corrected tonnage $= 31.96$ tons

Suction line pressure loss in °F

$$= \frac{\text{Actual equiv. length}}{50} \times \left(\frac{\text{Actual tons}}{\text{Table tons}}\right)^{1.8}$$

$$= \left(\frac{54}{50}\right) \times \left(\frac{40}{31.96}\right)^{1.8}$$

$$= (1.08)(1.5)$$

$$= 1.62°\text{F}$$

From the chart at the bottom of Table 19-2, the pressure loss in psi corresponding to 1.62°F at a 20°F suction temperature is approximately 1.1 psi.

When the suction piping is sized on the basis of a one or two degree drop in the saturated suction temperature, as in the preceding example, the resulting vapor velocity will ordinarily be sufficiently high to insure the return of oil up a suction riser during periods of minimum loading. However, exceptions to this are likely to occur in any system where the evaporator temperature is low, where the suction line is excessively long, and/or where the minimum system loading is less than 50% of the design load. When any of the above conditions exist, the vapor velocity should be checked for the minimum load condition to be sure that it will be above the minimum required for successful oil entrainment in risers.

The widespread use of automatic capacity control on modern compressors, in order to vary the capacity of the compressor to conform to changes in the system load, tends to complicate the design of all of the refrigerant piping. Through the use of automatic capacity control, single compressors are capable of unloading down to as little as 25% of the maximum design capacity. When two or more such compressors are connected in parallel, the system can be designed to unload down to as little as 10% of the combined maximum design capacity of the compressors. Obviously, when the system capacity is varied over such a wide range, any suction piping sized small enough to insure vapor velocities sufficiently high to carry oil up a riser during periods of minimum loading will cause a prohibitively high refrigerant pressure drop during periods

TABLE 19-5 Refrigerant Line Capacities for Refrigerant-717 (Ammonia) (Single or High Stage Applications) Tons of Refrigerant Resulting in a Line Friction Drop per 100 Ft Equivalent Pipe Length Corresponding to 1°F (ΔT) Change in Saturation Temp

*Suction Lines**							*Discharge Lines*	*Liquid Lines*			
Line Size		*Suction Temperature F*						*Line Size*		*Condenser to Receiver Velocity*	*Receiver to System*
IPS	*SCH*	*−40* $\Delta P=0.32$	*−20* $\Delta P=0.52$	*0* $\Delta P=0.78$	*20* $\Delta P=1.08$	*40* $\Delta P=1.48$	$\Delta P=3.3$	*IPS*	*SCH*	*=100 fpm*	$\Delta P=3.3$
3/8	80										
1/2	80						3.63	1/2	80	13.5	29.7
3/4	80				2.58	3.75	7.98	3/4	80	24.9	66.7
1	80		2.11	3.46	5.14	7.50	15.9	1	80	41.5	130.0
1 1/4	40	3.24	5.57	8.90	13.4	19.4	41.2	1 1/4	40	86.2	281.0
1 1/2	40	4.83	8.75	13.70	20.2	29.4	57.5	1 1/2	40	117.2	439.0
2	40	9.34	16.4	26.2	39.4	57.3	118.9	2	40	193.5	1004.0
2 1/2	40	15.0	26.0	42.2	62.5	91.2	187.2	2 1/2	40	276.0	1599.0
3	40	26.9	46.0	73.9	111.0	162.0	338.2	3	40	425.0	2341.0
4	40	56.1	94.5	151.0	226.0	327.0	676.0	4	40	736.0	5750.0
5	40	102.0	172.0	272.0	408.0	592.0	1228.0	5	40		
6	40	160.0	280.0	445.0	662.0	958.0	1986.0	6	40		
8	40	338.0	570.0	908.0	1355.0	1960.0	4120.0	8	40		
10	40	605.0	1030.0	1640.0	2430.0	3555.0		10	40		
12	ID	975.0	1660.0	2640.0	3940.0	5680.0		12	ID		

NOTES:

(1) *Basis of Table: 100 F Condensing Temperature, 1 F ΔT per 100 ft equivalent length. Discharge and liquid lines based on 0 F suction.*

(2) *For other ΔT's and Equivalent Lengths,*

$$\text{Line Capacity (tons)} = \text{Table Tons} \times \left(\frac{50 \times \text{Actual } \Delta T \text{ Loss Desired, F}}{\text{Actual Equiv. Length, ft}}\right)^{0.55}$$

(3) *For other Tons and Equivalent Lengths,*

$$\Delta T \text{ for a given pipe size} = \frac{\text{Actual Equiv. Length, ft}}{50} \times \left(\frac{\text{Actual Tons}}{\text{Table Tons}}\right)^{1.8}$$

(4) *Values based on 100 F condensing temp. For capacities at other condensing temp, multiply table value by line capacity multiplier:*

Line	*Condensing Temperature F* 70	80	90	100
Suction Lines	*1.0*	*1.0*	*1.0*	*1.0*
Discharge Lines	*0.70*	*0.80*	*0.90*	*1.0*

(5) *For the Equivalent Change in saturation temp for various line pressure drops in psi, refer to Fig.*
* (6) *Taken from chapter 9 of the 1957–58 ASRE Design Data Book. Initially developed from ARI preliminary data.*

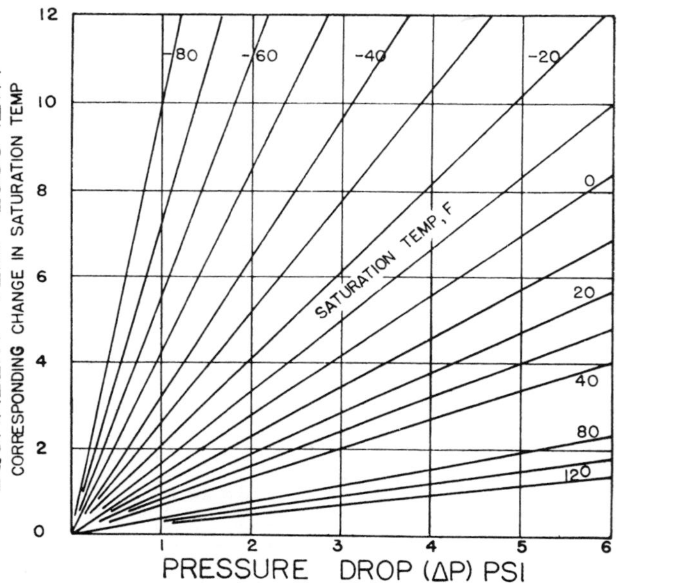

Temperature Equivalent
(ΔT) of Pressure Drop
(Ammonia)

TABLE 19-6 Refrigerant Line Capacities for Intermediate or Low Stage Duty (Tons) for Refrigerants 12, 22, and Ammonia

*Refrigerant and ΔT Equivalent of Friction Drop**	*Line Size Type L Copper OD*	*Suction Lines**							*Dis-charge Lines**	*Line Size Type L Copper OD*	*Liquid Lines*	
		Suction Temp F									*Condenser to Receiver V=100 fpm*	*Receiv-er to System**
		−90	*−80*	*−70*	*−60*	*−50*	*−40*	*−30*				
Refrigerant 12	1/2									1/2	See Table 19-2	2.1
	5/8									5/8		3.9
	7/8				0.2	0.3	0.4	0.5	0.9	7/8		11.0
	1 1/8	0.17	0.24	0.3	0.4	0.6	0.8	1.0	1.8	1 1/8		21.5
	1 3/8	0.30	0.42	0.6	0.8	1.1	1.4	1.7	3.2	1 3/8		37.0
	1 5/8	0.47	0.67	0.9	1.2	1.7	2.2	2.7	5.0	1 5/8		60.0
	2 1/8	1.00	1.40	1.9	2.5	3.5	4.6	5.7	10.5	2 1/8		125.0
2 F ΔT Per 100 ft Equiv. Length	2 5/8	1.7	2.4	3.3	4.5	6.1	8.0	10.0	18.5	2 5/8		220.0
	3 1/8	2.8	3.9	5.4	7.3	10.0	13.0	16.2	30.0	3 1/8		350.0
	3 5/8	4.1	5.9	8.2	10.8	15.0	19.5	24.3	44.0	3 5/8		
	4 1/8	6.0	8.5	11.7	15.6	21.5	28.0	35.0	65.0	4 1/8		
	5 1/8	10.6	15.1	20.8	27.8	38.5	50.0	62.5	113.0	5 1/8		
	6 1/8	18.1	25.8	35.4	47.2	65.4	85.0	106.0	180.0	6 1/8		
Refrigerant 22	1/2									1/2	See Table 19-3	3.6
	5/8								0.6	5/8		7.0
	7/8	0.16	0.23	0.31	0.44	0.57	0.75	0.94	1.5	7/8		18.0
	1 1/8	0.34	0.48	0.65	0.91	1.19	1.55	1.93	3.0	1 1/8		36.0
	1 3/8	0.59	0.81	1.12	1.59	2.07	2.7	3.4	5.2	1 3/8		63.0
	1 5/8	0.93	1.34	1.8	2.5	3.3	4.3	5.4	8.5	1 5/8		100.0
	2 1/8	1.9	2.8	3.7	5.2	6.8	8.9	11.1	17.5	2 1/8		210.0
2 F ΔT Per 100 ft Equiv. Length	2 5/8	3.5	5.0	6.6	9.4	12.3	16.0	20.0	31.0	2 5/8		375.0
	3 1/8	5.5	8.0	10.6	15.0	19.6	25.5	32.0	50.0	3 1/8		
	3 5/8	8.4	12.0	16.0	22.6	29.5	38.5	48.0	75.0	3 5/8		
	4 1/8	12.0	17.2	22.9	32.3	42.3	55.0	68.8	105.0	4 1/8		
	5 1/8	21.2	30.6	41.0	57.5	75.0	98.0	122.0	190.0	5 1/8		
	6 1/8	34.8	50.0	66.5	94.0	123.0	160.0	200.0	305.0	6 1/8		
	Steel IPS SCH				*−60*	*−50*	*−40*	*−30*		*Steel IPS SCH*		
										3/8 80	See Table 19-4	17.0
	1/2 40				0.26	0.38	0.50	0.62	1.0	1/2 80		34.0
	3/4 40				0.55	0.76	1.05	1.30	2.1	3/4 80		75.0
	1 40				1.05	1.53	2.00	2.50	4.1	1 80		150.0
Refrigerant 717 (Ammonia)	1 1/4 40				2.15	3.15	4.10	5.10	8.5	1 1/4 80		305.0
	1 1/2 40				3.4	5.0	6.5	8.1	12.5	1 1/2 80		490.0
	2 40				6.3	9.2	12.0	15.0	25.0	2 40		
	2 1/2 40				10.3	15.0	19.5	24.3	40.0	2 1/2 40		
1 F ΔT Per 100 ft Equiv. Length	3 40				18.4	26.8	35.0	43.7	71.0	3 40		
	3 1/2 40				27.3	39.8	52.0	65.0	105.0	3 1/2 40		
	4 40				37.8	55.2	72.0	90.0	145.0	4 40		
	5 40				68.3	100.0	130.0	162.0	260.0	5 40		
	6 40				110.0	161.0	210.0	262.0	425.0	6 40		
	8 40				258.0	376.0	490.0	610.0		8 40		

NOTES:

* *(1) Values in this table are tons of refrigeration resulting in a line friction drop per 100 ft of equivalent pipe length corresponding to the (ΔT) change in saturation temp indicated in the left hand column under the refrigerant designation.*

(2) Values based on 0 F saturated discharge temp. For capacities at other saturated discharge temp, multiply table value by proper line capacity multiplier:

Sat. Dis-charge Temp, F	*Line Capacity Multipliers*				
	Refrigerant 12		*Refrigerant 22*		*Ammonia*
	Suction	*Discharge*	*Suction*	*Discharge*	*Discharge*
−30	1.12	0.55	1.09	0.58	
−20	1.07	0.70	1.06	0.71	
−10	1.03	0.85	1.03	0.85	0.77
0	1.00	1.00	1.00	1.00	1.00
10	0.96	1.25	0.97	1.20	1.23
20	0.93	1.50	0.94	1.45	1.45
30	0.90	1.80	0.90	1.80	1.67

(3) For other ΔT's and Equivalent Lengths,

$$\text{Line Capacity (Tons)} = \text{Table Tons} \times \left(\frac{100}{\text{Actual Equiv. Length, ft}} \times \frac{\text{Actual } \Delta T \text{ Loss Desired, F}}{\text{Table } \Delta T \text{ Loss, F}} \right)^{0.55}$$

(4) For other Tons and Equivalent Lengths in a given pipe size,

$$\Delta T(F) = \text{Table } \Delta T \times \frac{\text{Actual Equiv. Length, ft}}{100} \times \left(\frac{\text{Actual Tons}}{\text{Table Tons}} \right)^{1.8}$$

(5) Values obtained from Carrier Corp. data.

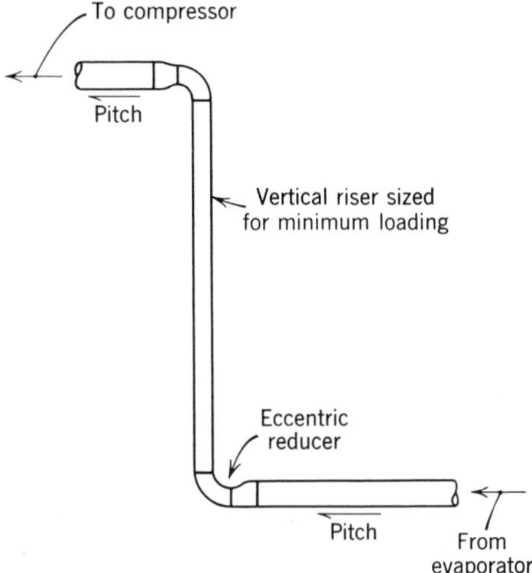

Fig. 19-2 Illustrating method of reducing the size of a vertical suction riser. (Courtesy York Corporation.)

of maximum loading. On the other hand, sizing the pipe for a low pressure drop at maximum loading will result in riser velocities too low to return oil. Fortunately, the vapor velocity in horizontal piping is not critical and the problem is mainly one of riser design. In most cases, when the minimum system load is not less than 25% of the design load, the problem can be solved by reducing the size of the riser only, with the balance of the suction piping being sized for a low pressure drop at maximum loading.

Figure 19-2 illustrates the proper method of reducing the line size at a vertical riser. An eccentric reducer with its flat side down should be used at the bottom connection before entry to the elbow. This is done to prevent forming an area of low gas velocity which could trap a layer of oil extending the length of the horizontal line. At the top of the riser, the line size is increased beyond the elbow with a standard reducer, so that any oil reaching this point cannot drain back into the riser.

19-7. Double-Pipe Risers

As a general rule, when the suction riser is reasonably short and the minimum system load does not fall below 25% of the maximum design load, undersizing of the riser to provide adequate vapor velocity during periods of minimum loading will not cause a significant increase in the overall suction line pressure drop, particularly if the horizontal portion of the piping is liberally sized. On the other hand, when the suction riser is quite long and/or when the minimum system loading is less than 25% of the design loading, undersizing of the riser to conform to the requirements of minimum loading will ordinarily result in an excessive pressure loss in the suction piping during periods of maximum loading, especially in low temperature installations. In such cases, the double-pipe riser, shown in Fig. 19-3, should be employed.

The small diameter riser is sized for the minimum load condition, whereas the combined capacity of the two pipes is designed for the maximum load condition. The larger riser is

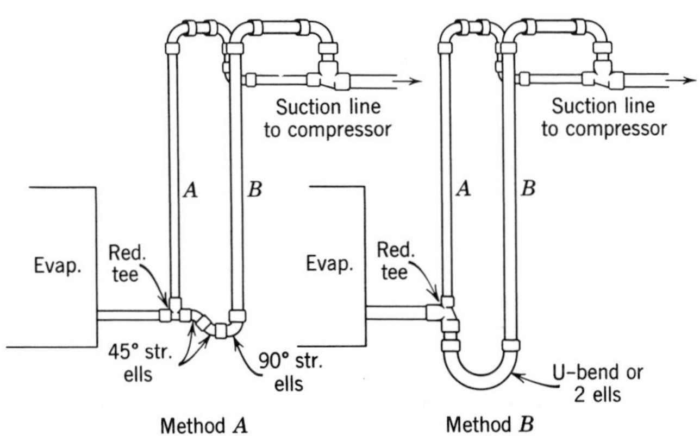

Fig. 19-3 Double suction riser construction. (Courtesy Carrier Corporation.)

trapped slightly below the horizontal line at the bottom. During periods of minimum loading, oil will settle in the trap and block the flow of vapor through the larger riser, thereby increasing the flow rate and velocity in the smaller riser to a level high enough to insure oil return up the riser. As the system load increases, the velocity increases in the small riser until the pressure drop across the riser is sufficient to clean the oil out of the trap and permit flow through both pipes.

Notice that the trap at the bottom of the large riser is made up of two 45° elbows and one 90° elbow. This is done to keep the volume of the trapped oil as small as possible. Notice also that inverted loops are used to connect both risers to the top of the upper horizontal line, so that oil reaching the upper line cannot drain back into the risers.

19-8. General Design of Suction Piping

The suction piping should always be so arranged as to eliminate the possibility of liquid refrigerant (or large slugs of oil) entering the compressor during either the running or off cycles, or during compressor start-up. Generally, unless the system is operated on a pump-down cycle, it is good practice to install a liquid-suction heat exchanger in the suction line of all systems employing dry-expansion evaporators. The reason for this is that thermostatic expansion valves frequently do not close off tightly during the compressor off cycle, thereby permitting off cycle leakage of liquid refrigerant into the evaporator from the liquid line. When the compressor starts, the excess liquid often slops over into the suction line and is carried to the compressor unless a liquid-suction heat exchanger is employed to trap the liquid and vaporize it before it reaches the compressor. The liquid-suction heat exchanger also serves to trap and vaporize any liquid which may carry over into the suction line because of overfeeding of the expansion valve during start-up or during sudden changes in the evaporator loading.

Ordinarily, the liquid-suction heat exchanger can be safely omitted if the system is operated on a pump-down cycle, in which case the liquid refrigerant will be pimped out of the evaporator before the compressor cycles off, and the liquid line solenoid installed ahead of the refrigerant control will prevent liquid refrigerant from entering the evaporator from the liquid line, even though the expansion valve itself may not close off tightly.

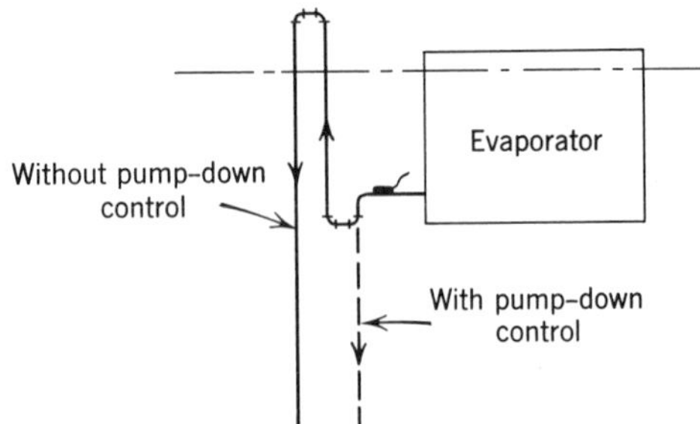

Fig. 19-4 Evaporator located above compressor.

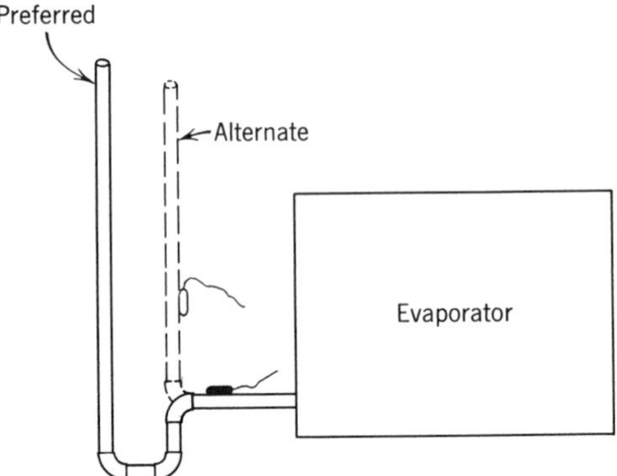

Fig. 19-5 Evaporator below compressor.

When the evaporator is located above the compressor, and the system is not operated on a pump-down cycle, the suction line should be trapped immediately beyond the expansion valve bulb, as shown in Fig. 19-4, so that liquid refrigerant cannot drain by gravity from the evaporator to the compressor during the off cycle. If the system is operated on a pump-down cycle, the trap may be omitted and the piping arranged for free draining (dotted lines in Fig. 19-4), since all the liquid is pumped from the evaporator before the compressor cycles off.

When the evaporator is located below the compressor and the suction riser is installed immediately adjacent to the evaporator, the riser should be trapped as shown in Fig. 19-5 to prevent liquid refrigerant from trapping at the thermal bulb location. In the event that trapping of the line is not practical, the trap may be omitted and the thermal bulb moved to a position on the vertical riser approximately 12 to 18 in. above the horizontal header, as shown by the dotted line in Fig. 19-5.

When a multiple of evaporators is to be connected to a common suction main, each evaporator (or separately fed evaporator segment) should be connected to the main with an individual riser, as shown in Fig. 19-6. Since thermostatic expansion valves do not perform properly when the load on the evaporator falls below 50% of the design capacity of the valve, the flow rate through the individual risers should never drop below 50% of the design flow rate. Therefore, the use of individual risers for each evaporator (or separately fed segment) should eliminate the problem of oil return at minimum loading. Figures 19-7 through 19-9 illustrate some of the various methods of connecting multiple evaporators to a common suction main when it is not practical to use individual risers.

The suction piping at the compressor should be brought in above the level of the compressor suction inlet. The piping should be designed without liquid traps and so arranged that the oil will drain by gravity from the suction line into the compressor. When multiple compressors are connected to a common suction header, the piping should be designed so that oil return to the several compressors is as nearly equal as possible. The lines to the individual compressors should always be connected to the side

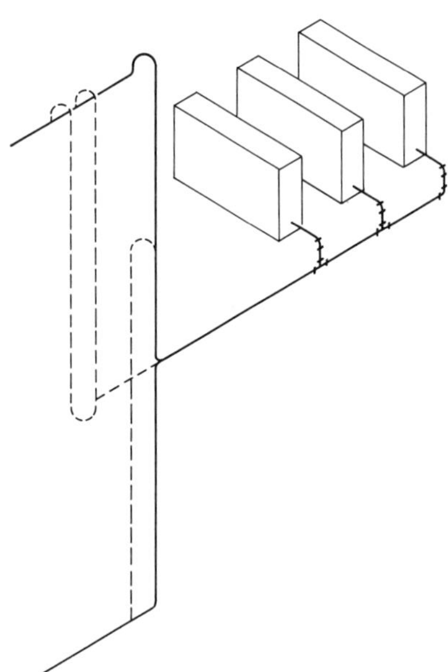

Fig. 19-7 Multiple evaporators, common suction line.

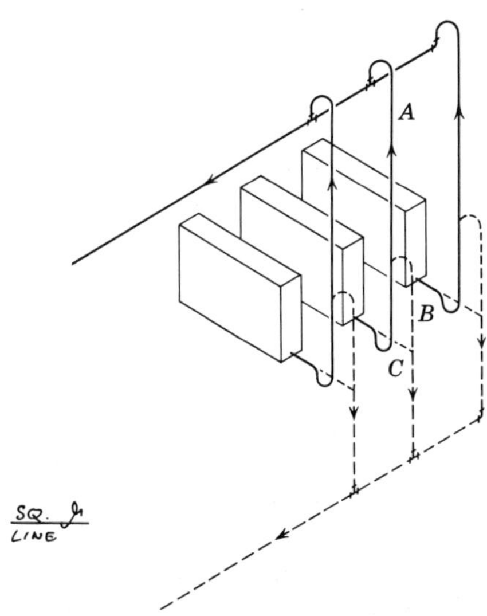

Fig. 19-6 Multiple evaporators, individual suction lines.

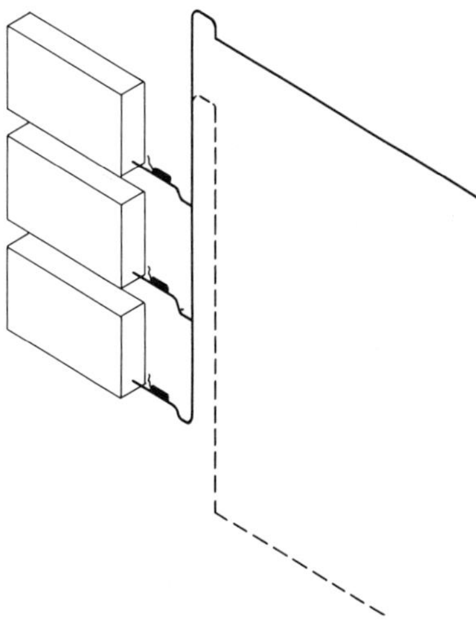

Fig. 19-8 Evaporators at different levels connected to a common suction riser.

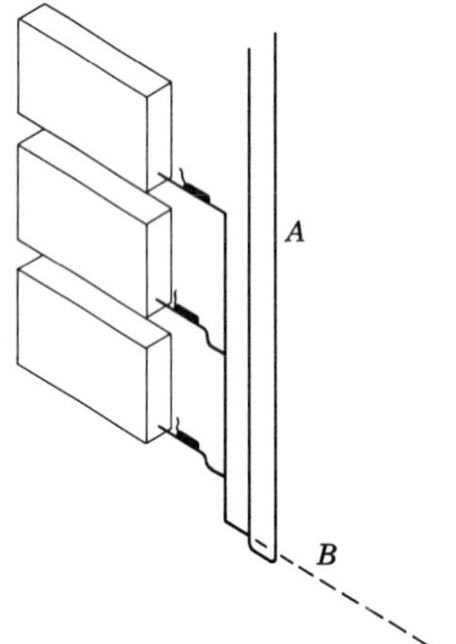

Fig. 19-9 Evaporators at different levels connected to a double suction riser.

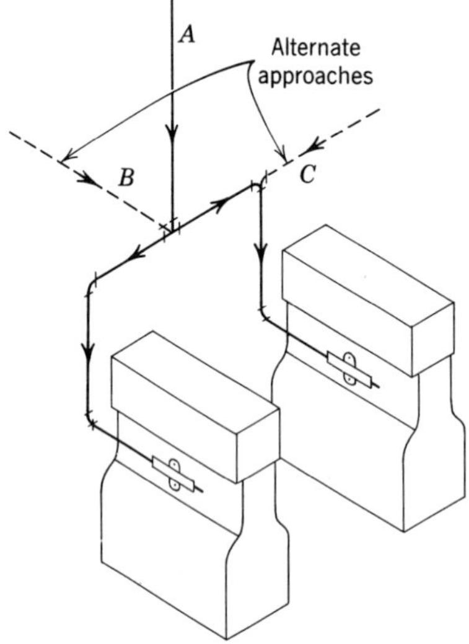

Fig. 19-10 Suction piping for compressors connected in parallel.

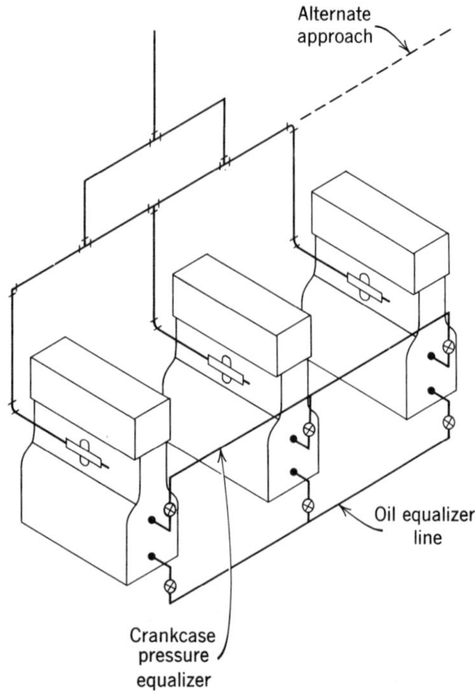

Fig. 19-11 Suction piping for compreesors connected in parallel.

of the header. Some typical piping arrangements for multiple compressors are shown in Figs. 19-10 and 19-11.

19-9. Discharge Piping

Sizing of the discharge piping is similar to that of the suction piping. Since any refrigerant pressure drop in the discharge piping tends to increase the compressor discharge pressure and reduce the capacity and efficiency of the system, the discharge piping should be sized to provide the minimum practical refrigerant pressure drop. Tonnage capacities for various sizes of discharge pipes are given in Tables 19-2, 19-3, 19-4, and 19-5. The values listed in the tables are based on an overall refrigerant pressure drop per 100 ft (30 m) of equivalent length corresponding to a 2°F (1°C) drop in the saturation temperature of Refrigerants-12 and 22, and for a 1°F loss in saturation temperature for R-502 and ammonia. The procedure for sizing the discharge piping is the same as that used for the suction piping.

All horizontal discharge piping should be pitched downward in the direction of the refrigerant flow so that any oil pumped over from the compressor into the discharge line will drain toward the condenser and not back into the compressor head. Although the minimum vapor velocity in horizontal discharge piping is not ordinarily critical, special attention must be given to the vapor velocity in discharge line risers. As in the case of a suction riser, the discharge line riser must be designed so that

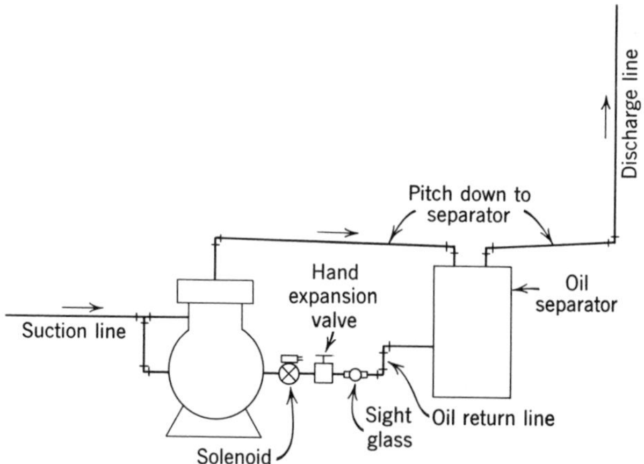

Fig. 19-12 Arrange for preventing liquid return to compressor crankcase.

the vapor velocity in the riser under minimum load conditions is sufficiently high to entrain the oil and carry it up the riser (see Table 9-1B). When the system capacity varies over a wide range, a double-pipe riser may be necessary, unless a discharge line oil separator is used. When an oil separator is installed in the discharge line, the vapor velocity in the discharge riser is not critical and the riser should be sized for a low pressure drop, since any oil which is not carried up the riser during periods of minimum loading will drain back into the separator (Fig. 19-12).

When the compressor is not operating, the oil adhering to the inside surface of a discharge riser tends to drain by gravity to the bottom of the riser. If the riser is more than 8 to 10 ft long, the amount of oil draining from the riser may be quite large. Therefore, the discharge line from the compressor should be looped to the floor to form a trap so that oil cannot drain from the discharge piping into the head of the compressor. Since this trap will also collect any liquid refrigerant which may condense in the discharge riser during the compressor off cycle, it is especially important when the discharge piping is in a cooler location than the liquid receiver and/or condenser. Additional traps, one for 25 ft (7.5 m) of vertical rise, should be installed in the discharge riser when the vertical rise exceeds 25 ft (7.5 m), as shown in Fig. 19-13. The horizontal width of the traps should be held to a minimum and can be constructed of two standard 90° elbows. The depth of the traps should be approximately 18 in. (0.5 m).

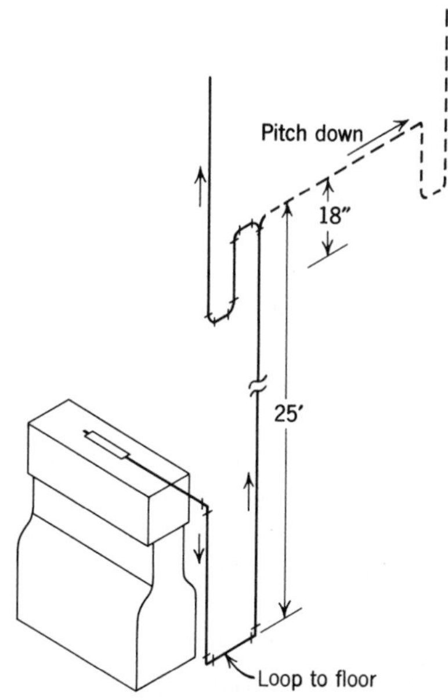

Fig. 19-13 Piping of discharge riser.

The traps may be omitted if an oil separator is used, since any oil or liquid refrigerant draining from the vertical riser during the off cycle will drain into the oil separator. However, certain precautions must be taken to eliminate the possibility of liquid refrigerant passing from

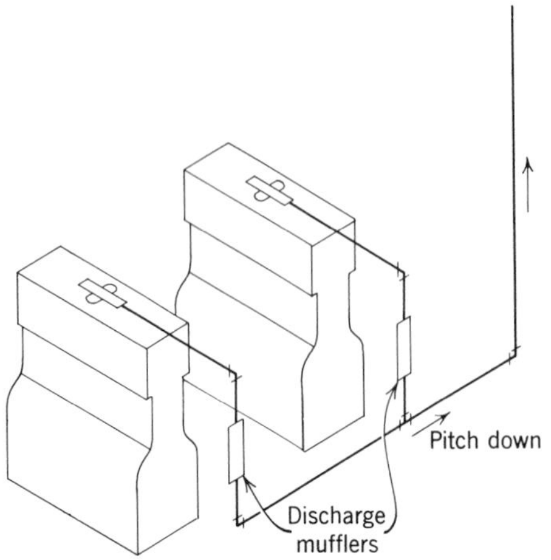

Fig. 19-14 Discharge piping of multiple compressors connected in parallel.

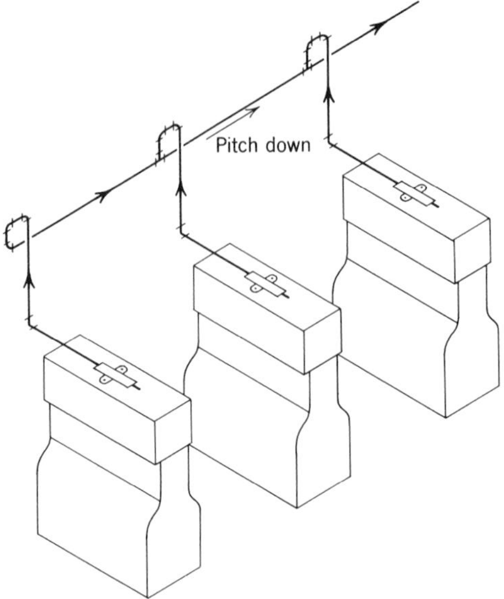

Fig. 19-15 Discharge piping for multiple compressors connected in parallel.

the oil separator to the compressor crankcase during the off cycle (see Section 19-12).

A purge valve, to permit purging of noncondensible gases from the system, should be installed at the highest point in the discharge piping or condenser.

When two or more compressors are connected together for parallel operation, the discharge piping at the compressors must be arranged so that the oil pumped over from an active compressor does not drain into an idle one. Under no circumstances should the piping be arranged so that the compressors discharge directly into one another. As a general rule, it is good practice to carry the discharge from each compressor nearly to the floor before connecting to a common discharge main (Fig. 19-14). With this piping arrangement, a discharge line trap is not required at the riser, since the lower horizontal header serves this purpose.

In the event that the discharge header must be located above the compressors, the discharge from the individual compressors should be connected to the top of the header, as shown in Fig. 19-15, so that oil cannot drain from the header into the head of an idle compressor.

In order to reduce the noise and vibration created by the compressor discharge pulsations, discharge mufflers are recommended for all multiple compressor installations and for a single compressor installation where the noise of the discharge pulsations may become objectionable. Discharge mufflers must be installed for free draining, either in a horizontal line or in a down-comer, as shown in Fig. 19-14, but never in a riser.

19-10. Liquid Lines

The function of the liquid line is to deliver a solid stream of subcooled liquid refrigerant from the receiver tank to the refrigerant flow control at a sufficiently high pressure to permit the latter unit to operate efficiently. Since the refrigerant is in the liquid state, any oil entering the liquid line is readily carried along by the refrigerant to the evaporator, so that there is no problem with oil return in liquid lines. For this reason, the design of the liquid piping is somewhat less critical than that of the other refrigerant lines, the problem encountered being mainly one of preventing the liquid from flashing before it reaches the refrigerant control.

Flash gas in the liquid line reduces the capacity of the refrigerant control, causes erosion of the valve pin and seat, and often results in erratic control of the liquid refrigerant to the evaporator. To avoid flashing of the liquid in the liquid line, the pressure of the liquid in the

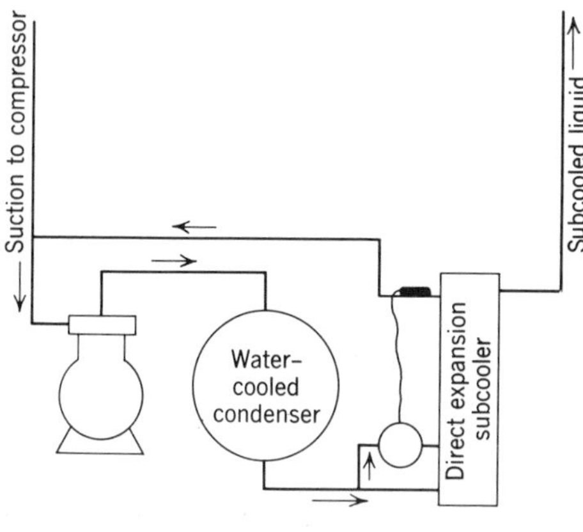

Fig. 19-16

line must be maintained above the saturation pressure corresponding to the temperature of the liquid.

Since the liquid leaving the condenser is usually subcooled 5° to 10°F, flashing of the liquid will not ordinarily occur if the overall pressure drop does not exceed 5 to 10 psi. However, if the pressure drop is much in excess of 10 psi, it is very likely that some form of liquid subcooling will be required if flashing of the liquid is to be prevented. In most cases, a liquid-suction heat exchanger and/or a water-cooled subcooler will supply the necessary subcooling. In extreme cases, a direct-expansion subcooler may be required. (Fig. 19-16.)

The amount of subcooling required in any individual installation can be determined by computing the liquid line pressure drop. Pressure drop in the liquid line results not only from friction losses but also from the loss of head due to vertical lift. Tonnage capacities of various sizes of liquid pipes are listed in Tables 19-2, 19-3, 19-4, and 19-5. The static pressure loss in pounds per square inch per foot of vertical lift is found by dividing the density of the liquid by 144 in.^{2}/ft^2.

Example 19-3 A Refrigerant-12 system with a condensing temperature of 100°F has a capacity of 35 tons. The equivalent length of the liquid line including fittings and accessories is 60 ft. If the line contains a 20-ft riser, determine:

(a) the size of the liquid line required,
(b) the overall pressure drop in the line,
(c) the amount of subcooling (°F) required to prevent flashing of the liquid.

Solution

(a) From Table 19-2, $1\frac{3}{8}$ in. OD copper tubing has a capacity of 35.1 tons based on a 1.8 psi pressure drop per 100 equivalent feet of pipe.

(b) For 60 ft equivalent length, the friction loss in the pipe = 1.8 psi × 0.6 = 1.00 psi

From Table 16-3, the density of 100°F liquid = 78.8 lb/ft^3

Pressure loss per foot of lift = 78.8/144 = 0.55

Static pressure loss = (0.55 psi/ft)(20 ft) = 11.0 psi

Overall pressure loss in liquid line = 1 psi + 11 psi = 12.0 psi

(c) Assuming the condensing temperature to be 100°F, the pressure at the condenser is 131.6 psia. The pressure at the refrigerant control is 119.6 psia, which corresponds to a saturation temperature of approximately 93°F. The amount of subcooling required is approximately 7°F (100° − 93°).

19-11. Condenser to Receiver Piping

Since the amount of refrigerant in the evaporator and condenser varies with the loading of the system, a liquid receiver tank is required on all systems employing hand expansion valves, automatic expansion valves, thermostatic expansion valves, or low pressure float valves. Some exceptions to this are installations using water-cooled condensers, wherein the water-cooled condenser also serves as the liquid receiver, and also those air-cooled condenser installations where the condenser has special built-in liquid subcooling circuits. In addition to accommodating fluctuations in the refrigerant charge, the receiver tends to keep the condenser drained of liquid, thereby preventing the liquid level

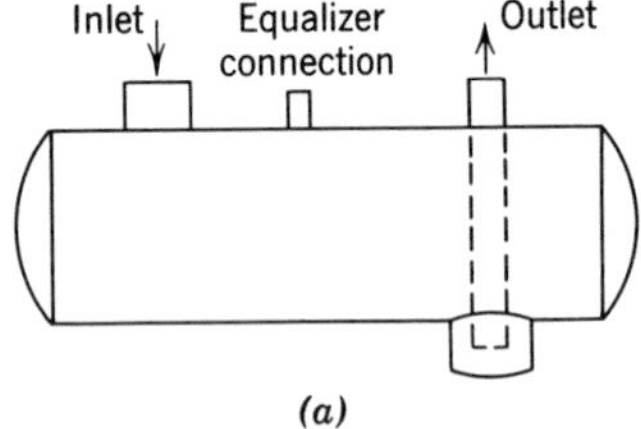

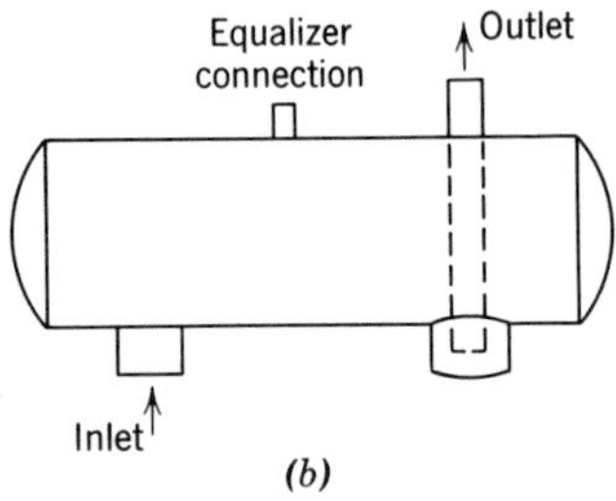

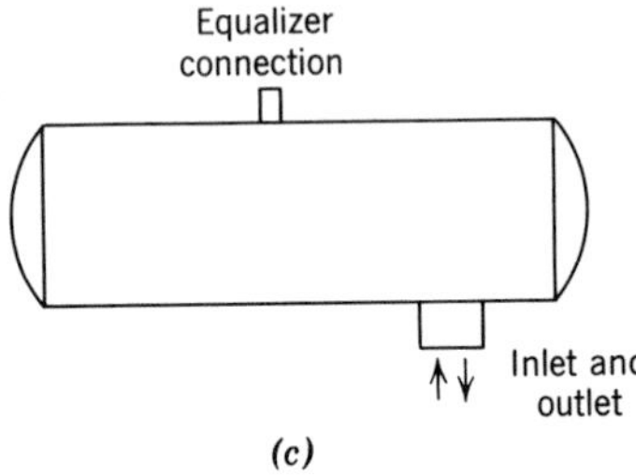

Fig. 19-17 (*a*) Top inlet through-flow receiver. (*b*) Bottom inlet through-flow receiver. (*c*) Surge-type receiver.

from building up in the condenser and reducing the amount of effective condenser surface. The liquid receiver serves also as a pump-down storage tank for the liquid refrigerant.

In general, the condenser to receiver piping must be so designed and sized as to allow free draining of the liquid from the condenser at all times. If the pressure in the receiver is permitted to rise above that in the condenser, vapor binding of the receiver will occur and the liquid refrigerant will not drain freely from the condenser. Vapor binding of the receiver is likely to occur in any installation where the receiver is so located that it can become warmer than the condenser. The problem of vapor binding is more acute in the wintertime and during periods of reduced loading.

Although the exact preventative measures which can be taken to eliminate vapor binding of the receiver depends somewhat on the type of condenser, in every instance it involves proper equalization of the receiver pressure to the condenser.

Basically, there are two types of liquid receivers: the through-flow and the surge (Fig. 19-17). The through-flow type may be either bottom inlet or top inlet. With the through-flow type receiver, all the liquid from the condenser drains into the receiver before passing into the liquid line. The surge type

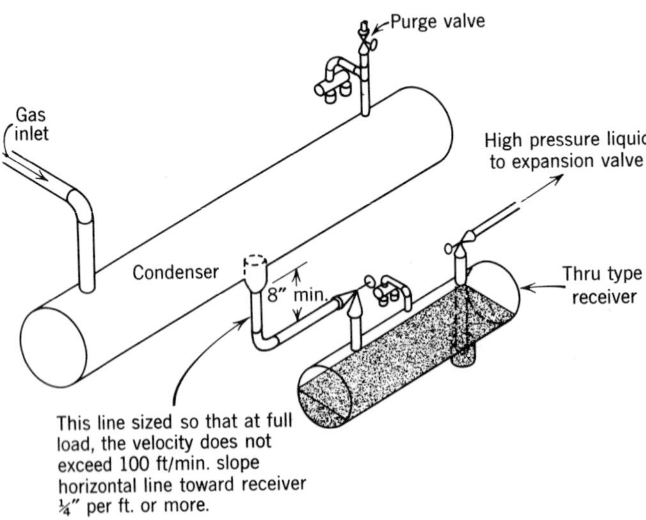

Fig. 19-18 Top inlet through type receiver hookup. (Courtesy York Corporation.)

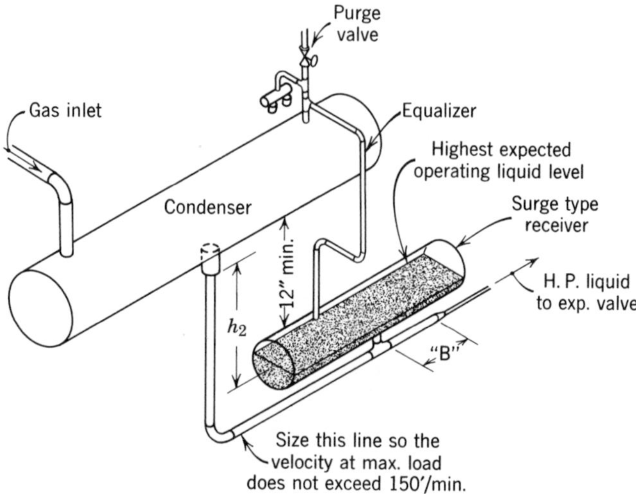

Fig. 19-19 Surge-type receiver hookup. (Courtesy York Corporation.)

differs from the through-flow in that only a part of the liquid from the condenser, that part not required in the evaporator, enters the receiver. With the surge-type receiver, the refrigerant liquid enters and leaves the receiver through the same opening.

When a top-inlet, through-flow type receiver is used, equalization of the receiver pressure to the condenser can be accomplished directly through the condenser to receiver piping, provided that the piping is sized so that the refrigerant velocity does not exceed 100 fpm and that the line is not trapped at any point (Fig. 19-18). All horizontal piping should be pitched toward the receiver at least $\frac{1}{4}$ in. per foot. When a stop valve is placed in the line, it should be located a minimum distance of 8 in. below the liquid outlet of the condenser and should be installed so that the valve stem is in a horizontal position. In the event that a trap in the line is unavoidable, a separate equalizing line must be installed from the top of the receiver to the condenser, as shown in Fig. 19-19. When an equalizing line is used, the condenser to receiver piping may be sized for a refrigerant velocity of 150 fpm (0.76 m/s).

All bottom inlet receivers of both the through-flow and surge types must be equipped with equalizing lines. The minimum vertical distance (h_2 of Fig. 19-19) between the outlet of the condenser and the maximum liquid level in the receiver to prevent the back-up of liquid in the condenser is listed at the bottom of Fig. 19-19. This value increases as the pressure loss in the condenser to receiver line increases, but should never be less than 12 in (0.3 m).

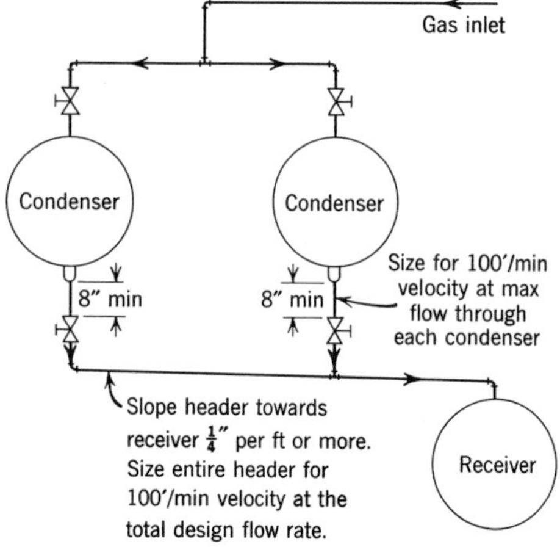

Fig. 19-20 Parallel shell-and-tube condensers with top inlet receiver. (Courtesy York Corporation.)

Figure 19-20 illustrates a satisfactory piping arrangement for multiple condensers connected in parallel to a top inlet receiver. A piping arrangement for multiple condensers with a bottom inlet receiver is shown in Fig. 19-21. The vertical distance h_2 is determined from the bottom of Fig. 19-19.

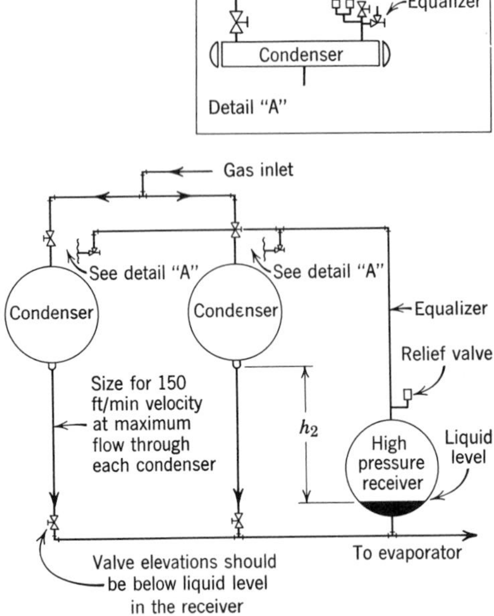

Fig. 19-21 Parallel shell-and-tube condensers bottom inlet receiver. (Courtesy York Corporation.)

19-12. Oil Separators

As a general rule, discharge line oil separators should be employed in any system when oil return is likely to be inadequate or difficult to accomplish and/or when the amount of oil in circulation is apt to be excessive or to cause an undue loss in the efficiency of the various heat transfer surfaces. Specifically, discharge line oil separators are recommended for (1) all systems employing nonmiscible refrigerants, (2) low temperature systems, (3) all systems employing nonoil-returning evaporators, such as flooded liquid chillers, when oil bleeder lines or other special provisions must be made for oil return, and (4) any system where capacity control and/or long suction or discharge risers cause serious piping design problems.

Discharge line oil separators are of two basic types: (1) impingement and (2) chiller. The impingement-type separator (Fig. 19-22) consists of a series of screens or baffles through which the oil-laden refrigerant vapor must pass. On entering the separator, the velocity of the refrigerant vapor is considerably reduced because of the larger area of the separator with relation to that of the discharge line, whereupon the oil particles, having a greater momentum than that of the refrigerant vapor, are caused to impinge on the surface of the screens or baffling. The oil then drains by gravity from the screens or baffles into the bottom of the separator, from where it is returned through a float valve to the compres-

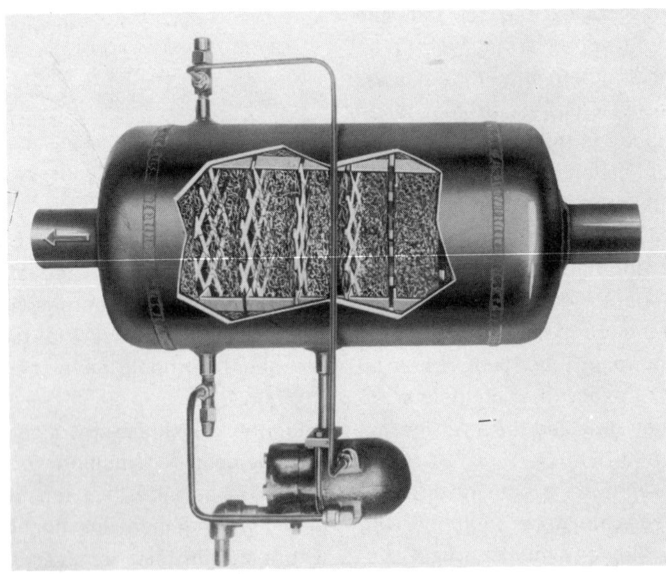

Fig. 19-22 Impingement-type oil separator with float drainer. (Courtesy York Corporation.)

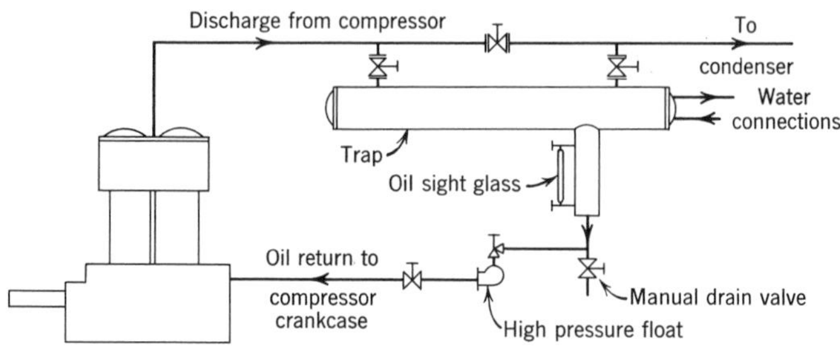

Fig. 19-23 Application of chiller-type oil separator. (Courtesy York Corporation.)

sor crankcase or, preferably, to the suction inlet of the compressor (Fig. 19-12).

The water-cooled, chiller-type separator, sometimes called an oil chiller, is similar in construction to the water-cooled condenser. Water is circulated through the tubes while the discharge vapor passes through the shell. The oil is separated from the vapor by precipitation on the cold water tubes, from where it drains into a drop-leg sump. The oil may be drained manually from the sump or returned automatically to the compressor through a float valve. (Fig. 19-23.) The water flow rate through the separator must be carefully controlled so that the refrigerant vapor is not cooled below its condensing temperature, in which case liquid refrigerant could be condensed in the separator and passed to the compressor crankcase through the float valve.

In some low temperature applications, a direct-expansion, chiller-type separator is installed in the liquid line. The operation and installation of the direct-expansion oil separator are similar to those of the direct-expansion liquid subcooler described in Section 19-10. The liquid refrigerant passing through the separator is chilled to a temperature below the pour point of the oil, thereby causing the oil to congeal on the chiller tubes. The oil is drained from the separator periodically by taking the separator out of service and allowing it to warm up to a temperature above the pour point of the oil.

Although properly applied oil separators are usually very effective in removing oil from the refrigerant vapor, they are not 100% efficient. Therefore, even when an oil separator is used, some means should be provided for removing the small amount of oil which will always pass through the separator and find its way into other parts of the system. Some of the various methods of accomplishing this are described in the following sections.

The principal hazard associated with the use of a discharge line oil separator is the possibility of liquid refrigerant passing from the oil separator to the compressor crankcase when the compressor is idle. The liquid refrigerant may drain into the separator from the discharge piping or it may condense in the separator itself during the off cycle.

While the compressor is operating, the temperature of the oil separator is relatively high and the possibility of liquid refrigerant condensing in the oil separator is rather remote, particularly if the separator is located reasonably close to the compressor. However, after the compressor cycles off, the separator tends to cool to the condensing temperature, at which time some of the high pressure refrigerant vapor is likely to condense in the separator. This raises the liquid level in the separator and causes the float to open and pass a mixture of oil and liquid refrigerant to the compressor crankcase. Condensation of the refrigerant in the separator is most apt to occur when the oil separator is installed in a cooler location than the condenser, in which case the liquid will boil off in the condenser and condense in the separator.

To eliminate the possibility of liquid refrigerant draining from the oil separator into the compressor crankcase during the off cycle, the oil drain line from the separator should be connected to the suction inlet of the compressor, rather than to the crankcase, and the line should be equipped with a solenoid valve, a sight glass, a hand expansion valve, and a manual shut-off valve (see Fig. 19-12). With the help of the sight glass, the hand throttling valve can be adjusted so that the liquid (oil and refrigerant) from the separator is bled slowly into the suction inlet of the compressor. The solenoid valve is interlocked with the compressor motor starter so that it is energized (open) only when the compressor is operating. This arrangement prevents the liquid refrigerant and oil in the separator from draining to the compressor during the off cycle, but permits slow draining into the suction inlet of the compressor when the compressor is operating.

To minimize the condensation of refrigerant vapor in the oil separator during the off cycle, oil separators should be installed near the compressor in as warm a location as possible. The separators should also be well insulated in order to retard the loss of heat from the separator after the compressor cycles off.

In some instances, the oil from the separator is drained into an oil receiver where it is stored until needed in the compressor crankcase (Fig. 17-39). The oil receiver contains a heating element which boils off the liquid refrigerant to the suction line. Oil from the oil receiver is admitted to the compressor crankcase as needed through a float valve located in the compressor crankcase.

Oil receivers cannot be used with compressors equipped with crankcase oil check valves. Since the oil receiver is at the suction pressure, the higher crankcase pressure could force all the oil out of the crankcase into the oil receiver. When oil receivers are employed, oil check valves must be removed and crankcase heaters installed.

19-13. Ammonia Piping

Since ammonia is a nonmiscible refrigerant, the oil pumped over into the discharge line from the compressor is not readily carried along through the system by the refrigerant. Therefore, an oil separator should be installed in the discharge line of all ammonia systems to reduce to a minimum the amount of oil that passes into the system. Provisions must be made also to remove from the system and return to the crankcase the small amount of oil that gets by the separator.

Oil, being heavier than liquid ammonia, tends to separate from the ammonia and settle out at various low points in the system. For this reason, oil sumps are provided at the bottom of all receivers, evaporators, accumulators, and other vessels in the system containing liquid ammonia, and provisions are made for draining the oil from these points either continuously or periodically. Since the amount of oil involved is small, when an operator is on duty the draining is usually done manually and the oil discarded. Of course, this requires that the crankcase oil be replenished periodically.

Since the lubricating oil is not returned to the compressor through the refrigerant piping, the minimum vapor velocity in ammonia piping is of no consequence and the piping is sized for a low pressure drop without regard for the minimum vapor velocity.

19-14. Non-Oil-returning Halocarbon Evaporators

The design of some halocarbon evaporators is such that the oil reaching the evaporator cannot be entrained by the refrigerant vapor and carried over into the suction line and back to the compressor. The more common of these evaporators are flooded liquid chillers and certain types of air-cooling evaporators that are operated semiflooded by bottom-feeding with a thermostatic expansion valve. In both cases, the problem with oil return results from the lack of sufficient refrigerant velocity and turbulence in the evaporator to permit entraining the oil and carrying it over into the suction line.

With the semiflooded evaporator, oil return from the evaporator is usually accomplished by adjusting the expansion valve for a low superheat and slightly overfeeding the evaporator so that a small amount of the oil-rich liquid refrigerant in the evaporator is continuously carried over into the suction line. This arrangement will ordinarily keep the oil concentration in the evaporator within reasonable limits. As shown in Fig. 19-24 a liquid-suction heat exchanger is installed in the suction line to evaporate the liquid refrigerant from the oil-

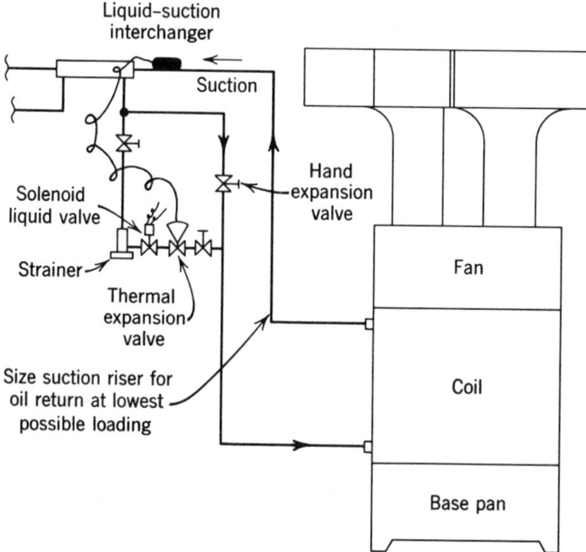

Fig. 19-24 Forced circulation air cooler with direct expansion feed of flooded-type coil. (Courtesy Carrier Corporation.)

refrigerant mixture before the latter reaches the compressor.

Flooded liquid chillers are usually equipped with oil bleeder lines which permit a measured amount of the oil-rich liquid in the chiller to be bled off into the suction line or, in some instances, into an oil receiver. A throttling valve is installed in the bleeder line so that the flow rate through the line can be adjusted. The bleeder line also contains a solenoid valve which is wired to open only when the compressor is operating, so that flow through the bleed line does not occur when the compressor is idle.

Since Refrigerant-12 is completely oil miscible at all temperatures above the pour point of the oil, the oil bleeder connection on Refrigerant-12 chillers may be located at any point below the liquid level in the chiller. Refrigerant-22, on the other hand, is partially oil miscible at evaporator temperatures. Therefore, when refrigerant turbulence in the evaporator is relatively low, there is a tendency for the oil-refrigerant mixture in the evaporator to separate into two layers, with the upper layer containing the greater concentration of oil. For this reason, the oil bleeder connection of Refrigerant-22 chillers should be located just above the midpoint of the liquid level in the chiller.

As shown in Fig. 17-30, when the bleed off is directly to the suction line, a liquid-suction heat exchanger is required in order to evaporate the liquid refrigerant from the bleed mixture before it reaches the compressor. Figures 17-39 and 17-41 illustrate typical piping arrangements when bleed off is to an oil receiver.

19-15. Crankcase Piping for Parallel Compressors

When two or more compressors are operating in parallel off a common suction line it is very unlikely that the oil returning through the suction line will be evenly distributed to the several compressors. Too, it is very unlikely that the amount of oil pumped over by any two compressors will be exactly the same even when the compressors are alike in both design and size. For these reasons, when compressors are piped for parallel operation, it is necessary to interconnect the crankcases of the several compressors both above and below the crankcase oil level, as shown in Fig. 19-11.

This requires that the compressors be so placed on their respective foundations that the oil tappings of the individual compressors are all at exactly the same level. The crankcase oil equalizing line may be installed either level with or, preferably, below the level of the crankcase oil tappings. Under no circumstances should the oil equalizing line be allowed to rise above the level of the crankcase tappings.

Since any small difference in the crankcase pressure of the several compressors will cause a difference in the crankcase oil levels, it is necessary also to equalize the crankcase pressures. This is done by interconnecting the crankcases above the crankcase oil level with a crankcase pressure equalizing line. This line may be installed level with, or above, the crankcase tappings of the compressors. The crankcase pressure equalizing line must not be allowed to drop below the level of the crankcase tappings and it must not contain any liquid traps of any kind.

Both the oil equalizing line and the crankcase pressure equalizing line should be the same size as the crankcase tappings. Manual shut-off valves should be installed in both lines between the compressors so that individual compressors can be valved off for maintenance or repairs without the necessity of shutting down the entire system.

19-16. Liquid Indicators (Sight Glasses)

A liquid indicator or sight glass installed in the liquid line of a refrigerating system provides a means of determining visually whether or not the system has a sufficient charge of refrigerant. If the system is short of refrigerant, the vapor bubbles appearing in the liquid stream will be easily visible in the sight glass. The sight glass should be installed as close to the liquid receiver as possible, but far enough downstream from any valves so that the resulting disturbance does not appear in the sight glass. When liquid lines are long, an additional sight glass is frequently installed in front of the refrigerant control (or liquid line solenoid, when one is used) to determine if a solid stream of liquid is reaching the refrigerant control. Bubbles appearing in the sight glass at this point indicate that the liquid is flashing in the liquid line as a result of excessive pressure drop, in which case the bubbles can be eliminated only by reducing the liquid line pressure drop or by further subcooling of the liquid refrigerant. Typical sight glasses are shown in Fig. 19-25.

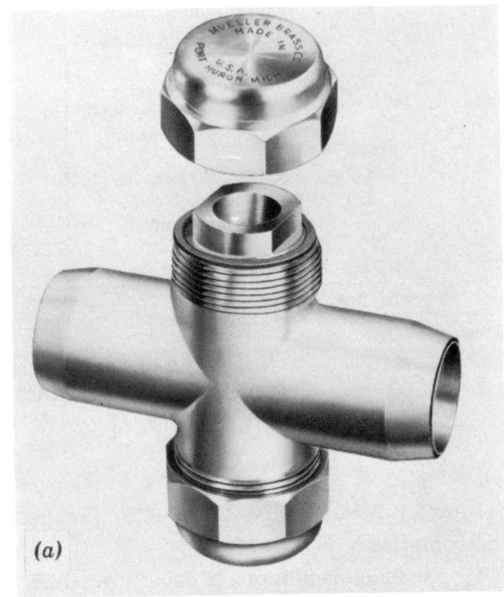

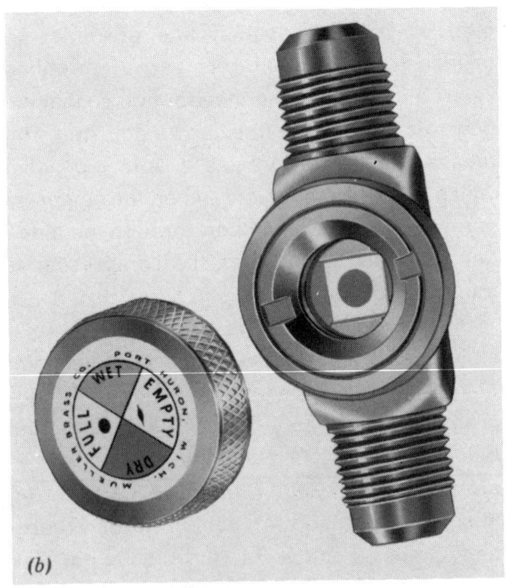

Fig. 19-25 Typical liquid indicators or sight glasses. Notice moisture indicator incorporated in single port sight glass. The color of the moisture indicator denotes the relative moisture content of the system. (*a*) Double port sight glass. (*b*) Single port sight glass. (Courtesy Mueller Brass Company.)

19-17. Refrigerant Dehydrators

Refrigerant driers (Fig. 19-26) are recommended for all refrigerating systems employing a halocarbon refrigerant. In small systems the drier is usually installed directly in the liquid line. In larger systems the by-pass arrangement shown in Fig. 19-27 is employed. With the latter method of installation, the drier cartridge can be removed and reinstalled without interrupting the operation of the system. Too, the drier can be used intermittently as needed. When the drier is not being used valves *A* and *B* are open and valve *C* is closed. When the drier is in service valves *B* and *C* are open and valve *A* is closed. Under no circumstances should valves *B* and *C* ever be closed at the same time, except when the drier cartridge is being changed. With valves *B* and *C* both closed, cold liquid could be trapped in the drier, which, upon warming, could create tremendous hydraulic pressures and burst the drier casing.

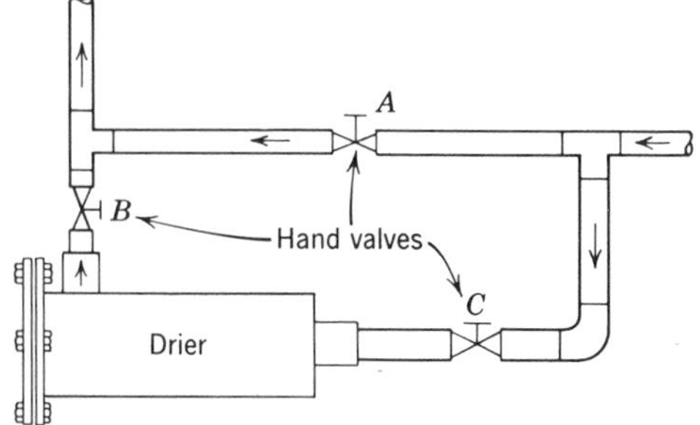

Fig. 19-27 Side outlet drier installed in bypass line.

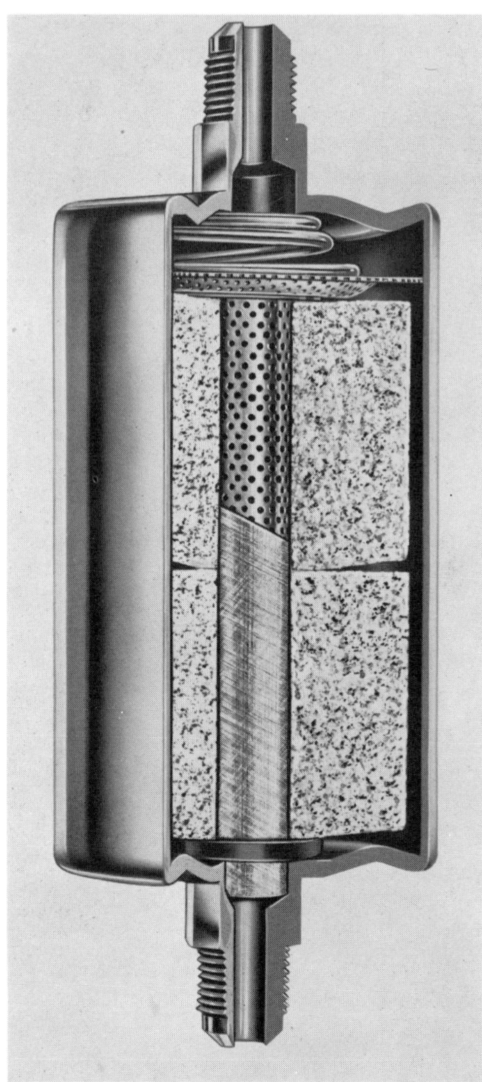

Fig. 19-26 Straight-through type, nonrefillable drier. (Courtesy Mueller Brass Company.)

19-18. Strainers

Strainers should be installed immediately in front of all automatic valves in all refrigerant lines. When two or more automatic valves are installed close together, a single strainer, placed immediately upstream of the valves, may be used. In all cases, the strainer should be amply sized so that the accumulation of foreign material in the strainer will not cause an excessive refrigerant pressure drop.

Most refrigerant compressors come equipped with a strainer in the suction inlet chamber. When installing the refrigerant piping, care should be taken to arrange the suction piping at the compressor so as to permit servicing of this strainer.

19-19. Pressure Relief Valves

Pressure relief valves are safety valves designed to relieve the pressure in the system to the atmosphere, or to the outdoors through a vent line, in the event that the pressure in the system rises to an unsafe level for any reason. Most refrigerating systems have at least one pressure relief valve (or fusible plug) mounted

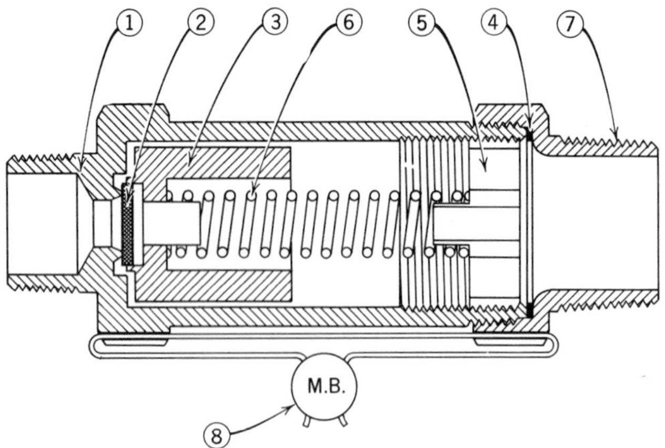

1. Valve body
2. Seat disc
3. Disc holder
4. Gasket
5. Spring retainer
6. Spring
7. Outlet connection
8. Lead seal and locking wire (prevents alteration of factory setting)

Fig. 19-28 Typical pressure relief valve. (Courtesy Mueller Brass Company.)

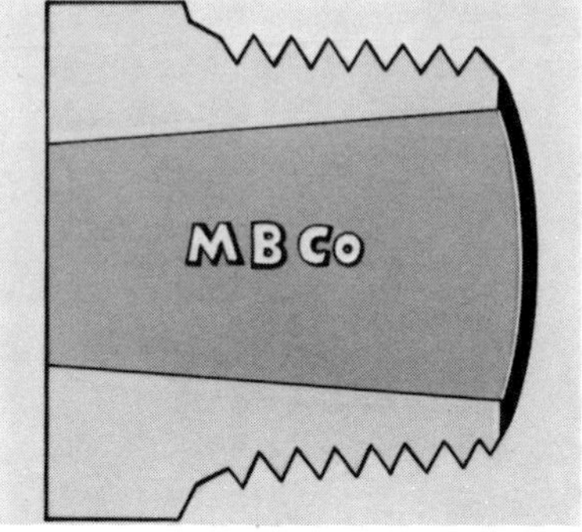

Fig. 19-29 Fusible plugs. (Courtesy Mueller Brass Company.)

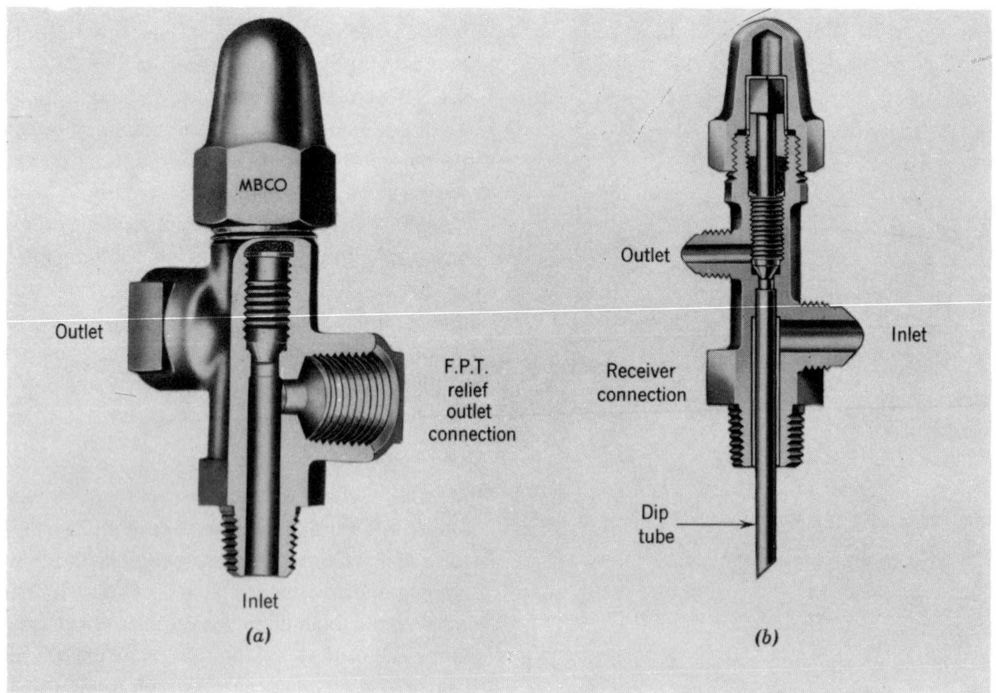

Fig. 19-30 Receiver tank valves. (*a*) Angle type with pressure relief outlet (nonbackseating). (*b*) Angle type with dip tube (nonbackseating). (Courtesy Mueller Brass Company.)

on the receiver tank or water-cooled condenser. In many instances, additional relief valves are required at other points in the system. The exact number, location, and type of relief devices required are set forth in the American Standard Safety Code for Mechanical Refrigeration, and depends for the most part on the type and size of the system. Since local codes vary somewhat in this respect, they should always be considered when designing an installation. A typical pressure relief valve is illustrated in Fig. 19-28.

A fusible plug is sometimes substituted for the pressure relief valve. A fusible plug is simply a pipe plug which has been drilled and filled with a metal alloy designed to melt at some predetermined fixed temperature (Fig. 19-29). The design melting temperature of the fusible plug depends on the pressure-temperature relationship of the refrigerant employed in the system.

19-20. Receiver Tank Valves

Receiver tank valves are usually of the packed type, equipped with a cap seal and designed for direct installation on the receiver tank. (Fig. 19-30.) When designed for installation on top of the receiver tank, the valves must be provided with dip tubes so that the liquid refrigerant can be drawn from the bottom of the receiver. Some valves also have tappings to accommodate a relief valve or fusible plug.

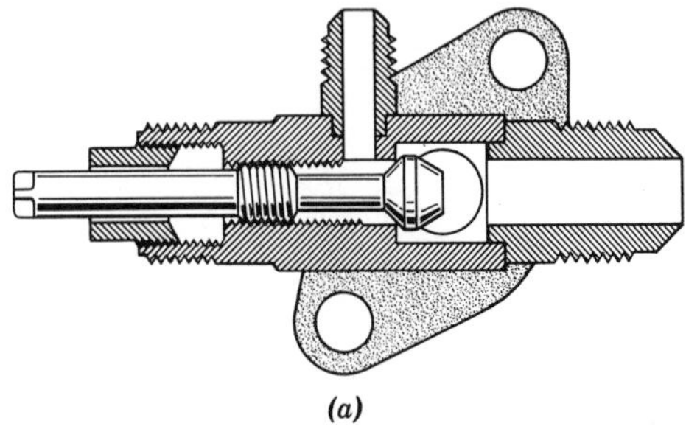

(a)

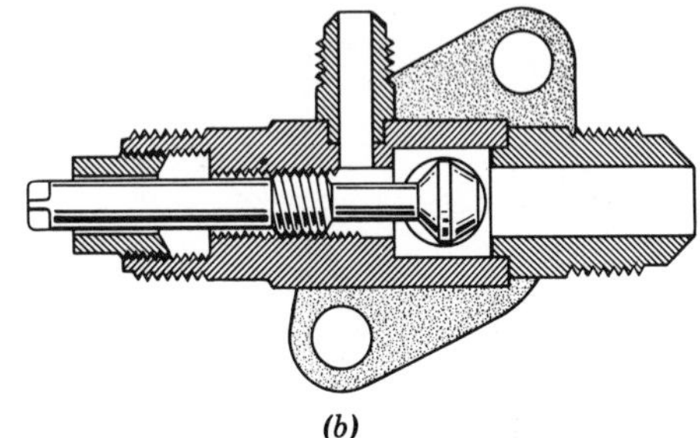

(b)

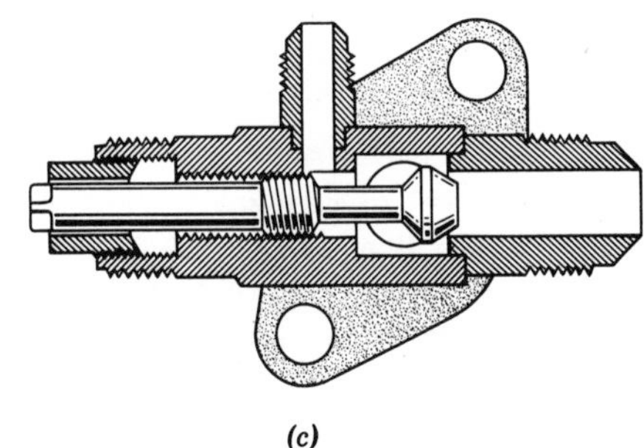

(c)

Fig. 19-31 Compressor service valve. (*a*) Backseated. (*b*) Intermediate position. (*c*) Frontseated.

19-21. Compressor Service Valves

Compressor service valves are usually designed to bolt directly to the compressor housing. As shown in Fig. 19-31, they have both "front" and "back" seats. The "front seat" controls the flow between the refrigerant lines and the compressor, and the "back seat" controls the gage port of the valve. When the valve stem is "back-seated," the gage port is closed and the refrigerant line is open to the compressor. When the valve is "front-seated," the gage port is open to the compressor and the refrigerant line is closed to the compressor and gage port. With the valve stem in an intermediate position between the seats, both the refrigerant line and the gage port are open to the compressor and, of course, to each other.

19-22. Manual Valves

Manual valves used for refrigeration duty may be of the globe or angle types. Since the piping code prohibits the use of gate valves in refrigerant lines, except in large installations where an operating attendant is on duty, they are used primarily in water and brine lines. Gate valves have a very low pressure drop, but do not permit throttling and therefore can be employed only where full-flow or no-flow conditions are desired. Both globe and angle

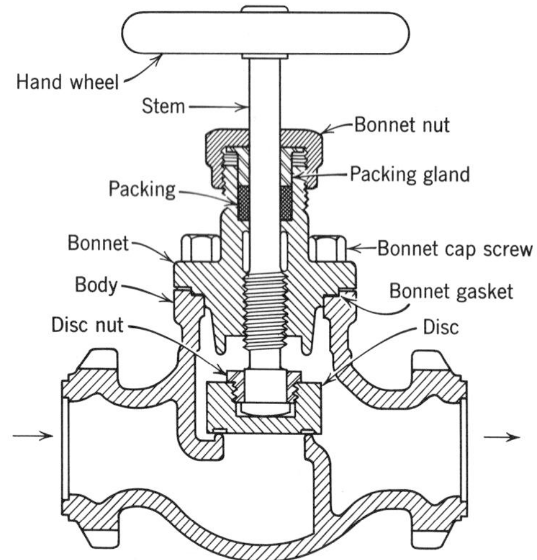

Fig. 19-32 Packed-type manual valve. (Courtesy Vilter Manufacturing Company.)

valves are suitable for throttling. Since the angle valve offers the least resistance to flow, its use is recommended whenever practical.

Either the "packed" or "packless" type valve is suitable for refrigeration duty, provided the valve has been designed for that purpose. Packed valves should be of the back-seating type in order to permit packing under pressure and to reduce the possibility of leakage through the packing in the full-open position (Fig. 19-32.) Many packed valves are equipped with cap seals which completely cover and seal the valve stem, thereby eliminating the possibility of leakage when the valve is not in use.

Customary Problems

19-1 Determine the minimum size of a suction riser that will ensure oil return at minimum loading for a 50-ton R-502 system that is equipped with a compressor having capacity steps of 25%, 50%, 75%, and 100%, if the design suction temperature at minimum loading is −20°F and the temperature of the liquid refrigerant approaching the refrigerant control is 80°F.

19-2 A 30-ton Refrigerant-22 system has an evaporator temperature of 40°F and a condensing temperature of 110°F. The suction line is 40 ft long and contains three standard 90° elbows. Determine: (a) the OD size of type L copper tube that will result in a loss in saturated suction temperature of less than 2°, (b) the total equivalent length of the suction pipe, (c) the actual loss in saturation temperature in degrees Fahrenheit, and (d) the actual pressure loss is pounds per square inch under full load conditions.

19-3 Rework Problem 19-2 using R-12 as the refrigerant.

19-4 A 30-ton ammonia system has an evaporator temperature of −20°F and a condensing temperature of 100°F. The suction line is 40 ft long and has five standard 90° elbows. Determine: (a) the size of Schedule 40 iron pipe that will result in a loss in saturated suction temperature of less than 1°F at full load conditions, (b) the total equivalent length of the suction pipe, (c) the actual loss in saturated suction temperature, and (d) the actual pressure loss in the suction pipe in pounds per square inch.

19-5 Rework Problem 19-4 using R-502 as the refrigerant and type L copper tube for the suction line.

19-6 The discharge pipe for the system described in Problem 19-2 is 40 ft long and contains four standard 90° elbows and an 18-ft riser. If the compressor is equipped with capacity control steps of 25%, 50%, 75%, and 100%, determine: (a) the OD size of type L copper tube that should be used for the discharge line, giving consideration to oil return under minimum loading, (b) the total equivalent length of the discharge pipe, (c) the loss in saturation temperature in the pipe, and (c) the pressure loss in pounds per square inch.

19-7 The pressure at the bottom of a 30-ft R-12 liquid riser is 115 psi and the temperature of the liquid is 90°F. Determine the pressure and approximate saturation temperature of the liquid at the top of the riser.

19-8 Assume that the refrigerant in the riser is saturated liquid ammonia at 90°F and rework Problem 19-8.

19-9 An R-12 system has a capacity of 25 tons. The equivalent length of the liquid line is 60 ft, and the line has a 20-ft riser and a solenoid valve selected for a 2 psi pressure drop. Assuming that saturated liquid at 100°F enters the liquid line, determine: (a) the OD size of type L copper pipe required, (b) the overall pressure loss in the line, and (c) the amount of subcooling in degrees Fahrenheit required to prevent flashing of the liquid in the line.

20

DEFROST METHODS—LOW TEMPERATURE, MULTIPLE TEMPERATURE, AND ABSORPTION REFRIGERATION SYSTEMS

20-1. Defrosting Intervals

The necessity for periodically defrosting air-cooling evaporators which operate at temperatures low enough to cause frost to collect on the evaporator surface has already been established. How often the evaporator should be defrosted depends on the type of evaporator, the nature of the installation, and the method of defrosting. Large, bare-tube evaporators, such as those employed in breweries, cold storage plants, etc., are usually defrosted only once or twice a month. On the other hand, finned blower coils are frequently defrosted as often as once or twice each hour. In some low temperature installations defrosting of the evaporator is continuous by brine spray or by some antifreeze solution.

In general, the length of the defrost period is determined by the degree of frost accumulation on the evaporator and by the rate at which heat can be applied to melt off the frost. For the most part, the degree of frost accumulation will depend on the type of installation, the season of the year, and the frequency of defrosting. As a general rule, the more frequently the evaporator is defrosted the smaller is the frost accumulation and the shorter is the defrost period required.

20-2. Methods of Defrosting

Defrosting of the evaporator is accomplished in a number of different ways, all of which can be classified as either "natural defrosting" or "supplementary-heat defrosting" according to the source of the heat used to melt off the frost. Natural defrosting, sometimes called "shut-down" or "off-cycle" defrosting, utilizes the heat of the air in the refrigerated space to melt the frost from the evaporator, whereas supplementary-heat defrosting is accomplished with heat supplied from sources other than the space air. Some common sources of supplementary heat are water, brine, electric heating elements, and hot gas from the discharge of the compressor.

All methods of natural defrosting require that the system (or evaporator) be shut down for a period of time long enough to permit the evaporator temperature to rise to a level well above the melting point of the frost. The exact temperature rise required and the exact length of time the evaporator must remain shut down in order to complete the defrosting vary with the individual installation and with the frequency of defrosting. However, in every case, since the heat to melt the frost comes from the space air, the temperature in the space must be allowed to rise to whatever level is necessary to melt off the evaporator frost, which is usually about 37° to 40°F. For this reason, natural defrosting is not ordinarily practical in any installation when the design space temperature is below 34°F.

The simplest method of defrosting is to shut the system down manually until the evaporator warms up enough to melt off the frost, after which the system is started up again manually. When several evaporators connected to the same condensing unit are located in different spaces or fixtures, the evaporators can be taken out of service and defrosted one at a time by manually closing a shut-off valve located in the liquid line of the evaporator being defrosted. When defrosting is completed, the evaporator is put back into service by opening the shut-off valve.

If automatic defrosting is desired, a clock timer can be used to shut the system down for a fixed period of time at regular intervals. Both the number and the length of the defrost periods can be adjusted to suit the individual installation. As a general rule, natural convection evaporators are defrosted only once a day, in which case the defrost cycle is usually started around midnight and lasts for several hours. On the other hand, unit coolers should be defrosted at least once every 3 to 6 hr. Since it is usually undesirable to keep the system out of service for any longer than is necessary, the length of the defrost period should be carefully adjusted so that the system is placed back in service as soon as possible after defrosting.

In one variation of the time defrost, the defrost cycle is initiated by the defrost timer and terminated by a temperature or pressure control that is actuated by the evaporator temperature or pressure. With this method, the defrost period is automatically adjusted to the required length, since the evaporator temperature (or pressure) will rise to the cut-in setting of the control as soon as defrosting is completed.

The most common method of natural defrosting is the "off-cycle" defrost. As described in an earlier chapter, off-cycle defrosting is accomplished by adjusting the cycling control so that the evaporator temperature rises to 37°F or 38°F during every off cycle. If the system has been properly designed, the evaporator will be maintained relatively free of frost at all times, since it will be completely defrosted during each off cycle.

20-3. Water Defrosting

For evaporator temperatures down to approximately minus 40°F, defrosting can be accomplished by spraying water over the surface of the evaporator coils. For evaporator temperatures below minus 40°F, brine or some antifreeze solution should be substituted for the water. A typical water defrost system is illustrated in Fig. 20-1.

Although water defrosting can be made automatic, it is often designed for manual operation. Ordinarily, the following procedure is used to carry out the defrosting:

1. A stop valve in the liquid line is closed and the refrigerant is evacuated from the evaporator, after which the compressor is stopped and the evaporator fans are turned off so that the water spray is not blown out into the refrigerated space. If the evaporator is equipped with louvers, these are closed to isolate further the evaporator and prevent fogging of the refrigerated space.
2. The water sprays are turned on until the evaporator is defrosted, which requires approximately 4 to 5 minutes. After the sprays are turned off, several minutes are allowed for draining of the water from the evaporator coils and drain pan before the evaporator fans are started and the system put back in operation.

To eliminate the possibility of water freezing in the drain line, the evaporator should be located close to an outside wall and the drain line should be amply sized and so arranged that the water is drained from the space as rapidly as possible. A trap is installed in the drain line outside the refrigerated space to prevent warm air being drawn into the space through the drain line during normal operation. In some instances, a float valve is employed in the drain pan to shut off the water spray and prevent overflowing into the space in the event that the drain line becomes plugged with ice.

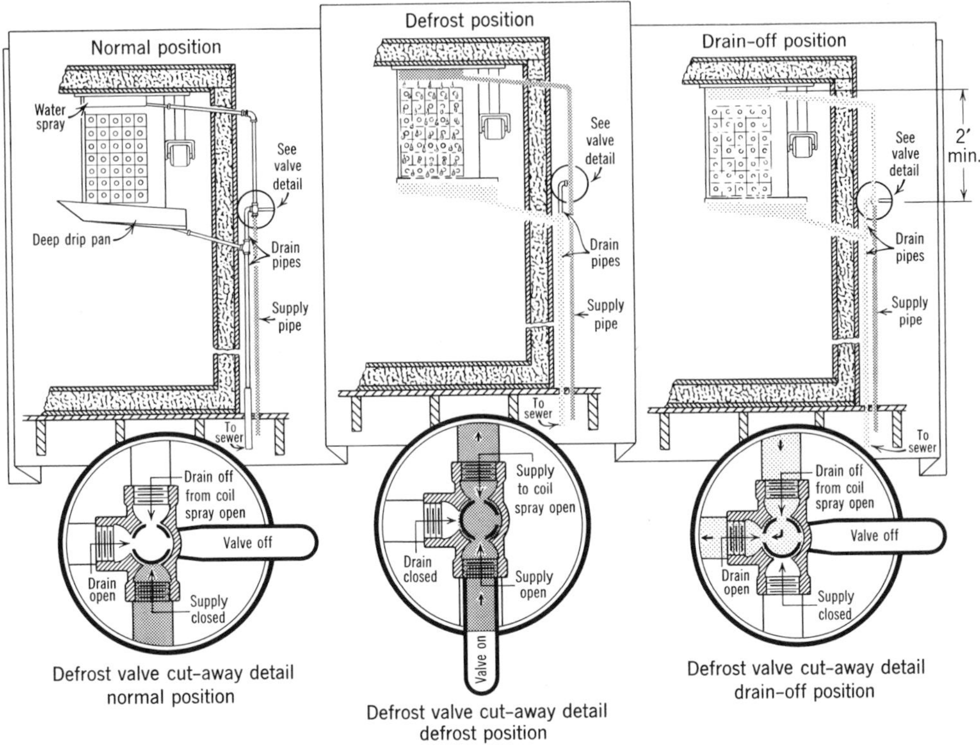

Fig. 20-1 Typical water defrost system. (Courtesy Dunham-Bush, Inc.)

When brine or an antifreeze solution replaces the water spray, the defrosting solution is returned to a reservoir and recirculated, rather than wasted. Unless the reservoir is large enough so that the addition of heat is not required, some means of reheating the solution in the reservoir may be necessary. Since the water from the melting frost will weaken the solution, the defrost system is equipped with a "concentrator" to boil off the excess water and return the solution to its initial concentration.

One manufacturer circulates a heated glycol solution through the inner tube of a double-tube evaporator coil. The principal advantage gained is that the glycol is not diluted by the melting frost.

20-4. Electric Defrosting

Electric resistance heaters are frequently employed for the defrosting of finned blower coils. An evaporator equipped with defrost heaters is shown in Fig. 20-2. Ordinarily, the drain pan and drain line are also heated electrically to prevent refreezing of the melted frost in these parts.

The electric defrost cycle can be started and stopped manually or a defrost timer may be used to make defrosting completely automatic. In either case, the defrosting procedure is the same. The defrost cycle is initiated by closing a solenoid valve in the liquid line causing the evaporator to be evacuated, after which the compressor cycles off on low pressure control. At the same time, the heating elements in the evaporator are energized and the evaporator fans turned off so that the heat is not blown out into the refrigerated space. After the evaporator is defrosted, the heaters are deenergized and the system put back in operation by opening the liquid line solenoid and starting the evaporator fans.

20-5. Hot Gas Defrosting

Hot gas defrosting has many variations, all of which in some way utilize the hot gas discharged from the compressor as a source of

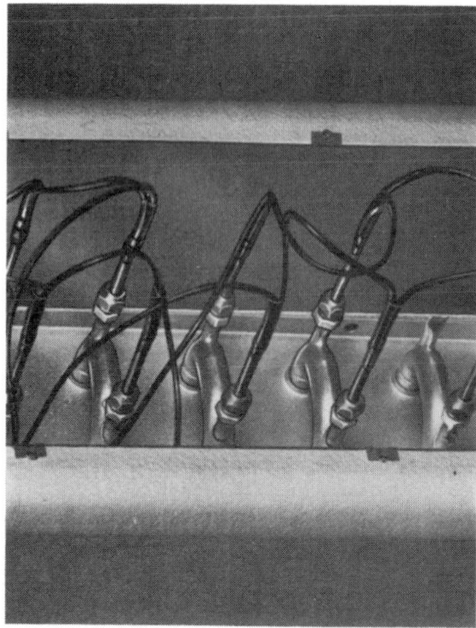

Fig. 20-2 Evaporator equipped for electric defrosting. Heater elements are installed through the center of the tubes. Inset shows details of mechanical sealing of the heater elements. (Courtesy Dunham-Bush, Inc.)

heat to defrost the evaporator. One of the simplest methods of hot gas defrosting is illustrated in Fig. 20-3. A bypass equipped with a solenoid valve is installed between the compressor discharge and the evaporator. When the solenoid valve is opened, the hot gas from the compressor discharge bypasses the condenser and enters the evaporator at a point just beyond the refrigerant control. Defrosting is accomplished as the hot gas gives up its heat to the cold evaporator and condenses into the liquid state. Some of the condensed refrigerant stays in the evaporator while the remainder returns to the compressor where it is evaporated by the compressor heat and recirculated to the evaporator.

This method of hot gas defrosting has several disadvantages. Since no liquid is vaporized in the evaporator during the defrost cycle, the amount of hot gas available from the compressor will be limited. As defrosting progresses, more liquid remains in the evaporator and less refrigerant is returned to the compressor for recirculation, with the result that the system tends to run out of heat before the evaporator is completely defrosted.

Another, and more serious, disadvantage of this method is the possibility that a large slug of liquid refrigerant will return to the compressor and cause damage to that unit. This is most likely to occur either at the beginning of the defrost cycle or immediately after defrosting is completed.

Fortunately, both these weaknesses can be overcome by providing some means of re-evaporating the liquid which condenses in the evaporator before it is returned to the com-

Hot gas solenoid
By-pass line

Fig. 20-3 Simple hot gas defrost system.

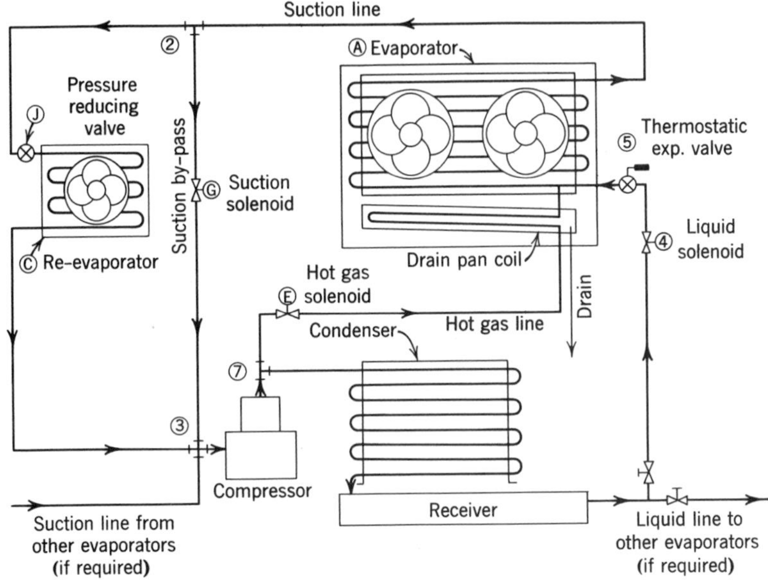

Fig. 20-4 Hot gas defrost system employing reevaporator coil. (Courtesy Kramer-Trenton Company.)

pressor. The particular means used to reevaporate the liquid is the principal factor distinguishing one method of hot gas defrosting from another.

20-6. Reevaporator Coils

One common method of hot gas defrosting employs a reevaporator coil in the suction line to reevaporate the liquid, as shown in Fig. 20-4. During the normal running cycle the solenoid valve in the suction line is open and the suction vapor from the evaporator bypasses the reevaporator coil in order to avoid an excessive suction line pressure loss. At regular intervals (usually 3 to 6 hr) the defrost timer starts the defrost cycle by opening the solenoid in the hot gas line and closing the solenoid in the suction bypass line. At the same time, the evaporator fans are stopped and the reevaporator fan is started. The liquid condensed in the evaporator is reevaporated in the reevaporator coil and returned as a vapor to the compressor, where it is compressed and recirculated to the evaporator. When defrosting is completed, the defrost cycle may be terminated by the defrost timer or by an evaporator-temperature actuated temperature control. In either case, the system is placed back in operation by closing the hot gas solenoid, opening the suction solenoid, stopping the reevaporator fan and starting the evaporator fans.

20-7. Defrosting Multiple Evaporator Systems

When two or more evaporators are connected to a common condensing unit, the evaporators may be defrosted individually, in which case the operating evaporator can serve as a reevaporator for the refrigerant condensed in the evaporator being defrosted. A flow diagram of this arrangement is illustrated in Fig. 20-5.

20-8. Reverse Cycle Defrosting

By employing the reverse cycle (heat pump) principle, the condenser can be utilized as a reevaporator coil to reevaporate the refrigerant that condenses in the evaporator during the defrost cycle. An automatic expansion valve is used to meter the liquid refrigerant into the condenser for reevaporation. Flow diagrams for defrosting and normal operation are shown in Figs. 20-6*a* and *b*, respectively. Modern practice replaces valves *A*, *B*, *C*, and

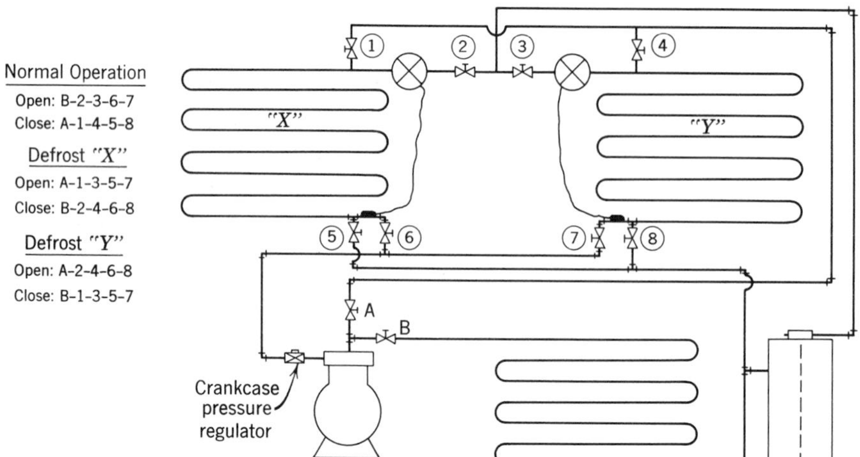

Fig. 20-5 Hot gas defrost—multiple evaporator system.

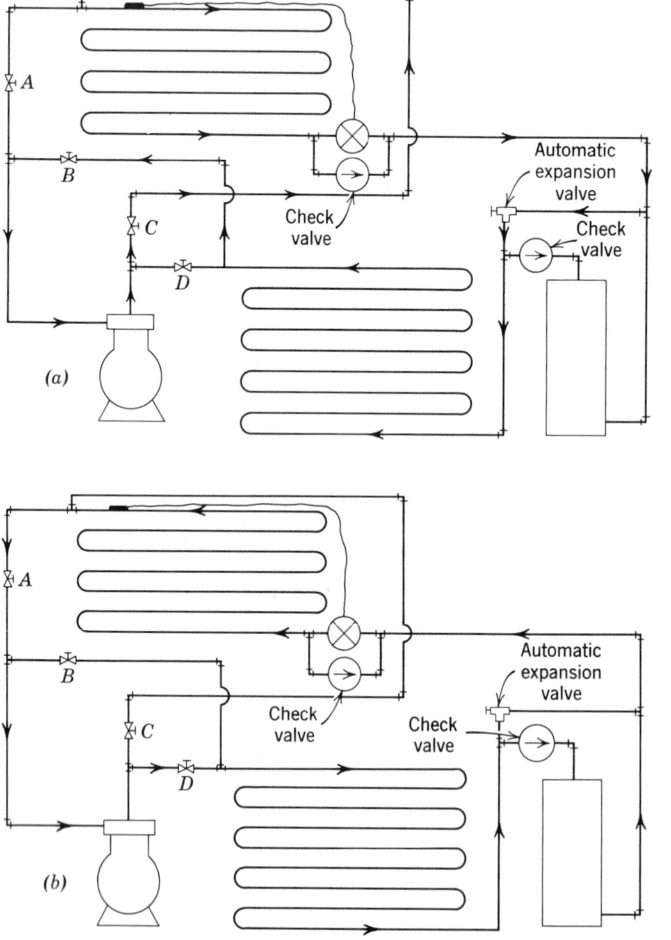

Fig. 20-6 (*a*) Reverse cycle hot gas defrost system (defrost cycle). (*b*) Reverse cycle hot gas defrost system (normal operation).

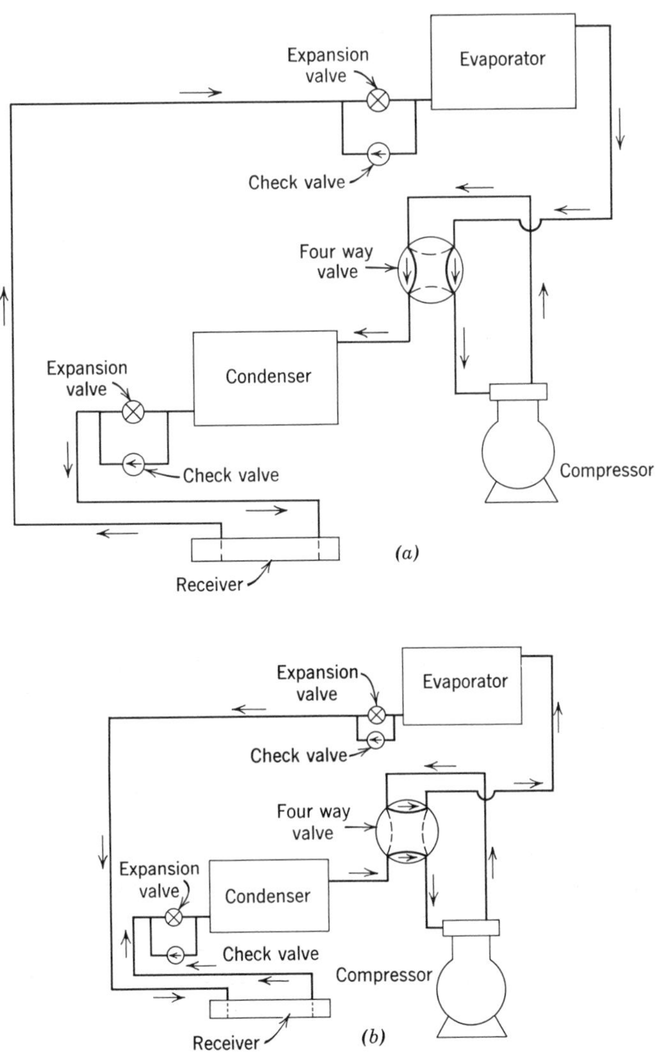

Fig. 20-7 (*a*) Reverse cycle hot gas defrost—normal operation. (*b*) Reverse cycle hot gas defrost—defrost cycle.

D of Fig. 20-6 with a single four-way valve as illustrated in Fig. 20-7.

20-9. Heat Bank Defrosting

The Thermobank* method of hot gas defrosting employs a water bank to store a portion of the heat ordinarily discarded at the condenser when the evaporator is being refrigerated. During the defrost cycle, the heat stored in the water bank is used to reevaporate the refrigerant condensed in the defrosting evaporator.

During normal operation (Fig. 20-8*a*), the discharge gas from the compressor passes through the heating coil in the water bank first and then goes to the condenser, so that a portion of the heat ordinarily discarded at the condenser is stored in the bank water. Notice that the suction vapor bypasses the holdback valve and bank during the refrigerating cycle in order to avoid unnecessary suction line pressure loss and superheating of the suction vapor by the bank water. Also, to control the maximum

* A proprietary design of the Kramer-Trenton Company.

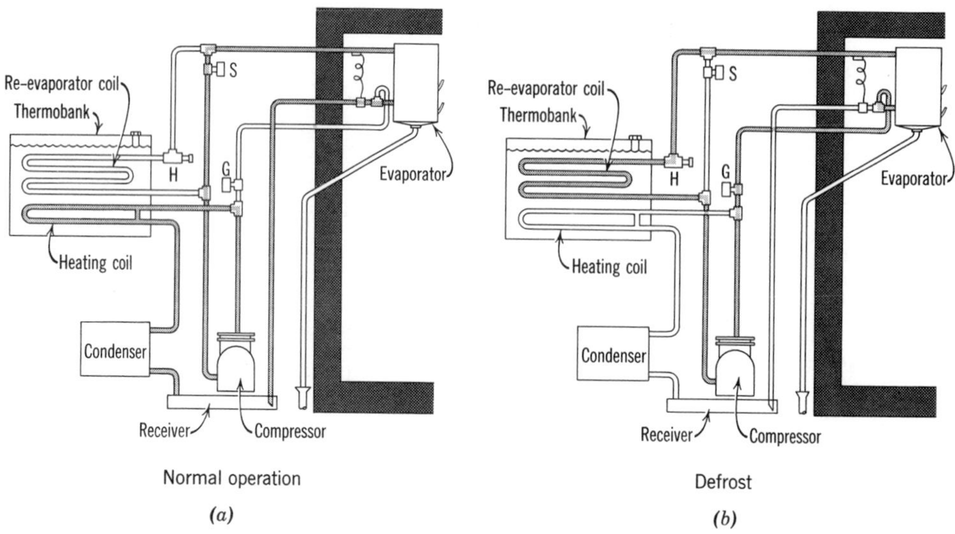

Fig. 20-8 Thermobank hot gas defrosting. (Courtesy Kramer-Trenton Company.)

water bank temperature, a bypass is built into the water bank heating coil. The bypass is so sized that a greater portion of the discharge gas bypasses the heater coil and flows directly to the condenser as the temperature of the bank water increases.

When the frost reaches a predetermined thickness, the defrost cycle (Fig. 20-8*b*) is initiated by an electric timer which opens the hot gas solenoid valve, closes the suction solenoid valve, and stops the evaporator fans. Hot gas is discharged into the evaporator where it condenses and defrosts the coil. The condensed refrigerant flows to the holdback valve which acts as a constant pressure expansion valve and feeds liquid to the reevaporator coil immersed in the bank water. In this process, the bank water actually freezes on the outside of the reevaporator coil. The heat stored in the bank is transferred to the refrigerant which evaporates completely in the reevaporator coil. Thus both sensible and latent heat are abstracted from the bank water, making available vast heat quantities for fast defrost and the refrigerant returns to the suction inlet of the compressor completely evaporated.

Defrost is completed in approximately 6 to 8 min. This is followed by a postdefrost period lasting a few minutes after the closing of the hot gas solenoid valve. During postdefrost any liquid refrigerant in the coil and suction line is reevaporated. The timer then returns the system to normal operation. When normal operation is resumed, the bank water is promptly restored to its original temperature by the hot gas passing through the heating coil.

20-10. Vapot Defrosting

A schematic diagram of the Vapomatic defrosting system is shown in Fig. 20-9, along with an enlarged view of the Vapot,* which is the heart of this hot gas defrost system. The Vapot, which is actually a specially designed suction line accumulator, traps the liquid refrigerant condensed in the evaporator and, by means of a carefully sized bleed tube, continuously feeds a measured amount of the liquid back to the compressor with the suction vapor. The small amount of liquid feeding back to the compressor is vaporized by the heat of compression and returned to the evaporator. In this way, the Vapot provides a continuous source of latent heat for defrosting the evaporator and at the same time eliminates the possibility of large slugs of liquid returning to the compressor. The heat exchanger in the Vapot has no significance in the defrost cycle.

* Proprietary designs of Refrigeration Engineering, Inc.

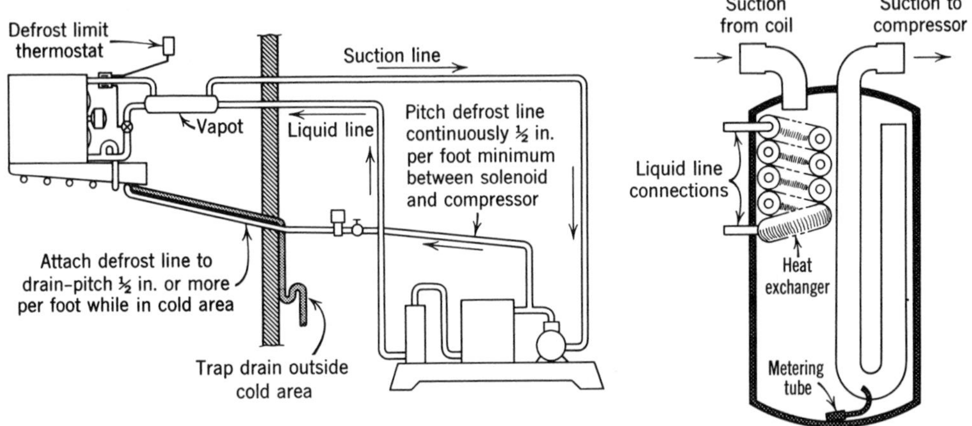

Fig. 20-9 (Left) Typical application of Vapot. (Right) Enlarged view of Vapot. (Courtesy Recold Corporation.)

The defrost cycle is initiated by a defrost timer which opens the hot gas solenoid valve and stops the evaporator fans. An evaporator temperature control terminates the defrost and restores the system to normal operation.

20-11. Multistage (Booster) Compression

It has already been established that the capacity and efficiency of any refrigerating system diminish rapidly as the difference between the suction and condensing temperatures is increased by a reduction in the evaporator temperature. The losses experienced are due partially to the rarification of the suction vapors at the lower evaporator temperatures and partially to the increase in the compression ratio. Since any increase in the ratio of compression is accompanied by a rise in the discharge temperature, discharge temperatures also tend to become excessive as the evaporator temperature is reduced.

Whereas conventional single-stage systems will usually give satisfactory results with evaporator temperatures down to −40°F, provided that condensing temperatures are reasonably low, for evaporator temperatures below minus 40°F, some form of multistage compression must be employed in order to avoid excessive discharge temperatures and to maintain reasonable operating efficiencies. In larger installations, multistage operation should be considered for any evaporator temperature below 0°F.

All methods of accomplishing multistage compression can be grouped into two basic types: (1) direct staging and (2) cascade staging. The direct staging method employs two or more compressors connected in series to compress a single refrigerant in successive stages. A flow diagram of a simple, three-stage, direct staged multicompression system is shown in Fig. 20-10. Notice that the pressure of the refrigerant vapor is raised from the evaporator pressure to the condenser pressure in three increments, the discharge vapor from the lower stage compressors being piped to the suction of the next higher stage compressor.

Cascade staging involves the use of two or more separate refrigerant circuits which employ refrigerants having progressively lower boiling points (Fig. 20-11). The compressed refrigerant vapor from the lower stage is condensed in a heat exchanger, usually called a cascade condenser, which is also the evaporator of the next higher stage refrigerant.

Both methods of multistaging have relative advantages and disadvantages. The particular method which will produce the best results in a given installation depends for the most part on the size of the installation and on the degree of low temperature which must be attained. In some instances, a combination of the cascade and direct staging methods can be used to an advantage. In these cases, the compound compression (direct staging) is usually applied to the lower stage of the cascade.

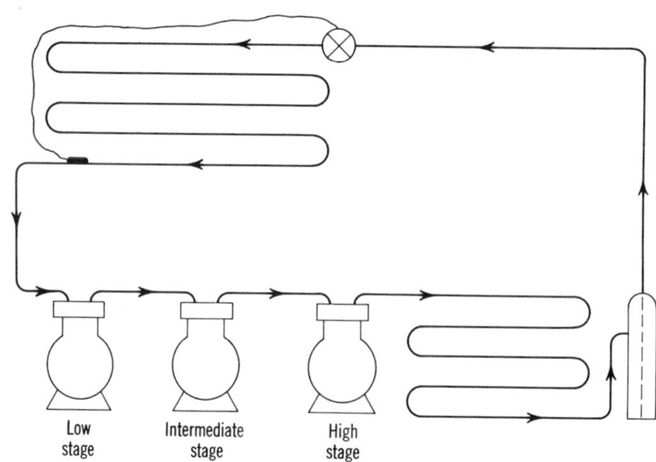

Fig. 20-10 Three-stage, direct staged compression system.

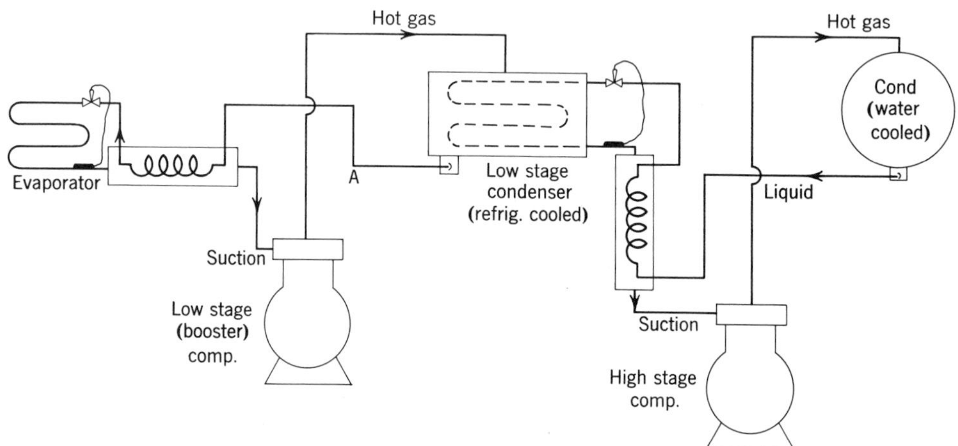

Fig. 20-11 Cascade system (two-stage). (Courtesy Carrier Corporation.)

20-12. Intercoolers

With direct staging, cooling of the refrigerant vapor between the several stages of compression (desuperheating) is necessary in order to avoid overheating of the higher stage compressors. Since the refrigerant gas is superheated during the compression process, if the gas is not cooled before it enters the next stage compressor, excessive discharge temperatures will result with subsequent overheating of the higher stage machines.

Because of the large temperature differential between the condenser and the evaporator, cooling of the liquid refrigerant is also desirable in order to avoid heavy losses in refrigerating effect because of excessive flashing of the liquid in the refrigerant control, and the accompanying increase in the volume of vapor which must be handled by the low stage compressor.

Three common methods of gas desuperheating and liquid cooling for direct staged systems are illustrated in Fig. 20-12. The intercooler shown in Fig. 20-12*a* is an "open" or "flash" type intercooler. The liquid from the condenser is expanded into the intercooler where its temperature is reduced by flashing to the saturation temperature corresponding to

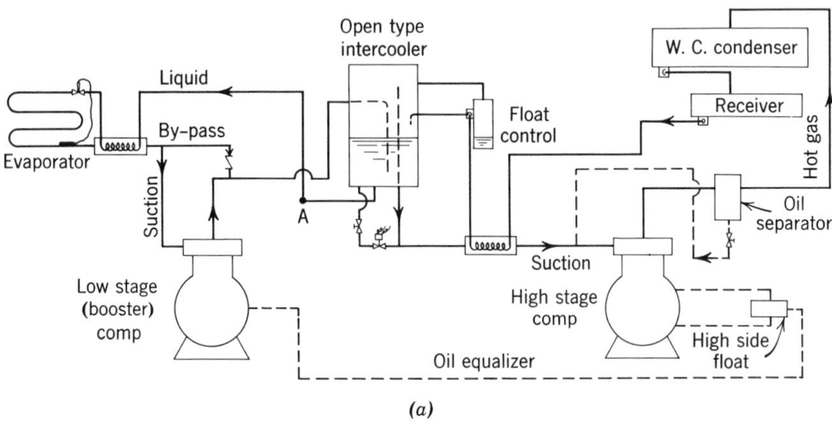

(a)

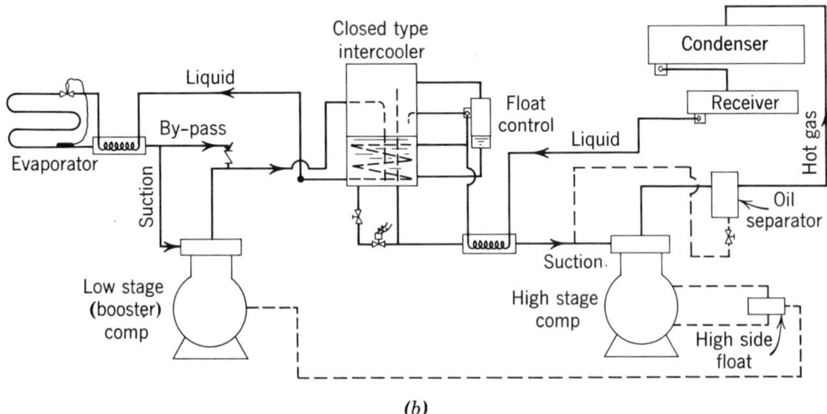

(b)

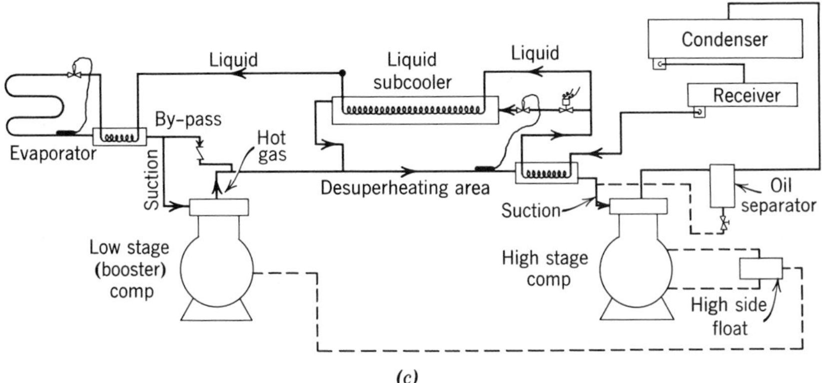

(c)

Fig. 20-12 Various types of gas and liquid intercoolers. (a) Direct staged system—open, flash-type intercooler. (b) Direct staged system—closed, shell-and-coil type intercooler. (c) Direct staged system—dry-expansion intercooler. (Courtesy Carrier Corporation.)

the intercooler pressure. Since suction from the intercooler is taken into the high stage compressor, the temperature of the liquid leaving the intercooler to go to the low temperature evaporator is the saturation temperature corresponding to the intermediate pressure (pressure between stages). Since the refrigeration expended in cooling the liquid to the intermediate temperature is accomplished much more economically at the level of the high stage suction than at that of the low stage suction, cooling of the liquid in the intercooler has the effect of reducing the horsepower per ton as well as the displacement required for the low stage compressor.

The discharge gas from the low stage compressor is desuperheated by causing it to bubble up through the liquid in the intercooler, after which the discharge gas passes to the suction of the high stage compressor along with the flash gas from the intercooler.

The principal advantages of the flash type intercooler are in its simplicity and low cost, and the fact that the temperature of the liquid refrigerant is reduced to the saturation temperature corresponding to the intermediate pressure. The chief disadvantage is that the pressure of the liquid going to the evaporator is reduced to the intermediate pressure in the intercooler. This reduces the pressure drop available at the expansion valve and necessitates oversizing of the valve, which often results in sluggish operation. Too, since the low-temperature, low-pressure liquid leaving the intercooler is saturated, there is a tendency for the liquid to flash in the liquid line between the intercooler and the evaporator. For this reason, the liquid line should be designed for the minimum possible pressure drop.

A shell-and-coil intercooler, sometimes called a "closed type" intercooler, is illustrated in Fig. 20-12*b*. This type of intercooler differs from the flash type in that only a portion of the liquid from the condenser is expanded into the intercooler, whereas the balance, that portion going to the evaporator, passes through the coil submerged in the intercooler liquid. Therefore, with the shell-and-coil type intercooler, the pressure of the liquid is not reduced to the intermediate pressure, that is, the liquid cooling occurs in the form of subcooling rather than as a reduction in the saturation temperature. The advantages gained by this method, of course, are the higher liquid pressures made available at the expansion valve and the elimination of flash gas in the liquid line. With good intercooler design the liquid can be cooled to within 10 to 20 degrees of the saturation temperature corresponding to the intermediate pressure.

Vapor velocities in both types of flooded intercoolers should be limited to a maximum of 200 fpm and ample separation area should be allowed above the liquid level in the intercoolers in order to prevent liquid carryover into the high stage compressor.

Another arrangement, employing a dry-expansion intercooler, is shown in Fig. 20-12*c*. This type of intercooler is not suitable for ammonia systems, but is widely used with Refrigerants-12 and 22. The liquid from the condenser is subcooled as it passes through the coil in the intercooler. Desuperheating is accomplished by overfeeding of the intercooler so that a small amount of liquid is carried over into the desuperheating area where it is vaporized by the hot gas from the discharge of the low stage compressor. The gas is cooled by vaporizing the liquid and by mixing with the cold vapor from the intercooler.

Since ammonia has a very high latent heat value, liquid cooling is not as important in ammonia systems as in systems employing fluorocarbon refrigerants. For this reason, liquid cooling is sometimes neglected in ammonia systems, in which case the discharge vapor from the low stage compressor is usually desuperheated by injecting a small amount of liquid ammonia directly into the line connecting the low and high stage compressors (Fig. 20-13). The vaporization of the liquid ammonia in this line provides the necessary gas cooling.

In some ammonia systems, the discharge gas is cooled in a water-cooled intercooler that is similar in design to the shell-and-tube or shell-and-coil water-cooled condensers. The effectiveness of this type of intercooler depends on the temperature of the available water and increases as the available water temperature decreases. As a general rule, water-cooled intercoolers will not produce

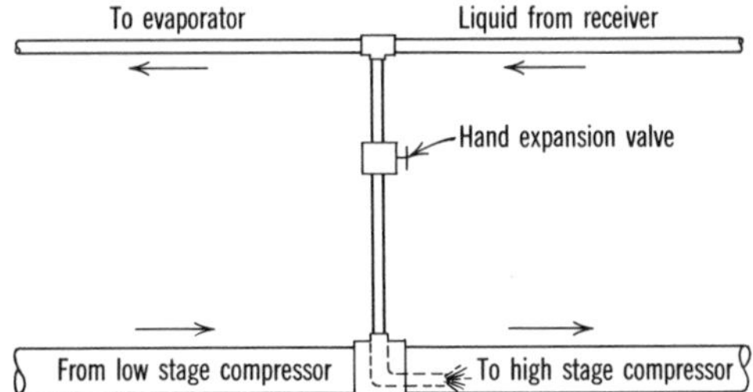

Fig. 20-13 Liquid injection gas intercooling.

enough gas cooling to lower the power requirements, but will usually provide sufficient cooling to keep the discharge temperature within the maximum limit and thereby prevent overheating of the high stage compressor.

In direct-staged multiple compression systems the intermediate pressure is usually selected so that the compression ratios for the several stages are approximately equal. An exception to this is when additional fixtures are to be maintained at the intermediate temperature, in which case the intermediate pressure will usually be determined by the temperature requirements of these additional fixtures. The intermediate pressure, which will provide equal compression ratios for a two-stage, direct-staged system, is found by taking the square root of the product of the absolute evaporating and condensing pressures, that is,

$$p_i = \sqrt{(p_e)(p_c)} \tag{20-1}$$

Example 20-1 An R-12 system is to operate with saturated evaporating and condensing temperatures of −50°F and 100°F. Determine the intermediate pressure that will result in equal compression ratios for the low and high stages of compression.

Solution From Table 16-3, the saturation pressures corresponding to −50°F and 100°F are 7.125 psia and 131.6 psia. Applying Equation 20-1,

$$p_i = \sqrt{(7.125)(131.6)} = 30.62 \text{ psia}$$

20-13. Direct Staging versus Cascade Staging Direct staging requires the use of refrigerants that have boiling points low enough to provide the low temperatures desired in the evaporator and which, at the same time, are condensable under reasonable pressures with air or water at normal temperatures. This requirement tends to limit the degree of low temperature that can be attained by the direct staging method. The practical low limit is approximately −125°F with either Refrigerants-12 or 502 and −90°F with ammonia. Below these temperatures cascade staging is ordinarily required, with some high-pressure, low boiling point refrigerant, such as methane, ethane, ethylene, R-13, R-13BI, or R-503, being used in the lower stage. Because of their extremely high pressures at normal condensing temperatures and/or their relatively low critical temperatures, these high pressure refrigerants must be condensed at rather low tempratures and therefore are cascaded with R-12, R-22, 502, or propane.

Also, to avoid the development of excessively high system pressures during periods when the low-temperature stage is inoperative and the temperature of the refrigerant in the system rises to the ambient air temperature, a pressure relief valve relieves the refrigerant into an expansion tank when the system pressure rises to some predetermined maximum saturation pressure. Usually, no receiver tank is employed, so that the system refrigerant charge is limited to the minimum. The expansion tank is sized so that the total system volume, including the expansion tank, is sufficient to hold the entire refrigerant charge in the vapor state at the predetermined maximum saturation temperature. As long as any liquid is present in the system, the pressure exerted by the refrigerant will be the saturation pressure corresponding to the temperature of the refrigerant. However, with all the refrigerant in the vapor state, any further increase in system pressure as the temperature increases will be approximately in accordance with Charles' law. Recall that this same pressure fade-out principle is used with gas-charged, pressure-limiting expansion valves.

The size of the expansion tank can be determined by the following equation:

$$V_f = \frac{(V_s - m_s v_2)}{(v_2/v_1) - 1} \tag{20-2}$$

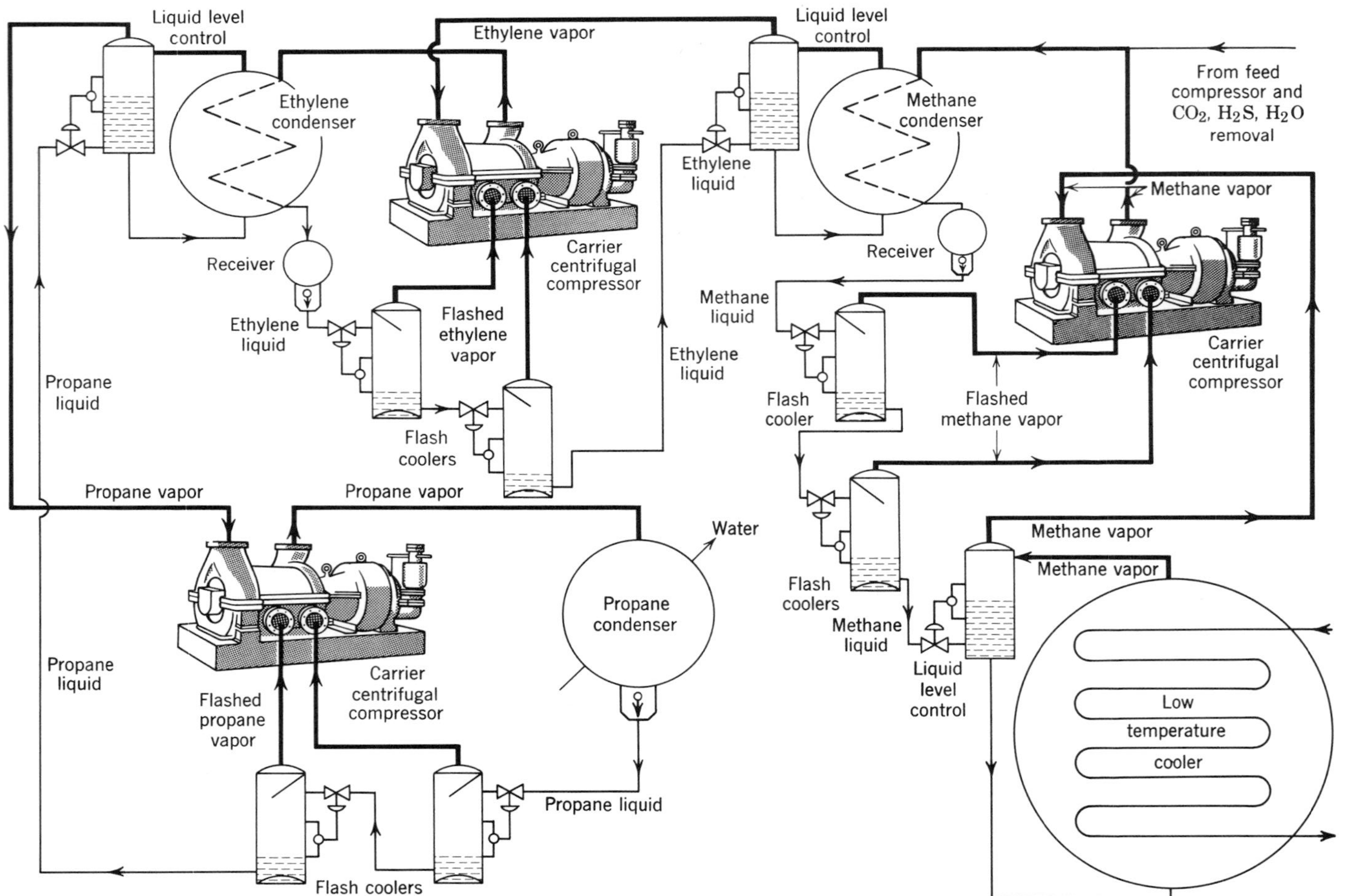

Fig. 20-14 Three-stage cascade system employing methane, ethylene and propane in the low, intermediate, and high stages, respectively. (Courtesy Carrier Corporation)

where m_s = the mass of the system charge in pounds.

V_f = the volume of the expansion tank in cubic feet.

V_s = the volume of the system in cubic feet.

v_1 = the specific volume of the refrigerant at the operating pressure in cubic feet per pound.

v_2 = the specific volume of the refrigerant at the maximum shutdown pressure in cubic feet per pound.

A three-stage cascade system employing methane, ethylene, and propane in the low, intermediate, and high stages respectively is illustrated in Fig. 20-14.

The chief disadvantage of cascade staging is the overlap of refrigerant temperatures in the cascade condenser, which tends to reduce the thermal efficiency of the system somewhat below that of the direct staged system. On the other hand, cascade staging makes possible the use of high density, high pressure refrigerants in the lower stages, which will usually result in a considerable reduction in the displacement required for the low stage compressor. The use of high pressure refrigerants also simplifies the design of the low stage evaporator in that higher refrigerant pressure losses through the evaporator can be permitted without incurring excessive losses in system capacity and efficiency. Too, since the refrigerants in the several stages do not intermingle, and each stage is a separate system within itself, the problem of oil return to the compressors is somewhat less critical than in the direct staged system.

A single stage Refrigerant-12 system and a two-stage, direct staged Refrigerant-12 system operating between the same temperature limits are compared on pressure-enthalpy coordinates in Fig. 20-15. If the volumetric and compression efficiences of the compressors were considered, the difference between the single and multistage compression systems would be approximately twice as great as that indicated by the values because of the difference in the compression ratios.

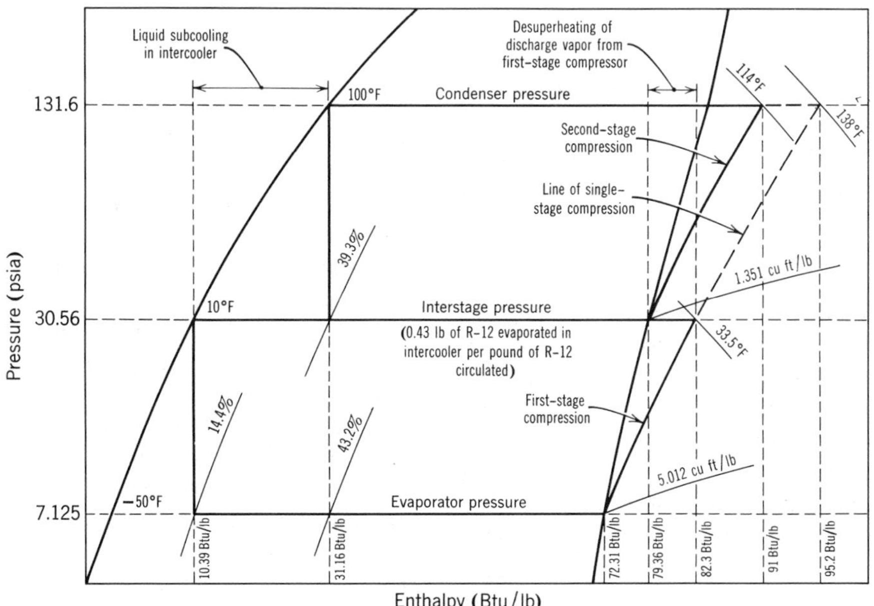

Fig. 20-15 *ph* Diagram of two-stage direct-staged R-12 system with flash intercooler. First-stage compressor displacement—16.19 cfm/ton. Second-stage compressor displacement—6.24 cfm/ton including vapor from intercooler. Compression ratio for each stage is approximately 4.3 to 1.

20-14. Oil Return in Multistage Systems

Since the several stages of the cascade system are actually separate and independent systems, oil return is accomplished in the individual stages in the same manner as in any other single stage system operating under the same conditions. This is not true of the direct staged system. Whenever two or more compressors are interconnected, either in parallel or in series, there is no assurance that oil return to the individual compressors will be in equal amounts. Therefore, some means of insuring equal distribution of the oil among the several compressors must be provided. When the compressors are connected in parallel, the oil can be maintained at the same level in all the compressors by interconnecting the crankcases as shown in Fig. 19-11. However, this simple method of oil equalization requires that the crankcase pressures in the several compressors be exactly the same (Section 19-15) and therefore is not practical when the compressors are connected in series, since the higher pressures existing in the crankcase of the high stage compressors would force the oil through the equalizing lines into the lower pressures existing in the crankcases of the low stage compressors.

One common method of equalizing the oil levels in the crankcases of compressors connected in series is shown in Fig. 20-12. An oil separator installed in the discharge line of the high stage compressor separates the oil from the discharge gas and returns it to the suction inlet of that machine. High side float valves maintain the desired oil level in the high stage compressors by continuously draining the excess oil returning to these compressors to the next lower stage compressor through the oil transfer lines. For manual operation, hand stop valves (normally closed) can be substituted for the float drainers, in which case the hand valves are opened periodically to adjust the oil levels by bleeding oil from the higher stage compressors to the lower stage compressors.

20-15. Pull-Down Loads and Capacity Control

Although the problem of pull-down loads is not limited to low-temperature systems, it is apt to be more severe in the low-temperature system because of the greater temperature differential involved. In general, corrective measures are the same as those prescribed for the higher temperature operations. Some authorities recommend that the compressor driver be sized for 150% of the design power requirements and that the compressor suction pressure be limited by a crankcase pressure regulator or a pressure-limiting expansion valve so that the compressor power requirements during pull-down do not exceed this limit.

Notice in Fig. 20-12 that the bypass around the low-stage compressor permits use of the high-stage compressor to reduce the evaporator temperature (load) to the interstage condition before starting the low-stage compressor.

Capacity control methods for multistage systems are the same as those employed with single-stage systems.

20-16. Multiple Temperature System

A multiple temperature system is one wherein two or more evaporators operating at different temperatures and located in different spaces or fixtures (or sometimes in the same space or fixture) are connected to the same compressor or condensing unit. The chief advantages gained by this type of operation are a savings in space and a reduction in the initial cost of the equipment. However, since higher operating costs will usually more than offset the initial cost advantage, a multiple temperature system is economically justifiable only in small capacity installations where operating costs in any case are relatively small. One obvious disadvantage of the multiple temperature system is that in the event of compressor breakdown all spaces served by the compressor will be without refrigeration, thereby causing the possible loss of product which otherwise would not occur.

A typical three-evaporator multiple temperature system is illustrated in Fig. 20-16. An evaporator pressure regulator valve is installed in the suction line of each of the warmer evaporators in order to maintain the pressure, and therefore the saturation temperature of the refrigerant, in these units at the desired high level. A check valve is installed in the suction line of the lowest temperature evaporator to prevent the higher pressure from the warmer

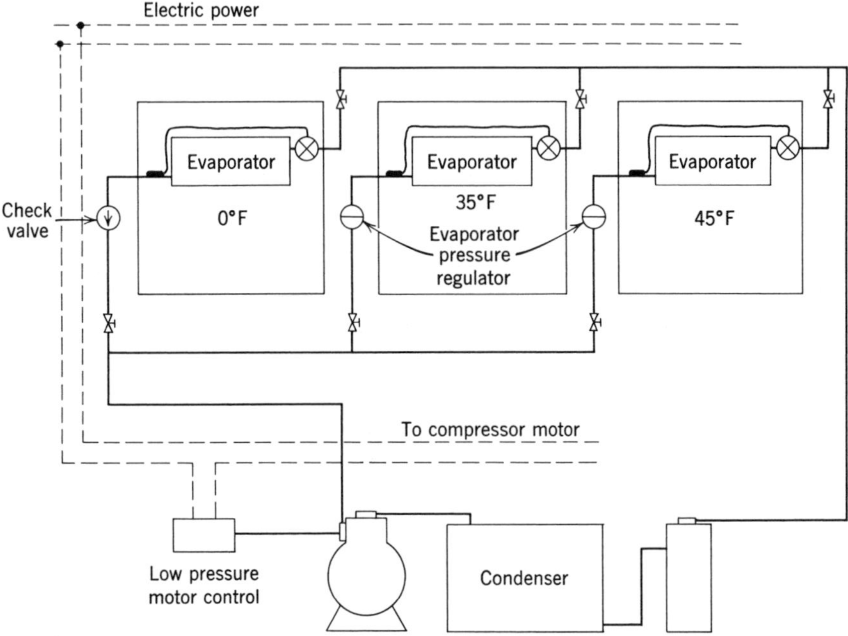

Fig. 20-16 Three-evaporator multiple temperature installation. Manual stop valves in suction and liquid lines permit isolation of the individual evaporators for maintenance.

evaporators from backing up into the cold evaporator when the former are calling for refrigeration. Since the check valve will remain closed as long as the pressure in the suction main is above the pressure in the low temperature evaporator, it is evident that the low temperature evaporator will receive little, if any, refrigeration until the refrigeration demands of the high temperature evaporators are satisfied.

For this reason, if a multiple temperature system is to perform satisfactorily, the load on the lowest temperature evaporator must account for at least 60%, and preferably more, of the total system load. When the high temperature evaporator(s) constitute more than 40% of the total load, the refrigeration demands of the high temperature evaporator(s) will cause the compressor to operate a greater portion of the time at suction pressures too high to permit adequate refrigeration of the low temperature evaporator, with the result that temperature control in that unit will be erratic.

The evaporator pressure regulators installed in the suction line of the warmer evaporators may be either the throttling or snap-action types, the latter being employed whenever "off-cycle" defrosting of the high temperature evaporator is desired (Section 17-24).

In multiple temperature systems, a low pressure motor control is ordinarily used to cycle the compressor on and off, the cut-in and cut-out pressures of the control being adjusted to suit the conditions required in the low temperature evaporator.

When the compressor is operating, the pressure at the sucton inlet of the compressor will depend on the rate at which vapor is being generated in all the evaporators. If the combined load on the several evaporators is high, the suction pressure will also be high. When the refrigeration demands of one of the higher temperature evaporators is satisfied, the evaporator pressure regulator will close (or throttle) so that little or no vapor enters the suction main from that unit, thereby causing a reduction in the suction pressure. As previously mentioned, whether or not the low temperature evaporator is being refrigerated at any given time depends on whether the suction pressure is below or above the pressure in that evaporator. However, the low temperature evaporator will always be open to the compressor (refrigerated) at any time that the demands of the high temperature evaporators are satisfied and the regulator valves are closed (or throttled). When the demands of the low temperature evaporator are also satisfied, the suction pressure will drop below the cut-out setting of the low pressure control and the compressor will cycle off.

With the compressor on the off cycle, any one of the evaporators is capable of cycling the compressor on again. For the low temperature evaporator, a gradual rise in pressure in that unit to the cut-in pressure of the low pressure control will start the compressor, whether or not either of the high temperature evaporators also requires refrigeration.

Since the pressure maintained in the high temperature evaporators, even at the lower limit, is always above the cut-in setting of the low pressure control, if either of the evaporator pressure regulators opens one of the high temperature evaporators to the suction main, the suction pressure will immediately rise above the cut-in setting of the low pressure control and the compressor will cycle on.

The fact that the pressure in the high temperature evaporators is always above the cut-in setting of the low pressure control poses somewhat of a problem when throttling-type evaporator pressure regulators are employed on the high temperature evaporators. Since this type of regulator is open any time the evaporator pressure is above the pressure setting of the regulator, it will often cause the compressor to cycle on when the evaporator it is controlling does not actually require refrigeration. As soon as the compressor starts, the pressure in the evaporator is immediately reduced below the regulator setting and the regulator closes, causing the compressor to cycle off again on the low pressure control.

To prevent short cycling of the compressor on the low pressure control when throttling type evaporator pressure regulators are employed, it is usually necessary to install a solenoid stop valve, activated by a space or evaporator temperature control, in the suction lines of the higher temperature evaporators in order to obtain positive shut-down of these evaporators during the compressor off-cycle. With pilot operated evaporator pressure regulators, positive close-off of the regulator during the compressor off cycle can be obtained by controlling the regulator with a temperature or solenoid pilot.

In small systems, a surge tank installed in the suction main will usually prevent short cycling of the compressor. Because of the relatively large volume of the surge tank, it is capable of absorbing reasonable pressure increases occurring in the high temperature evaporator and thereby preventing the suction pressure from rising prematurely to the cut-in pressure of the low pressure control. The size of the surge tank required depends on the ratio of the high temperature load to the low temperature load and on the temperature differential between the high and low temperature, the size of the tank required increasing as each of these factors increases. Obviously, this method of preventing short cycling is best applied to systems where the high temperature load is only a small portion of the total load and/or where the difference in temperature between the high and low temperature evaporators is relatively small.

20-17. Solenoid Controlled Multiple Temperature Systems

Thermostatically controlled solenoid stop valves, installed in either the liquid or the suction lines of the high temperature evaporators, are frequently used to obtain multiple temperature operation. A typical installation employing solenoid valves in the liquid line is shown in Fig. 20-17. The operation of this type of system is similar to that of systems employing snap-action regulators in the suction line, except that there is no control of the evaporator pressure and temperature. A space-temperature actuated thermostat controls the solenoid valves. When the space temperature rises, the thermostat contacts close, energizing the solenoid coil and opening the liquid line to the evaporator. The pressure in the evaporator rises as the liquid enters the evaporator, causing the low pressure control to cycle the compressor on, if the latter is not already running. If the compressor is already running, the entrance of liquid into the evapo-

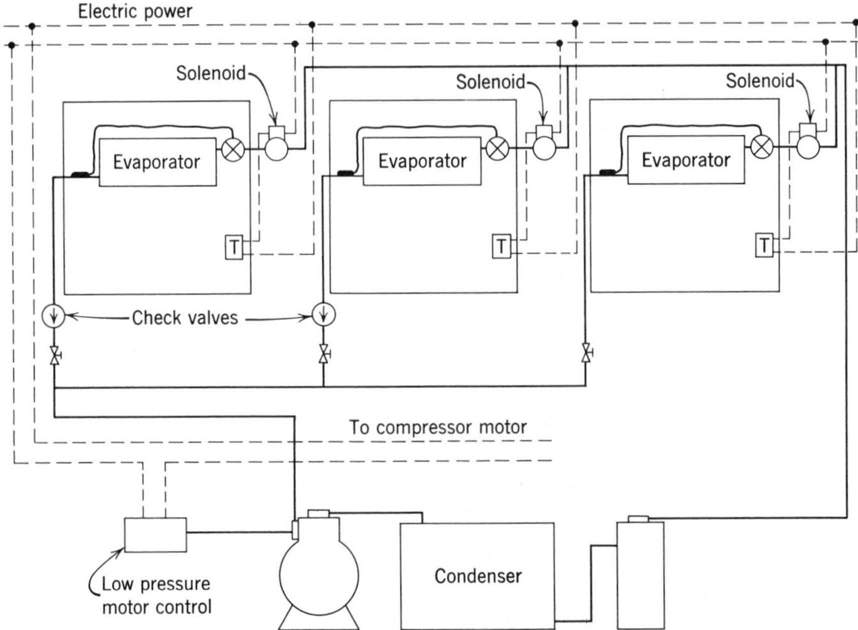

Fig. 20-17 Multiple unit installation employing thermostat-solenoid control. Manual stop valves in suction lines permit isolation of individual evaporators for maintenance.

rator will cause a rise in the operating suction pressure.

When the temperature in the space is reduced to the desired low level, the thermostat contacts open, deenergizing the solenoid coil and closing the liquid line, whereupon the evaporator pumps down to the operating suction pressure, or to the cut-out pressure in the event that none of the other evaporators is calling for refrigeration. Since all the evaporators pump-down as they are cycled out, the system receiver tank must be large enough to hold the entire system refrigerant charge. This is not true, however, if the solenoids are intalled in the suction line rather than in the liquid line.

Placing the solenoid valves in the suction line has the disadvantage of requiring larger, more expensive valves. Too, in the event of a leaky refrigerant control, there is always the danger that liquid will accumulate in the evaporator while the suction solenoid is closed and flood back to the compressor when the solenoid is opened.

When solenoids are employed to obtain multiple temperature operation, there is no direct control of the evaporator pressure and temperature, since the pressure in the evaporator at any given time will depend upon the number of evaporators open to the compressor. Obviously, with this type of operation it is very difficult to maintan a balanced relationship between the space and evaporator temperatures, and therefore humidity control in the refrigerated space becomes very indefinite. For this reason, when humidity control is important, it is usually necessary to install a throttling-type evaporator pressure regulator in the suction lines of the high temperature evaporators in order to maintain the pressure and temperature in these evaporators at the necessary high level. However, evaporator pressure regulators should be used only with suction line solenoids. Short cycling of the compressor may result if they are employed in conjunction with liquid line solenoids.

With either suction line or liquid line solenoids, check valves should be installed in the suction lines of the lower temperature evaporators to avoid excessive pressures and temperatures in these units when the warmer evaporators are open to the suction line.

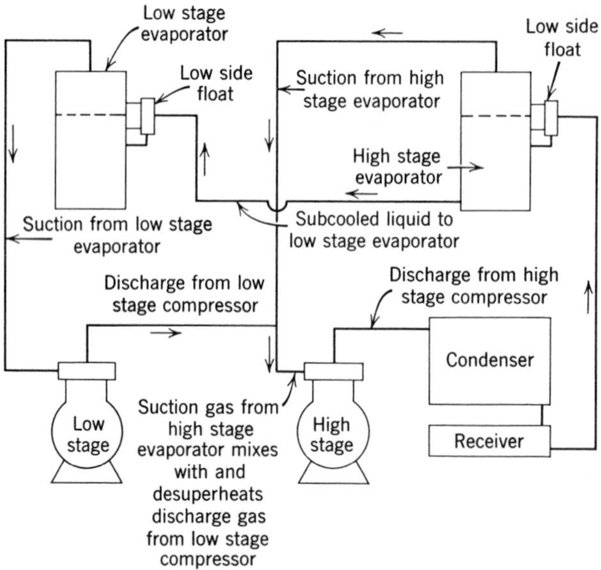

Fig. 20-18 A multiple temperature system employing two-staged direct staged compression.

20-18. Multiple Temperature Operation in Staged Systems

A multiple temperature system employing direct staged compression is illustrated in Fig. 20-18. Notice that the high temperature evaporator serves also as the gas and liquid cooler for the lower stage. The high stage compressor must be selected to handle the high temperature load in addition to the load passed along by the low stage compressor. A three-stage, multiple temperature, cascade-staged system employing Refrigerant-12 in all three stages is shown in Fig. 20-19.

20-19. Absorption Refrigeration Cycle

The absorption refrigeration cycle is similar to the vapor-compression cycle in that it employs a volatile refrigerant, usually either ammonia or water, which alternately vaporizes under low pressure in the evaporator by absorbing latent heat from the material being cooled and condenses under high pressure in the condenser by surrendering the latent heat to the condensing medium.

The principal difference in the absorption and vapor-compression cycles is the motivating force that circulates the refrigerant through the system and provides the necessary pressure differential between the vaporizing and condensing processes. In the absorption cycle, the vapor compressor employed in the vapor-compression cycle is replaced by an absorber and generator, which perform all the functions performed by the compressor in the vapor-compression cycle. In addition, whereas the energy input required by the vapor-compression cycle is supplied by the mechanical work of the compressor, the energy input in the absorption cycle is in the form of heat supplied directly to the generator. The source of the heat supplied to the generator is usually low-pressure steam or hot water, although in smaller systems the heat is usually supplied by the combustion of an appropriate fuel, such as natural gas, propane, or kerosene, directly in the generator or by an electric resistance heater installed in the generator.

A simple absorption system is illustrated in Fig. 20-20. Notice that the system consist of four basic components: an evaporator and an absorber, which are located on the low-pressure side of the system, and a generator and a condenser, which are located on the high-pressure side of the system. Two working fluids are employed, a refrigerant and an absorbent. The flow cycle for the refrigerant is

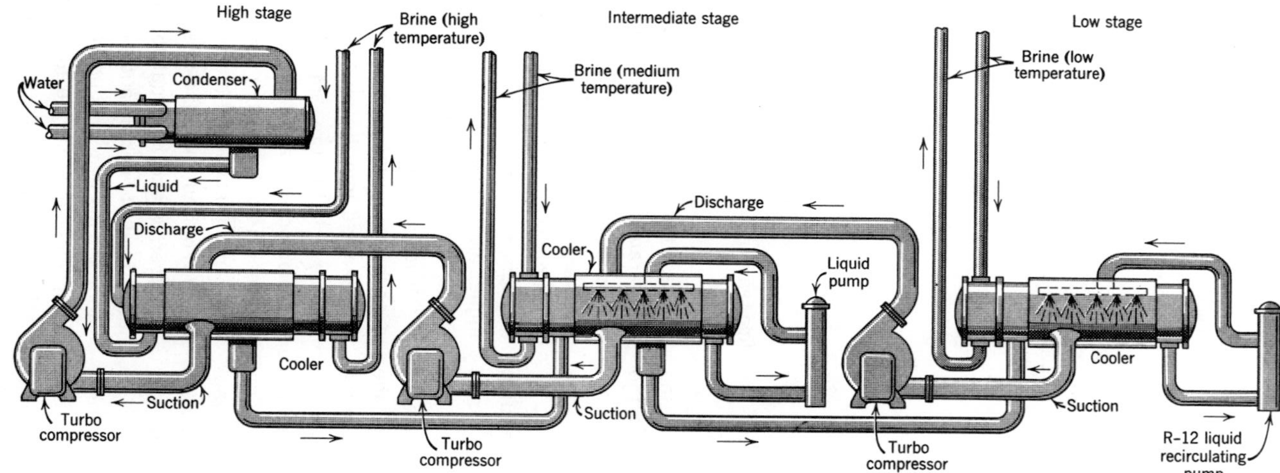

Fig. 20-19 Three-stage, multiple temperature system employing Refrigerant-12 with centrifugal compressors. (Courtesy York Corporation.)

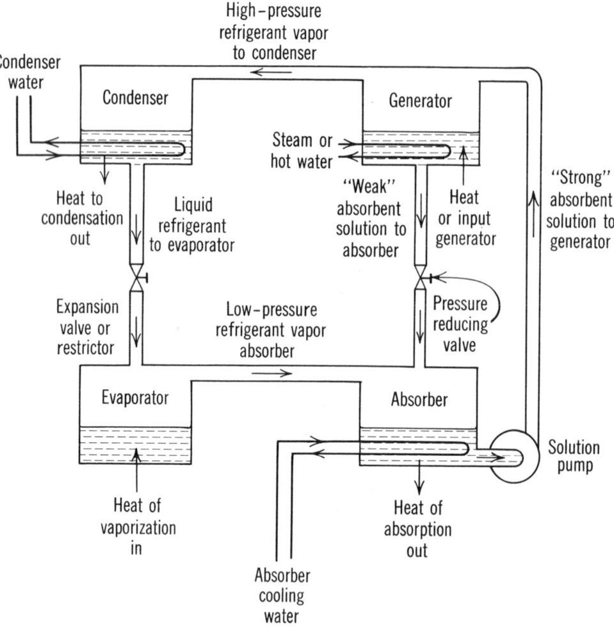

Fig. 20-20 Basic absorption refrigeration cycle.

from the condenser to the evaporator to the absorber to the generator and back to the condenser, while the absorbent passes from the absorber to the generator and back to the absorber.

With reference to Fig. 20-20. the operating sequence is as follows. High-pressure liquid refrigerant from the condenser passes into the evaporator through an expansion device or restrictor that reduces the pressure of the refrigerant to the low pressure existing in the evaporator. The liquid refrigerant vaporizes in the evaporator by absorbing latent heat from the material being cooled, and the resulting low pressure vapor then passes from the evaporator through an unrestricted passage to the absorber, where it is absorbed by, and goes into solution with, the absorbent.

The refrigerant flows from the evaporator to the absorber because the vapor pressure of the absorbent-refrigerant solution in the absorber is lower than the vapor pressure of the refrigerant in the evaporator. It is the vapor pressure of the absorbent-refrigerant solution in the absorber that determines the pressure on the low-pressure side of the system and, consequently, the vaporizing temperature of the refrigerant in the evaporator. In turn, the vapor pressure of the absorbent-refrigerant solution depends on the nature of the absorbent, its temperature, and its concentration. The lower the temperature of the absorbent and the higher the concentration (the smaller the percentage of refrigerant in the solution), the lower is the vapor pressure of the solution.

As the refrigerant vapor from the evaporator is dissolved into the absorbent solution, the volume of the refrigerant is decreased (compression occurs) and the heat of absorption is released. In order to maintain the temperature and vapor pressure of the absorbent solution at the required level, the heat released in the absorber, which is equal to the sum of the latent heat of condensation of the refrigerant vapor and the heat of dilution of the absorbent, must be discarded to the surroundings, usually to the same heat sink or condensing medium used for the heat discard at the condenser.

In order to transfer heat from the absorber to the sink, the temperature of the absorbent must be higher than that of the available sink. Inasmuch as the efficiency of the absorber increases as the temperature of the absorbent solution is reduced, it follows that the efficiency of the absorber depends in part on the temperature of the available coolant.

Since the refrigerant vapor dissolving into the absorbent solution increases the strength (percentage refrigeration) and vapor pressure of the solution, it is necessary to continuously reconcentrate the solution in order to maintain the vapor pressure of the solution at a level low enough to provide the required low pressure and temperature in the evaporator. The reconcentration is accomplished by continuously removing the "strong" absorbent solution from the absorber and recirculating it through the generator, where most of the refrigerant vapor is "boiled off" by the application of heat, and the resulting "weak" solution is returned to the absorber to absorb more refrigerant vapor from the evaporator.

Since the absorber is on the low-pressure side of the system and the generator is on the high-pressure side, the strong solution must be pumped from the absorber to the generator and the weak solution returned to the absorber through a pressure reducing valve or restrictor. While the pressure of the absorbent solution is increased from the low side pressure to the high side pressure as the solution is pumped from the absorber to the generator, no compression of the refrigerant occurs in the process, since compression of the refrigerant has already been accomplished in the absorber. Consequently, the power required by the solution pump is relatively small.

In the generator, the refrigerant is separated from the absorbent by heating the solution and vaporizing the refrigerant. The resulting high-pressure refrigerant vapor then passes to the condenser, where it is condensed by giving up latent heat to the condensing medium, after which it is ready for recirculation to the evaporator.

The weak absorbent solution left in the generator is returned to the absorber through the absorber return pipe as previously described. The relative strength of the weak solution is controlled by the amount of heat supplied to the generator.

To realize maximum system efficiency, the pressure differential between the low-pressure and high-pressure sides of the system should be maintained as small as possible by keeping the low side pressure as high as possible consistent with the refrigeration requirements and the high side pressure as low as possible with the available condensing medium. Recall that the low side pressure is determined primarily by the vapor pressure of the absorbent solution, which in turn depends on the temperature and concentration of the solution. Since control of the solution temperature is limited by the temperature of the available coolant, control of the low side (evaporator) temperature and pressure is usually obtained by varying the concentration of the absorbent solution.

An improvement in system efficiency can be achieved by providing for a heat exchange between the strong solution going to the generator and the high-temperature weak solution returning from the generator to the absorber. Since the temperature of the strong solution going to the generator is increased, while the temperature of the weak solution going to the absorber is decreased, the heat exchange results in a reduction in both the heat supplied to the generator and the cooling required by the absorber.

Also, as in the vapor-compression cycle, flashing of the liquid refrigerant, with the resulting loss of refrigerating effect, occurs as the high-pressure liquid refrigerant from the condenser undergoes a drop in pressure as it passes through the expansion device into the evaporator. Consequently, subcooling of the liquid refrigerant going from the condenser to the evaporator by heat exchange, with the low-temperature refrigerant vapor going from the evaporator to the absorber, will increase the refrigerating effect and improve system efficiency.

20-20. Refrigerant-Absorbent Combinations

To be suitable for use in an absorption refrigeration system, there are certain criteria that the refrigerant-absorbent combination should meet, at least to some degree. Obviously, the absorbent must have a strong affinity for the refrigerant vapor, and the two must be mutually soluble over the desired range of operating conditions. The two fluids should be safe, stable, and noncorrosive, both individually and in combination. Ideally, the absorbent should have a low volatility so that the refrigerant vapor leaving the generator will contain little or no absorbent, and working pressures should be reasonably low and preferably near atmospheric pressure to minimize equipment weight and leakage into and out of the system. The refrigerant should have a reasonably high latent heat value so that the required refrigerant flow rate is not excessive.

At present, there are two refrigerant-absorbent combinations that are in common use. The oldest in terms of usage is ammonia and water, wherein the ammonia is the refrigerant and water is the absorbent. A more recent combination is water and lithium bromide, wherein the water is the refrigerant and the lithium bromide, a hydroscopic salt, is the absorbent. The relative advantages and disadvantages of the two systems are discussed in the following sections.

20-21. Ammonia-Water Systems

Ammonia-water systems are widely used in domestic refrigerators and in commercial and industrial systems where the evaporator temperature is maintained close to or below 32°F. The ammonia-water combination meets some of the more important criteria exceptionally well but falls somewhat short in others. The water absorbent has a very strong affinity for the ammonia vapor, and the two are mutually soluble over a wide range of operating condi-

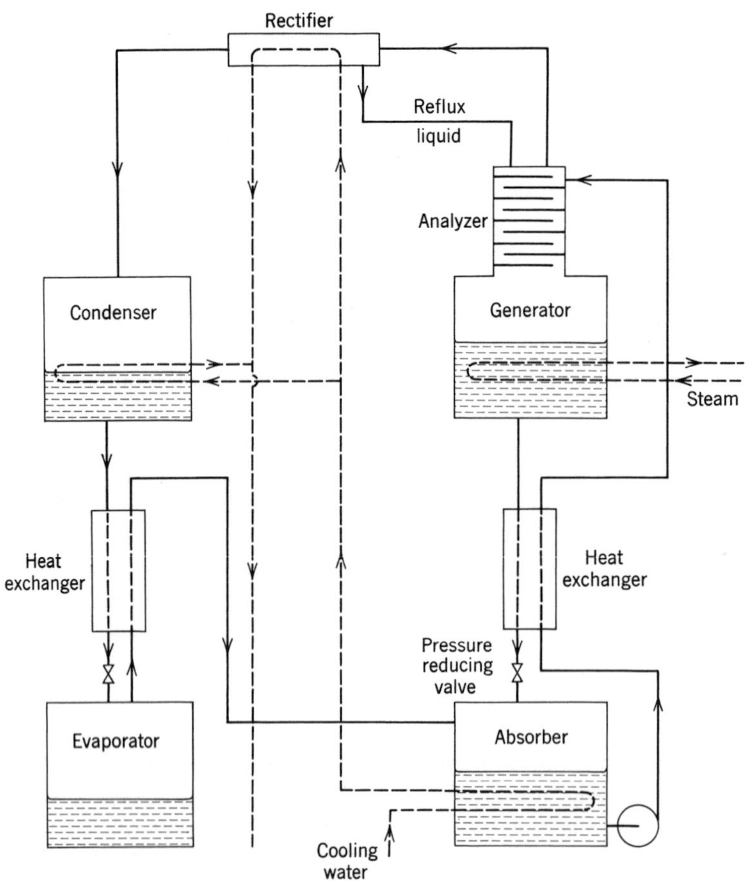

Fig. 20-21 Ammonia-water absorption system.

tions. Both fluids are highly stable and are compatible with most materials found in refrigeration systems. One notable exception is copper and its alloys, which, as previously mentioned, are not suitable for use in any ammonia system. The ammonia refrigerant has a high latent heat value but is slightly toxic, which limits its use in air-conditioning applications, and operating pressures are relatively high.

Probably the major disadvantage of the ammonia-water system is the fact that the absorbent (water) is reasonably volatile, so that the refrigerant (ammonia) vapor leaving the generator will usually contain appreciable amounts of water vapor, which, if allowed to pass through the condenser and go into the evaporator, will raise the evaporator temperature and reduce the refrigerating effect by carrying unvaporized refrigerant out of the evaporator.

For this reason, the efficiency of the ammonia-water system can be improved by the use of an analyzer and a rectifier which function to remove the water vapor from the mixture leaving the generator before it reaches the condenser. As shown in Fig. 20-21, the analyzer is essentially a distillation column that is attached to the top of the generator. As the ammonia and water vapors coming off the generator rise up through the analyzer, they are cooled, and the water vapor, having the highest saturation temperature, condenses and drains back to the generator, while the ammonia vapor continues to rise and leaves at the top of the analyzer. The ammonia vapor then passes to the rectifier, or reflux condenser, where the remaining water vapor, and a small amount of ammonia vapor, condenses and drains back through the analyzer in the form of a weak reflux solution, the latter being required for the analyzer to function properly. Cooling for the reflux condenser is usually accomplished with a portion of the condenser water and is limited in order to control the amount of reflux liquid going to the analyzer.

20-22. Water-Lithium Bromide Systems

Water-lithium bromide systems are used extensively in air conditioning and other high-temperature applications, but with water as the refrigerant, they are not suitable for use in any application where the evaporator temperature is below 32°F.

The lithium bromide, when not in solution, is a hydroscopic salt, and its brine has a great affinity for water vapor. However, one disadvantage of the water-lithium bromide combination is that the absorbent is not completely soluble in water under all the conditions likely to occur in the system, and special precautions must be taken in the design and operation of these systems to avoid conditions that will allow precipitation and crystallization of the absorbent.

One of the principal advantages of the water-lithium bromide system is that the absorbent is nonvolatile, so that there is no absorbent mixed with the refrigerant (water) vapor leaving the generator, and consequently, no analyzer or rectifier is required in the system.

Since water is the refrigerant, operating pressures are very low, being well below atmospheric. For example, assuming a 40°F evaporator temperature and a 100°F condensing temperature, the evaporator and condenser pressures are 0.248 in. Hg absolute and 1.93 in. Hg absolute, respectively. With the small pressure differential between the low and high side pressures, pressure-reducing valves between the high and low pressure sides are not ordinarily required, since the pressure losses through the connecting piping and spray nozzles will usually provide the necessary pressure differential.

It has already been shown that the ability of an absorbent to absorb refrigerant vapor depends on the relative concentration of the absorbent. For this reason, controlling the concentration of the absorbent solution provides a convenient means of varying the capacity of the system in response to varying refrigeration loads. Some of the methods employed to control solution concentration include control of the steam or hot water supply to the generator, control of the flow of the condenser water, and direct control of the reconcentrated solution from the generator.

A typical water-lithium bromide system is illustrated in Fig. 20-22. In addition to the

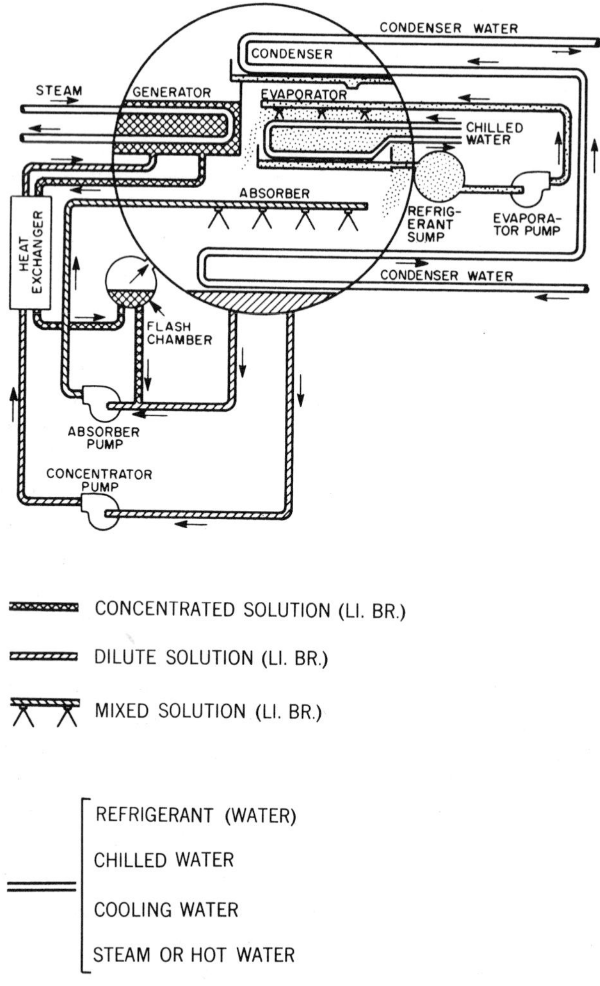

Fig. 20-22 Water-lithium bromide absorption system. (Courtesy York Corporation.)

equipment shown, a purge unit is required to purge nonconsiderable gases from the system and maintain the evaporator pressure at the required low level.

20-23. Domestic Absorption Systems

By adding a third fluid, an inert gas such as hydrogen, to the absorption system to balance the low and high side pressures of the system, an absorption cycle can be made to function without a solution pump or any other moving parts. Such systems, employing ammonia and water, have found widespread use in domestic refrigerators, particularly for mobile homes and recreational vehicles, in which case the heat is supplied to the generator by the direct combustion of propane, or by an electric heating element, or both.

The cycle operates on the principle defined by Dalton's law, which states that the total pressure of any mixture of gases and vapors is the sum of the individual partial pressures exerted by each of the gases or vapors in the mixture. With this type of system, the total pressure exerted by the gases and vapors in the system is virtually the same in all parts of the system. However, because of the presence of the hydrogen and the partial pressure that it exerts on the "low-pressure" side of the sys-

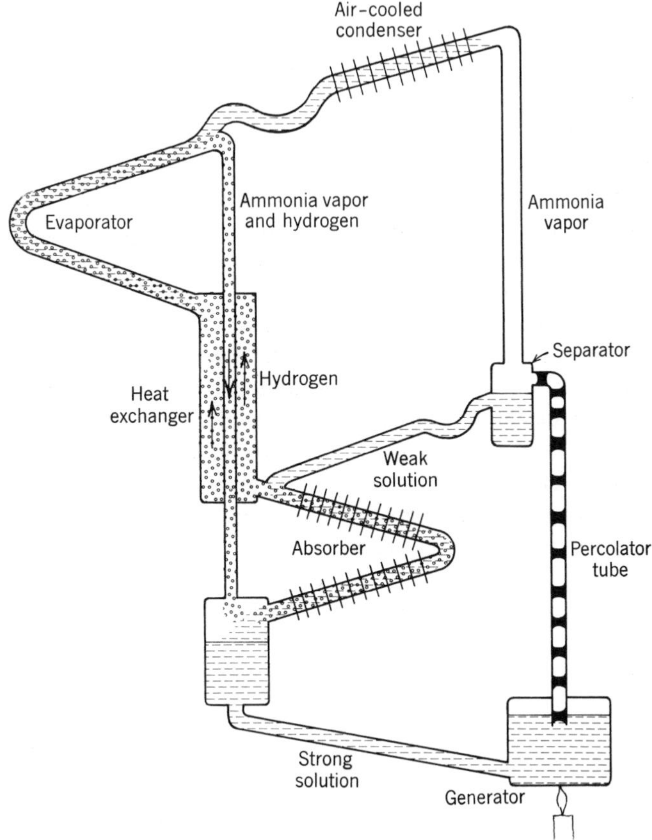

Fig. 20-23 Simplied domestic absorption system employing ammonia and water in conjunction with hydrogen gas to balance the low and high side pressures.

tem (evaporator and absorber), the partial pressure exerted by the ammonia vapor in these parts will be lower than that exerted by the ammonia vapor in the generator and condenser, where no hydrogen gas is present. As a result, the ammonia can evaporate at a low pressure and temperature in the evaporator and at the same time condense at high pressure and temperature in the condenser. Circulation of the absorbent solution is accomplished through the use of a percolator or bubble tube (as in a coffee percolator), which replaces the liquid absorbent pump employed in larger systems.

With reference to the simplified cycle illustrated in Fig. 20-23, heat supplied to the generator boils off the ammonia vapor, and the action of the percolator tube is such that slugs of liquid absorbent are carried along with the ammonia vapor up the percolator tube to the separator. From the separator, the liquid absorbent drains by gravity through the *U*-tube liquid trap into the air-cooled absorber, while the ammonia vapor passes directly to the air-cooled condenser, where it is condensed. The ammonia liquid then drains from the condenser by gravity through the *U*-tube liquid trap into the evaporator, where it is vaporized by the absorption of latent heat from the refrigerated space. The ammonia vapor, along with some of the hydrogen gas, passes from the evaporator to the absorber where the ammonia vapor goes into solution with the absorbent, while the hydrogen gas, which has no affinity for the absorbent, passes through the absorber and returns to the evaporator. The purpose of the two U-tube liquid traps is to contain the hydrogen gas in the evaporator and absorber, and thereby prevent its migration to the generator, separator, and condenser.

20-24. Absorption Systems Versus Vapor-Compression Systems

The absorption equipment itself is much simpler and less expensive to manufacture than vapor-compression equipment of comparable tonnage. Moreover, since there are few moving parts, absorption systems operate with less noise and with less maintenance than vapor-compression systems. However, the thermodynamic coefficient of performance is much lower for the absorption machine, being less than 1 as compared to 4 or more for a vapor-compression system operating under the same conditions. However, since energy supplied directly in the form of heat is less expensive and more efficient than when it must undergo several transformations, the foregoing cannot be used directly as a basis for comparing base energy costs or requirements.

In general, absorption systems are not usually economical when a boiler must be installed and maintained for refrigeration only. However, where the boiler capacity is already available, as when steam is used for winter heating or when waste heat is available from various industrial processes, the absorption system should be given consideration. Absorption machines are frequently used also in conjunction with turbine-driven centrifugal machines, in which case exhaust steam from the turbine is employed to drive the absorption equipment.

21

ELECTRIC MOTORS AND CONTROL CIRCUITS

21-1. Electric Motors

Single-phase and three-phase alternating current motors of various types are employed in the refrigerating industry as drives for compressors, pumps, and fans. A few two-phase and direct current motors are also used on occasion.

Single-phase motors range in size from approximately $\frac{1}{20}$ hp up through 10 hp, whereas three-phase motors are available in sizes ranging from approximately $\frac{1}{3}$ hp on up, although the latter are seldom employed in sizes below 1 hp. When three-phase power is available, the three-phase motor is usually preferred to the single-phase type in integral horsepower sizes because of its greater simplicity and lower cost.

Practically all power in the United States is generated as 60-hertz, three-phase alternating current and is supplied at the point of use as single-phase and/or three-phase alternating current. Voltages available at the point of use depend somewhat on the type of transformer connection. In areas where power consumption is predominantly single-phase low voltage, transformers are "*Y*" connected so that the low voltage load can be distributed evenly among the three transformers. As shown in Fig. 21-1*b*, voltages supplied from a "*Y*" connected transformer bank are 120 V and 208 V single-phase and 208 V three-phase. When the power load is predominantly three-phase, the transformers are delta (Δ) connected as shown in Fig. 21-1*a*, in which case the supply voltages are 115 V and 230 V single-phase and 230 V three-phase. With the open delta arrangement shown in Fig. 21-1*c*, three-phase power for isolated users can be supplied with only two transformers.

Power is frequently delivered to commercial establishments at 460 V and is available to large industrial users at much higher voltages by special arrangement with the power company. Naturally, all motors must be selected to conform to the characteristics of the available power supply. Power companies guarantee to maintain voltages within plus or minus 10% of the design voltage. Most motors will operate satisfactorily within these voltage limits. Many motors are designed so that they can be operated on either low or high voltage by reconnecting external motor leads.

In addition to the type of power supply available, the following are some of the other more important factors that must be taken into account in order to select the proper type motor:

1. The conditions prevailing at the point of installation with respect to the ambient temperature and to the presence of dust, moisture, or explosive materials.

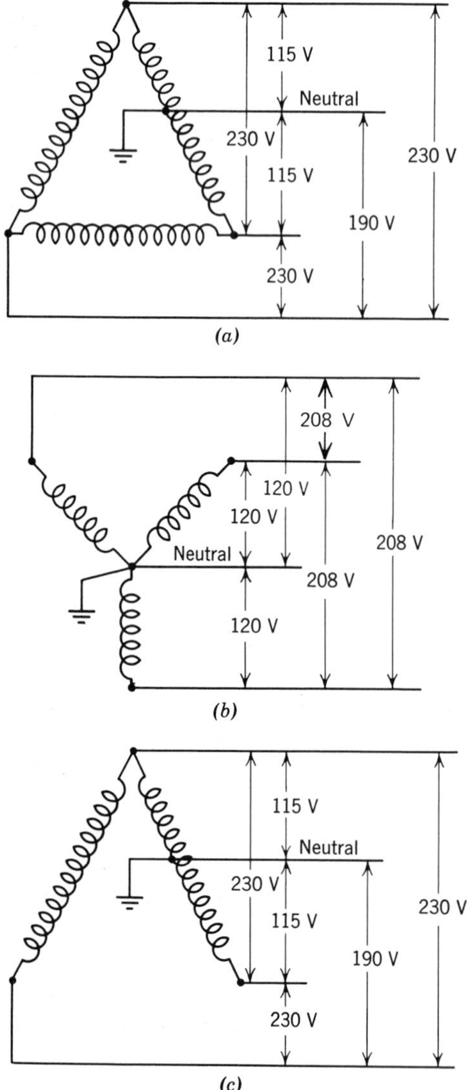

Fig. 21-1 (*a*) Delta connected transformers. (*b*) "Y" connected transformers. (*c*) Transformers connected open delta.

2. Starting torque requirements (loaded or unloaded starting).
3. Starting current limitations.
4. Single or multispeed operation.
5. Continuous or intermittent operation.
6. Efficiency and power factor.

All motors generate a certain amount of heat due to power losses in the windings. If this heat is not dissipated to the surroundings, motor temperature will become excessive and breakdown of the winding insulation will result. Open type (ventilated) motors are designed to operate at temperatures approximately 40°C (72°F) above the ambient temperature under full load, whereas totally enclosed motors are designed for a 55°C temperature rise at full load. Both types are guaranteed by the manufacturer to operate continuously under full load conditions without overheating when the ambient temperature does not exceed 40°C. When higher ambient temperatures are encountered, it is sometimes necessary to employ a motor designed to operate at a proportionally lower temperature rise.

Most motors can be operated with small overloads for reasonable periods of time without damage. However, since the heat generated in the motor increases as the load on the motor increases, continuous operation of the motor under overload conditions will cause excessive winding temperatures and materially shorten the life of the insulation.

Motors may be classified according to the type of enclosure as (1) open, (2) totally enclosed, (3) splash-proof, and (4) explosion-proof. Open-type motors are designed so that air is circulated directly over the windings to carry away the motor heat. This type of motor can be used in any application where the air is relatively free of dust and moisture, the motor is not subject to wetting, and the hazards of fire or explosion do not exist. Splash-proof motors are designed for outdoor installation or in any other location where the motor may be subject to wetting. Totally enclosed motors are designed for use where dust and moisture conditions are severe. These motors are unventilated and must dissipate their heat to the surrounding air through the motor housing. Explosion-proof motors are designed for installation in hazardous locations, as when explosive dusts or gases are contained in the air.

For the most part, motor starting torque requirements depend on the load characteristics of the driven machine. Low starting torque motors may be used with any machine that starts unloaded. On the other hand, high starting torque motors must be used when the driven machine starts under load. Since motor starting (locked rotor) currents often exceed five to six times the full load current of the motor, where large motors are employed, a low starting current characteristic is desirable in order to reduce the starting load on the wiring, transformers, and generating equipment.

Alternating current motors may be classified according to their principle of operation as either induction motors or synchronous motors. An induction motor is one wherein the magnetic field of the rotor is induced by currents flowing in the stator windings. A synchronous motor is one wherein the rotor magnetic field is produced by energizing the rotor directly from an external source.

21-2. Three-Phase Induction Motors

Three-phase induction motors are of two general types: (1) squirrel cage and (2) wound rotor (slip ring). The two types are similar in construction except for rotor design. As shown in Fig. 21-2, each has three separate stator

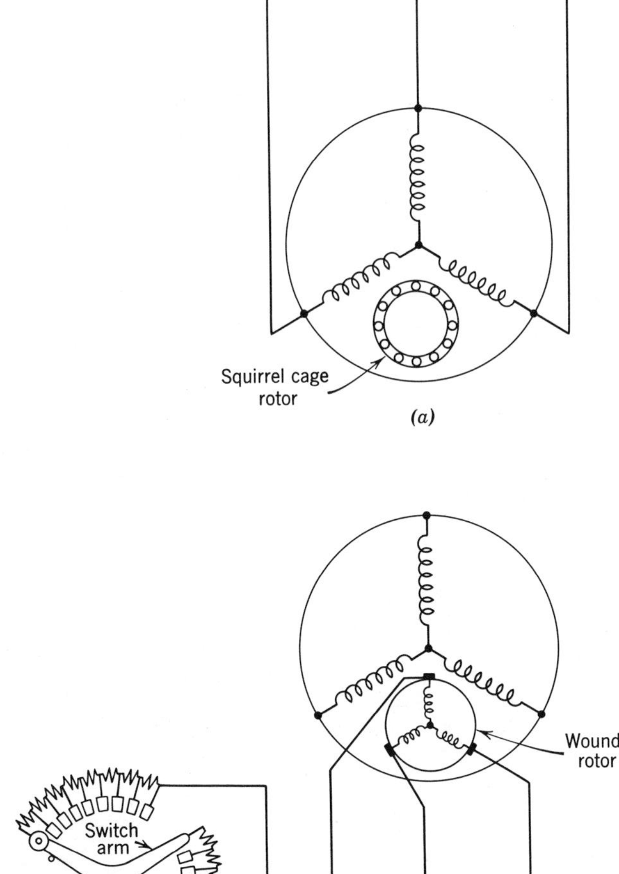

Fig. 21-2 (*a*) Squirrel cage polyphase motor. (*b*) Wound rotor (slip ring) polyphase motor.

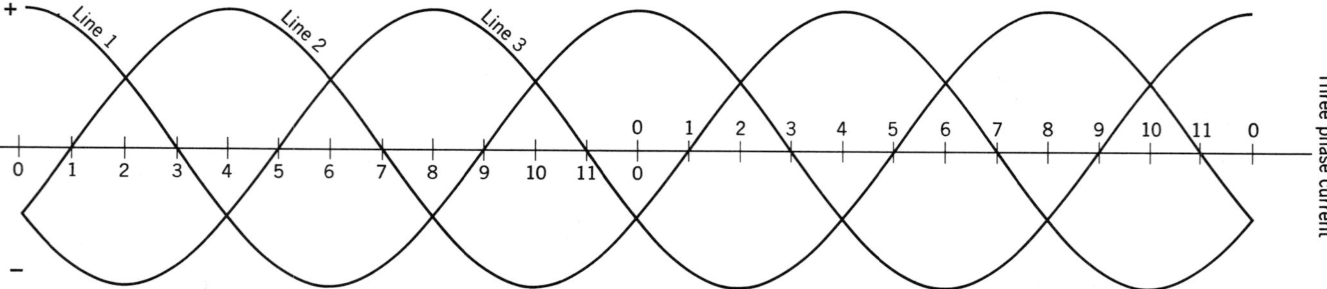

This diagram of three-phase voltages covers two complete cycles. The numbers on it refer to the numbers on the diagrams below. Each diagram shows the condition in the armature at the instant indicated by the corresponding number on this curve. The action of the magnetic field is smooth and regular; the rise and fall of currents in the conductors are also smooth and regular.

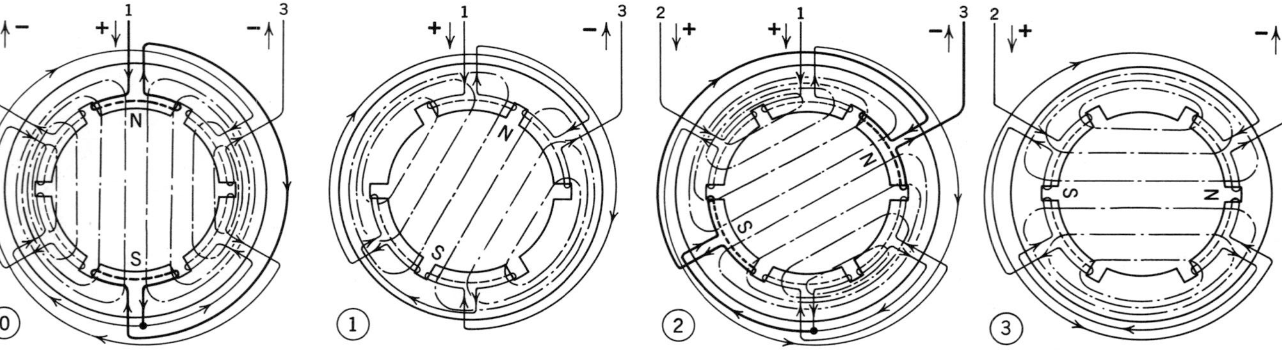

The current entering the motor on line 1 divides equally and leaves the motor on line 2 and line 3.

Now the current in line 2 is zero and that flowing in at line 1 leaves at line 3. The magnetic field revolves clockwise.

The current in line 1 is small and joining that from line 2 flows out in line 3 which carries a maximum negative current.

This and the following diagrams show how the magnetic field continues to rotate throughout the remainder of the cycle.

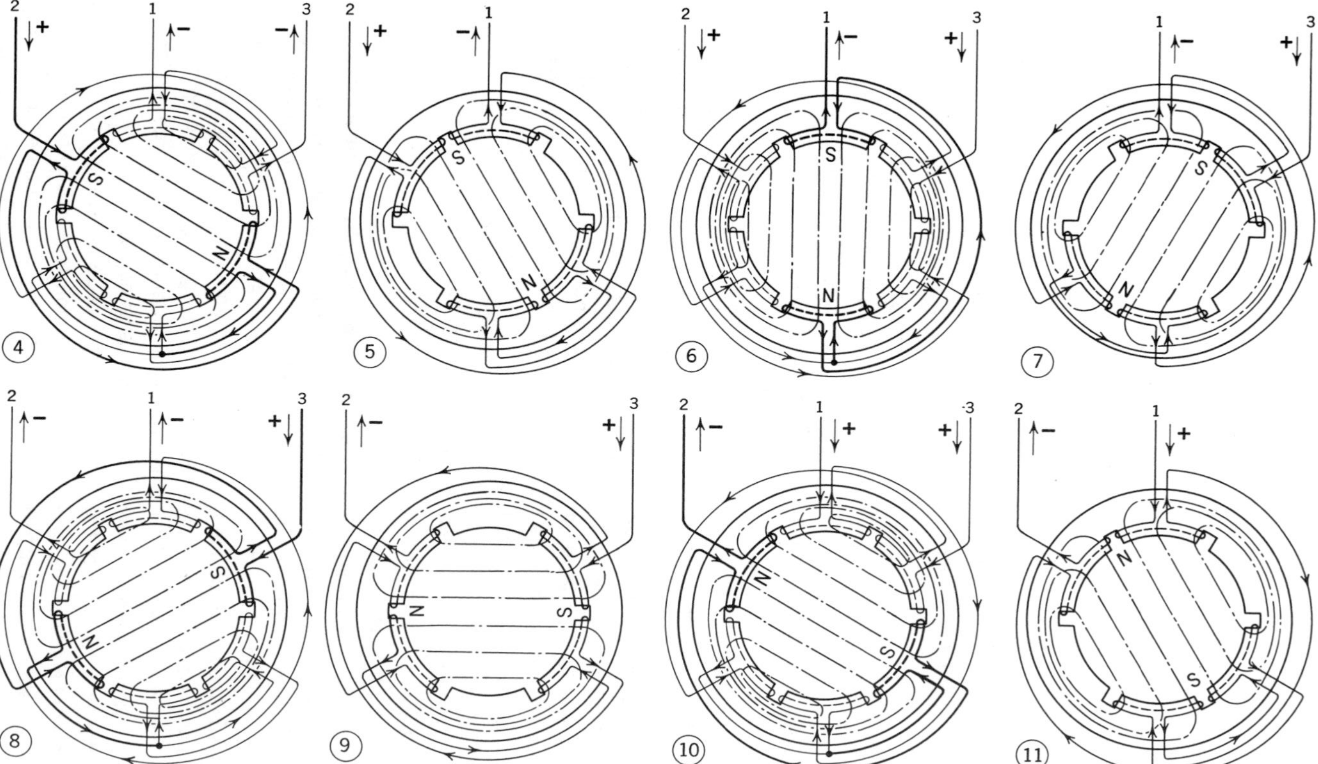

Fig. 21-3 This series of 12 diagrams shows the electric and magnetic conditions in a two-pole three-phase motor at the end of 12 equal parts of one cycle.

windings, one for each phase, which are evenly and alternately distributed around the stator core to establish the desired number of poles. A four-pole, three-phase motor will have twelve poles, four poles for each of the three phases.

When the stator is energized, three separate currents, each 120 electrical degrees out-of-phase with the other two, flow in the stator windings and produce a rotating magnetic field in the stator. At the same time, the currents induced in the rotor windings establish a magnetic field in the rotor. The magnetic poles of the rotor field are attracted by, and tend to follow, the poles of the rotating stator field, causing the rotor to rotate as shown in Fig. 21-3.

The rotor of an induction motor always rotates at a speed somewhat less than that of the rotating stator field. If the speed of the rotor were the same as that of the field, the conductors of the rotor winding would be standing still with respect to the rotating stator field rather than cutting across it, in which case no voltage would be induced in the rotor and the rotor would have no magnetic polarity. Therefore, it is necessary that the rotor turn at a speed slightly less than that of the stator field so that the conductors of the rotor winding continuously cut the flux of the stator field as the latter slips by. The difference between the speed of the rotor and that of the stator field is called the "magnetic slip" or "rotor slip." The greater the load on the motor, the greater is the amount of rotor slip. However, since the amount of slip changes only slightly as the load increases or decreases, and is very small even at full load, three-phase induction motors are usually considered to be constant speed motors.

Neglecting rotor slip, the speed of any alternating current motor is a function of the frequency and the number of stator poles. The synchronous speed of an alternating current motor can be determined by the following equation:

$$\text{Motor speed (rpm)} = \frac{\text{Frequency} \times 120}{\text{Number of poles}} \tag{21-1}$$

For a four-pole alternating current motor operating on 60-hertz power, the synchronous speed is

$$\frac{60 \times 120}{4} = 1800 \text{ rpm}$$

Rotor slip is usually expressed as a percentage of the synchronous speed. For instance, for a motor having a synchronous speed of 1800 rpm and operating at 1750 rpm, the percentage slip is

$$\frac{1800 - 1750}{1800} = 2.78\%$$

Rotor slip is also a measure of the power losses in the motor. In this instance, 2.78% of the total power input to the motor is converted to heat in the motor. Hence, there is a definite relationship between the amount of slip and the efficiency of the motor. The higher the slip, the lower is the efficiency of the motor.

The starting torque of a motor is the turning effort or torque that the motor develops at the instant of starting when full voltage is applied to the motor terminals. Starting torque is usually expressed as a percentage of full load torque and depends to some extent on the resistance of the rotor winding. An increase in rotor resistance increases the starting torque, but also increases the amount of rotor slip and decreases motor efficiency.

21-3. Three-Phase Squirrel Cage Motors

The rotor winding of a squirrel cage motor consists of bar-type copper conductors embedded in a laminated iron core and connected together (short-circuited) at the ends with heavy endrings, giving the winding the appearance of a squirrel cage, from which the motor derives its name.

The squirrel cage induction motor is by far the most common type of three-phase motor and is available in a number of designs which provide a variety of starting torque-starting current characteristics. Two designs frequently employed with refrigerating equipment are designs B and C. Design B motors develop a locked rotor or starting torque between 125% and 275% of full load torque with relatively low starting currents. The normal torque-low starting current characteristic of this design makes it ideal for use as a drive for blowers, fans, and pumps, and for compressors which are started unloaded. Design C motors have a high-starting torque-low starting current characteristic which makes them suitable as drives for compressors which must start under load. Design C motors develop a starting torque between 225% to 275% of full load torque, but are slightly less efficient than the design B motor.

Multispeed operation of squirrel cage induction motors can be obtained by proper design of the stator windings. Since motor speed decreases as the number of poles increase, it follows that by doubling the number of stator poles, the speed of the motor can be reduced by one-half. For a single-winding motor, the number of poles can be changed in a two to one ratio by bringing extra leads outside of the motor. When more than two speeds are desired, the motor is wound with two or more separate windings for each phase. Two speeds are available with each separate winding.

21-4. Wound Rotor (Slip Ring) Motors

Wound rotor motors are employed in applications where the excessive starting current of a large squirrel cage motor would be objectionable and/or where a number of operating speeds are desired in the range between one-half and maximum speed.

A slip ring or wound rotor motor induction motor differs from the squirrel cage type only in the rotor winding. The rotor winding consists of insulated coils, grouped to form definite pole areas so that the rotor has the same number of poles as the stator. The terminal connections of the rotor windings are brought out to slip rings. The leads from the brushes on the slip rings are connected to external resistors, as shown in Fig. 21-2*b*. The operating principle of the slip ring motor is the same as that of the squirrel cage motor, except that by inserting external resistance in the rotor circuit when starting, high-starting torque can be developed with low values of starting current. As the motor accelerates, the resistance is gradually cut out of the rotor circuit until at full speed the rotor windings are short-circuited. With the rotor winding short-circuited, the motor operates with low slip and high efficiency.

The speed of the wound rotor motor can be varied from maximum down to approximately 50% of maximum by inserting resistance in the rotor circuit. The starting resistors can be used for this purpose provided they are designed for a large enough current capacity to prevent excessive heating in continuous service. At reduced speeds, the wound rotor motor tends to lose its constant speed characteristic and the speed of the motor will vary somewhat with the load.

21-5. Synchronous Motors

The synchronous motor is so named because the field (rotor) poles are synchronized with the rotating poles of the armature (stator) windings. Therefore, the speed of the synchronous motor depends only on the frequency of the power supply and the number of poles, and is independent of motor load. The armature winding of the synchronous motor is similar to those of the squirrel cage and wound rotor motors. The field (rotor) winding consists of a series of coils which make up the field poles. The field coils are connected through slip rings to a direct current power source and are so connected together that alternate north and south poles are formed when the field winding is energized with direct current. The direct current is usually supplied by a small direct current generator, called an exciter, which is mounted on the motor shaft. The rotor is also equipped with a squirrel cage winding, called the "damper" winding, which is used to start the motor. The damper winding can be designed for a variety of starting torque-starting current characteristics.

When polyphase power is applied to the armature winding, the rotating magnetic field acts on the squirrel cage damper winding and causes the rotor to rotate. Since the motor starts as a squirrel cage motor, the speed will be slightly less than synchronous speed. After the motor comes up to speed, direct current is applied to the field windings. This produces alternate north and south poles on the rotor which lock the rotor into synchronization with the rotating armature field.

By adjusting the flow of direct current to the field coil, the synchronous motor can be operated at unity power factor. Therefore, synchronous motors can be applied to an advantage in any large installation where constant speed and high efficiency are desired. However, the big advantage of the synchronous motor is that it can be used to correct the low power factor that results from heavily inductive loads. By increasing the flow of direct current through the field winding (overexciting the field), the synchronous motor is operated with a leading power factor which can be adjusted to offset exactly the lagging power factor produced by inductive loads. Power factor correction will in no way affect the load-carrying capacity of the synchronous motor.

21-6. Single-Phase Motors

Single-phase motors commonly used in the refrigerating industry are of the following types: (1) split-phase, (2) capacitor start, (3) capacitor start and run, (4) permanent split capacitor, and (5) shaded pole. All these motors are induction motors and all employ a squirrel cage rotor. The principal factor which distinguishes one type from another is the particular method used to produce a starting torque.

When a single-phase stator winding is energized, current flow is simultaneous in all the stator poles and no rotating stator field is produced. Furthermore, the current induced in the squirrel cage rotor winding is such that the magnetic field set up in the rotor is exactly in line with the magnetic field of the stator. The condition occurring can be compared to the "dead center" condition of a single piston engine. Therefore, there is no tendency for the rotor to rotate. However, if the rotor is started to rotate by some means, the current induced in the rotor winding will lag slightly behind the current in the stator winding. This causes the rotor field to lag the stator field and produces a torque that keeps the rotor turning. Hence, once the rotor of a single-phase motor is started, a rotating field is produced and the motor operates in a manner similar to that of the three-phase squirrel cage motor.

21-7. Split-Phase Motors

In order to produce a starting torque in the single-phase motor and make the motor self-starting, a second stator winding, called the "starting" or "auxiliary" winding, is employed in addition to the phase winding, the latter winding being referred to as the "main" or "running" winding. The relative position of the two windings in the stator of a four-pole, single-phase motor is shown in Fig. 21-4. Notice that the starting and running windings are connected in parallel directly across the single-phase line. In the split-phase type motor, the starting winding is wound with small wire so that the winding has a high resistance and low inductance, whereas the running winding is wound with large wire to have a low resistance and a high inductance. Both windings are energized at the instant of starting. However, because of the higher inductance of the running winding with relation to that of the starting winding, the current flow in the running winding lags the current flow in the starting winding by approximately 30 electrical degrees. Since the currents flowing in the two windings are 30 degrees out-of-phase with each other, the single phase is "split" to give the effect of two phases and a rotating field is set up in the stator which produces a starting torque and causes the rotor to rotate. When the rotor has accelerated to approximately 70% of maximum speed, which is a matter of a second or two, a centrifugal mechanism mounted on the rotor shaft opens a switch in the starting winding. With the starting winding disconnected, the motor continues to operate on the running winding alone. Since the starting winding is wound with relatively small wire, it will heat very quickly and, if allowed to remain in the circuit for an appreciable length of time, will be destroyed by overheating.

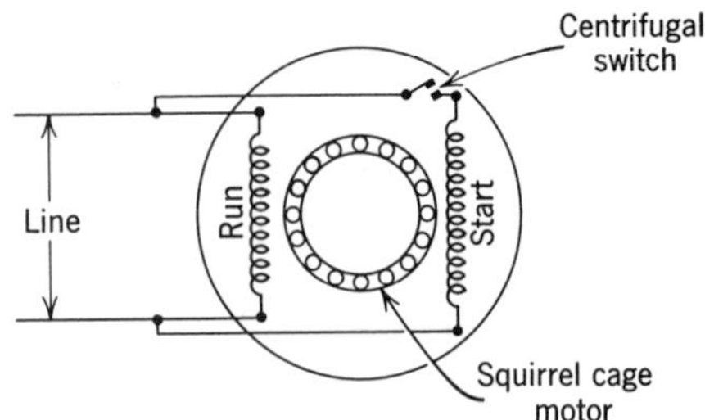

Fig. 21-4 Split-phase motor.

Since the maximum phase split that can be achieved with the split-phase motor is approximately 30 electrical degrees, the split-phase motor has a relatively low starting torque and can be used only with machines which start unloaded. These motors are generally available in sizes ranging from $\frac{1}{20}$ to $\frac{1}{3}$ hp for both 115 V and 230 V operation. They are used primarily as drives for small fans, blowers, and pumps.

21-8. Capacitor Start Motors

The capacitor start motor is identical to the split-phase motor in both construction and operation, except that a capacitor is installed in series with the starting winding, as shown in Fig. 21-5. Too, the starting winding of the capacitor start motor is usually wound with larger wire than that used for the starting winding of the split-phase motor. The use of a capacitor in series with the starting winding causes the current in this winding to lead the voltage, whereas the current in the running winding lags the voltage by virtue of the high inductance of that winding. With this arrangement, the phase displacement between the two windings can be made to approach 90 electrical degrees so that true two-phase starting is achieved. For this reason, the starting torque of the capacitor start motor is very high, a circumstance which makes it an ideal drive for small compressors that must be started under full load.

As in the case of the split-phase motor, the starting winding of the capacitor start motor is taken out of the circuit when the rotor approaches approximately 70% of maximum speed, and thereafter the motor operates on the running winding alone. Capacitor start motors are generally available in sizes ranging from $\frac{1}{6}$ through $\frac{3}{4}$ hp for both 115 V and 230 V operation.

21-9. Capacitor Start-and-Run Motors

Construction of the capacitor start-and-run motor is identical to that of the capacitor start motor with the exception that a second capacitor, called a "running" capacitor, is installed in series with the starting winding but in parallel with the starting capacitor and starting switch, as shown in Fig. 21-6. The operation of the capacitor start-and-run motor differs from that of the capacitor start and split-phase motors in that the starting or auxiliary winding remains in the circuit at all times. At the instant of starting, the starting-and-running capacitors are both in the circuit in series with the auxiliary winding so that the capacity of both capacitors is utilized during the starting period. As the rotor approaches 70% of rated speed, the centrifugal mechanism opens the starting switch and removes the starting capacitor from the circuit, and the motor continues to operate with both main and auxiliary windings in the circuit. The function of the running capacitor in series with the auxiliary winding is to correct power factor. As a result, the capacitor start-and-run motor not only has a high starting torque but also an excellent running efficiency. These motors are generally available in sizes ranging from approximately $\frac{1}{2}$ through 10 hp, and are widely used as drives for refrigeration compressors in single-phase applications.

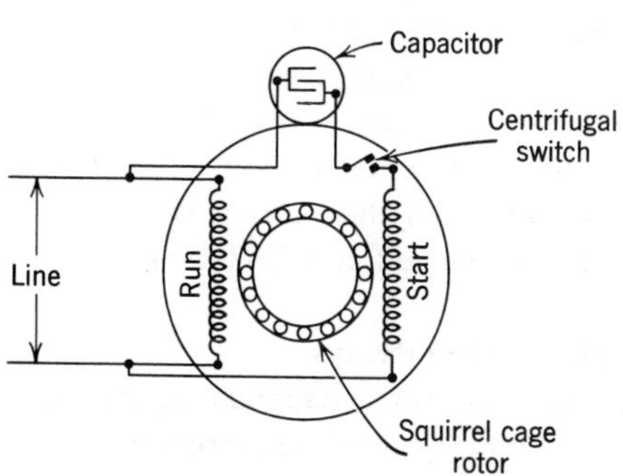

Fig. 21-5 Capacitor start motor.

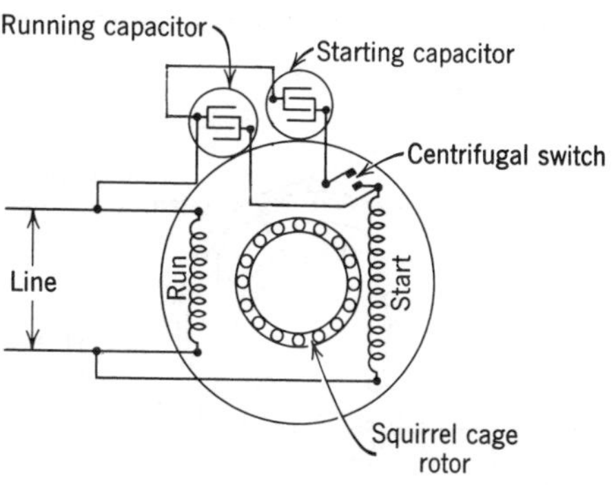

Fig. 21-6 Capacitor start-and-run motor.

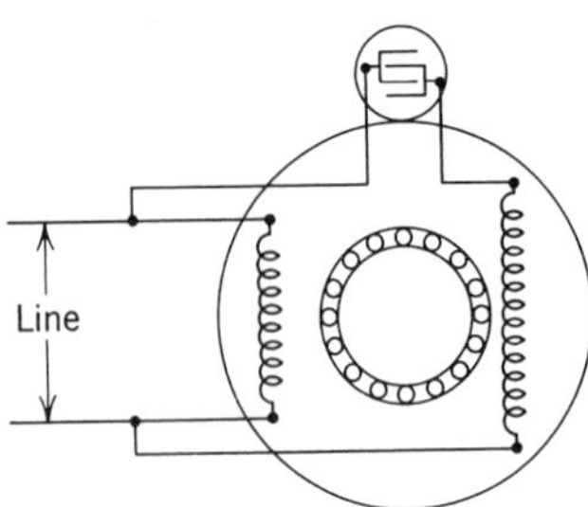

Fig. 21-7 Permanent capacitor motor.

21-10. Permanent Capacitor Motors

Construction of the permanent capacitor motor is similar to that of the capacitor start-and-run motor, except that no starting capacitor or starting switch is used. The capacitor shown in series with the auxiliary winding in Fig. 21-7 remains in the circuit continuously. The capacitor is sized for power factor correction but is used also as a starting capacitor. However, since the capacitor is too small to provide a large degree of phase displacement, the starting torque of the permanent capacitor is very low. These motors are available only in small fractional horsepower sizes. They are used mainly as drives for small fans which are mounted directly on the motor shaft. The chief advantage of this type of motor is that it lends itself readily to modulating speed control. Also, it does not require a starting switch.

21-11. Shaded Pole Motors

Construction of the shaded pole motor differs somewhat from that of the other single-phase motors in that the main stator winding is arranged to form salient poles, as shown in Fig. 21-8. The auxiliary winding consists of a shading coil, which surrounds a portion of one side of each stator pole. The shading coil usually consists of a single turn of heavy copper wire which is short-circuited and carries only induced current. In operation, the flux produced by the induced current in the shading coil distorts the magnetic field of the stator poles and thereby produces a small starting torque. Shaded pole motors are widely used as drives for small fans which are mounted directly on the motor shaft. They are available in sizes ranging from $\frac{1}{125}$ through approximately $\frac{1}{20}$ hp. In addition to its ready adaptability to modulating speed control, the main advantages of the shaded pole motor are its simple construction and low cost.

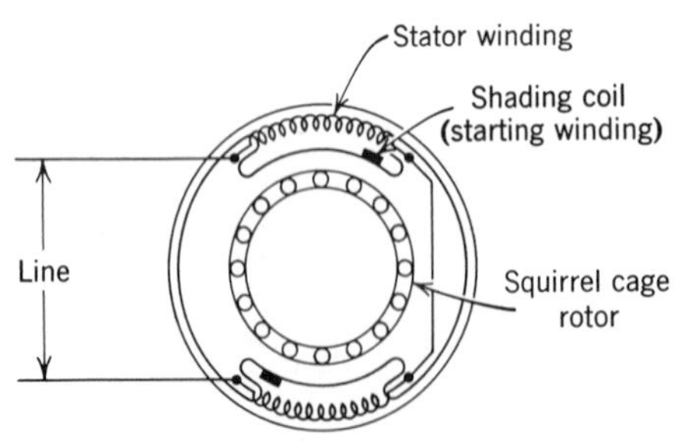

Fig. 21-8 Shaded pole motor.

21-12. Hermetic Motors

Motors frequently employed in hermetic motor-compressor units are three-phase squirrel cage motors and split-phase, capacitor start, and capacitor start-and-run single-phase motors. Whereas the split-phase and capacitor start motors are limited to small fractional horsepower units, the capacitor start-and-run motor is used in sizes from $\frac{1}{2}$ through 10 hp. Three-phase squirrel cage motors are employed from 3 hp up.

Although air, water, oil, and liquid refrigerant are sometimes used as cooling mediums to carry away the heat of hermetic motors, the large majority of hermetic motors are suction vapor cooled. For this reason, hermetic motor-compressor units should never be operated for any appreciable length of time without a continuous flow of suction vapor through the unit.

In the single-phase hermetic motor, a specially designed starting relay replaces the shaft-mounted centrifugal mechanism as a means of disconnecting the starting winding (or starting capacitor) from the circuit after the motor starts. Three types of starting relays have been used, namely: (1) the hot wire or timing relay, (2) the current coil relay, and (3) the voltage coil or potential relay.

21-13. Hot Wire Relays

The hot wire relay depends on the heating effect of the high-starting current to cause the thermal expansion of a special alloy wire, which in turn acts to open the starting contacts and remove the starting winding from the circuit. As shown in Fig. 21-9, the hot wire relay contains two sets of contacts, "S" and "M," which are in series with the starting and running windings, respectively. Both sets of contacts are closed at the instant of starting so that both windings are connected to the line (Fig. 21-9*a*). The high-starting current heat the wire and causes it to expand sufficiently to pull contacts "S" open and remove the starting winding from the circuit. After the starting winding is out of the circuit, the normal running current through the running winding will generate enough heat to maintain the "S" contacts in the open position, but not enough to cause additional expansion of the wire and open contacts "M" (Fig. 21-9*b*). However, if for any reason the motor draws a sustained overcurrent, the wire will expand further and pull open contacts "M," removing the running winding from the circuit (Fig. 21-9*c*). Contacts "M" are actually overload contacts which act as overcurrent protection for the motor. The mechanical arrangement of the two sets of contacts is such that contacts "M" cannot open without also opening contacts "S."

Since the action of the hot wire relay depends on the amount of current flow through the alloy

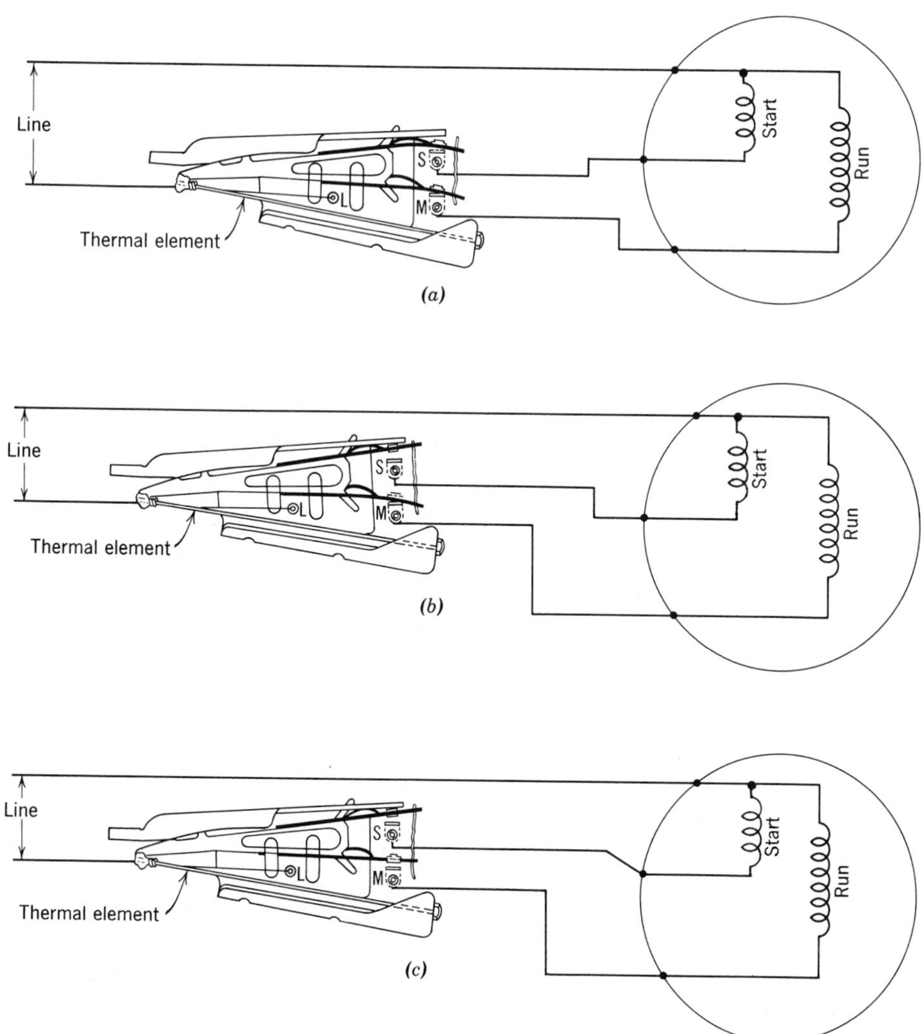

Fig. 21-9 Hot wire starting relay. (*a*) Starting position. (*b*) Run position. (*c*) Overload position.

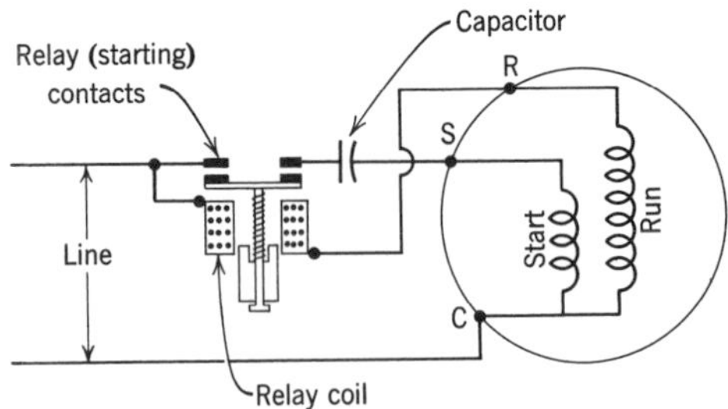

Fig. 21-10 Current-coil type starting relay.

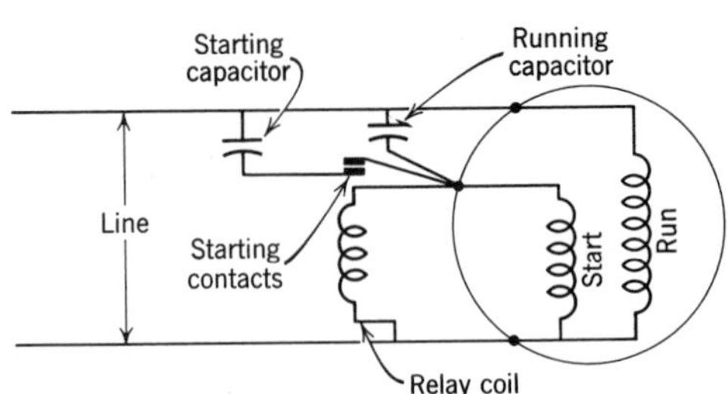

Fig. 21-11 Potential type starting relay.

wire, these relays must be sized to fit the current characteristics of the motor. They are best applied to the split-phase type motor.

21-14. Current Coil Relays

The current coil relay is used primarily with capacitor start motors. It is a magnetic type relay and is actuated by the change in the current flow in the running winding during the starting and running periods. The coil of the relay, which is made up of a relatively few turns of large wire, is connected in series with the running winding. The relay contacts, which are normally open, are connected in series with the starting winding, as shown in Fig. 21-10.

When the motor is energized, the high locked rotor current passing through the running winding and through the relay coil produces a relatively strong magnet around the coil and causes the relay armature to "pull-in" and close the starting contacts energizing the starting winding. With the starting winding energized, the rotor begins to rotate and a counter emf is induced in the stator windings which opposes the line voltage and reduces the current through the windings and relay coil. As the current flow through the relay coil diminishes, the coil field becomes too weak to hold the armature, whereupon the armature falls out of the coil field by gravity (or by spring action) and opens the starting contacts. The motor then runs on the running winding alone.

21-15. Potential Relays

Potential or voltage coil relays are employed with capacitor start and capacitor start-and-run motors. The potential relay differs from the current coil type in that the coil is wound with many turns of small wire and is connected in parallel with (across) the starting winding, rather than in series with the running winding, as shown in Fig. 21-11. The relay contacts are connected in series with the starting capacitor and are closed when the motor is not running. When the motor is energized, both the starting and running windings are in the circuit. As the motor starts and comes up to speed, the voltage in the starting winding increases to a value considerably above that of the line voltage (approximately 150%), as a result of the action the capacitor(s) in series with this winding.*

The high voltage generated in the starting winding produces a relatively high current flow through the relay coil and causes the coil armature to pull in and open the starting contacts. With the capacitor start motor, opening the relay contacts disconnects both the starting winding and starting capacitor from the circuit. With the capacitor start-and-run motor, only the starting capacitor is disengaged. With either type of motor, the starting winding voltage decreases somewhat when the starting contacts open, but remains high enough to hold the coil armature in the field and keep the starting contacts open until the motor is stopped.

21-16. Thermal Overload Protection for Hermetic Motors

All hermetic motor-compressors should be equipped with some type of thermal device which will protect the motor against overheating regardless of the cause. Thermal overload devices of this type are usually designed to be fastened directly to, and in good thermal contact with, the motor-compressor

* The vector sum of the voltages across the starting winding and capacitor(s) is equal to the line voltage.

housing, so that they are sensitive not only to motor overcurrent but also to overheating resulting from high discharge temperatures and other such causes.

Although some types of motor starting relays contain built-in overcurrent protection, these are usually sensitive only to motor overcurrent and do not provide protection against overheating from other causes.

21-17. Motor Operating Devices

For fractional horsepower motors the motor starting equipment sometimes consists only of a direct acting (line voltage) manual switch, thermostat, or low pressure control installed in the motor circuit between the motor and the power source (Fig. 21-12*a*). The control acts to open and close the motor circuit to stop and start the motor, respectively. Safety controls, such as high pressure cut-outs, overcurrent

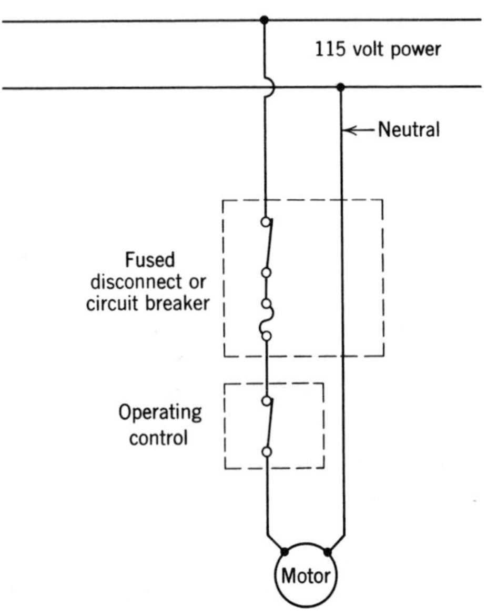

Fig. 21-12*a* Direct acting, line voltage controls with 115 V single-phase motor.

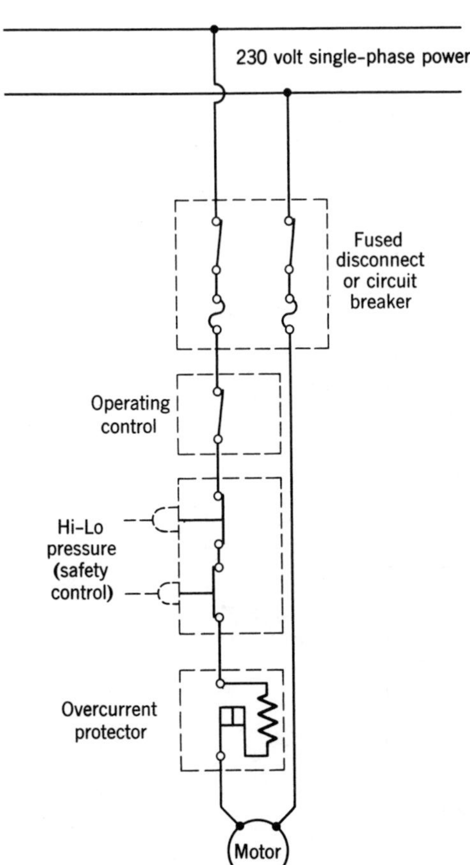

Fig. 21-12*b* Direct acting, line voltage controls used with 230 V single-phase power.

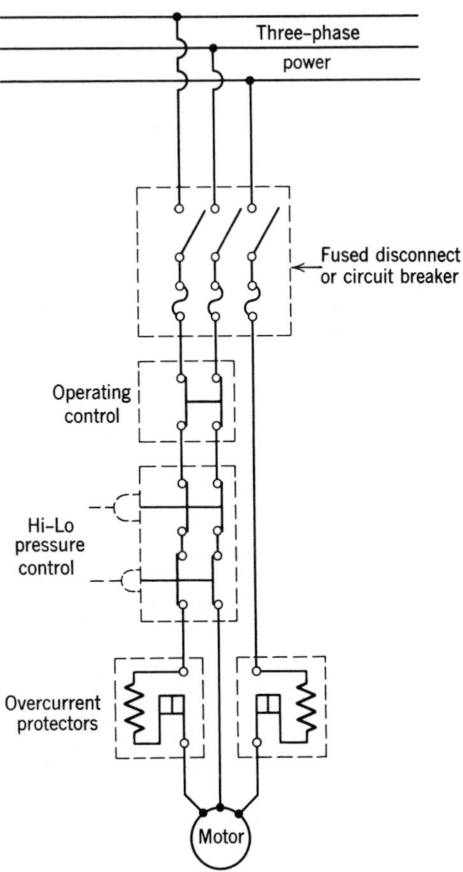

Fig. 21-12*c* Direct acting, line voltage controls used with 230 V three-phase power.

protective devices, etc., are connected in series with the operating or "cycling" control, as shown in Fig. 21-12*b*. The contacts of the safety controls are normally closed and do not open to break the circuit unless called on to perform their protective function.

With low voltage, single-phase power, the line voltage controls are installed in the "hot" line, never in the neutral (Fig. 21-12*a*). With high voltage, single-phase power, the controls may be installed in either one or both of the power lines (Fig. 21-12*b*). In the case of three-phase power, at least two of the three power lines must be opened to disconnect the motor from the power source. This requires the use of double-pole controls, as illustrated in Fig. 21-12*c*. However, in all cases, regardless of the type of power supplied, all "hot" lines must be protected individually with a properly sized fuse or circuit breaker.

Since the contacts of direct acting controls must be heavy enough to carry the full load current of the motor(s) they are controlling, these controls tend to become unwieldy when the full load current of the motor exceeds 15 or 20 amperes. Therefore, general practice is to control larger motors indirectly through a magnetic contactor.

A magnetic contactor or motor starter is essentially an electrical relay which in its simplest form consists of a coil of insulated wire, called a holding coil, and an armature to which the electrical contacts are attached (Fig. 21-13). The operation of the magnetic contactor is similar to that of the solenoid valve described in Chapter 17. When the holding coil is energized, the armature is pulled into the coil magnetic field, thereby closing the electrical contacts and connecting the motor to the power source. When the holding coil is deenergized,

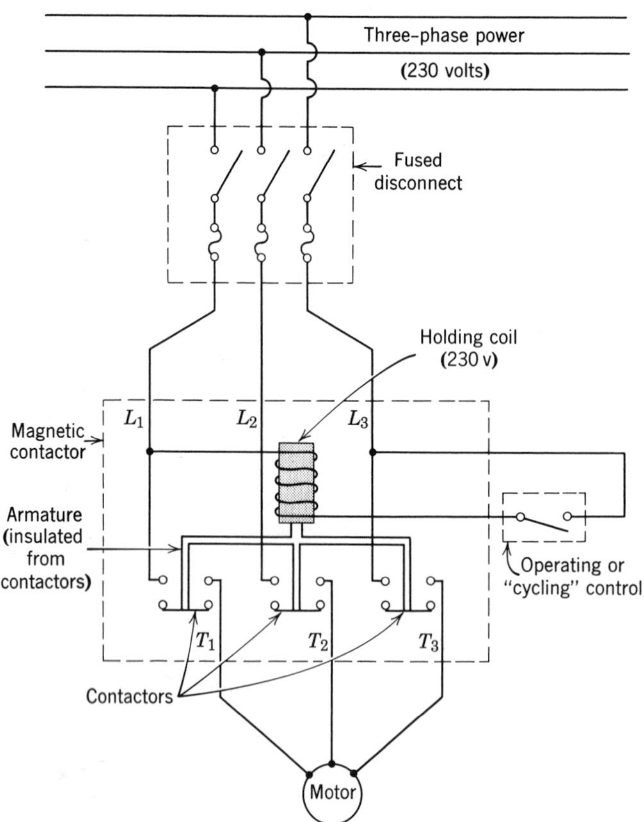

Fig. 21-13 Motor controlled indirectly through magnetic contactor.

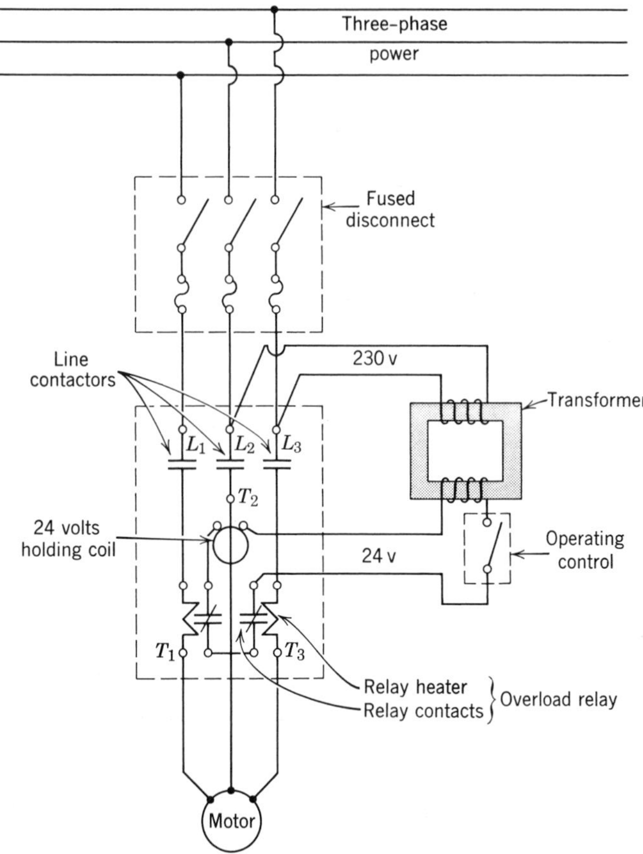

Fig. 21-14 Magnetic contactor with low voltage control circuit.

the armature drops out of the coil field, causing the contacts to open and disconnect the motor from the power source. When a magnetic contactor is employed, the motor is controlled indirectly by controlling the contactor holding coil. Therefore, the operating control is installed in series with the holding coil in the holding coil circuit rather than directly in the motor circuit.

The advantages gained by employing magnetic contactors to connect motors to the power source are several. First, since the current required to energize the holding coil is small, the contacts of the operating and safety controls can be of relatively light construction, which results in a reduction in both the size and the cost of the controls. Second, since the holding coil circuit is electrically independent of the motor circuit, the holding coil circuit voltage may be different from that of the motor circuit. This permits the use of low voltage (usually 24 V) control circuits, which are safer and generally less expensive to buy and install. A magnetic contactor employing a low voltage control circuit is shown in Fig. 21-14.

Holding coils for magnetic contactors are manufactured for all standard voltages and frequencies and are readily interchangeable in the field. The holding coil voltages most commonly used are 24, 115, 230, and 460.

21-18. Reduced Voltage Starters

The magnetic contactor described in the preceding section is called an "across-the-line" starter, and is so named because it connects the motor directly across the line at full voltage immediately when the holding coil is energized. This type of motor starter is suitable for motors up to 20 or 25 hp and is more widely used than any other type. However, in order to

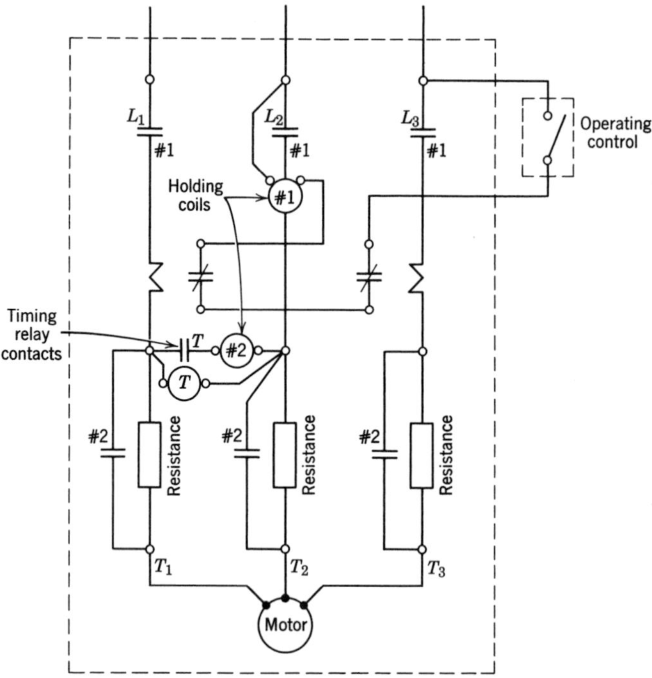

Fig. 21-15 Resistance type reduced voltage starter.

prevent excessive current surges in the power lines during the starting period, general practice is to start squirrel cage motors above 25 hp under reduced voltage. Reduced voltage starting is accomplished through the use of reduced voltage starters which introduce resistors or auto-transformers into the motor circuit during the starting period. A resistance type reduced voltage starter is illustrated in Fig. 21-15. When the operating control closes, the #1 holding coil is energized and the main contacts (#1) close, thereby connecting the motor to the power source through the resistors. This allows the motor to start under reduced voltage and, at the same time, energizes the timing relay. After a predetermined time interval, the timing relay closes and energizes the #2 holding coil, which in turn closes the #2 contactors and shunts the resistors out of the motor circuit.

21-19. Motor Overcurrent Protection

It is important to recognize that line fuses and circuit breakers are designed to protect the circuit only and do not provide overcurrent protection for the motor. Therefore, unless the motor is equipped with a built-in thermal overload, separate overcurrent protection must be provided in the circuit of each motor. To satisfy the need for overcurrent protection, many magnetic motor starters come equipped with overload relays.

The "overload relay" consists essentially of two parts: (1) a heater element installed in the motor circuit and (2) a set of contacts installed in the holding coil circuit. In the event that the motor is subjected to a sustained overcurrent, the temperature of the heater element increases above normal and the excess heat given off by the heater causes warping of a bimetal element (or melting of a special alloy metal) which opens the overload contacts in the holding coil circuit. This deenergizes the holding coil which in turn disengages the motor from the power source. A time delay action built into the overload relay prevents tripping of the overload during the motor starting period and during momentary overloads.

Fig. 21-16

21-20. Oil Pressure Failure Control

Another safety control frequently encountered in the control circuits of refrigeration equipment is the oil pressure failure control. The function of this control is to cycle the compressor off when the useful oil pressure developed by the oil pump falls below a predetermined minimum, or in the event that the oil pressure fails to build up to the minimum safe level within a predetermined time interval after the compressor is started. An external view of the oil pressure failure control is shown in Fig. 21-16.

In studying the operating characteristics of the oil pressure failure control it is important to recognize that the total oil pressure, as measured by an oil pressure gage, is the sum of the crankcase (suction) pressure and the pressure developed by the oil pump, and therefore is not the true or useful oil pressure. To determine the useful oil pressure, the suction pressure must be subtracted from the total oil pressure, the difference between the two being the useful oil pressure developed by the oil pump.

To be effective, the oil pressure failure switch must be actuated by the useful oil pressure rather than by the total oil pressure. This is accomplished by using two pressure bellows opposed to each other, as shown in Fig. 21-17. One bellows is connected to the crankcase and reflects crankcase pressure, whereas the other bellows is connected to the discharge of the oil pump and reflects total oil pressure. The pressure differential between the two bellows

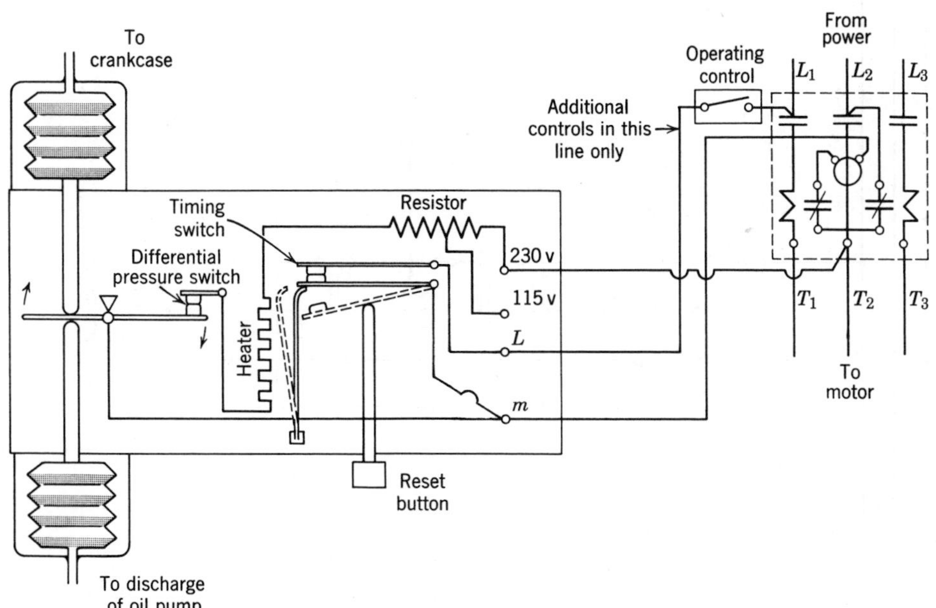

Fig. 21-17 Oil pressure failure control.

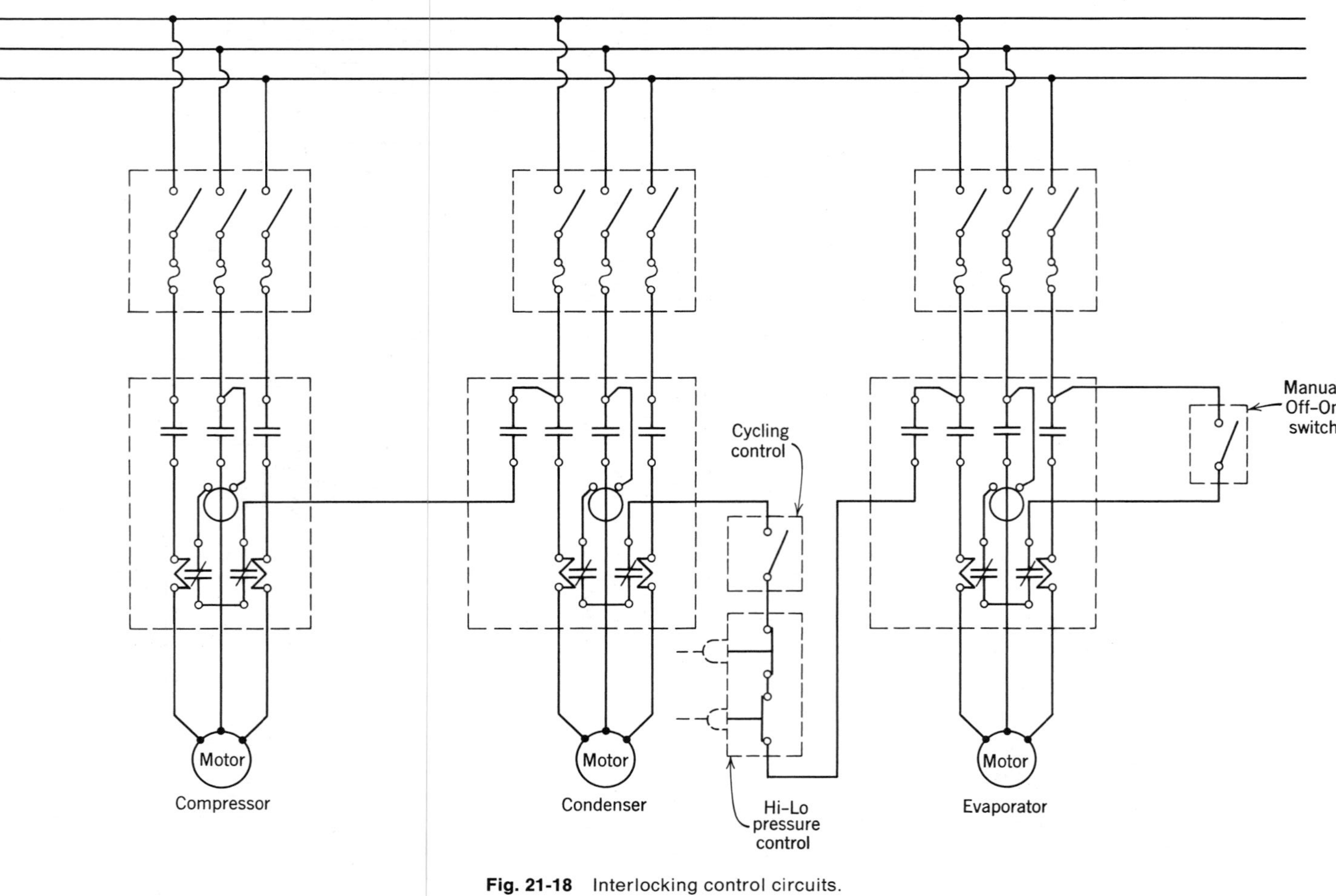

Fig. 21-18 Interlocking control circuits.

pressures is equal to the useful oil pressure and is utilized to actuate the pressure differential switch of the oil pressure failure control.

A time delay relay incorporated into the oil pressure failure control allows the compressor to operate 90 to 120 sec with the oil pressure below the safe level. This permits the compressor to start with zero oil pressure and also prevents unnecessary shut-down of the compressor in the event that the oil pressure momentarily falls below the minimum safe limit. However, if the oil pressure does not rise to the safe level within the alloted time, the oil pressure failure control will shut-down the compressor. Before the compressor can be restarted, the oil pressure failure control must be reset manually.

Referring to Fig. 21-17, notice that the timing relay consists of a timing switch and a heater element. The timing switch is connected in series with the holding coil of the magnetic starter, and the heater is connected in parallel with the holding coil. The pressure differential switch is connected in series with, and controls the operation of, the relay heater. The resistor in series with the relay heater limits the current flow through the heater and makes the oil pressure failure control adaptable to both 115 V and 230 V control circuits.

Since the oil pump operates only when the compressor is operating, the total oil pressure will be exactly equal to the crankcase pressure during the compressor off cycle, so that both the timing relay heater and the holding coil are energized when the compressor is started. If, after the compressor starts, the useful oil pressure builds up to the cut-in pressure of the oil pressure safety control, the differential pressure switch will open and remove the relay heater from the circuit. This action will allow the compressor to continue normal operation. On the other hand, if the useful oil pressure does not build up to the cut-in pressure of the control within the alloted time, the differential pressure switch will not open and the heater is left in the circuit. Continued operation of the relay heater will cause the bimetal of the timing switch to warp and open the timing contacts. This breaks the holding coil circuit and stops the compressor.

If the useful oil pressure falls below the cut-out point of the oil pressure failure control while the compressor is operating, the differential pressure switch closes and energizes the relay heater. If the oil pressure does not build up to the cut-in pressure again within the alloted time interval, continued operation of the heater will open the timing switch and stop the compressor.

As indicated in the foregoing, the oil pressure failure control has both a cut-in pressure and a cut-out pressure. These should be set in accordance with the compressor manufacturer's instructions whenever such data are available. In the absence of these data, general practice is to set the cut-in point of the control for a pressure approximately 5 psi below the useful oil pressure when the compressor is in operation. The cut-out point is usually set for a pressure approximately 5 psi below the cut-in pressure. For example, assume that the crankcase pressure is 37 psig and the total oil pressure is 72 psig, so that the useful oil pressure is 35 psi (72 − 37). The cut-in pressure should be set at approximately 30 psi (35 − 5) and the cut-out pressure at approximately 25 psi (30 − 5).

21-21. Interlocking Controls

As a general rule, a refrigerating system employs at least three motors: the compressor motor, the evaporator blower motor, and the condenser fan (or pump) motor. Good design practice requires that the controls of these motors be so interlocked that the compressor cannot operate unless the evaporator blower and the condenser fan or pump are operating. One of the more common methods of achieving the desired interlocking is illustrated in Fig. 21-18. In this instance, the evaporator blower motor is permitted to operate continuously and is controlled with a manual off-on switch. With this particular control arrangement, the fan control is the lead control and as such may be used to start and stop the entire system.

The holding coil of the condenser starter is wired through an auxiliary contact in the evaporator blower starter. Since the auxiliary contact will be closed only when the holding coil

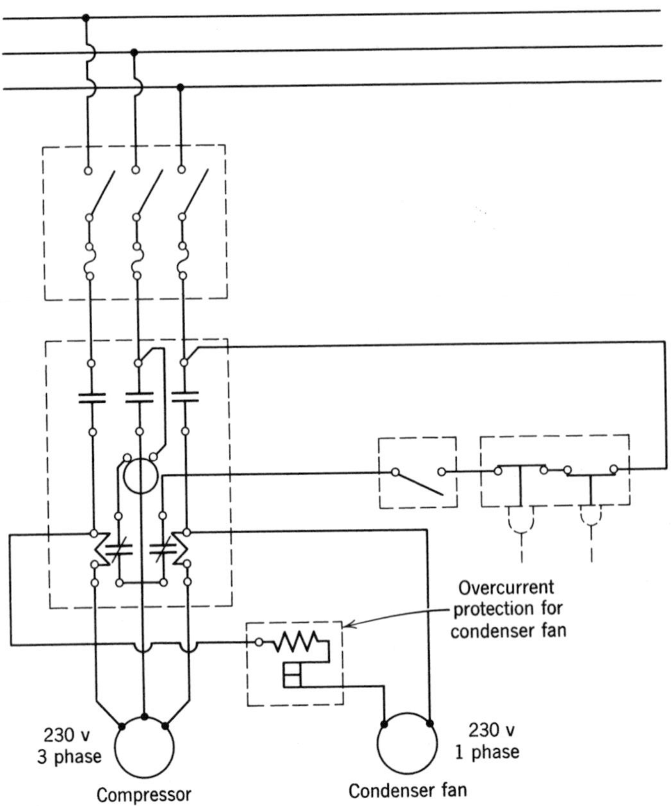

Fig. 21-19 Two motors operating through one magnetic contactor.

of the blower starter is energized, the condenser fan or pump cannot be started without first starting the evaporator blower. Likewise, the holding coil of the compressor starter is connected through an auxiliary contact in the condenser starter so that the compressor cannot start unless the condenser and evaporator starters are energized. Notice also that the cycling control (thermostat) controls the condenser starter rather than the compressor starter. This arrangement permits the condenser fan or pump to cycle off and on with the compressor.

Another common method of accomplishing the same result is to operate the compressor and condenser motors through the same magnetic contactor, as shown in Fig. 21-19. This method is usually confined to small, packaged equipment and requires that separate overcurrent protection be provided for each motor.

A wiring diagram for a simple pump-down system with interlocking control is illustrated in Fig. 21-20. Notice particularly the method of interlocking low voltage and high voltage control circuits.

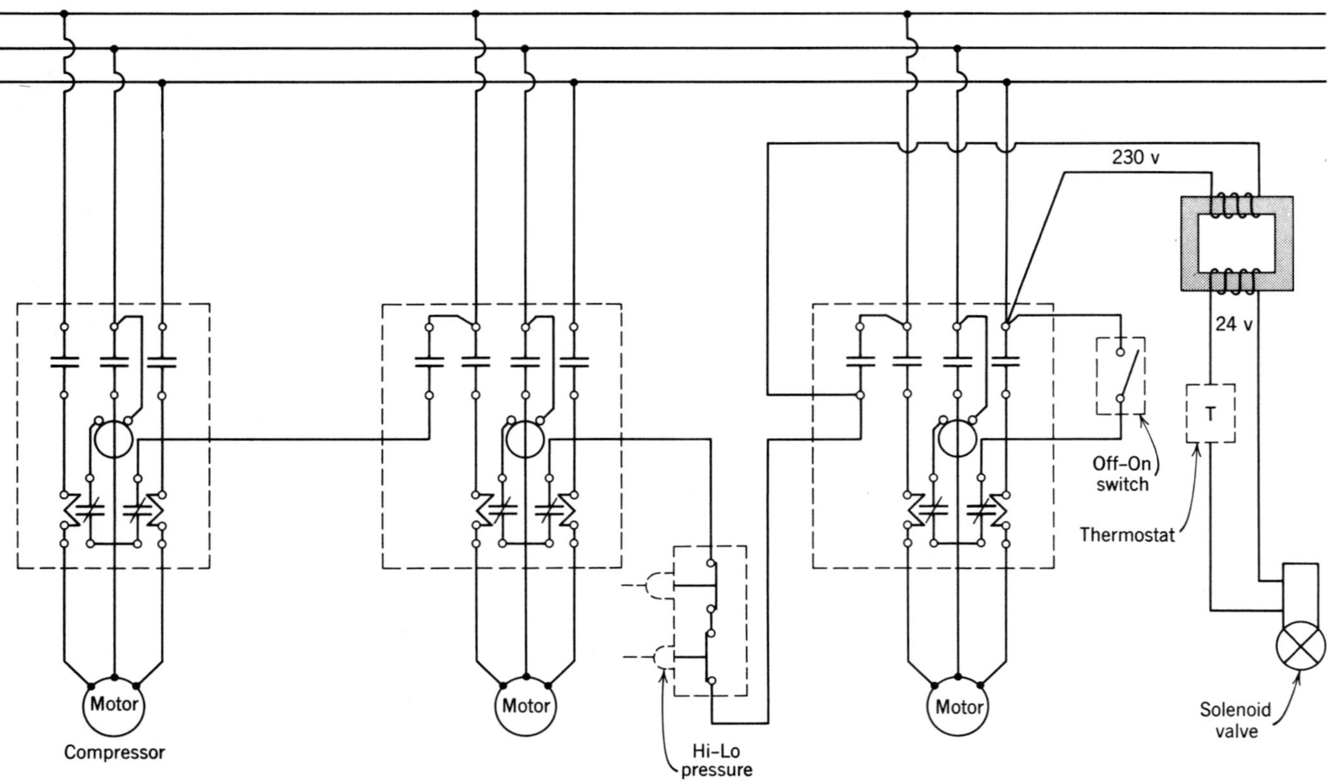

Fig. 21-20 Pump-down system. A time delay relay should be installed in solenoid circuit to prevent solenoid energizing at the same time that the evaporator blower is started.

Tables

TABLE R-1 Natural Convection Cooling Coils Capacities Btu × hr/in. Finned Length

Model	(7″ less than overall) Width	No. Tubes	Fin Spac.	Surface Sq Ft per in.	Btu per Hr per in. 1° TD	Btu per Hr per in. 15° TD
		Single Row High Coils				
J-14	13″	4	1/3	1.26	1.47	22.1
			1/2	0.86	1.21	18.2
K-16	18½″	6	1/3	1.89	2.21	33.1
			1/2	1.29	1.82	27.3
L-18	24″	8	1/3	2.52	2.94	44.1
			1/2	1.72	2.44	35.5
N-1.10	29½″	10	1/3	3.15	3.68	55.2
			1/2	2.15	3.04	45.9
J-18	30″	8	1/3	2.52	2.94	44.2
			1/2	1.72	2.44	36.6
K-1.12	42″	12	1/3	3.78	4.42	66.3
			1/2	2.58	3.65	54.8
L-1.16	54″	16	1/3	5.04	5.88	88.4
			1/2	3.44	4.87	73.2
PJ-14	13″*	8	1/3	2.52	2.94	44.2
			1/2	1.72	2.44	36.6
PK-16	18½″*	12	1/3	3.78	4.42	66.2
			1/2	2.58	3.65	54.8
PL-18	24″*	16	1/3	5.04	5.88	88.3
			1/2	3.44	4.87	73.0
PN-1.10	29½*	20	1/3	6.30	7.33	110.0
			1/2	4.30	6.10	91.1
		Two Row High Coils				
J-24	13″	8	1/3	2.52	2.65	39.8
			1/2	1.72	2.23	33.5
K-26	18½*	12	1/3	3.78	3.96	59.4
			1/2	2.58	3.34	50.1
L-28	24″	16	1/3	5.04	5.30	79.5
			1/2	3.44	4.45	66.8
N-2.10	29½″	20	1/3	6.30	6.62	99.3
			1/2	4.30	5.57	83.6
J-28	30″	16	1/3	5.04	5.30	79.5
			1/2	3.44	4.45	66.8
K-2.12	42″	24	1/3	7.56	7.92	118.8
			1/2	5.16	6.68	100.2
L-2.16	54″	32	1/3	10.08	10.60	159.0
			1/2	6.88	8.90	133.5
PJ-24	13″*	16	1/3	5.04	5.30	79.5
			1/2	3.44	4.45	66.8
PK-26	18½″*	24	1/3	7.56	7.92	118.8
			1/2	5.16	6.68	100.2
PL-28	24″*	32	1/3	10.08	10.60	159.0
			1/2	6.88	8.90	133.5
PN-2.10	29½″*	40	1/3	12.60	13.25	198.8
			1/2	8.60	11.12	166.8

* Width of each section.
Courtesy Dunham-Bush, Inc.

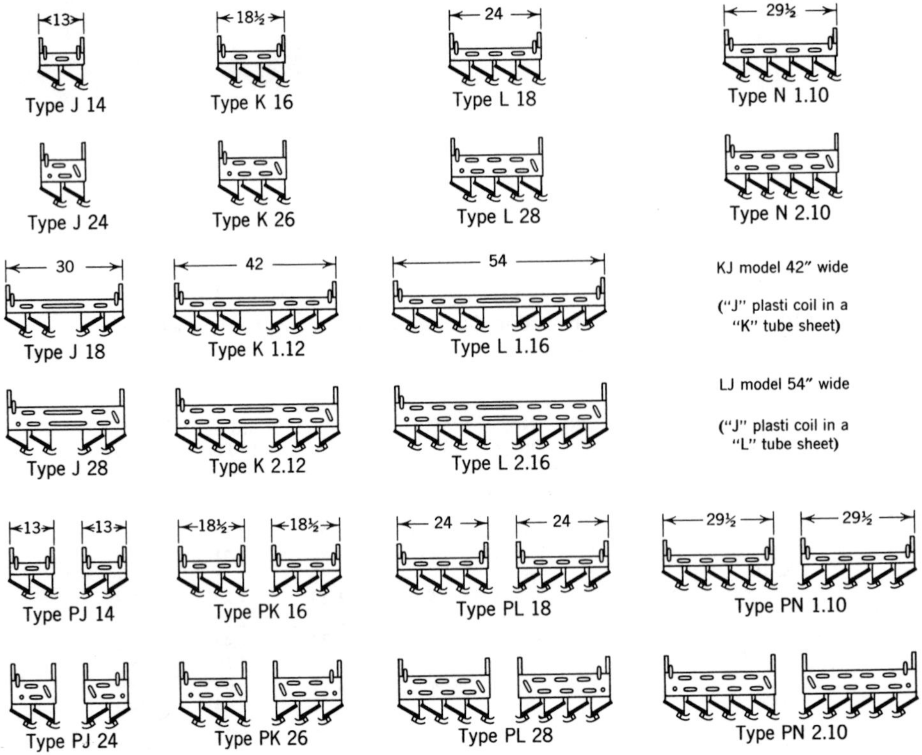

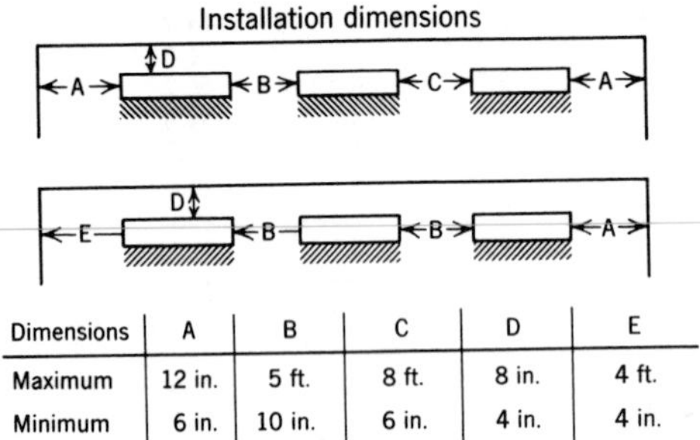

Dimensions	A	B	C	D	E
Maximum	12 in.	5 ft.	8 ft.	8 in.	4 ft.
Minimum	6 in.	10 in.	6 in.	4 in.	4 in.

Note:
If the width of the box requires additional plasti-units, care should be taken to see that they are installed in a similar manner as shown above.

Fig. R-1

TABLE R-2 Single Plate Evaporators

Catalog Number	Size, Inches	No. Plates	Feet of Pass	No. of Expansion Valves— R-12	No. of Expansion Valves— Ammonia	Total Btu's per Hour at 15° TD	
						Below 32°	Above 32°
4-1248-B	12 × 48	4	87	1	1	1260	1440
5-1248-B	12 × 48	5	109	1	1	1575	1880
6-1248-B	12 × 48	6	131	1	1	1890	2160
4-1260-B	12 × 60	4	111	1	1	1560	1800
5-1260-B	12 × 60	5	139	1	1	1950	2250
6-1260-B	12 × 60	6	166	1	1	2340	2700
4-1272-B	12 × 72	4	136	1	1	1860	2160
5-1272-B	12 × 72	5	170	1	1	2325	2700
6-1272-B	12 × 72	6	204	2	1	2790	3240
4-1284-B	12 × 84	4	160	1	1	2220	2520
5-1284-B	12 × 84	5	200	2	1	2775	3150
6-1284-B	12 × 84	6	240	2	1	3300	3780
4-12108-B	12 × 108	4	207	2	1	2880	3240
5-12108-B	12 × 108	5	258	2	1	3600	4050
6-12108-B	12 × 108	6	310	2	1	4320	4860
4-12144-B	12 × 144	4	279	2	1	3780	4320
5-12144-B	12 × 144	5	348	2	1	4725	5400
6-12144-B	12 × 144	6	418	2	1	5670	6480

Courtesy Kold-Hold Division, Tranter Manufacturing Co.

TABLE R-3 Plate Banks

Catalog Number	Width, Inches	Length, Inches	Feet of Pass	Sq Ft Surface	Total "K" per Plate		Total Btu's per Hour at 15° TD	
					Below 32°	Above 32°	Below 32°	Above 32°
1224	12	24	9.9	4.18	10	13	150	195
1236	12	36	15.8	6.26	16	18	240	270
1248	12	48	21.8	8.35	21	24	315	360
1260	12	60	27.8	10.45	26	30	390	450
1272	12	72	33.8	12.55	31	36	465	540
1284	12	84	39.8	14.65	37	42	555	630
12108	12	108	51.8	18.83	48	54	720	810
12144	12	144	69.8	25.2	63	72	945	1080
2230	22	30	25.7	9.64	24	28	360	420
2236	22	36	31.6	11.58	29	33	435	495
2248	22	48	43.6	15.4	38	44	570	660
2260	22	60	55.6	19.3	48	55	720	825
2272	22	72	67.6	23.2	58	66	870	990
2284	22	84	79.6	27.24	68	79	1020	1185
22108	22	108	103.6	34.9	87	100	1305	1500

Courtesy Kold-Hold Division, Tranter Manufacturing Co.

TABLE R-4 "K" Factors for Bare Pipe Coils in Liquid* Btu/(hr)(ft²)(°F)

Desired Liquid Temp.	Refr. Temp.	"K"	Desired Liquid Temp.	Refr. Temp.	"K"
65° (*a*)	38°	15.7	35° (*c*)	19°	13.5
60 (*a*)	38	15.5	30 (*c*)	15	13.0
55 (*a*)	38	15.2	25 (*c*)	11	12.5
50 (*a*)	36	15.0	20 (*c*)	7	12.0
45 (*a*)	32	14.5	15 (*c*)	3	11.2
40 (*a*)	28	14.0	10 (*c*)	−1	10.5
35 (*a*)	24	13.5	5 (*c*)	−5	9.8
35 (*b*)	19	12.5	0 (*c*)	−9	9.0
35 (*b*)	15	10.8	−5 (*c*)	−12	8.2
35 (*b*)	11	10.0	−10 (*c*)	−16	7.5
35 (*b*)	7	9.0			

(*a*) Water cooling. (*b*) Water cooling, ice formation on coils. (*c*) Brine cooling.

* For dry expansion tubing or pipe submerged in water or brine without agitation. (Courtesy Vilter Manufacturing Company.)

TABLE R-5 "K" Factors for Bare Pipe Coils in Air* Btu × (hr)(ft²)(°F)

Refrig. Temp.	Room Temperature Degrees Fahrenheit										
°F	−20°	−10°	0°	10°	20°	30°	36°	40°	44°	50°	60°
32										2.30	2.49
28									2.11	2.50	2.52
24								2.11	2.49	2.51	2.52
20							2.11	2.49	2.49	2.47	
14						1.79	2.50	2.52	2.48	2.52	
12						1.80	2.49	2.49	2.52		
9					1.40	1.79	2.50	2.49	2.49		
6					1.39	2.01	2.48	2.51			
3					1.40	1.99	2.48	2.53			
0				1.39	1.59	1.99	2.51				
−4				1.39	1.59	1.99	2.50				
−8			1.30	1.49	1.80	2.01					
−13			1.39	1.60	1.74	1.98					
−17			1.50	1.70	1.79						
−25		1.50	1.60	1.80							
−30	1.39	1.70	1.79	1.80							
−40	1.59	1.80	1.80								
−50	1.79	1.80									

* For iron pipe coils with gravity air circulation. (Courtesy Vilter Manufacturing Company.)

TABLE R-6 Lineal Foot of Pipe per Square Foot of External Surface

Pipe Size	Lineal Feet
$\frac{1}{2}''$	4.55
$\frac{3}{4}''$	3.64
1″	2.90
$1\frac{1}{4}''$	2.30
$1\frac{1}{2}''$	2.01
2″	1.61

Courtesy Vilter Manufacturing Company.

TABLE R-7 Prestfin Capacities in Btu/(hr)(ft²)(°TD)

Refrigerant Temperature °F	Natural Air Circulation		Forced Air 300–500 fpm Face Vel.	
	Flooded or Brine	Thermal Valve	Flooded or Brine	Thermal
Refrig. above 32°	2.3	2.1	3.0	2.7
Refrig. 32° to 0°	1.6	1.4	2.0	1.8
Refrig. −1° and lower	1.2	1.1	1.6	1.5

One lineal foot of Prestfin pipe provides 8.1 ft² of surface area.

TABLE R-8 Unit Cooler Capacity Ratings and Specifications

	Btu/Hr Rating		Core			Motor and Fan					
Model	10° TD	15° TD	Surf. FT²	Circuits	Tube	HP	Motor Heat Btu/24 Hr	Fan	Rpm	Cfm	Air Throw
UC25	2,500	3,750	67	1	$\frac{1}{2}''$	$\frac{1}{25}$	7,600	10″	1500	390	20
UC35	3,500	5,250	93	1	$\frac{1}{2}''$	$\frac{1}{25}$	8,000	12″	1500	510	20
UC45	4,500	6,750	156	1	$\frac{1}{2}''$	$\frac{1}{20}$	11,500	14″	1500	700	17
UC65	6,500	9,750	210	1	$\frac{5}{8}''$	$\frac{1}{12}$	12,600	16″	1140	1,000	23
UC85	8,500	12,750	266	1	$\frac{3}{4}''$	$\frac{1}{12}$	13,350	16″	1140	1,480	27
UC105	10,500	15,750	328	Split	$\frac{3}{4}''$	$\frac{1}{6}$	15,100	18″	1140	1,730	25
UC120	12,000	18,000	378	Split	$\frac{3}{4}''$	(2) $\frac{1}{20}$	18,000	(2) 14″	1140	1,950	25
UC180	18,000	27,000	566	Split	$\frac{3}{4}''$	(2) $\frac{1}{12}$	25,200	(2) 16″	1140	2,550	20
UC240	24,000	36,000	755	Two	$\frac{3}{4}''$	(2) $\frac{1}{6}$	34,000	(2) 18″	1140	4,050	27
UC320	32,000	48,000	1030	3	$\frac{3}{4}''$	(2) $\frac{1}{4}$	75,000	(2) 22″	1140	6,000	28

Courtesy Dunham-Bush Inc.

TABLE R-9 Heavy-duty Product Coolers

Model No.	CFM at 600 FPM	Sq. Ft. Face	Rows Deep	CAPACITY – BTUH / F° TD								Total Coil Surface Sq. ft.		Int. Coil Vol. Cu. Ft.	Centrifugal Blower Wheel Data		GPM Req'd. for Water Defr. at 7 PSIG Nozzle Press.
				Above + 32° F Evap. Temp.				Below + 32° F Evap. Temp.									
				3 fins/inch		4 fins/inch		3 fins/inch		4 fins/inch							
				DX	Flood. Recir.	DX	Flood. Recir.	DX	Flood. Recir.	DX	Flood. Recir.	3 fins/inch	4 fins/inch		No.	Size, In.	
BL-1674			4	1395	1672	1540	1850	1240	1490	1360	1630	321	406	.43			16.2
BL-1676	4020	6.7	6	1925	2310	2100	2520	1740	2080	1890	2270	482	610	.65	1	13	16.2
BL-1678			8	2370	2840	2570	3080	2160	2590	2330	2800	631	812	.86			16.2
BL-2114			4	2290	2750	2530	3030	2040	2450	2230	2680	540	684	.90			24
BL-2116	6600	11.0	6	3160	3790	3470	4160	2850	3420	3100	3720	811	1025	1.34	2	13	24
BL-2118			8	3880	4650	4230	5075	3550	4260	3830	4600	1080	1368	1.80			24
BL-3174			4	3530	4240	3910	4690	3240	3880	3450	4140	812	1028	1.27			37.8
BL-3176	10200	17.0	6	4880	5860	5350	6420	4400	5280	4800	5750	1215	1535	1.91	2	14-5/8	37.8
BL-3178			8	6000	7200	6520	7820	5490	6580	5920	7100	1622	2050	2.54			37.8
BL-31710			10	6960	8350	7480	8960	6400	7680	6860	8230	2028	2565	3.18			36
BL-4254			4	5200	6240	5750	6900	4630	5550	5075	6090	1192	1505	1.87			46
BL-4256			6	7180	8620	7860	9440	6480	7780	7050	8450	1790	2260	2.81			46
BL-4258	15000	25.0	8	8825	10600	9600	11500	8075	9700	8700	10450	2380	3010	3.76	2	19-1/2	46
BL-42510			10	10250	12300	11000	13200	9420	11300	10100	12100	2980	3770	4.70			46
BL-42512			12	11380	13650	12150	14550	10600	12700	11300	13550	3580	4520	5.65			46
BL-5364			4	7360	8840	8150	9775	6550	7850	7150	8580	1690	2140	2.59			55.2
BL-5366			6	10150	12200	11150	13400	9160	11000	10000	12000	2535	3210	3.86			55.2
BL-5368	21300	35.4	8	12500	15000	13600	16300	11400	13700	12300	14750	3380	4280	5.16	2	22-3/4	55.2
BL-53610			10	14500	17400	15600	18700	13330	16000	14300	17150	4225	5350	6.46			55.2
BL-53612			12	16100	19300	17200	20600	15000	18000	16050	19250	5070	6420	7.75			55.2
BL-6464			4	9560	11500	10600	12700	8500	10200	9325	11200	2198	2785	3.36			80.2
BL-6466			6	13200	16850	14500	17400	11930	14330	13950	16750	3298	4180	5.04			80.2
BL-6468	27600	46.0	8	16250	19500	17700	21250	14880	17850	16000	19200	4396	5570	6.22	2	26	80.2
BL-64610			10	18900	22700	20320	24350	17400	20900	18600	22300	5498	6970	8.40			86.4
BL-64612			12	21000	25200	22300	26700	19600	23500	20800	25000	6596	8360	10.10			86.4
BL-7736			6	20900	25100	23100	27700	18900	22700	20600	24700	5220	6610	7.73			118.8
BL-7738	43800	72.9	8	25600	30800	28000	33600	23500	28200	25400	30500	6960	8820	10.30			118.8
BL-77310			10	29900	35900	32200	38700	27600	33100	29400	35300	8700	11000	12.85	3	26	118.8
BL-77312			12	33200	39800	35500	42500	30900	37100	33000	39600	10420	13200	15.40			118.8

DX LOW TEMPERATURE CAPACITY CORRECTION TABLE FOR REFRIGERANTS R12, R22, R502	
SUCTION	FACTOR
– 20°	1.0
– 30°	.9
– 40°	.8
– 50°	.7
– 60°	.6

Finned Coil Capacity Correction Factors for Various Rates of Air Flow				
ROW DEPTH DIRECTION OF AIR FLOW	Coil Face Velocity, FPM			
	400	500	600	700
4	.82	.91	1.0	1.07
6	.80	.90	1.0	1.08
8	.79	.89	1.0	1.09
10	.77	.88	1.0	1.10
12	.75	.87	1.0	1.10

A. Capacity is based on sensible heat removal only and a 600 FPM face velocity. Temperature difference is the temperature of the air entering the coil and the coil evaporating temperature.

B. Direct expansion ratings are applicable for evaporating temperatures of −20°F and above. For evaporating temperatures below −20°F, apply the capacity correction multiplier for the temperature required. (Example: The direct expansion rating for the BL-31710-3 at a −40° evaporator temperature using R-502 will be 6400 BTU's/HR/°FTD X .8 = 5120 BTU's/HR/°TD.)

Note: **Do not use direct expansion ammonia below 0°F evaporating temperatures unless the compressor system is protected and designed to handle the overfed liquid by the use of a suction trap or accumulator.**

C. For Brine Systems, consult the factory for the correct rating. Give capacity required, type of brine, room and brine temperature and GPM available. Warning: **Do not attempt to use the flooded rating for brine systems.**

D. To re-rate the base CFM ratings for other than 600 FPM, use the multiplier shown and the fin coil capacity correction factor table. (Example: A BL-31710-4-DXR12 operating at a +35° coil temperature has a basic rating or 7480 BTU's/HR/°TD; however, the air requirements will be 6800 CFM of air and 400 FPM face velocity. The revised rating would then be 7480 BTU's/HR/°TD x .77 = 5760 BTU's/HR/°TD.)

TABLE R-10 Water Chillers

Acme Model No.*	Effective Tube Area Sq Ft	Capacity Range Tons	Std. No. of Circuits	Number of Tubes
DXH-805	41			
806	50			
807	59			
808	67			
809	76	5.4 to 19	1	44
810	85			
811	94			
812	102			
813	111			
814	119			
DXH-1005	64			
1006	77			
1007	91			
1008	104			
1009	118			
1010	131	8 to 30	1	68
1011	145			
1012	158			
1013	171			
1014	184			
1015	197			
1016	211			
DXH-1206	105			
1207	123			
1208	141			
1209	159			
1210	177			
1211	195	11 to 40	2	92
1212	213			
1213	231			
1214	249			
1215	267			
1216	285			
DXH-1406	136			
1407	159			
1408	184			
1409	207			
1410	231			
1411	255	14 to 52	2	120
1412	278			
1413	302			
1414	325			
1415	349			
1416	372			
DXH-1606	187			
1607	219			
1608	251			
1609	283			
1610	315			
1611	348	20 to 71	2	164
1612	380			
1613	412			
1614	444			
1615	477			
1616	509			
DXH-2006	286			
2007	335			
2008	384			
2009	437			
2010	487			
2011	535			
2012	583	31 to 111	2	252
2013	633			
2014	683			
2015	733			
2016	782			
2017	830			
2018	880			
2020	976			

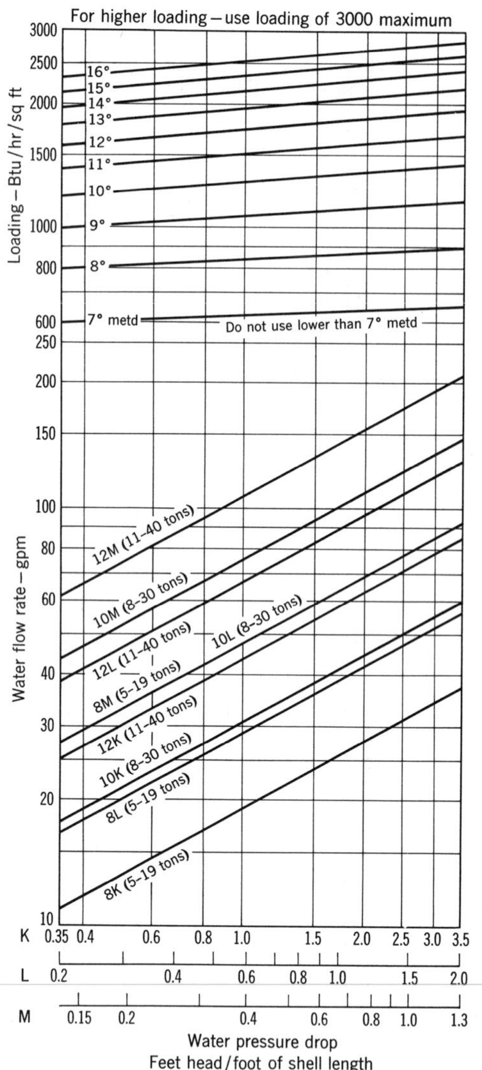

Fig. R-10 (Courtesy Acme Industries.)

TABLE R-11 Open Compressor—Refrigerant R-12

Capacity (extraction) ratings are based on: suction line temperature 65°F except where a lower limit for suction line temperature is indicated.
No liquid subcooling

1740 RPM

Evap. Temp. °F	Cond. Temp. °F	E027				E036				E050			
		Cap (Tons)	Shaft BHP	THR (Tons)	Max. Suct. Line Temp.	Cap (Tons)	Shaft BHP	THR (Tons)	Max. Suct. Line Temp.	Cap (Tons)	Shaft BHP	THR (Cap)	Max. Suct. Line Temp.
					°F				°F				°F
−40	80	2.50	7.01	3.5	65	3.3	9.3	4.2	55	4.5	13.7	5.8	58
	100	2.00	8.2	3.0	49	2.7	11.0	3.7	48	3.7	15.0	5.1	48
	120	1.60	9.5	2.5	30	2.2	12.5	3.3	31	2.9	16.4	4.3	30
−30	80	3.76	9.0	4.8	65	5.0	12.0	6.2	65	6.8	16.3	8.4	65
	100	3.04	10.0	4.6	65	4.1	13.0	5.5	65	5.4	18.3	7.2	65
	120	2.40	11.4	4.4	45	3.3	15.0	4.7	45	4.5	20.6	6.4	40
−20	80	4.75	10.3	6.2	90	6.3	13.5	7.7	90	8.8	19.6	11	90
	100	4.10	11.4	5.6	85	5.3	15.0	7.1	73	7.4	22.6	9.6	88
	120	3.40	12.8	4.9	72	4.5	17.2	6.2	68	6.3	25.2	8.6	70
−10	80	6.40	11.4	8.1	90	8.5	15.2	10.1	90	11.2	21.6	13.3	90
	100	5.60	13.0	7.5	88	7.5	18.0	9.5	90	10.2	24.9	12.9	90
	120	4.70	14.7	6.7	85	6.2	20.0	8.2	86	9.5	28.8	12.7	88
0	80	8.40	12.8	10.2	90	10.71	17.7	12.6	90	14.7	24.5	17.2	90
	100	7.34	14.4	9.6	90	9.8	20.0	12.0	90	13.4	28.1	16.6	90
	120	6.30	16.7	9.6	90	8.4	22.9	11.0	90	11.8	32.1	15.4	90
+10	80	10.6	13.9	12.9	90	13.8	20.0	15.9	90	18.8	26.5	21.4	90
	100	9.42	15.5	12.1	90	12.0	22.6	14.6	90	16.8	31.0	20.5	90
	120	8.16	17.9	11.3	90	11.0	25.8	14.0	90	14.8	36.6	18.8	90
	140	7.01	20.4	10.7	90	9.2	29.4	12.4	90	13.4	38.9	17.8	90

TABLE R-11 *(Continued)*

Evap. Temp. °F	Cond. Temp. °F	E027				E036				E050			
		Cap (Tons)	Shaft BHP	THR (Tons)	Max. Suct. Line Temp.	Cap (Tons)	Shaft BHP	THR (Tons)	Max. Suct. Line Temp.	Cap (Tons)	Shaft BHP	THR (Cap)	Max. Suct. Line Temp.
					°F				°F				°F
+20	80	13.0	14.5	14.8	90	17.3	22	19.9	90	25.4	28.6	29.0	90
	100	11.8	17.2	14.3	90	15.8	24.8	19.1	90	23.9	32.7	28.7	90
	120	10.4	19.4	13.1	90	14.1	28.6	17.8	90	22.2	38.3	28.0	90
	140	9.3	22.2	12.2	90	12.4	32.4	16.4	90	20.8	41.5	27.2	90
+30	80	16.3	15.7	18.5	90	21.3	23.5	24.3	90	31.3	30.4	35.4	90
	100	14.5	18.1	17.5	90	19.5	27.5	23.2	90	28.7	35.7	33.9	90
	120	12.8	20.4	16.0	90	17.0	28.9	21.4	90	26.5	40.9	32.9	90
	140	11.1	23.7	14.5	90	14.5	35	18.9	90	23.8	44.5	30.7	90
+40	80	19.8	16.3	22.4	90	27.5	25.2	31.4	90	38.2	32.4	42.8	90
	100	18.7	19.6	22.0	90	25.0	28.6	29.5	90	36.0	36.9	42.5	90
	120	17.4	22.0	21.5	90	23.2	32.7	28.8	90	34.1	43.5	41.9	90
	140	16.0	25.7	20.5	90	21	37.3	27.0	90	32.2	47.11	41.2	90
+50	80	25.2	17.2	28.2	90	35.6	26.5	37	90	47.4	33.4	53.1	90
	100	22.3	20.1	26.3	90	31.9	29.8	35.2	90	42.8	38.9	50.1	90
	120	19.8	23.2	24.5	90	28.5	34.0	33.0	90	38.5	45.5	47.3	90
	140	17.4	27.3	22.3	90	25.7	38.9	30.7	90	34.8	49.0	44.5	90

(1) Rated suction gas temperature of 65°F except where a lower limit is indicated.

Rating Basis Factor

Difference between actual suction temp. and rated suction temp. °F	0	10	20	30	40	50	60	70	80	90
Capacity multiplier	1.00	0.99	0.98	0.97	0.96	0.95	0.94	0.93	0.915	0.90

The above factors apply to R-12 and R-502. For R-22 the capacity figures in the tables should be used unchanged as variation in suction superheat (provided it is usefully obtained) has a negligible effect on evaporator duty.

(2) No liquid subcooling.
For every 2°F of liquid subcooling at the condensing outlet, 1% may be added to capacity figure. When adjusting for subcooling, power input does not change.

(3) Rated RPM of 1740.

Multiplying factors for other RPM

RPM	1620	1450	1260	1050
Capacity factor	0.93	0.85	0.72	0.60
BHP factor	0.96	0.90	0.78	0.65

(4) Capacities shown on Table are full capacities.

Capacity Control Steps

Capacity Step % of full capacity	100	83	67	50
Number of active cylinders	.6	5	4	3
Power input % of full cap. power	100	85–87	70–74	55–60

TABLE R-12 Air Cooled Condensing Units

$\frac{1}{2}$ hp

Application and Speed	Sat. Suction Temp. (F)	Sat. Suction Press. (Psig)	Ambient Air Temperature (F) 80 F Btu/Hr	80 F Disch. Press. Psig	80 F Kw	90 F Btu/Hr	90 F Disch. Press. (Psig)	90 F Kw	100 F Btu/Hr	100 F Disch. Press. (Psig)	100 F Kw
High temp. 1725 rpm	45	41.7	6,760	149	0.73	6,280	169	0.76	5,810	187	0.80
	40	37.0	6,300	143	0.71	5,830	162	0.74	5,480	181	0.78
	35	32.6	5,810	137	0.68	5,370	154	0.71	4,950	175	0.75
	30	28.5	5,340	133	0.65	4,930	148	0.68	4,520	169	0.73
	25	24.6	4,900	126	0.62	4,490	142	0.64	4,100	163	0.70
Med. temp. 1725 rpm	25	24.6	5,780	141	0.73	5,360	157	0.75	4,960	179	0.76
	20	21.1	5,300	136	0.71	4,900	151	0.72	4,500	172	0.74
	15	17.7	4,800	130	0.68	4,420	145	0.69	4,040	165	0.71
	10	14.7	4,320	125	0.66	3,960	141	0.67	3,600	160	0.69
	5	11.8	3,900	119	0.63	3,560	136	0.64	3,220	154	0.66
	0	9.2	3,490	113	0.60	3,190	131	0.61	2,880	148	0.63
Low temp. 1725 rpm	0	9.2	4,020	118	0.60	3,670	134	0.61	3,320	153	0.63
	−5	6.7	3,560	115	0.57	3,250	129	0.58	2,940	149	0.60
	−10	4.5	3,140	112	0.54	2,840	125	0.55	2,600	145	0.57
	−15	2.5	2,740	108	0.51	2,450	120	0.52	2,180	141	0.54
	−20	0.6	2,360	107	0.48	2,090	119	0.49	1,830	138	0.51
	−25	2.28*	2,000	105	0.44	1,750	117	0.45	1,520	135	0.48

TABLE R-12 *(Continued)*

$\frac{3}{4}$ hp

Application and Speed	Sat. Suction Temp. (F)	Sat. Suction Press. (Psig)	80 F Btu/Hr	80 F Disch. Press. (Psig)	80 F Kw	90 F Btu/Hr	90 F Disch. Press. (Psig)	90 F Kw	100 F Btu/Hr	100 F Disch. Press. (Psig)	100 F Kw
High temp. 1725 rpm	45	41.7	9,860	148	1.10	9,160	170	1.11	8,480	186	1.16
	40	37.0	9,130	143	1.02	8,470	162	1.05	7,780	181	1.10
	35	32.6	8,460	137	0.95	7,820	154	0.99	7,200	175	1.04
	30	28.5	7,850	132	0.90	7,220	148	0.95	7,610	169	0.99
	25	24.6	7,250	126	0.85	6,650	142	0.90	6,060	162	0.95
Med. temp. 1725 rpm	25	24.6	7,960	140	1.10	7,380	155	1.10	6,830	175	1.13
	20	21.1	7,350	134	0.98	6,780	149	1.04	6,250	169	1.07
	15	17.7	6,610	128	0.95	6,090	142	0.98	5,560	162	1.00
	10	14.7	6,000	123	0.90	5,500	139	0.93	5,000	157	0.96
	5	11.8	5,450	117	0.85	4,980	133	0.88	4,500	151	0.91
	0	9.2	4,900	111	0.80	4,490	127	0.83	4,050	145	0.86
Low temp. 1725 rpm	0	9.2	5,720	118	0.89	5,200	138	0.91	4,730	154	0.92
	−5	6.7	5,100	116	0.83	4,640	133	0.85	4,200	151	0.87
	−10	4.5	4,540	114	0.78	4,090	129	0.80	3,720	148	0.82
	−15	2.5	4,030	111	0.72	3,600	124	0.74	3,200	144	0.76
	−20	0.6	3,550	109	0.67	3,140	122	0.69	2,770	141	0.71
	−25	2.28*	3,090	106	0.62	2,710	120	0.64	2,360	138	0.66

* Inches of Mercury Vacuum.

NOTES:

1. Refrigeration effect is given in Btu per hour. To obtain tons of refrigeration effect, divide by 12,000.

2. Refrigeration effect values given are based upon an actual suction gas temperature of 80°. To obtain this gas temperature usually requires the use of a liquid-suction interchanger.

3. Selection of condensing unit should be made on the basis of the maximum air temperature surrounding the condensing unit. When condenser is not connected directly to outdoors, or where poor ventilation and heat dissipation exists, an ambient temperature higher than the maximum outdoor temperature should be the basis for selection.

4. Operation at suction temperatures lower than those shown is permissible. Operation at suction temperatures higher than those shown will result in overloading of the compressor motor. When low or medium temperature range unit selections are made, it is usually necessary to use some form of suction pressure control to prevent overloading of the compressor motor during pull-down or other abnormal conditions producing a high suction pressure.

5. Power input to motor is given in kilowatts, and includes the power required by the condenser fan. Kilowatt values given are for single phase 60 cycle a-c and d-c motors. To obtain approximate Bhp, divide Kw by 1.04.

TABLE R-12 *(Continued)*

	Sat. Suction		1 hp Ambient Air Temperature (F)								
			80 F			90 F			100 F		
Application and Speed	Temp. (F)	Press. (Psig)	Btu/Hr	Disch. Press. (Psig)	Kw	Btu/Hr	Disch. Press. (Psig)	Kw	Btu/Hr	Disch. Press. (Psig)	Kw
High temp. 1725 rpm	45	41.7	13,430	151	1.50	12,500	172	1.55	11,580	190	1.61
	40	37.0	12,450	145	1.41	11,540	164	1.46	10,660	184	1.51
	35	32.6	11,680	139	1.32	10,790	156	1.36	9,920	177	1.40
	30	28.5	10,750	133	1.23	9,860	151	1.27	9,060	171	1.31
	25	24.6	9,900	127	1.13	9,080	143	1.17	8,300	164	1.22
Med. temp. 1725 rpm	25	24.6	10,930	142	1.52	10,140	158	1.57	9,400	179	1.62
	20	21.1	10,100	137	1.41	9,340	152	1.46	8,570	172	1.51
	15	17.7	9,250	131	1.30	8,500	145	1.35	7,753	164	1.40
	10	14.7	8,320	125	1.21	7,620	140	1.26	6,920	159	1.31
	5	11.8	7,410	119	1.11	6,790	135	1.16	6,120	154	1.21
	0	9.2	6,360	113	1.01	5,830	130	1.06	5,260	149	1.11
Low temp. 1725 rpm	0	9.2	8,000	120	1.21	7,300	138	1.24	6,620	157	1.27
	−5	6.7	7,220	118	1.14	6,580	134	1.17	5,960	153	1.20
	−10	4.5	6,400	116	1.07	5,790	130	1.10	5,200	149	1.13
	−15	2.5	5,730	113	0.99	5,120	125	1.02	4,550	145	1.05
	−20	0.6	5,000	110	0.92	4,410	123	0.95	3,880	142	0.98
	−25	2.28*	4,300	106	0.86	3,770	120	0.89	3,290	138	0.92

TABLE R-12 *(Continued)*

$1\frac{1}{2}$ hp

Application and Speed	Sat. Suction Temp. (F)	Sat. Suction Press. (Psig)	80 F Btu/Hr	80 F Disch. Press. (Psig)	80 F Kw	90 F Btu/Hr	90 F Disch. Press. (Psig)	90 F Kw	100 F Btu/Hr	100 F Disch. Press. (Psig)	K
			Ambient Air Temperature (F)								
	45	41.7	20,820	148	1.98	19,370	166	2.07	17,920	188	2.18
	40	37.0	19,460	144	1.90	17,900	160	1.98	16,660	182	2.10
High temp. 1725 rpm	35	32.6	18,110	139	1.82	16,720	154	1.90	15,400	175	2.01
	30	28.5	16,670	133	1.75	15,360	148	1.83	14,080	169	1.94
	25	24.6	15,220	126	1.68	13,970	142	1.76	12,760	163	1.87
	25	24.6	15,220	126	1.68	13,970	142	1.76	12,760	163	1.87
	20	21.1	13,700	122	1.62	12,630	137	1.70	11,630	158	1.81
Med. temp. 1725 rpm	15	17.7	12,300	118	1.56	11,310	132	1.64	10,330	152	1.74
	10	14.7	11,050	115	1.51	10,140	128	1.59	9,210	147	1.68
	5	11.8	9,800	111	1.45	8,960	123	1.53	8,090	142	1.62
	0	9.2	8,700	107	1.40	7,870	139	1.48	7,010	137	1.56
	0	9.2	11,300	120	1.73	10,240	135	1.78	9,360	154	1.85
	−5	6.7	10,100	116	1.61	9,200	131	1.67	8,320	149	1.73
Low temp. 1725 rpm	−10	4.5	9,040	113	1.50	8,070	128	1.56	7,330	145	1.62
	−15	2.5	7,970	109	1.39	7,120	124	1.45	6,330	140	1.51
	−20	0.6	6,940	107	1.31	6,150	122	1.37	5,420	137	1.43
	−25	2.28*	5,900	104	1.22	5,170	119	1.28	4,510	134	1.34

TABLE R-12 *(Continued)*

			2 hp Ambient Air Temperature (F)								
	Sat. Suction		80 F			90 F			100 F		
Application and Speed	Temp. (F)	Press. (Psig)	Btu/Hr	Disch. Press. (Psig)	Kw	Btu/Hr	Disch. Press. (Psig)	Kw	Btu/Hr	Disch. Press. (Psig)	Kw
High temp. 1725 rpm	45	41.7	27,260	153	2.69	25,370	173	2.83	23,500	194	2.96
	40	37.0	25,490	148	2.60	23,430	167	2.72	21,830	188	2.85
	35	32.6	23,720	142	2.50	21,880	160	2.61	20,160	181	2.74
	30	28.5	21,860	136	2.40	20,100	154	2.51	18,450	175	2.64
	25	24.6	20,000	130	2.30	18,320	147	2.40	16,750	168	2.53
Med. temp. 1725 rpm	25	24.6	20,000	130	2.30	18,320	147	2.40	16,750	168	2.53
	20	21.1	17,960	126	2.20	16,610	142	2.30	14,300	162	2.42
	15	17.7	16,160	122	2.10	14,870	137	2.20	13,580	156	2.31
	10	14.7	14,490	119	2.00	13,310	133	2.10	12,100	151	2.22
	5	11.8	12,820	115	1.90	11,750	128	2.00	10,600	145	2.12
	0	9.2	11,300	112	1.80	10,400	124	1.90	9,300	140	2.02
Low temp. 1725 rpm	0	9.2	15,870	122	2.43	14,650	138	2.51	13,360	158	2.60
	− 5	6.7	14,220	118	2.28	12,960	134	2.36	11,720	153	2.44
	−10	4.5	12,690	115	2.14	11,410	130	2.22	10,290	148	2.30
	−15	2.5	11,160	111	1.99	9,970	126	2.07	8,850	143	2.15
	−10	0.6	9,750	109	1.87	8,640	124	1.95	7,610	140	2.03
	−25	2.28*	8,340	107	1.75	7,310	121	1.83	6,360	136	1.91

TABLE R-12 *(Continued)*

Application and Speed	Sat. Suction Temp. (F)	Sat. Suction Press. (Psig)	3 hp Ambient Air Temperature (F) 80 F Btu/Hr	80 F Disch. Press. (Psig)	80 F Kw	90 F Btu/Hr	90 F Disch. Press. (Psig)	90 F Kw	100 F Btu/Hr	100 F Disch. Press. (Psig)	100 F Kw
	45	41.7	43,050	152	4.02	40,000	173	4.20	37,050	191	4.40
	40	37.0	30,230	147	3.86	37,000	165	4.03	34,430	185	4.24
High temp. 1750 rpm	35	32.6	37,400	141	3.70	34,560	158	3.86	31,800	179	4.07
	30	28.5	34,500	136	3.56	31,750	152	3.72	29,150	173	3.93
	25	24.6	31,600	130	3.42	28,950	146	3.58	26,450	166	3.78
	25	24.6	31,600	130	3.42	28,950	146	3.58	26,450	166	3.78
	20	21.1	28,350	126	3.30	26,250	141	3.46	24,120	160	3.66
Med. temp. 1750 rpm	15	17.7	25,550	122	3.17	23,500	136	3.34	21,460	154	3.53
	10	14.7	22,900	119	3.07	21,000	131	3.23	19,100	149	3.43
	5	11.8	20,200	115	2.96	18,510	127	3.13	16,700	143	3.32
	0	9.2	17,000	112	2.50	15,500	123	3.03	13,900	138	3.20
	0	9.2	20,080	120	3.08	18,480	138	3.20	16,800	159	3.40
	−5	6.7	17,980	116	2.86	16,380	133	2.98	14,810	152	3.10
Low temp. 1750 rpm	−10	4.5	16,020	113	2.66	14,410	129	2.78	13,000	146	2.90
	−15	2.5	14,050	110	2.46	12,550	125	2.58	11,170	141	2.70
	−20	0.6	12,240	109	2.29	10,850	123	2.42	9,560	138	2.54
	−25	2.28*	10,420	107	2.13	9,150	120	2.25	7,960	135	2.37

* Inches of Mercury Vacuum.

NOTES:

1. Refrigeration effect is given in Btu per Hour. To obtain tons of refrigeration effect, divide by 12,000.

2. Refrigeration effect values given are based upon an actual suction gas temperature of 80°. To obtain this gas temperature usually requires the use of a liquid-suction interchanger.

3. Selection of condensing unit should be made on the basis of the maximum air temperature surrounding the condensing unit. When condenser is not connected directly to outdoors, or where poor ventilation and heat dissipation exists, an ambient temperature higher than the maximum outdoor temperature should be the basis for selection.

4. Operation at suction temperatures lower than those shown is permissible. Operation at suction temperatures higher than those shown will result in overloading of the compressor motor. When low or medium temperature range unit selections are made, it is usually necessary to use some form of suction pressure control to prevent overloading of the compressor motor during pull-down or other abnormal conditions producing a high suction pressure.

5. Power input is given in kilowatts, and includes the power required by the condenser fan. Kilowatt values given are for three phase 60 cycle a-c motors. To obtain approximate Bhp, divide Kw by 0.93.

TABLE R-13 Water-Cooled Condensing Units

Usage	Sat. Suction Temp. (F)	Sat. Suction Press. (Psig)	$\frac{1}{3}$ hp Btu/Hr	$\frac{1}{3}$ hp Disch. Press. (Psig)	$\frac{1}{3}$ hp Kw	$\frac{1}{2}$ hp Btu/Hr	$\frac{1}{2}$ hp Disch. Press. (Psig)	$\frac{1}{2}$ hp Kw	$\frac{3}{4}$ hp Btu/Hr	$\frac{3}{4}$ hp Disch. Press. (Psig)	$\frac{3}{4}$ hp Kw	1 hp Btu/Hr	1 hp Disch. Press. (Psig)	1 hp Kw
	45	41.7	4,300	130	0.43	7,580	129	0.80	10,890	131	1.25	14,810	133	1.49
High temp.	40	37.0	3,890	127	0.42	6,860	126	0.77	9,860	128	1.20	13,410	130	1.45
75° Water In	35	32.6	3,550	125	0.41	6,250	124	0.75	8,980	126	1.16	12,230	127	1.40
95° Water Out	30	28.5	3,260	123	0.40	5,740	122	0.72	8,250	123	1.01	11,230	125	1.36
	25	24.6	3,000	120	0.39	5,280	120	0.70	7,580	121	0.97	10,320	123	1.32
	25	24.6	3,970	117	0.44	6,400	118	0.70	8,720	119	0.95	12,030	120	1.31
Med. Temp.	20	21.1	3,630	115	0.43	5,860	116	0.68	7,990	117	0.92	11,010	118	1.27
75° Water In	15	17.7	3,300	113	0.42	5,310	113	0.66	7,250	114	0.89	10,000	116	1.23
90° Water Out	10	14.7	2,970	112	0.41	4,770	112	0.64	6,510	112	0.86	8,980	114	1.19
	5	11.8	2,630	110	0.40	4,230	110	0.62	5,770	110	0.83	7,960	112	1.15
	0	9.2	2,290	108	0.39	3,680	109	0.60	5,030	108	0.80	6,940	110	1.11
	0	9.2	2,790	107	0.42	4,360	108	0.62	5,950	111	0.90	8,510	110	1.24
Low temp.	− 5	6.7	2,390	105	0.41	3,740	106	0.60	5,100	108	0.87	7,290	108	1.19
75° Water In	−10	4.5	2,030	103	0.40	3,180	104	0.59	4,330	105	0.80	6,200	106	1.14
85° Water Out	−15	2.5	1,720	101	0.39	2,700	102	0.57	3,670	102	0.76	5,260	103	1.09
	−20	0.6	1,440	99	0.38	2,250	100	0.56	3,060	100	0.71	4,380	101	1.04
	−25	2.28*	1,180	96	0.37	1,850	97	0.54	2,520	97	0.66	3,600	98	0.99

TABLE R-13 *(Continued)*

Usage	Sat. Suction Temp. (F)	Sat. Suction Press. (Psig)	1½ hp Btu/Hr	1½ hp Disch. Press. (Psig)	1½ hp Kw	2 hp Btu/Hr	2 hp Disch. Press. (Psig)	2 hp Kw	3 hp Btu/Hr	3 hp Disch. Press. (Psig)	3 hp Kw
	45	41.7	23,000	131	1.65	29,980	134	2.15	47,300	135	3.25
High temp.	40	37.0	20,800	129	1.60	27,150	131	2.08	42,850	132	3.10
75° Water In	35	32.6	18,970	126	1.55	24,740	128	2.00	39,080	129	2.95
95° Water Out	30	28.5	17,410	124	1.50	22,700	126	1.93	35,850	127	2.82
	25	24.6	16,020	122	1.45	20,850	124	1.85	32,940	124	2.68
	25	24.6	16,020	122	1.45	20,850	124	1.85	32,940	124	2.68
Med. temp.	20	21.1	14,630	119	1.40	19,080	122	1.72	30,120	122	2.56
75° Water In	15	17.7	13,300	116	1.35	17,330	119	1.60	27,350	119	2.43
90° Water Out	10	14.7	11,930	115	1.30	15,670	117	1.52	24,550	117	2.30
	5	11.8	10,580	114	1.25	13,800	115	1.45	21,780	115	2.18
	0	9.2	9,220	113	1.20	12,010	113	1.38	18,980	113	2.06
	0	9.2	12,020	109	1.58	16,870	109	2.09	21,080	112	2.95
Low Temp.	− 5	6.7	10,300	107	1.50	14,460	106	1.96	18,060	109	2.80
75° Water In	−10	4.5	8,740	105	1.43	12,460	104	1.83	15,620	106	2.65
85° Water Out	−15	2.5	7,550	102	1.35	10,660	101	1.70	13,320	103	2.50
	−20	0.6	6,490	100	1.29	9,090	99	1.57	11,410	101	2.35
	−25	2.28*	5,430	97	1.23	7,690	97	1.44	9,610	99	2.20

* Inches of Mercury Vacuum.

NOTES:

1. Refrigeration effect is given in Btu per Hr. To obtain tons of refrigeration, divide by 12,000.

2. Refrigeration effect values given are based upon an actual suction gas temperature of 65° F. To obtain this gas temperature usually requires the use of a liquid-suction interchanger.

3. Operation at suction temperatures lower than those shown is permissible. Operation at suction temperatures higher than shown will result in overloading of the compressor motor. When low or medium temperature range unit selections are made, it is usually necessary to use some form of suction pressure control to prevent overloading of the compressor motor during pull-down or other abnormal conditions producing high suction pressure.

4. For each 10° lower entering water temperature, increase above capacities 6%
For each 10° higher entering water temperature, decrease above capacities 6%

5. Power input to motors is given in kilowatts. To obtain approximate Bhp divide Kw by the factor from the table below:

Motor hp	Factor	Motor hp	Factor
⅓	1.09	1½	0.96
½	1.07	2	0.94
¾	1.04	3	0.92
1	1.02		

6. Condenser water quantity (gpm) at full load $= \dfrac{\text{Btu/hr} \times F}{(\text{leaving water temp.} - \text{entering water temp.}) \times 500}$

Where F = 1.2 for high temperature usage
F = 1.3 for medium temperature usage
F = 1.4 for low temperature usage

TABLE R-14A Air Cooled Condenser Ratings

		Total Heat Rejection at 30°F Temp. Difference					
		R-12 Capacity		R-22 Capacity		R-502 Capacity	
Unit Size	No. Circuits Avail.	Total Unit MBH	BTUH per Circuit	Total Unit MBH	BUTH per Circuit	Total Unit MBH	BTUH per Circuit
3	2	45.0	22,500	48.0	24,000	47.1	23,550
5	2	75.0	37,500	78.0	39,000	76.5	38,250
8	2	120.0	60,000	126.0	63,000	123.0	61,500
9	6	120.0	20,000	126.0	21,000	123.0	20,500
11	8	147.0	18,375	153.0	19,125	150.0	18,750
16	12	216.0	18,000	228.0	19,000	222.0	18,500
18	16	255.0	15,937	267.0	16,687	261.0	16,312
21	16	291.0	18,187	306.0	19,125	300.0	18,750
23	16	330.0	20,625	348.0	21,750	342.0	21,375
27	24	375.0	15,625	396.0	16,500	387.0	16,125
30	21	429.0	20,428	450.0	21,428	441.0	21,000
36	31	510.0	16,452	534.0	17,226	525.0	16,935
41	28	573.0	20,464	603.0	21,536	591.0	21,107
45	(1) 15 (1) 16	633.0	20,419	666.0	21,484	654.0	21,097
53	(2) 21	756.0	18,000	795.0	18,928	780.0	18,571
58	(1) 15 (1) 16	795.0	25,645	834.0	26,903	816.0	26,323
66	(2) 21	945.0	22,500	993.0	23,643	972.0	23,143
82	(2) 21	1,128.0	26,857	1,185.0	28,214	1,161.0	27,642
95	(2) 26	1,476.0	28,385	1,551.0	29,827	1,521.0	29,250
110	(2) 32	1,650.0	25,781	1,731.0	27,047	1,698.0	26,531
115	(2) 39	1,746.0	22,385	1,833.0	23,500	1,797.0	23,038

TABLE R-14B TD Correction Factor

Design TD	Correction Factor
10	3.00
15	2.00
20	1.50
25	1.20
30	1.00
35	.86
40	.75

TABLE R-15 Refrigerant Condensers—Capacity and Engineering Data

Shell O.D. Inches	No. of Tubes	Model Number ▲	Total Effective Sq. Ft. of Surface	Nominal Rating—Tons*								Pump Down Capacity Pounds Freon-12 **	Nominal Operating Charge lbs. F-12 ***
				75°—95° Water 102° Cond. Temp.				85°—95° Water 105° Cond. Temp.					
				Capacity Tons*		Water P.D. PSI		Capacity Tons*		Water P.D. PSI			
				4-Pass.	2-Pass.	4-Pass.	2-Pass.	4-Pass.	2-Pass.	4-Pass.	2-Pass.		
		STF 84	73.5	5.41	‡	.24	‡	9.93	‡	2.4	‡	61	8.2
		STF 85	93.4	9.38	‡	.72	‡	14.4	8.7	5.1	.35	77	10
		STF 86	113	13.6	‡	1.50	‡	19.3	12.7	9.2	.74	93	13
		STF 87	133	17.9	‡	2.65	‡	§	16.8	§	1.3	109	15
		STF 88	153	22.4	12.2	4.2	.20	§	21.1	§	2.1	125	17
		STF 89	173	26.9	16.0	6.3	.40	§	25.5	§	3.0	141	19
8⅝	40	STF 810	193	31.7	20.1	8.7	.64	§	30.2	§	4.3	158	21
		STF 811	213	36.6	24.3	12.2	.95	§	35.1	§	6.0	174	24
		STF 812	232	41.6	28.5	15.9	1.30	§	40.2	§	7.8	190	26
		STF 813	252	§	32.8	§	1.78	§	§	§	§	206	28
		STF 814	272	§	37.2	§	2.30	§	§	§	§	222	30
		STF 815	292	§	41.6	§	2.95	§	§	§	§	238	32
		STF 816	312	§	46.1	§	3.65	§	§	§	§	254	34
		STF 104	110	8.1	‡	.24	‡	14.9	‡	2.4	‡	98	15
		STF 105	140	14.1	‡	.72	‡	21.6	13.0	5.1	.35	124	18
		STF 106	170	20.4	‡	1.50	‡	29.0	19.1	9.2	.74	150	22
		STF 107	200	26.9	‡	2.65	‡	§	25.2	§	1.3	175	26
		STF 108	229	33.6	18.3	4.2	.20	§	31.7	§	2.1	201	30
		STF 109	259	40.4	24.0	6.3	.40	§	38.2	§	3.0	226	34
10¾	60	STF 1010	289	47.5	30.2	8.7	.64	§	45.3	§	4.3	252	38
		STF 1011	319	54.9	36.4	12.2	.95	§	52.6	§	6.0	278	42
		STF 1012	349	62.5	42.7	15.9	1.30	§	60.3	§	7.8	304	46
		STF 1013	378	§	49.2	§	1.78	§	§	§	§	330	50
		STF 1014	408	§	55.8	§	2.30	§	§	§	§	355	53
		STF 1015	438	§	62.5	§	2.95	§	§	§	§	381	57
		STF 1016	468	§	69.2	§	3.65	§	§	§	§	407	61
		STF 124	169	12.4	‡	.24	‡	22.8	‡	2.4	‡	136	20
		STF 125	215	21.6	‡	.72	‡	33.1	20.0	5.1	.35	172	26
		STF 126	260	31.3	‡	1.50	‡	44.4	29.2	9.2	.74	207	31
		STF 127	306	41.2	‡	2.65	‡	§	38.7	§	1.3	243	36
		STF 128	352	51.5	28.0	4.2	.20	§	48.6	§	2.1	279	42
		STF 129	397	61.9	36.8	6.3	.40	§	58.7	§	3.0	315	47
		STF 1210	443	73.0	46.3	8.7	.64	§	69.5	§	4.3	351	53
		STF 1211	489	84.3	55.9	12.2	.95	§	80.7	§	6.0	386	58
12¾	92	STF 1212	535	95.8	65.6	15.9	1.30	§	92.5	§	7.8	422	63
		STF 1213	580	§	75.5	§	1.78	§	§	§	§	458	69
		STF 1214	626	§	85.6	§	2.30	§	§	§	§	494	74
		STF 1215	672	§	95.7	§	2.95	§	§	§	§	530	80
		STF 1216	718	§	106	§	3.65	§	§	§	§	566	85
		STF 1217	763	§	117	§	4.5	§	§	§	§	602	90
		STF 1218	809	§	127	§	5.5	§	§	§	§	638	96
		STF 1219	854	§	138	§	6.6	§	§	§	§	674	101
		STF 1220	900	§	149	§	7.8	§	§	§	§	710	107

TABLE R-15 *(Continued)*

Shell O.D. Inches	No. of Tubes	Model Number ▲	Total Effective Sq. Ft. of Surface	Nominal Rating—Tons* 75°—95° Water 102° Cond. Temp. Capacity Tons* 4-Pass.	2-Pass.	Water P.D. PSI 4-Pass.	2-Pass.	85°—95° Water 105° Cond. Temp. Capacity Tons* 4-Pass.	2-Pass.	Water P.D. PSI 4-Pass.	2-Pass.	Pump Down Capacity Pounds Freon-12 **	Nominal Operating Charge lbs. F-12 ***
14	120	STF 144	220	16.2	‡	.24	‡	29.8	‡	2.4	‡	159	26
		STF 145	280	28.1	‡	.72	‡	43.2	26.1	5.1	.35	202	33
		STF 146	339	40.8	‡	1.50	‡	57.9	38.1	9.2	.74	244	40
		STF 147	399	53.7	‡	2.65	‡	§	50.4	§	1.3	287	46
		STF 148	458	67.2	36.6	4.2	.20	§	63.3	§	2.1	329	53
		STF 149	518	80.7	48.0	6.3	.40	§	76.5	§	3.0	372	60
		STF 1410	578	95.1	60.3	8.7	.64	§	90.6	§	4.3	414	67
		STF 1411	637	110	72.9	12.2	.95	§	105	§	6.0	457	74
		STF 1412	697	125	85.5	15.9	1.30	§	121	§	7.8	499	81
		STF 1413	757	§	98.4	§	1.78	§	§	§	§	542	88
		STF 1414	816	§	111.7	§	2.30	§	§	§	§	584	94
		STF 1415	876	§	125	§	2.95	§	§	§	§	626	101
		STF 1416	936	§	138	§	3.65	§	§	§	§	669	108
		STF 1417	995	§	152	§	4.5	§	§	§	§	712	115
		STF 1418	1055	§	166	§	5.5	§	§	§	§	754	122
		STF 1419	1115	§	180	§	6.6	§	§	§	§	797	129
		STF 1420	1174	§	195	§	7.8	§	§	§	§	839	136
16	164	STF or SRF 164	301	22.2	‡	.24	‡	40.7	‡	2.4	‡	208	35
		STF or SRF 165	382	38.5	‡	.72	‡	59.0	35.7	5.1	.35	264	44
		STF or SRF 166	464	55.8	‡	1.50	‡	79.1	52.0	9.2	.74	319	53
		STF or SRF 167	545	73.4	‡	2.65	‡	§	68.9	§	1.3	374	62
		STF or SRF 168	627	91.8	50.0	4.2	.20	§	86.5	§	2.1	430	72
		STF or SRF 169	708	110	65.6	6.3	.40	§	105	§	3.0	485	81
		STF or SRF 1610	790	130	82.5	8.7	.64	§	124	§	4.3	541	
		STF or SRF 1611	871	150	99.7	12.2	.95	§	144	§	6.0	596	99
		STF or SRF 1612	953	171	117	15.9	1.30	§	165	§	7.8	652	108
		STF or SRF 1613	1034	§	135	§	1.78	§	§	§	§	707	118
		STF or SRF 1614	1116	§	153	§	2.30	§	§	§	§	763	127
		STF or SRF 1615	1197	§	171	§	2.95	§	§	§	§	818	136
		STF or SRF 1616	1278	§	189	§	3.65	§	§	§	§	874	145
		STF or SRF 1617	1360	§	208	§	4.5	§	§	§	§	929	155
		STF or SRF 1618	1441	§	227	§	5.5	§	§	§	§	983	164
		STF or SRF 1619	1523	§	246	§	6.6	§	§	§	§	1038	173
		STF or SRF 1620	1604	§	266	§	7.8	§	§	§	§	1092	182

TABLE R-15 (*Continued*)

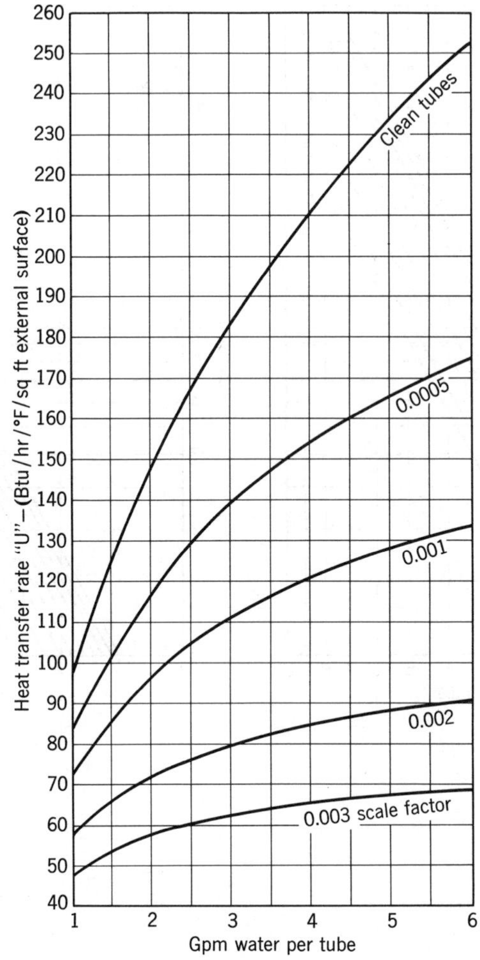

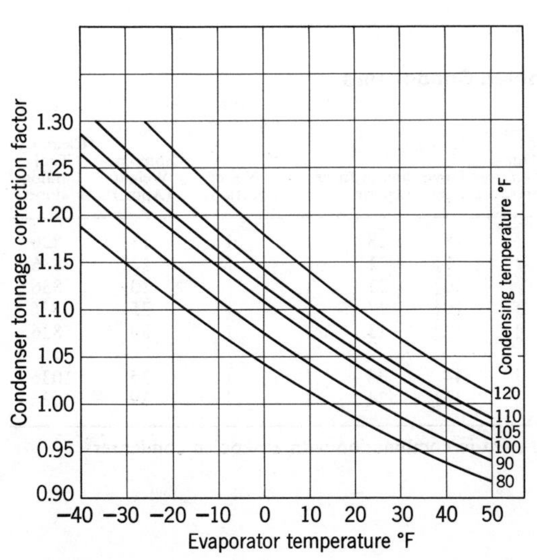

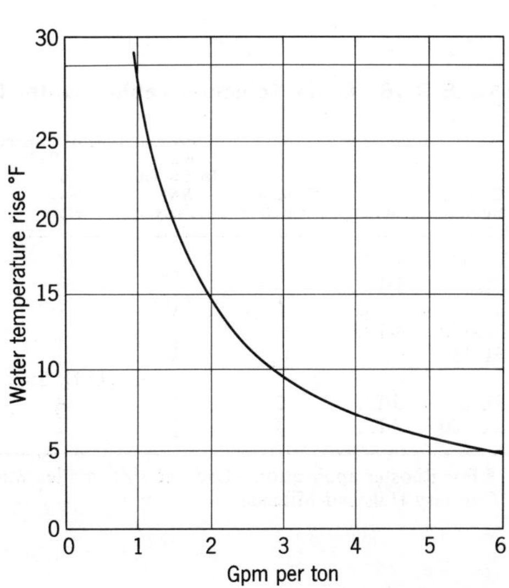

TABLE R-15 *(Continued)*

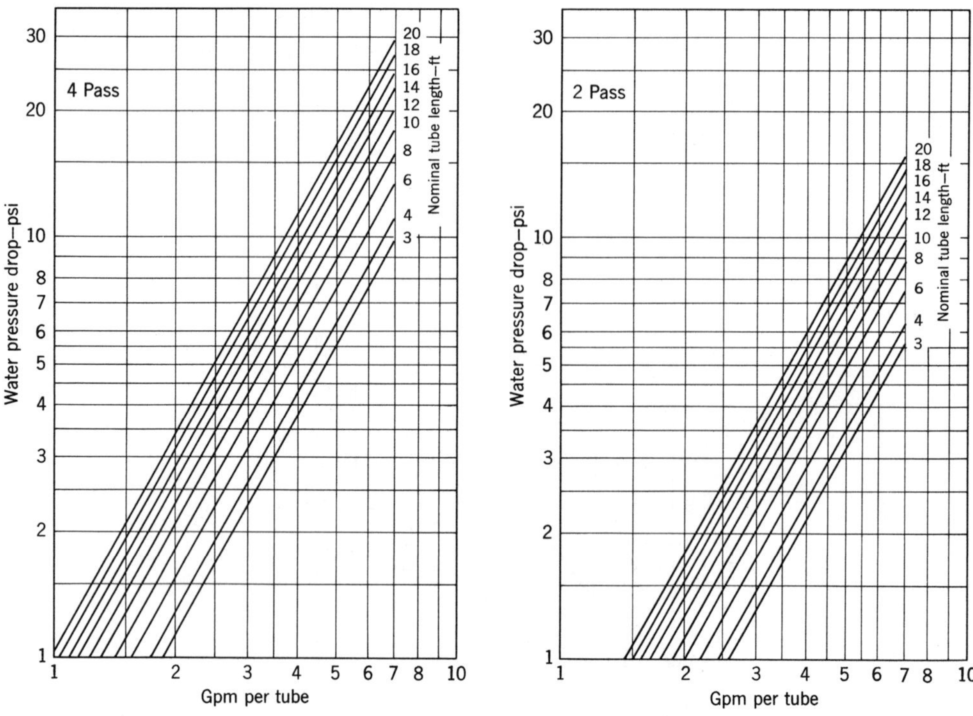

TABLE R-16 Quick Selection Table—Water-Cooled Condensers

Catalog Number	Stock Number	*Nominal HP Rating	Size and Type of Connections: Water In and Out SAE Flare	Refrigerant In SAE Flare	Refrigerant Out SAE Flare	Dimension in Inches: Height	Length	Depth	No. of Sections	Shipping Weight (Appr.)	Cleaning Tool Catalog Number
EL-33	1-EL	1/3	1/2″	1/2	3/8	8 3/4	18	1	1	13	836
EL-50	2-EL	1/2	1/2″	1/2	3/8	8 3/4	21	1	1	18	836
EL-75	3-EL	3/4	1/2″	1/2	3/8	10 3/4	21	1	1	20	836
EL-100	4-EL	1	1/2″	5/8	3/8	10 3/4	27	1	1	25	836
EL-150	5-EL	1 1/2	1/2″	5/8	3/8	12 3/4	33	1	1	30	836
				O.D. Swt.							
EL-200	6-EL	2	1/2″	7/8	1/2	14	34	1	1	35	1036
EL-300	7-EL	3	1/2″	7/8	1/2	16 1/4	34	1	1	39	1036

* For Booster application—Use one size smaller when used in combination with air-cooled condensers.
Courtesy Halstead Mitchell.

TABLE R-17 Ratings for Atmospheric Cooling Tower Capacities based on 3 mph wind velocity

Capacities based on 3 mph wind velocity

		Refrigeration						Tons			20° Range—14° Approach			
	Tower Basin W.B.	5 Gpm Per Ton		4 Gpm per Ton				3 Gpm per Ton			Gas and Gasoline Engine	Diesel Engine	Steam Condensing	Comp. Air 100 lb
		91 85 78	93 87 80	90½ 83 75	93½ 86 78	92½ 85 78	94½ 87 80	90 80 70	95 85 75	96 86 78				
Tower No.	Nom. Gpm						(Nom.)				HP	HP	lb/hr	Cfm
CSA 23	12	2.7	2.6	2.8	3.1	2.9	3.0	3.4	3.8	3.4	20	33	100	670
CSA 33	20	4.5	4.4	4.8	5.1	4.8	5.0	5.7	6.4	5.8	30	50	150	1,000
CSA 34	24	5.4	5.3	5.7	6.1	5.7	6.0	6.8	7.7	6.9	40	67	200	1,330
CSA 44	32	7.2	7.0	7.6	8.2	7.7	8.0	9.1	10.2	9.2	50	83	250	1,670
CSA 45	40	9.0	8.8	9.5	10.2	9.6	10.0	11.4	12.8	11.5	60	100	300	2,000
CSA 55	50	11.2	11.0	11.8	12.8	12.0	12.5	14.2	16.0	14.4	80	133	400	2,670
CSA 66	72	16.2	15.8	17.1	18.4	17.2	18.0	20.5	23.0	20.7	110	183	550	3,670
SA 33	22	5.0	4.8	5.2	5.6	5.3	5.5	6.3	7.0	6.3	40	67	200	1,330
SA 34	30	6.7	6.6	7.1	7.7	7.2	7.5	8.6	9.6	8.6	50	83	250	1,670
SA 44	40	9.0	8.8	9.5	10.2	9.6	10.0	11.4	12.8	11.5	60	100	300	2,000
SA 45	50	11.2	11.0	11.8	12.8	12.0	12.5	14.2	16.0	14.4	80	133	400	2,670
SA 46	60	13.5	13.2	14.2	15.3	14.4	15.0	17.1	19.2	17.1	100	167	500	3,330
SA 56	75	16.9	16.5	17.8	19.2	18.0	18.8	21.4	24.0	21.6	120	200	600	4,000
SA 58	100	22.5	22	23.7	25.5	24	25.0	28.5	32.0	29	160	267	800	5,330
SA 68	120	27	26	28	31	29	30.0	34	38	34	200	333	1,000	6,670
SA 610	150	34	33	36	38	36	37.5	43	48	43	250	410	1,250	8,200
SA 612	180	41	40	43	46	43	45.0	51	57	52	300	500	1,500	10,000
SA 615	225	50	49	53	57	54	56.3	64	72	64	350	580	1,750	11,600
SA 616	240	54	53	57	61	57	60.0	68	77	69	400	670	2,000	13,400
SA 618	270	61	60	65	70	65	67.5	77	87	78	450	750	2,250	15,000
SA 620	300	67	66	71	77	72	75.0	86	96	86	500	830	2,500	16,600
SA 624	360	81	79	85	92	86	90.0	103	115	104	600	1,000	3,000	20,000
SA 824	400	90	88	95	102	96	100.0	114	128	115	700	1,167	3,500	23,300
SA1224	450	102	100	107	115	108	112.5	129	144	130	800	1,330	4,000	26,700
SA1230	550	124	121	131	140	132	137.5	157	176	158	1,000	1,670	5,000	33,300
SA1236	650	146	143	154	166	156	162.5	186	208	187	1,200	2,000	6,000	40,000
SA1242	750	169	165	178	192	180	187.5	214	240	216	1,400	2,330	7,000	46,700
SA1248	850	192	187	202	218	204	212.5	243	272	245	1,600	2,670	8,000	53,300
SA1254	950	213	210	226	243	228	237.5	271	305	274	1,800	3,000	9,000	60,000
SA1260	1100	247	240	261	280	263	275.0	314	350	315	2,000	3,330	10,000	66,700
SA1266	1200	270	260	280	310	290	300.0	340	380	340	2,200	3,670	11,000	73,300

Courtesy Star Cooling Towers, Inc.

TABLE R-18 Evaporative Condenser Ratings

TABLE A (In terms of evaporator load at + 40°F evaporator)

Model No.	Cond. Temp. (° F)	Wet Bulb Temperature of Entering Air (°F) 60°	65°	70°	75°	78°	80°	85°
	90	2.6	2.2	1.9	1.5	1.2	1.1	0.6
	95	3.2	2.9	2.5	2.1	1.9	1.7	1.2
E-80F	100	3.8	3.5	3.2	2.8	2.5	2.3	1.9
	105	4.5	4.2	3.9	*3.5	3.3	3.1	2.6
	110	5.3	5.1	4.8	4.4	4.1	4.0	3.5
	115	6.3	6.0	5.7	5.3	5.1	4.9	4.5
	90	4.4	3.9	3.3	2.6	2.1	1.8	0.9
	95	5.5	5.0	4.3	3.7	3.3	2.9	2.1
E-135F	100	6.6	6.1	5.5	4.8	4.4	4.1	3.2
	105	7.8	7.3	6.7	*6.0	5.6	5.4	4.5
	110	9.4	8.9	8.3	7.6	7.2	6.9	6.1
	115	11.0	10.5	9.9	9.3	8.9	8.6	7.8
	90	8.8	7.8	6.6	5.2	4.2	3.6	1.8
	95	11.0	10.0	8.6	7.4	6.6	5.8	4.2
E-270F	100	13.2	12.2	11.0	9.6	8.8	8.2	6.4
	105	15.6	14.6	13.4	*12.0	11.2	10.8	9.0
	110	18.8	17.8	16.6	15.2	14.4	13.8	12.2
	115	22.0	21.0	19.8	18.6	17.8	17.2	15.6

* ASRE standard rating conditions.

TABLE B Evaporator Temperature Correction Factors

Evaporator Temp. (° F)	Correction Factor	Evaporator Temp. (° F)	Correction Factor
50	0.97	0	1.11
40	1.0	−10	1.16
30	1.03	−20	1.20
20	1.05	−30	1.26
10	1.09		

Courtesy McQuay, Inc.

TABLE R-19 Water Valve Selection Table

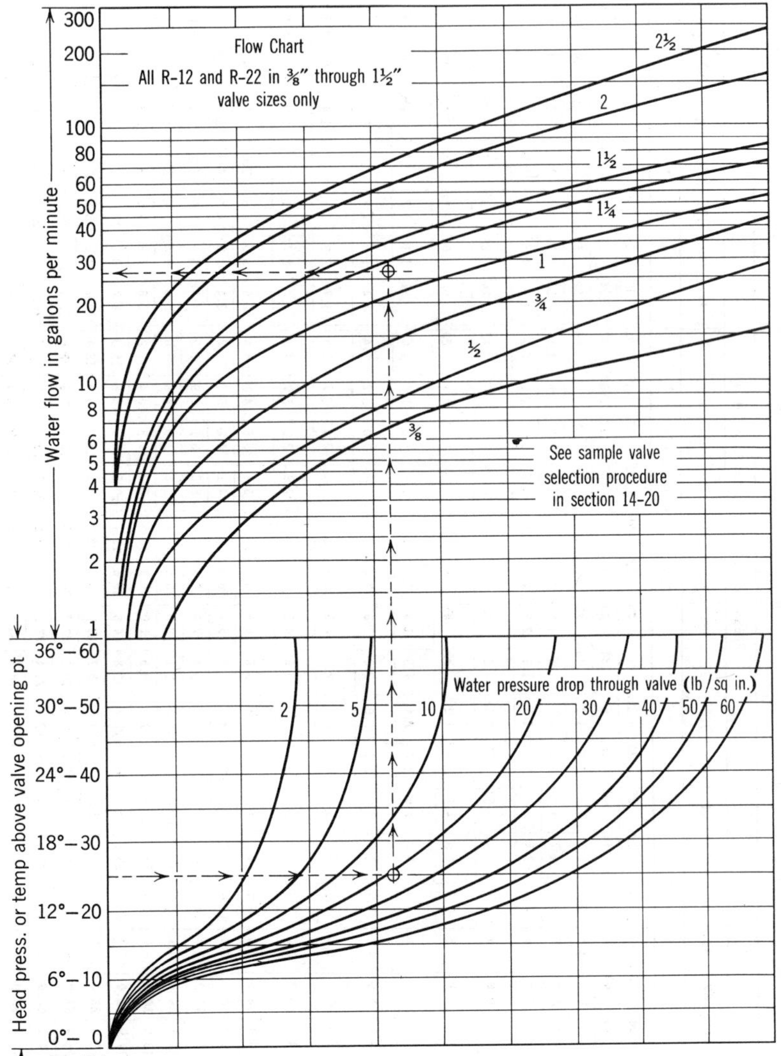

TABLE R-20 Centrifugal Pump Capacity Table

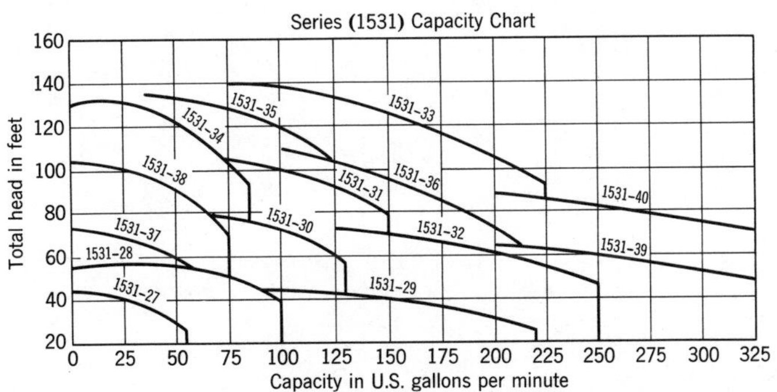

TABLE R-21 Thermostatic Expansion Valve Selection Valve Selection Table

New Type No.	EVAPORATOR TEMPERATURE °F.																							
	40°				20°				0°				−10°				−20°				−30°			
	PRESSURE DIFFERENCE ACROSS VALVE, PSI																							
	40	**60**	80	100	60	80	100	120	75	100	125	150	75	100	125	150	75	100	125	150	75	100	125	150
	TONS OF REFRIGERATION																							
TK25F	.24	**.30**	.35	.39	.22	.25	.28	.31	.17	.20	.22	.25	.15	.17	.19	.21	.12	.14	.16	.18	.10	.12	.14	.16
TK50F	.49	**.60**	.70	.78	.44	.50	.56	.62	.34	.40	.44	.50	.30	.34	.38	.42	.24	.28	.32	.36	.20	.24	.28	.32
TK100F	1.1	**1.3**	1.5	1.7	1.0	1.1	1.2	1.3	.75	.86	.97	1.1	.63	.72	.81	.89	.52	.60	.67	.73	.42	.49	.55	.60
TK200F	1.9	**2.2**	2.5	2.9	1.7	1.9	2.1	2.2	1.3	1.5	1.7	1.9	1.1	1.2	1.4	1.5	.90	1.0	1.1	1.2	.71	.83	.93	1.1
TK300F	2.8	**3.3**	3.8	4.3	2.5	2.8	3.0	3.3	1.9	2.2	2.5	2.8	1.6	1.8	2.0	2.2	1.3	1.5	1.7	1.9	1.1	1.2	1.4	1.5
TL50F TCL50F	.49	**0.6**	.70	.78	.44	.50	.56	.62	.34	.40	.44	.50	.30	.34	.38	.42	.24	.28	.32	.36	.20	.24	.28	.32
TL100F TCL100F	1.1	**1.3**	1.5	1.7	1.0	1.1	1.2	1.3	.75	.86	.97	1.1	.63	.72	.81	.89	.52	.60	.67	.73	.42	.49	.55	.60
TL200F TCL200F	1.9	**2.3**	2.6	3.0	1.7	1.9	2.2	2.4	1.3	1.5	1.7	1.9	1.1	1.3	1.4	1.6	.91	1.1	1.2	1.3	.75	.86	.97	1.1
TL300F TCL300F	2.9	**3.5**	4.0	4.5	2.6	3.0	3.3	3.6	2.0	2.3	2.6	2.9	1.7	2.0	2.2	2.4	1.4	1.6	1.8	2.0	1.1	1.3	1.5	1.6
TL400F TCL400F	3.5	**4.3**	5.0	5.6	3.2	3.6	4.0	4.5	2.5	2.9	3.2	3.5	2.1	2.4	2.7	2.9	1.7	2.0	2.2	2.4	1.4	1.6	1.8	2.0
TL600F TCL600F	4.9	**6.0**	7.0	7.8	4.4	5.0	5.7	6.2	3.5	4.0	4.5	4.9	2.9	3.3	3.7	4.1	2.4	2.7	3.1	3.4	1.9	2.2	2.5	2.7
TAL650F TDL650F	5.3	**6.5**	7.5	8.4	4.7	5.5	6.1	6.7	3.8	4.4	4.9	5.4	3.1	3.6	4.1	4.4	2.6	3.0	3.3	3.6	2.1	2.4	2.7	3.0
TJL800F	6.9	**8.5**	9.8	11.0	6.2	7.2	8.0	8.8	4.9	5.7	6.4	6.9	4.1	4.8	5.3	5.8	3.4	3.9	4.4	4.8	2.8	3.2	3.6	3.9
TJL1100F	9.0	**11.0**	12.7	14.2	8.1	9.3	10.4	11.4	6.4	7.4	8.3	9.1	5.3	6.1	6.9	7.5	4.4	5.0	5.6	6.2	3.6	4.1	4.6	5.0
TEL1400F	11.4	**14.0**	16.1	18.1	10.3	11.8	13.2	14.6	8.1	9.3	10.5	11.4	6.8	7.8	8.8	9.6	5.6	6.4	7.2	7.9	4.5	5.2	5.9	6.4
TEL1600F	13.5	**16.5**	19.0	21.3	12.1	13.9	15.6	17.2	9.5	11.0	12.3	13.5	8.0	9.2	10.4	11.2	6.6	7.6	8.5	9.3	5.3	6.2	6.9	7.6
TEL2200F	17.9	**22.0**	25.4	28.4	16.1	18.6	20.8	22.7	12.8	14.8	16.6	18.1	10.6	12.3	13.8	15.0	8.7	10.0	11.3	12.3	7.1	8.2	9.2	10.1
TIL2700F	22.0	**27.0**	31.2	34.8	19.8	22.8	25.4	27.9	15.7	18.2	20.3	22.2	13.1	15.0	16.9	18.4	10.7	12.3	13.8	15.1	8.8	10.1	11.3	12.3
TIL3300F	26.9	**33.0**	38.0	42.6	24.3	27.9	31.2	34.3	19.0	22.0	24.7	26.9	16.0	18.5	20.8	22.6	13.1	15.1	17.0	18.5	10.7	12.3	13.9	15.1
THL4200F	34.3	**42.0**	48.5	54.2	30.8	35.5	39.6	43.4	24.4	28.2	31.6	34.6	20.3	23.4	26.2	28.7	16.6	19.1	21.4	23.5	13.6	15.7	17.6	19.2
THL5000F	40.8	**50.0**	57.6	64.6	36.8	42.3	47.3	52.0	28.9	33.3	37.4	40.8	24.3	28.0	31.5	34.3	19.9	22.9	25.7	28.1	16.2	18.7	21.0	22.9

*New valve size.

Based on 100° F. condensing temperature, 1° subcooled liquid, and a maximum superheat change of 4° F. (Temperature rise of remote bulb required to move valve pin from closed to rated open position.) For each 10° subcooling, the capacities are increased approximately 6%.

SUBJECT INDEX

TABLE AND CHART INDEX

TABLES

CHARTS

EQUIPMENT RATING TABLES